PLANT ECOLOGY

PLANT ECOLOGY

Edited by Michael J. Crawley

Department of Biology,
Imperial College of Science, Technology and Medicine
Silwood Park, Ascot, Berks

SECOND EDITION

**Blackwell
Science**

© 1986, 1997 by
Blackwell Science Ltd
Editorial Offices:
Osney Mead, Oxford OX2 0EL
25 John Street, London WC1N 2BL
23 Ainslie Place, Edinburgh EH3 6AJ
238 Main Street, Cambridge
  Massachusetts 02142, USA
54 University Street, Carlton
  Victoria 3053, Australia

Other Editorial Offices:
Arnette Blackwell SA
  224, Boulevard Saint Germain,
  75007 Paris, France

Blackwell Wissenschafts-Verlag GmbH
  Kurfürstendamm 57
  10707 Berlin, Germany

  Zehetnergasse 6
  A-1140 Wien
  Austria

First published 1986
Reprinted 1989, 1991, 1994
Second edition 1997

Set by Semantic Graphics, Singapore
Printed and bound in Great Britain
at the University Press, Cambridge

The Blackwell Science logo is a
trade mark of Blackwell Science Ltd
registered at the United Kingdom
Trade Marks Registry

DISTRIBUTORS

Marston Book Services Ltd
PO Box 269
Abingdon
Oxford OX14 4YN
(*Orders:* Tel: 01235 465500
        Fax: 01235 465555)

USA
Blackwell Science, Inc.
238 Main Street
Cambridge, MA 02142
(*Orders:* Tel: 800 215-1000
        617 876-7000
        Fax: 617 492-5263)

Canada
Copp Clark Professional
200 Adelaide Street, West, 3rd Floor
Toronto, Ontario
Canada, M5H 1W7
(*Orders:* Tel: 416 597 1616
        800 815 9417
        Fax: 416 597 1617)

Australia
Blackwell Science Pty Ltd
54 University Street
Carlton, Victoria 3053
(*Orders:* Tel: 03 9347 0300
        Fax: 03 9349 3016)

A catalogue record for this title
is available from the British Library

ISBN 0 632 03639 7

Library of Congress
Cataloging-in-publication Data

Plant ecology / edited by
Michael J. Crawley. — 2nd ed.
        p.   cm.
  Includes bibliographical references
  and index.
  ISBN 0-632-03639-7
  1. Plant ecology.
  I. Crawley, Michael J.
QK901.P57   1997
581.5—dc20   96-23347
                              CIP

Contents

Colour plates fall between pp. 366 and 367

List of Contributors

Mike Ashmore *Department of Environmental Technology, Imperial College of Science, Technology and Medicine, Silwood Park, Ascot, Berkshire, SL5 7PY*

Michael J. Crawley *Department of Biology, Imperial College of Science, Technology and Medicine, Silwood Park, Ascot, Berkshire, SL5 7PY*

James R. Ehleringer *Department of Biology, University of Utah, Salt Lake City, Utah, 84112, USA*

Alastair Fitter *Department of Biology, University of York, York, YO1 5DD*

John Grace *Institute of Ecology and Resource Management, University of Edinburgh, Edinburgh, EH9 3JU*

J. Philip Grime *Unit of Comparative Plant Ecology, University of Sheffield, Sheffield, S10 2TN*

James P. Grover *Department of Biology, University of Texas, Arlington, PO Box 19498, Texas, 76019, USA*

Jeffrey B. Harborne *Department of Botany, University of Reading, Whiteknights, Reading, Berkshire, RG6 6AS*

Susan E. Hartley *Institute of Terrestrial Ecology, Banchory Research Station, Hill of Brathens, Glassel, Banchory, Kincardineshire, AB31 4BY*

Robert D. Holt *Museum of Natural History, Department of Systematics and Ecology, University of Kansas, Lawrence, Kansas, 66045-2454, USA*

Henry F. Howe *Department of Biological Sciences, University of Illinois, 845 W. Taylor Street, Chicago, Illinois, 60607, USA*

Michael J. Hutchings *School of Biological Sciences, University of Sussex, Falmer, Brighton, Sussex, BN1 9QG*

Clive G. Jones *Institute of Ecosystem Studies, Box AB, Milbrook, NY 12545, USA*

Harold A. Mooney *Department of Biological Sciences, Stanford University, Stanford, CA 94305, USA*

Stephen W. Pacala *Department of Ecology and Evolutionary Biology, Princeton University, Princeton, New Jersey, 08544, USA*

Mark Rees *Department of Biology, Imperial College of Science, Technology and Medicine, Silwood Park, Ascot, Berkshire, SL5 7PY*

David Tilman *Department of Ecology, Evolution and Behaviour, 1987 Upper Burford Circle, University of Minnesota, St. Paul, Minnesota, 55108, USA*

Andrew R. Watkinson *Schools of Biological and Environmental Sciences, University of East Anglia, Norwich, NR4 7TJ*

Lynn C. Westley *Department of Biology, Lake Forest College, Lake Forest, Illinois, 60045, USA*

Preface to the Second Edition

Our aim in the first edition was to produce a stimulating set of papers that might form the basis of tutorial or small group teaching as part of a general ecology course. The fact that the book was used more generally than originally intended has highlighted the need for a comprehensive textbook of plant ecology which covered theory, experimental and field observations, and which dealt with individuals, populations and communities from both pure and applied perspectives. We hope you will feel that the second edition succeeds in fulfilling this broader remit.

It is worth recalling just how much has happened during the 10 years since the first edition appeared in 1986. The Berlin Wall has tumbled, the Cold War has ended and the world now has one superpower instead of two. The arrival of AIDS has changed the way that human beings think about themselves. When the first edition was produced, scientists still had typewriters on their desks, rather than personal computers. Genetic engineering has transformed the way that plant breeders go about their craft and has given the environmental movement another issue to worry about. Sequencing has produced the first molecular phylogeny of plants, and genetic fingerprinting is set fair to revolutionize the way that ecologists study gene-flow in the field. New kinds of plant pathogen have been discovered (e.g. the mycoplasma-like organisms). In ecology, theoreticians have moved on from studying point-equilibrium (mean-field) models to the analysis of spatially explicit models. Such models can give diametrically opposite answers to identical questions; for example, where once a point-equilibrium model might predict competitive exclusion, a spatially explicit version, incorporating exactly the same ecological assumptions, might predict coexistence. Previously, field ecologists were preoccupied with observational studies; now they combine field experiments with theory and observation.

The first edition derided the 'Many, Complex and Interacting' school of ecology, in an attempt to foster a more critical, question-focused approach to the subject. It is interesting that several of the most important recent advances in population dynamics have involved substantial increases in the complexity of the models employed. It is worth recalling Albert Einstein's famous remark: 'A model should be as simple as possible. But no simpler'. Modern theory is often substantially more complex (more realistic, some might say) than equivalent models from just 10 years ago. For example, instead of concentrating on single-factor explanations (food, predators), a modern investigation of population cycles in voles might include generalist predators, specialist predators, food and seasonality (Hanski *et al.* 1993). It is a study of the interaction between factors, analysed using the simplest possible models, that is likely to produce the most productive insights into ecological function.

The second edition reverts to the traditional sequence of things, with the physiology introduced first, then the populations, then the communities and, finally, the applied ecosystem-level processes. Material that was not covered by the first edition, but which is dealt with here, includes water relations, secondary chemistry, sex, seeds, plant defences, herbivory, the theory of community dynamics, pollution, climate change and biodiversity. There is much more of an evolutionary emphasis on trade-offs and plant life history throughout the text. The level of the book, and I hope its tone, are the same. The idea is to present something of the excitement of current plant ecology and to highlight the critical issues rather than to gloss over them or to pretend that we have all of the answers.

Since the first edition, most people working with plants have adopted the new family names and old favourites like Gramineae and Leguminosae are gone from current usage. Here is a complete list of the changes as they affect this book: Apiaceae replaces Umbelliferae, Arecaceae replaces Palmae, Asteraceae replaces Compositae, Brassicaceae replaces Cruciferae, Clusiaceae replaces Guttiferae, Fabaceae replaces Leguminosae, Lamiaceae replaces Labiatae, Poaceae replaces Gramineae. The family name is given in parentheses after the generic and specific names of an example are first used in any particular chapter. Exceptions are when the family name and the generic name come from the same root (e.g. *Pinus* (Pinaceae) or *Brassica* (Brassicaceae)), or when the Latin binomial is preceded by an explanatory phrase (e.g. the grass *Agrostis capillaris*, or the orchid *Ophrys apifera*), when the family name is omitted. Good ecologists know their organisms, and so in addition to courses in mathematics, statistics, experimental design, geology, soil science, molecular genetics, physiology, quantitative genetics, evolution, behavioural ecology and biogeography, make sure that you learn how to identify plant families in the field. The number of people who can identify plants continues to decline year by year. Do not be put off from learning plant taxonomy. It's fun. Enjoy it.

Michael J. Crawley
Ascot

Preface to the First Edition

This book is about the factors affecting the distribution and abundance of plants. It aims to show how pattern and structure at different levels of plant organization (communities, populations and individuals) are influenced by abiotic factors like climate and soils, and by biotic interactions including competition with other plants, attack by herbivorous animals and plant pathogens, and relationships with mutualistic organisms of various kinds. One further aim has been to convey something of the excitement and dynamism of modern plant ecology. For too long the subject has been regarded as the poor relation of animal ecology. It has been transformed over the last two decades into a veritable growth industry, thanks chiefly to the inspiration of John Harper, and to the wide dispersal of the students he trained.

The present work differs from other textbooks on plant ecology in stressing dynamics rather than statics, and by espousing an experimental rather than a descriptive methodology. Throughout the book, the patterns of plant abundance which we see in the field are interpreted as the outcome of dynamic processes involving gains and losses. For example, we view the species richness of communities as the resolution of immigration and extinction, population density as the balance of recruitment and mortality within single species, and the size of individual plants as the result of births and deaths of modular component parts.

This book is based on a plant-centred view of ecology rather than the traditional, quadrat-centred view. Traditional quadrat-based measures like 'percentage cover' consign individual plants to oblivion, and discourage thinking about the evolutionary ecology of individuals. Quadrat based tests of association may detect positive statistical associations between species, even when the individual plants of the two species never grow next to one another, and never influence one another's population dynamics! The plant-centered view of ecology is intended to rectify some of these defects by focusing attention directly on the interactions between a plant and its immediate neighbours, and between plants and their mycorrhizal associates, pollinators and natural enemies. Throughout the book we are at pains to stress that our ultimate objective is to measure the impact of the processes we describe on the *fitness* of individual phenotypes (difficult though we recognize this task to be).

If there is a single philosophical strand running through the following chapters, it is this: 'seek simplicity but distrust it' (Lagrange). Our approach to ecology is to develop theory by the recognition of patterns in the field and the development of simple models to describe them. As Lewontin and Levins have said, only slightly tongue-in-cheek: 'Things are similar – this makes science possible. Things are different – this makes science necessary'! The role of theory in ecology is rather different than in some other sciences. It is not, for example, intended to make accurate predictions, as in astronomy or physics. Rather it attempts to separate the expected from the unexpected, the possible from the impossible and the surprising from the unsurprising (Lewontin). In the past, however, plant ecologists have tended to emphasize the complexity of ecological systems and to decry simple, theoretical models as being naive and unrealistic. So prevalent, indeed, is this viewpoint amongst practising ecologists that it has now been elevated to the position of a formal school of thought. During a recent botanizing trip to a species-rich area of serpentine soils, we had recourse to the local botanists' guide book. It promised enlightenment of the most profound kind. It began: 'the factors responsible for the richness of the serpentine flora are' . . . we held our breath in anticipation . . . 'many, complex and interacting'. On that day the 'MC&I School' of ecology was born. Many and various have been its subsequent publications!

Of course this is not the only modern school of plant ecologists. Other recognizable disciplines include the 'Nothing's Happening School' (who believe in null models, random processes and non-equilibrium communities), the Biochemical Ecologists (the 'Find 'em and Grind 'em School'), the Mathematical Modellers (quantifying the bleeding obvious), the Agricultural Ecologists (spray and pray), and the Phytosociologists (ignore that plant, it shouldn't be here). There are as many schools of thought represented in this book as there are different contributors to it. Your editor, however, is a confirmed adherent to the Experimentalist School who 'suck it and see'!

The layout of the chapters is upside down in terms of the conventional atom-to-universe textbook. I have chosen to begin with communities and work progressively downwards through populations and individuals for three reasons. First, it is with communities that we gain our initial impressions of plants and of plant ecology. Second, the theory of community ecology is relatively straightforward, so the theoretical material becomes more rather than less challenging as the book goes on. Third, it makes a change!

It is worth pointing out explicitly what the book is *not* about. First and foremost, it is not about ecological methods. There is a vast literature on methodology that we have made no attempt to précis. Students would do well to study carefully the problems of sampling plant populations, and to acquaint themselves with the statistical analysis and presentation of data. Only then, perhaps, should the more rarified quantitative methods like pattern analysis and the plethora of multivariate techniques be addressed. Three particularly good introductory papers for those about to embark on ecological experimentation are Lewontin (1974), Hurlbert (1984) and Bender, Case & Gilpin (1984). Other important aspects of plant ecology not covered by this book are micrometeorology, soil science, plant anatomy, plant physiology, plant evolution, plant geography and ecological biochemistry. Not only do these fields provide vital background information, but they also afford a variety of different perspectives from which plant ecology can be viewed.

Finally, I should like to thank the many people who have helped in the preparation and production of this book.

Michael J. Crawley
Ascot, 1985

1: Photosynthesis

Harold A. Mooney and James R. Ehleringer

1.1 Introduction

The growth of plants depends upon their capacity to incorporate atmospheric carbon into organic compounds through the use of light energy absorbed during photosynthesis. This is a two-step process: (i) an initial photochemical reaction traps light energy in absorbing pigments (chlorophyll and accessory pigments), producing a reductant (NADPH) plus ATP; (ii) subsequently, atmospheric CO_2 is reduced and biochemically incorporated into carbohydrates (Fig. 1.1). The CO_2 to fuel this reaction diffuses from the atmosphere to the site of fixation within the chloroplast. Limitations to the overall rate of photosynthesis may occur through restrictions at either of these steps; the nature of these limitations is discussed in detail in the following sections.

1.2 Background

1.2.1 Photochemical reactions

A portion of the light energy impinging on a leaf is absorbed by chlorophyll pigments and then transferred to specialized reaction centres where electrons are moved along an energy gradient (Fig. 1.2). The free energy released in a series of subsequent electron transfers (known as the 'light reactions' of photosynthesis) is utilized to phosphorylate ADP and to reduce NADP (Barber & Baker 1985). Both molecules are

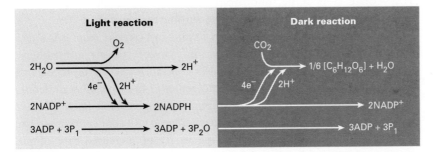

Fig. 1.1 Coupling of photochemical and biochemical reactions in photosynthesis. Since the biochemical reactions are not dependent on light they are termed 'dark reactions'. (From Miller 1979.)

essential in the biochemical reduction cycle (the 'dark reactions') described below.

1.2.2 Biochemical reactions

In most plants, the initial reduction of CO_2 in photosynthesis depends upon the activity of a single enzyme, ribulose bisphosphate carboxylase oxygenase (Rubisco). A fundamental feature of this enzyme is that both CO_2 and O_2 are competitive substrates for the active site. In the carboxylation reaction, atmospheric CO_2 is coupled with the five-carbon acceptor molecule, ribulose bisphosphate (RuP_2), to form two molecules of a three-carbon product, 3-phosphoglycerate (3-PGA).

$$
\begin{array}{c}
CH_2OP \\
| \\
CO \\
| \\
CHOH \\
| \\
CHOH \\
| \\
CH_2OP
\end{array}
\;+\; CO_2 \;\xrightarrow{\text{Rubisco}}\;
\begin{array}{c}
COOH \\
| \\
CHOH \\
| \\
CH_2OP
\end{array}
\;+\;
\begin{array}{c}
COOH \\
| \\
CHOH \\
| \\
CH_2OP
\end{array}
\qquad (1.1)
$$

In the oxygenation reaction, O_2 and RuP_2 combine to form a two-carbon molecule, phosphoglycolate, and PGA.

$$
\begin{array}{c}
CH_2OP \\
| \\
CO \\
| \\
CHOH \\
| \\
CHOH \\
| \\
CH_2OP
\end{array}
\;+\; O_2 \;\xrightarrow{\text{Rubisco}}\;
\begin{array}{c}
COOH \\
| \\
CHOH \\
| \\
CH_2OP
\end{array}
\;+\;
\begin{array}{c}
COOH \\
| \\
CH_2OP
\end{array}
\qquad (1.2)
$$

The glycolate is broken down in a series of reactions (called photorespiration) that involve the generation of CO_2. Photosynthesis and photorespiration are linked as shown in Fig. 1.3. The photorespiratory cycle serves no known function, except to salvage as much of the carbon fixed as glycolate as possible; photorespiration appears to be an inevitable consequence of Rubisco activity in an aerobic environment.

The balance between the oxygenase and carboxylase reactions shifts according to atmospheric composition. Under today's atmospheric conditions (~350 ppm CO_2 and 21% O_2), about seven molecules of CO_2 react with RuP_2 for every two molecules of O_2. One CO_2 molecule is subsequently lost during photorespiration to give a net fixation of six molecules of CO_2. At an increased atmospheric CO_2 concentration, the CO_2 reaction predominates, resulting in (i) an increased rate of net photosynthesis, and (ii) a decrease in the compensation point (the concentration at which net CO_2 uptake balances O_2 evolution). Certain kinds of plants, the so-called C_4 species, possess mechanisms for main-

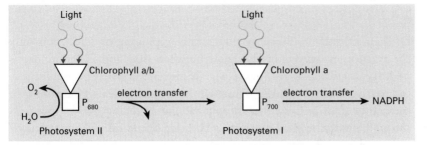

Fig. 1.2 Light reactions involved in the production of NADPH and ATP. Light is absorbed by 'antennae' chlorophyll pigments and transferred to specialized reaction centres and then to photosystems I and II. The photosystems are responsive to different wavelengths, with photosystem I absorbing light of 680 nm and photosystem II absorbing wavelengths longer than this. The electrons removed from the splitting of water are moved along this pathway obtaining reducing potential through the light reactions and eventually producing stored energy as ATP and the strong reductant NADPH. These, in turn, are utilized to reduce CO_2 to carbohydrate during non-light-requiring reactions. Plants from sun and shade habitats have leaves with different proportions of pigments, electron transfer chains and reaction centres.

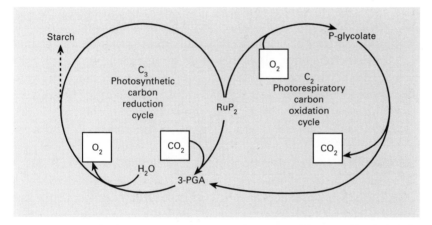

Fig. 1.3 Linkage of photosynthesis and photorespiration. (Modified from Osmond *et al.* 1981.)

taining the site of carboxylation at high CO_2 concentrations, so that carbon is not lost in photorespiration (see section 1.6).

1.3 Environmental influences on photosynthetic capacity

With this brief sketch of the process of photosynthesis as background, we are in a position to examine the influences of the physical environment on photosynthetic rates. We begin by considering the bulk of the world's plant species, which utilize RuP_2 as the primary carbon acceptor in photosynthesis. These are called C_3 plants because the initial product of photosynthesis is a three-carbon compound, PGA.

1.3.1 Light

Of the total solar and terrestrial radiation impinging on a leaf, it is only the fraction lying within the band between 400 and 700 nm that is photosynthetically active (Fig. 1.4). This is referred to as photosynthetically active radiation (PAR). Photosynthetic rates of leaves increase with increasing PAR because the supply rate of reducing power increases through photochemical reactions. The rate levels off as limitations of carboxylating capacity and diffusion begin to predominate. The stomatal conductance to CO_2 greatly influences the maximum photosynthetic rates achieved at high light intensities. At very low intensities, there is no net uptake of CO_2 since the rate of CO_2 uptake through photosynthesis is less than the rate of CO_2 evolution from mitochondrial respiration.

The photosynthetic light response may differ considerably between species and among leaves on the same individual plant (Fig. 1.5). The nature of these differences has been studied most thoroughly in leaves of

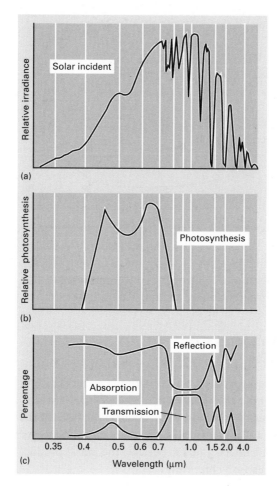

Fig. 1.4 Incoming solar radiation (a), photosynthetically active radiation, PAR (b), and leaf spectral characteristics (c). Note that the leaf has a high absorptance to PAR and a low absorptance to longer solar radiation wavelengths.

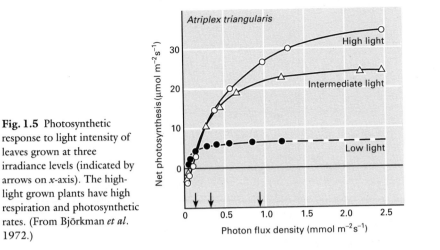

Fig. 1.5 Photosynthetic response to light intensity of leaves grown at three irradiance levels (indicated by arrows on *x*-axis). The high-light grown plants have high respiration and photosynthetic rates. (From Björkman *et al.* 1972.)

the same plant that have been produced under either low or high light intensities ('shade leaves' or 'sun leaves'). Under these contrasting conditions, the sun leaves almost always have higher saturated photosynthetic rates and higher light compensation points (Boardman 1977). The mechanisms underlying these differences are complex, and involve both morphological and biochemical components. Sun leaves generally have a greater density of stomata and hence a greater conductance to gas transfer. Further, they have a greater capacity for photochemical electron transport, higher mitochondrial respiration rates, and a higher content (and activity) of carboxylating enzymes than occurs in shade leaves.

Moving a shade leaf directly into the sun can cause damage to the photosynthetic system (photoinhibition), because the light energy trapped cannot be used fully in photosynthesis and these leaves do not have the capacity to dissipate this excess energy through the xanthophyll cycle (Demming-Adams & Adams 1992). In this case, there is inactivation of the photosystem II reaction centres. Over the course of several days, however, a shade leaf may be able to adjust (acclimate) to sun conditions, by increasing its photosynthetic enzyme content (dark reactions) and light reaction components, thereby enhancing its capacity to utilize light energy.

1.3.2 Carbon dioxide

Carbon dioxide is a primary substrate for photosynthesis and Rubisco activity is often the rate-limiting step in photosynthesis. Rates of photosynthesis increase linearly with increasing intercellular CO_2 concentrations (c_i) at low intercellular CO_2 levels, because RuP_2 levels are not limiting (Fig. 1.6). At higher intercellular CO_2 concentrations, the photosynthetic rate begins to level off as the capacity to regenerate RuP_2 fails to keep pace with the increased CO_2 supply. The regeneration of

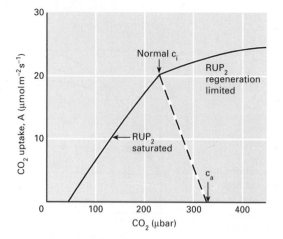

Fig. 1.6 The solid line represents the response of an intact leaf to differing intercellular CO_2 concentrations. This is the overall capacity of a leaf to take up CO_2 at high light levels for a given intercellular CO_2 concentration. The arrow on the curve indicates the transition between Rubisco-limited and RuP_2 regeneration limitations on overall CO_2 uptake rates. The arrow on the *x*-axis indicates the atmospheric CO_2 concentration. The dashed line emerging from this point has a slope equal to that of the leaf conductance (but negative in sign). The intersection of the solid and dashed lines is the operational intercellular CO_2 concentration. Under non-limiting light and water conditions, leaves often operate at the break-point (indicated by the arrow at the intersection) (Adapted from Farquhar & Sharkey 1982.)

RuP_2 is dependent on photochemical activity (electron transport and photophosphorylation), which means that photosynthetic limitations at low light intensities and high CO_2 concentrations are due to similar causes. Most leaves appear to operate at a stomatal conductance that maintains the c_i value in the vicinity of the break-point between Rubisco limitation and RuP_2 regeneration limitation (Farquhar & Sharkey 1982). This means that carboxylating capacity and electron transport capacity are co-limiting, and results in an economical investment by the leaf in the biochemical components of these processes.

Analyses of the carbon isotope discrimination (Δ) that continuously occurs during photosynthesis provides a long-term estimate of the intercellular CO_2 concentration in C_3 plants (Farquhar *et al.* 1989). There are two stable isotopes of carbon, ^{12}C and ^{13}C. While ^{13}C constitutes only about 1% of the total pool of these two isotopes, small changes in the isotopic composition of leaves can be determined through mass spectroscopy. Discrimination occurs whenever the isotopic composition of the source carbon (R_{air}) is different from that of the product (R_{plant}). We express this discrimination in parts per thousand (‰) as

$$\Delta = \left(\frac{R_{air}}{R_{plant}} - 1 \right) \times 1000‰ \qquad (1.3)$$

Where $R = {}^{13}C/{}^{12}C$. Because of mass differences, the two CO_2 mole-

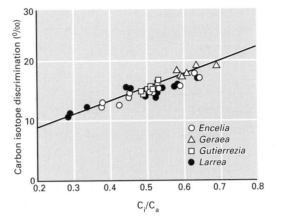

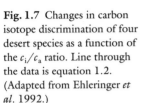

Fig. 1.7 Changes in carbon isotope discrimination of four desert species as a function of the c_i/c_a ratio. Line through the data is equation 1.2. (Adapted from Ehleringer *et al.* 1992.)

cules will have different diffusivities in air, with a, the discrimination associated with slower diffusivity of $^{13}CO_2$, equal to 4.4‰. A second discrimination step (b) of 27‰ occurs during photosynthesis because Rubisco favours $^{12}CO_2$ over $^{13}CO_2$. However, the overall discrimination during photosynthesis depends on c_i as

$$\Delta = a + (b - a)\frac{c_i}{c_a} \tag{1.4}$$

where c_a is the atmospheric CO_2 concentration. The result of this discrimination provides valuable information on plant–environment relationships. For example, as stomata regulate inward CO_2 diffusion during drought or other stress conditions, c_i values will change and this is faithfully recorded in the organic material produced as shown in Fig. 1.7.

Water stress will induce a stomatal closure, impeding the inward CO_2 diffusion and therefore reducing c_i values; as a consequence photosynthesis is also decreased (Fig. 1.6). Water stress can develop slowly as a soil dries out or may exist as steep gradients along a transect (Fig. 1.8). In either situation, the closure of stomata reduces photosynthesis first by decreasing c_i values; secondarily, there may be a downward regulation of photosynthetic capacity (Farquhar & Sharkey 1982).

1.3.3 Temperature

Photosynthesis increases with temperature because an increase in enzymatic activity leads to an enhanced capacity to bind CO_2. At high temperatures, diffusion of CO_2 and photorespiration become limiting and the temperature response levels off. Finally, at extreme temperatures, the integrity of the photosynthetic system begins to break down and rates begin to decrease. At the highest temperatures, the decline may be irreversible. The nature of the decline depends on the structure of the membrane lipids, and the composition of these lipids varies from species to species (Berry & Björkman 1980).

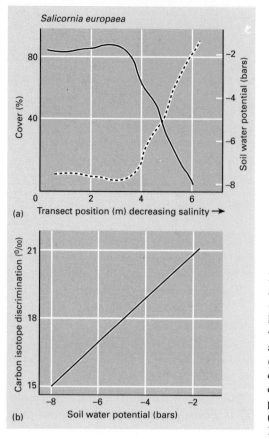

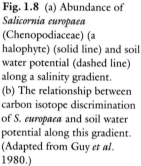

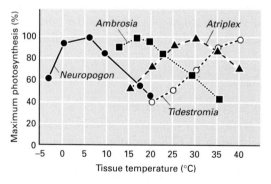

Fig. 1.8 (a) Abundance of *Salicornia europaea* (Chenopodiaceae) (a halophyte) (solid line) and soil water potential (dashed line) along a salinity gradient. (b) The relationship between carbon isotope discrimination of *S. europaea* and soil water potential along this gradient. (Adapted from Guy *et al.* 1980.)

Plants vary greatly in their photosynthetic response to temperature, depending on the kind of conditions they experience in their natural environments (Fig. 1.9). For example, some desert perennials have thermal optima of more than 40 °C, whereas Antarctic lichens have their optima close to freezing point. Plants that occur in the same

Fig. 1.9 Temperature photosynthetic response curves of plants from dissimilar habitats. Curves from left to right are for *Neuropogon acromelanus*, an Antarctic lichen (Lange & Kappen 1972); *Ambrosia chamissonis* (Asteraceae), a cool coastal dune plant (Mooney *et al.* 1983); *Atriplex hymenolytra* (Chenopodiaceae), an evergreen desert shrub and *Tidestromia oblongifolia* (Amaranthaceae), a summer active desert perennial (Mooney *et al.* 1976).

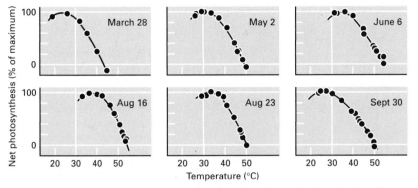

Fig. 1.10 Changes in the photosynthetic temperature optima of irrigated apricot *Prunus armeniaca* (Rosaceae) with seasons. (Adapted from Lange *et al.* 1974.)

habitat may also have different seasonal growth patterns which are associated with differences in temperature-related photosynthesis (Kemp & Williams 1980).

Many plants have been shown to have leaves that adjust their photosynthetic capacity with the changing seasons in an apparently adaptive manner (acclimation). A good example of this is the changing optimum temperature of photosynthesis in leaves of a Negev Desert plant, which experiences temperatures that fluctuate widely throughout the year (Fig. 1.10). Those species that show the most complete acclimation are those which have no reduction in photosynthetic capacity through the year (although their thermal optimum may shift).

1.3.4 Photosynthesis with respect to water use

An inevitable consequence of the expansion of plants onto land was the trade-off between controlling water loss while enhancing CO_2 uptake. In order to fix carbon, plants must lose water, simply because water vapour and CO_2 diffuse through the same stomatal-pore pathway. While stomata are open and CO_2 is diffusing inward to the sites of fixation, water vapour is diffusing outward into the drier atmosphere. This loss of water is directly proportional to the water vapour concentration gradient between the leaf and the atmosphere expressed as a water vapour mole fraction gradient (V), multiplied by the stomatal (g) and boundary layer (g_b) conductances (just as CO_2 uptake is proportional to the CO_2 concentration gradient across the stomata multiplied by similar conductances; see below). The water vapour gradient across the stomata is about 200-fold greater than the CO_2 gradient and usually changes during the course of the day, due largely to changing temperature and the associated effects on vapour pressure. The CO_2 gradient also changes during the day as stomata open at sunrise, partially closed during the driest midday periods, and then completely closed at sunset.

Water loss from stomata is regulated in the short term in response to

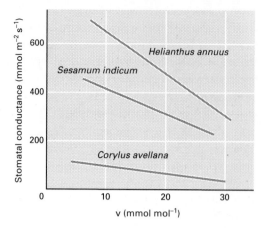

Fig. 1.11 Short-term responses of stomata to changes in the evaporative gradient between the leaf and air by three species, representing a range of life forms. *Helianthus* (Asteraceae) is an annual herb, *Sesamum* (Pedaliaceae) is a herbaceous oil-seed plant and *Corylus* (Betulaceae) is a small deciduous tree. (Adapted from Schulze & Hall 1982.)

diurnal changes in the water vapour gradient between the leaf and the atmosphere (Fig. 1.11), and in the long term in response to changes in available soil moisture levels (Fig. 1.12). These changes in stomatal conductance appear related to the loss of bulk leaf water (reduced leaf water potential) and to hormonal signals received from the roots (Davies & Zhang 1991). As a result of these combined responses, plants appear to be able to regulate the rates of water loss during dry or stressful periods; this has led to the notion of water-use efficiency, which is defined as the ratio of photosynthesis to transpiration rates.

There are times of the day, especially mornings, when the amount of water lost per unit of carbon fixed is rather low (i.e. water-use efficiency is high). It has been proposed that plants 'manage' stomatal conductance in such a way as to optimize this relationship (Cowan 1977), and, particularly during periods of drought, plants may close their stomata during the midday period when the water loss to carbon gain would be most unfavourable (Tenhunen *et al.* 1981).

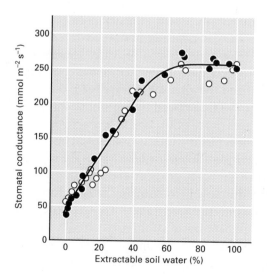

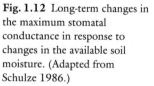

Fig. 1.12 Long-term changes in the maximum stomatal conductance in response to changes in the available soil moisture. (Adapted from Schulze 1986.)

Water-use efficiency can be measured at any given moment by determining photosynthesis (A) and transpiration (E) simultanously. A more important ecological measure, however, may be the lifetime amount of carbon gained versus water lost. Such measures are difficult to make, but there is an indirect way of evaluating water-use efficiency based on carbon isotope analysis. This arises because

$$A = (c_a - c_i)g/1.6 \qquad (1.5)$$

and

$$E = vg \qquad (1.6)$$

with water-use efficiency (A/E) then equal to

$$A/E = \frac{c_a\left(1 - \dfrac{c_i}{c_a}\right)}{1.6v} \qquad (1.7)$$

If we assume equivalent leaf temperatures among the plants under comparison, any denominator differences disappear and all differences in water-use efficiency are based on differences in c_i/c_a, which is measured by Δ as previously shown in Fig. 1.7.

When variations in carbon isotope discrimination among species in a habitat are examined, there are distinct patterns associated with life history. Longer-lived plants exhibit lower discrimination values than shorter-lived plants (Ehleringer & Cooper 1988). Within a semi-arid grassland community, annual plants discriminate more than do perennials; among the perennials, forbs discriminate more than do grasses (Fig. 1.13). These patterns of carbon isotope discrimination may relate to overall differences among the different life forms with respect to water stress tolerance and the nature of xylem architecture (especially susceptibility to cavitation; see Chapter 2).

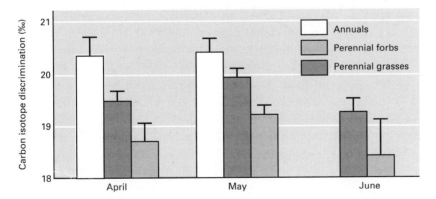

Fig. 1.13 Differences in the carbon isotope discrimination by annual plants, perennial grasses and perennial forbs in a semi-arid g... ... s... in three different months. The annuals had died by June. (Adapted from Smedley *et al.* 1991.)

1.3.5 Energy balance considerations

Transpirational cooling is an important mechanism for reducing leaf temperatures. As noted in section 1.3.1, leaves generally absorb a high fraction of the usable radiation between 400 and 700 nm, and reflect a high fraction of the non-photosynthetic radiation of wavelengths greater than 700 nm (Fig. 1.4). Under these conditions, a high transpiration rate will cool the leaves by offsetting the high solar radiation load on the leaf. In environments where radiation is abundant but water is limited, leaf reflectance may also be high in the visible wavebands especially as water stress develops (Fig. 1.14). This change in leaf absorptance is an advantage since stomatal closure under drought conditions can lead to detrimentally high leaf temperatures. In such arid environments, the loss of potential absorbed radiation is of little consequence to carbon gain, because photosynthesis is generally water-limited rather than light-limited.

There are alternative mechanisms for reducing leaf temperature under periods of water deficit other than through the use of a reduced leaf absorptance. Increased vertical leaf orientation occurs in many species under long-term drought periods and midday 'leaf wilting' occurs in many plants such as cassava and sunflower. Active leaf movements, whereby leaf lamina are oriented parallel to the sun's rays through the day, is called paraheliotropism and commonly occurs among legume species. All three mechanisms serve to reduce the heat load on the leaf by reducing the incident solar radiation. To reduce the leaf boundary and increase convectional heat exchange, leaves produced under periods of water deficit or in high solar radiation environments often are smaller and/or deeply lobed.

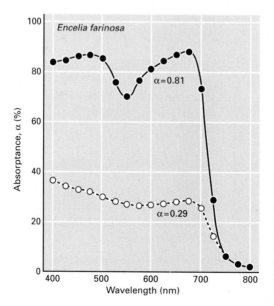

Fig. 1.14 Leaf absorptances of *Encelia farinosa* (Asteraceae), a desert shrub, collected early (high absorptance) and late (low absorptance) in the growing season. (Adapted from Ehleringer & Björkman 1978.)

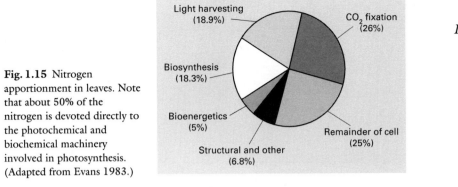

Fig. 1.15 Nitrogen apportionment in leaves. Note that about 50% of the nitrogen is devoted directly to the photochemical and biochemical machinery involved in photosynthesis. (Adapted from Evans 1983.)

1.3.6 Nutrients

Nutrients can affect photosynthetic performance in a relatively direct fashion, since nitrogen and phosphorus are both involved in the photosynthetic reactions. Alternatively, nutrients may act indirectly through their effects on the overall metabolic environment. Direct effects are most conspicuous for nitrogen. Of the nitrogen found in a leaf, a large fraction is contained in the carbon-fixing enzyme Rubisco (Fig. 1.15). It is not surprising, therefore, that there is generally a strong positive correlation between photosynthetic capacity and leaf nitrogen content (Fig. 1.16). This relationship only holds, however, if other factors such as solar radiation are not limiting. As nitrogen content is reduced, so the amount of carbon fixed per molecule of nitrogen present in the leaf is reduced, since a fraction of the nitrogen is involved in processes other than carbon fixation (Fig. 1.17).

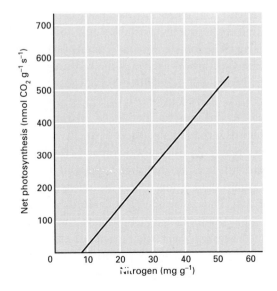

Fig. 1.16 Light-saturated photosynthetic rates as a function of leaf nitrogen content for leaves of plants from a variety of habitats. (From Field & Mooney 1983.)

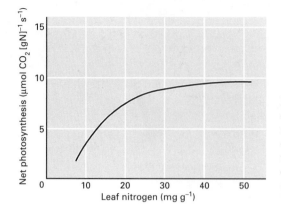

Fig. 1.17 Nitrogen-use efficiency (carbon gained/unit nitrogen) versus leaf nitrogen content for a variety of plant types. At low nitrogen contents, the efficiency is low. (From Field & Mooney 1986.)

1.3.7 Atmospheric pollutants

Many atmospheric pollutants affect the photosynthetic rate of leaves because they enter directly into the leaf mesophyll by the very same pathway as CO_2. The nature of their effects may be complex, and can differ from one pollutant to another. Often, as in the case of SO_2, the effect of the pollutant is to reduce photosynthesis through its effect on reducing leaf conductance. Photosynthesis is reduced because of a reduced c_i value and the impacts of varying pollutant concentrations can be measured over extended time periods through either leaf or tree-ring analyses (see Chapter 17).

1.4 Seasonality of photosynthesis

1.4.1 Individual leaves

The amount of photosynthesis that a leaf performs during its lifetime depends on: (i) its intrinsic photosynthetic capacity; (ii) the amount of limiting resources available; and (iii) how long the leaf stays on the plant. The longer the leaf remains productive, the greater its return (in terms of carbon fixed) on the resources invested to build and maintain it. Plants will generally maintain leaves as long as they are providing a positive carbon input.

In addition to the direct relationship between leaf duration and photosynthetic gain, there is an inverse relationship between leaf longevity and photosynthetic capacity (Fig. 1.18). Leaves of plants with fast growth rates do not live long, but they have a high photosynthetic capacity and a low water-use efficiency as discussed earlier in section 1.3.4. Thus, plants with different leaf-durations (e.g. an evergreen tree and a deciduous tree) could have similar integrated carbon gains.

The photosynthetic capacity of a leaf changes with age as well as with the seasons. The leaves of most plants go through a predictable change in their photosynthetic capacity as they age. The highest rates are

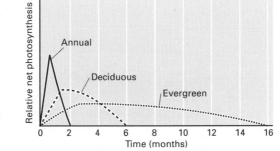

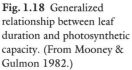

Fig. 1.18 Generalized relationship between leaf duration and photosynthetic capacity. (From Mooney & Gulmon 1982.)

attained prior to, or near, the period of maximal leaf expansion, after which time fixation capacity begins to decline. It is interesting that leaves with basal growth, such as grasses, do not appear to show this decline (Sestak 1985).

In slower growing plants, the life of individual leaves is prolonged, and leaves may remain active throughout different seasons or over several years. Such leaves have a changing photosynthetic capacity through time, and exhibit reduced (or no) capacity during winter or in periods of drought (Fig. 1.19). Evergreen leaves in severely cold climates, for example, may completely lose their capacity for photosynthesis in the winter, even under temporally favourable conditions (Larcher & Bauer 1981). Evergreen leaves of plants in less severe environments, however, maintain their competence, and thus have the potential to fix carbon throughout the year.

1.4.2 Whole plants

The productive potential of a plant through the seasons depends on the behaviour of its entire population of leaves. There is remarkably little information available on the seasonal pattern of change in leaf number on plants of different life histories. In certain perennial plants, one crop of leaves is produced per year, and all leaves appear and are lost more or less synchronously. In other species, leaves may be produced and lost throughout the growing season, with mean leaf longevity changing from one season to another. In the case of evergreen plants whose leaves last several years, leaf production may occur continuously or in simultaneous bursts. The net result of these diverse patterns in the birth and death rates of leaves, and of their changing photosynthetic capacity, is that whole plant photosynthesis shows pronounced seasonal changes (Constable & Rawson 1980).

The net photosynthesis of entire leaf canopies can be conveniently described in terms of the leaf area index (LAI); this is the area of photosynthetic surface per unit area of ground. Early in the growing season of an annual crop, for example, LAI will be much less than 1 and there is substantial bare ground. Later in the season LAI may reach

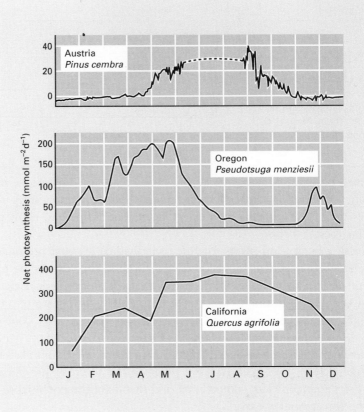

Fig. 1.19 Seasonal course of photosynthesis of evergreen trees from three different climatic regions. Top, A timberline tree, *Pinus cembra*, growing at 1980 m in the Swiss Alps. The needles have no photosynthetic capacity during the winter months. (From Hassler, upublished data.) Centre, Simulated net photosynthesis of *Pseudotsuga menziesii* (Pinaceae) in Oregon. Summer drought limits photosynthesis. (From Emmingham 1982.) Bottom, Simulated photosynthesis of the evergreen oak *Quercus agrifolia* (Fagaceae) in central California. These plants have adequate water even during the summer drought because of a deep root system. (From Hollinger 1983.) Simulations are based on site climatic features in combination with plant physiological responses to environmental parameters.

values of 6 or more. In both agronomic and natural canopies, the distribution of LAI is not random, but rather leaves are distributed in the upper portions of the canopy with those leaves at the top having the steepest leaf angles (Fig. 1.20). This allows more of the lower leaves to remain above their light-compensation points. Leaves lower in the canopy are in a decreased light environment and nitrogen is allocated in leaves through the canopy in such a way as to maximize canopy productivity (Fig. 1.20). Of course, water can play a significant role in determining canopy development and the amount of water available for transpiration places an upper limit on the realized maximum LAI for any canopy (Fig. 1.21).

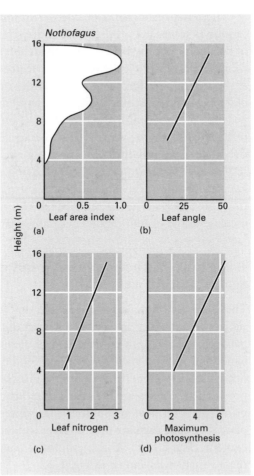

Fig. 1.20 Integration of canopy architecture to maximize canopy-level photosynthesis in *Nothofagus solandri* (Fagaceae) forests of New Zealand. (a) The distribution of leaf area index by layer. (b) Distribution of leaf angle from the horizontal (degrees). (c) Distribution of leaf nitrogen (% dry weight). (d) Maximum photosynthesis by layer ($\mu mol\ m^{-2}s^{-1}$). (Based on data from Hollinger 1989.)

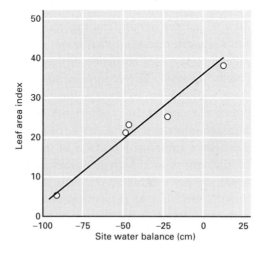

Fig. 1.21 Relationship between site water balance and leaf area index for five forest zones in western Oregon. (Adapted from Gholz 1982.)

There is a reallocation of nitrogen resources among leaves within a canopy as leaves get progressively more shaded at the bottom. The rate of decline of photosynthesis in individual leaves is correlated with many factors, but is most clearly related to the growth rate of the plant. Leaves at the top of a fast-growing plant rapidly overtop the older leaves. The older leaves thus become light-limited and the repayment in carbon gain per unit of nitrogen invested declines. Limiting nitrogen may then be reallocated to the newer leaves at the apex. Photosynthetic capacity of the older leaves is reduced by exporting nitrogen and their maintenance respiratory losses are decreased by having reduced leaf protein levels. Eventually, the older, shaded leaves form an abscission layer and fall from the canopy. The consequence of this reallocation of nitrogen is an increase in the overall canopy net photosynthetic rate.

1.5 Photosynthetic capacity and defence against herbivores

The photosynthetic capacity of leaf and the probability of it being subject to herbivory are related through the strong positive correlations between leaf nitrogen content and photosynthetic capacity on the one hand, and leaf nitrogen content and food quality for herbivores on the other (see Chapter 10). A specific illustration of this relationship is provided by the shrub *Diplacus aurantiacus* (Scrophulariaceae) and its principal herbivore, the butterfly *Euphydryas chalcedona*. Because the photosynthetic capacity of sun leaves is directly related to their nitrogen content (Fig. 1.16), we can employ a common axis to describe both these variables, as shown in Fig. 1.22.

On an artificial medium (and, presumably, in nature) larval growth is positively related to nitrogen content. Leaves of *Diplacus*, however, contain a phenolic resin called diplacol that can constitute as much as 30% of leaf dry weight, and this resin effectively inhibits larval growth. Thus one can readily see that leaves could discourage herbivores in two different ways: by reducing nitrogen content, or by increasing the content of resin. The economics of these two options can be assessed from a knowledge of the amount of photosynthate (or other limiting resources) needed to produce the resin. If a plant maintains a low nitrogen content it will suffer a reduced photosynthetic capacity and it will take a substantial fraction of its lifetime carbon gain in order to produce the resin in high quantities. In contrast, with high leaf nitrogen levels, the payback time is relatively short. From this kind of analysis one might conclude that the optimum leaf would have high leaf nitrogen and resin levels, since both photosynthetic capacity and herbivore protection would be greatest under these conditions.

The interrelationships between photosynthetic capacity and the direct and indirect costs of defence can be evaluated in more general terms

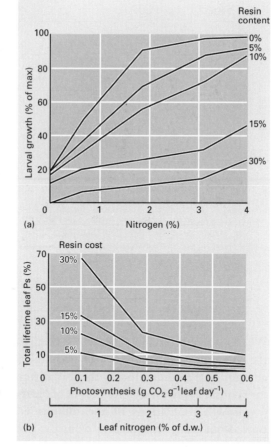

Fig. 1.22 (a) Nitrogen–resin interaction in relation to the growth of the larvae of *Euphrydras chalcedona*, the principal herbivore of the shrub *Diplacus aurantiacus*. (From Lincoln *et al.* 1982.) (b) The fractional cost of leaf lifetime-acquired carbon. (From Mooney & Gulmon 1982.)

by calculating the direct costs (in carbon units) of making a particular defensive compound, and the indirect costs in terms of loss of future carbon gain (the opportunity cost of investing in that compound). The increase in dry weight of a leaf is determined by: (i) its CO_2 fixation rate (A); (ii) the conversion efficiency of making dry matter from CO_2 (k) (iii) the allocation of acquired assimilate to new leaf material (L); and (iv) the initial amount of leaf material present (W_1) as

$$\frac{dW_1}{dt} = W_1 \cdot A \cdot L \cdot k \qquad (1.8)$$

The direct cost of making a defensive compound (C) can be subtracted directly from the photosynthesis term to indicate the impact of defence cost on future carbon gain as

$$\frac{dW_1}{dt} = W_1 \cdot (A - C) \cdot L \cdot k \qquad (1.9)$$

The direct costs are then calculated as

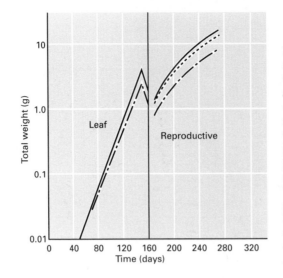

Fig. 1.23 Growth simulations for leafy tissue of the California annual *Hemizonia luzulaefolia* (Asteraceae) assuming no production of resin (solid line), production throughout the life of the plant (dash/dot), or production during reproduction only (dashed). The reduction in dry weight at initial reproduction is due to basal mesophytic leaf loss associated with the annual drought. (From Gulmon & Mooney 1986.)

$$C(t) = SC\left(\frac{\mathrm{d}\,\frac{S(t)}{W(t)}}{\mathrm{d}t}\right) \qquad (1.10)$$

Where SC is the specific cost of the compound expressed as weight of the compound per weight of CO_2 required to produce it, and $S(t)$ is the amount of the compound in the leaf at time t. Specific costs appear to vary considerably for different types of leaves (Table 1.1).

Growth simulations of the indirect costs of defence have been carried out using information from the herbaceous annual *Hemizonia luzulaefolia* (Asteraceae) (Fig. 1.23). This plant produces a resinous compound on cauline leaves at the beginning of reproduction, at which time the basal leaves are lost with the onset of the dry season. During the annual drought *Hemizonia* is one of the few plants with green tissue, and it suffers enhanced herbivory in consequence. The production of

Table 1.1 Costs of construction of leaves and various leaf constituents presumed effective in herbivore defence. (From Gulmon & Mooney 1986.) See text for definitions.

Type	Species example	Compound	Formula	Specific cost (SC) $(\text{g } CO_2\,\text{g}^{-1})$	Content (% leaf d.w.) $\left(\frac{S}{W}\right) \times 100$
Phenolic resin	*Diplacus aurantiacus*	Diplacol	$C_{22}H_5O_7$	2.58	29
Cyanogenic glucoside	*Heteromeles arbutifolia*	Prunasin	$C_{14}H_{17}NO_6$	2.79	6
Alkaloid	*Nicotiana tabacum*	Nicotine	$C_{10}H_{14}N_2$	5.00	0.2–0.5
Long-chain hydrocarbon	*Lycopersicon hirsutum*	2-Tridecanone	$CH_3(CH_2)_{10}$	4.78	0.9–1.7
Terpene array	*Salvia mellifera*	Camphor (50%) + others	$C_{10}H_{16}O$ + others	4.65	1.3
Whole leaves	Various shrub species			1.93–2.69	

resin has a different impact on biomass accumulation depending on whether it is produced throughout the life of the plant, or only during the reproductive period (as happens in nature). Clearly, if the probability of herbivory is low during the early growth stages, the plant might increase its fitness by delaying the elaboration of the defensive resins.

1.6 Variations on the basic photosynthetic pathway

So far, we have considered the basic mode of carbon fixation via C_3 photosynthesis. Several important variations on this basic metabolic theme allow improvements in photosynthetic efficiency in certain kinds of habitats. The most common variant is C_4 photosynthesis, so called because the initial products of CO_2 fixation are four-carbon organic acids, rather than the usual three-carbon PGA (Fig. 1.24). The carboxylating enzyme in C_4 photosynthesis is phosphoenolpyruvate carboxylase (PEP carboxylase). C_4 plants possess specialized cells that surround their vascular bundles, known as bundle sheaths, where chloroplasts operate in normal C_3 mode (Kranz anatomy). Four-carbon products of the initial CO_2 fixation are transported from the surrounding mesophyll cells to the bundle sheath cells where the CO_2 is released and then refixed by the C_3 pathway. The refixation of CO_2 in isolation in the bundle sheath acts as a CO_2 'pump', and overcomes the oxygenation reaction of Rubisco.

Another photosynthetic pathway, termed crassulacean acid metabolism (CAM), occurs in certain desert plants and tropical epiphytes. It is similar to C_4 photosynthesis, except that separate carboxylations take place within the same cells, displaced in time rather than in space as in C_4 photosynthesis. In typical CAM plants, the stomata open at night rather than during the heat of the day. CO_2 diffusing into the leaf at night is fixed in four-carbon organic acids through the use of stored energy (Fig. 1.24). During the day, while the stomata are closed, the stored CO_2 is refixed via the C_3 pathway using light energy. An essential

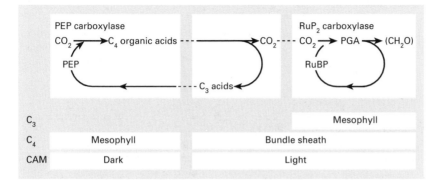

Fig. 1.24 Comparative features of C_3, C_4 and CAM photosynthesis. (Modified from Jones 1983.)

feature of this pathway is succulence or the possession of large vacuoles in which the organic acids can be temporarily stored. Large, columnar cacti in the deserts and orchids in the tropics are typical of the kind of plants that employ CAM photosynthesis.

Other CAM plants, particularly leafy succulents of the family Crassulaceae, exhibit flexibility in their mode of photosynthesis. During wet periods they fix carbon directly through the C_3 mode during the day, and at night they use the CAM mode. As drought sets in, they shift entirely to the CAM mode. This adjustment has become highly evolved in some South African plants such as *Frerea indica* (Asclepidiaceae), which has deciduous C_3 leaves and an evergreen succulent CAM stem.

1.7 Ecological consequences of different photosynthetic pathways

1.7.1 Water-use efficiency

The biochemical dissimilarity of C_3, C_4 and CAM plants results in dissimilar physiological behaviour and this, in turn, leads to different ecological performance (Table 1.2). Because they possess an effective CO_2 pumping mechanism, C_4 plants are able to saturate net photosynthesis at lower internal CO_2 concentrations than C_3 plants. This means that stress-induced stomatal closure tends to have a much greater effect on photosynthesis in C_3 compared with C_4 plants. While photosynthesis of C_3 and C_4 plants is differentially affected by stomatal closure, transpiration rate is not. This simple relationship has a profound influ-

Table 1.2 Comparative characteristics of the different photosynthetic pathways. (Modified from Jones 1983.)

	Pathway		
	C_3	C_4	CAM
Initial carboxylating enzyme	RuP_2	PEP	PEP, RuP_2
Tissue isotope range ($\delta^{13}C$, ‰)	-20 to -35	-10 to -20	Spans C_3–C_4 range depending on fraction of daytime versus night-time fixation
Anatomy	Normal	Kranz	Succulent
Water-use efficiency	Low	Medium	High
Photosynthetic capacity	Medium	High	Low
Oxygen inhibition of photosynthesis	Yes	No	Yes in day, no at night
Growth form occurrence	All	Shrubs and herbs	Succulents
Principal geographic range	Everywhere	Open tropical areas or arid habitats	Arid regions or habitats

PEP, phosphoenolpyruvate; RuP_2, ribulose bisphosphate.

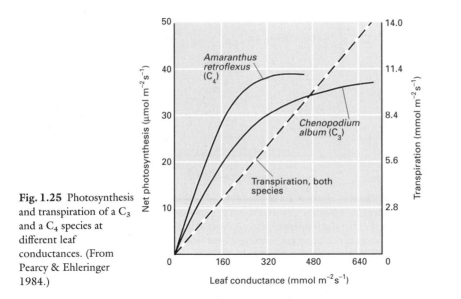

Fig. 1.25 Photosynthesis and transpiration of a C_3 and a C_4 species at different leaf conductances. (From Pearcy & Ehleringer 1984.)

ence on the instantaneous water-use efficiency (i.e. on the amount of carbon fixed per unit water lost). At an equivalent stomatal conductance, C_3 and C_4 plants will lose identical amounts of water but the C_4 plant will fix more carbon and so will have higher water-use efficiency (Fig. 1.25). One consequence of this is that C_4 species tend to become more common in saline habitats, where conservative water use may be of significant selective advantage.

CAM plants have an even higher water-use efficiency than C_4 plants because they only open their stomata at night, when evaporative water loss is minimal. CAM plants are typically found in the driest environments, such as coastal and interior deserts or as rootless epiphytes in mesic environments.

1.7.2 Significance of temperature

Another important physiological difference between C_3 and C_4 plants is their differential efficiency of fixing carbon at different temperatures when light intensities are low. As we saw in section 1.3.1, at low light intensities photosynthetic rate is directly proportional to the amount of light absorbed by the leaf (the quantum yield). C_4 plants have an intrinsically lower quantum yield because of the extra costs associated with this pathway (they require two additional molecules of ATP in order to regenerate PEP). At low temperatures (10–20 °C), the quantum yield of C_3 plants is as much as 30% greater than that of C_4 plants. However, this difference is offset in favour of C_4 plants at high temperatures, because C_4 plants lack photorespiration which reduces the quantum yield of C_3 plants at high temperatures. The result is a reduction of net photosynthesis in C_3 plants at higher temperatures because previously fixed CO_2 is lost.

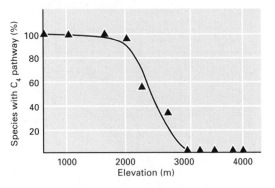

Fig. 1.26 Relative distribution of C_3 and C_4 grasses in relation to elevation on an African tropical mountain. The vegetation at the lowest elevations has all C_4 grass species and at the high elevations all C_3. (From Tieszen *et al*. 1979.) Similar distribution patterns along temperature gradients have been described in other parts of the world.

Although individual leaves on a plant may not be light-limited, whole plants in natural environments generally *are* light-limited because of carbon losses through respiration of non-photosynthetic tissues such as stems and roots. Thus, differences in quantum yields take on considerable significance in overall plant performance. From the quantum yield considerations just described, we would predict that C_4 plants would be most abundant in habitats where summer temperatures are high and moisture is available to support growth. This would include regions such as tropical and subtropical grasslands and summer-monsoon deserts. Similarly, we would predict that along elevational transects C_4 plants would be more frequent at lower elevation grasslands and C_3 plants more common at higher elevations (Fig. 1.26).

1.8 Climate change and photosynthesis

1.8.1 Photosynthesis in the recent past and near-future CO_2 environments

Atmospheric CO_2 concentrations have been measured continuously since the late 1950s (Fig. 1.27). These data show a clear trend of increasing CO_2 concentration in response to fossil fuels burned by humans. There is an annual oscillation in these data, reflecting natural ecosystem shifts between photosynthesis (reducing c_a) and decomposition (increasing c_a) in the summer and winter periods. In 1960, the mean atmospheric CO_2 concentration was 316 ppm. By 1990, it had reached 354 ppm, and by 2020 the atmospheric CO_2 concentration is expected to be approximately 600 ppm. From the air bubbles trapped during snow deposition in Arctic regions, we can examine the ice layers and reconstruct the history of atmospheric CO_2 concentration over a period of 160 000 years. This record shows fluctuations between 180 and 260 ppm, depending on whether it was a glacial or interglacial

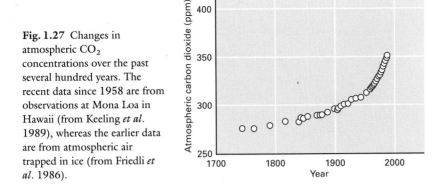

Fig. 1.27 Changes in atmospheric CO_2 concentrations over the past several hundred years. The recent data since 1958 are from observations at Mona Loa in Hawaii (from Keeling *et al.* 1989), whereas the earlier data are from atmospheric air trapped in ice (from Friedli *et al.* 1986).

period, but the concentration was always lower than present-day levels. In recent times, it is only since the Industrial Revolution that the earth has seen an increase in atmospheric CO_2 (Fig. 1.27).

There is geological evidence, however, which indicates that atmospheric CO_2 concentrations were much higher in previous geological periods, and that it is only since the Miocene that atmospheric CO_2 concentrations reached levels as low as they are today (Berner 1991). Indications are that atmospheric CO_2 concentrations may have been as high as 1200–2800 ppm during the Cretaceous. By comparison, the CO_2 levels during the Pliocene and Pleistocene were so much lower (180–280 ppm) that we may consider that plants today are relatively CO_2 starved.

What happens to photosynthesis under these different ambient CO_2 levels and what ecological changes are likely to occur in the future? Several common patterns are observed when plants are grown under elevated CO_2 conditions. First, plants invest significantly less nitrogen in leaves, which results in a CO_2 dependence curve (Fig. 1.6) that has a lower slope under elevated CO_2 conditions (Stitt 1991). However, since CO_2 concentrations are much higher, the photosynthetic rate is increased. Second, there is an increase in water-use efficiency. Since nutrient levels are not expected to increase under elevated CO_2 conditions, it is not clear how much overall canopy growth rates will increase (Norby *et al.* 1992). One consequence of a reduced nitrogen allocation to leaves is that they become less palatable to insect herbivores. Therefore animals would need to eat more leaf tissues to acquire the equivalent amount of nitrogen they obtain under present atmospheric conditions (Lincoln *et al.* 1993).

1.8.2 Climate change and the evolution of photosynthetic pathways

The C_3 pathway is ancestral, with the CAM and C_4 pathways having evolved after plants invaded land (Ehleringer & Monson 1993). CAM

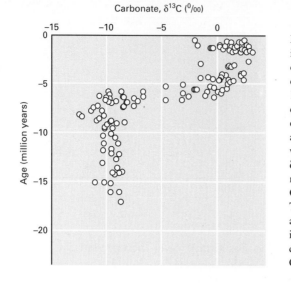

Fig. 1.28 The carbon isotope composition of carbonate nodules are enriched approximately 14–16‰ above that of the carbon isotope composition of plants growing in the soil at the time that carbonate was formed. The negative $\delta^{13}C$ values prior to 7.5 million years ago indicate a C_3-dominated vegetation. The positive $\delta^{13}C$ values after 7.5 million years ago indicate a switch to a C_4-dominated flora. (From Cerling *et al.* 1989.)

photosynthesis is found in diverse taxa in all the major extant groups and appears to have evolved in response to water-limited habitats, a condition that has existed since plants invaded land (Monson 1989).

The taxonomic and geographic distribution of C_4 plants suggests that this pathway evolved independently numerous times and over a short time period. This pathway is found in nearly two dozen diverse families, and intermediates on the evolutionary road to becoming C_4 are known from numerous other families. It has been suggested that C_4 photosynthesis evolved in response to the greatly reduced atmospheric CO_2 conditions that prevailed from the Miocene onwards (Ehleringer *et al.* 1991). At low CO_2 levels, reduced diffusion rates and increased photorespiration rates would have greatly decreased the carbon-gaining capacity of C_3 plants. The C_4 pathway has what amounts to a PEP carboxylase-driven CO_2 pump, and this would help to overcome the growth-limiting effects of a reduced CO_2 environment. Soil carbonate evidence indicates that C_4 plants first appeared about 7 million years ago, when atmospheric CO_2 levels were low, and that C_4 plants have persisted since that time (Fig. 1.28). As global atmospheric CO_2 increases in response to anthropogenic activity, CO_2 levels will soon reach a level where the C_4 pathway is no longer of advantage.

1.9 Conclusions

Photosynthesis is a central process in the functioning of all green plants. It provides the carbon skeletons and energy required to build biomass and to synthesize the wide variety of products utilized by plants in their metabolism. The basic chemistry of photosynthesis does not vary greatly among plant species, although there are three fundamentally different biochemical pathways for the process, each with different ecological consequences. The vast majority of the world's plants operate with the

C_3 pathway, where CO_2 is incorporated initially into the three-carbon product, phosphoglycerate. Plants utilizing a second pathway, CAM, are able to fix CO_2 into organic acids during the night, refixing it during the day into carbohydrate, and utilizing light energy even though their stomata are closed. CAM results in a high ratio of carbon gained to water lost, and is found typically in water-limited regions such as deserts and epiphytic habitats. A third pathway, C_4 photosynthesis, also results in efficient water use and may have evolved in response to a global reduction in atmospheric CO_2 levels beginning in the Miocene. In C_4 plants, high photosynthetic rate is gained through both anatomical and biochemical features that result in the maintenance of high photosynthetic capacity even whilst the stomata are partially closed. The C_4 pathway is found commonly in tropical and subtropical grasses and in plants of saline regions.

Photosynthesis is very sensitive to variations in the supply rates of light and CO_2, the principal resources utilized in the process. Photosynthesis is further influenced by a wide array of environmental factors including temperature, nutrients, tissue water status, and atmospheric pollutants. These factors have influences on different time scales because of their different rates of change in natural environments. For example, leaf temperature and the quantity of radiation absorbed by a leaf change greatly during the course of a single day, whereas tissue water and nutrient status change over longer time spans.

Although plants do not differ greatly in the basic machinery utilized in photosynthesis, they do differ radically in the ways they acquire the resources needed for the process. Dissimilarities exist among plants in the amounts, display and duration of their leaves, and these affect the total amount of light intercepted and hence photosynthate accumulated. They further differ in their photosynthetic responses to various environmental factors both in the short and long term. For example, species may differ in the amount of photosynthesis they perform at a given light level, as well as in the way they respond (acclimate) to a long-term change in the light environment. These differences are often the result (and may be the cause in some cases) of dissimilar patterns of resource acquisition by plants. Such differences between species presumably play a role in permitting their coexistence.

The photosynthetic capacity of plants is directly linked to their ability to acquire water, light and nutrients, and the process itself serves as an integrator of 'success' in a given habitat. Photosynthetic capacity, however, may also be related directly to potential rates of herbivory, since leaves that have high photosynthetic rates generally have high leaf protein contents, and this makes them additionally attractive to herbivores.

2: Plant Water Relations

John Grace

2.1 Introduction: water and life

2.1.1 Water as a physical and chemical medium

Water constitutes some 85–90% of the growing tissues of plants, and 5–15% of the mass of seeds. It is a solvent and the medium for biochemical reactions, and also the means of transport of many materials within the plant. It differs from other small molecules (e.g. NH_3, CH_4, NO_2) in being a liquid as opposed to a gas at 'normal' temperatures. This is because the H_2O molecule is strongly polarized, making groups of molecules cling together through weak hydrogen bonding (Fig. 2.1). The energy required to separate the molecules so that they can escape by evaporation is the heat of vaporization, λ. Its value for water is one of the highest known (2.48 MJ kg^{-1} at 20 °C), and as a consequence evaporation has a powerful cooling effect.

The hydrogen bonding between the oxygen and hydrogen of adjacent molecules of water accounts for the high tensile strength of water in capillaries, which can be as much as one-tenth that of copper wires of

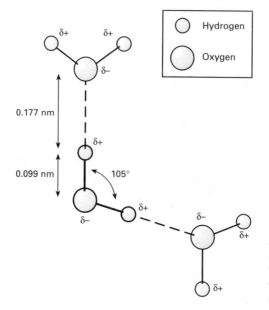

Fig. 2.1 Adjacent molecules clinging together as a result of electrostatic attraction between the positively charged hydrogen and the negatively charged oxygen. Many molecules are associated in this way as 'clusters' in liquid water and they form crystals in ice. (Redrawn from Jones 1992.)

similar diameter if the water is pure (Smith 1994). It enables water to exist as a continuum in the plant, and to be drawn through the roots, stem and leaves when evaporation at the leaf surface occurs.

Water is an excellent solvent for mineral salts. Ions such as Na^+ and Cl^- are attracted to each other. When placed in water, their attraction is reduced. This capacity of a solvent to influence the attraction between ions is expressed as the dielectric constant. The value for water is 80 whilst that of non-polar organic solvents is less than 2. Thus mineral ions needed for plant growth, and originating from the soil, accompany the transpiration stream to all parts of the plant, and become available for metabolic uptake in the growing regions of all organs. The solutes in the transpiration stream are not only the mineral elements. Most biochemically important substances in plants are charged, and readily soluble in water. Water molecules become associated with polar groups of organic biochemical constituents, including proteins, polysaccharides and other macromolecules as water of hydration.

The non-compressible nature of water makes it a useful hydroskeleton; leaves of many land plants owe their rigidity to the water pressure inside them, which is often in the range 0.3–1 MPa (3–10 atmospheres). This pressure is vital during cell expansion.

Other properties of water seem less useful. Firstly, it is more viscous than most other solvents as a result of the attraction between its molecules, which tends to prevent layers of the liquid sliding over each other. Thus the tensions during transpiration, when water is drawn through the fine capillaries that constitute the cellulose cell walls and xylem tissues of the plant, are considerable. Secondly, the rates at which oxygen and carbon dioxide diffuse in water are very low indeed, some 10 000 times slower than that in air, although both gases are sparingly soluble in water (Table 2.1), and this may have imposed limitations on the maximum thickness of leaves. However, vascular plants have

Table 2.1 Some properties of water in the range 0–40 °C.

		0 °C	10 °C	20 °C	30 °C	40 °C
S (m^3 gas m^{-3} water)	NH_3	1130	870	680	530	400
	CO_2	1.7	1.2	0.85	0.65	0.52
	H_2S	4.5	3.3	2.5	2.0	1.6
	N_2	0.023	0.018	0.015	0.013	0.011
	O_2	0.047	0.037	0.03	0.026	0.022
	SO_2	80	57	39	27	19
D ($mm^2 s^{-1}$)	CO_2	0.001 58	0.001 69	0.001 8	0.001 9	0.002
	O_2	0.001 76	0.001 88	0.002	0.002 1	0.002 2
	heat	0.127	0.135	0.144	0.153	0.162
ρ (kg m^{-3})		1000*	999.7	998.2	995.6	992.2
λ (MJ kg^{-1})			2.50	2.48	2.45	2.43	2.41
η (mPa m^{-1})			1.79	1.30	1.00	0.80	0.65

*Water reaches its maximum density at 3.98 °C.
S, solubility of gases; D, diffusivity of gases and heat; ρ, density; λ, latent heat of vaporization; η, viscosity.

evolved intercellular spaces in their leaves so that gases do not need to diffuse in water by more than the length of about one cell.

Finally, water participates directly and fundamentally in biochemical processes. In photosynthesis the water molecule is split, the H^+ being used to reduce CO_2 to carbohydrates $(CH_2O)_n$ whilst O_2 is evolved as oxygen gas. This is the origin of the oxygen in the earth's atmosphere (see Chapter 1).

2.1.2 State of water in the plant

The water content of a leaf or other plant organ is measured as the relative water content ($\Re$), which is the water content stated as a percentage of the maximum water content that the tissue is capable of holding.

$$\Re = 100 \, (M_f - M_d)/(M_t - M_d) \qquad (2.1)$$

where M_f is the mass of the plant material fresh from the plant, M_t is the mass when the material is fully hydrated by being placed in water in the dark until no further water can be absorbed (such a leaf is said to be fully turgid) and M_d is the mass after drying by removing all water in an oven at a reference temperature, often 80 °C. This index of tissue hydration is generally more useful than the water content stated as a percentage of the dry mass, as the latter is more sensitive to the varying amount of structural tissue, and the transient nature of storage materials such as starch. $\Re$ was originally devised for reporting the water content of leaves, but can also be used to report the water content of woody tissues (Sobrado *et al.* 1992).

The state of water in the plant is measured as the water potential (ψ), which is the difference in free energy ($J \, mol^{-1}$ or $J \, m^{-3}$) between the water under consideration and that of pure water at sea level. It is the work that would be required to move water from where it is, to the pure state at sea level. The water potential tells us about the tendency of the water to move one direction or another. Water always moves from high potential to low potential. For historical reasons, the units used are those of pressure, pascals (Pa), which are dimensionally the same as $J \, m^{-3}$ (see Kramer & Boyer 1995). Water potential of pure water at sea level is arbitrarily set to zero, and the water potentials in plant leaves are nearly always negative, often by as much as − 1 or − 2 MPa. In leaves, the water potential tends to be reduced by the presence of solutes, and increased by the force of the cell walls tending to squeeze the water from the cells. The cellulose walls are not rigid but elastic, and they exert their greatest pressure on the protoplast when the tissue is fully hydrated, and a declining pressure as water is lost from the system. Total water potential ψ_t is the sum of the solute potential ψ_s and the pressure potential ψ_p brought about by the wall pressure:

$$\psi_t = \psi_s + \psi_p \qquad (2.2)$$

The relationship between the water potential and the water content is very important. As the leaf loses water, the cells reduce in volume and the solutes become more concentrated (ψ_s declines). At the same time, the pressure exerted by the walls declines (ψ_p declines). The form of the relationship between water potential and water content is described as the Höfler–Thoday diagram (Fig. 2.2). In the one illustrated, a fully hydrated leaf has a total water potential of zero, composed of a pressure potential of 2 MPa exactly balanced by a solute potential of -2 MPa. The leaf wilts when the wall pressure is zero (by definition) and in this case wilting occurs at a relative water content of about 0.75.

The relationship between the water potential and the water content differs markedly between species, and may influence the ecological range of the species. For example, a tomato plant (an example of a mesophyte, a plant unable to grow in dry places) may show a small decline in water potential for a particular decline in relative water content, but an acacia (a xerophyte, normally growing in dry places) shows a relatively large decline whilst still maintaining a positive turgor. Thus, the xerophyte is more able to extract water from the soil, by virtue of its highly negative water potential, and thus is well suited for survival in dry soils.

Water potentials are routinely measured using a pressure chamber (Turner 1988). Leaves are cut from the plant with a sharp blade and placed inside a pressurized vessel with their cut petioles protruding (Fig. 2.3). On cutting, the water meniscus retreats into the cut end of the xylem and the cut surface appears very dry when viewed with a hand-lens. Pressure is applied by adding nitrogen gas to the chamber, squeezing the leaf until water begins to exude from the cut surface of the petiole. This pressure (the balancing pressure) is equal and opposite to the water potential. An alternative technique using a thermocouple psychrometer gives very similar readings, and both support the classical

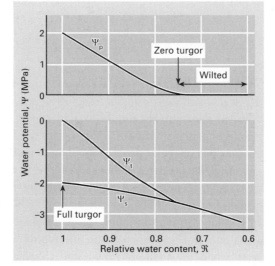

Fig. 2.2 Example of a Höfler–Thoday diagram. As the leaf loses water ($\Re$ declines), the pressure on the protoplasts from the cell wall (ψ_p) declines, finally reaching zero when the leaf is observed to be in a wilted state. The solute potential ψ_s declines as a result of the shrinkage in the protoplast. The total water potential ψ_t is the sum of ψ_p and ψ_s. (Redrawn from Jones 1992.)

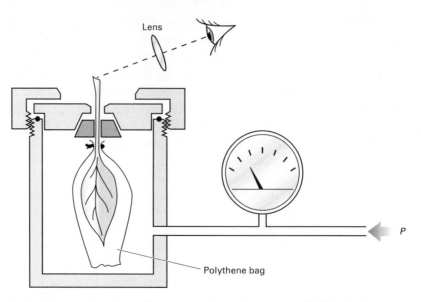

Fig. 2.3 Pressure chamber used to measure the total water potential. The chamber is pressurized until water exudes from the cut surface of the petiole (the balancing pressure). (Reproduced with permission from Jones 1992.)

view (the cohesion theory of water transport) that water in the stem is under considerable tension when the plant is actively transpiring.

The range of water potentials usually found in plants varies on a diurnal cycle. Immediately before dawn, vascular plants are in a relatively hydrated state, and typically ψ_t is between zero and -0.05 MPa. At dawn, transpiration begins and ψ_t falls to a minimal value soon after noon. The relationship between transpiration rate and water potential is usually almost linear (Kozlowski 1981). Minimum water potentials recorded in vascular plants vary from -1.0 MPa in wetland herbs to -6.0 MPa in some desert shrubs (Scholander *et al.* 1965; Larcher 1995).

2.1.3 Acquiring and conserving water on land

Life on land poses great challenges to plants. In average conditions prevailing at the earth's surface in the summer, evaporation rates from a free water surface are in the range 5–8 mm daily, and so a wet thallus dries up very quickly unless it is in touch with moist ground. Many primitive plants are described as poikilohydric, and simply tolerate large fluctuations in their hydration. These include some bacteria, fungi, terrestrial algae, most lichens, and the gametophytes of mosses and ferns. A small number of higher plants are also poikilohydric for much of their life cycle (Tuba *et al.* 1993), and most are poikilohydric as seeds and pollens. In poikilohydric leaves the photosynthetic capacity declines as water is lost from the tissues, and recovers after days, months or even years when the tissue is rehydrated (Bewley 1979; Tuba *et al.* 1993;

Larcher 1995). Poikilohydric vegetation is important over much of the land surface, for example the lichen crusts of hot deserts (Lange *et al.* 1994) and the lichen mats of the tundra. In former times, before the evolution of vascular plants, such mats may have had much higher rates of photosynthesis than they do today, as the CO_2 concentration was higher then. Poikilohydric photosynthetic mats are among the oldest plant communities known: the cyanobacterial colonies known as stromatolites were the first photosynthetic organisms some 3.5×10^9 years ago (Raven 1995). Much later, about 0.4×10^9 years ago in the Silurian, vascular plants began widespread colonization of the land, mainly as a result of having evolved superior control of water use.

The most important evolutionary innovation enabling plants to avoid internal drought was the cuticle. The leaves and stems of most land plants are covered by a waterproof film of chitinous material containing wax layers and coated with wax crystals. The structure is exuded over the leaf surface during leaf expansion, and in its final form is relatively impervious to water and CO_2, very resistant to chemical attack and moderately resistant to mechanical damage.

To enable some control over gas exchange between leaf and air, and especially to provide a port of entry for CO_2 during photosynthesis, the plant epidermis has pores, called stomata, typically $50–400\,\mu m^{-2}$. Stomata are variously arranged on the leaf surface, often sunk in pits or sheltered by a chimney (Meidner & Mansfield 1968; Fig. 2.4). Each stomatal pore has a diameter when fully open of about $20\,\mu m$. The apertures of the stomata are determined by pressure differences between the cells (guard cells) surrounding the pore and those in the rest of the epidermis. Stomatal pore size is under physiological control, allowing regulation of carbon gain and water loss and, in some cases, coarse control of surface temperature. Stomata are found in the sporophytes of mosses, pteridophytes, gymnosperms and angiosperms.

Natural selection on land has favoured tall plants; they outgrow their neighbours by shading them, and are thus more effective in acquiring resources during the struggle for existence. Moreover, they disperse their spores or seeds to the atmosphere more effectively. However, the penalty of being a tall plant is the associated cost of transporting water and nutrients from the soil, and moving assimilates from the leaves to wherever they are needed. Water transport requires pipework made from specialized xylem cells; these are cells that have died, leaving only the hard exterior walls. Primitive land plants from the Silurian, such as *Rhynia* and *Cooksonia*, had a vascular system made of stiff lignified tubes (tracheids), which provided both water transport and the mechanical support for a tall structure. Tracheids are spindle shaped and joined through valves called pits (Fig. 2.5). Angiosperms have a mixture of tracheids and vessel elements, the latter being broader and wider, offering less resistance to water flow (Esau 1965; Zimmermann 1983). Tracheids and vessel elements have an extremely high elastic modulus

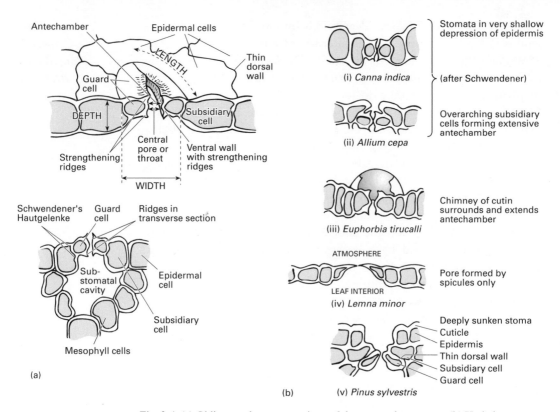

Fig. 2.4 (a) Oblique and transverse views of the stomatal apparatus. (b) Variations between species, all except the aquatic plant (iv) showing some increase in the pathway for water diffusion brought about by stomata being sunk or having a chimney. (Reproduced with permission from Meidner & Mansfield 1968.)

(i.e. they are very stiff) to avoid collapse under tensions during water transport.

Land plants living in regions of water shortage avoid desiccation and possess a large number of structural and life-history adaptations that are believed to conserve water (xeromorphic traits), a few of which are listed here:

1 highly reflective leaf surfaces brought about by wax deposits or reflective hairs;

2 stomata sunk into wells, or positioned within invaginations of the leaf itself;

3 reductions in leaf area per mass of plant;

4 leaves reduced in size;

5 deep roots, or a large mass of root per mass of shoot;

6 modifications of stem or root to form water-storage organs;

7 ability to shed leaves during the driest periods to avoid water deficits;

8 the annual life-style, where the plant survives dry periods as a seed, or perennial life-style, where the plant survives as a subterranean or otherwise much-reduced structure.

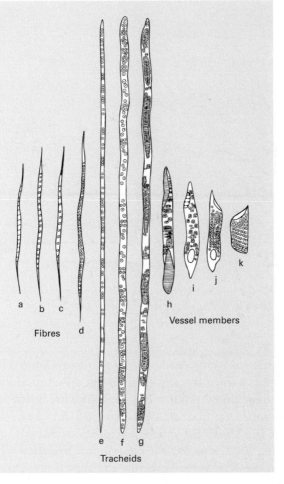

Fig. 2.5 Some cell types found in xylem: (a–d) are fibres; (e–g) are long *tracheids* from primitive woods showing the circular pits, through which water flows from one tracheid to another; (h–k) shows the postulated evolutionary sequence towards short broad *vessels* in angiosperm wood, which operate like drainpipes with water flowing mainly through the broad ends; the sequence (d–a) shows the evolution of *fibres* from tracheids. (From Esau 1965.)

For some of these characters (commonly 1, 3, 4 and 5) individuals of any single species display plasticity in response to drought, developing the characteristic or not depending on the water supply during growth.

2.1.4 Water as a limiting resource

Water and temperature are the two most important determinants of the vegetative cover of the land surface. The most productive biomes, and the ones with the largest biomass per area and the tallest plants, are those that are both warm and wet (Fig. 2.6). The hot dry deserts cover as much as 30% of the land surface, and are very unproductive unless irrigated.

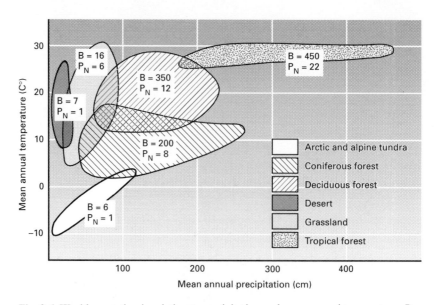

Fig. 2.6 World vegetation in relation to precipitation and mean annual temperature. *B*, biomass in dry mass ha^{-1}; P_N, net primary productivity in t ha^{-1} year^{-1}. Compiled from several sources.

On a world scale, crop production is frequently limited by the supply of water, a fact known to early civilizations, who introduced irrigation to enhance agricultural production. Arid lands are distributed in broad latitudinal zones centred at 30 °N and 30 °S, where the potential evaporation far exceeds the annual precipitation, and it is in these zones that irrigation has historically been developed. Irrigation was used 5000 years ago in the Nile Valley, utilizing the annual flood and the associated rich alluvial soil. The Bible records the consequent reliability of Egypt's irrigated agriculture in contrast to Israel's rain-fed agriculture (Deuteronomy 11: 10–17). Today, irrigation is important over 271 million ha of the earth's surface, distributed as follows: Europe, 29; Asia, 184; Africa, 13; North America, 34; South America, 9; Australasia, 2 million ha.

Even in northern climates, plant growth responds to irrigation. For example, 20-year-old Scots pine in Sweden showed a significant response to irrigation, and a substantial response in plots that had received fertilizer treatment (Linder & Axelsson 1982). Some degree of water deficit is certain to occur in all photosynthesizing land plants during the day, because of the frictional resistance offered by the path of water transport in the soil, root and stem.

The effects of water deficits on plants have been widely studied, and many are documented in Kramer (1980) and Kozlowski (1981). When water is withheld from higher plants, the pressure potential of the cells declines over a period of hours and days, and rates of cell expansion are consequently reduced. New leaves are thus smaller, with smaller cells.

Even a slight reduction in the hydration of the leaves is likely to influence biochemical processes such as protein synthesis, and thus influence cell division and growth rate. Rates of photosynthesis may decline, mainly as a result of stomatal closure. Over periods of weeks there is usually a considerable degree of acclimation, which tends to offset the impact of drought. For example, there is a profound change in the pattern of allocation, with an enhanced translocation to roots (Khalil & Grace 1992). In forest species, it is possible to observe the result of such effects acting over several years (Mencuccini & Grace 1995).

At the ecosystem scale, the vegetation of drought-stricken areas is typically sparse and sclerophyllous, with selection for drought-avoiding strategies as in ephemerals that complete their life cycle in a short growing season during the rains, and perennial species that shed leaves in the dry season. Competition for water often limits the extent of development of leaf area at a community level (Ehleringer 1984).

2.2 Transpiration rate

2.2.1 Energetics

Rates of transpiration E can be expressed as flux of water vapour per area of surface (mmol H_2O m^{-2} s^{-1} or mg H_2O m^{-2} s^{-1}), or as λE the energy flux accompanying the change in state from liquid to vapour, where λ is the latent heat of vaporization expressed per mole or per mass. For hydrological purposes, especially when dealing with stands of vegetation, it useful to express transpiration as an equivalent amount of rainfall (in mm).

The loss of water by evaporation is closely related to the flux of heat, as both are transferred between the surface and the atmosphere by turbulent diffusion through the air layers over the surface. In bright sunshine a dry surface will be warmer than a comparable wet surface. Vegetation presents a surface that is neither dry nor wet, and its energy balance in a particular environment, its temperature and rate of water loss, will depend on the extent of stomatal opening.

In the energy balance of leaves or stands of vegetation, λE is a major component along with the convective heat transport to the atmosphere C and the net radiation R:

$$R = C + \lambda E + P + S \qquad (2.3)$$

Photosynthesis P captures only a few percent of the energy available in sunlight, and the heat storage term S is important only when considering massive plant organs. In the day, R is the main *heat gain*, and C, λE, P and S are (usually) *heat losses*. Equation 2.3 thus simply states that energy is conserved. More rigorously, if we adopt a sign convention and say that heat gains to the plant are deemed positive, and heat losses are negative, the sum of all the terms is zero:

$$R + C + \lambda E + P + S = 0 \tag{2.4}$$

We can simplify the expression to say that P and S are negligible and the energy gain obtained by radiative transfer is partitioned between convection and evaporation:

$$R + C + \lambda E = 0 \tag{2.5}$$

The convection term C depends on the difference between the temperature of the surface and the air $(T_s - T_a)$ and also on the aerodynamic conductance (g_a), a measure of the ease with which heat can be lost from the leaf by turbulent diffusion and roughly proportional to the average length of the pathway for diffusion through the air layers that tend to cling to the surface.

$$C = g_a \, pC_p \, (T_s - T_a) \tag{2.6}$$

Where p is the density of air and C_p is the specific heat capacity of air at constant pressure. For leaves, the aerodynamic conductance g_a is a function of wind speed u and leaf dimension d (van Gardingen & Grace 1991):

$$g_a = f(u, d) \tag{2.7}$$

In an analogous fashion, the evaporative term λE depends on the difference between the concentration of water in the internal and external atmosphere $(\chi_i - \chi_a)$, and also on $g_a + g_s$ where g_s is the stomatal conductance, a measure of the ease with which water molecules can diffuse through the stomata. Usually, g_s is measured with a porometer, but it can also be calculated from the number and size of the stomata (van Gardingen *et al.* 1989).

$$E = (g_a + g_s)(\chi_i - \chi_a) \tag{2.8}$$

χ_a, the humidity of the air, is measured in the field with a psychrometer or hygrometer. The internal atmosphere of the leaf is assumed to be water saturated, and the water content of water-saturated air is a well-known function of temperature (see Monteith & Unsworth 1990 for an empirical equation):

$$\chi_i = f(T_s) \tag{2.9}$$

The energy balance equation has been used in many ways, and is the basis for the derivation of the Penman–Monteith equation frequently used by hydrologists and micrometeorologists. The approach can be applied at various levels of organization from leaves to ecosystems. At the ecosystem scale it is assumed that vegetation acts like a 'big leaf' characterized by appropriately scaled values of g_a and g_s. At the leaf scale it is possible to estimate the transpiration rate and temperature of the leaf in any environment by assuming typical values for g_a and g_s (van

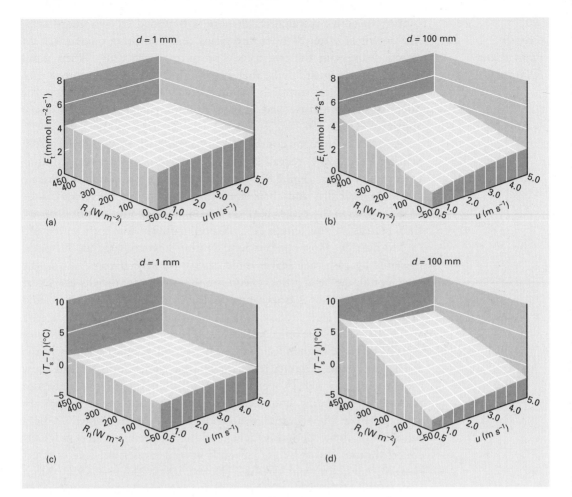

Fig. 2.7 Solution of the energy balance to explore the effect of leaf size (d) and environmental variables on the transpiration rate (a,b) and the difference between the surface and air temperature ($T_s - T_a$, c,d). R_n and u are the net radiation and wind speed respectively. In the calculation the stomatal conductance was 500 mmol m^{-2} s^{-1}, the air temperature was 15 °C and the saturation deficit of the air was 8 mmol mol^{-1}. This corresponds to a temperate environment. Other examples are given in van Gardingen & Grace (1991) from which these figures were obtained.

Gardingen & Grace 1991). This requires an iterative solution of the equations, in which the value of T_s is adjusted until the heat lost ($C + E$) is equal to the heat gained in radiation transfer (R). Some of the results of such calculations are presented in Fig. 2.7, showing a remarkably different response for large ($d = 100$ mm) versus small ($d = 1$ mm) leaves.

2.2.2 Stomatal conductance

Stomatal conductance is under physiological control, though the underlying mechanisms of this are still imperfectly understood despite over a century of active research. Each pore is bounded by two guard cells. Movement of water between the guard cells and other epidermal cells causes changes in the hydrostatic pressure of the cells. When water moves into the guard cells, they become inflated and the pore opens. Damage to the leaf surface by wind, particles or pathogens reduces the integrity of the cuticle and impairs the competence of the stomatal apparatus (Grace 1977).

A survey of stomatal conductance (Körner *et al.* 1979; Körner 1994; Schulze *et al.* 1994) reveals some major correlations between conductance and habitat (Fig. 2.8). Many of the highest conductances are found in ruderals, and their domesticated cousins, the crop plants. Many species in dry places have their stomata sunk into pits or grooves, thus increasing the length of diffusion pathway for both water and carbon dioxide. Most conifers have a rather low stomatal conductance, probably because their leaves must survive for several years, and in the boreal or alpine habitat they encounter seasonal water deficits every year when the soil water is frozen but the leaves are exposed to dry air and bright sunlight.

In the field, stomatal conductance is near zero at night, increases sharply over a few hours following sunrise, and declines abruptly in the afternoon. Some examples from the tropical rain forest of Brazil show these trends very well (Roberts *et al.* 1990). Conductance is highest at the top of the canopy, and declines in the shade within the canopy. In the laboratory the stomatal conductance is normally shown to be a function of light, humidity deficit of the air, temperature and CO_2 concentration, and there are currently several models relating conductance to environmental variables (Dewar 1995; Leuning 1995; Monteith 1995). It is also possible to demonstrate 'sleep movements' in which the stomata continue to show opening and closing responses for

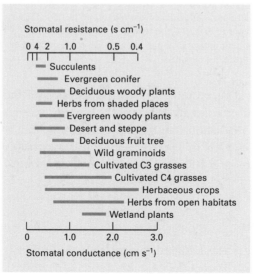

Fig. 2.8 Range of stomatal conductance in the leaves of vascular plants (and its reciprocal, stomatal resistance). (From Körner *et al.* 1979.)

several days even though exposed to continuous light or continuous darkness (Meidner & Mansfield 1968).

Afternoon closure of stomata, and sometimes midday closure and then reopening in the afternoon, is more pronounced in plants growing in dry soil, and was formerly considered to be the response to a decline in the water potential in the leaves. This is now believed to be less likely, as stomatal conductance in drought-affected plants is usually more closely related to the soil water potential than to the leaf water potential. It is now considered that stomata respond to chemical signals passing from the roots to the shoots (Gollan *et al.* 1985; Zhang & Davies 1989; Khalil & Grace 1993). Abscisic acid has been identified as the likely chemical signal, both in experimental studies indoors and in the natural environment (Gowing *et al.* 1993; Jackson *et al.* 1995a).

There is good evidence that mechanisms have evolved for the coarse control of leaf temperature by utilizing transpirational cooling. For example, the desert cucurbit *Citrullus colocynthis* has very deep roots and an exceptionally large number of stomata per area of leaf. It is one of a class of desert plants known as 'water spenders' because it spends water to keep cool (Lange 1959; Althawadi & Grace 1986). In this species the leaves are held near to the ground, where the air temperatures can rise above 50 °C at noon, which is well above the lethal temperature for the species. When the temperature of the leaf reaches 42 °C the leaves move to a vertical posture, thus reducing the heat load on them, and the stomata open very widely, thus cooling the leaf and maintaining it several degrees below its lethal point (Fig. 2.9). In other cases, leaf movement seems to be more passive and is more likely to be associated with stomatal closure. Many species show midday wilting, which effectively reduces R, the net absorption of energy (Chiariello *et al.* 1987).

The stomata of cacti and many succulent plants behave quite differently. The stomata stay closed during the day to conserve water. They open at night, and nocturnal CO_2 fixation into malic acid occurs through the activity of phosphoenolpyruvate carboxylase (PEP carboxylase). The malic acid is stored in the vacuoles of cells, and utilized during the day to synthesize sugar (Nobel 1991). This special adaptation to the desert environment (crassulacean acid metabolism, or CAM)

Fig. 2.9 Transpirational cooling in the desert plant *Citrullus colocynthis*. Graph shows air temperature (heavy line), leaf temperature (thin line) and temperature of another leaf which was excised to stop transpiration (broken line). The arrow shows when excision was carried out. (From Lange 1959.)

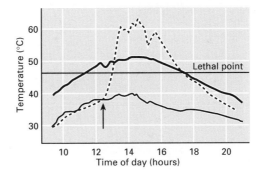

enables succulent plants to survive and grow in remarkably hot and dry conditions, but they do not usually grow fast (see Chapter 1).

2.3 Soil–plant–atmosphere continuum

2.3.1 Pathway

The detailed pattern of water movement in plant tissues, especially the root and leaf, remains an important subject for investigation. The extent to which movement in these tissues is symplastic (i.e. through the cytoplasm from cell to cell via the plasmodesmata, which are holes in the wall where strands of cytoplasm pass) or apoplastic (through the cell walls) is much debated in the literature, but will depend on which pathway offers the least resistance (Kramer 1983; Boyer 1985; Canny 1990, 1995). However, the general pathway is well established (Fig. 2.10). Water moves vertically upwards in the plant because the total water potential in the leaves becomes negative when the leaf loses water by transpiration, whilst the water potential in moist soil is close to zero. Water moves from the soil to the plant through root hairs and/or mycorrhizas, apoplastic and symplastic pathways in the root tissues (movement through the cytoplasm must occur at the endodermis because cell wall pathways are blocked by suberization), vertical movement in the xylem to the leaves, and both apoplastic and symplastic movement to the substomatal cavities.

The rate at which water moves cannot be predicted from the difference in water potential, unless the hydraulic resistance is known. This is a measure of the frictional resistance imposed by the water-conducting pathway, and depends on the numbers and diameters of water-conducting cells in the leaves and stems, and also on the root

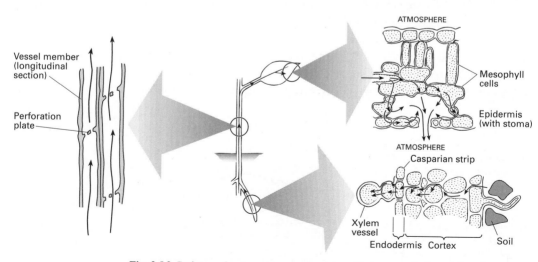

Fig. 2.10 Pathway of water movement in plants. (Redrawn and adapted from Jones 1992.)

structure. The concept of water moving from the soil to the atmosphere down a gradient of water potential by passing through a series of resistances was first proposed by van den Honert (1948) and is still widely used in plant ecophysiology and related fields such as micrometeorology (Monteith & Unsworth 1990; Jones 1992).

2.3.2 Pipe model of hydraulic architecture

The pipe model, first proposed by Leonardo da Vinci, is a very attractive representation of the plant as a hydraulic system. He observed that 'all the branches of a tree at every stage of its height, when put together, are equal in thickness to the trunk below them, (Richter 1970). This places important constraints on the form and carbon economy of plants, and has some biomechanical implications.

In the pipe model, a unit of leaf area is supplied by a unit area of water-conducting tissue. The relationship may vary for different species, because of differences in the microstructure and therefore the hydraulic conductivity of stem tissues. However, the relationship has been demonstrated many times and is linear (Fig. 2.11).

In Scots pine, *Pinus sylvestris*, the ratio of the total cross-sectional area of the stem and branches immediately above the first branch (A_1) to the stemwood area below the first branch (A_2) was examined in forests at different places in Europe. The fraction A_1/A_2 was not always 1.0 as Leonardo da Vinci had suggested, but depended on the availability of water at the site, estimated crudely as the potential evapotranspiration (Berninger *et al.* 1995). For wetter sites, A_1/A_2 was up to 1.8, and at dryer sites it was as low as 0.75. Berninger *et al.* (1995) concluded that the architecture of the tree acclimates to the site conditions, providing a larger hydraulic pathway in the canopy as the conditions become drier. Similar conclusions were derived from a related study examining the leaf area (A_L). The fraction A_L/A_2 varied markedly between Scots pine in the dry south-east of England and a population from the same seed origin growing at a wetter site in Scotland (Mencuccini & Grace 1995). At the drier site, A_L/A_2 was reduced and hence the tendency for leaves

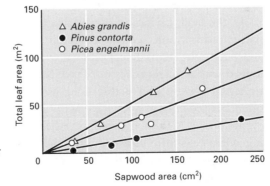

Fig. 2.11 Relationship between sapwood area and total leaf area for three conifer species. (Adapted from Waring *et al.* 1982.)

to become water stressed under the drier conditions was reduced. Indeed, leaves at the dry site exhibited water potentials no lower than those of leaves at the wet site (Mencuccini & Grace 1995).

2.3.3 How vulnerable is the pipeline?

A mechanical vacuum pump is capable of lifting water from a well to only 10 m as the difference between the pressure in the pump (assuming a vacuum) and the well cannot exceed 1 atmosphere (0.1 MPa). The leaf as a water pump does not face this limitation, as it relies on the capacity of an aqueous solution of solvents in the leaves to attract water (ψ_s). Trees in humid tropical forests grow to 50 m, but much less in the dry tropics and not at all in very arid regions. The limitation on the height of trees is probably imposed by the difficulty in maintaining such a long water-conducting pathway, or by some other constraint relating to transport over long distances (Mencuccini & Grace 1996).

As we have seen, the tensions that occur in the transpiration stream are considerable, frequently more than – 1 MPa, and in some cases – 2 MPa. Under these tensions water ought to boil spontaneously, but inside the xylem it is in a metastable state. Indeed, water columns confined to Z-shaped tubes and spun in a centrifuge are easily capable of withstanding these tensions if the water is pure and without dissolved gases, as a result of the attractive forces between water molecules (see Smith 1994). The first indication that water columns inside plants may break came from Milburn and Johnson (1966) who showed that water-stressed stems and petioles emitted sound as 'clicks', which they inter-preted as breakages with the release of tension and a consequent shock wave. The phenomenon was already known in the engineering sciences as cavitation. Their work, which was controversial at the time, is now widely accepted on the basis of several independent lines of evidence: the hydraulic conductivity of water-stressed plants declines after acous-tic emissions have been detected; gravimetric analysis of stems from water-stressed woody plants suggests a considerable gas-filled fraction; such stems are more transparent to X-rays and γ radiation, and cavitated regions of stems may be mapped using γ-ray scanning, X-rays or nuclear magnetic resonance (Borghetti *et al.* 1993).

In the field, cavitation can be detected over the course of a day by means of ultrasonic acoustic emission (Fig. 2.12). Suitable transducers attached to the stems detect acoustic events that are counted and usually logged with a microprocessor (Tyree & Sperry 1989). Generally, acous-tic signals occur after a threshold water potential has been reached, often between – 1.0 and – 2.0 MPa. It is thought that most acoustic events originate from 'air-seeding' where air is drawn into the conduct-ing xylem cells from outside, rather than cavitation occurring inside the mass of water held by the cell (Zimmermann 1983). Cavitation is apparently a feature of most species under summer conditions, and

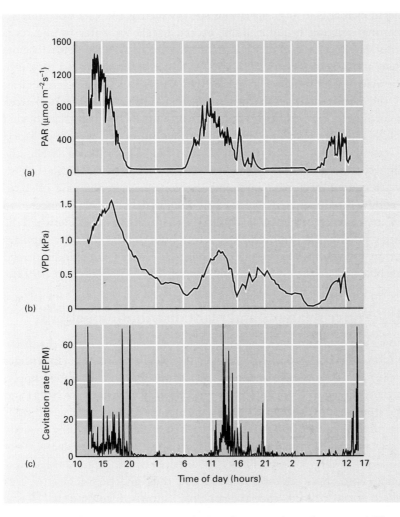

Fig. 2.12 Cavitation detected by ultrasonic acoustic sensors clamped to a mature *Pinus sylvestris* at Thetford Forest, England. (a) PAR = Photosynthetically active radiation; (b) VPD = vapour pressure deficit; (c) cavitation rate estimated from acoustic emissions per minute (EPM). (From Jackson *et al.* 1995c.)

probably occurs in winter also when the soil water is frozen or very viscous at low temperature (Tyree & Sperry 1989; Borghetti *et al.* 1993).

When cavitation occurs, it leads to a reduction in hydraulic conductivity, which in turn predisposes the plant to further cavitation. Consequently, attention has been drawn to the possibility of 'runaway' cavitation (Tyree & Sperry 1989). It seems that in trees and shrubs, cavitation is more likely to occur in the small twigs than in the main stem, so loss of leaves and twigs occurs whilst the main 'highways' of water transport in large stems are preserved.

Is cavitation reversible? After it has occurred it is presumed that the embolized xylem cells contain gas and water vapour. In the case of large xylem elements in angiospermous species such as the vine *Vitis*, there is seasonal refilling brought about by the uptake of water by the roots (root pressure) and the consequent increase in water potential of the stem above zero (Sperry *et al.* 1987). Water is simply forced back into the xylem. The situation in gymnospermous species is much less clear, as the stem undergoes changes in water content from week to week according to the weather (Waring & Running 1978). It is thought that refilling occurs after rain, but the mechanism is not well understood (Grace 1993a,b). Water potentials measured before dawn (when the highest water potentials occur) are usually − 0.3 to − 0.5 MPa, and so water cannot be forced into embolized cells as in the case of *Vitis*. Refilling experiments, in which partly dehydrated stems are connected to a water reservoir at a lower level, suggest that refilling can occur when the water potential is slightly below zero, about − 0.04 MPa, by capillary action involving the traces of water in the ends of the spindle-shaped tracheid cells of the xylem (Borghetti *et al.* 1991; Sobrado *et al.* 1992).

2.4 Water relations and plant distribution patterns

There are many instances in which water apparently controls the distribution of species at a local level, not only in the desert (Ehleringer 1984) but also in environments with a more favourable water supply, such as the South African fynbos vegetation (Richards *et al.* 1995) and the alpine zone (Dawson 1990). In the alpine regions of the world, soils are often shallow, topography is uneven and the water-holding capacity varies sharply from place to place. Dawson (1990) describes patterns of variation in the Canadian Rocky Mountains, where exposed ridge crest grades into moist alpine meadow, and then a depressed wetter area where snow collects and a distinctive 'snowbed community' forms (Fig. 2.13). Three dwarf willow species are found: *Salix reticulata* is an exceedingly dwarf (< 50 mm) species with small leaves and a generally sclerophyllous appearance, and is found on the very shallow soil of the plateau; *S. barrattiana* is a low alpine bush (0.2–1.3 m) with large leaves and deep roots, found in the snowbeds; and *S. arctica* grows in an intermediate situation on the exposed ridges (Fig. 2.13). The species showed markedly different responses of stomatal conductance to humidity. At high humidity (low leaf–air water vapour pressure gradient, Δw) *S. barrattiana* showed a much higher conductance than the other two species, but the stomata closed when Δw became large (> 20 mmol mol^{-1}). *Salix reticulata* had the lowest conductance of the three species, but its stomata remained open at the highest values of Δw. Dawson (1990) also reported the corresponding tissue water relations using a pressure chamber as outlined in (section 2.1.3). The pressure–volume curves were markedly different, irrespective of whether the tissues were rehydrated for a day or not. The difference between the rehydrated and non-rehydrated curves provides some information on the ability of the species to acclimate over a short period.

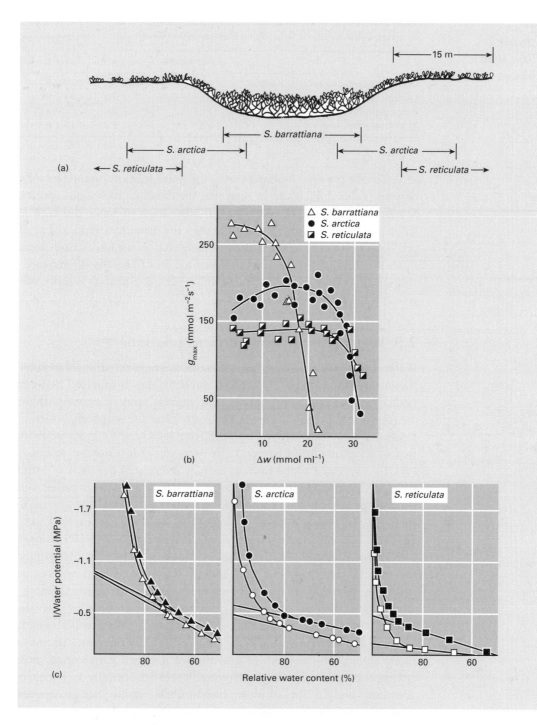

Fig. 2.13 Distribution and physiology of co-occurring species of *Salix* in the Canadian Rocky Mountains: (a) distribution on plateaux, ridges and snowbeds of three species; (b) maximal stomatal conductance in relation to leaf-to-air water vapour pressure deficit; (c) pressure–volume curves in the rehydrated (open symbols) and non-rehydrated cases (solid symbols). In these curves the reciprocal of balancing pressure is plotted against the relative water content of the leaves using data from a pressure chamber. (From Dawson 1990.)

Table 2.2 Comparison of some water relation parameters (units are MPa) for three co-occurring alpine species of *Salix*. (From Dawson 1990.)

	S. barrattiana	*S. arctica*	*S. reticulata*
Osmotic potential near full hydration	– 1.21	– 1.35	– 2.37
Osmotic potential near turgor loss	– 1.55	– 1.73	– 2.91
Bulk elastic modulus	2.65	8.30	17.31
Minimum water potential	– 0.8	– 1.3	– 1.6

From the pressure–volume curves the water relation parameters were derived (Table 2.2). It can be seen that the most sclerophyllous species, *S. reticulata*, does show physiological adaptation to water shortage. It had a much lower osmotic potential than the other two species, and its cell walls were considerably stiffer (as measured by the bulk tissue elastic modulus, one of the parameters obtained from analysis of pressure–volume curves). As a result of these differences, *S. reticulata* is more able to extract water from drying soil.

2.5 Water, carbon and nutrient relations

Water loss is an inevitable consequence of opening the stomata to allow the inward diffusion of CO_2 and so a relationship is expected between photosynthesis and transpiration. The highest rates of photosynthesis are found in C_4 tropical grasses like sugar cane and sorghum, which in bright sunlight can achieve up to $50\,\mu\text{mol}\ CO_2\ \text{m}^{-2}\text{s}^{-1}$, about twice that of C_3 plants (see Chapter 1). It turns out that water-use efficiency (WUE), expressed as the mass of carbon assimilated per mass of water transpired, varies greatly between species and with the weather but depends principally on whether the plant has C_3, C_4 or CAM photosynthesis. For C_3 plants WUE is 1–$3\,\text{g}\ CO_2$ per kg H_2O, whereas for C_4 the figure is 2–5 and up to 10–40 for CAM plants (Nobel 1991; Jones 1992). WUE, rather than assimilation rate *per se*, is presumably selected for in arid environments, and indeed CAM photosynthesis occurs mainly in desert plants (but also, intriguingly, in some aquatic plants).

We must also expect mineral nutrient supply to be related to the carbon and water relations, because high rates of photosynthesis imply high rates of nutrient supply (see Chapter 3). Any of several nutrients might limit photosynthesis or growth, but nitrogen has received most attention because quite large amounts (leaves are typically 1–4% nitrogen) are needed for ribulose bisphosphate carboxylase oxygenase (Rubisco), the enzyme responsible for the first stage of carbon assimilation. Only recently has it been possible to assemble sufficient data to search for such relationships. Schulze *et al.* (1994) showed that the maximum surface conductance of vegetation (the stomatal conductance for the canopy as a whole, which depends on the stomatal conductance of the leaf, and the leaf area index) is closely related to the assimilation

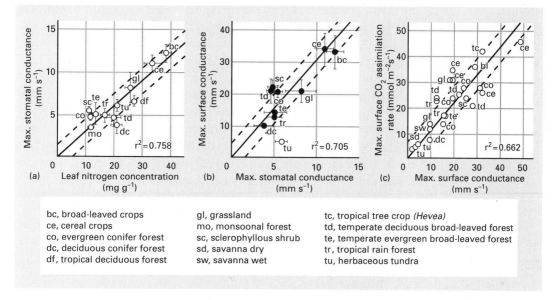

bc, broad-leaved crops gl, grassland tc, tropical tree crop (Hevea)
ce, cereal crops mo, monsoonal forest td, temperate deciduous broad-leaved forest
co, evergreen conifer forest sc, sclerophyllous shrub te, temperate evergreen broad-leaved forest
dc, deciduous conifer forest sd, savanna dry tr, tropical rain forest
df, tropical deciduous forest sw, savanna wet tu, herbaceous tundra

Fig. 2.14 Relationships between (a) leaf nitrogen and maximum stomatal conductance; (b) maximum stomatal conductance and whole-ecosystem surface conductance; and (c) surface conductance and whole-ecosystem assimilation rate. (From Schulze *et al.* 1994.)

maximal rate (Fig. 2.14), the data showing a highly significant linear trend. This is like the trend that others have reported at the scale of leaves. The nitrogen content of leaves is weakly related to the leaf stomatal conductance, the relationship being highly influenced by the crop plants, which tend to have high nitrogen and a high conductance.

2.6 Concluding remarks

The colonization of the land by plants about 400 million years ago was arguably the most significant event in the earth's history since the evolution of photosynthesis around 3.5×10^9 years ago, because it permitted the development of a multilayered canopy with a relatively high rate of carbon assimilation and oxen evolution, without which the development of the current oxygen-rich atmosphere might not have been possible. New life forms evolved that avoided desiccation by developing a cuticle, stomatal apparatus, lignified cell walls, xylem and a parallel transport system for organic materials synthesized in the leaves.

An important evolutionary step was the development of xylem walls, which did not collapse under great tensions and yet did not offer undue resistance to the flow of water. Both tracheids and vessels are intricate structures, made of lignin and cellulose, in which these materials are utilized economically for both water transport and mechanical strength. The diversion of carbon into wood involves considerable cost in the form of lost opportunity for carbon investment in photosynthetic and metabolic machinery for growth. This has long been presumed to be the main reason for the lower growth rates of trees compared to herbaceous species (Jarvis & Jarvis 1964). The benefit of woody stems for water transport and mechanical support has however been well worth the cost. Trees have dominated terrestrial ecosystems since the appearance of the conifers in the Carboniferous. They are excluded from the very dry places, where they have been defeated by the problem of extraction and transport of water, probably because embolized tracheids and vessels cannot be refilled. They are also excluded from the cold places, where tissue temperatures are adequate only for the dwarf plants that can utilize the warmer microclimates near the ground.

3: Nutrient Acquisition

Alastair Fitter

3.1 Availability of nutrients

Plants require about 15 essential elements, and with a few important exceptions these are obtained as ions dissolved in soil water. In physiological experiments, it is easy to induce a deficiency of any of these elements, even of those such as Zn, Cu or Mo that are only required in minute quantities, and on some soils these micronutrients may be naturally limiting. For example, peat soils are wholly organic and lack a mineral reservoir; they often induce copper deficiency. To plants growing in most field situations, however, the two elements that most commonly determine plant performance are nitrogen and phosphorus. The ways in which these two elements becomes available to plants differ strikingly. Nitrogen is an abundant element: air contains around 79% dinitrogen gas (N_2), but this is unusable by eukaryotes, which must have it 'fixed' into ionic form, either as ammonium (NH_4^+) or nitrate (NO_3^-) ions. Plants then convert these to organic forms, and these organic N compounds are eventually returned to the soil as litter, where they are acted upon by microbes (Fig. 3.1). Decomposition results in the reappearance of NH_4^+ ions in the soil and these may experience one of four fates:

1 some may be adsorbed onto clay minerals in soil;
2 much is often leached out of the root zone into ground water, but this is more likely if;
3 it is converted to nitrate by nitrifying bacteria, which may be a major pathway;
4 finally it may be taken up by plant roots or microbes, depending on their activity.

Importantly, there is no large mineral reserve of N in soil, both because the principal ions of nitrogen (NO_3^- and NH_4^+) are soluble in soil solutions and because they are rapidly converted by plants and micro-organisms to organic forms. Virtually all the N in soil, therefore, is present as organic N in soil organic matter. Its release as inorganic ions (the process known as mineralization) depends on the activity of decomposers, and that in turn depends on a range of soil characteristics, such as pH, temperature, oxygen concentration, moisture concentration, and so on. The speed at which mineralization occurs mainly determines the fertility of the soil in terms of N. If organic N is

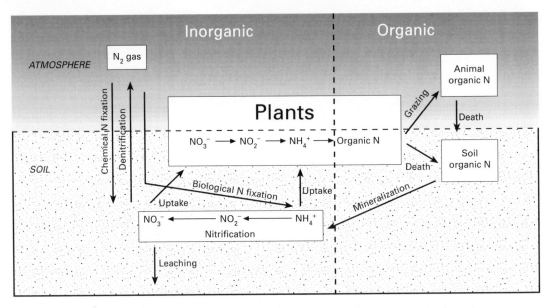

Fig. 3.1 A schematic representation of the nitrogen cycle, showing the major transformations that determine plant availability. The rate-determining step is typically mineralization of organic N. The cycle is dominated by organic compounds.

mineralized, little of the resulting inorganic N remains in solution as ions. NH_4^+ ions follow the pathways listed above; if converted to NO_3^- by nitrifying bacteria then leaching is more likely, since NO_3^- ions are very weakly adsorbed by soil particles, and hence easily moved in soil. Under anaerobic conditions, NO_3^- ions may be converted to N_2 gas by bacteria such as some pseudomonads, which gain energy by using NO_3^- ions as an alternative electron acceptor.

Most plants can take up N as either NH_4^+ or NO_3^- ions, though some species such as heathers, characteristic of acid peats where nitrifying bacteria are scarce, cannot cope with NO_3^- because they lack the enzyme system (nitrate and nitrite reductase) that converts NO_3^- to NH_4^+ (Dirr *et al.* 1973). N assimilation in these plants requires that the inorganic N be either taken up as NH_4^+ ions or converted to them.

In contrast to N, phosphorus is a rare element. As with most nutrient elements (N is the important exception), there is no atmospheric reservoir. Rocks typically contain much less than 1% P, and all types of phosphate ions are extremely insoluble in combination with the dominant cations in soils, such as aluminium, iron and calcium. The phosphorus cycle is therefore dominated by a large pool of insoluble inorganic P (Fig. 3.2), although as soils age an increasing amount is found in the soil organic matter (Fig. 3.3). The availability of P is therefore a function of soil chemistry, whereas that of N depends more on soil biology. Other nutrients fall somewhere between these two extremes, although potassium is different again since it is not meta-

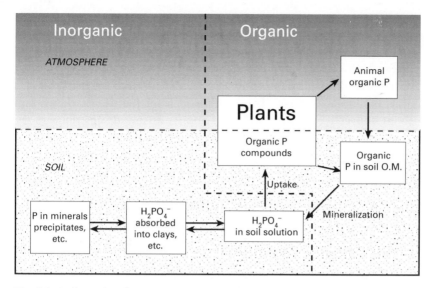

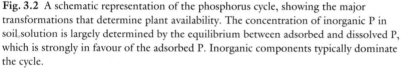

Fig. 3.2 A schematic representation of the phosphorus cycle, showing the major transformations that determine plant availability. The concentration of inorganic P in soil solution is largely determined by the equilibrium between adsorbed and dissolved P, which is strongly in favour of the adsorbed P. Inorganic components typically dominate the cycle.

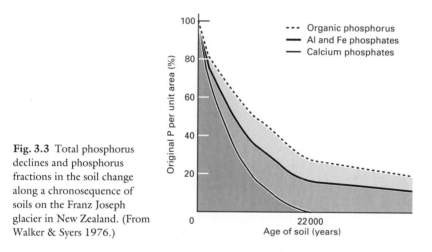

Fig. 3.3 Total phosphorus declines and phosphorus fractions in the soil change along a chronosequence of soils on the Franz Joseph glacier in New Zealand. (From Walker & Syers 1976.)

bolized by organisms, never forms organic compounds and is therefore found only in the inorganic phase of the soil.

3.2 Nutrient uptake by root systems

3.2.1 Transport through the soil

All the resources used by plants reach the plant by physical transport processes: radiation in the case of photosynthetically active radiation

(PAR); diffusion in the case of CO_2 and many ions in the soil (notably $H_2PO_4^-$); and convection in the case of water and other ions in the soil, particularly those present at high concentrations. With the exception of PAR, which is simply intercepted by plant tissues, the supply of all other resources is created by absorption. It is the removal of water from soil, a process ultimately driven by evaporation from the leaves, that brings about convective transport of water and its dissolved ions through the soil. Where this convective supply of ions is less than the rate of absorption at the root surface, depletion there will create concentration gradients in the soil and initiate diffusion down those gradients (Fitter & Hay 1987). For all resources other than PAR, it is *use* of a resource that creates supply.

Since mechanisms of interception of PAR and absorption CO_2 are dealt with in Chapter 1, only absorption of water and minerals are discussed here. Uptake of water is entirely passive and depends on the existence of a water potential gradient between root cells and soil, generated by evaporation from leaves (see Chapter 2). Since the root cell walls are freely permeable to water as far as the suberized endodermis (and as far as the xylem in very young, unsuberized roots), the surface area for uptake is equivalent to that of all the root cortical cells. Actual transport into the symplasm may occur anywhere in the cortex, although hydraulic considerations make transport through the cell walls (apoplastic transport) the more likely (Newman 1974). This water movement produces convective transport of dissolved ions to the root surface, where the cations may either be absorbed by negatively charged cell wall materials or move passively into the symplasm down the electrochemical gradient (root cortical cells have a standing potential of -60 to $-150\,mV$); anions are actively absorbed against that gradient.

Since soil water permeates the whole cortex and can therefore be taken up into the symplasm by any of the cortical cells, whereas many ions tend to move into the symplasm only at or near the epidermis, the effective surface area for absorption of ions is much smaller than that for water. The convective ion flux F is given simply by the water flux at the root surface, V, and the concentration of ions in the soil solution C_1:

$$F = V \cdot C_1 \tag{3.1}$$

When, for a particular ion, F is less than the rate of absorption by root cells, depletion at the root surface will lead to diffusion of the ion from the bulk soil. The diffusion coefficient for this depends on the moisture content of the soil (θ), the tortuosity of the diffusion pathway (f) and the reactivity of the ion with the soil (soil buffer power, dC/dC_1):

$$D^* = D \cdot \theta \cdot f \cdot \frac{dC_1}{dC} \tag{3.2}$$

where D^* and D are the diffusion coefficients for the ion in soil solution and free solution respectively (Nye & Tinker 1977). Note that buffer power enters the equation as its reciprocal. The effect of these relationships is that the availability of a particular ion in soil is determined by the following factors.

1 *Soil buffer power* (dC/dC_1): this controls the soil solution concentration (C_1) in relation to total soil concentration (C) and so the level of convective supply. In consequence ions that are strongly buffered in any soil will tend to have low values of C_1 (unless the total concentration, C, is very high) and hence low rates of both convective and diffusive supply. The extreme example is phosphate, where uptake is nearly always diffusion-limited. If buffer power is very weak, as is the case for nitrate (where $C = C_1$), convection may be inadequate if C_1 is low. In this event, rapid diffusion may lead to exhaustion of soil nitrate stocks.

2 *Soil water content*: the demand for water by leaves, largely to maintain energy balances, interacts with soil water content to control the rate of water uptake and hence convective supply of ions (Equation 3.1). Where soil water content is sufficiently low to limit convective supply, rates of diffusion are also depressed, because of the effect on diffusion coefficients (Equation 3.2).

3 *Soil structure*: the compaction of the soil and the nature of the soil aggregates determine both the size and distribution of soil pores, affecting both the hydraulic conductivity of the soil and the length of the diffusion pathway (f in Equation 3.2).

These soil factors control water and ion availability per unit root area or volume. Thus the actual levels of resources captured are strongly influenced by the total amount of root and its three-dimensional distribution within the soil.

3.2.2 Transport across the root

The membranes of plant root cells contain transport proteins that can move ions across the membrane, with great precision and at very high rates. The details of the transport process are described in several reviews (see, for example, Clarkson & Lüttge 1991). From an ecological standpoint, some of the important features of such mechanisms are:

1 they frequently involve movement against concentration gradients, so that the concentration of an ion inside the root cells can be many times that outside, and yet uptake may continue;

2 uptake frequently results in the loss of protons (H^+) to the external medium, with consequent changes in the acidity of the soil around roots;

3 uptake requires the expenditure of energy, and there is therefore an energetic cost to ion uptake, which is responsible for part of the respiration in roots.

The potential rate of ion uptake by roots (especially excised roots, which are often used in physiological experiments) under ideal conditions is usually very high, and typically much higher than is observed in intact plants, and even more so than is measurable by plants growing in the field. When lettuce plants were grown in carefully controlled conditions and the nitrate concentration in solution was varied over a range from 0.5 to 10 mmol l^{-1}, their nitrate uptake rate was constant (Blom-Zandstra & Jupijn 1987); if excised roots are given such a range of concentrations, the uptake rate would show a classic asymptotic response, increasing initially as external concentration increased and then saturating. This tells us that the rate of nitrate uptake is not controlled by the uptake mechanism, but by the plant itself: in other words, there is regulation.

This is an immensely important discovery. In effect, we can regard the uptake system as capable of acquiring any ions that the plant needs and that are available at the surface of the root cell membranes. Plant 'need' is determined by, for example, growth rate. When the plant is growing rapidly, it has a greater demand for nutrients, because metabolic processes consume nutrients already available. Consequently, the concentrations of either the ions themselves (as appears to be the case for phosphate and potassium) or a product of their metabolism (possibly amino acids in the case of N) change in the phloem and hence in the root cells. This seems to send a signal to the ion transporters and down-regulates the rate of uptake. There is currently great research activity aimed at discovering more about the molecular nature of these transporters, and this should help to reveal more about the control mechanism.

However, much of this work is done on plants that are capable of high rates of growth, typically crop plants (and increasingly nowadays the tiny annual crucifer *Arabidopsis thaliana*, Brassicaceae). Many wild plant species are characterized by very low rates of growth, even under optimal conditions. When presented with high concentrations of an ion such as NO_3^- at the root surface, these species do not down-regulate uptake to the same extent, and may consequently accumulate higher concentrations of N in tissues than they can use. They thus appear superficially to be very inefficient: one measure of this is photosynthetic nitrogen-use efficiency (PNUE), i.e. photosynthetic rate expressed per unit N, rather than leaf area or biomass. Pons *et al.* (1992) grew four graminoids (three grasses and a sedge) at both low and high N supply rates. Two species (*Carex flacca*, Cyperareae and *Briza media*, Poaceae) were inherently slow growing; two (*Brachypodium pinnatum*, Poaceae and, especially, *Dactylis glomerata*, Poaceae) were capable of fast growth under good conditions. At the low rate of N supply, all species grew equally slowly, with a relative growth rate of around 0.05 day^{-1}. When N was supplied at an optimum rate, all species grew faster: *Carex* and *Briza* approximately doubled their growth rate, but *Brachypodium* tripled it and *Dactylis* increased it fourfold. PNUE was higher for all species under N-poor compared with N-rich growth conditions. However, when N supply increased, *Dactylis* showed the smallest reduction in PNUE because it had the smallest increase in leaf N (Fig. 3.4). The other species, to varying extents appeared to accumulate N in the leaves, but not to increase photosynthetic rate proportionately.

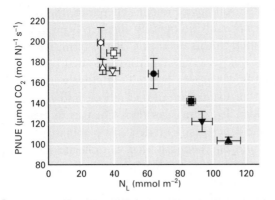

Fig. 3.4 Photosynthetic nitrogen-use efficiency (PNUE, $\mu mol\ CO_2\ (mol\ N)^{-1}\ s^{-1}$) is the rate at which a plant fixes CO_2 as a function of its N content. PNUE declines as leaf N concentration increases, when plants are fertilized: solid symbols on the graph represent plants grown at high N. The effect is much greater, however, for slow-growing species such as the grasses *Brachypodium pinnatum* ($\triangle$) and *Briza media* ($\triangledown$) and the sedge *Carex flacca* ($\square$), than for the fast-growing grass *Dactylis glomerata* ($\bigcirc$) which shows the smallest increase in leaf N concentration and little or no decline in PNUE. (From Pons *et al.* 1994.)

It would be a mistake to equate high PNUE with fitness. In fast-growing plants, the bulk of N in the leaf is present in the photosynthetic enzymes (ribulose bisphosphate carboxylase oxygenase, or Rubisco, in C_3 plants; see Chapter 1); they are fast-growing because they have such a large photosynthetic capacity. Slow-growing plants may use N in different ways: for survival in saline habitats, for defence against herbivores, or as a store. The reasons why some plants have the potential to grow rapidly and some do not are poorly understood but there may well be trade-offs between maximum growth rate and ability to survive at low rates of resource supply (see Chapter 4).

3.3 Responses to nutrient deficiency

3.3.1 Modifying the rhizosphere

If plants are deficient in nutrients, important changes in their physiology occur and these alter the interaction that roots have with the surrounding soil. All roots are surrounded by a zone of soil called the rhizosphere that has been chemically altered by root activity. Many root activities affect concentrations within the rhizosphere: for example, evolution of carbon dioxide from root respiration will increase local concentrations, whereas removal of water and ions due to uptake will reduce them (Dunham & Nye 1974; Hainsworth & Aylmore 1986). Most importantly, roots continually lose organic compounds, by active secretion, passive exudation from epidermal cells or by the death of these and other cells. This supply of organic carbon fuels an active microbial community, which itself transforms other compounds and further alters the soil environment. Therefore, the rhizosphere is physically, chemically and biologically distinct from the bulk soil. It is important to stick to this definition of the term, which is increasingly and incorrectly used as a synonym for the rooted zone in a soil horizon.

Nutrient-deficient plants have quantitatively and qualitatively different patterns of exudation and secretion, and this alters the rhizosphere microflora. In some cases these changes in exudation are direct responses to the deficiency. Iron-deficient plants may increase the secretion of siderophores, chemicals that can chelate Fe ions and prevent them being immobilized physicochemically by soil (Marschner *et al.* 1989). In consequence the Fe ions are more available for uptake by the root. However, various microbes in the rhizosphere can both produce these siderophores themselves (Crowley *et al.* 1991) and degrade them (von Wiren *et al.* 1993), so that there is uncertainty about the role played by siderophores in soil-grown plants. Similarly, phosphate-deficient plants may increase the activity of their extracellular acid phosphatase enzymes (Boutin *et al.* 1981). It has been suggested that this is an adaptive response that enables the plant to decompose organic phosphates to release inorganic ions that can be taken up by the root, but there is little evidence to support this.

The most dramatic impact of the root on the rhizosphere is related to N nutrition. Since N can be taken up as either NO_3^- or NH_4^+, the balance between these two ions has large implications for the electrochemical balance of the root cells. If the anion is the main form, the plant excretes OH^- ions to balance this; if the cation, then it is H^+ ions that are lost. Where NH_4^+ is the main source of N, therefore, there can be marked acidification of the rhizosphere, by several pH units, a phenomenon that is easily visualized using pH indicators (Plate 1, facing p. 366). This acidification may have profound effects on the microbiology and chemistry of the rhizosphere and hence on plant community structure (see Chapter 14).

3.3.2 Resource allocation

The best characterized response to deficiency of soil-based resources is a change in the pattern of growth, favouring root growth over shoot growth. Most plants grown in ample light but with inadequate nutrient supply increase root growth relative to shoot growth, producing a high root weight fraction (RWF):

$$RWF = \frac{\text{root dry weight}}{\text{total plant dry weight}}$$

conversely plants grown in shade have low RWF. This is an intuitively adaptive response and can be shown to represent optimal behaviour under idealized conditions (Iwasa & Roughgarden 1984).

However, increased root growth will not necessarily result in increased nutrient uptake, and hence alleviation of the deficiency. Since nutrient uptake depends to a great extent on the geometry of the root system, the greatest return on this investment will be achieved if root length is maximized. This implies that the production of fine roots will be favoured, since they achieve the greatest root length for a given weight. This can be measured as the ratio of length to weight, the specific root length (SRL). Generally, plants grown in low nutrient environments have higher SRL than those well supplied with nutrients (Fig. 3.5), and this is apparently due to the production of thinner roots (Christie & Moorby 1975).

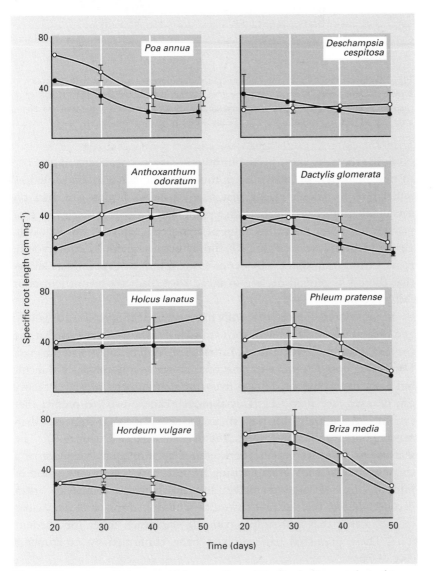

Fig. 3.5 Roots generally become thicker overall as soil fertility increases as shown by changes in specific root length (SRL: length per unit weight) In a set of eight grass species. Time-trends of SRL were very variable, roots becoming thicker with age in some species and thinner in others, but in almost all cases roots of plants at low fertilizer addition rates were finer than those grown at high rates. Open symbols represent low fertility, solid symbols, high fertility (from Fitter 1985).

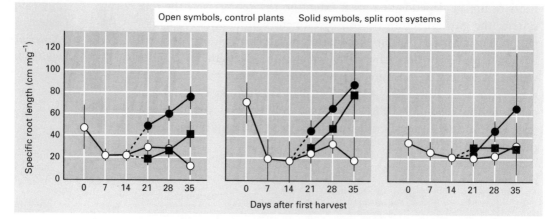

Fig. 3.6 Specific root length increases in split-root experiments in those parts of root systems supplied with N in each of three grass species, but the behaviour of the part of the root system deprived of N differs between species. Circles describe roots in low N after day 14, squares those in high N. (From Robinson & Rorison 1983.)

Some roots are short-lived, but others survive for longer and give rise to new branches. In most species this development is accompanied by radial growth, which means that newly produced roots are finer on average than the existing roots on a plant. Therefore, an increase in root growth following the reallocation of resources caused by nutrient deficiency will result in a fall in SRL simply because there are more young roots in the root system. Evidence from split-root experiments shows that there is a general decline in mean root diameter in nutrient-poor soils (Fig. 3.6).

If root systems in infertile soils have different proportions of young and old roots compared with those in richer soils, this suggests that the overall form or *architecture* of the root system may also change. Describing complex three-dimensional patterns is notoriously difficult but a set of techniques derived from the mathematics of graph theory can be used (see Box 3.1). This approach can be used to predict that certain types of branching pattern are more expensive to construct than others, but are also more efficient at exploring soil (Fig. 3.7). The increased cost arises because herringbone patterns, which consist of a single main axis with laterals arising from it, have a high proportion of interior links of large diameter, and the increased efficiency arises because, despite their greater cost, they tend not to have links very close to each other that will simply compete with each other for nutrient ions. These predicted patterns have been confirmed by experiments (Fig. 3.7b).

Adaptive responses to nutrient deficiency based on changes in resource allocation are, therefore, more complex than might initially be imagined, since they involve not only increased root growth in terms of mass, but also changes in the overall structure of the root system. A

Box 3.1 The architecture of root systems

In an architectural analysis, root systems are treated as sets of links or edges, which are segments between branching points, nodes or vertices. Links can be exterior (ending in a meristem) or interior. A system has a magnitude, which is the number of exterior links or root tips, and its branching pattern or topology is described by the relationship between magnitude and altitude, which is the greatest number of links that can be found in a single path from any root tip to the base of the root system.

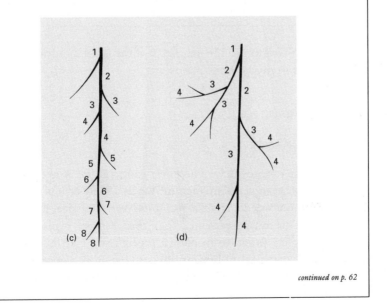

The two root systems (a) and (b) represent extreme topologies. Each has magnitude 8: that is to say, each has eight exterior links. The magnitude of an interior link is the number of exterior links it feeds.

continued on p. 62

Box 3.1 *Continued.*

The branching pattern (topology) can be characterized by calculating the path length of each link, which is the number of links in the path that connects it to the base. In (c) the longest path length is 8; in (d) it is 4. This is called the altitude of the system. If altitude (α) is plotted against magnitude (μ), a simple relationship emerges (e): systems with a single main axis and side branches (herringbone) have the largest possible altitude for a given magnitude, while (d) represents the minimum possible altitude, found in a dichotomous system.

Most root systems have intermediate root branching patterns, represented by the middle line on the graph. The slope of the line describes the branching pattern and reflects the rules by which the root system develops. Fitter (1995) gives a fuller account of this approach.

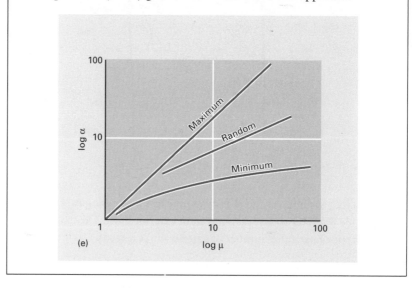

more radical alteration to the geometry of the plant's nutrient gathering system occurs in mycorrhizal plants.

3.3.3 Symbioses

3.3.3.1 Mycorrhizas

Root diameter varies widely. The finest roots can be as great as 1.5 mm (e.g. in *Drimys winteri* (Winteraceae); Baylis 1975) or less than 100 µm in the common weed *Capsella bursa-pastoris* (Brassicaceae) (Fitter & Peat 1994). This must represent one of the least well explained patterns of morphological variation in plants, though some at least is probably related to mycorrhizal colonization (see Fig. 3.8). Since the cost of constructing unit length of fine root is proportional to volume (assuming constant density), it is also proportional to the square of its

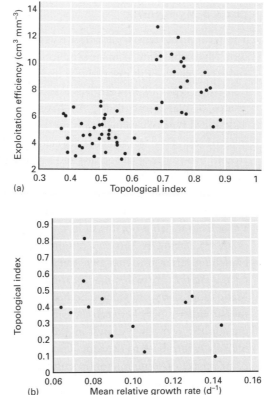

Fig. 3.7 (a) The efficiency with which a root system exploits soil in a simulation analysis increases as the root becomes more herringbone in architecture (high values of topological index (see Box 3.1)). (b) The topological index of plant root systems are highest in plants of low relative growth rate, which are typically adapted to growth on nutrient-poor soils and would benefit most from high exploitation efficiency. (From Fitter *et al.* 1991.)

diameter. As we have seen, roots vary by nearly two orders of magnitude in diameter, and so by nearly four orders of magnitude in specific cost (tissue mass per unit length).

An economic approach to understanding this variation (Fitter 1991) must offset the obvious benefit of increased SRL in fine roots by some disbenefit. Coarse roots may be favoured because they have greater disease or herbivore resistance, because they are longer-lived or because they have greater growth potential. In addition, coarse-rooted species may increase their nutrient acquisition ability by being mycorrhizal. Mycorrhizal fungi are symbionts that live in or on plant roots and extend their mycelium into the soil. There are several different types of mycorrhiza, but the most widespread is the arbuscular (sometimes known as the vesicular-arbuscular or VA) mycorrhiza (see Box 3.2 and Plate 3, facing p. 366). In this association the fungus is able to obtain phosphate ions, the most immobile of all plant nutrient ions, from soil outside the depletion zone in which the root is itself able to forage, and transport it back to the root. The fungus obtains its carbon from the root, and is obligately dependent on the plant for its energy requirements.

The economic explanation for this symbiosis lies in the dimensions of the fungal hyphae. They are typically less than 5 μm in diameter, at least an order of magnitude finer than the finest plant roots. In conse-

Box 3.2 Mycorrhizas

The term *mycorrhiza* comes from two Greek words meaning fungus-root. Mycorrhizas are intimate symbiotic associations of fungi and roots, and most plants form mycorrhizas under natural conditions. There are numerous types of mycorrhizal associations, each involving distinct groups of plants and fungi. All have evolved separately. The most widespread type of mycorrhiza is the arbuscular or vesicular-arbuscular mycorrhiza. This is formed between most plant species (probably around two-thirds of all species) and a restricted group of fungi that are members of the Zygomycotina (order Glomales); about 120–150 species of these fungi have been described, but there may be many more. The fungus is an obligate symbiont that can only survive when linked to a root from which it obtains all its organic food sources. Inside the root (Plate 3), facing p. 366 the fungus forms characteristic arbuscules, branching structures that penetrate the root cortical cell walls and invaginate the plasmalemma, creating a huge surface area of contact. Many fungi also form storage bodies called vesicles. The hyphae of the fungus either ramify between the cells or pass directly from cell to cell; outside the root they ramify in the soil, from where they transport phosphate ions back to the root. These fungi are also known to increase plant uptake of some micronutrients, provide protection against some toxic ions, fungal pathogens and grazing insects, and to affect plant water relations. The exact balance of all these various benefits to a particular plant growing in the field remains unclear.

The best known type of mycorrhiza is the ectomycorrhiza, formed between some forest trees, especially in boreal and cool-temperate forests, and large fungi ('toadstools') that are mainly members of the Basidiomycotina. Here the fungus forms a sheath around the outside of the root and spreads between the cells of the cortex, but does not penetrate cell walls. Most obviously, mycorrhizal roots usually have a different growth form, often stubby and dichotomous, Plate 2, facing p. 366 and this is the only type of mycorrhiza that can be recognized by the unaided eye. Ectomycorrhizas probably involve less than 10% of plant species, but more fungal species are known than in the case of arbuscular mycorrhizas.

The other main mycorrhizal types are taxonomically restricted, in one case to heathers and allies (Ericales: ericoid mycorrhizas) and in the other to orchids. Ericoid mycorrhizas involve a few fungi in the Ascomycotina and form dense infections in very fine 'hair-roots'. Orchid mycorrhizas are found only in the orchids. The fungi are mainly members of the Basidiomycotina, including some species that are parasitic on trees, but many of the fungi are permanently sterile and cannot be reliably classified. The fungi form coils in root cells, but what physiological interactions occur there is almost completely unknown. Intriguingly, the same coils are found in some arbuscular mycorrhizal associations.

quence, at least 100 times the length of hyphae can be constructed for a given investment of resources, as compared to roots. The hyphae extend the zone that can be exploited for immobile nutrients (especially phosphate) by several millimetres. The benefit to the plant is directly proportional to the diameter of its roots, since this determines the differential in cost between growing more roots to obtain P and subcontracting the task to fungal hyphae. Thick-rooted plants are therefore more likely to be mycorrhizal than fine-rooted species (Fig. 3.8).

Arbuscular mycorrhizas (AM) are the most widespread type, found in around two-thirds of all vascular plants. The other, perhaps better known type is the ectomycorrhiza, an association between trees and a quite different set of fungi, mostly members of the Basidiomycotina, many of which are familiar woodland toadstools and often characteristically found under a particular species of tree. *Boletus elegans*, for example, is typically found under larches *Larix* spp., with which it forms a specific symbiosis. Not all fungi are specific, however; many, like the AM fungi, have wide host ranges. Ectomycorrhizas have marked effects on the root systems of the plant: often the colonized roots have much reduced growth and they may fork dichotomously. The fungus does not penetrate the root cells, forming instead a sheath around the outside and then penetrating between the cells of the cortex, forming a structure known as a Hartig net.

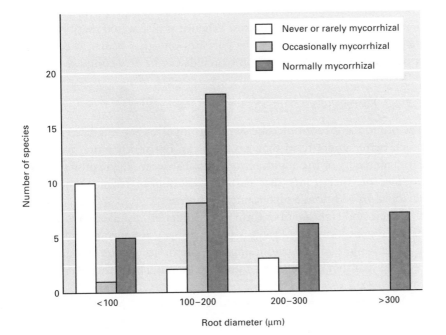

Fig. 3.8 Plants that are never or rarely mycorrhizal have particularly fine roots, confirming theoretical predictions. Data are for British plant species and come from the Ecological Flora Database. (From Peat & Fitter 1993.)

Although ectomycorrhizal associations are certainly involved in P transfer from soil via fungus to the roots, their nutritional role is certainly wider. Some fungal species can decompose organic matter and transport the N back to the tree, effectively closing the nutrient cycle, since the N is never free in soil and therefore much less likely to be lost by leaching or denitrification. Ectomycorrhizas are found in rather few ecosystems, most notably the boreal forest, and it seems that they are favoured in those ecosystems where N deficiency is a major controlling factor (Read 1991).

A third type of mycorrhiza is different again. Ericoid mycorrhizas are found in the roots of members of the order Ericales (the heathers and their allies). These fungi belong to the Ascomycotina and there is no evidence for P transport to the plant. Decomposition of organic N is probably important, and these fungi appear to be involved in protecting the plant against toxic metal ions in soil – heathers typically live in extreme, often acid soils.

The fourth mycorrhizal type emphasizes the point that the concept of mycorrhiza is a functional not a taxonomic one. All orchids form mycorrhizal associations, but in this case the orchid appears to be parasitic on the fungus, since even in green orchids it is not possible to detect any carbon movement from plant to fungus. Some orchids have no chlorophyll at all, so they clearly cannot supply fixed carbon to the fungus. Since they have no root system either, it is not obvious that these orchids are capable of acquiring any resources independently. These non-green orchids are usually referred to as saprophytes, but it is better to classify them as parasites. When it is discovered that the fungi they parasitize can themselves be parasites on other plants, it becomes apparent that the life-style of some orchids can be very complex.

3.3.3.2 Root nodules

The other major root–microbe symbiosis involves N-fixing bacteria; it is much better understood than mycorrhizas, but much more restricted. Most members of the angiosperm family Fabaceae (the legumes), form nodules on their roots that harbour a group of bacteria known as rhizobia. In addition, a taxonomically diverse group of species in about 17 genera and eight families form rather different nodules with another bacterial genus, *Frankia* (Table 3.1). Rhizobia were once thought to be a rather simple group taxonomically, but molecular techniques have revealed that they do not form a single taxonomic unit (there are at least four genera).

Wild plants frequently appear to be limited by N and so, at first sight, the selective advantage of possessing bacteria in roots capable of fixing atmospheric N_2, which is extremely abundant, and converting it to NH_4^+, which is not, would seem very large. It is surprising therefore that so few plant species have acquired this trait, and those that have are

Table 3.1 The genera of flowering plants that can form nitrogen-fixing associations with the actionmycete *Frankia*. The distribution is haphazard taxonomically, occuring in a diverse group of families, in a few genera within those families and only regularly in some of those.

Family	Genera in which most species normally form nitrogen-fixing symbioses	Genera in which some species may form nitrogen-fixing symbioses
Betulaceae	*Alnus*	
Casuarinaceae		*Casuarina*
Coriariaceae	*Coriaria*	
Datiscaceae	*Datiscus*	
Elaeagnaceae	*Hippophae, Shepherdia*	*Elaeagnus*
Ericaceae		*Arctostaphylos*
Myricaceae	*Myrica, Comptonia*	
Rhamnaceae	*Ceanothus*	*Trevoa, Discaria*
Rosaceae	*Pershia*	*Rubus, Dryas, Cercocarpus*

not more successful than they are. Although legumes are extremely widespread in agriculture, natural communities dominated by legumes or by *Frankia*-associated species are much less common. Species of alder *Alnus* may form almost pure woods and some tropical leguminous trees (e.g. *Tetra-berlinia*) can be important canopy dominants, but many N-fixers are rare or uncommon in the wild.

Part of the explanation for this may lie in the cost of the symbiosis. The fixation of N is an energy-demanding process and typically around 10% of the carbon fixed in photosynthesis is diverted to the nodule and to the fixation process in an actively fixing, nodulated plant. How serious this transfer of carbon is to the plant must depend on the extent to which photosynthesis is limited by sink activity (which seems often to be the case) or by the direct impact of environmental factors, notably lack of light (see Chapter 1). Very few N-fixing plants grow in shade. Plants growing in sun, however, are usually photosynthesizing below the maximum rate that they could achieve, because photosynthetic rate is then determined by the metabolic activity of growing points and other plant processes, much in the way that ion uptake is controlled (see above).

Another reason why N fixation may not be more widespread in plants is that the benefit can only be large in N-deficient soils; yet, by its very existence, the N-fixing plant must enrich the soil with fixed N as its tissues contribute to soil organic matter, and so it will reduce the benefit it receives. Nitrogen fertilization of a community rich in legumes quickly reduces their abundance, as shown clearly by the famous Park Grass experiment at Rothamsted Experimental Station (Fig. 3.9, Plate 12, facing p. 366 and see Chapter 14). N-fixing plants themselves achieve the same effect over a longer period of time by adding N to soil. This is shown in the successional sequence at Glacier Bay, Alaska, where the community is dominated by the N-fixing alder *Alnus crispa* from

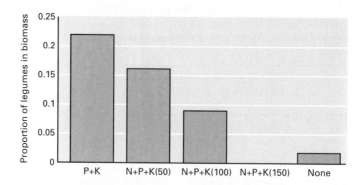

Fig. 3.9 The relative abundance of legumes, measured as the proportion of shoot biomass in the June hay crop made up by *Trifolium pratense*, *Lathyrus pratensis* and *Lotus corniculatus*, declines with increasing nitrogen, N, addition on those plots which receive both potassium, K, and phosphorus, P, fertilizers (the figure in parenthesis is the N application rate in kg/ha/yr). The control plots, which receive no nutrient fertilizers at all, have more legumes than the plots receiving the highest rates of nitrogen, but less than other fertilized plots; legumes on the control plot are P limited. Data from Crawley & Silvertown, unpublished results.

about 30 to 80 years after glacial retreat. During this period, soil N concentrations increase rapidly, greatly improving the competitive ability of spruce *Picea sitchensis*, which then replaces the alder (Fig. 3.10).

3.4 Heterogeneity

3.4.1 Patchiness

Because soil is a solid medium, transport processes within it are slow. Nutrients are therefore rarely evenly distributed, either in time or in space. To understand how plants acquire nutrients from soil, it is important to consider the distribution of nutrient availability and how plants respond to that.

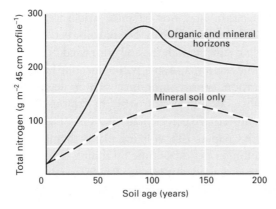

Fig. 3.10 Nitrogen builds up rapidly in the developing soils on newly exposed glacial moraines at Glacier Bay, Alaska, especially during the period from about 30 to 80 years after colonization corresponding to the time when N-fixing alder (*Alnus crispa*) is dominant. (From Crocker & Major 1955.)

Heterogeneity in time arises because of the seasonal nature of many environments. This imposes a regular pattern on many processes, including decomposition. In temperate ecosystems there is a pulse of organic matter deposition in autumn and often a peak of decomposition in spring or summer. This leads to a clear pattern in the production of inorganic N (Fig. 3.11), irregularity is imposed by weather variation.

Heterogeneity in space is caused by the pattern in which organic matter is deposited: leaves and shoot material arrive on the surface and tend to produce a regular gradient in nutrient availability from the surface downwards. Roots, in contrast, though they may be concentrated at the surface, occur as discrete patches throughout the soil profile, providing local hotspots of organic matter and decomposition activity. The same effect is achieved when earthworms drag leaves down into soil and when individual soil animals die. The net result of these processes, and of inherent variation in the distribution of soil minerals, is often pronounced spatial pattern, which can be revealed by careful and stratified sampling procedures (Fig. 3.12).

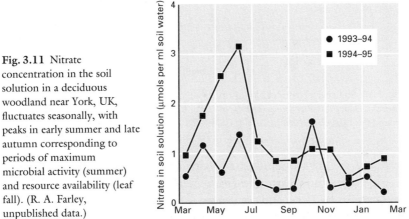

Fig. 3.11 Nitrate concentration in the soil solution in a deciduous woodland near York, UK, fluctuates seasonally, with peaks in early summer and late autumn corresponding to periods of maximum microbial activity (summer) and resource availability (leaf fall). (R. A. Farley, unpublished data.)

It is possible to classify all this variation in terms of the attributes of the patches (Fitter 1994). Patch attributes are distribution, extent and number. Patches may be distributed randomly, regularly or be clumped, in both space and time; they may be small or large, short- or long-lived; they may be abundant or rare in space or occur frequently or rarely in time. When these attributes are combined it is obvious that each soil may have its own unique patchiness, and plants need to be able to respond to these patches effectively.

3.4.2 Response to patches

When roots encounter nutrient-rich patches they may display a proliferative response that seems intuitively adaptive (Robinson 1994). This

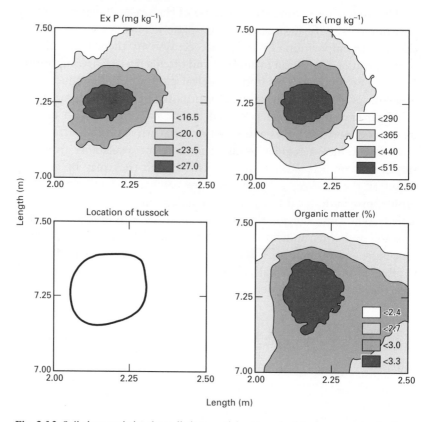

Fig. 3.12 Soil characteristics show distinct spatial pattern, in this case around a single tussock of the grass *Pseudoroegneria spicata*. Soil P, K and organic matter are all greater around the tussock. (From Jackson & Caldwell 1993.)

response is easy to demonstrate with species such as cereals (barley, wheat, maize) and with numerous other species. In a classic series of experiments by Drew and coworkers (e.g. Drew 1975), one section of a barley root axis was exposed to $10 \, \text{mmol L}^{-1}$ of NO_3^-, NH_4^+ or phosphate ions, while the rest experienced only $0.1 \, \text{mmol L}^{-1}$. The segment receiving the high concentration showed greatly enhanced growth of primary laterals and consequently much denser development of secondary laterals. Intriguingly, the roots failed to respond to potassium ions, perhaps because potassium ions are not metabolized in the root; in the case of the other ions, increased uptake would lead to a demand for more metabolites for their utilization and this increased supply of fixed carbon may have stimulated growth.

This proliferative response is so clear and apparently so adaptive that it has often been assumed that it is universal, but this is not so. In a neat experiment, Campbell *et al.* (1991) allowed nutrient patches to develop in sand culture by dripping solutions onto the surface of a hemispherical bowl. This created four quadrants with no barriers between but with very well-maintained differences in nutrient concentration. They then allowed plants of a range of uncultivated species to explore this hetero-geneous environment. Strikingly, whereas some species did, as expected, concentrate root growth in the nutrient-rich areas, others grew their roots almost randomly through the bowls. The species that were most precise in their root placement, developing roots predominantly in the nutrient-rich patches, were the smallest and slowest growing. Those that ignored the patches and grew roots randomly were the largest, fastest growing and most competitive (Fig. 3.13). The authors pointed out that this suggested that a contrast should be drawn between the scale and precision of foraging.

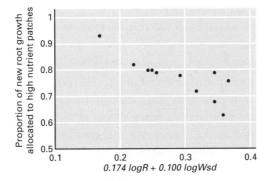

Fig. 3.13 The precision with which plants allocated new root growth to nutrient-rich patches in the experiments of Campbell *et al.* (1991) was greatest in small plants and least in large ones. The *x*-axis in this graph is a composite variable derived from multiple regression of precision of allocation on relative growth rate (R) and seed weight (W_{sd}); high values represent large and/or fast-growing plants. (From Fitter 1994.)

3.4.3 Turnover

The fact that plants do not always exploit nutrient-rich patches by proliferation highlights the complex economics of patch exploitation. For proliferation to be an advantageous response, the benefit from the increased root density must exceed the costs of the new roots. If a patch is short-lived, the gain may in fact be small, but the cost may be large and the need for maintaining the roots will continue for as long as the roots survive. The coarser the roots, the less likely it will be that the construction costs will be recouped. Coarse-rooted species might there-fore be expected to be less likely to exploit patches, and this appears to be what occurs.

Another important distinction between coarse and fine roots seems to lie in their longevity. Fine roots are often remarkably short-lived: in rhizotron studies, where roots are observed against glass walls or the sides of tubes inserted into soil, the half-life of a population of fine roots can be less than 10 days (Fig. 3.14). In other words, half of a set of roots initiated at a particular time may be dead 10 days later. This rapid

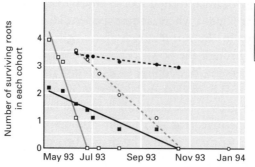

Fig. 3.14 Survival of cohorts of roots under a grassland sward growing either at the summit of Great Dun Fell in northern England (alt. 845 m) or at Newton Rigg (alt. 170 m). Mean survival time is always lower for spring cohorts compared with summer cohorts, and for low compared with high altitude cohorts. (G. K. Self & T. K. Brown, unpublished data.)

turnover of roots has enormous implications for studies of carbon cycling, since it means that estimates of below-ground productivity based on biomass measurements are almost always too low. For studies of resource acquisition, it suggests the possibility of responding to spatial and temporal heterogeneity in nutrient supply by the maintenance of a labile root system, in which roots are continuously developing in areas of high nutrient supply and then dying.

What appears to be unknown is whether these roots senesce in the sense that leaves do, with the nutrients being withdrawn and re-used, or whether they just die or are killed by soil organisms. If there is no nutrient recycling the adaptiveness of rapid turnover is less obvious. This is an area where new research is urgently needed. Nevertheless, heterogeneity of nutrient supply and the nature of plant response to it are too often neglected in studies of mineral nutrition that have their intellectual springs in the controlled environment chamber and the hydroponic culture. An ecological approach must be more aware of the nature of the soil environment.

3.5 Summary

The acquisition of nutrients from soil by plant roots cannot be understood by reference to the physiology of ion uptake alone. The movement of ions from soil to roots represents two processes in series and the limiting step is frequently that through soil. Plant responses to this limitation include direct modifications of the soil around the roots, and complex changes in the architecture of root systems. In addition, most plants are symbiotic, typically with mycorrhizal fungi; N-fixing bacteria also promote nutrient acquisition in some species. The responses that plants make to nutrient limitation are complicated by the variability of nutrient supply in time and space.

4: Life History and Environment

Michael J. Crawley

4.1 Introduction

Suppose that you had to design a plant to do well in a particular environment. Where would you start? I suspect that you would begin by thinking about the abiotic conditions that are likely to prevail: maximum and minimum temperatures, soil acidity, the amount and seasonal distribution of rainfall, the prevalence of fires, the frequency and intensity of storm-force winds, the likelihood of inundation by flood, the presence of toxic ions in the soil, the kinds and abundances of herbivorous animals, and so on. Your ideal plant design would also depend strongly on what the other plant species were doing. Under conditions of moderate temperatures, on well-irrigated, neutral, non-toxic soils, away from severe exposure, the other plant species would almost certainly be trees and, if you want your plant to do well, then it had better be a tree as well. But what kind of tree? Will it be broad-leaved or coniferous? Evergreen or deciduous? How long will it live, and how long will it wait before flowering for the first time? How tall will it grow and how much will it invest in shoot versus root, and in support structures like trunk and branches versus productive tissues like leaves and flowers? What kind of seeds will it produce, and how are they to be dispersed? Will they germinate at once, or will they exhibit some form of protracted dormancy?

These are questions of plant life history, and each of the 250 000 living vascular plant species possesses a more or less unique combination of traits such as these, reflecting both their phylogeny and their recent ecological circumstances. Some environments are characterized by plants that show a remarkable degree of convergence in some of their conspicuous life-history traits (e.g. most of the shrubs in deserts have grey foliage, and have small leaves with inrolled margins; the majority of trees in tropical rain forest have glossy, dark-green, oval leaves with drip-tips). Other habitats contain a wide range of coexisting plant species with markedly different life histories (e.g. in a temperate forest you could find both winter-green and winter-deciduous members in most of the plant guilds: trees, shrubs, ferns, vines, herbs and grasses all have green and leafless representatives in winter). While it is possible to make lists of plant life histories that appear to be precluded by certain combinations of conditions (e.g. you do not find annual grasses in tropical rain forest, or geophytes in arctic heathland), it is generally the diversity of coexisting life forms which catches our attention.

It is obvious, therefore, that for most plant communities there is no such thing as an optimal life history. Indeed, given the fact that the Darwinian struggle is never-ending (remember the Red Queen in *Alice Through the Looking Glass*, who had to run as fast as she could, just to stay where she was) there never could be an optimum; evolution would always come up with an improvement that would competitively exclude its predecessor. Theoretical models may predict that there is only one mathematically optimum life history for a given environment, but history and circumstance will have thrown together many species which exhibit a variety of solutions to the problems posed by survival and reproduction in any one place (Cohen 1966; Stearns 1976, 1977). The mathematical optimum may also be quite different from the best *practical* solution. Plants have life histories that 'work' in the sense that they are capable of persistence (all species show the ability to increase when rare; the 'invasion criterion'; see Chapter 19), but they are bundles of traits cobbled together from countless compromises. They do not represent multidimensional optima. Their major traits may have evolved millions of years ago in quite different environments, surrounded by a totally different suite of competitors, herbivores and pollinators. All of their traits are likely to be trade-offs between conflicting costs and benefits. These trade-offs and compromises are the subject of this chapter.

4.2 Neighbourhoods

A fundamental theme running through this book is that processes like growth, mortality, pollination and seed dispersal all occur in the context of very small numbers of individuals, and that the spatial arrangement of plants is of vital importance in determining their fitness. Unlike animal ecology, where the individuals are assumed to blunder about at random, like molecules obeying the ideal gas laws, each individual plant does not have the potential to meet and interact with every other. For example, plants compete strongly with only the six or so individuals in their immediate vicinity (their 'first order neighbours'). Furthermore, this competition is highly asymmetric, with the larger plants exerting a much greater influence on the fitness of the smaller individuals than vice versa (see Chapter 11). Plants exchange pollen over very limited distances, even in wind-pollinated species (see Chapter 6), and most seeds fall in the immediate vicinity of their maternal parents (see Chapter 9). Most of the important interactions between plants are extremely local; where there is large-scale uniformity in plant communities, this speaks of large-scale uniformity of substrate (e.g. soil nutrients or water conditions), or of uniformly severe disturbance (e.g. fire, or defoliation by large, mobile vertebrate herbivores). This spatial limitation matters, because the behaviour of spatially explicit theoretical models is much richer than that of point-equilibrium, mean-field models, especially when there are trade-offs between competitive ability and dispersal ability (see Chapters 14 & 15).

4.3 Life history

A good deal of attention has been directed by theorists to the question of which patterns of life history are likely to prosper in different kinds of environments. One of the first distinctions in life-history theory was between *r*-selection and *K*-selection (MacArthur 1972). If superior fitness at low population density is incompatible with superior fitness at high density, then a range of different plant types may evolve. In Fig. 4.1 type A increases faster than type B at low density, but type B increases faster than type A at high density. Type A is *r*-selected and type B is *K*-selected. From these traits, Pianka (1970) predicted that at high densities (where competition is assumed to be more severe) the optimum strategy is (i) to allocate fewer resources to reproduction and more to adult maintenance; and (ii) to produce fewer, larger offspring (Gadgil & Solbrig 1972; Schaeffer & Gadgil 1975). It soon became clear, however, that this dichotomy of traits was very crude, because fitness is affected by factors influencing development rate, survivorship *and* fecundity, and these may be influenced in quite different ways by population density. Grime (1979) extended the *r* and *K* classification to

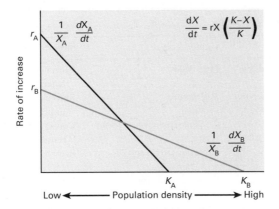

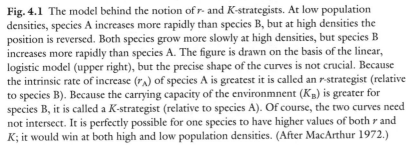

Fig. 4.1 The model behind the notion of *r*- and *K*-strategists. At low population densities, species A increases more rapidly than species B, but at high densities the position is reversed. Both species grow more slowly at high densities, but species B increases more rapidly than species A. The figure is drawn on the basis of the linear, logistic model (upper right), but the precise shape of the curves is not crucial. Because the intrinsic rate of increase (r_A) of species A is greatest it is called an *r*-strategist (relative to species B). Because the carrying capacity of the environmnent (K_B) is greater for species B, it is called a *K*-strategist (relative to species A). Of course, the two curves need not intersect. It is perfectly possible for one species to have higher values of both *r* and *K*; it would win at both high and low population densities. (After MacArthur 1972.)

account for the intensity of disturbance (physical damage to vegetation), competition (the attempt by neighbours to capture the same unit of resource, be it light, water or nutrient) and stress (defined as any constraint on production but, in practice, the stress axis is usually related to soil nutrient availability). Although Grime often presents his three strategies (ruderals, competitors and stress tolerators) as a two-dimensional triangle (the empty space (the upper triangle) is defined by high disturbance and high stress, and is assumed to be uninhabitable), it is better to think of the strategies as defining the first three principal axes of a multidimensional statistical ordination of species. The principal axis for herbaceous plant species in northern England turns out to be nutrient stress (see Table 4.1). Other authors have added a disturbance axis directly to the r–K axis; for example Southwood (1977) described a 2×2 contingency table with durational stability and adversity as its two axes (high durational stability = K, high adversity = stress tolerance). The major problem with these classifications is that 'stress' and 'adversity' are not measurable. Conditions that would be extremely stressful for one species may be optimal for another. Remember that there is no such thing

Table 4.1 Co-occurring suites of traits associated with Grime's three principal dimensions of plant life history. Although Grime's triangle is obviously a two-dimensional (flat) ordination, the attributes are best seen as defining the first three dimensions of a multidimensional ordination of species. In a statistical ordination of herbaceous plants of the Sheffield region using all of the traits measured during the Integrated Screening Programme (Hendry & Grime 1993), the stress axis turned out to be the most important, ranking species from those requiring high levels of all the macronutrients (e.g. *Urtica dioica*) to those found on chronically nutrient-poor soils (e.g. *Festuca ovina*; J.P. Grime and K. Thompson, personal communication). RGR = relative growth rate (see Box 11.3).

Ruderals	Competitors	Stress tolerators
High RGR of seedlings	High RGR	Chronically low productivity habitats
Early onset of flowering	Tall stature	Resource availability low and unpredictable
Self-pollination	Consolidated growth form	Conservation of captured resources
Heavy allocation to reproduction	Vigorous lateral spread	Evergreen leaves
	Underground storage organs	Plants long-lived
	Large peak shoot biomass	Low RGR
	High plasticity of root and shoot	Delayed onset of reproduction
	Short leaf life span	Strongly defended against herbivores
	Flexible canopy distribution	
	Active foraging of roots and shoots	
	Delayed seed production	
	Weak anti-herbivore defence	

as a 'benign' environment; the struggle for existence is intense in *all* environments (e.g. if low nutrients don't get you, then pathogens will).

An important cause of semantic confusion arises from the difference between the use of the word 'competitor' by Grime (1979) and Tilman (see Chapter 8); to Grime a competitor is typified by a fast-growing renascent herbaceous perennial which is good at accumulating resources, whereas to Tilman the superior competitor is the species that can reduce the rate of resource supply to the lowest level ($R*$), and hence cause the competitive exclusion of the inferior competitor. In Grime's terminology, Tilman's competitor is a stress tolerator (e.g. a shade-tolerant tree like *Tsuga canadensis* (Pinaceae), which will eventually replace faster-growing deciduous trees). In Tilman's terminology, Grime's competitor is nothing special; it is only a moderate competitor that can outcompete annuals and low-growing perennials by shading them, but whose high light and nutrient requirements mean that it is destined to be outcompeted by shade-casting or nutrient-depleting species later on in succession (e.g. tall herbs like *Chamerion angustifolium* (Onagraceae) or *Urtica dioica*).

Further sophistications to life-history theory have developed from the 'principle of allocation' in which organisms with finite resources must allocate them between competing demands (Cody 1966). Thus, for example, there must be a trade-off between traits promoting fecundity and those promoting survival (Williams 1966; Stearns 1976). The aim of life-history theory is to compare the *fitness* of phenotypes that differ in their development rates or age-specific fecundity or survivorship schedules in specified ecological circumstances. Fitness is measured as the intrinsic rate of increase, r, of the phenotype, p, in environment, E according to the Euler equation (pronounced 'oiler'):

$$1 = \int e^{-rt} l_t(p, E) b_t(p, E) dt \qquad (4.1)$$

where t is the age of the plant in years, $l_t(p, E)$ is the fraction of plants which survive from seed disperal to age t, and $b_t(p, E)$ is the fecundity of phenotype p at age t in environment E. This is an adequate measure of fitness in constant, density-independent environments, but also holds for density-dependent populations (Charlesworth 1980). Indeed, maximizing the intrinsic rate of increase in density-dependent populations is equivalent to maximizing the carrying capacity of the environment. This equivalence is not widely appreciated by ecologists, many of whom consider maximizing the rate of increase and maximizing the carrying capacity as *alternatives*, as in the dichotomy between r- and K-selection (Sibly & Calow 1983).

Fisher's (1930) concept of *reproductive value* V_x is an important element of life history that shows the 'value' of individuals of different ages x in terms of their discounted future production of offspring. Seeds have low reproductive value, because they are highly likely to die before they ever begin to reproduce. Senescent individuals have low reproductive value because they have so few years left to live. Somewhere in between individuals peak in reproductive value, and it is at this age that we might expect plants to be best defended against their competitors and natural enemies (Crawley 1983). The formula is:

$$V_x = \frac{e^{rx}}{l_x} \int_x^\infty e^{-rt} l_t m_t \, dt \qquad (4.2)$$

and the terms are defined in Box 4.1. The important thing to remember is that future reproduction $\int_x^\infty l_t m_t \, dt$ is *discounted* at the intrinsic rate of increase r, which draws attention to the importance of early reproduction in determining fitness (see below). For clonal growth, Sibly and Calow (1982) show that more, smaller offspring should be produced in conditions that are good for individual growth. For sexually reproducing forms, and if there are no interactions between variables, then the number of offspring per breeding should be maximized, survival until first (or next) breeding should be maximized, and time to first (or next) breeding should be minimized. If, as would seem likely, there are trade-offs between variables, then the optimum trade-off will be that which maximizes fitness. In cases where juvenile mortality is low, there should be selection towards semelparity with many offspring per brood, while if juvenile mortality is high there should be selection towards iteroparity with few offspring per brood (Sibly & Calow 1983). Further trade-offs between reproduction, survivorship and development time are explained below.

One of the most severe practical problems in plant ecology derives from the fact that there is often pronounced polymorphism in life-history traits within species, and even within a single clutch of offspring produced by an individual parent. At best, this means that very large samples of plants must be analysed in order to determine representative mean values for life-history parameters. At worst, it means that average values of life-history parameters are meaningless as characterizations. These difficulties are clearly illustrated by the polymorphism for cyanogenesis shown by the legume *Lotus corniculatus* (Fabaceae); not only do individual plants differ from one another in their production of HCN, but the same plant may vary its phenotype, sometimes appearing cyanogenic, at other times acyanogenic (Ellis *et al.* 1977). Also, what happens to an individual plant depends very much upon when, exactly, recruitment occurs. The fates of plants in different cohorts can be radically different depending upon the phenology of the species and the timing of recruitment, due to differences in weather conditions, natural enemy activity, and so on. From their detailed study of the grass *Bromus tectorum*, Mack and Pyke (1984) found that the probability of a seed germinating, or a seedling dying varied dramatically both within and between years, so that population dynamics were determined by 'the chronology of environmental events'. Clearly, then, generalizations

Box 4.1 Rates of increase and life table parameters

The key concept is the net multiplication rate λ (lambda) which is related to the intrinsic rate of increase r ('little r') by the equation

$$\lambda = e^r \tag{4.1.1}$$

which can also be written so that

$$r = \ln(\lambda) \tag{4.1.2}$$

The rate of increase in population density is determined by the balance between births and deaths; in other words by *survivorship* l_x and *fecundity* m_x. In discrete time, the value of r can be found from the Euler equation as follows:

$$1 = \sum_{x=0} e^{-rx} l_x m_x \tag{4.1.3}$$

where the value of r is found numerically by computer (an initial value is adjusted until the right-hand side is equal to 1). Note that this is not exactly the same as a Leslie Matrix (see Chapter 12) where the seeds produced this year become the first age class of next year; here the seeds produced this year become *this years* first age class (see Caswell 1989 for details). An example should show how the process works:

x	l_x	m_x	$l_x m_x$	$x l_x m_x$	$e^{-rx} l_x m_x$
0	1	0	0	0	0
1	0.1	5	0.5	0.5	0.385
2	0.06	15	0.9	1.8	0.533
3	0.018	10	0.18	0.54	0.082
Totals			1.58	2.84	1

The extreme right-hand column shows the exact numerical solution of $r = 0.2618$ from Equation 4.1.3. A good approximation can be obtained by using the important equation:

$$r \approx \frac{\ln R_0}{T} \tag{4.1.4}$$

where R_0 ('r nought') is the average number of offspring per individual per lifetime, defined as

$$R_0 = \Sigma l_x m_x \tag{4.1.5}$$

(1.58 in this example) and T is the cohort generation time, which can be thought of as the average time between the birth of an individual and the birth of one of its offspring (1.797 in this case):

$$T = \frac{\Sigma x l_x m_x}{R_0} \tag{4.1.6}$$

s, or $\approx 0.457/1.797 \approx 0.255$

continued on p. 80

Box 4.1 *Continued.*

This is a *density-independent model* (survivorship and fecundity are assumed to be constants) that allows only two outcomes of interest: when $r > 0$ ($\lambda > 1$) the population grows exponentially and when $r < 0$ ($\lambda < 1$) the population declines exponentially (to eventual extinction). The case where $r = 0$ ($\lambda = 1$) would only occur in equilibrium, and since this is a density-independent model *it can have no equilibrium* (see Chapter 12). In field studies, the measured value of λ is often close to 1 (see Enright *et al.* 1995) but it is seldom clear whether this means that the population was close to equilibrium or that (especially with long-lived plants like trees) the duration of the study was simply too short to estimate the true rate of increase. In order to do this properly, it is vital to *measure the empty quadrats* to assess recruitment (Crawley 1990a), because in many species recruits are not produced close to the parent plant (see Chapters 9 & 15). Life histories of plants can be described in terms of their age-specific reproductive and mortality schedules (as here), but bear in mind that plant size at a given age is usually a much better predictor of its likely seed production and risk of death than its age alone (see Chapter 11). Also remember that l_x and m_x are interrelated, so that increased investment in reproduction at age x will usually cause reduced survivorship in subsequent years (see text).

about the life-history attributes of 'the average individual' must be treated with great caution.

4.3.1 The growth forms of plants

Amongst the vascular plants there is a marvellous variety of growth forms, from tiny annual plants that produce 10 seeds in only a few weeks, to gigantic coniferous trees that produce huge crops of seeds periodically over the course of many centuries. Several classification systems have been proposed to describe the range of plant growth forms, but easily the most valuable was devised by the Danish ecologist P. Raunkiaer in the 1920s. His system is based on the precise position in which the plant maintains its perennating buds for the duration of the unfavourable season (see Fig. 14.10). If there is no unfavourable season (e.g. in the continually moist tropics), then most plant species of closed communities are tree-like: they maintain their perennating buds on aerial, usually woody, shoots. Raunkiaer called these plants 'phanerophytes', distinguishing between large (> 30 m), medium (8–29 m), small (2–7 m) and tiny (< 2 m) species as mega-, meso-, micro- and nano-phanerophytes. When the unfavourable season is extremely cold, plants with their perennating buds very close to the soil surface are favoured (chamaephytes), while in extremely arid conditions, perennials that die back to subterranean storage tissues may prevail (geophytes, and other cryptophytes). Despite its anachronistic terminology, the classification system has survived because it is thorough and is based on ecologically significant characteristics (see Chapter 14).

Leaf size and shape vary widely both within and between different life forms of plants. Different forms of venation, degrees of division, serration of the leaf edge, thickness and kinds of surface covering occur in countless combinations, each influencing light interception, heat balance, temperature regulation, water balance and CO_2 diffusion (Parkhurst & Loucks 1972; Givnish 1979). There is a broad correlation, for example, between water availability and leaf size; the largest leaves are found in tropical rain forest, medium-sized leaves in temperate forests, and small leaves in deserts, tundra or heathland communities that are dry or cold. The longevity of leaves also differs markedly both between species within a habitat, and between environments of different kinds (Chabot & Hicks 1982; see Chapters 1–3).

4.3.2 Annual plants

In terms of fitness, annual plants appear to have two great advantages: (i) they reproduce early, so they have the potential for very high intrinsic rates of increase; and (ii) they can survive adverse conditions as dormant seeds in the soil. Weighed against these attributes, however, are two rather severe constraints: (i) it is difficult for annual plants to grow sufficiently large in one season that they can compete with tall, perennial plants; and (ii) annual plants rely upon the availability of establishment microsites in *every* generation – if there are no microsites, there can be no recruitment. The balance between these opposing forces is determined by the severity of the unfavourable season, the density of competing vegetation, and the frequency of disturbance.

Put formally, if F is the average fecundity of plants surviving to reproduce, and s is the fraction of seeds that survives to reproduce, then the annual rate of increase, λ, is:

$$\lambda = s.F \qquad (4.3)$$

When λ is less than 1, the population declines; when it is larger than 1, the population increases exponentially. Static populations would have λ exactly equal to 1 (see Chapter 12). It is a mistake, however, to equate an annual life history with r-selected demographic behaviour, since some annual species have extremely low intrinsic rates of increase (e.g. *Vulpia fasciculata* (Poaceae); Watkinson 1978). The inclusion of seed dormancy allows the development of a seed bank; now the equation becomes:

$$\lambda = (1 - d - g) + g.s.F \qquad (4.4)$$

where d is the fraction of seeds that dies in 1 year in the soil, g is the germination rate and s is the survival of germinated seeds to fruiting. The first term on the right-hand side incorporates the stabilizing effect of a seed bank; it means that $\lambda > 0$ even if either F or s were zero in a given year because of environmental stochasticity (details in Chapter 7).

If recruitment only occurs at intervals of more than 1 year because of intermittent microsite availability, then the mortality suffered by seeds in the bank must be taken into account, and the rate of increase must be discounted in calculating the annual average value. For example, if recruitment occurred every other year, and the survival of seeds in the bank was b during the first year, then $b.s.F$ seeds would be produced every 2 years, so the annual rate of increase would be much lower, namely:

$$\lambda = (b.s.F)^{1/2} \qquad (4.5)$$

Note that we take the square root of the 2-yearly rate (*not* half; Fig. 4.2a), so λ is a *decelerating* function of fecundity. If gaps appear on average only every x years, then:

$$\lambda = (b^{x-1}.s.F)^{1/x} \qquad (4.6)$$

where b is the per-annum survival of seeds in the soil (for details, see Kelly 1985). The model can be adapted to account for temporal variation (Cohen 1966, 1968) and spatial variation (Venable & Brown 1988) in recruitment probability.

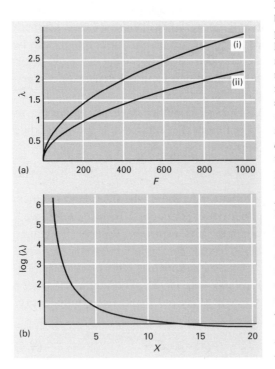

Fig. 4.2 The relationship between the net rate of population increase λ and life-history attributes in monocarpic plants. (a) The relationship between λ and fecundity F in a biennial plant, showing the decelerating (square root) relationship; as net survival bs declines from (i) $bs = 0.01$ to (ii) $bs = 0.005$, so the fecundity threshold for population increase ($\lambda = 1$) increases from $F = 100$ to $F = 200$. (b) The relationship between $\log(\lambda)$ and the age of first reproduction x in a long-lived monocarpic plant, showing the hyperbolic decline in $\log(\lambda)$ as reproduction is delayed. The population declines ($\lambda < 1$) once the delay x exceeds $\ln(1/F)/\ln(s)$; in this case with $F = 100$ and $s = 0.7$ the threshold is at $x = 12.9$ years.

Clearly, the potential advantages of an annual life history are re-duced considerably when recruitment is limited by the intermittent availability of microsites, especially if survival in the seed bank is low. Under these circumstances, a long-lived life history, with resources devoted to high adult survivorship, and with only intermittent repro-duction (triggered, ideally, by the same factor that makes microsites available) might confer greater fitness.

In habitats characterized by moist winters and dry summers, many of the annual species germinate in autumn and produce a rosette of overwintering leaves (so called 'winter annuals'). Rapid growth begins very early in the year, flowering occurs in early spring and seeds are ripened before the onset of the summer drought. This kind of life history is common in temperate sand dune communities (e.g. many small crucifers and grasses) and garden weeds (e.g. *Cardamine hirsuta* (Brassicaceae), *Lamium purpureum*). The phenology of winter annuals ensures that the vegetative stage escapes the attentions of most inverte-brate herbivores. However, the plants are vulnerable to vertebrate herbivores that forage throughout the winter (e.g. rabbits and voles), because they offer young, green foliage at a time of year when many other species are leafless, or carry only old leaves of low nutritional value. Summer annuals overwinter as dormant seeds, then germinate in spring and grow rapidly to ripen their seeds in summer or early autumn. A few summer annuals are able to grow sufficiently large in a single season that they are capable of forming closed, monospecific stands (e.g. the alien *Impatiens glandulifera* (Balsaminaceae) on muddy river banks in England). Others, like *Galium aparine* (Rubiaceae), are able to persist in dense, mesic vegetation composed of woody and herbaceous peren-nials, by virtue of large seeds, early germination, rapid growth and a scrambling habit.

Because recruitment from seed is so vital to the perpetuation of the genes of annual plants, it comes as no surprise to find that annuals display a variety of traits which ensure that all their eggs are not put in one basket (so called 'bet-hedging' tactics; Stearns & Crandall 1981). It would be suicide, for example, for an annual plant to have synchronous germination of its entire seed crop in an environment, or at a time, where conditions were likely to deteriorate so badly that all the seedlings would die. Thus, for example, we find that individual plants produce seeds which, within a single clutch, display a wide range of dormancy-breaking and germination requirements (Leon 1985).

By altering the relative costs and benefits of dispersal, environmental uncertainty also determines whether it is better for the plant to stay where it is, or to move on (De Angelis *et al.* 1979; Venable & Lawlor 1980). (Note, however, that selection is likely to favour dispersal even in stable environments; Hamilton & May 1977). Many annual plants adopt a mixed strategy by equipping some seeds for long-range dispersal and others for staying put. This trait is particularly common amongst members of the Asteraceae, where fruits of the outermost flowers are large and have no parachute, while the more numerous, smaller fruits from the centre of the inflorescence have a pappus of hair and are wind dispersed (Zohary 1950; Baker & O'Dowd 1982). The question of how many large, non-dispersing seeds and how many small, far-dispersing seeds to produce is a problem for analysis as an evolutionary stable strategy (ESS) (see Chapter 6), since the production of large numbers of non-dispersing seeds would lead to competition between the sibs. The ESS is to produce an increasing proportion of far-dispersed seeds as total seed production increases. Where plants have been found to show variable morph ratios with clutch size (e.g. plants producing both subterranean and aerial fruits), they conform to the ESS, producing larger, near-dispersed fruit first, and changing the ratio in favour of smaller, far-dispersed fruit as clutch size increases (Silvertown 1985).

In variable environments, selection will tolerate some loss of average (arithmetic mean) fitness, if there is a corresponding reduction in the *variability* of fitness and this, too, will lead to the evolution of bet-hedging, risk-avoiding strategies (Bulmer 1985). Ritland and Jain (1984) found just such patterns in two contrasting annual species of *Limnanthes* which differed in their seed dormancy, vegetative growth rate, timing of reproduction, allocation to reproduction and seed size.

4.3.3 Monocarpic perennials

Plants that spend one or more years in a vegetative condition before flowering once then dying have been called 'big bang' strategists. At one end of the spectrum are strict biennial plants like *Melilotus alba* (Fabaceae) that flower and die in their second year of life. At the other extreme are long-lived species like the sword-leaved *Lobelia wollastonii* (Campanulaceae) from the mountains of Uganda, or the silversword *Argyroxiphium sandwicense* (Asteraceae) from Hawaii, which grows until it achieves a rosette diameter of about 60 cm, then flowers and dies; this takes about 7 years. *Puya raimondii* (Bromeliaceae) from the Bolivian Andes does not flower until it is over 100 years old. Between these extremes are many kinds of plants that are more or less long-lived, depending upon the opportunities for growth, and upon their histories of defoliation and crowding.

There has been considerable debate about the relative fitness of annual, biennial and perennial monocarpic life histories (Hart 1977; Silvertown 1983; Kelly 1985), but, as we saw earlier, these comparisons are of limited value in the absence of considerations about the frequency and reliability of microsite availability. Indeed, for a great many so-called biennials, the populations themselves are ephemeral, and, as Harper (1977) wryly observes, most populations of biennials are on their way to extinction by the time an ecologist decides to study them. Be that as it may, if the plant spends *x* years in the vegetative stage prior to flowering, its average annual rate of increase will be:

$$\lambda = (s^x.F)^{1/x} \qquad (4.7)$$

where s is the annual survival rate. For a given fecundity, F, the rate of increase falls rapidly as the delay in reproduction, x, is increased (Fig. 4.2b). Weighed against this are two potential advantages for big bang reproduction: (i) fecundity of the larger plants may be sufficiently high to compensate for the delay; or (ii) the large seed crop may satiate local seed predators, so that seedling recruitment is assured. Furthermore, the long gap between seed crops may reduce the risk of mortality from specialist seed-feeding species because the herbivores cannot increase in abundance during the intervening period (see Chapter 13).

4.3.4 Herbaceous perennial plants

Herbaceous perennial plants comprise an extremely diverse array of different growth forms (Table 4.2); each type will have more or less unique aspects of their population dynamics and life history. For example, the main difference in life-style between trees and the renascent herbaceous perennials is in their occupancy of the aerial environment. Both maintain tenure of a piece of ground for many years, but the shoots of renascent herbs die back to ground level at the end of every growing season. The advantage of this growth form is that the shoots are not vulnerable to drought, cold or natural enemies during the unfavourable season. The main disadvantages, compared to a woody habit, is that the individual's position in the canopy of plants competing for light must be re-established every year. Weighed against this, however, is the fact that herbaceous perennials need not divert resources into the production or maintenance of permanent, woody, supporting structures.

A further advantage of the herbaceous perennial habit (shared by a few woody species) is the ability to 'move about' by virtue of rhizomes, stolons or rooting shoots. These plants are capable of spreading radially to form large, clonal stands (e.g. *Holcus mollis* (Poaceae)), of exploiting patchy habitats by 'jumping over' barriers of uninhabitable substrate (e.g. the long, tough runners of *Potentilla reptans* (Rosaceae)), or of insinuating themselves through a dense matrix of taller plants (e.g. the 'guerrilla' growth form of *Trifolium repens* (Fabaceae)).

In a study of rhizome development in patchy environments differing in soil salinity, Salzman (1985) found that *Ambrosia psilostachya* (Asteraceae) showed a distinct tendency to spread into less saline (and presumably more favourable) soils. There was a strong correlation between the salt tolerance of a clone and its preference for non-saline patches; while the tolerant clones were almost indifferent, the intolerant clones put over 90% of their shoots from new rhizomes in the non-saline patches. This could have been due to: (i) direct suppression of growth or increased rhizome mortality in saline soils; (ii) increased

Table 4.2 Growth forms of herbaceous perennials.

Renascent herbs	*Sendentary*	
	Multicapital rhizomes	*Vincetoxicum*
	Mat geophytes with	
	(a) stem tubers	*Crocus*
	(b) root tubers	*Ophrys*
	(c) bulbs	*Lilium*
	(d) tuberous stems	*Cyclamen*
	Mobile	
	Rhizome geophytes on	
	(a) sand	*Ammophila*
	(b) woodland	*Anemone*
	(c) wet mud	*Phragmites*
	(d) grassland	*Cirsium arvense*
Rosette plants	*Sedentary*	
	Short internodes and close-set leaves	
	(a) alpine fell fields	*Draba*
	(b) elongate leaves	*Taraxacum*
	(c) succulent leaves	*Sedum*
	(d) leathery leaves	*Saxifraga*
	(e) *Musa*-form	
	(f) branched tuft trees	*Yucca*
	(g) unbranched tuft trees	palms
	Mobile	
	Stoloniferous plants	*Fragaria*
Creeping plants	*Mobile*	
	Prostrate, rooting, leafy shoots of	
	(a) woods	*Lysimachia nummularia*
	(b) grasslands	*Trifolium repens*
	(c) wetlands	*Menyanthes trifoliata*
Cushion plants	*Sedentary*	
	Richly branched densely packed shoots	*Silene acaulis*
Undershrubs	*Sedentary*	
	Flowering shoots die after blooming	*Lavandula*
	Weakly lignified tropical forest spp.	Acanthaceae
	Rhizomatous	*Vaccinium*
	Canes	*Rubus idaeus*
Succulent-stemmed plants	*Sedentary*	
	Cactus-form	*Opuntia*
Aquatic herbs	*Free-floating*	
	Surface leaves	*Lemna*
	Submerged leaves	*Utricularia*
	Rooted	
	Surface leaves	*Nuphar*
	Submerged leaves	*Litorella*

investment in rhizome growth in less saline patches; or (iii) both of
these. While there is some evidence to suggest selective investment
(plants produced fewer shoots at a given salinity when given the 'choice'
of a less saline patch in which to grow than when grown in a habitat of
uniform salinity), it is not yet possible to distinguish critically between
these processes under field conditions (Salzman 1985).

The Achilles heel of a herbaceous species is its perennating organ (rhizome, corm, bulb, tuber or fleshy rootstock). If the plant is to survive for a protracted period, these structures must be resistant to herbivores and pathogens, as well as being capable of surviving the rigous of the unfavourable season. One of the reasons that we know so little about the ecology of most herbaceous perennials is because it is so difficult to study this subterranean world. For example, we know virtually nothing about the extent and importance of feeding by herbivores on (or in) rhizomes, corms and tubers in the population ecology of wild plants.

Grasses and sedges are amongst the most important groups of herbaceous perennials. They display a remarkable range of form and life history in different environments: they may be evergreen or summergreen, rhizomatous or tussock-forming, prostrate or erect, cryptophyte or chamaephyte, outbreeding, inbreeding or not breeding. Perhaps their greatest claim to fame, however, lies in their ability to withstand grazing. Those species with prostrate growth form, buried meristems, rapid tillering and substantial, inaccessible reserves of carbohydrate achieve dominance under continuously high levels of vertebrate grazing (Johnson & Parsons 1985). Not all grasses are grazing tolerant, however, and species with vulnerable meristems or storage organs are quickly eradicated under grazing (Holmes 1980). Grasses also possess a range of chemical defences against defoliation (especially in the early seedling stages; Bernays & Chapman 1978) and many species have sharp or coarse silica-rich tissues that are unpalatable to vertebrates.

4.3.5 Trees and shrubs

In order to buy themselves a place in the sun, trees delay investment in reproduction until they have grown sufficiently large to ensure their survival over a protracted period. Then they can devote resources to reproduction as and when they are available (e.g. in years of exceptionally good weather for pollination, in times of unusually high productivity, or in years when herbivores are scarce). Plants unable to obtain a place in the canopy will usually die without leaving any progeny at all.

With presently available data it is virtually impossible to calculate the annual rate of increase of tree species, because we have no accurate data on the fraction of seeds that survives to reproductive age (this is hardly surprising given the longevity of most ecologists). We can make some rough estimates, however, by considering two extreme cases. First, we assume that only the first bout of seed production is important (i.e. the trees have approximately the same rate of increase as monocarpic perennials). Second, we assume that the surviving plants reproduce at an average rate F each year for ever, once they have begun to reproduce at age x (i.e. they are essentially immortal). In this case

$$(l_x F)^{1/x} < \lambda < \exp(l_x F) \qquad (4.8)$$

where l_x is the fraction of seeds reaching reproductive age x, and F is the average number of seeds produced per plant per year. Suppose that an oak tree produces its first acorns at age $x = 25$ years, that an acorn has a 1 in 1000 chance of surviving to 25 years (most of the mortality will occur in the acorn stage) and that a 25-year-old oak tree produces 1500 acorns in its first bout of reproduction. This means that $l_x F = 0.001 \times 1500 = 1.5$ so, applying Equation 4.8, we find that λ must lie between 1.016 and 4.482. This range is so large as to be virtually useless. The important information is that $l_x F > 1$, so we know that the population will increase. In practice, the range of values will often include $\lambda = 1$, so we cannot be sure whether the population will persist or not.

In making calculations of the rate of increase of long-lived plants we must resist the temptation of estimating survivorship by dividing the numbers of mature plants found per unit area by the numbers of seeds falling on the same area, since this procedure implicitly assumes that the population is stable; i.e. we are trying to estimate λ, but assuming that $\lambda = 1$. For instance, if we found 100 mature oak trees ha^{-1} and 50 acorns m^{-2} we might estimate that survivorship is 0.0002 but this is true *only if* the population is stable. Thus, we cannot use this information to calculate the actual rate of increase (or the potential rate of increase) of the oak population (see Chapter 12).

4.4 Trade-offs

The original trade-off (and still one of the most interesting and important) had to do with colonizing ability and competitive ability: r-strategists were good dispersers but poor competitors while K-strategists were poor dispersers but good competitors (see above). Since the pioneering work of MacArthur and the expansion of behavioural ecology, a great many aspects of plant performance have been investigated from the viewpoint of trade-offs (Stearns 1989).

4.4.1 Colonization/competitive ability

This is the essential trade-off of the r–K continuum. The inferior competitor (the r-strategist) persists by dint of its superior colonizing ability (e.g. it rapidly produces large numbers of small, wind-dispersed seeds). Whenever it meets the K-strategist it loses out in a strongly asymmetric (contest) competition (e.g. Crawley & May 1987). In a community where the maintenance of high species richness depends on the colonization/competitive ability trade-off (Tilman 1994), then seed-addition experiments should show that recruitment of the potential dominant species is seed-limited, but that recruitment of the inferior competitor is not seed-limited.

4.4.2 Root growth/shoot growth

Individual plants show a refined ability to adjust their root–shoot ratio according to prevailing conditions of light availability (see Chapter 1), water availability (see Chapter 2), nutrient availability (see Chapter 3; Ericsson 1995) and the rigours of above- and below-ground herbivory (see Chapter 13). Over evolutionary time, plant species have adapted to a trade-off in investment in root or shoot biomass that tends to lock them into a particular, limited range of the successional continuum from high light/low nutrients in early primary succession, to low light/high nutrients of late succession (see Chapters 8 & 14). Plants specializing in early succession need large root systems relative to their shoot mass, whereas late successional forest species need to invest in large shoot systems to obtain their place in the sun at the top of the canopy.

4.4.3 Palatability/competitive ability

This trade-off lies at the heart of herbivore effects on plant community structure. If the most palatable plants were also the least competitive, then they could not persist in any environments, whatever the herbivore density. If herbivory is to allow increased plant species richness by fostering plant coexistence, then there must be a *positive* correlation between palatability and competitive ability (note that selective herbivory is a necessary but not a sufficient condition for coexistence; see Chapter 13). This trade-off is consistent with the growth/defence trade-off (see below and Chapter 10).

4.4.4 Seed size/seed number

This is the classic fitness trade-off from theoretical models of plant life history. For a fixed investment in reproduction, the plant must trade off the benefits of large seeds (more competitive seedlings, see below) against the costs of large seeds (higher risk of predation, lower dispersal distance, less chance of survival in the seed bank) and the evident benefit of producing a larger number of small seeds (each of which is probably less likely to be eaten by granivores and more likely to be dispersed a substantial distance away from the parent plant).

4.4.5 Seed size/seedling performance

Bigger seeds make more competitive seedlings. Thus, it might be worth while for a plant to trade off seed number for increased seed size. But the benefits of larger seeds are likely to saturate quite quickly (the graph of fitness against seed size is likely to be asymptotic; indeed, it might even have a hump if the very largest seeds were irresistible to herbivores;

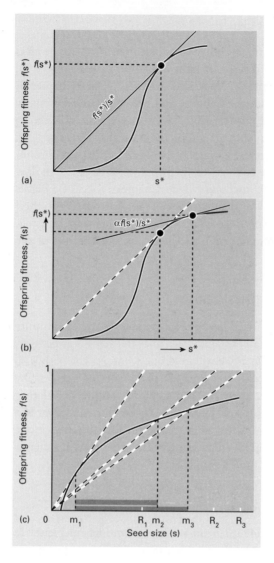

Fig. 4.3 Seed size and plant fitness: the development of optimality ideas. (a) The original optimality model of Smith and Fretwell (1974). The optimal seed size *s** maximizes the slope of the tangent passing through the origin. (b) Venable's (1992) modification showing that optimal seed size is greater under sibling competition. (c) Geritz's (1995) ESS model allows a wide range of seed sizes to coexist (inside the shaded bars) on the basis of considerations about seed predators and asymmetric competition between seedlings; a single seed size was never found to be optimal.

Leishman *et al.* 1995). Figure 4.3 shows the derivation of the optimal seed size under various assumptions (Smith & Fretwell 1974; Venable 1992, Geritz 1995). Armstrong and Westoby (1993) found that seedlings from larger seeds were more tolerant of defoliation. The fact that real plant communities support plants with such a wide range of seed sizes suggests that the trade-off between competitive ability and dispersal ability may be widespread in spatially structured communities.

4.4.6 Seed size/dormancy

The largest seeds tend to germinate immediately. This has a range of advantages (microsite pre-emption, obtaining a head start on other seedlings in the cohort, root development before the onset of the unfavourable season), but the plant is unlikely to prosper in habitats where seedling herbivory is predictably intense (escape in space through wind-dispersed small seeds might be favoured under these circumstances). It is difficult (and presumably expensive) to defend a large dormant seed in the soil from herbivores and pathogens for a protracted period; we would expect that seeds with protracted dormancy would be small (see Chapter 7). The presence of seed dormancy can also affect the evolution of post-germination traits (e.g. germination in drier years would be expected to select for more xerophytic traits; Evans & Cabin 1995).

4.4.7 Dormancy/dispersal

Plants with adaptations for long-distance dispersal by wind or water tend to show little dormancy, while plants with no specialized dispersal mechanisms sometimes show high rates of dormancy (e.g. in environments like deserts with unpredictable rainfall) or low rates of dormancy (e.g. in temperate coastal sand dunes with reliably wet winters). See Chapter 7 for details.

4.4.8 Longevity/growth rate

This trade-off is the botanical equivalent of the tortoise and the hare. The plants that grow tall most quickly (e.g. bulky herbaceous perennials; Grime's 'competitors') tend to be overtopped and eventually outcompeted by slower-growing, shade-tolerant species (e.g. evergreen trees). Note, however, that some plant species combine great longevity with relatively high growth rates as juveniles, so this trade-off is clearly not inevitable (e.g. *Quercus robur* (Fagaceae) can establish during the first year of a secondary succession, yet could survive for more than 1000 years, but it is not shade tolerant). The longevity/growth rate trade-off is related to the growth rate/defence trade-off (see below), and to the fact that plants from poor soils tend to grow slowly and to have long-lived, well-defended, evergreen leaves (Kerner 1894).

4.4.9 Longevity/reproductive output

The key to the outcome of this trade-off depends upon the *costs of reproduction* in which increased investment in reproduction causes a reduction in subsequent survival (Ashman 1994). This cost might be measured in terms of nutrient allocation or meristem allocation. For example, it is commonly observed that the longevity of monocarpic herbaceous plants can be increased by removing their flowers to stop them setting seed (ragwort *Senecio jacobaea* (Asteraceae) deflorated by cinnabar moths *Tyria jacobaeae*; Fig. 4.4; lives much longer than than plants allowed to set seed; Gillman & Crawley 1990).

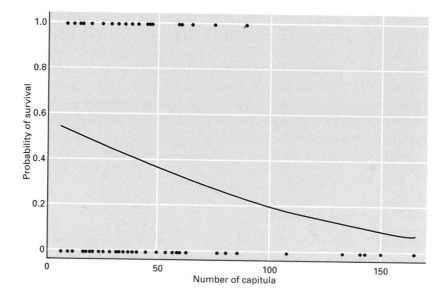

Fig. 4.4 Logistic regression of probability of survival of ragwort as investment in reproduction increases. The data consist of 1s (alive) and 0s (dead) and the curve is the fitted logistic (see Crawley 1993a). In this experiment the size of the plant's rootstock also influenced the risk of death; for a given production of flowerheads (capitula) a plant with a larger rootstock was less likely to die over winter. (From Gillman & Crawley 1990.)

4.4.10 Resource extraction/growth rate

This trade-off involves the ability to extract and retain nutrients under conditions of chronically low nutrient availability versus the ability to grow rapidly when nutrients are readily available (Mamolos *et al.* 1995; Ryser & Lambers 1995). The suite of traits exhibited by stress tolerators is listed in Table 4.1; these are traded off against traits exhibited by ruderals on the one hand and by competitors on the other. It reflects the principal axis (the *stress* axis) of Grime's ordination of temperate herbaceous vegetation.

4.4.11 Defence/growth rate

Should plants put limited resources into vegetative growth or into defence of the tissues they already have? This trade-off is discussed in detail in Chapter 10 (see Mole 1994).

4.4.12 Growth/reproduction

Because there is usually a 'cost of reproduction' in terms of reduced subsequent survival, there is typically a trade-off between investment in vegetative growth and reproduction, and the trade-off is likely to be most severe in resource-poor habitats (Biere 1995). In annual plants, this trade-off is generally resolved in favour of a simple switch; 100% invest-ment in vegetative growth is followed by a switch to 100% investment in reproduction (see above). In long-lived, repeat-flowering perennials, in-vestment in reproduction is typically not attempted until the require-ments of survival (growth, storage and defence) have been met (see above). There are a few examples in which hand pollination has increased reproduction, and this had led to increased rather than reduced survivor-ship (e.g. *Primula veris*); Lehtila & Syrjanen 1995a), but it is more usual to find that increased reproduction is correlated with shorter lifespan (e.g. in populations of the short-lived perennial *Arabis fecunda* (Brassi-caceae) in Montana; Lesica & Shelly 1995).

4.4.13 Male/female reproductive function

Hermaphrodites can put a varying proportion of their reproductive effort into male (pollen) and female (seeds) function and they can vary the relative timing of production of male and female flowers (see Chapter 6). In dioecious species, the conflicts between male and female investment can lead to the evolution of sexual dimorphism (Meagher 1992), and females are expected to show greater trade-offs with other life-history traits than males because of their higher investment in reproduction (Delph & Meagher 1995).

4.4.14 Shade growth rate/shade death rate

In models of forest regeneration, Pacala (see Chapter 15) discovered an important axis (second only to the dispersal/competitive ability axis, see above) in which the saplings and seedlings of different tree species appear to trade off their growth rates in shade against their survival in shade. One group of species grew slowly but could live in shade for a long time, while another group were capable of relatively rapid growth in the shade but had low survivorship unless a canopy gap opened up above them.

4.4.15 Gap/forest regeneration niche

Seedlings of some tropical tree species trade off high tolerance of direct sunlight against tolerance of varying degrees of shade. Regeneration of the light-tolerating species is restricted to occasional light gaps in which the saplings grow very rapidly; the shade tolerators grow much more slowly but can regenerate over a much larger fraction of the forest floor (Clark & Clark 1992; Condit *et al.* 1992, 1995).

4.4.16 Sun leaves/shade leaves and water/light

There is a trade-off in leaf construction between tolerance of water limitation and tolerance of light limitation; leaves that are effective at photosynthesis in low-light environments tend to be poor at water retention in dry atmospheres (see Chapters 1 & 2; Barton 1993; Berkowitz *et al.* 1995; but see Schmitt 1993).

4.4.17 Growth rate/nutrient retention

Plant traits leading to low nutrient loss rates (long leaf lifespan and low nutrient concentrations in leaves) are correlated with low rates of dry matter production (e.g. low leaf nitrogen contents; see Chapter 1). Thus, evergreen plants are favoured in low-nutrient environments because the benefits of low nutrient loss rates outweigh the costs of reduced maximum photosynthetic rate; low nutrient loss rate leads to higher equilibrium biomass despite slower initial growth rates (Fig. 4.5; see Aerts 1995).

4.4.18 Fruit weight/seed weight

If a plant invests too heavily in the pulp of a fruit, it will be highly attractive to frugivores, but the payload (seeds equipped to create competitive seedlings) will be unprofitably low. With too heavy a seed, however, the whole fruit is likely to be dropped directly beneath the parent plant (see Chapter 9).

4.4.19 Pollen quantity/pollen quality

The plant might invest in such elaborate flowers that it guarantees the attentions of just one species of pollinating insect (e.g. *Ophrys* (Orchidaceae) described in Box 6.4). This ensures that pollen quality will be high (it is bound to have come from a different individual of the same species) but the quantity of pollen received could be extremely low if the year happens to be bad for that particular insect. Alternatively, the plant could produce flowers that were attractive to most flying insects (e.g.

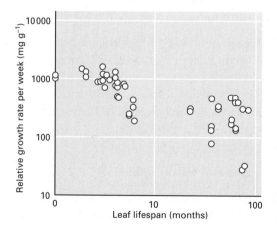

Fig. 4.5 Relative growth rate and leaf longevity. Evergreen plants grow more slowly than deciduous plants, partly because they have lower tissue nitrogen concentrations and partly because they invest more heavily in defence of leaves (see Chapter 10).

Apiaceae). Now the quantity of pollen would be assured, but the quality is likely to be low (the stigmas could easily become clogged by useless pollen from unrelated plant species; see Chapter 6).

4.4.20 Flammability/competitive ability

Attributes that influence flammability (volatile oils, retaining copious dead wood in the canopy) may have other functions (e.g. volatile oils may protect the plant from herbivores). Given that it is possible to construct models suggesting that individual-based selection can favour flammability and self-immolation (Bond & Midgley 1995, and see below), there may well be trade-offs between traits increasing flammability (resources stored in above-ground fuels) and traits favouring survival after fire (resources stored in fireproof, underground tissues like lignotubers; Possingham *et al.* 1995).

4.5 Canopy architecture

All terrestrial plants must fulfil the basic requirements of survival, growth and reproduction:
1 they must intercept sunlight efficiently;
2 they must be capable of thermoregulation and control over water loss;
3 they must sustain the weight of the above-ground parts;
4 they must exchange gases with the external environment for photosynthesis and respiration;
5 they must absorb and conduct liquids;
6 their roots must be securely anchored and capable of extracting mineral nutrients from the soil;
7 each of their successive growth stages must be competent in all of these functions; and
8 they must reproduce as adults (Niklas 1988).
Plant form is therefore a compromise between resource gathering (leaves and roots), sex (pollination and seed dispersal) and the physical demands of the abiotic and biotic environments (protection against wind, ultraviolet (UV) radiation, desiccation, pathogens and herbivores). This compromise is resolved in different ways by the species coexisting in any given environment, but all of these compromises in morphological expression represent the interaction between physical and developmental constraints on growth.

The literature on canopy architecture of trees draws attention to the wide variety of different ways in which woody plants can exploit the aerial environment. For example, there are 23 developmentally distinct architectural models in tropical trees, based on characters like periodicity of growth flushes, orientation on primary shoots and the position of inflorescences (e.g. terminal or axillary; Hallé *et al.* 1978). One important way that tree canopies differ from one another is in their 'branch order' as defined by the 'Strahler technique'. The outermost twigs of this year's growth are defined as first order; the meeting of two first-order shoots produces a second-order branch, the meeting of two second-order branches produces a third-order branch, and so on. Some trees like ash *Fraxinus excelsior* (Oleaceae) have very low branch orders (e.g. order 3–6) while others like oak *Quercus robur* are much more intricately branched (e.g. order 6–9). The three-dimensional arrangement of shoots and leaves within the canopy is another important trait (Waller & Steingraeber 1986).

Control of plant shape is highly localized and occurs at a very small scale. For example, branches and twigs that grow vigorously this year tend to produce more and bigger twigs next year, while stressed sub-units frequently die back or remain dormant. This allows the plant to respond opportunistically to favourable patches and to minimize investment in areas that are shaded or otherwise unsuitable for growth. Thus the asymmetric illumination of shoots affects the interconversion of the two forms of phytochrome with the result that auxin production on the shadier side is cut, which stimulates elongation of the shady side and causes the shoot to bend towards the light (Waller 1986). Branches and parts of branches that receive more light grow and divide more prolifically than shaded branches, with the result that trees planted in a free-standing group only have branches on the outward-facing sides of their trunks (Plate 4, facing p. 366).

Clearly, there are problems with simple generalizations like these: (i) three-dimensional branching patterns are difficult to quantify; (ii) it is difficult to separate genetic and microenvironmental determinants of branching; and (iii) the selective advantage of different branching patterns between species and between habitats is by no means clear (Niklas 1988). With field work and theoretical modelling, however, it may be possible to distinguish between alternative hypotheses of canopy architecture. We could ask, for example, whether plants are selected: (i) to maximize cover per unit biomass; (ii) to overtop and shade-out other individuals; (iii) to maximize photosynthetic area per unit volume; (iv) to minimize self-shading of branches, and optimize leaf area index; (v) to optimize canopy microclimate; or (vi) to maximize the efficiency of pollination and/or fruit dispersal (Cody 1984). We need to bear in mind that botanical branched structures are complex systems that perform many tasks simultaneously, and their design is constrained by the history of their development from germination to senescence, by resource availability and by phylogenetic constraints. It is most unlikely, therefore, that the designs are in any sense 'optimal': it is much more likely that they represent *efficient designs* given their historic, genetic and environmental constraints (Fig. 4.6; Farnsworth & Niklas 1995).

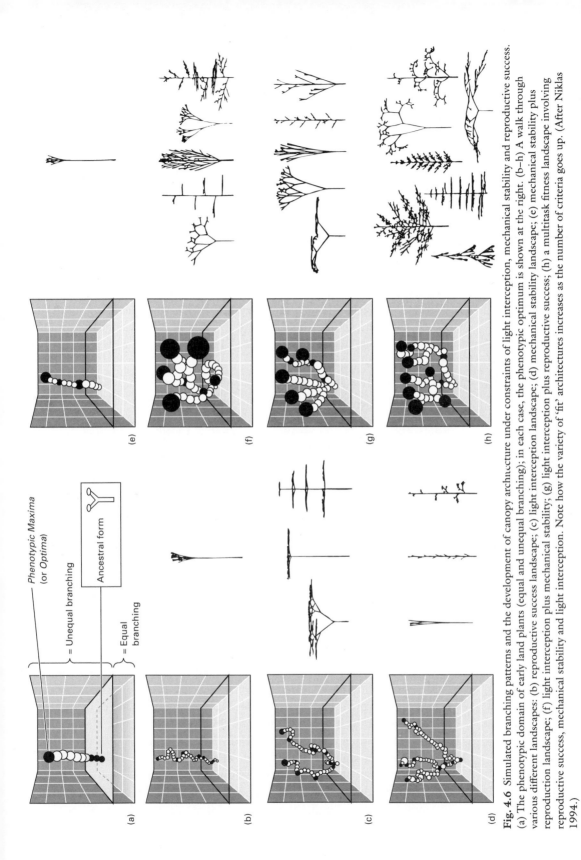

Fig. 4.6 Simulated branching patterns and the development of canopy archuecture under constraints of light interception, mechanical stability and reproductive success. (a) The phenotypic domain of early land plants (equal and unequal branching); in each case, the phenotypic optimum is shown at the right. (b–h) A walk through various different landscapes: (b) reproductive success landscape; (c) light interception landscape; (d) mechanical stability landscape; (e) mechanical stability plus reproduction landscape; (f) light interception plus mechanical stability; (g) light interception plus reproductive success; (h) a multitask fitness landscape involving reproductive success, mechanical stability and light interception. Note how the variety of 'fit' architectures increases as the number of criteria goes up. (After Niklas 1994.)

4.5.1 Modular growth

Growth in most vascular plants is indeterminate, with meristems continuing to produce new shoots until the opportunities for further local growth are exhausted. The modules are made up of smaller units ('metamers' *sensu* White 1984) consisting of a leaf, its internode and the associated axillary bud. The iterated modular nature of growth gives the potential for indefinite growth and relative autonomy among plant parts, giving great plasticity in final size. Despite this, most plants have an easily recognizable overall growth form (what naturalists call their 'jizz', a phrase derived from wartime aircraft recognition, where GIS was the 'general impression of shape'). The existence of jizz implies that there are characteristic, and probably rather simple, rules of growth.

Plant shape is important underground as well. In a range of herbaceous species, Bell *et al.* (1979) recognized two major patterns of rhizome arrangement: (i) near-hexagonal forms well suited to filling space efficiently; and (ii) linear forms, capable of rapid expansion into new areas (what Lovett Doust (1981) called 'phalanx' and 'guerrilla' growth forms).

4.5.2 Integration of plant growth

Roots and shoots are obviously integrated to the extent that they depend on one another for carbohydrates, proteins, soil nutrients and water transported though phloem and xylem. Sources (mature expanded leaves) supply photosynthate for sinks (growing points, roots, storage), but the costs of mobilization and active transport mean that it is more economic to use local sinks rather than more distant ones (Watson & Casper 1984). Also, vascular integration tends to be local (Pitelka & Ashmun 1985) and integration works both by vascular connections and communication via hormones (e.g. this could occur through root grafts; Chiariello *et al.* 1982). The impact of herbivores on plant shape through the alteration of apical dominance is discussed in Chapter 13.

4.5.3 Allometry

Larger organisms have a larger surface area to volume ratio, which means that the plant body must devote an increasing fraction of its total mass to support as it grows larger. The strength of a column or a beam scales with the square of its radius, while its mass increases with volume (i.e. with radius squared times length). For example, the diameter of a trunk tends to enlarge faster than its length, generally scaling with length to the power $3/2$ (McMahon 1975). This allometry preserves dynamic or elastic similarity, making a limb of any size bend to a constant deflection per unit length, and highlighting the importance of dynamic stresses in the design of tree canopies. The number, length and angle of divergence of branches may all vary. These allometric relationships indicate important constraints which determine those plant architectures that are possible and those that are likely to be profitable. The performance of the canopy needs to by integrated over the daily and seasonal course of the sun and this, in turn, is affected by the location, size and the specific identity of neighbouring plant canopies.

4.5.4 Plant height

Light intensity declines roughly exponentially with distance beneath the surface of the canopy, but the costs of structural tissue scale less rapidly than exponentially with height. This means that increasing competition for light favours increased investment in height growth (Givnish 1982). However, plants tend to grow only as tall as is necessary to compete; there is no fitness payoff for being taller than necessary (benefits from increased visibility to pollinators and increased seed dispersal distances are likely to be offset by an increased likelihood of being blown over during storms). Large plants like trees have to grow through a series of stages from tiny seedling to giant mature plant; each one of these stages must be physiologically and ecologically competent, and each stage must be attainable by simple growth rules from the preceding stage (Waller 1986).

4.5.5 Leaf arrangement

Photosynthesis saturates in about 25% full sunlight (see Chapter 1) so that a plant growing in the open can increase its total productivity by producing several diffuse layers of leaves with a certain amount of self-shading, rather than producing a single continuous layer of leaves (multilayer canopy architecture, *sensu* Horn 1971). In shade, however, it is best to avoid self-shading, and a single layer of shade-adapted leaves is preferable (monolayer canopy; Horn 1971). Other things being equal, monolayers should have higher rates of *net* photosynthesis at very low light intensities (because there is no self-shading), while multilayers should be more drought resistant than monolayers (because they experience reduced heat loads; see Horn 1971 for details). This reasoning led Horn to predict that multilayer canopy architecture would be common amongst early successional trees and monolayer architecture amongst late succession, shade-adapted tree species. Another adaptation for rapid tree growth in shade is the use of 'throw-away branches' in the form of long, tough petioles that fall to leave a pencil-like trunk with a wide, diffuse crown without the need for investment in permanent support tissue (Givnish 1978).

4.5.6 Phyllotaxis

In addition to leaf shape, the geometry of angular arrangement of leaves on stems (phyllotaxy) can influence the efficiency of light interception. Niklas (1988) points out that in rosette-forming species with short internodes phyllotaxy is important in reducing self-shading of leaves, but in taller plants with longer internodes factors like leaf-flutter, elongated petioles and solar tracking, coupled with the daily and seasonal movements of the sun, mean that phyllotaxis is of lesser importance. For prostrate rosette species, however, light interception is maximized by a phyllotactic pattern that produces a leaf divergence angle close to 137.5° (e.g. 8/21, 21/55 or 34/89). This angle represents the limit approached by the series of Fibonacci fractions that characterizes the various phyllotactic patterns commonly seen in plants (1/2, 1/3, 2/5, 3/8, 5/13, etc.; see Niklas 1988 for examples).

4.5.7 Switch from growth to reproduction

For annual plants growing in relatively predictable seasonal environments, the optimal strategy is to put 100% of effort into growth, and then to switch and put 100% of subsequent effort into reproduction (Paltridge & Denholm 1974; Macevicz & Oster 1976); this is the so-called 'bang-bang' model of reproductive investment. If the end of the growing season is unpredictable, or the plants face strong competition favouring increased height growth, then the optimal strategy is a little more complex; Cohen (1971) shows that under these circumstances the plant should show a period of partial investment in reproduction before finally switching to all-out commitment to seed production. For longer-lived plants like polycarpic trees, it appears that survival is much more important than reproduction in any given year. Oak *Quercus robur* only invests resources in acorn production that are surplus to its requirements of wood growth, storage and provisioning of buds for the following spring. Part of the evidence for this hierarchical investment by the plant is that exclusion of insect herbivores using pesticides increases acorn production without increasing shoot length or girth increment (Crawley 1985; Crawley & Long 1995).

4.5.8 Ageing and senescence

Since meristems are capable of continuing their growth indefinitely, why is it that plants show ageing? For example, all the leaves on an ancient oak tree in summertime are just a few months old. It may be that traits favouring fitness early in life have an inevitable cost in senescence (*antagonistic pleiotropy*, Williams 1957; Hamilton 1966; Finch & Rose 1995). Alternatively, the gradual deterioration in plant performance may have to do with the accumulation of deleterious somatic mutations or of plant viruses (Gill 1986). The cause might be environmental (e.g. the depletion of certain trace nutrients, or the accumulation of allelopathic chemicals in the soil). The mystery of senescence is much more profound for clonal herbs and grasses that have no really long-lived structures than it is for woody plants (Orive 1995; Pedersen 1995). For example, in the case of trees, senescence might be nothing more than the collapse of the woody support structures.

4.6 Environmental factors affecting plant performance

Life-history patterns and growth forms place constraints on the survival and reproduction of plants, which mean that they are competitive with other species only under a *limited* range of environmental conditions (Orians & Solbrig 1977). Some species may be 'Jacks of all trades' but no single species can be 'master of all'. In this section I describe some of the morphological, physiological and phenological traits of plants that appear to be associated with particular environmental factors. Remember, however, that without careful, long-term experimentation, it is all too easy to misinterpret the 'adaptive significance' of a plant's features.

Throughout this section, a distinction is made between the *avoidance* and *tolerance* of extreme conditions. Similarly, we should distinguish between those responses that reflect constraints imposed by the plant's genetic programme, and those that reflect phenotypic plasticity in response to local environmental conditions (Clausen *et al.* 1948). Within a species there will be both genotypic variation and phenotypic plasticity. Indeed there may often be genotypic variation *for* phenotypic plasticity (Bradshaw 1973). The ecological range occupied by a species is determined by the interplay between all these factors.

Finally, it must be stressed that single factor explanations will usually prove to be inadequate as descriptions of a plant's environment. In most communities, plant performance is affected by the interaction of several kinds of factors. So, for instance, many Australian ecosystems are extremely arid, extremely nutrient-poor and very prone to fires. Similarly, high alpine plant communities suffer extremely short growing seasons, extremely intense sunlight and extreme exposure to high winds.

4.6.1 Fire

Fire is one of the most important factors in many of the world's plant communities (Kozlowski & Ahlgren 1974; Mooney *et al.* 1981). The likelihood of fire is determined by the interplay between the weather (how dry it is), the flammability of the plant species (both living and dead) and the amount of fuel present. This, in turn, depends upon the time since the last fire. Communities differ in the intensity of fires (Glitzenstein *et al.* 1995), in their frequency (Morrison *et al.* 1995) and in the area burnt during a typical outbreak (especially the availability of unburned patches to act as sources of propagules; Turner *et al.* 1994). For example, some communities may suffer frequent surface or 'grass fires', while others experience much less frequent (but more devastating) canopy fires. Peatlands may experience 'ground fires' in which the peat itself is burned, with severe consequences for the seed bank and for below-ground perennating organs (Mallik *et al.* 1984).

The results of fire include: (i) release of nutrients from accumulated standing dead biomass; (ii) breakdown of hydrophobic plant litter, which leads to improved soil wettability and opens up establishment microsites; (iii) breaking dormancy in many fire-adapted species (Went *et al.* 1952); and (iv) removal of inhibiting chemicals including allelochemicals from the soil surface. The impact of a fire on the plant community is determined by the season in which it occurs (because species vary seasonally in their vulnerability) and in the area burnt (as this influences the proximity of surviving plants that can act as seed parents). The great majority of species are killed by fires of even moderate intensity, and frequent fires greatly restrict the pool of species capable of persisting in a community (Naveh 1975; Mooney & Conrad 1977; Kruger 1982).

The four main kinds of adaptations to fire are:
1 resistance, where plants have thick, fireproof bark, as in *Pinus resinosa*;
2 regeneration by sprouting from root stocks or surviving stems, common in many deciduous trees like *Populus* (Salicaceae) and *Betula*;
3 possession of specialized underground organs like the lignotubers of certain *Eucalyptus* (Mytraceae) species;
4 specialized, long-lived fruits that accumulate on the plant over a number of years, only opening to release their seeds after the passage of a fire (as in the serotinous cones of *Pinus banksiana* and the woody follicles of Australian *Hakea* (Proteaceae).

Life-history traits associated with flammability and susceptibility to fire are related to time since the last fire; models sugest that two species of fire-adapted tree are likely to coexist with a late successional tree if their flammabilities are very different and if the most flammable species is more susceptible to fire but less likely to die from other (non-fire) causes (Possingham *et al.* 1995).

Why should a plant want to burn? Surely it would make sense for plants to evolve towards being fireproof? Yet where climatic conditions favour fire (e.g. shrublands in North American chaparral, Australian mallee or South African fynbos), plants seem to possess traits like high volatile oil content and retention of copious quantities of fine dead branches within the canopy that encourage fire spread and increase fire intensity (Zedler 1995). In 1970, Mutch proposed that natural selection had favoured the characteristics that make plants from fire-prone regions more flammable, but this argument was considered too group selectionist to be credible. In a cellular automation model of competition between flammable 'torches' and non-flammable 'damps', Bond & Midgley (1995) showed how individual selection could favour an increase in flammability so long as the benefits outweigh the costs. The case was considered for two strategies: (i) for species that regenerated from seed following fire, the 'torch' strategy prospered so long as it had superior fecundity and/or superior seedling establishment than the 'damp'; while (ii) for species that regenerated from woody stem bases or lignotubers (sprouters) the 'torch' had to have a lower probability of being killed by the fire than the 'damp'. In both cases, the 'torch' strategy was more likely to be favoured as the radius of neighbours killed by immolation of the 'torch' increased.

Frequent grass fires and infrequent canopy fires exert quite different selection pressures, a fact exemplified by the contrasting life histories of different species of pines. Seedlings of most pines are killed by grass fires, but species like *Pinus montezumae* have grass-like seedlings that do not produce a woody aerial shoot for several years after germination. Instead, they produce herbaceous 'grass foliage' which dies back to the ground each year, and the seedlings build up an underground reserve in a woody rootstock over a period of years. When the reserve is large enough, the plants suddenly produce a shoot that grows tall enough in a single season to take it out of the range of grass fires. In contrast, pines from communities domainated by canopy fires (like bishop pine *Pinus monticola* are not severely harmed by grass fires. They grow for 60 years or so, accumulating serotinuous cones on their branches and shedding no seed. After a canopy fire kills the mature trees, the accumulated seeds of 40 or more years of reproduction are released into the ash in a single burst, where they germinate free of interspecific competitors.

Survival of heather *Calluna vulgaris* (Ericaceae) following fire depends on the intensity of the fire. Cool fires do not kill heather plants, which then resprout from surviving stem bases. Alternatively, where survival is poor, there is recruitment from the seed bank. Severe fires, however, cause the upper organic soil horizons to burn, killing most of the seed bank; fires that burn the peat to a depth of > 10 cm are likely to kill the entire seed bank. The lichen/algal crust that eventually develops over the surface of exposed organic soil after severe moorland fires further hinders heather regeneration by reducing germination of heather seeds by about 40% (Legg *et al.* 1992).

4.6.2 Drought

The availability of soil water is a prime determinant of plant community structure. Communities growing where water is in short supply (i.e. where potential evaporation greatly exceeds actual evapotranspiration) tend to be dominated by plants showing one or more adaptations to drought tolerance ('xeromorphic' features), or by annual plants which escape the drought as dormant seeds in the soil (the annuals only germinate after rainfall, and grow rapidly in conditions of relatively high water availability).

Drought-tolerant plants improve their water relations by both increasing their efficiency in extracting and storing water, and reducing the rate at which they lose water through evapotranspiration (Slatyer 1967). Water extraction is improved by modifications to the root system, and in places where the water table lies far beneath the surface, long-lived plants may develop very deep root systems (e.g. mesquite, *Prosopis juliflora* (Fabaceae), may have roots reaching down to 12 m). The young, shallow-rooted individuals of such species are particularly vulnerable to death through desiccation, and frequently exhibit novel means of water gathering (e.g. mesquite seedlings can absorb dew through their leaves). Where the water table is beyond the range of even the deepest-rooted plants, and where rainfall occurs as brief, light showers, then plants with rapidly growing, extensive, but shallow roots may be dominant (e.g. saguaro cactus, *Carnegia gigantea)*.

Water loss through leaves is reduced in a wide variety of different ways including:

1   reduced leaf area;
2   abscission of leaves during the dry season (drought-deciduous plants);
3   modifications of the stomata (e.g. setting them in deep pits in the epidermis);
4   possession of a thick, waxy epidermis;
5   positioning of the stomata on the inner surface of tightly inrolled leaves;
6   orienting the leaves vertically to minimize the direction of radiation they receive;
7   investing the leaf surface with a clothing of hairs.

It is well known, for example, that species that are normally glabrous become hairy in dry places, and that hairy species become more hairy (e.g. *Ranunculus bulbosus*). Further improvements in the efficiency of water use are brought about by a range of physiological modifications to photosynthesis and respiration (see Chapter 2), by tolerance of high levels of tissue dehydration (Kozlowski 1972) and by the possession of a deep, highly branched root system (Chapin 1995) and large seeds (Fagouri *et al.* 1995).

The second problem faced by plants in sunny, arid environments is overheating. Unable to cool their leaves by increased evapotranspiration, they show a range of morphological features that reduce heat loads, and physioloical adaptations which permit the tolerance of higher temperatures. Overheating is avoided by: (i) increasing the surface area for heat dissipation; (ii) protecting the plant surface from direct sunlight; and (iii) increasing the reflectance of the surface. All of these can be accomplished by covering the surface with a dense felt of hairs. The hairs also produce a boundary layer of still air next to the stomata, which traps moisture and reduces evaporation (note that the hairs may also have a role in defence against herbivorous insects). Not all xerophytic plants are hairy, however, and some desert species have completely hairless, smooth surfaces. In these cases, water loss is restricted by a thick covering of wax, or by modifications to the walls of the epidermal cells (Levitt 1972).

Water shortages are not restricted to plants of desert regions. Xeromorphic features are also exhibited by plants growing in habitats where water is unavailable because it is frozen or too salty to extract (this is sometimes called 'physiological drought'). Even temperate species from mesic environments may suffer water shortage during one or more of the summer months. In tropical rain forests, many of the canopy tree species show xeromorphic leaf features on the mature individuals that emerge into the intense sunlight above the surface of the leaf canopy, while saplings and shaded mature individuals of the same species have larger, thinner shade leaves lacking xeromorphic features.

Indirect consequences of drought on plant performance include an increased susceptibility to damage by pathogens (Boyer 1995) and insect herbivores (White 1993). Many insects, for example, show a preference for wilted rather than turgid tissues (Lewis 1979), and aphids feeding on irrigated crops do far less damage than on unwatered plots (Cammell & Way 1983). Water shortage alone leads to an increase in free amino acids in the foliage and in the phloem, which may increase the quality of these tissues as food for insect herbivores (see also Chapter 10).

Experimental studies on competition intensity along natural gradients of productivity tend to support Grime's contention that competition is more intense on relatively productive sites, whereas results from experimentally created gradients tend to suggest that competition intensity is similar across a range of productivities (Goldberg & Barton 1992; Wilson & Tilman 1993). In an experiment on plant competition along an experimental gradient in productivity produced by a range of watering treatments, Kadmon (1995) manipulated densities of the desert annual grass *Stipa capensis* by neighbour removal. He measured absolute competition intensity as $\Upsilon_E - \Upsilon_C$ and relative competition intensity as $(\Upsilon_E - \Upsilon_C)/\Upsilon_E$, where Υ_E is mean per capita fitness (seed production) of experimentally isolated plants, and Υ_C is the performance of control plants subject to competition. Not surprisingly, absolute competition intensity was more sensitive to changes in productivity compared with relative competition intensity (Fig. 4.7), but both measures of competition increased with standing crop (Kadmon 1995).

Experimental treatments that indirectly modify root size and architecture can have important effects on drought tolerance. For example, nitrogen-fertilized grasses suffered most from drought and recovered from drought most slowly (Tilman & Downing 1994) and the acidified,

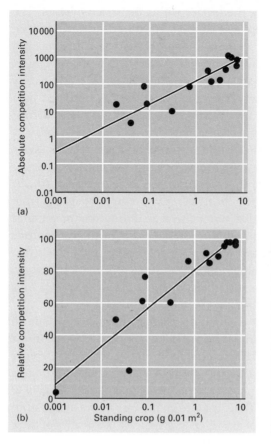

(a)

(b)

Standing crop (g 0.01 m²)

Fig. 4.7 Competition experiments using watering and plant removal: the difference involved in measuring competition as absolute competition intensity (the difference in biomass) or relative competition intensity (the proportional difference in biomass). In both cases, the intensity of competition increases with standing crop (see text for details). (After Kadmon 1995.)

high-nitrogen plots at Park Grass, with their shallow-rooting domainants, showed much greater variation in hay yield in response to fluctuating spring and summer rainfall (see Chapter 14 and Fig. 14.5).

4.6.3 Waterlogging

Waterlogging of the soil by flooding is common in many habitats following snow melt or after very heavy rains. This has both direct effects on plant root systems and indirect effects on soil structure and function. Physical attributes of the soil are altered, including restriction of gas exchange (depletion of molecular O_2, and accumulation of soil gases like N_2, CO_2, CH_4 and H_2), thermal effects (altered radiation absorption and reflectance, modified heat flux, and so on), and alterations to soil structure (increased soil plasticity, breakdown of crumb structure, and swelling of soil colloids, especially in sodic soils with high clay content).

Within hours of flooding, normal soils are virtually devoid of O_2 and aerobic soil organisms are replaced by facultative anaerobes. Prolonged flooding leads to the replacement of these forms by obligate anaerobes. For example, fungi and actinomycetes are dominant in freely drained, aerobic soils, but are largely replaced by bacteria under waterlogged conditions, leading to a reduction in the average rate of decomposition. Not all waterlogged soils are uniformly anaerobic, however, because the roots of many vascular marsh plants contain special air-conducting cells (aerenchyma) and give off oxygen, making their rhizospheres aerobic (Jackson & Drew 1984). Flooding also has profound effects on the electrochemistry of the soil solution; there is an immediate dilution of the solution and a great reduction in redox potential, usually accompanied by an increase in pH (Gebauer *et al.* 1995; Huang *et al.* 1995; Joly & Brandle 1995).

Many plant species are killed by even brief exposure to waterlogging. Others are capable of a variety of responses that enable them to survive flooding of their root systems. The most frequent responses are swelling of the stem base (hypertrophy), wilting, reduced leaf growth, abscission, epinastis, root death and rotting, with adventitious rooting higher up the stem. As a consequence, waterlogged plants often show signs of nutrient or hormone deficiency. There is often an alteration in the pattern of hormone distribution within the plant, with increased levels of ethylene, auxin and abscisic acid in the shoot, and decreased levels of gibberellins and cytokinins. Flooding tends to increase the rate of flower and fruit abortion, and prolonged waterlogging leads to increased mortality, even amongst somewhat flood-tolerant spieces. In a comparison of the performance of three species of *Veronica* (Scrophulariaceae) Dale & Causton (1992) found that waterlogging had a bigger negative effect on relative growth rate and unit leaf rate compared with drought. Plants in waterlogged soils are also more susceptible to disease (Kozlowski 1984).

Certain species, of course, inhabit permanently wet soils. Specialized 'flood trees' are capable of growing with their root systems permanently immersed in fresh water (*Taxodium distchum, Nyssa aquatica*) or salt water (*Rhizophora mangle*). In these plants there are specialized root morphologies that allow oxygen transportation to the growing tips. These include 'knee roots' (pneumatophores) and surface-sprouting adventitious roots, coupled with hypertrophied lenticels, which allow gas exchange and toxin release to the surface waters (Keeley 1979). The seeds of flood trees are tolerant of long submersion, and often germinate immediately the floodwaters recede. Some, like *Populus deltoides* (Salicaceae), can even germinate underwater, and their seedlings can survive considerable periods of submersion (Clark & Benforado 1981).

Herbaceous vascular plants from permanently wet sites show a range of morphological and physiological traits including: (i) anatomical features allowing oxygen transportation to the roots (Arenovski & Howes 1992); (ii) the ability to exclude or tolerate soil toxins such as Fe^{2+} and H_2S; and (iii) biochemical features that allow prolonged fermentation (anaerobic glycolysis) in the roots (Fitter & Hay 1981). Salt marsh plants like *Salicornia* (Chenopodiaceae) have to cope with both waterlogging and salinity; the frequency of flooding is greater at lower elevations on the marsh but salinity is greater at higher elevations. The interaction between waterlogging and salinity tolerance affects interspecific competitive ability and accounts for the zonation of plant species within the marsh (Pennings & Callaway 1992). Waterlogging of the marsh grass *Sporobolus virginicus* caused initiation of adventitious surface roots, increased internal aeration, induced alcohol dehydrogenase activity and decreased below-ground biomass (Naidoo & Mundree 1993).

4.6.4 Shade

The relationship between plant performance and light intensity is discussed in detail in Chapter 1, and here we are concerned with the features exhibited by plants living in more or less permanent shade (see Boardman 1977 for a review). In temperate, deciduous forest, the most typical class of shade plants exhibit 'phenological escape'. They grow rapidly and flower in the spring before the forest floor is shaded (e.g. *Hyacinthoides non-scripta* (Liliaceae) and *Anemone nemorosa* (Ranunculaceae)). They show few morphological modifications for shade tolerance and their leaves die back to ground level when the canopy closes in summer. It is interesting that plants with this life history are absent from deciduous woodlands on very infertile soils, where the ground flora tends to consist of a rather stable community of slow-growing, evergreen herbs (Grime 1979).

Plants living in permanent shade include many ferns, mosses and lichens, as well as the (usually evergreen) vascular plants of the floor of evergreen forests. These species must maximize their photosynthetic gain from the low levels of energy they receive, by means of:

1 a reduced respiration rate – this lowers the compensation point so that the plant can maintain positive rates of net photosynthesis at lower light intensities;

2 increased unit leaf rate (the photosynthetic rate per unit energy per unit leaf area);

3 increased chlorophyll per unit leaf weight;

4 increased leaf area per unit weight invested in shoot biomass (by having thinner leaves, or arranging leaves in flat tiers to minimize self-shading; see above).

Interesting case histories are described by Björkman (1968) for sun and shade ecotypes of *Solidago virgaurea* (Asteraceae) and by Solbrig (1981) who compared the performance of three species of *Viola* in reciprocal transplant experiments in different degrees of shade.

Leaf morphology is highly plastic, and many tree species in tropical forests possess large, thin shade leaves as juveniles, but smaller, thicker sun leaves when they emerge into the full sunlight of the upper canopy.

Other differences include:

1 a single-layered upper and lower epidermis in young shade leaves, but a multilayered epidermis in sun leaves;

2 short, broad palisade cells in shade leaves, but long, thin cells in leaves of the adult tree;

3 fewer, larger stomata in the shade leaves of young trees;

4 almost hairless leaves in the shade but dense indumentum on sun leaves; and

5 low frequency of vascular bundles (each with weak bundle sheaths) in the shade leaves, but high frequency (with very strong bundle sheaths) in sun leaves of the adult tree (Roth 1984).

Leaf weight per area (LWA) and leaf size in 85 species of woody plants were investigated by Niinemets and Kull (1994). They found that LWA increased with relative light availability (shade-tolerant leaves were thinner) and that leaf size decreased with species light demand and increased with total plant height. In stoloniferous clonal plants, both leaf area and petiole extension are responsive to light intensity. In a fenland species *Hydrocotyle vulgaris* (Apiaceae) interconnected ramets in patches of different light supply developed very different morphologies, but in a woodland species *Lamiastrum galeobdolon* (Lamiaceae) physiological intergration amongst the ramets meant that local responses were evened out and morphology was similar in light and shade (Dong 1995). Other factors influencing leaf morphology are discussed in Chapters 1–3.

One of the more subtle adaptations to light is the ability of seedlings to detect neighbours growing to their sides, and hence to anticipate subsequent competition for light. When green light is played sideways onto these seedlings they etiolate more rapidly than seedlings irradiated by light of different colours. This phytochrome-mediated stem elongation is cued by the lowered red : far-red ratio of light reflected off or transmitted through neighbouring plants, and the youngest internodes tend to be more responsive than older internodes (Dudley & Schmitt 1995).

4.6.5 Disturbance

The word 'disturbance' is used in two different ways in the literature of plant ecology. The first use covers various accidental injuries suffered by mature or seedling plants (e.g. landslide, burial by sand, fire, or storm damage), defoliation by herbivores, or intervention by humans (e.g. ploughing arable land, road construction, etc.). Grime (1979) defines this usage of disturbance as 'the mechanisms which limit the plant biomass by causing its partial or total destruction'. The second use covers processes leading to the creation of bare ground, loose soil, or light gaps, which constitute microsites in which recruitment can occur. Clearly, the two meanings overlap, but I concentrate on the second meaning in this section – disturbance as a creator of establishment microsites (Mooney & Godron 1983; Pickett & White 1985).

The most important aspects of the disturbance regime are:

1 its spatial *scale* (the characteristic patch size of disturbance (e.g. a single plant canopy in a gap-regenerating forest, or a 10-km swathe of forest felled by a tropical storm);

2 the *frequency* of disturbance (e.g. annual snow-melt floods in an arctic river, or once in 500 year storms in southern England); and

3 the *intensity* of the disturbance (i.e. the fraction of the biomass destroyed within the characteristic patch size; this could vary from a worm cast destroying 0.01% of the biomass in a 1×1 m quadrat to a mud slide that removes 100% of the biomass).

In most communities, individual plants have tenure over pieces of real estate for protracted periods. This 'site pre-emption' means that other plants are kept out, unless or until the owner of the site dies naturally, or succumbs to accident or attack by its natural enemies. The degree to which the occupant of a site must debilitated before its space is invasible by other individuals is poorly understood, but is clearly of vital importance in studies of long-term vegetational dynamics (e.g. successional studies; see Chapters 14 & 15).

Specific adaptations related to disturbance and microsite creation can be summarized in two categories: (i) morphological, phenological or physiological attributes that increase the probability of getting into (and staying in) a newly created microsite; and (ii) traits that enable a plant to resist disturbance, and to keep other individuals out (McIntyre *et al.* 1995).

Seed and fruit morphology are amongst the most important determinants of microsite colonizing ability. In general, the larger the seed, the less the disturbance necessary for successful establishment. Thus oaks *Quercus robur* with their large acrons can produce vigorous seedlings in dense vegetation, whereas small-seeded trees like birch *Betula pendula* are light-demanding. In addition to their direct role in creating microsites (see Chapter 13) herbivores may also influence plant recruitment if defoliation leads to the production of smaller seeds, which are capable of recruitment only under a more restricted range of conditions than can be exploited by the larger seeds from undefoliated individuals (Crawley & Nachapong 1985). Species with wind-dispersed seeds (dust-like seeds of orchids, the pappus fruits of many Asteraceae, or the winged fruits of trees like *Acer* or *Fraxinus*) are especially prevalent in areas subject to erratic, large-scale disturbance (Grime 1979). Similarly, in habitats where there is frequent, severe disturbance, many of the short-lived species retain a bank of dormant seeds in the soil (Hastings 1980; see Chapter 7).

In addition to attributes of seed dispersal, the probability of obtaining tenure of a microsite is greatly increased by such *seedling* traits as shade tolerance, increased resistance to fungal attack and low palatability to herbivores. Seedlings capable of prolonged survival can exploit ephemeral resources (e.g. temporarily improved light or temperature conditions), whereas seed germination is an all-or-nothing process, irreversible if conditions deteriorate.

The second set of disturbance-related traits concern the ability of an individual plant to keep others out. Tolerance of factors such as fire, grazing, trampling and drought all enable a plant to maintain its tenure of a site in the face of severe disturbance. Other traits may (perhaps fortuitously) enable an established individual to exclude potential invaders either by: (i) casting a dense shade; (ii) allelochemical effects; (iii) the production of deep, persistent leaf litter; (iv) harbouring high densities of invertebrate herbivores that kill seedlings beneath the canopy; (v) desiccating the soil surface by virtue of a shallow, but highly competitive root system. For example, apparently large gaps in shoot cover in arid grassland are typically exploited by the root system of the dominant grass *Bouteloua gracilis*; disturbance within these gaps is necessary to remove root competition and to make the microsites invasible (Hook *et al.* 1994), and recruitment was only seed-limited following disturbance (Aguilera & Lauenroth 1995).

Communities can be seed-limited at one scale and disturbance (microsite)-limited at smaller scales. For example, Crawley and Brown (1995) studied recruitment of oilseed rape *Brassica napus* (Brassicaceae) on the M25 motorway around London. They found that the perennial grassland which dominated most of the motorway verge was impervious to invasion by seed (i.e. recruitment was microsite-limited). Thus, sowing seeds in a small-scale study (e.g. 100×100 m) would be unlikely to detect seed limitation. However, at a bigger scale (e.g. hundreds of kilometres), sufficient disturbances are encompassed that seed addition does cause increased recruitment (e.g. rape seed spilled from trucks in transit from the farm to the oil-processing plant; Fig. 4.8).

The intermediate disturbance hypothesis (see Chapter 14) predicts that plant species richness will be greatest in communities with moderate levels of disturbance and will occur at intermediate time spans following disturbance. Data from two long-term field experiments on periodic burning in tallgrass prairie vegetation in the USA demonstrated a monotonic decline in species richness with increasing disturbance (fire) frequency, with no evidence of an optimum. However, species richness did peak at an intermediate time period since the last disturbance (Collins *et al.* 1995). During a long-term study of exclusion of gophers from serpentine grassland, the abundance of perennial species increased greatly but then subsequently declined; it is clear that disturbance history is a major factor controlling local community variation in these serpentine grasslands (Hobbs & Mooney 1995).

4.6.6 Low nutrient availability

The principles of uptake, translocation and use of nutrients are de-

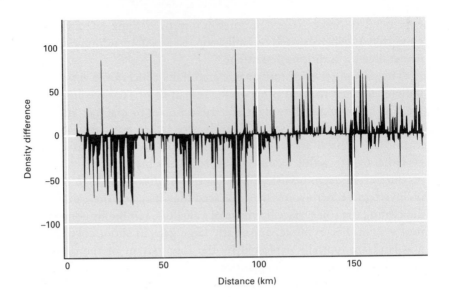

Fig. 4.8 Oilseed rape on the M25. The difference in mean oilseed rape population density per 100 m quadrat (average of 10 samples each of 100 m over 1 km for 1993 and 1994) between the clockwise and anticlockwise verges. From km 115 upwards there is an excess of positive density differences, while from km 0 up to km 115 there is a highly significant surfeit of negative differences. The cause of the difference appears to be oil seeds spilled from full lorries carrying their loads to Erith in Kent, close to km 0. The data suggest that rape recruitment is seed-limited at this scale, even though sowing experiments at smaller scales show no evidence of seed limitation in undisturbed vegetation. (From Crawley & Brown 1995.)

scribed in Chapter 3, and the implications of the relative availability of different nutrients on the outcome of competition between plant species are discussed in Chapter 8 (see Berendse 1994). Here we look at the features of plants found in environments where certain soil nutrients are available at chronically low rates of supply.

Perhaps the most conspicuous features of plants from low-nutrient environments are their small size, their tendency to have small, leathery, long-lived leaves, and their high root : shoot ratios (Chapin 1980; Vitousek 1982). Since the individuals have relatively small shoot systems, plant communities on nutrient-poor soils tend to be open, with bare gaps between the plants. Nutrient-use efficiency might be expected to be high in low-nutrient environments, as indicated by: (i) the production of a given amount of plant tissue from a smaller amount of nutrients; (ii) a lower rate of leaching of essential nutrients from live foliage, minimizing uptake needs; (iii) reabsorption of a larger proportion of foliar nutrients prior to leaf fall (Boerner 1984; Vitousek & Mattson 1984). The root systems of species tolerant of low soil nutrients often consist largely of storage tissues, which means that their typically high root : shoot ratios do *not* necessarily reflect extensive systems of nutrient-absorbing roots.

Physiological traits of plants from nutrient-poor soils include:

1 slow growth rates;

2 high investment in anti-herbivore defences;

3 selectivity (e.g. ability to take up calcium in the presence of high concentrations of Mg on serpentine soils; see below);

4 low saturation rates of nutrient uptake coupled with failure to respond to fertilizer application;

5 storage of nutrients for use when the supply rate drops ('luxury uptake');

6 efficient nutrient utilization;

7 flexible allocation patterns (e.g. plasticity within the same root system, so that there is proliferation of lateral roots in localized zones of higher nutrient availability);

8 high investment in mycorrhizal development;

9 efficient mechanisms of internal nutrient recycling to ensure minimal losses through leaf fall, exudation or leaching (Clarkson & Hanson 1980).

Perhaps the most fascinating adaptation to low-nutrient soils is found amongst the insectivorous plants. Some species exhibit movements in the capture of their prey, as in the Venus fly trap, *Dionaea* (Droseraceae); others possess traps or pitfalls to ensnare unwary insects, like the pitcher plants, *Nepenthes*; a third group traps insects on sticky leaves (e.g. butterworts, *Pinguicula* (Lentibulariaceae); or with special, sticky glands (e.g. sundews, *Drosera*). A full description of these plants is given by Kerner (1894; see Fig. 4.9). Experiments using radiolabelled amino acids have shown that the plants can take up intact amino acids from

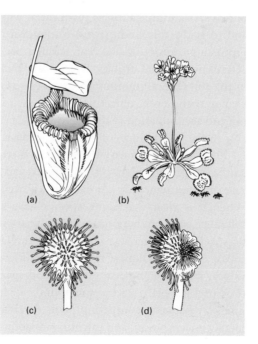

Fig. 4.9 Insectivorous plants. In nutrient-poor environments, certain species supplement their nitrogen and phosphorus requirements by trapping and digesting small, invertebrate animals. (a) the pitcher plant, *Nepenthes*; (b) the Venus fly trap, *Dionaea*; the leaf tip of a sundew, *Drosera*, showing its open sticky tentacles (c), and when inflected over a captured insect (d). (After Kerner 1894.)

their prey, and also that valuable supplies of phosphorus are obtained by this means (Chandler & Anderson 1976).

The flora of serpentine soils has attracted a great deal of attention from plant ecologists chiefly because of the number of rare plant species found there. The communities are open, and support low densities of rather small plants. Many of the constituent species are found only on the highly nutrient-poor and magnesium-rich serpentine soils (Proctor & Woodell 1975). The question that has occupied ecologists is whether these 'serpentine endemics' have a *requirement* for some specific factor provided only by serpentine soils (e.g. high magnesium), or whether the serpentine endemics are simply poor competitors in closed vegetation. Although there is some evidence for both processes (e.g. *Poa curtifolia* has a high Mg requirement), the weight of evidence points to the low competitive ability of the serpentine endemics on normal soils. When serpentine ecotypes of *Plantago erecta* were sown experimentally in cleared plots on normal soil they were healthy and produced good crops of seed, but in competition with native vegetation on uncleared plots they were suppressed and failed to set any seed at all. Their inability to compete on normal soils is thought to be a consequence of their intrinsically slow growth rate (see Chapter 12).

4.6.7 Soil acidity

The hydrogen ion concentration of the soil (pH) depends on the balance of basic and acid substances, and on the degree of dissociation of the acids. Several processes contribute to acidification of the soil, including (i) mineral weathering, which releases relatively large amounts of bases; (ii) humification of plant litter, which gives rise to organic acids and to acid-stable humus; (iii) respiration by roots and soil organisms producing carbon dioxide that acidifies the soil solution; (iv) uptake of certain nutrients as cations, which leads to an increase in hydrogen ions in the soil (e.g. ammonium-nitrogen, see Chapter 3). Soil pH tends to change with depth, reflecting differences in the balance between leaching and accumulation (Etherington 1982). The role of industrial pollutants in the acidification of soils and water is described by Paces (1985).

Because pH is so easy to measure, there has been a vast number of studies correlating plant distribution to soil acidity (e.g. Table 4.3). This ease of measurement is not matched, however, by ease of interpretation. While it is clear that some species have a very narrow range of pH tolerance, and are therefore good indicators of soil acidity (these tend to be species characteristic of either extremely acidic or extremely basic soils), others, including most species found in the mid range of pH values, can grow under a rather broad span of soil acidity. Away from the extremes, plant stature and abundance tend to be determined by the interplay between competing plant species, nutrient availability, heavy metal concentration, and calcium availability, rather than by pH as such.

Table 4.3 The relation between soil pH and distribution in Denmark of some meadow species. The figures represent the percentage distribution in the various pH ranges. Note the increase in species richness with increasing pH. (After Olsen 1923.)

Species	pH range								
	3.5 to 3.9	4.0 to 4.4	4.5 to 4.9	5.0 to 5.4	5.5 to 5.9	6.0 to 6.4	6.5 to 6.9	7.0 to 7.4	7.5 to 7.9
Deschampsia flexuosa	54	31	15						
Calluna vulgaris	31	23	31	15					
Molinia coerulea	25	25	25	11		4	4	7	
Potentilla erecta	18	18	18	21	10	5	3	8	
Hieracium pilosella			20	30	30	20			
Anthoxanthum odoratum	7	7	15	17	26	17	7	2	2
Deschampsia caespitosa				3	30	27	12	15	12
Galium palustre				11	29	32	14	11	4
Veronica chamaedrys				7	29	14	36	7	7
Scirpus sylvaticus							63	12	25
Succisa pratensis			8	17	8	8	17	42	
Tussilago farfara						11	11	44	33
Agrostis stolonifera							14	29	57
Total species	5	5	7	9	7	9	10	10	7

In general, high pH tends to be associated with 'unfavourability' (like deficiencies of iron or phosphorus), whereas low pH is associated with toxicity (e.g. high concentrations of dissolved aluminium and iron; Quist 1995). Hydrogen ion concentration of the soil solution exerts its most important ecological effects by influencing nutrient availability and the concentration of potentially toxic ions. High pH may lead to a decrease in potassium, phosphorus and iron availability due at least in part to competition for sites on soil ion-exchange complexes. However, increased pH may also lead to increased rates of nitrogen fixation by symbiotic bacteria (Andrew 1978), and to a change in availability from predominantly ammonium-nitrogen at low pH to nitrate-nitrogen at high pH. Hydrogen ions have a twofold effect on nutrient uptake; (i) they have a direct effect by displacing cations; and (ii) an indirect effect by favouring bicarbonate ions, which compete with phosphate ions (Olsen 1953).

The solubilities of many metallic cations vary markedly with pH, and aluminium, for example, may be released in soluble form at toxic concentrations at pH values of about 4 (Jones 1961). Soluble aluminium interferes with nutrient uptake and can inhibit root growth, particularly that of seedlings (Clymo 1962). Certain species that appear to perform best on acid soils (calcifuges) may, in fact, simply be more competitive than other species when soil levels of Al^{3+} are high (e.g. the grass *Deschampsia flexuosa* which has its optimum pH for growth under experimental conditions at 5.5–6.0, yet is found in the field on much more acid soils; Hackett 1965).

In addition, the structure of the soil itself is influenced by pH, and factors such as coherence, swelling, porosity, aeration and water-holding capacity may all be altered; the more a species is affected by any of these, the more it is affected by soil acidity (Stalfelt 1972). Soils may also show a striking degree of small-scale heterogeneity in pH (e.g. from pH 4 to 7 over only a few metres of dune heath; Ranwell 1972; van der Maarel & Leertouwer 1967).

Where the role of pH has been investigated critically, it has been found that the interaction between acidity and calcium ion concentration can be the vital factor in determining plant performance. For example, many *Sphagnum* mosses suffer disproportionately severe depression of growth in the presence of high pH and high calcium availability together, but suffer rather little under conditions of high pH or high calcium availability alone (Clymo 1973).

The biology of the so-called calcicole (lime-loving) and calcifuge (lime-hating) plants is closely linked with soil acidity (Tyler 1994). Botanists have long recognized that the floras of chalk and limestone regions are exceptionally rich in species, and have sought to understand the properties of those plants that appear to be restricted to calcareous or to very acid soils (Table 4.3 and Fig. 4.10). Away from chalk and limestone substrates, soils tend to be calcium deficient because of weathering and leaching. However, calcium deficiency tends to be associated with low pH, simply because calcium is the most abundant base cation. Because of this correlation, it is difficult by field observation alone to distinguish between plants that actually *require* high calcium, and others which are simply not competitive on soils of low pH.

Most of the ecological effects of calcium on plant performance appear to be indirect.

1 Calcium modifies soil nutrient availability, and excess calcium can bring about deficiencies of iron, magnesium or trace elements (e.g. 'lime-induced chlorosis'; Grime & Hodgson 1969).

2 Calcium may render phosphoric acid unavailable, so that plants with high phosphate requirements appear to be calcifuges.

3 Calcium-rich soils tend to be porous and to have low soil water content, with the consequence that soil temperatures are relatively high and aeration is good (e.g. the grass *Bromus erectus* is restricted to calcareous soils in the moist, temperate parts of Europe, but occurs on non-calcareous soils in warm, dry Mediterranean regions; Salisbury 1921).

4 Calcium modifies the relative competitive abilities of plants for different nutrients (see Chapter 8). Young seedlings of limestone plants exude considerably more dicarboxylic and tricarboxylic acids than calcifuges, which mainly exude monocarboxylic acids; the tricarboxylic citric acid is a powerful extractor of iron and the dicarboxylic oxalic acid is a very effective extractor of phosphate from limestone soils (Tyler & Strom 1995). Clones of *Ditrichia viscosa* (Asteraceae) differ in their ability to control potassium and calcium balance (calcifuge genotypes

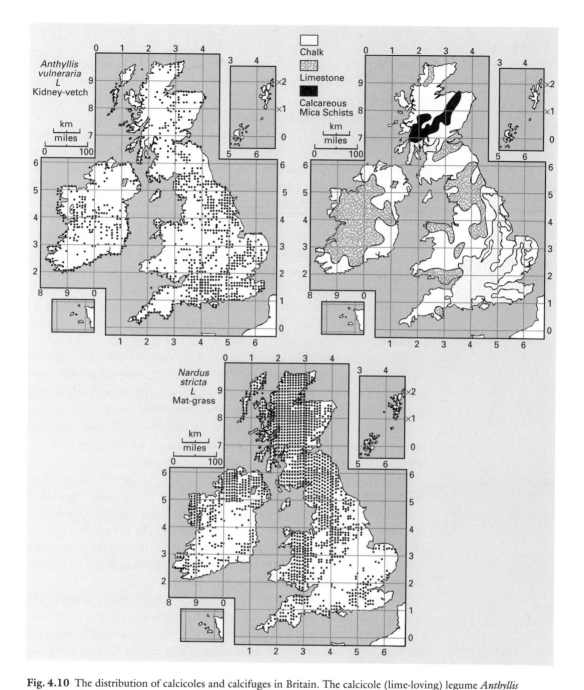

Fig. 4.10 The distribution of calcicoles and calcifuges in Britain. The calcicole (lime-loving) legume *Anthyllis vulneraria* follows closely the distribution of the chalk in southern England. It is more scattered on the limestones of northern England and Ireland, and scarce on the high mountains of the Scottish Highlands. It is also found in coastal habitats where there are calcareous dunes of shell sand. The calcifuge (lime-avoiding) grass *Nardus stricta* is found on acid soils throughout upland Britain. It is absent from the chalk, but occurs in limestone regions where acid glacial drift overlies the calcareous rocks. Note that species distributions cannot be explained by reference to *single* environmental factors. For example, *Nardus* requires high rainfall and cannot tolerate heavy defoliation, while *Anthyllis* is absent from exposed, high-altitude calcareous rocks, and from wet substrates generally. Also, different environmental factors tend to be correlated with one another, so that chalky soils, for example, are usually thin, relatively warm and freely drained, while acid soils are often peaty, cool and wet. (Maps by kind permission of the Biological Records Centre, Monks Wood.)

accumulated proportionately more K and less Ca; Wacquant & Picard 1992).

Calcicoles, therefore, tend to be plants which do not require high calcium as such, but which are sensitive to other environmental factors *influenced* by calcium, such as acidity, nutrient availability, aluminium toxicity, soil temperature or aeration.

A detailed study of calcicoles and calcifuges was carried out in 'chalk heath' by Grubb *et al.* (1969). This is an unusual plant community where calcifuges like heather, *Calluna vulgaris*, grow alongside chalk grassland species like *Asperula cynanchica* (Rubiaceae). The classical explanation of this phenomenon was based on an alleged stratification of the root systems: the calcifuges were supposed to be shallow-rooted plants living in an acid, upper soil layer overlying the chalk, while the calcicoles were deep-rooted species with their root systems in contact with the chalk. This hypothesis failed to explain how the seedlings of the lime-loving plants were supposed to survive in the acid, upper layer. It was also at odds with the observation that many of the calcicoles were actually *shallow* rooted. By concentrating on the dynamics of plant recruitment, Grubb *et al.* (1969) were able to show that both kinds of plants form healthy root systems at pH 5–6, and that soils with this pH were relatively frequent amongst a mosaic of more acid and more calcareous patches. However, unless heavy grazing kept rank-growing grasses and acidifying shrubs at bay, the lime-loving species were unable to regenerate. In short, what set the calcicoles aside was: (i) their intolerance of soil pH < 5; (ii) some resistance to drought; and (iii) some degree of sensitivity to competition with tall grasses.

In this same area, relaxation of rabbit grazing after the introduction of myxomatosis led to very rapid acidification of the soil, because of vigorous growth by *Calluna vulgaris* and *Ulex europaeus* (Fabaceae), both of which produce strongly acidifying litter. The pH of the top 1 cm of mineral soil fell from about 5.5 to about 4.0 in only 10 years. This example illustrates the important point that plants may affect pH just as much as pH affects plants (Grubb *et al.* 1969).

Soil acidity interacts with low nutrient availability by altering the relative availability of nitrate (NO_3^-) and ammonium (NH_4^+) to plant roots, and this has consequences for allocation to root and shoot (e.g. the root weight ratio (RWR) of the grass *Deschampsia flexuosa* increased as NO_3^- increased from 0 to 100% of the available nitrogen; Troelstra *et al.* 1995), and by reducing the ability of plants to obtain nutrient-gathering benefits from their mycorrhizae (Habte & Soedarjo 1995). Brief episodes of elevated H^+ or Al^{3+} concentrations may be decisive for plant performance, and it is important to consider the extreme, rather than the long-term average concentrations to which plants are exposed (e.g. 2 weeks' exposure of *Galium odoratum* (Rubiaceae) roots to pH < 4.0 caused irreversible damage in the presence of dissolved aluminium; Quist 1995).

4.6.8 Heavy metals in soil

A great deal of work has been carried out on the ecology of heavy metals in soils, particularly in relation to the revegetation of mineral spoil heaps and the reclamation of derelict land (Bradshaw & Chadwick 1980). The term 'heavy metals' covers a diverse range of elements like lead (Mesmar & Jaber 1991; Brown 1995), cadmium (Palit *et al.* 1994), zinc (Davis-Carter & Shuman 1993) and copper (Lidon & Henriques 1992) that are classed as environmental pollutants, and a number of others like nickel (Gasser *et al.* 1995), cobalt (Palit *et al.* 1994), chromium (Bishnoi *et al.* 1993), arsenic (Reuther 1992) or mercury that are less important pollutants but may exhibit important effects on the growth of certain plants. Because the range of elements is so broad, there are few generalizations about the responses of plants to heavy metals as a group, and detailed responses vary considerably from one plant species to another (Antonovics *et al.* 1971; Hughes *et al.* 1980).

The effects of heavy metals on plants are felt in three main ways: (i) through direct toxicity, leading to stunting and chlorosis: (ii) through antagonism with other nutrients, often leading to symptoms of iron deficiency; and (iii) through inhibition of root penetration and growth. Plants have both constitutive (present in most phenotypes) and adaptive (present only in tolerant phenotypes) mechanisms for dealing with the perturbations of cell homeostasis that occur as a result of elevated metal concentrations (Meharg 1994). The impact of heavy metals on mycorrhizal performance and the role of mycorrhizae in protecting plants from the deleterious consequences of heavy metals is receiving increasing attention (Gadd 1993; Bucking & Heyser 1994; Hetrick *et al.* 1994; Leyval *et al.* 1995).

There is no doubt that high concentrations of heavy metals can act as potent agents of natural selection, and where soil levels are high a specialized, metal-tolerant flora may arise. Some species are so substrate specific that they can be used in prospecting for commercial ore deposits (e.g. the Rhodesian copper plant *Becium homblei* (Lamiaceae) is used by copper and nickel prospectors in Central Africa; Howard-Williams 1970; Brooks *et al.* 1992). The metal-tolerant plants exhibit one or more processes for containing the otherwise toxic metals: (i) the metals may be excluded by modification to the root's uptake mechanisms; (ii) the metals may be complexed internally to render them unchanged, but innocuous; (iii) they may be degraded to innocuous by-products; or (iv) the plant may possess resistant enzymes that can function in the presence of elevated levels of the pollutant (Bradshaw 1976).

In species that exhibit polymorphism in heavy metal tolerance, the tolerant genotypes are usually little different in fitness from the non-tolerant genotypes when grown on normal soils (e.g. *Agrostis capillaris* (Poaceae)). There are exceptions, however. Zinc-tolerant *Anthoxanthum odoratum* (Poaceae) is distinctly less fit than normal genotypes, and zinc-tolerant *Armeria maritima* (Plumbaginaceae) will not grow on normal soils without an application of zinc. These heavy metal-tolerant species have contributed greatly to our understanding of the evolutionary ecology of plants (see Chapter 6). In particular, the sharpness of the boundary between tolerant and normal populations has emphasized that the strength of natural selection under field conditions can be sufficiently great to maintain polymorphism even in the face of strong gene flow (e.g. in wind-pollinated grasses; Bradshaw & McNeilly 1981).

Interactions between the effects of heavy metals occur commonly. For example, high levels of cobalt induce iron deficiency in plants and suppress the uptake of cadmium by roots. Beneficial effects of cobalt include retardation of leaf senescence, increased drought resistance in seeds, regulation of alkaloid accumulation and the inhibition of ethylene biosynthesis (Palit *et al.* 1994).

4.6.9 Salinity

Soils dominated by Na^+ and Cl^- ions are found in coastal salt marshes, and ionic dominance by Na^+ and SO_4^{2-} occurs in 'salt deserts' (arid, inland environments where freshwater drainage is impeded, and evaporation exceeds precipitation). Although plants growing in salty soils (halophytes) do not normally *require* high salinity, some likely *Halogeton glomeratus* (Chenopodiaceae), will not survive in non-saline culture solution, while others, like *Salicornia* (Chenopodiaceae), grow only poorly (Wainwright 1984). Some halophytes can tolerate extraordinarily high salt concentrations: *Suaeda maritima* (Chenopodiaceae) can grow in $500 \, mol \, m^{-3}$ NaCl (Greenway & Munns 1980).

The principal effects of salinity are felt via the plant's altered osmotic balance. The low external solute potential (e.g. – 20 bar for seawater of roughly 3% NaCl) means that in order to take in water the plant must achieve even lower intracellular potential, which can lead to: (i) reduced growth; (ii) depressed transpiration rate; (iii) reduced water availability; and (iv) an excessive accumulation of ions and reduced uptake of essential mineral nutrients (Slatyer 1967; Ranwell 1972).

Salt-tolerant plants tend to display one or more of the following traits:

1   ion selection (e.g. the ability to absorb ions like potassium in the presence of high concentrations of sodium in the external medium);
2   ion extrusion (e.g. the possession of specialized 'salt glands' by species like *Spartina* (Poaceae));
3   ion accumulation in tissues away from active metabolic sites (e.g. *Agropyron elongatum* (Poaceae) accumulates chlorides in roots which, as in most grasses, are shed annually along with their accumulated ions);
4   ion dilution (as in succulent halophytes like *Aster*, where succulence has the effect of increasing ion dilution by increasing the volume : surface area ratio of the plant; see Chapman 1977 for details).

Yield reductions on saline soils tend to be associated with reduced root growth, increased potassium loss from leaves stems and roots (Cachorro *et al.* 1995; Khan *et al.* 1995), and reduced uptake of nitrate (Cramer *et al.* 1995). The result is that leaf tissue concentrations of nitrate and potassium are reduced in salt-stressed plants (see Chapters 1 & 2).

The Dead Sea in Israel is one of the lowest and most saline lakes on earth (it lies at 400 m below sea level and contains a salt concentration of 340 g per litre). As a result of water extraction from the R. Jordan, the lake level is falling and exposing hypersaline sea shores which are initially completely sterile. The first perennial plants to colonize this new substrate survive the extreme salinity by making use of occasional floodwater, which is distinct from the bulk of the hypersaline soil water found in their root zone (Yakir & Yechieli 1995). Water sharing between ramets may allow clonal salt marsh grasses to overcome the water stress associated with hypersaline but patchy environments; salinity conditions sufficient to preclude seedling recruitment do not prevent invasion from the edge of the patch by physiologically integrated ramets (Shumway 1995).

4.6.10 Atmospheric pollutants

Atmospheric pollution with gases such as SO_2, O_3, hydrogen fluoride, peroxyacetyl-nitrate (PAN), or oxides of nitrogen (NO_x) can have dramatic effects on plant growth and community structure (Grace *et al.* 1981; Koziol & Whatley 1984). For example, along a 60-km transect downwind from a smelter in Ontario, Canada, there were no trees or shrubs at all in the first 8 km, and there was high mortality of mature trees as far as 25 km away. The species richness of the ground flora was reduced for up to 35 km downwind (Gordon & Gorham 1963). A heated controversy has raged for many years over the role of 'acid rain' in the destruction of coniferous forests in Germany and Scandinavia (see references in van Breemen 1985 and Paces 1985). These issues are discussed in depth in Chapter 17.

The gases responsible for air pollution are typically produced by large-scale industrial processes and from burning fossil fuels, although substantial air pollution can arise from volcanic eruptions and from natural forest fires. These pollutants affect plants both directly and indirectly. The direct, phytotoxic effects influence net photosynthesis, stomatal resistance, and metabolic and reproductive activity (Treshow 1984). For example, 8 hours of exposure to SO_2 at $785\,g\,m^{-3}$ produces symptoms such as chlorosis and bleached spotting of the leaves, and 4 hours of exposure to O_3 at $59\,g\,m^{-3}$ causes flecking of the leaves and necrosis of conifer needle tips. Other pollutants take longer to act, but are effective at very low concentrations. Exposure to hydrogen fluoride at concentrations of only $0.08\,g\,m^{-3}$ for 5 weeks causes tip- and margin-burn to leaves, dwarfing and leaf abscission (Stern *et al.* 1984). This visible damage is due to the death of mesophyll and/or epidermal cells. Whether or not it is associated with yield reductions, or with impaired plant survival or fecundity, depends on a host of other factors, including weather conditions, the intensity of plant competition and the abundance of herbivorous animals. Substantially impaired performance may result from air pollution levels far lower than those that produce visible damage symptoms (Ashenden & Mansfield 1978).

Some species are so sensitive to air pollutants that their presence can be used to produce quite accurate maps of mean levels of air pollution. Lichens have been particularly useful in this kind of study since they demonstrate such a wide range of pollution sensitivity, from species like *Lobaria pulmonaria*, which are exceptionally sensitive and can survive only in the cleanest air, to others like *Lecanora conizaeoides*, which thrive even in the most polluted urban environments (Hawksworth 1973; Seaward 1977). The indirect effects of air pollutants act via the alterations in plant biochemistry that they induce. One of the most interesting side-effects involves the plant's invertebrate herbivores. It has been known for some time that plants which are exposed to extreme conditions (e.g. drought, pollutants, physical disturbance) exhibit altered nitrogen metabolism; they tend to show increased tissue concentrations of nitrogen and altered patterns of amino acid composition (White 1974; Jager & Grill 1975). It now appears that these increased levels of amino acid availability can represent substantial improvements in food quality for the insects feeding on these plants, with the result that air pollution may even lead to insect outbreaks (Port & Thompson 1980; Edmunds & Alstad 1982). In a detailed study of blackfly *Aphis fabae* on broad beans *Vicia faba* (Fabaceae), Dohmen *et al.* (1984) suggest that the increased pest status of blackfly in Essex, downwind from London, is due to alterations in the concentration and composition of amino acids in the bean plant, induced by SO_2 and NO_2 in the London air. Experiments with these gases, and with filtered and unfiltered London air, show that the gases have no effect on the insects directly, and that the increase in aphid growth which they observe in polluted air is mediated entirely through induced changes in host plant chemistry.

4.6.11 Exposure

In habitats like mountain ridges and sea cliffs exposed to strong, persistent winds, or in aquatic communities subject to strong wave action, the shear mechanical battering to which plants are subject imposes severe limits to the growth forms that can survive there. Indeed, one of the arguments put forward to explain why vascular plants have been so conspicuously unsuccessful at colonizing the rocky intertidal zone is that the algal holdfast is greatly superior to the vascular plant root system as a means of anchorage under these conditions.

The most severely exposed terrestrial habitats (e.g. dry, vertical rock faces) are characterized by life forms such as endolithic cryptogams which actively bore into the surface of the rock itself (e.g. certain lichens, cyanobacteria and blue-green algae). Vascular plants of exposed places tend to dwarfed ('nanism'), prostrate (e.g. *Salix herbacea*), or to form dense, hemispherical cushions that nestle in relatively sheltered crevices amongst the rocks (Bliss 1956). Tall plants can only survive in exposed habitats if they are very securely anchored, have snap-resistant stems, and leaves that do not tatter. The plants of sea cliffs have to contend with both strong winds and salt spray, while mountain plants are exposed to both the wind and prolonged coverage by snow.

The main effects of strong winds are felt through their drying action, and the direct damage they cause to leaves, shoots and buds (Grace 1977). Many plants of exposed places show pronounced xeromorphic features (see Chapter 2). The wind-sculptured canopies of krumholtz trees growing at the timberline on mountains, and the close-cropped, umbrella-like crowns of seaside coniferous trees, are vivid testimony to the pruning action of the wind. Over the course of many years, long-lived plants in exposed habitats may actually move downwind (e.g. 'islands' of *Picea engelmannii* and *Abies lasiocarpa* (both Pinaceae) moved at an average rate of about 2 cm per year, due to rooting of low, horizontal branches on the sheltered side, while the exposed, windward edge died back through freezing and desiccation of the shoots; Benedict 1984).

Some extraordinary plants are found in exposed habitats. For example, the Adriatic coast of the former Yugoslavia is one of the stormiest in the world, and is buffeted throughout the year by very strong winds known as the bora. This area has served as a refuge for numerous bizarre plants, which are relics of the Palaeo-Mediterranean flora. Many of the characteristic endemics belong to the Brassicaceae (e.g. *Brassica cazzae*) and Apiaceae (e.g. *Seseli palmoides*), and form columnar, palmoid treelets with huge, monocarpic inflorescences. Other endemics include shrubby, bottle-like succulents ('hyperpachycauls'), while further species look like miniature baobabs (e.g. *Astragalus dalmaticus* (Fabaceae) and *Centaurea lungensis* (Asteraceae)). A common feature of these plants is that they have persistent petioles which form a collar around the shoot apex protecting it against desiccation and freezing (Lovric & Lovric 1984). Species from sheltered, lowland habitats are often less productive when grown in windy environments because of: (i) direct tattering and abrasion of their leaves; and (ii) the diversion of limited resources into extra supporting tissue.

4.6.12 Trampling

The contrast between the vegetation of footpaths through grassland and that of the surroundings is often extremely marked, especially in winter when the green of the footpath stands out in vivid contrast to the dead remains of plant life on either side. It is also noticeable that livestock prefer to graze on the footpath rather than the surrounding grassland (Bates 1935). Trampling by humans and domestic animals creates a characteristic plant community dominated by species whose morphology gives them a certain tolerance of bruising, compression and other physical abuse. For example, in mesic grasslands in Britain, the central part of the path is dominated by grasses like *Poa pratensis* and *Lolium perenne* and by herbs such as *Trifolium repens* (Fabaceae) and *Plantago major*. Chamaephytes (plants with their perennating buds held above the ground surface) are especially susceptible to trampling, while matted and rosette hemicryptophytes and geophytes are relatively tolerant of trampling (Cole 1995).

Trampling tolerance in *Poa* and *Lolium* derives from the unusual structure of their leaf sheaths, with their conduplicate stem and folded leaf section (compared with the normal rolling of the leaf in other grasses). The leaves thus offer a flat surface to the crushing action of the foot. Further, the grasses of the footpath are cryptophytes, with their perennating buds buried just below soil level, whereas the grasses of the surrounding area tend to be hemicryptophytes or chamaephytes with their buds at or above the surface. Trampling tolerance in the herbs derives from their prostrate growth habit. In *Plantago major*, the tough leaves are held in a ground-hugging rosette, with the upper leaves protecting the lower. On the path this species grows as a cryptophyte, whereas in untrodden areas nearby it is a hemicryptophyte with semi-erect leaves. The trampling tolerance of *Trifolium repens* derives from its tough, rather flattened, prostrate runners, but its leaves and perennating buds can be damaged by treading. It tends to be found on the edge of the path, and its presence may be due to reduced competition for light, rather than to trampling tolerance as much.

Another characteristic plant community occurs in gateways and on farm tracks. Annual plant species like *Matricaria discoidea* (Asteraceae) and *Polygonum aviculare* and perennials like *Potentilla anserina* (Roaceae) and *Plantago major* are particularly abundant in these places, owing to an ability to withstand the pressure of wheels and the puddling action of animal hooves. The annual species germinate late in the year, and because of this they may be unable to establish in closed vegetation (Bates 1935); indeed, most of the species of tracks and gateways appear to be light-demanding. *Potentilla* persists by sending out copious runners over the disturbed ground during the summer, but the plant does not tolerate much direct treading. Treading resistance in *Polygonum aviculare* derives from its tough, wiry, prostrate stems, whereas the normally upright *Matricaria discoidea* possesses an extremely pliable stem, which is tough, fibrous and does not snap if bent double. Trampling tolerance in Australian grasses is positively correlated with their tiller production rate, and is more dependent upon stem flexibility than leaf strength (Sun & Liddle 1993). Trampling by large herbivores like elephant, buffalo and mountain gorilla was found to affect roughly

0.01% of vegetation cover per day over a 12.2-km^2 study area; although the effects of elephant trampling could last for well over a year, regeneration occurred sufficiently rapidly that trampling by these large animals had no significant effect on available biomass (Plumptre 1994). Trampling by sheep on heavily burned heather *Calluna vulgaris* moorland is a major factor preventing seedling establishment (Legg *et al.* 1992).

4.6.13 Extremes of heat

4.6.13.1 Low temperatures

Since biochemical reaction rates are temperature dependent, the most obvious effects of reduced temperature are felt through reduced net photosynthesis and reduced growth rates. The consequence of this is that plants take longer to reach their threshold size for flowering, or fail to accumulate sufficient resources during the growing season to flower at all (Went 1953). Thus, in arctic and alpine environments where the growing season is very short, annual plants are extremely uncommon and many of the perennial plants have foregone the production of seed in favour of entirely vegetative reproduction (Bliss 1985).

The physiological effects of low temperature are complex and still rather poorly understood (see Fitter & Hay 1981). Some cold-tolerant species, however, can be chilled to – 38°C without damage. Chilling of tropical plants brings about a phase change (liquid to solid) in their membrane lipids, and inactivation of membrane-bound enzymes in the mitochondria. Temperate plants, however, do not suffer until ice begins to form, leading to mechanical damage, progressive dehydration and eventually to cell death. Slow cooling is much less harmful than rapid cooling to the same temperature, and rapid thawing may also have lethal effects (Levitt 1978). Several genes whose expression is induced by cold have been cloned and characterized; transgenic plants express enzymes of lipid, carbohydrate and protein metabolism, structural proteins and various putative cryoprotectants that protect against intracellular dehydration ('antifreeze genes', Howarth & Ougham 1993).

Plants from extremely cold environments exhibit a number of common traits.

1 If any seeds are produced they show marked dormancy, which can only be broken by cold treatment.

2 The plants often possess carbohydrate storage organs, which allow them to grow very rapidly in the spring and also to accumulate resources over several brief growing seasons before investing in a burst of seed production.

3 Flower buds can be formed one or more years before flowering.

4 Leaves tend to be small, long-lived (investment in leaves is paid off over a long period), and dark coloured (an accumulation of anthocyanin

pigments in the leaves allows them to heat up rapidly because dark-coloured leaves absorb more radiation).

5 The tissues can accommodate intercellular ice without damage.

6 They are tolerant of the cell dehydration caused by freezing.

7 They are able to acquire freezing resistance by 'hardening' through exposure to temperatures below 5 °C for several days once growth has stopped and dormancy has been established. The degree of cold tolerance often depends on the temperature of hardening; e.g. herbaceous plants hardened at 0 °C were freezing tolerant to −12 °C while plants hardened at −10 °C survived down to −24 °C; woody plants hardened at 0 °C survived down to −15 to −36 °C while those hardened at −10 °C survived to −32 to −80 °C (Bauer *et al.* 1994).

Xylem embolism (breakage of the water column) occurs during freeze–thaw cycles. In ring porous woods (e.g. *Quercus*) it is necessary to produce new xylem tissues to restore hydraulic conductance, but in diffuse porous woods (e.g. *Betula*) or conifers (which possess tracheids rather than vessels) cavitation can be restored by refilling existing xylem (Sperry *et al.* 1994). Plants grown experimentally on high-nitrogen soils tend to be less frost hardy than plants grown on nitrogen-poor soil (Caporn *et al.* 1994).

Greater exposure of plants to clear night skies increases susceptibility of leaves to radiant frost, and at high altitudes, these frosts can occur throughout the summer. Plants with broad, horizontal leaves are particularly susceptible to freezing at night. For example, leaf size in *Taraxacum officinale* (Asteraceae) decreased with increasing elevation and a corresponding decrease in infrared radiation from the night sky (Jordan & Smith 1995).

4.6.13.2 High temperatures

Plant temperatures in excess of 40 °C are usually associated with failure of the cooling system, when stomatal closure in response to water shortage cuts off evaporation and its consequent transpirational cooling (see Chapter 2). The result of high tissue temperatures is that cell metabolism is severely disrupted, possibly by protein denaturation, membrane damage or the production of toxic substances (Levitt 1972).

Plants from hot environments show one or more of the following traits:

1 small, dissected leaves, which increase the rate of convective heat loss;

2 leaves that reflect a high proportion of the incident radiation (e.g. some plants possess a thick, white, reflective pubescence at times of the year when tempeatures are highest, but less hairy, green leaves in cooler seasons; Ehleringer & Mooney 1978);

3 a C_4 photosynthetic system that continues to function up to 45–60 °C rather than a C_3 system which operates only up to 35–45 °C (see Chapter 1);

4 physiological tolerance of very high tissue temperatures, especially in those species like succulents and sclerophylls which cannot cool their tissues by high rates of transpiration during the daytime (the maximum recorded plant temperature is 65 °C from an *Opuntia* cactus).

As with frost and drought resistance, the mechanisms of heat resistance are not yet clear, but resistance can be induced by 'thermal hardening', which may be related to the stability of protein structure (Fitter & Hay 1981).

The occurrence of heat-shock proteins (HSPs) in response to high temperatures appears to be a universal phenomenon in higher plants. Transgenic plants of *Arabidopsis thaliana* (Brassicaceae) expressing a heat-shock transcription factor showed 20% of maximum heat-inducible HSPs at normal temperatures (20 °C) and this significantly increased the thermotolerance of the transgenic plants (Lee *et al.* 1995).

In many desert plants, transpiration occurs throughout the hottest periods of the day and appears to be essential for leaf cooling; there is a threshold leaf temperature above which photosynthesis is severely impaired (Laurie *et al.* 1994). Plants with lower stomatal densities exhibit higher leaf temperatures (Tan & Buttery 1995). Heat stress increases accumulation of foliar sucrose and decreases starch accumulation. In stem succulent plants, the increased rate of water loss in cooling is not matched by an immediate increase in water uptake through roots. Instead, quick and considerably enhanced transpiration is buffered by internal water reserves which are refilled over the following 24 hours, once the plant is relieved of heat and drought stress (Flach *et al.* 1995).

4.6.14　Mutualists

Almost all kinds of mutualisms are of concern to plant ecologists, from the intimate symbioses of algae and fungi living together as lichens (Seaward 1977) to the loose, facultative mutualisms involved in the dispersal of fruit and the stimulation of germination (Krefting & Roe 1949; Boucher 1985). Many aspects of mutualism are covered elsewhere in the book: pollination in Chapters 6 and 9; fruit dispersal and germination enhancement by animals in Chapter 9; nodulation of legume roots in Chapter 8; mycorrhizae in Chapter 3; and plant defence by ants later in this chapter.

Because nodulation and mycorrhizae occur underground, their study has suffered from the 'out of sight, out of mind' syndrome, and it is only recently that the study of these vital mutualisms had begun in earnest (Alexander 1983; Harley & Smith 1983; Sprent 1983). Root nodulation by *Rhizobium* bacteria occurs in most species of legumes, and nodulation has also been reported in 158 species from 14 genera of non-legumes, including *Alnus* (Betulaceae), *Myrica*, *Dryas* (Rosaceae), *Casuarina* and *Hippophae* (Eleagnaceae) (Bond 1976). Not surprisingly, nodulation is particularly prevalent on nitrogen-deficient soils, and during the early stages of primary successions (Sprent 1983). Furthermore, nitrogen fertilization of soils tends to reduce the incidence of nodule formation (see Chapter 8).

Mycorrhizal development is prevalent in most habitats and most plant taxa, but can be classified into three main types: (i) many of the dominant tree species found in low-diversity or monospecific stands have sheathing (ectotrophic) mycorrhizae; (ii) the great majority of herbaceous plants support vesicular-arbuscular (VA, or endotrophic) mycorrhizae; (iii) heaths have their own, distinct 'ericaceous' mycorrhizae.

In a mycorrhizal association, the fungus obtains carbohydrate from the root system of the plant, and the plant obtains nutrients from the fungus (Bowen 1980; Fogel 1980; Malloch *et al.* 1980). (The relationship between germinating orchid seeds and their fungi is untypical, in that the vascular plant parasitizes carbohydrate from the fungus; Summerhayes 1951.) While the principal advantage of mycorrhizal association is usually said to be an enhanced ability to gather immobile soil nutrients like phosphates (Caldwell *et al.* 1985), Alexander (1983) emphasizes its potential importance in the nitrogen economy of plants, and Newsham *et al.* (1994) suggest that mycorrhizae play an important role in protecting roots from attack by fungal pathogens (see Chapter 13). Studies of nutrient transfer between plants whose root systems are bridged by mycorrhizae have demonstrated exchange of labelled phosphorus but, as stressed by Ritz and Newman (1984), this does not imply *net* movement from one plant to another. Just as nitrogen fertilization reduces nodulation, so phosphate fertilization tends to reduce the development of mycorrhizal fungi.

The amounts of carbohydrate paid to the nodulating bacteria for the nitrogen they fix and to the mycorrhizal fungi for the services they provide are unknown. It is clear that large amounts of carbohydrate 'leak' from the root system into the soil in nutrient-rich systems (e.g. agricultural soils; Coleman *et al.* 1984; Whipps 1984), but little of this material is payment to soil mutualists. The likely functions of secreted carbohydrates include lubrication of root tips penetrating beween soil particles, defence of roots against soil pathogens, and the production of allelopathic compounds. In nutrient-poor soils, however, carbohydrates passing out of the roots are more likely to be payment for the services of mycorrhizal fungi or nodulating bacteria (Newman 1985; Read *et al.* 1985). These topics are discussed more fully in Chapters 3 and 13.

4.6.15 Enemies

Plants are protected against their enemies by a wide range of structural, biochemical, phenological and ecological features (see Chapters 5 & 10; for an introduction to the earlier literature see Rosenthal & Janzen 1979; Harborne 1982; Strong *et al.* 1984; Rhoades 1985). The protective function of many of the surface features of plants, such as hooked or glandular hairs, spines, thorns and thick waxes, can be seen by comparing the damage done to genetic strains of plants that lack these traits when exposed to vertebrate or invertebrate herbivores (Maxwell & Jennings 1980). Plants also possess many morphological and physiological features that increase their *tolerance* of defoliation:

1 inaccessible meristems and rapid refoliation (many pasture grasses);

2 reserves of dormant buds (many trees);

3 readily mobilized reserves of carbohydrate and proteins which allow regrowth;

4 plasticity in the distribution of their photosynthate between roots and shoots and between reproductive and vegetative functions within the shoot; and

5 the ability to increase the rate of photosynthesis per unit area of surviving leaf tissue (Fig. 4.11 and see Chapter 13).

Defence against insects through 'phenological escape' appears to be rather common, especially in seasonal environments. Individuals that leaf out very early will escape herbivore damage because their leaves are mature and unpalatable by the time the insect herbivores emerge from their overwintering stages. Weighed against this, early-leafing plants are extremely vulnerable to late frosts which could kill all their leaves. Individuals that produce their leaves very late may escape herbivore damage because the emerging insects starve in the absence of young foliage on which to feed. However, late flushing trees are prone to

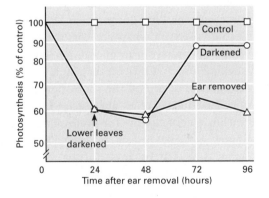

Fig. 4.11 The source–sink hypothesis states that the rate of photosynthesis in the leaves and other green parts (the 'source') is controlled (at least in part) by the availability of 'sinks' for the carbohydrate produced (e.g. rapidly growing or respiring tissues). The effect is shown in this classic experiment by King *et al.* (1967). In wheat, the flag leaf exports most of its photosynthate to the developing ear. When the ear is experimentally removed at time 0, sink strength is reduced and photosynthetic rate falls to only 60% of control levels. However, if a new sink is created by shading the lower leaves at 24 hours, the photosynthetic rate of the flag leaf recovers to levels not significantly different from the controls. Feeding by herbivorous animals can lead to increases in the photosynthetic rate when they increase the sink strength (e.g. sucking insects like aphids), or when they reduce the size of the source (e.g. defoliating insects and grazing vertebrates).

attack by larger caterpillars emigrating from nearby trees later in the season. Late flushing trees also 'waste' a certain amount of potentially interceptible sunlight by remaining leafless until relatively late in the season, and would be at a competitive disadvantage in years when herbivore attack was slight. Clearly, the precise timing of bud burst is a delicate compromise depending, at least in part, on what other plants in the population are doing.

Another ecological means of escape from enemies involves life-history characteristics. Very short-lived plants may be able to escape the attentions of specialist herbivores by developing so quickly that their enemies are unlikely either to discover them or to build up in damaging numbers on them. Such plants are said to be 'unapparent' to their enemies, in contrast to 'apparent' plants (such as trees) which are 'bound to be found' by their herbivores during the course of their long lives (Feeny 1976; Rhoades & Cates 1976). A second kind of ecological defence is 'associational resistance' (Root 1973). This is not an evolved characteristic of individual plants, but occurs when the rate of attack suffered by an individual plant is influenced by the structure and species composition of the surrounding vegetation. Plants may simply be hidden from their enemies by taller vegetation, or the surrounding plants may mask the chemical cues herbivores use in locating their hosts. Other protective effects of complex vegetation may include alterations to microclimate that favour the plant more than its enemies, and increased predator or parasite densities in the mixed community (Bach 1980; Risch 1980).

Finally, some plants pay 'protection money' to ants in order to secure defence from their herbivorous enemies or from neighbouring plants attempting to invade their air space. Ants may be attracted to the plants to feed from specialized extra-floral nectaries, or to collect honeydew from sucking insects. Once on the plant, the ants kill or expel foliage-feeding caterpillars, or nip off the shoot tips of invading plants. In other, more specialized cases (e.g. 'ant acacias'), the plants harbour their protectors within special hollow (or hollowable) stems where the ants culture colonies of scale insects or other Homoptera (Janzen 1985). Plant species defended by ants do not invest in such costly chemical defences as their related non-ant species (see Chapter 10).

4.6.16 Nurse plants

It is a mistake to think of neighbouring plants as always having a negative effect on plant performance (the competitors of Chapters 12 & 15). In many environments like deserts and exposed mountain slopes, plant establishment only occurs directly beneath established perennial plants, which afford shelter from desiccation, wind-blast or herbivore attack. Shade-demanding species are soon lost if the shade-casting plants below which they live are cut down (e.g. the loss of 'ancient woodland' species following felling of British woodlands; Peterken 1981). Hacker and Bertness (1995) described the morphological and physiological consequences of the positive interaction between a shrub *Iva frutescens* (Asteraceae) and the dominant rush *Juncus gerardi* in coastal salt marsh in New England. Experimental removal of the *Juncus* in the lower marsh caused increased soil salinity and anoxia, resulting in lower growth, biomass and survival of the *Iva*. The positive effect of *Juncus* on *Iva* resulted from shading of the marsh surface, which minimized salt accumulation, coupled with increased soil oxygen levels as a result of radial oxygen loss from the rhizosphere of the rush.

4.7 Conclusions

The principal message of this chapter is that the 'ecological optimum' and the 'physiological optimum' of plants are not one and the same (Ellenberg's rule). The conditions under which plants occur at their maximum abundance in the field (the so-called ecological optimum) are often quite different from the levels of those same environmental conditions under which the plants perform best in single-factor, single-species experiments (the physiological optimum). The reasons for this discrepancy are many, complex and interacting, but they can be summarized as follows. Plant abundance under a given set of field conditions is determined by both its *relative competitive ability* with other plants and its *relative susceptibility* to herbivores and pathogens. As often as not, therefore, plant distributions in the field are refuges from competitors or enemies, rather than places that present the plant with 'ideal environmental conditions'.

Trade-offs between plant physiological functions are universal and they have implications at both the population and community levels, especially when they affect plant dispersal. As we shall see in Chapters 14 and 15, there is a fundamental difference in the behaviour of mean-field versus spatially explicit models of population dynamics, and trade-offs lie at the heart of this difference. For example, mean-field models of species richness predict competitive exclusion and the development of a monoculture of the competitive dominant. For exactly the same system, spatially explicit models incorporating the competitive ability/dispersal ability trade-off predict that there is no limit to species richness because a randomly generated mosaic of patches provide refugia for the inferior competitor.

5: Plant Secondary Metabolism

Jeffrey B. Harborne

5.1 Introduction

A major difference between plants and animals is the ability of plants to accumulate a wide variety of low molecular weight constituents, the so-called products of secondary metabolism. Indeed, over 80% of the organic compounds found in the natural world are of plant origin. In animals, secondary metabolites are mainly produced as defensive secretions, as in the arthropods, or as pheromones, released from special glands for signalling purposes. By contrast in plants, all species have a wealth of phenolic and terpenoid constituents, while as many as one-third of species also contain nitrogen-based metabolites, such as alkaloids, cyanogens or glucosinolates. Any one plant may have more than 100 such constituents, although only a few of them are likely to accumulate in any quantity.

The purpose of this confusing profusion of secondary metabolites in plants has long been debated. Earlier suggestions that they are simply 'an overflow' from primary metabolism or are 'waste products', accumulating because of the absence of an effective waste disposal system, are no longer accepted by most plant scientists. Some of these compounds have been shown to have an essential role in growth and development. This is true of the 102 gibberellins so far recorded and also of molecules such as abscisic acid, auxin, the cytokinins and the brassinosteroids. A few may be produced, such as the glycinebetaine of salt-tolerant plants (halophytes), to protect them from the consequences of environmental stress. Yet others, notably the anthocyanins, carotenoids and the volatile oils of floral tissues, have an important role as colours and scents in attracting animal pollinators to plants for reproductive purposes. Fruit pigments again are needed to encourage animals to eat the fruits and disperse the seeds.

However, there is still the majority of secondary constituents that must be produced for some other benefit to the plant. The most favoured theory today is that many, perhaps most, are involved in chemical defence systems, which protect plants from herbivory and from microbial infections. This view is supported by the fact that many secondary metabolites are toxic in varying degrees to other forms of life, if not to mammals, then to insects and molluscs or to pathogenic bacteria and fungi. Moreover, it is known that some animals, such as ants, mimic plants in the production of identical defensive chemicals and also that other animals, such as butterflies, borrow poisonous chemicals from plants and secrete them to deter their own predators. Other evidence supporting the defensive theory of secondary metabolism is mentioned later. First of all, some consideration must be given to the different secondary constituents and their production within the plant.

5.2 Secondary metabolites

Secondary constituents represent the accumulation of end-products from relatively lengthy pathways of enzyme-catalysed steps in biosynthesis. These products are then channelled within the plant, or organ, to a site of storage (e.g. to the cell vacuole) or are excreted and laid down at the surface. These chemicals have a slow turnover within the plant, compared to primary metabolites, and biosynthesis may continue throughout the life of the plant. Secondary metabolism is costly and depends on primary metabolism for the sources of energy and the necessary precursors (e.g. amino acids, carbohydrates or acetyl coenzyme A). The cost of synthesis of alkaloid is estimated at 5 g of photosynthetic CO_2 per gram of toxin, while a comparable figure for a phenolic would be 2.6 g (Gershenzon 1994). Secondary metabolism is regulated within the plant according to the availability of primary precursors, which are also needed to feed the primary pathways of growth and respiration. Secondary synthesis has to be balanced against the cost of new growth; plants may have the dilemma of the choice between producing new leaves or defending with chemistry those leaves that are already laid down (Herms & Mattson 1992).

The three main classes of secondary products are the terpenoids, nitrogen-containing compounds and phenolics. The terpenoids, or iso-prenoids, are formed from acetyl coenzyme A via mevalonic acid and built up from five-carbon units into molecules of regularly increasing molecular size. The largest group of nitrogen-containing metabolites are the alkaloids, which are variously derived from one or other of the 20 protein amino acids. Phenolic metabolites, including the well-known anthocyanin pigments and the astringent-tasting tannins, are derived from the aromatic amino acid phenylalanine and also from malonyl coenzyme A, a precursor also required in fatty acid and lipid biosynthesis.

Although secondary metabolism is the major process by which materials toxic to animals accumulate in plant tissues, products of primary metabolism are occasionally involved. Examples are known in the case of organic acid and amino acid synthesis. For example, soil fluoride can be taken up into a plant where it may be bound as the fluoracetate anion, $CH_2FCO_2^-$. This happens in *Dichapetalum cymosum* (Dichapetalaceae) and in species of *Gastrolobium* and *Oxylobium* (Fabaceae). Fluoracetate is highly toxic to mammals, since it is incorporated into the Krebs respiratory tricarboxylic cycle, causing a block at the fluorocitrate stage. Cattle poisoning in South Africa and Australia occurs when these plants are eaten. However, some members of the native fauna in Australia, such as the grey kangaroo (*Macropus fuliginosus*), are adapted to this toxin and can feed on fluoracetate-containing plants with impunity.

Another simple plant organic acid dangerous to animals is oxalate. The potential threat depends on which cation the oxalate anion is associated with in the plant. Calcium oxalate (e.g. *Rumex* spp. (Polygonaceae)) is relatively safe to eat, but potassium oxalate (e.g. in *Setaria sphacelata* (Poaceae)) may be toxic. Additionally, plant organic acids such as citric acid, may play a role in sequestering heavy metal cations, which accumulate in certain plants growing on appropriate soils. This chelation probably occurs in the hyperaccumulation of nickel in a plant such as *Streptanthus polygaloides* (Brassicaceae) growing on serpentine soils; as a result this plant is quite toxic to lepidopteran larvae (Martens & Boyd 1994).

While protein amino acids in plants are essential dietary components, they can occasionally cause problems to animals. This occurs in plants growing on selenium-rich soils, which adapt to the selenium toxicity by storing selenium-containing amino acids in their tissues in place of the sulphur-containing amino acids methionine and cysteine. Species of *Astragalus* (Fabaceae) are among the best known of these selenium-accumulating plants, and they may contain up to 5000 ppm selenium. Since selenium is poisonous to cattle at levels of 5 ppm, these plants are clearly dangerous to livestock.

The effectiveness of secondary metabolites in deterring feeding is closely connected with their localization in the plant. If they are 'advertised' to the herbivore (e.g. as a crystalline exudate on the leaf surface or as a malodorous smell issuing from leaf trichomes), they are likely to arrest feeding at an early stage. Secondary metabolism is often channelled towards the external surface and harmful substances do accumulate in the waxes or in leaf hairs. Alternatively, damage to the plant may release noxious materials that limit grazing. Some 12 000 plant species belonging to 22 families secrete a milky, sticky latex in their tissues and this may pour forth when tissue is damaged. Typically, the latex may contain bitter-tasting metabolites, as in the case of chicory (*Cichorium intybus* (Asteraceae)), where intensely bitter sesquiterpene lactones are present (Rees & Harborne 1985). Similarly, trees may release a copious oleoresin, consisting of monoterpenes mixed with diterpene acids, when the bark on the trunk is damaged by browsing. Even where the secondary metabolite is located inside the plant (e.g. as a bound toxin stored in soluble form in the cell vacuole), it may still deter animal feeding through taste. A significant number of plant metabolites are bitter or otherwise unpleasant to taste. This is true not only of alkaloids but also of many cyanogenic glycosides, iridoids, cardenolides, limonoids and quassinoids.

The concentration of a defensive toxin within the plant may well be a crucial factor in determining whether a leaf is eaten or not. Environmental stress, e.g. drought, can often significantly increase toxin production so that an otherwise palatable species becomes unpalatable to the herbivore. This is true of essential oil production in Mediterranean plants, where the levels of oil may double as a result of drought stress. Likewise, plants of camphorweed (*Heterotheca subaxillaris* (Asteraceae)) grown under nitrate-limiting conditions accumulate 50% more oil in the leaves compared with control plants. Similarly, in *Cleome serrulata* (Capparidaceae) the glucosinolate (mustard oil glycoside) content in drought-stressed plants is three or four times greater than that of normal plants (Louda *et al.* 1987). (Note, however, that some species are *more* palatable to their herbivores when the plant is drought stressed; see Chapter 13.)

A most remarkable feature of secondary metabolism in plants is that increases in concentration may be induced dynamically in response to either mechanical damage (punching a hole in the leaf) or insect or mollusc feeding. Typically, leaves originally palatable to feeding by a generalist insect such as the armyworm *Spodoptera uridania* become unpalatable a few days after such damage (Edwards & Wratten 1983). In a few cases, increases in alkaloid or furanocoumarin synthesis have been recorded. For example, in *Atropa acuminata* (Solanaceae), tropane alkaloid production increases to 153% of control levels after leaf damage, to 164% after mollusc feeding and to 186% on repeating the leaf damage (Khan & Harborne 1991). In *Nicotiana sylvestris* (Solanaceae), nicotine alkaloid production increases to 400% of control levels after leaf damage and 286% after lepidopteran feeding (Baldwin 1988). It is important to point out that for every plant that responds positively to this induction, there is another plant where no detectable change in secondary chemistry occurs and there is still much to discover about the process of induced chemical response to herbivory (Tallamy & Raupp 1991; see Chapter 10).

While it is easy to propose that the secondary metabolites of a particular plant are able to protect it from herbivory, it is much more difficult to establish a defensive role by ecological experiments. The interaction between a plant and its many potential herbivores is a complex one and is likely to change with time. Plants can almost always survive a significant amount of herbivory and leaf loss so that chemical protection may be concentrated in the more vulnerable tissues. There is also the well-recognized ability of animals to modify and detoxify the poisonous compounds present in plants. Most wild animals are well equipped to detoxify many of the poisons that are present in the foliage within their habitat and it is only in exceptional circumstances that a plant may gain complete protection from feeding. Milkweeds, such as *Asclepias curassavica* (Apocynaceae), are normally completely protected from mammalian herbivory, due to the bitter-tasting, emetic cardiac glycosides secreted in their latex. Even such a plant, however, is used as a food plant by the larvae of the monarch butterfly, *Danaus plexippus*, since these insects are able to deal with their heart-poisoning effects.

While detoxification biochemistry remains an important adaptive response in animals to harmful dietary chemicals, it can occasionally produce the wrong result. For example, the mammalian detoxification of pyrrolizidine alkaloids, such as senecionine, can produce through *in vivo* dehydrogenation a more toxic product, a pyrrole base that binds to DNA in the liver. The protection that the ragwort (*Senecio jacobaea* (Asteraceae)) enjoys from cattle and sheep grazing lies in the fact that animals die as a result of forming this so-called 'detoxification' product. Insects by contrast appear to be able to avoid the toxic effects of pyrrolizidine alkaloids, since certain Lepidoptera actually collect these compounds from plant sources and use them as male pheromones (cf. Harborne 1993).

The defensive roles of secondary metabolites are now considered in separate accounts of the terpenoid, nitrogen-containing and phenolic constituents of plants.

5.3 Terpenoid metabolites

The terpenoids are distinguished from other classes of secondary metabolite by their biosynthetic origin from isopentenyl and dimethylallylpyrophosphates and their broadly lipophilic properties. Chemically, they are mainly cyclic unsaturated hydrocarbons, with varying degrees of oxygenation in the substituent groups attached to the basic carbon skeleton. The terpenoids are classified according to the number of five-carbon (C_5) units that are present, from the monoterpenoids (C_{10}) through the sesquiterpenoids (C_{15}) and diterpenoids (C_{20}) to the triterpenoids (C_{30}) and tetraterpenoids (C_{40}). Their biosynthetic relationships are illustrated in Fig. 5.1.

In the monoterpene and sesquiterpene series, many lactones are known and these are often dealt with separately from the related hydrocarbons. The monoterpene lactones are also known as iridoids. In the case of the triterpenoids, biosynthesis can extend to the partial degradation of the C_{30} skeleton first formed. Thus, there are a series of nortriterpenoids, such as the C_{26} limonoids, the C_{23} cardenolides (or cardiac glycosides), and the C_{20} and C_{19} quassinoids. A further structural refinement encountered mainly in the triterpenoid series is the attachment of sugar residues. Two well-known classes of triterpenoid glycosides are the cardenolides and the saponins. Terpenoids are broadly colourless, apart from the yellow to red carotenoid pigments, with 40 carbon atoms, formed by the union of two geranyl-geranyl phosphate units, as indicated in Fig. 5.1. The number of fully characterized terpenoids is somewhere around 20 000 and exceeds that of any other group of plant product. Recent studies on the biochemisty and function of many of these terpenoids have been reviewed (see Harborne & Tomas-Barberan 1991).

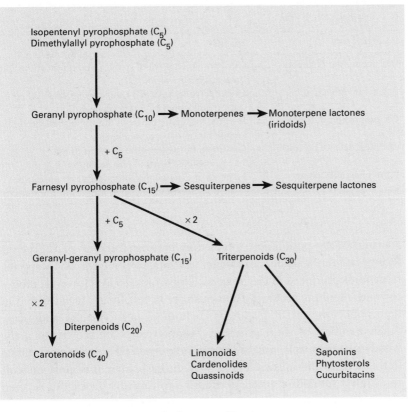

Fig. 5.1 Biosynthetic relationships in the terpenoid series.

Much is known of the natural functions of plant terpenoids, particularly the sesquiterpenoid abscisins and the diterpenoid gibberellins, which are growth regulators in higher plants. Plant terpenoids have also dominated the subject of chemical ecology and terpenoid molecules have been implicated in almost every conceivable interaction between plant and animal or plant and plant. They are regularly recognized as phytoalexins, insect antifeedants, pheromones, defence agents, allelochemicals or signal molecules. Some are highly toxic to animals while others have the ability to interfere hormonally with animal growth and reproduction. Their contributions to plant defence mechanisms is now considered for three of the major classes of terpenoid molecule, the monoterpenoids, the sesquiterpenoids and the triterpenoids.

5.3.1 Monoterpenoids

Monoterpenes occur widely in the plant kingdom, and some of the ecological roles that have been assigned to them are summarized in Table 5.1. Their presence in flower scents, usually in small quantities, can be clearly related to pollinator attraction, and bees, moths and butterflies in particular are sensitive to terpenoid odours. The special

Table 5.1 Some ecological roles of plant monoterpenoids. For references see Harborne & Tomas-Barberan (1991).

Ecological role	Example
Defence against herbivory	Camphor protects white spruce foliage from snowshoe hare grazing
Pollination attraction	Orchid terpenes such as γ-cadinene attract male bees for pseudocopulation (see Chapter 6)
Pheromonal production	Myrcene from pine oleoresin utilized by bark beetles as aggregation/sex pheromone
Chemical defence borrowing	*Euphydryas* butterflies store *Plantago* iridoids (e.g. aucubin) to avoid bird predation
Plant–plant allelopathy	Terpenes of *Calamintha* shrub, such as borneol and carvone, inhibit grasses in Florida scrub community

coevolutionary pollination interaction between orchids of the genus *Ophrys* and the male bees of the genus *Andrena* is reinforced by specific terpenoids produced in the flowers (Bergström 1991). However, monoterpenes regularly accumulate in quantity in leaf tissue, being stored in high concentration in glandular trichomes, as in *Mentha* and many other members of the Lamiaceae. Monoterpenes are present in many woody plants as well, notably in gymnosperms, where they accumulate in the needles and in the oleoresins of the bark. Here it is more difficult to assign a functional role. A defensive role against herbivory is often assumed to account for such storage but the evidence is still largely circumstantial. A defensive role assumes: (i) the compounds are toxic or deterrent to a range of herbivores; (ii) they represent a real barrier to feeding; (iii) that their production protects plant tissue most vulnerable to herbivory; and (iv) the concentration increases in response to herbivory. These and related points are now considered.

There is much evidence that monoterpenes in quantity are toxic and/or repulsive to mammals and especially rodents. Pine oil is a mixture of monoterpenes such as α- and β-pinene, limonene and myrcene, and several monoterpene alcohols. It is an effective feeding repellent to voles and snowshoe hares since these animals do not become habituated to the pine odour (Bell & Harestad 1987).

In the case of insects, the evidence for toxicity is less compelling, since some monoterpenes can act as feeding attractants to aphids, silkworms and pinebark beetles. Furthermore, a species of sawfly *Neodiprion sertifer* is able to ingest dietary monoterpenes and store them in a special gland for later use as a toxic discharge against ant predators. Nevertheless, monoterpenes are probably toxic to most insects, other than those species adapted to them. For example, when a number of monoterpenes were tested against locusts, the majority proved to be feeding deterrents at concentrations around 0.01% dry weight (Bernays & Chapman 1978).

The clear-cut ecological evidence of a defensive role for monoterpenes is still limited. Here it is useful to consider their effects on mammalian feeding, before moving onto their effects on insect feeding. Mammals have a very well developed detoxification system, based on cytochrome P450 oxidases, for dealing with dietary terpenoids, so that terpenes are only likely to be effective in reducing the level of feeding when they are present in quantity and when the animal concerned is restricted to feeding on terpenoid-containing foliage. On the other hand, if terpenes are absorbed during feeding by ruminants, they may inhibit the microorganisms of the rumen and hence have an antinutritional effect. Adaptation to dietary terpene has been established in marsupials such as the greater glider (*Petauroides volans*) and the brushtail possum (*Trichosurus vulpecula*). These Australian animals feed mainly on eucalyptus leaves, which are rich in monoterpenes such as limonene, cineole, terpineol, piperitone and β-phellandrene. They are able to avoid any deleterious nutritional effects because the microbial populations in their hindguts are protected by absorption of the terpenes from the stomach and small intestine and detoxification via the liver (Foley *et al.* 1987).

Adaptation to dietary terpenes appears to be unusual, since many other mammals that have been tested show aversion to terpenoid odours and to plants rich in terpene constituents. Red deer calves (*Cervus elaphus*), for example, reject food contaminated with monoterpenes such as borneol or with terpenoid-containing tissue from Sitka spruce and lodgepole pine. Rejection is based primarily on concentration rather than on the odour quality of a particular monoterpene (Elliott & Loudon 1987).

Caecal digestors such as snowshoe hares and voles are also sensitive to monoterpenes in feeding trials (Bell & Harestad 1987) and there is much evidence that gymnosperms resistant to deer or snowshoe hare damage (e.g. some strains of Douglas fir) owe this resistance to the high levels of monoterpenoids present. More detailed studies have been carried out on the resistance of white spruce (*Picea glauca* (Pinaceae)) to browsing by snowshoe hares and it is apparent that the monoterpene camphor is a specific antifeedant to these hares (Sinclair *et al.* 1988). Camphor is only produced in quantity to protect the juvenile growth from hare herbivory, since the mature foliage is out of reach of this animal. This is reflected in the fact that the camphor content of juvenile twigs and leaves of white spruce is four times that of the mature tissues. The antifeedant nature of camphor was confirmed in feeding trials with the snowshoe hare.

The effects of plant monoterpenes on insect grazing are varied and complex, but not neutral. It is a reasonable assumption that they are broadly defensive, even if specialist insect herbivores have become adapted to dealing with the toxic effects present in their host plants. Even insects that depend on plant monoterpenes to attract them to their food plant and which utilize dietary terpenoids for pheromonal purposes (such as the pinebark beetles) are sensitive to variations in terpenoid vapours released by the tree and may become inhibited from feeding. Although these beetles have a well-developed detoxification system for dealing with dietary monoterpenes, they may be repelled by particular oil components. The pinebark beetle *Dendroctonus brevicomis*, for example, does not colonize all *Pinus ponderosa* trees in a given population. Some trees are resistant and resistance is due to a change in the balance of the five major terpenes in the oleoresin. Limonene increases at the expense of α- and β-pinene, myrcene and 3-carene, and this change in concentration is sufficient to inhibit the beetles from feeding. It is even possible to show that host monoterpenes can have toxic effects on these beetles, e.g. *Dendroctonus frontalis* and *Ips calligraphus* (Cook & Hain 1988).

The pinebark beetle *Dendroctonus brevicomis* is also sensitive to the pheromonal signals of related bark beetles and some of these, e.g. ipsdienol and verbenol, are terpenoid in nature. Release of such pheromones prevents interspecific competition among beetle species for feeding sites on pine trees. The pine shoot beetle, *Tomiscus piniperda*, which feeds on *Pinus sylvestris* is able while in flight to recognize a host tree that is unsuitable for colonization from the release of verbenone by a colonized tree. Verbenone is released in increasing concentration as a given infestation increases and it inhibits completely the attractiveness of the host monoterpenes to the beetle (Byers *et al.* 1989).

Other groups of animal herbivore may be deterred from feeding by terpene constituents, although evidence for this is still limited. Geese apparently reject plants with essential oils, while feeding on those with alkaloids or amines (Wink *et al.* 1993). Slugs are more variable in their responses. *Ariolimax dolichophallus* avoids feeding on chemotypes of *Satureja douglasii* (Lamiaceae), which are rich in pulegone or carvone (Rice *et al.* 1978), while *Arion ater* seems to consume all terpene-rich plants with impunity (Mølgaard 1986).

5.3.2 Sesquiterpenoids

Sesquiterpenoids are derived biosynthetically from three isoprene units and share farnesyl pyrophosphate as a common biosynthetic intermediate. There are many more sesquiterpenoids known in nature than there are monoterpenoids. True sesquiterpenes accompany monoterpenes in plant essential oils, although because of their higher molecular weight they are in the less volatile fraction. The sesquiterpene lactones, of which some 5000 are known, are lipophilic substances secreted in leaf trichomes, waxes and in latexes. These lactones are more restricted in their natural occurrence than sesquiterpenes, being confined principally to one very large plant family, the Asteraceae, and to the related Apiaceae.

Some of the different ecological roles that have been assigned to sesquiterpenoids are indicated in Table 5.2. The biological activities

Table 5.2 Some ecological roles of plant sesquiterpenoids. For references see Harborne & Tomas-Barberan (1991).

Ecological role	Example
Defence against herbivory	Sesquiterpene lactones of chicory against insects and mammals
Pollination attraction	Orchid floral sesquiterpenes attract male bees for pseudocopulation (see Chapter 6)
Pheromonal mimicking	*E*-β-farnesene, an aphid alarm pheromone, produced in *Solanum* trichomes to ward off aphid attack
Antifeedant activity	Caryophyllene epoxide protects plants against *Atta cephalotes* leaf cutting ant attack
Phytoalexin defence	Rishitin and related sesquiterpenes in members of the Solanaceae, including the tomato and potato

associated with sesquiterpenes are many and varied, from plant growth regulation (abscisic acid) to interference with insect metamorphosis. The sesquiterpene lactones likewise have a wide range of demonstrated biological activities, as cytotoxic compounds, vertebrate poisons, insect feeding deterrents, schistosomicidal substances and allergenic agents. Here, attention is concentrated on their defensive role in nature; the sesquiterpenes are considered first and then the lactones.

Individual sesquiterpenes occurring in plants can have direct toxic effects on potential insect herbivores. The sesquiterpene zingiberene, originally reported from the ginger plant *Zingiber officinale* has been identified as a major component of the glandular trichomes of the wild tomato plant *Lycopersicon hirsutum* f. *hirsutum* (Solanaceae). Furthermore, it appears to be responsible for the toxicity of this plant to the Colorado potato beetle. Pure zingiberene is poisonous to the beetle when applied at a concentration of 12–25 µg per larva and this is significantly less than the amount (160–250 µg per 2-cm square) present in wild tomato leaflets. A more effective plant defence than simple synthesis of a sesquiterpene toxin like zingiberene may be to vary the chemical armoury within the plant. This appears to happen in *Ageratina adenophora* (Asteraceae) (Proksch *et al.* 1990). While the first leaves of the plant contain three insecticidal chromenes, the later leaves lack these compounds completely, having instead chlorogenic acid and sesquiterpene diketone. Bioassay against generalist insect feeders confirmed that this sesquiterpene is a useful defence agent, showing both larval growth inhibition and contact toxicity.

There is much indirect evidence that sesquiterpene lactones cause insect herbivores to avoid plants containing them, because of their antifeedant, growth inhibitory or toxic effects. However, ecological data confirming the defensive role of these lactones in the plants where they occur are still relatively limited. In 1978, Burnett *et al.* described such experiments with glaucolide A, the toxic and bitter-tasting major lactone of *Vernonia glauca* and *V. gigantea* (both Asteraceae). Although laboratory experiments showed that the compound caused severe growth inhibition and antifeedant activity against six lepidopterous larvae, there was little evidence in the field that the plants were so protected from insect herbivory. Nevertheless, glaucolide A was of considerable value under natural conditions in significantly limiting deer and rabbit grazing on the two *Vernonia* species.

Rees and Harborne (1985) examined the protective value of the sesquiterpene lactones lactupicrin and 8-deoxylactucin to chicory *Cichorium intybus* (Asteraceae). In this plant, the repellency of these intensely bitter lactones is reinforced by secretion in the latex, which is distributed throughout the leaf and root. In two-choice and no-choice feeding experiments with borosilicate discs, the lactones significantly reduced feeding of the locust *Schistocerca gregaria* at levels (0.2% dry weight) comparable to those present in the plant. Furthermore, the levels of lactones remained high throughout the growing season, being highest (0.45% dry weight) in the most exposed top-stem leaves. The only organs lacking lactone are the flowers, but since these only last 24 hours their need of protection from herbivory is very limited. Here again, the protective value of the bitter lactones to wild chicory plants probably extends beyond limiting insect herbivory to include the deterrence of mammalian (e.g. deer and rabbit) browsing.

These experiments are strongly suggestive that sesquiterpene lactones have an important role in plants of the Asteraceae as chemical defence agents. They are often present in high concentration and relatively complex mixtures of different structures are found in many species. Usually bitter-tasting, they are obviously deterrent to many kinds of herbivore including humans. Their effects on insects are very damaging and they are also poisonous to livestock. However, many further experiments – monitoring their distribution within the plant, their seasonal variations and their interactions with animals – are needed to confirm their defensive role.

5.3.3 Triterpenoids

The triterpenoids are formed biosynthetically from six isoprene units and share in common the C_{30} precursor, squalene. Different types of ring closure in this acyclic intermediate can give rise to more than one type of triterpenoid and in the later stages of synthesis small carbon fragments may be removed to produce molecules with less than 30 carbon atoms. As a result of all the rearrangements that can occur during biosynthesis, there are more triterpenoids produced in plants than any other group of terpenoid. Conjugation is a common feature in certain classes of triterpenoid, with sugar in the case of the saponins and cardenolides, and with fatty acid in the case of the phytosterols.

Of the many types of triterpenoid the most significant ecologically are the cardenolides, cucurbitacins, limonoids, phytoecdysones, quassinoids and saponins. Their various biological activities include mammalian toxicity (cardenolides), molluscicidal activity (saponins), bitter taste (cucurbitacins, quassinoids), antifeedant activity (limonoids) and hormonal interference in insects (phytoecdysones) (see Table 5.3).

While there is much indirect evidence that accumulations of particular triterpenoids in plants are defensive against herbivory, experiments implicating triterpenoids definitively as feeding deterrents are relatively few. One of the first detailed studies of this type was the work of Reichardt *et al.* (1984) on the defensive role of papyriferic acid in protecting the winter-dormant Alaska paper birch against snowshoe hares *Lepus americanus*. In this case, the triterpenoid specifically accumulates on the surface of winter-dormant juvenile twigs, protecting the tree only when it is vulnerable to attack. The concentrations drop 25-fold in the mature internodes, when herbivory is no longer harmful to the survival of the paper birch tree. Furthermore, feeding experiments show that the triterpenoid is highly distasteful and that the concentrations present are more than sufficient to explain the absence of herbivory at the juvenile stages of growth. In older trees, the concentrations fall off, because much of the winter-dormant tissue is out of reach of the hares. Older trees are also able to withstand a greater degree of herbivory than young trees.

Cardenolides are C_{23} steroids that typically occur in plants with a sugar or sugars attached at the 3-hydroxyl group. The sugar is often an unusual one, specifically associated with these degraded triterpenoid derivatives. The ecological chemistry of the cardenolides centres around the interaction that occurs in nature between milkweed plants (*Asclepias* spp.) and the monarch butterfly *Danaus plexippus*. The milkweed–monarch butterfly interaction has been studied continuously by Brower and his associates since 1970 and much is now known about the biochemistry and ecology of the plants and insects involved. The key feature is the utilization by the butterfly of the plant cardenolides for its own protection from bird predation.

Over its natural range in North America, the female monarch

Table 5.3 Some ecological roles of plant triterpenoids. For references see Harborne & Tomas-Barberan (1991).

Ecological role	Example
Defence against herbivory	Papyriferic acid protects paper birch from snowshoe hare grazing
Hormonal defence	Phytoecdysones disrupt metamorphosis in grazing insects
Chemical defence borrowing	Monarch butterflies store *Asclepias* cardenolides to avoid bird predation
Antifeedant defence	Azadirachtins from neem tree inhibit insect feeding
Hatching factor for cyst nematodes	Glycinoeclipin from soyabean, *Glycine max*

butterfly may lay her eggs on several different *Asclepias* spp., and these vary considerably in both the concentration and type of cardenolide present (Seiber *et al.* 1984). By one-dimensional thin-layer chromatography of the cardenolides of host plants and adult butterflies, it is possible to determine which *Asclepias* sp. a particular adult butterfly fed upon in the larval state. Profiles have now been compared for five *Asclepias* spp. and each is distinctive in the cardenolide pattern of plant and butterfly (Lynch & Martin 1987). It is also clear from studying the concentrations of cardenolides accumulating in different butterfly individuals and populations, and comparing the concentrations in the different host plants, that there is an adjustment in the degree of absorption of cardenolide during feeding. Thus individuals feeding on plants with high levels of cardenolide are able to avoid absorbing too much cardenolide, while individuals feeding on plants low in cardenolide can increase their cardenolide intake accordingly. Variation in cardenolide content from plant to plant ranged from 1.02 to 9.19 mg of digitoxin equivalents in *Asclepias ericarpa*, with an average of 4.21 mg (Seiber *et al.* 1984).

The quality of cardenolide also varies from one species to another, in that more emetic/poisonous compounds may occur in one *Asclepias* sp. compared with another. Thus the cardenolides of *A. syriaca* and *A. speciosa* are of relatively low potency, whereas those of *A. curassavica* and *A. eriocarpa* are of high potency. Hence, some monarch butterfly populations will be better defended than others, depending on the species utilized as a larval food plant.

Although cardenolides are not generally toxic to either the larva or adult butterfly, there is nevertheless a cost to the insect in adopting a life-style based on a poisonous plant genus. There is evidence that the stem of the milkweed is not eaten because the aglycone uzarigenin is present and this is one of the very few cardenolides toxic to the caterpillar. There is also a restriction on milkweed feeding, due to the abundant terpenoid-based latex that exudes from cut leaf surfaces. Dussourd and Eisner (1987) have shown that the monarch larvae when feeding on milkweed plants cut the leaf veins before feeding distal to these cuts. Cutting the veins blocks the flow of latex to the feeding sites and represents a counter-adaptation by the insect to the plant's defence.

The cardenolides undergo metabolism and presumably also conjugation during the process of absorption and storage, and the profile of cardenolides in the food plant differs from that found in the insect. The biochemical changes that occur have yet to be fully documented, but there is evidence that uscharidin can be metabolized by tissue homogenates of monarch larvae to a mixture of calotropin and calactin. Other cardenolides, however, probably remain unchanged; for example aspecioside has been isolated from two food plants. *A. speciosa* and *A. syriaca*, and also from the butterfly.

A detailed study of the storage of cardenolides within the monarch's tissues has revealed that most of the cardenolides are tightly bound to the cuticle in the different organs. Presumably, this reduces the bioavailability of the toxins, keeps these cellular toxins away from the neuronal tissues, and hence avoids self-poisoning. Such storage in cuticles, however, has disadvantages for the insect, in that it allows some birds to overcome the emetic potential of the cardenolides, so that they can prey on the adult monarchs in their overwintering sites in central Mexico. The black-backed oriole *Icterus abeillei* and the black-headed grosbeak *Pheuctius melanocephalus* avoid an emetic dose by feeding only on the body contents, and by defecating the butterfly cuticular tissues where the cardenolide is mainly concentrated. The majority of bird predators, however, are still deterred by the emetic potential of these toxins.

The exploitation of the milkweed by *Danaus plexippus* and other insects raises the question of the long-term survival of these host plants. Why does the milkweed continue to synthesize these cardenolides if they no longer afford protection from insect herbivory? The answer clearly lies in the continued benefit to the plant in deterring mammalian herbivory through the manufacture of these highly poisonous constituents. It has been estimated that as little as 25 g of dry *Asclepias labriformis* leaf is fatal to a 50-kg sheep. *Asclepias* plants are likewise poisonous to cattle and goats. Generally, grazing animals find *Asclepias* spp. unpalatable and avoid them but occasionally, in heavily overgrazed pastures, they may be forced to eat milkweeds and poisoning nearly always results. Similarly, cut fodder containing *Asclepias* material can cause livestock deaths (Seiber *et al.* 1984). There is little doubt, therefore, that cardenolides provide a competitive advantage to *Asclepias* spp. in nature by limiting their consumption by large herbivores.

5.4 Nitrogen-containing metabolites

Alkaloids are the best known of the nitrogen-containing secondary metabolites of plants. They are organic bases that have a nitrogen atom as part of their structures and usually as part of a carbon cyclic system. They are also exceedingly numerous; some 10 000 structures have been recorded in a modern dictionary of alkaloids (Southon & Buckingham 1989). They can be conveniently subdivided according to their differing ring structures and natural distribution patterns. Thus, pyrrolizidine alkaloids such a retronecine characteristically occur in *Senecio* (Asteraceae) and *Borago* (Boraginaceae), whereas quinolizidine alkaloids like anagyrine are found mainly in lupins and other related members of the Fabaceae.

From the viewpoint of plant defence, alkaloids are less important than the phenolics, since they are only found in about 20% of angiosperm species. They are also generally absent from ferns, mosses and gymnosperms. Other classes of nitrogen-containing metabolites – the non-protein amino acids, cyanogenic glycosides and glucosinolates – are also of limited occurrence. Cyanogenic glycosides occur in around 2000 species from 100 plant families, but their distribution is relatively sporadic. They are common in some families (e.g. Rosaceae) but quite rare in others (Seigler 1991). Likewise, the glucosinolates are mainly restricted to Brassicaceae and four related families, although there are occasional records in six other families (Louda & Mole 1991).

One simple reason for the relatively restricted distribution in plants of secondary metabolites based on nitrogen is that the supply of this element to the plant is nearly always limiting, even in those plants such as legumes that fix their own nitrogen (see Chapter 3). Nitrogen-containing metabolites are all ultimately derived from the protein amino acids (Fig. 5.2), so that there is always competition for precursors which are required for more important processes, such as protein synthesis. Even when plants do develop secondary metabolites containing nitrogen in response to herbivore pressure, there are limits to the extent of this synthesis. With alkaloids the quantities produced are usually low (sometimes one-fifth or one-tenth of comparable phenolic production), but their low concentration is offset by their high potency. Alkaloid poisoning of livestock from lupin or *Senecio* plants requires only trivial amounts of plant tissue to be ingested.

There is also good evidence that alkaloids are efficiently used as defensive agents and they may be moved around the plant to those parts needing greater protection during growth and development. This happens, for example, in pyrrolizidine alkaloid-containing plants such as *Senecio vulgaris* (Asteraceae). In the stems of *Senecio*, the alkaloid (e.g. retronecine) is concentrated near the surface with 10 times more alkaloid in the epidermal than mesophyll cells. Likewise the inflorescence is especially well protected from grazing; the alkaloid flows upwards from

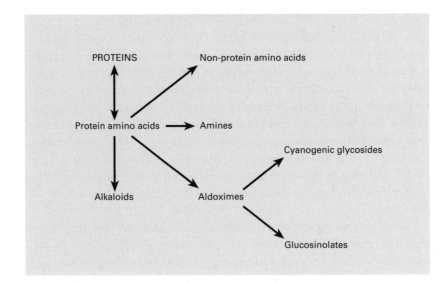

Fig. 5.2 Biosynthetic relationships of nitrogen-containing plant metabolites.

root and leaf during flowering, so that more than 85% of the total alkaloid becomes located in reproductive tissues (Hartmann *et al.* 1989).

In the coffee plant, *Coffea arabica* (Rubiaceae), the purine alkaloid caffeine is concentrated in those tissues vulnerable to herbivory. When no longer required for protection, the nitrogen in the alkaloid is recycled and utilized for protein synthesis. Caffeine is concentrated, for example, in the young, tender leaves where it reaches a level of 4% dry weight; the amount drops off exponentially as the leaf thickens and hardens. The expensive chemical protection of young leaves is replaced by mechanical defence in the older leaves. A similar change in metabolism occurs in the coffee bean, with the soft young bean containing about 2% alkaloid while the hard ripe bean contains only about 0.24%. Significantly, caffeine has a lethal effect on larvae of the tobacco hornworm *Manduca sexta* at a dietary concentration of 0.3%, while it causes sterility when fed to the beetle *Caliosobruchus chinensis* at a concentration of 1.5% (see Frischknecht *et al.* 1986). The protection of young leaves by high alkaloid levels and subsequent recycling of the nitrogen may occur regularly in alkaloid-containing plants; it has been observed in *Ipomoea parasitica* (Convolvulaceae), where young leaves have 10-fold the ergot alkaloid concentration at day 7 compared to mature leaves at day 20 (Amor-Prats & Harborne 1993).

A similar strategy for the maximal utilization of nitrogen locked up in secondary metabolites is employed in those legume species that protect their seeds with high levels of non-protein amino acid. Typically, seeds of *Mucuna* spp. may contain between 6 and 9% dry weight of the amino acid dopa, while seeds of *Canavalia ensiformis* have 4–6% dry weight of canavanine. These amino acids protect the seeds from being eaten by either bruchid beetles or mammalian feeders. The recycling of the nitrogen occurs during seed germination, since radioactive labelling experiments show that the dopa and canavanine are rapidly broken down and the nitrogen is transferred back into protein amino acid in the seedling plant.

Yet another strategy for nitrogen economy occurs in many of those plant species protected from herbivory by cyanogenic glycosides such as linamarin. In the legumes *Trifolium repens* and *Lotus corniculatus*, the production of cyanogenesis is a variable feature, such that one population may have only 5% of individuals cyanogenic while another may have 95% of individuals cyanogenic. The protective value of cyanogenesis has usually been assumed to lie in the enzymic release of hydrogen cyanide (Fig. 5.3), which is a respiratory toxin in the herbivore. More recent studies of cyanogenic glycosides and their effects on animals suggest that the non-nitrogenous aldehyde or ketone released during the enzymic breakdown may be equally effective in deterring herbivory. The substances involved, such as acetone or benzaldehyde, are also found in the defensive secretions of arthropods. This extra toxicity may be important in a defence agent where animal detoxification systems are

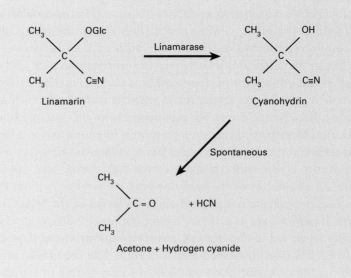

Fig. 5.3 Pathway of release of HCN and acetone from the cyanogenic glycoside linamarin.

readily evolved to deal with the primary toxin, namely the hydrogen cyanide (Jones 1988).

The cyanogenic glycoside linamarin is an example of a bound toxin, stored in the plant in a safe form and separated from the hydrolytic enzyme, linamarase, by compartmentation within the leaf. The free toxin, cyanide, is only released when tissue is damaged during grazing and the enzyme comes into contact with its substrate (Fig. 5.3). The glucosinolates (or mustard oil glycosides) represent another example of bound toxins found in plants. Sinigrin, a glucosinolate common in the cabbage family, gives rise by myrosinase hydrolysis to the acrid, volatile mustard oil allyl isothiocyanate (Fig. 5.4).

The protective role of glucosinolates against mammalian herbivores, as well as Lepidoptera and aphids, has been extensively studied. A high concentration is crucial for defence and this will even deter cows from

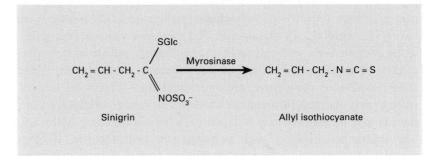

Fig. 5.4 Enzyme release of allyl isothiocyanate from the glucosinolate sinigrin.

feeding on wild cabbage. The interaction with specialist insects feeding on the Brassicaceae is a complex one, since at low concentrations glucosinolates are important feeding stimulants. Furthermore, they are used by the cabbage white butterfly as an oviposition signal. In the case of the cabbage aphid, *Brevicoryne brassicae*, feeding is limited by high sinigrin levels so that young plants are avoided. Nutritional factors also come into play and changes in the free amino acid balance stops the aphids from feeding on senescing plants (van Emden 1972).

An example, illustrating the changes that can occur in glucosinolate concentration during plant growth, is the study by Bodnaryk (1991) of sinalbin (*p*-hydroxybenzylglucosinolate) levels in white mustard *Sinapis alba* (Brassicaceae). The high concentrations in young cotyledons (20 mmol L^{-1}) and young leaves (up to 10 mmol L^{-1}) effectively deter both a specialist (the flea beetle, *Phyllotreta cruciferae*) and a generalist insect (the armyworm, *Mamestra configurata*). As the plant grows, the concentration drops so that older leaves have 2–3 mmol L^{-1}. These concentrations stimulate feeding by the flea beetle, but still deter the more generalist feeder.

The harmful effects of nitrogen-containing toxins on farm animals have been thoroughly documented, since fatalities occur regularly in livestock grazing on open pastures in Australia, South Africa and North America (James *et al.* 1992). Toxicity may be expressed directly through effects on the liver or brain or indirectly through damage to the offspring (teratogenicity). A number of plant alkaloids have been implicated, such as the hepatotoxic senecionine of ragwort and the teratogenic anagyrine from lupin. Less is known about the effects of such toxins on wild animals or how these animals may avoid the toxicity. More ecological studies are still needed to establish a defensive role for alkaloids and other nitrogen-containing metabolites in plants.

5.5 Phenolic metabolites

Phenolic compounds are aromatic structures bearing one or more hydroxyl groups. Most are polyphenols, having several hydroxyl substituents, one or more of which may be further substituted by methyl or glycosyl groups. They share a common biosynthetic origin from phenyalanine, one of the three amino acids formed from sedoheptulose via the shikimate pathway. The flavonoids are the best known group of polyphenols and comprise some 4000 structures (Harborne 1994). The total number of phenolics including all the non-flavonols must be near 8000. The great majority of these are of plant origin, although a few simple phenols are found in the pheromonal secretions of elephants and some flavonoids are stored in butterfly wings after dietary ingestion. Phenols range in molecular weight from simple compounds such as phenol itself, which occurs in the essential oil of *Pinus sylvestris*, through complex acylated anthocyanin pigments with molecular weights around

2000 to the polymeric condensed tannins where the molecular weights may reach 20 000.

Among the important properties of phenolics is their ability to ionize in the presence of bases. Many polyphenols have catechol (adjacent *o*-dihydroxy phenyl) groups and hence have the ability to chelate divalent or trivalent metal ions. Phenols with either *o*- or *p*-dihydroxy substitution (e.g. catechol or hydroquinone and their derivatives) are readily oxidized to the corresponding quinones; these are generally more active biologically than the original phenols. Such oxidations can occur enzymically through operation of the ubiquitous phenolases, which always accompany phenols in plant tissues, or non-enzymically by aerial oxidation.

Most phenolics are broadly toxic to life. Phytotoxicity is avoided in the plant by conjugation with sugar, sulphate or both, and storage in water-soluble form in the vacuole. The great majority of phenols in plants are glycosidic and many different glycosides have been recorded for the more widely occurring substances. For example, over 172 different glycosides of the common flavonol quercetin have been characterized. When free phenols are ingested orally by mammals, conjugation again takes place and they are converted to the corresponding sulphate or glucuronide. In insects, they are converted to glucosides. A central role in the formation of the major classes of plant phenols is occupied by *p*-hydroxycinnamic acid, formed from phenylalanine by deamination and *p*-hydroxylation (Fig. 5.5). The biosynthesis of flavonoids follows a regulated step-wise pathway from chalcones, the first C_{15} intermediates, to give the range of classes that are encountered in nature from flavones and flavonols to anthocyanins. The flavolans or condensed tannins are end-products of the pathway, while isoflavonoids are formed in a branching pathway (Fig. 5.5). Practically all the enzymes of these pathways have now been characterized.

Waterman and Mole (1994) have produced an excellent introduction to phenolic analysis especially for plant ecologists. Some of the ecological roles that have been suggested for phenolics are indicated in Table 5.4 and are now discussed in more detail. A major role may be related to the important antifungal and antibacterial activities that some structures possess. A number of such phenolics have been identified in the surface waxes of leaves or fruits in concentrations suitable to prevent germination of hostile fungal spores. One such example is the isopentenylisoflavone luteone, which occurs at the leaf surface of *Lupinus albus* (Fabaceae) and other lupins. Although lipophilic, it has sufficient water solubility to dissolve in the leaf surface moisture film and to arrest the germination of fungal spores *in situ* (Harborne *et al.* 1976). Not only are phenolics produced constitutively in this way to ward off microbial infection, but also they can be formed *de novo* in direct response to fungal elicitation. Indeed, some 50% of the 300 known phytoalexins are phenolic (Grayer & Harborne 1994). Such an example is the stilbene

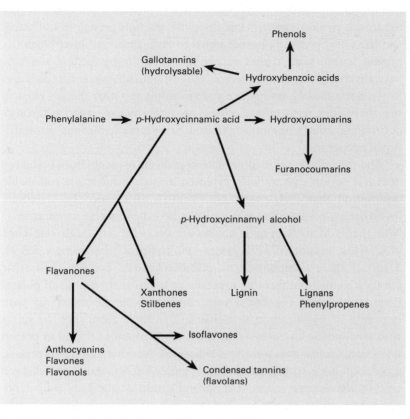

Fig. 5.5 Biosynthetic relationships of different phenolic classes.

resveratrol, identified as a phytoalexin in both the grapevine *Vitis vinifera* and the groundnut *Arachis hypogaea* (Fabaceae). Its effectiveness as an antifungal agent has been demonstrated by the introduction of the gene for resveratrol synthesis from the grapevine to the tobacco plant. The transgenic plant so produced showed the expected improved resistance to the common fungus *Botrytis cinerea* (Hain *et al.* 1993).

Plant phenolics are liable to interact with animals that eat plants throughout the whole process of food selection, eating and digestion.

Table 5.4 Some ecological roles for plant phenolics. For references see Harborne (1993).

Ecological role	Example
Defence against herbivory	Pinosylvin in alder inhibits snowshoe hare feeding in Alaska
Pollination attraction	Blue anthocyanin flower colours attract bee pollinators
Hormonal interaction	Clover oestrogen formononetin reduces fertility in sheep in Australia
Phytoalexin defence	Stilbene resveratrol produced on infection in grapevine and groundnut
Allelopathic interaction	Quinone juglone in walnut inhibits growth of neighbouring alder

Anthocyanin pigments in the skin of fruits provide a visual signal that the fruit is ripe for eating, while lipophilic phenols present in leaf waxes and hairs may provide a hostile signal to deter feeding. Other phenolics present within plant tissues will be absorbed during feeding and may have adverse effects. Some phenols may be immediately antifeedant and hence arrest feeding immediately after sampling. Other dietary phenols may have long-term effects, either on reproductive capacity through oestrogenic effects or on growth and development through antinutritional properties.

The role of allergenic substances in protecting plants from herbivory does not appear to have been explored in depth, but it is a reasonable assumption that such plants are likely to be avoided by vertebrate herbivores; we know that people learn by experience to avoid contact with them. Many allergenic toxins are phenolic, although one large group, the sesquiterpene lactones, are terpenes (see Section 5.3.2). Many of these compounds are photosensitizers, because their toxic effects are enhanced through exposure to light. Three groups of phenol deserve mention. The first group consists of simple molecules with isoprenyl or hydrocarbon side-chains. Urushiol, a catechol derivative with an aliphatic C_{13} side-chain, is the chief irritant in the oil of poison ivy, *Toxicodendron radicans* (Anacardiaceae). Another example is primin, a benzoquinone with methoxyl and pentanyl substitution, the allergen of *Primula obconica* leaves. The second group of allergens are furanocoumarins, in which the free hydroxyl groups are generally masked by methylation. These include angelicin, bergapten, psoralen and xanthotoxin, which occur in plants of the Apiaceae and cause contact dermatitis in humans who become sensitized to handling such plants. These furanocoumarins also interact unfavourably with the insects that feed specifically on umbelliferous plants and the chemical ecology of these interactions has been extensively explored (Berenbaum 1991). The third and last group of phenolic allergens are the polyhydroxy extended quinones, the harmful effects of which have been recognized in grazing farm animals. The best known example is hypericin, from species of *Hypericum* (Clusiaceae), which causes facial eczema when it is ingested by sheep, especially in Australia.

A group of phenols that may have long-term effects on grazing animals because of their oestrogenic properties are the isoflavones. The best known of these is formononetin (7-hydroxy-4′-methoxyisoflavone) which occurs in *Trifolium subterraneum* (Fabaceae) and other species of *Trifolium*. It is the cause of serious infertility in Australian ewes who ingest clover plants during feeding. Formononetin is metabolized *in vivo* to equol (7,4′-dihydroxyisoflavan). Equol interferes with normal oestrus and causes the so-called 'clover disease'. Isoflavonoids occur widely in legume pasture plants. These plants may benefit from long-term effects of isoflavonoids by reducing the fertility of grazing animals to limit the damage they do during feeding.

Phenolic metabolites may afford more immediate protection to plants by their antifeedant properties. The astringency of plant tannins must surely limit grazing and likewise a high content of low molecular weight phenols may be protective. Some of the most extensive antifeedant studies have been carried out with the snowshoe hare and mountain hare *Lepus timidus*. Several simple phenols are effective deterrents to hare feeding. Thus, snowshoe hares do not feed on *Populus balsamifera* (Salicaceae) twigs because of the presence of 2,4,6-trihydroxydihydrochalcone and do not eat *Alnus crispa* (Betulaceae) tissues due to the coating of pinosylvin and its methyl ether. Avoidance of apical twigs of birch (*Betula* spp.) by mountain hares in winter may be ascribed to the presence of a phenolic glycoside, identified as platyphylloside. Both species of hare avoid feeding on willow trees (*Salix* spp.) due to the phenolic glycosides based on salicin. These are all simple phenols, without isopentenyl substitution, and it is remarkable that they are so effective. The majority of woody gymnosperms and angiosperms tend to accumulate phenols in their tissues and these results suggest that they may enjoy significant protection from mammalian herbivores because of this. The meadow vole *Microtus pennsylvanicus*, for example, is deterred from feeding on conifer tree bark as soon as the phenolic content rises above 2.6% dry weight (Roy & Bergeron 1990).

Plant polyphenols that have attracted most attention as barriers to mammalian herbivory are the tannins. These are defined as those polyphenols that have the ability to bind with protein (i.e. they exhibit tanning activity on animal skins). Associated with tanning activity is astringency, a repellent taste property causing a puckering of the tongue and rejection. Chemically, there are two classes of tannin: the condensed tannins or oligomeric flavan-3-ols, which are widely present in all woody plants; and the hydrolysable tannins, sugar molecules esterified by a number of gallic (3,4,5-trihydroxybenzoic) acid moieties, which have a limited occurrence in certain angiosperm families. Since the chemical ecology of hydrolysable tannins has yet to be explored, the present remarks are restricted to the *raison d'être* of the condensed tannins.

Much has been written (Haslam 1989) about plant tannins, their quantitative measurement, effects on animal nutrition and defensive value to the plant. While there are still some controversial points, most authors agree that tannins do deter herbivory. For example, measurements of tannin levels in plants eaten and not eaten by colobus monkeys in Africa indicate that plants which have more than a certain level of tannin are avoided as food. Other chemicals, such as alkaloids, were measured in these plants but there is much less evidence that these are protective to the plant. Similar measurements on the tannins of plants browsed by kudu, impala and goat in the South African savannah show significant failure to feed when the plant species contains more than 5% dry weight of tannin in the leaves. Confirmatory experiments in which condensed tannins are added to otherwise nutritious diets of laboratory animals also support the hypothesis that ingestion of tannin above a certain level is harmful those animals. In fact, hamsters die if they are fed for more than 3 days with an artificial diet containing around 4% sorghum tannin.

The most impressive evidence confirming that tannins are a major feeding barrier is the discovery by Butler *et al.* (1986) that rats fed on a diet containing *Sorghum* (Poaceae) tannins are able to adapt to their adverse effects. In these rats there is enormously increased synthesis of a series of unique proline-rich proteins (PRPs) in the parotid glands. These salivary proteins have a high binding affinity for condensed tannins and remove them at an early stage in the digestive process. The high affinity of PRPs depends on the presence of up to 40% by weight of carbohydrate, the sugar units keeping the polypeptide chain in an open conformation so that it can bind strongly via hydrogen bonding to the tannin (Asquith *et al.* 1987). Adaptation in rats takes place within 3 days of commencing the dietary tannin and there is a 12-fold increase in PRPs in the parotid glands. The adapted animal can subsequently thrive on the tannin-containing diet without any adverse effects.

A survey of mammals (Mole *et al.* 1990) showed that rabbits and hares, like rats, are well endowed with the ability to produce PRPs when fed a tannin diet. As expected, these proteins are lacking in the saliva of carnivores such as dogs and cats. The response is weak in ruminants and there are many other animals where this adaptation is lacking. In insects, other modes of adaptation to tannin-rich food plants have occurred but, even so, there is still a strict limit on the ability of most insects to thrive on tannin-containing plant species.

5.6 Conclusions

It is suggested that the major functions of plant secondary metabolites are defensive. These substances assist the growing plant in its struggle for survival in a hostile environment, surrounded by animal herbivores of many different types, open to invasion by parasitic microorganisms and surrounded by competing plant species. This does not preclude a role for some secondary metabolites within the internal economy of the plant, since it has been known for some time that a number of secondary substances are involved in the processes of growth, cell division and plant development. Each of the major classes of secondary constituent – terpenoid, phenolic or alkaloid – have slightly different ecological roles, according to the biological properties of the molecules within each class. Thus, the intense and pervasive odours and scents of the volatile terpenoids provide valuable signals in angiosperm flowers for attracting pollinators, and similarly the brightly coloured anthocyanins and carotenoids are important floral attractants with the same purpose.

All three classes of secondary metabolite have a *raison d'être* in providing plants with defence against herbivory, although their effectiveness may vary. Some, like the alkaloids and cardenolides, are physiologically active at very low concentrations while others, such as the tannins, exert their deleterious effects at relatively high concentrations. Evidence for a defensive role is still quite limited and often depends on circumstantial clues from a variety of experimental approaches. The function of only a relatively few molecules has been explored in depth. The value to the plant of mixtures of related toxins has hardly been investigated, although this is a common feature of secondary metabolism production. Undoubtedly, the accumulation of the toxins at the site of feeding is crucial to the defensive role of secondary compounds but this is a subject where data are still relatively scarce.

Secondary metabolites can be excreted from plants, through leaching from the leaf or by exudation from the roots. They can thus interact with neighbouring plants, a phenomenon known as allelopathy (Rice 1984). How far secondary metabolities can limit the growth of neighbouring species before they are degraded by microbes in the soil is the subject of continued debate. It is not clear how significant allelopathy is in the normal competitive interactions that occur between plants growing in natural stands. It is easy to demonstrate in laboratory experiments some secondary compounds inhibit the germination of seeds of neighbouring species. More critical field experiments are needed to demonstrate that such inhibition has a controlling effect on the development of vegetation patterns in the natural environment.

A major subclass of phenolics, the monomeric flavonoids, has not been discussed here to any extent. There are the coloured flavonoids (anthocyanins, chalcones and aurones) which clearly contribute to plant pigmentation, but the majority of known flavonoids are colourless in daylight, i.e. the flavones and flavonols. That they have a role in providing plants with protection from damaging UV radiation has been postulated in recent years. Some evidence, such as their restricted presence in leaf epidermal cells, has accrued to support such a hypothesis. The biogenetically related hydroxycinnamic acid esters, which are also ubiquitous in angiosperm leaves, may also be involved in the same protective process and many more experiments are needed to test this idea.

Our present knowledge of the chemistry of secondary metabolites is immense. Furthermore, the new structures that are regularly uncovered in nature are relatively easily determined, following the advances in nuclear magnetic resonance spectroscopy of recent decades. Our knowledge of the biochemistry is reasonably well advanced; biosynthetic pathways have been outlined for many natural products and the enzymes of biosynthesis have been characterized for several major pathways. It is the ecological function of most of these molecules under field conditions that still eludes us. It is suggested that the major role is likely to be a defensive one but many more experiments are still needed to see whether this is true or not. This topic is taken up in detail in Chapter 10.

6: Sex

Michael J. Crawley

6.1 Introduction

Picture the scene. You are blind. You can't go anywhere. You are rooted to the spot. You've got limited resources. You don't know who your neighbours are and, even if you did, you couldn't talk to them. But you are seriously horny. You want to father as many seeds as you can, and you want to get as many of your seeds fathered by hunky pollen as you can. What's to be done? You could cast your pollen to the wind. Or you could advertise, and have the sex come to you.

You also need to decide what *kind* of sex you want. Do you want to transmit your genes mainly through paternity or maternity, or do you want to go for a balanced investment in both sexual functions? Finally, you need to decide how much to invest. Do you want to do it just the once, and go out in a blaze of glory (the big bang approach)? Or do you want to ration your sexual investment, and go for a strategy of little and often? Because you are hermaphrodite, then if the worst came to the worst, you could always have sex with yourself (as Woody Allen once said: 'don't knock it; it's sex with someone you love'). Or you might adopt a totally celibate life-style. A number of very successful plant species have foregone sexual reproduction altogether and persist by the production of asexual seeds that are carbon copies of their own genotype. But that's no fun at all. You want your progeny in their turn to be highly fecund mothers and prodigious fathers. Your offspring need to be competitive, disease resistant and herbivore tolerant. Your seeds need to be dispersed to the ideal habitats and to exhibit the correct degree of dormancy for the prevailing level of environmental uncertainty.

This chapter investigates these issues from a theoretical perspective based on evolutionary stable strategies (ESS) and optimality arguments, supported by selected case histories. The investigation of plant sexual success under field conditions is a young science, and is set to be completely revolutionized by the application of modern techniques from molecular genetics that will make it relatively straightforward to determine the male and female parentage of plants growing in the wild.

6.2 Sex: why bother?

There are several paradoxical questions raised by the prevalence of sexual reproduction in vascular plants. For example, a parthenogenetic mutant in a sexual population would have an immediate twofold advantage, so why is asexual reproduction so rare? Why allow recombination to reshuffle an obviously successful genotype? Why throw away half of an evidently well-adapted nuclear genome and replace it by a somewhat risky, unknown new half from a male? Presumably, the answer is that the benefits of sex outweigh the costs. However, it has to be admitted that the fundamental question of sex remains unresolved (Maynard Smith 1972, 1978, 1982, 1989, Williams 1966, 1975).

6.2.1 Costs of sex

1 The production of sons appears to be wasteful, and an asexual female could invest entirely in daughters and would therefore possess a twofold fitness advantage over a sexual mutant. This argument is much more compelling for animals than plants since only about 4% of the world's vascular plants are dioecious (the overwhelming majority of plant species are hermaphrodite; see Fig. 6.1).

2 The seeds produced by sexual reproduction might be inferior: it is a waste of resources to provision seeds that are less fit than the maternal plant (the so-called 'cost of meiosis'). However, given the existence of potentially strong competition between male gametes, the quality of paternal genes need not be any lower, on average, than the matching maternal genes.

3 Asexual mutants may be able to donate genes for asexuality to

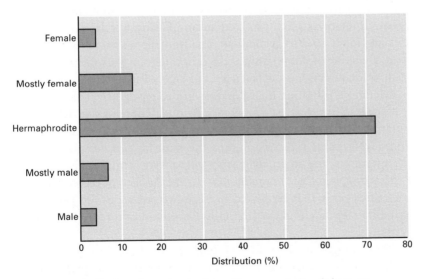

Fig. 6.1 The distribution of sex types in plants. The vast majority of plants are hermaphrodite, and the bulk of these have cosexual flowers (about 5% of plants are monoecious with single sex flowers on the same individual plant). Complete separation of the sexes on different individuals (dioecy) is rare in most habitats and for most plant groups.

co-occurring sexual individuals through pollen, but the reverse flow of sexual genes can obviously not occur.

4 Male-sterile asexuals will have more resources to allocate to seed than would sexual hermaphrodites, although this advantage cannot be spread, and could only win by long-term competitive exclusion of the less fecund, sexual seeds.

5 Sexual plants may suffer pollen-limited seed production; asexuals would produce more seeds in places or in years where pollinators were scarce, or where plant population density was low.

6.2.2 Benefits of sex

Sex might offer both short-term and long-term benefits to outweigh these costs.

1 *Generating variability.* There may be an immediate benefit to individual fitness in producing genetically variable rather than genetically uniform offspring. Given the spatial and temporal heterogeneity of real environments, there will almost always be a greater variety of recruitment opportunities for the variable offspring of a sexual parent than for the uniform offspring of an asexual parent. Computer simulations demonstrate that sexual strategies can persist in competition with the twofold advantage of asexual strategies under realistic levels of environmental heterogeneity (Maynard Smith 1978). Furthermore, a plant's competitors and natural enemies are constantly evolving, and species need to evolve constantly, just to keep up (see Chapter 4).

2 *Dissipating Muller's ratchet.* Recombination during sexual reproduction reverses the tendency of the whole genome to accumulate disadvantageous mutations over the long term (Charlesworth *et al.* 1993). Asexual clones tend to accumulate disadvantageous mutations because there is no way of eliminating them other than by back-mutation. As a result of demographic stochasticity, the class of individuals containing no deleterious mutation might happen not to reproduce at all, so in the next generation the fittest group of individuals contains one deleterious mutation (the ratchet has clicked round one notch). If that class fails to reproduce at some future date, then the most fit class has two deleterious mutations, and so on. This represents a long-term benefit of sex.

3 *Breaking up linkage disequilibrium.* In the absence of recombination, deleterious genes that are linked to advantageous genes can persist and accumulate by hitch-hiking (the Hill–Robertson effect; see Felsenstein 1974).

4 *The clean egg hypothesis.* Meiosis tends to weed out aberrant gametes, so that chromosomal abnormalities are reduced. Also, plant viruses appear to be unable to survive the process of meiosis, so sexual seeds are virus-free (the mechanism of this is not understood).

5 *Creation of linkage groups.* Sex can cause linkage by helping advantageously coadapted genes to come together in tight linkage on the same chromosome (see Maynard Smith 1989 for details).

6.2.3 Variable progeny and individual fitness

The fact that the grass *Dichanthium aristatum* can produce both apomictic and sexual offspring, with relative numbers under environmental control (Knox 1967), suggests that there must be a short-term advantage to sex, otherwise why would the sexuals not have been eliminated by the twofold advantage? Perhaps it is simply wrong to regard recombinations as a negative process that breaks up favourable, coadapted gene complexes. After all, recombination and directional selection are the tools of traditional plant breeding, and there is no reason to believe that similar processes do not work with natural selection. There are several possible mechanisms by which selection might favour sexual reproduction through the production of genetic variability within each brood of seeds.

Suppose that each parent contributes several seeds to one microsite. If there is strong sib competition, and only one seed can survive per microsite (i.e. selection is both intense and density dependent), then the seed that is genetically best adapted is likely to be the winner (assuming, of course, that it germinates at the same time as the others; a head start can be the difference between life and death for a seedling; see Chapters 4 & 7). So long as the environmental conditions per patch are sufficiently unpredictable (i.e. if there is low temporal autocorrelation), then a sexual genotype will increase in frequency.

Sexual reproduction is usually timed to coincide with seed dispersal into an uncertain environment, and this is the time when recombination is most likely to be advantageous (Williams 1975). Likewise, sex is more often lost in simple than in complex ecosystems (Levin 1975), and species that are sexual as natives often lose their sexuality if they become established as aliens in distant countries (see Chapter 19). However, perhaps the most persuasive argument is that continuing evolution in pathogens necessitates the flexibility of response that sexual recombination provides (Haldane 1949, Antonovics & Thrall 1994). Hamilton 1980 showed that in a model of host–pathogen coevolution, selection favouring sex can be strong enough to overcome the twofold advantage. Plants that are freed from their natural enemies by introduction to foreign countries often grow much larger than they ever did in their native habitats (see Chapter 19).

6.3 Mating systems

Mating systems can be divided into three broad classes: (i) autogamous (self-compatible); (ii) allogamous (entirely self-incompatible); and (iii) facultatively self-compatible. In this last case, it is often observed that the degree of selfing exhibited by a species increases towards the edge of a plant's geographic range. In the centre of the range, all of the plants are outbred, while at the extreme edge of the range all of the populations might be selfing. In terms of plant fitness there may be benefits to self-pollination (reliability at low population densities, matching of genes to the current local environment), but these will usually be outweighed by the disadvantages (inbreeding depression, loss of genetic variability). Which of the mating systems predominates in any given circumstances presumably reflects the interplay of these costs and benefits of outbreeding. I begin, however, by considering the way in which the genes are exchanged between individual plants during a single breeding season.

The textbook ideal of population genetics is *panmixis*. Under this assumption, every pollen grain has an equal probability of finding its way on to any stigma, and each stigma, is equally likely to receive pollen from any anther. It is as if all of the pollen grains from the entire population were to be put in a bag, shaken up vigorously, and then randomly dished out over the entire range of stigmas. It is immediately obvious that panmixis could never happen in a large population; there is no way that the probability of a pollen grain reaching a stigma 500 m away is the same as the probability of it reaching a stigma 0.5 m away.

There are two main processes at work in real plant populations that require more complex models of gene movement. First, even with the most efficient systems of wind pollination (see below), there is not random mixing within the pollen cloud; the pollen rain from individual plants is strongly leptokurtic, and close neighbours are much more likely to receive pollen from a given plant than are more distant individuals (Fig. 6.2). Second, the spatial distribution of genotypes of adult plants is unlikely to be random because of limited seed dispersal (see below), so that close neighbours are likely to have genotypes that are more similar to one another than the average for the population as a whole. Thus, individuals exchange genes with only a subset of the population (the 'genetic neighbourhood', see below) and these gene exchanges are more inbred than panmixis would predict. Understanding gene movements within plant populations is vastly improved by the use of spatially explicit models (see Chapter 14). These issues are expanded in subsequent sections.

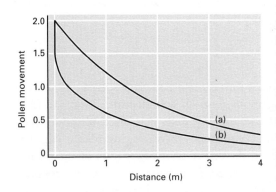

Fig. 6.2 Leptokurtic distributions of gene flow away from parents (located at $x = 0$). Such curves are observed for both pollen flow and for seed dispersal. The more gradual curve (a) has a higher mean dispersal distance; thus pollen flow by wind (a) is typically greater than by insects (b), and wind dispersal of seed (a) greater than by gravity alone (b). The scale of pollen movement is arbitrary.

Pollen movement

Distance (m)

6.4 Inbreeding and outbreeding

Inbreeding is the mating of individuals related by ancestry. It tends to increase or maintain homozygosity in which both alleles are identical by descent and to reduce heterozygosity. Inbreeding has consequences at both the individual level and population level, and plants display a wide range of features which reduce the likelihood of inbreeding (Grant 1958).

6.4.1 Population genetics of inbreeding

It is important to master two related definitions: (i) a *neighbourhood* is the spatial area in which mating is effectively random (note, however, that gene frequencies are assumed to vary continuously through space); and (ii) *effective population size*, N_e, which is the number of plants that would give rise to the observed level of inbreeding if there was random mating (local panmixis). At the population level, inbreeding causes loss of allelic diversity and increased homozygosity. The inbreeding coefficient, F, is the probability $(0 \leqslant F \leqslant 1)$ that two alleles at a locus in an individual are identical by descent. With inbreeding, the effective population size N_e is lower than the censused population N:

$$N_e = \frac{N}{1 + F} \tag{6.1}$$

This will lie somewhere between N and $N/2$, but in practice it is likely to be much lower than $N/2$ as a result of limited spatial mixing of alleles and, in very small populations, could lead eventually to the loss of all but one of the alleles at each locus. The effect of inbreeding is to reduce the frequency of heterozygotes, so that under predominant selfing the population tends to break up into essentially isolated lineages. Given long enough, a state of complete allelic identity by descent will be reached where all existing copies of a gene trace back to a single individual. This is an effective population of size $N_e = 1$ (Crow & Kimura 1970).

High inbreeding slows Muller's ratchet (see above) because the increased homozygous expression of mutant alleles in inbred populations has effects similar to stronger selection, and the fixation of deleterious mutations is accelerated in highly inbred populations (Charlesworth *et al.* 1993). Excessive inbreeding depression for a normally outbreeding species (defined as a change in F of more than 1% per generation) is likely to occur in populations with fewer than 50 individuals (Franklin 1980; Frankel & Soule 1981). For a habitually inbreeding species, the minimum population is much smaller than this. The longer-lived the species, the lower the minimum; for example, if the probability of an inbreeding perennial being replaced by at least one offspring during its lifetime is 0.75, then at least four individuals are required for $P > 0.99$ for population survival (Frankel *et al.* 1995).

6.4.2 Inbreeding depression

Inbreeding depression is any decrease in performance that arises because an individual's parents were more closely related than two randomly selected members of the population. The degree of inbreeding depression will be greater when mating occurs between brothers and sisters than between first cousins, and so on. New mutations are generally deleterious, but they may be protected from selection in the heterozygous condition. Homozygosity resulting from inbreeding leads to such mutations being expressed and the phenotypes suffer as a result of exposure to selection. The (usually deleterious) effects of inbreeding on the phenotype of individual offspring can be felt through reduced seed set, reductions in the numbers of seeds per fruit, reduced individual seed weights, poor germination, reduced seedling growth rates, reduced longevity, etc. These fitness reductions may be sufficient to offset any of the evolutionary advantages of sex discussed above. Some examples of the relative fitness of selfed versus outcrossed progeny are shown in Table 6.1.

The increased homozygosity that results from inbreeding can expose deleterious recessives and eliminate the benefits of heterozygosity (see below). The appearance of individuals that are homozygous for deleterious alleles is known as *partial dominance*, while the loss of hybrid vigour is known as *overdominance*. Charlesworth and Charlesworth (1987) argue that partial dominance is the most important cause of inbreeding depression in plants. While inbreeding depression of the order of 50% is common in normally outbreeding species, inbreeding depression of about 25% is not uncommon in normally inbreeding taxa. Given the persistence of inbreeding populations in the field, it is obvious that this level of inbreeding depression is not terminal and that the benefits of inbreeding are non-trivial (e.g. the potential for a single individual to colonize a new site). Also, if fecundity is sufficiently high and natural selection is sufficiently intense, then deleterious alleles might be elimi-

Table 6.1 Fitness values for self seeds or progeny relative to the value for outcrossed seeds or progeny.

Number of species and plant group	Character	Mean
12 Gymnosperm trees	% filled seed	0.43
7 Gymnosperm trees	Germination rate	0.81
12 Gymnosperm trees	Plant size	0.70
8 Angiosperm herbs	Seed production	0.63
6 Angiosperm herbs	Germination rate	0.88
4 Angiosperm herbs	Plant size	0.73
6 Angiosperm herbs	Fertility	0.58
6 Angiosperm herbs	Inferred viability	0.31
5 Angiosperm herbs	Fitness	0.46
2 *Lobelia* species	Two-year fitness	0.41
2 *Begonia* species	Total fitness	0.68
2 *Mimulus* species	Inferred viability	0.26

nated without causing local extinction. Inbreeding depression can be reduced if mutant allele frequencies are reduced due to selection at the pollen stage (Charlesworth & Charlesworth 1992). Mating between a small number of very large, long-lived plants is likely to increase the average rate of inbreeding.

The effect of self-pollination, open-pollination and controlled cross-pollination were compared for progeny of *Eucalyptus globosus* (Myrtaceae) by Hardner and Potts (1995). They followed progeny for 43 months after planting and found that selfing severely depressed seed set and field growth rate relative to outcrossing. Depressed height growth was not recorded until 8 months after planting and there was a general trend for the impact of inbreeding to increase with age.

6.4.3 Heterosis (hybrid vigour)

Heterosis is said to occur when progeny exhibit more vigorous growth, greater yield or increased disease resistance than *either* parent. It is typically associated with the accumulation of dominant alleles, each having additive effects on fitness. The dominant alleles can also mask the expression of deleterious recessive alleles. Suppose, for example, that a characteristic like seed size was affected by two genes X and Y and that the parents were $XXyy$ and $xxYY$, then the hybrid offspring $XxYy$ express *both* dominant alleles.

Heterosis can also occur through *pleiotropy* (genes that have more than one effect on phenotype) in cases where alleles affect different components of fitness in opposite directions. In such cases, both kinds of homozygote have relatively low fitness, but the heterozygote has high fitness (the classic example of overdominance from human genetics is sickle cell anaemia). Field studies on coniferous trees have shown that mature stands are much less inbred than the zygote or seed populations from which they were derived; mean heterozygosity of adults is much higher than in seedlings (Levin 1983), presumably as a result of selective mortality.

6.4.4 Outbreeding depression

This is said to occur when the pollen parent was 'too far away' from the maternal parent, so that the offspring were less well adapted to local conditions than if the pollen parent had been growing closer to the seed-bearing plant (Price & Wasser 1979; Wasser & Price 1983). Hand pollination studies in a large natural population of the herbaceous perennial *Gentiana pneumonanthe* found no effect of within-population crossing distance when selfing and interpopulation crosses were omitted from the analysis; selfing reduced fitness but crossing with pollen from a different (distant) population increased fitness (as measured by germination rate, seedling weight and adult weight; Oostermeijer *et al.* 1995). In contrast, pollen from long distance (2500 m) caused reduced seed set in *Agave schottii*, although the other patterns (local inbreeding depression and no midrange distance effect) were similar to the *Gentiana* example (Trame *et al.* 1995). There is no general agreement that outbreeding depression is either widespread or important under natural conditions (Barrett & Kohn 1991).

6.4.5 Kinds of self-pollination

Habitually inbreeding plant species show varying degrees of self-pollination and suffer inbreeding depression to varying degrees.

6.4.5.1 Autogamy

This is self-fertilization within the same hermaphrodite flower; pollen from the anther is moved by insects or through direct mechanical contact to the adjacent stigma. This is the commonest form of pollination in facultative autogamy (selfing may be a last resort if outcross pollen does not arrive).

6.4.5.2 Geitonogamy

This involves movement of pollen from one flower to another within the same individual plant (e.g. from branch to branch). It is extremely common in multiflowered clonal plants (Kiang 1972), and in animal-pollinated, mass-flowering trees (Gill 1986).

6.4.5.3 Cleistogamy

In cleistogamous species there is enforced selfing because the flowers never open. Some species adopt a mixed strategy, having some cleistogamous flowers (usually closer to ground level) and some open-pollinated flowers (generally closer to the top of the plant). It is normal for the seeds of the outcrossed flowers to be more widely dispersed than the self-fertilized seeds from cleistogamous flowers (occasionally, the closed flowers are produced underground, as in desert subspecies of the vetch *Vicia sativa* subsp. *amphicarpa* (Fabaceae)).

6.5 Sex types

Vascular plants display a wide spectrum of different sex types (Fig. 6.1). Some individuals are completely and permanently male, others are mostly male but partially (or sometimes) female, others are equally hermaphrodite, some are mainly female, and a few are completely (and permanently) female (Richards 1986). Botanists being what they are, each of these systems has been given its own long and profoundly unmemorable name (Box 6.1). Here, we are concerned mainly with the

Box 6.1 Glossary of terms used in the discussion of plant breeding systems.

Agamospermy asexual seed formation

Allogamy fertilization between pollen and ovules from different flowers

Androecy maleness

Apomixis asexual reproduction including vegetative reproduction and agamospermy

Autogamy fertilization within the same flower (selfing)

Cleistogamy flowers that stay closed and are always self-fertilized

Dichogamy separation in time within the same flower of pollen shedding and stigma receptivity (does not guarantee outbreeding because self-pollination from other flowers on the same plant (geitonogamy) may be possible)

Dicliny populations with male and/or female plants rather than just hermaphrodites

Dioecy where genets are either male or female (not hermaphrodite); rather uncommon in temperate floras but familiar examples include Salicaceae (willows and poplars), which have separate-sexed individuals

Geitonogamy fertilization between different flowers on the same individual plant (selfing; see also **autogamy**)

Genet a discrete genetic individual plant (in contrast to a ramet, which might be a structurally independent plant derived by lateral growth (vegetative reproduction), or part of a clonal plant with rhizome or stolon connections between the component ramets)

Gynodioecy female and hermaphrodite individuals occur in a population

Gynoecy femaleness

Herkogamy spatial separation of anthers and stigmas within an individual flower so that selfing does not occur without an insect visit

Hermaphrodite a plant (genet) with both male and female reproductive function (includes monoecious plants with separate male and female flowers as well as plants with hermaphrodite flowers)

Heterostyly coexistence of genetically controlled hermaphrodite floral types with different style lengths (usually with different anther positions as well); e.g. pins and thrums in *Primula*

Homogamy synchrony of anthesis and stigma receptivity within a flower

Inbreeding a non-panmictic breeding system in which heterozygotes are below Hardy–Weinberg expectation; commonly associated with low population densities, assortative mating or selfing.

continued on p. 166

Box 6.1 *Continued.*

Iteroparous polycarpic; repeat-flowering perennials

Monocarpic plants that flower once then die (also semelparous); opposite of polycarpic

Monoecy one individual plant supports separate male and female flowers (e.g. many temperate trees like oak and birch)

Outbreeding a panmictic breeding system where heterozygote frequencies might be expected to be close to Hardy–Weinberg equilibrium

Panmixis a theoretical ideal breeding system in which an infinitely large number of genets exchange genes with one another at random (all individuals are equally likely to cross with all others)

Protandrous a dichogamous condition in which pollen is shed before the stigma is receptive (opposite of protogynous, where the stigma is receptive before anthesis)

Pseudogamy asexual seeds are produced, but only after pollination and fertilization of the primary endosperm nucleus

Selfing (self-fertilization) fertilization of an ovule by pollen from the same individual plant (either the same flower (autogamy) or a different flower (geitonogamy) of the same genet)

Semelparous monocarpic

Tristyly populations in which short-, medium- and long-styled individuals (with reciprocal anther positions) coexist (Fig. 6.3)

Xenogamy fertilization of ovules by pollen from a different genet (a genetically distinct individual plant, rather than a different ramet of a clonal individual)

extremes of the distribution (where the individuals belong to separate sexes; dioecy) and the middle (hermaphrodite species). Within the hermaphrodites, some have single sex flowers on the same individual plant (monoecious species) while others have cosexual flowers that contain both male and female parts (stamens and pistil).

Until quite recently, the prevailing orthodoxy saw cosexual hermaphrodite flowers as essentially female in function. It is easy to measure female reproductive success by counting seeds and by determining the weight of limiting resources allocated to fruit and seed production. It is much harder to study paternity. A pollen grain is so much less substantial than a seed, and it is exceptionally difficult to discover where an individual pollen grain ends up, so it is easy to fall into the trap of assuming that the female reproductive function of a flower must be much more important than its male function. This view is simply wrong. Each plant receives half its genes from its father and half from its mother. In this fundamental sense, male reproductive function is just as important as female function. If anything, the data show that an individual hermaphrodite flower is more likely to be successful in its male function than in its female (see below).

If you were asked to design the ultimate plant breeding system, you might begin by thinking about the kind of conditions under which it had to operate. In some environments, you might favour flexibility in the face of environmental uncertainty. Under conditions where resources were locally plentiful for example, a desert annual might produce a few large, apomictic seeds with restricted dispersal, in order to take advantage of the conditions that proved so conducive for maternal development. Under poor growing conditions, however, the plant might switch to the production of many, small, outbred seeds, each provided with a means of wind dispersal, in the hope of finding better conditions somewhere else in the next generation.

To what extent can real plants be expected to conform to ideal models such as this? The success of any particular strategy depends very much on what the other individuals in the population are doing, so the problem requires a game theoretic solution (see Box 6.2). The question also arises as to the relative investment of limited resources in male and female reproductive functions, since it is obvious that plants cannot aspire simultaneously to maximize investment in both pollen and seeds (see below). Some advantages for hermaphroditism are:

1   it allows facultative self-fertilization, should outcross pollen fail to be delivered, or if it proved to be of inferior quality;
2   male and female function may be limited by different resources (e.g. protein or carbohydrate);
3   temporal separation within the year means that resources available to female function later in the summer are simply not available to male function at the beginning of spring (unless they are stored – at a cost – from the previous year);
4   in those cases where seed set is pollen-limited, then the single set of petals and nectaries increases both male and female function; or
5   it may be that pollen rather than nectar is the principal reward for the pollinators, in which case purely female flowers would simply not be visited.

With sequential hermaphroditism, the larger (older) individuals may be better at one sexual function than the other (e.g. attracting pollinators (male) or dispersing ripened seeds (female)). If reproduction is costly in terms of growth or survival, then reproducing through the more costly sex may have disadvantageous long-term implications (e.g. a reduction of the length of reproductive lifespan). In such a case, expression of the costly sex would be expected to begin later in life. It could be that economies of scale apply here; it may not be worth producing a small crop of fruits, because they are likely all to be destroyed by herbivores (see Chapter 13); the same investment in pollen might have a much better chance of getting the plant's genes into seeds of the next

Box 6.2 Evolutionary stable strategies (ESS).

The application of game theory (Maynard Smith 1989) allows a test of the fitness of a mutant plant by asking 'Could a plant with a slightly different reproductive strategy invade a population consisting of the hypothesized optimal strategy?' If the answer is yes, then either the strategy is not optimal or, because of frequency dependence, it is not globally optimal. In the second case, it may be that there is a balanced polymorphism, with a trait having greater fitness when it is rare than when it is common.

Suppose that the members of the population have one of two phenotypes *A* and *B*; we can call these *strategies*. Fitness (*W*) consists of a constant value *K* plus a *pay-off*, which is written $E(A,B)$ where *A* is the individual and *B* is its partner. The conditions for invasion are as follows: let *I* be the phenotype of most members of the population, and *M* is the rare mutant phenotype with frequency $p \ll 1$. Then

$$W(I) = K + (1-p)E(I,I) + pE(I,M)$$
$$W(M) = K + (1-p)E(M,I) + pE(M,M)$$

Strategy *I* is an ESS if, for all alternative strategies, *M*, $W(I) > W(M)$ when *p* is small. The theory allows for *pure strategies*, where a plant does one thing or another, and *mixed strategies* where a plant does one thing with a probability *q* and another thing with probability $(1-q)$. If individuals can adopt mixed strategies, then the population will come to consist of mixed strategists, each playing various strategies with differing probabilities. Alternatively, if only pure strategies are possible, then the population will become polymorphic, with some individuals following one strategy and others a different strategy.

Strategy one of a series of alternative courses of action (e.g. invest in extra ovules or extra pollen)

Strategy set the full range of options (this involves a consideration of what is biologically plausible plus a leavening of parsimony; the bigger the strategy set, the more difficult the analysis will be)

Pay-off value of a strategy in terms of Darwinian fitness (copies of the gene in subsequent generations). The difficulty arises because the pay-off of a given strategy often depends on what the other members of the population are doing (i.e. on frequency-dependent processes)

ESS a strategy is an ESS if, when adopted by most of the population, it cannot be invaded by the spread of any rare mutant exhibiting a different strategy. An ESS is a strategy that is robust against mutants playing alternative strategies (Maynard Smith 1972, 1989)

generation (Charnov 1982b, 1984; Charnov *et al.* 1977; Lloyd 1984). These issues are discussed in detail in the following sections.

6.6 Incompatibility systems

Most flowering plants are self-compatible, and this is thought to be the primitive condition. Prevention of inbreeding is a conspicuous feature of floral biology and has fascinated biologists for centuries; for example Darwin (1876) said that self-incompatibility was 'one of the most surprising facts which I have ever observed'. Prevention of inbreeding can be accomplished through the breeding system (e.g. dioecious individuals; see below) or through the possession of specific incompatibility traits within the flowers of hermaphrodite individuals.

There are two broad classes of genetically based self-sterility that prevent fertilization with other individuals carrying the same alleles: (i) *gametophytic* self-incompatibility (GSI) involves the pollen; and (ii) *sporophytic* self-incompatibility (SSI) involves the tissues of the pistil. Incompatibility works by one of three mechanisms: (i) prevention of pollen germination; (ii) retardation or distortion of the pollen tube; or (iii) failure of fertilization (nuclear fusion). Only one type of incompatibility is found in any one plant family: for example, GSI occurs in Solanaceae, Scrophulariaceae, Rosaceae, Fabaceae, Onagraceae and Papaveraceae and SSI occurs in Convolvulaceae and Brassicaceae (Clarke & Newbigin 1993; Sims 1993).

GSI is a single locus with many alleles; rejection occurs if the single *S* allele present in the haploid pollen grain matches *either* of the two alleles present in the diploid maternal tissue of the pistil. The pistil produces extracellular glycoproteins (surprisingly, these are related to extracellular ribonucleases of some fungi) that enter incompatible pollen tubes, where they act as a cytotoxin by degrading RNA (including ribosomal RNA). SSI also involves a single, multiallele locus, but rejection is controlled by interaction of the self-incompatibility genotype of the pistil with the genotype of the pollen parent and not with the haploid genotype of the pollen grain. Here, each pollen grain presents the products of two alleles and rejection occurs when *either* one of these matches *either* one of the *S* alleles in the pistil. Often there are complex dominant or codominant interactions between *S* alleles, affecting the outcome of particular crosses. One type produces an extracellular glycoprotein and another produces a membrane-associated protein able to phosphorylate serine/threonine residues. The classical view (Whitehouse 1950) was that other mechanisms for avoiding selfing like heteromorphic self-incompatibility and dioecy arose in secondarily self-compatible species in response to pressure favouring outbreeding, and that SSI (genes held by the pollen parent) evolved directly from GSI (genes held by individual pollen grains) (de Nettancourt 1977). However, at the nucleotide level, loci controlling SSI in *Brassica* bear no resemblance to loci controlling GSI in Solanaceae. They do, however, crop up in self-compatible *Brassica* spp. and in other plant families. The origin and evolution of SSI in Brassicaceae has been studied by Uyeno-yama (1995) using analysis of the nucleotide sequences that regulate the expression of self-incompatibility. In *Brassica* the S locus regulating sporophytic self-incompatibility shows homology to a multigene family present in self-compatible congeners and groups in which this form of self-incompatibility is not typical. It therefore appears that the age of the SSI system is four to five times greater than species divergence within the group. It looks as if both multigene families predate the function of self-incompatibility, and suggests multiple, independent origins of self-incompatibility through recruitment of pre-existing genes. Incompatibility presents a problem in small plant populations. In a large population, the greater diversity of S alleles means that the proportion of potential mates is relatively high. Simulations show that it is hard to maintain a high diversity of S alleles in populations less than about 50 individuals, and hence there is a decrease in the frequency of available mates (Byers & Meagher 1992). This process might drive a downward spiral towards local extinction in a strictly self-incompatible species.

6.7 Prevention of self-pollination

Herkogamy is spatial separation of the anthers and stigmas. The anthers and stigmas can be physically separated within the same cosexual flower, or the male and female flowers can be located on different parts of the plant (monoecy). *Heterostyly* typically involves reciprocal polymorphism in stamen and style lengths, physiological self-incompatibility, intra-morph incompatibility, and a set of associated polymorphisms of pollen and stigma characters (Eckert & Barrett 1994).

Distyly is the most familiar incompatibility mechanism, as exemplified by the 'pin and thrum' dimorphism of *Primula*. Populations contain two kinds of individuals. Pin plants have long, pin-like styles while others have short, thrum styles (Fig. 6.3a). Pin females can only be fertilized by thrum pollen and vice versa because of incompatibility (see above). This prevents self-fertilization and ensures crossing of a pin female by a thrum male. The pollen grains differ as well. The thrum grains that have to grow down the longer pin styles are bigger than the pin grains, which need only to grow through the shorter thrum styles (Ganders 1979).

Tristyly involves three morphs and individuals are found that possess long, medium or short styles. At equilibrium, populations should exhibit equal frequencies of the three style morphs, but historical accidents during post-glacial colonization are thought to account for the marked deficiency of mid-styled morphs of *Decodon verticillatus* (Lythraceae) in New England and central Ontario. Computer simulations show that skewed morph ratios can persist for > 10 000 years in

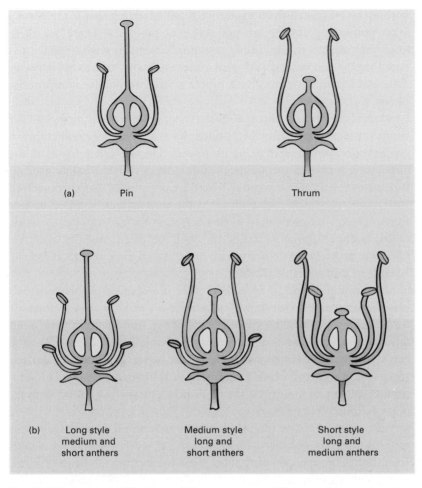

(a) Pin   Thrum

(b) Long style   Medium style   Short style
medium and   long and   long and
short anthers   short anthers   medium anthers

Fig. 6.3 Heterostyly. (a) Distyly, in which there are two self-incompatible morphs; pin has a long style and short filaments and thrum has a short style and long filaments. (b) Tristyly, in which there are three (usually equally abundant) morphs showing various permutations of long, medium and short styles and filaments; self-compatible mutants have stigmas and anthers at the same level.

long-lived clonal species like *Decodon* (Eckert & Barrett 1995). One of the best-known examples of tristyly involves the waterweed *Eichhornia paniculata* (Pontederiaceae) (Fig. 6.3b), which shows tristyly in its native range in tropical Brazil, but has secondarily evolved self-compatibility (smaller flowers, adjacent anthers and stigmas) where it occurs introduced on Caribbean islands (Manicacci & Barrett 1995; Barrett 1996).

Dichogamy is temporal separation of pollen shedding (anthesis) and stigma receptivity. The opposite trait (*homogamy*) involves simultaneous maturity of male and female parts and greatly increases the likelihood of selfing. When anthesis occurs first, the trait is known as *protandry*, and when the stigma becomes receptive before anthesis the trait is known as *protogyny*. There is no point in dispatching clouds of pollen to the winds if there are no receptive stigmas around at the time. Similarly, there is nothing to be gained from expensive floral advertisement if the plant's main pollinating insects are not active at the time. There are three important aspects to the timing of reproduction: (i) when to start; (ii) how long to go on; and (iii) what time-lag, if any, to exhibit between male and female function. Each aspect is likely to involve compromise of one sort or another.

Consider a wind-pollinated grass. There is something to be said for being amongst the first to shed pollen, because the dilution effect from competitors' pollen will be at its lowest. On the other hand, if the population is strongly protandrous, this strategy runs the risk of missing the first of the receptive stigmas. There are other risks to early flowering such as the increased probability that the flowers will be killed by late frosts. Thus, while we might expect selection for early flowering when pollen competition is intense, this will be attenuated by counter-selection to ensure synchrony with receptive stigmas and to avoid the hazards of unfavourable weather.

As to the duration of flowering, there is a continuum between two extreme strategies: to shed all of the pollen in a massive, single cloud; or to release a small amount of pollen every day throughout a protracted flowering season. The fact that most wind-pollinated plants adopt the first strategy suggests that this is clearly the better. For insect-pollinated species, however, the whole continuum is observed, from mass production of ephemeral flowers to the continuous production of small numbers of flowers (see below).

The duration of the time-lag between male and female function can vary from zero in obligate selfing species (e.g. cleistogamous species) to several weeks in species where there is no self-incompatibility but severe inbreeding depression. Brunet and Charlesworth (1995) present an ESS analysis of dichogamy in the context of male and female investment when there is directional movement of pollinators within inflorescences (see below). Floral longevity is reviewed by Primack (1985) who concludes that female flowers last longer than male flowers, and that flower longevity is higher in cooler, moister conditions.

6.7.1 Evolution of self-pollination from a cross-pollinating ancestor

All the things that you predict would happen with a switch from outcrossing to self-pollination have been observed in various permutations in a range of plant families: reduction in the number of flowers, reduction in peduncle length, reduction in petal and sepal size, change from rotate to funnel-shaped or closed corolla, less contrasting petal colour patterns, loss of floral scent and nectar, reduced anther size and increased proximity between anther and stigmas, reduced numbers of pollen grains, reduced length of style, synchrony of stigma receptivity and anther dehiscence, reduction in the numbers of ovules per flower (Ornduff

1969). These patterns are consistent with reduced costs of advertisement (fewer and smaller pedicels, petals and sepals) and reduced need to reward pollinators (no more nectar or scent need be produced), and increased efficiency of within-flower pollen transfer (increased proximity of anther and stigma, increased synchrony of pollen shed and stigma receptivity) and reduced waste (fewer pollen grains (and sometimes ovules) produced, with a trend towards the production of fewer pollen grains per ovule). These traits are discussed in detail below.

6.8 Limits to reproductive output

6.8.1 Resource-limited fecundity

In most vascular plant species, fecundity (total seed production per year) is a roughly linear function of plant size, and plant size reflects an individual's success in resource capture (see Chapter 4). To this extent, therefore, fecundity is typically resource-limited. For obligate outcrossing species, however, size is not everything; if pollen, or pollinating insects, are in short supply, then seed production will be lower than resources would otherwise permit. This is known as pollen-limited fecundity. A wide range of manipulative experiments show that seed production by outcrossing plants increases as resource availability is increased by the addition of mineral fertilizers (see Chapter 3), removal of herbivores (see Chapter 13) or removal of neighbouring plant competitors (see Chapter 12), consistent with the notion that seed production is typically resource-limited. Conventional wisdom about reproductive limitation is summarized succinctly in Bateman's principle (Box 6.3): female reproductive success is limited access to resources, while male reproductive success is limited by access to females.

6.8.2 Pollen-limited fecundity

Pollination limitation of female reproductive success appears to be rather more frequent than has been supposed (Bierzychudek 1981, 1984). In a study of published data on 258 species, Burd (1994a,b) found significant pollen limitation at some times, or in some sites, in 62% of cases. Perhaps the most extreme example of pollen limitation comes from Gill's (1989) study of slipper orchids *Cypripedium acaule* in deciduous woodlands of Virginia. Over a 10-year period, these ornate flowers appeared to be virtually useless at attracting insects. In only 1 year did more than 5% of the plants produce seed, and in 4 years out of 10 there was complete reproductive failure. The flowers cheat by producing neither nectar nor fragrance, and the insects appear to have learned to avoid them. Plant recruitment was demonstrably pollen-limited; cross-pollination by hand produced high seed set in all years, and increased seeds led, in turn, to dramatically increased seedling recruitment.

> **Box 6.3 Bateman's principle: female fitness is limited by access to resources; male fitness is limited by access to females.**
>
> Bateman (1948) studied the reproductive performance of male and female fruit-flies *Drosophila*. He found that most females produced offspring, but that male reproductive performance was much more variable; some males obtained many mating and other males fathered no offspring at all (in statistical terms, the variance in female reproductive performance was lower than the variance in male performance). The females needed to mate only once to obtain sufficient sperm to fertilize their entire batch of eggs and hence to realize their maximum reproductive potential. For males, however, reproductive performance was directly proportional to the number of matings they obtained.
>
> Trivers (1971) expressed the concept in terms of parental investment; he suggested that the parent which has invested least in the offspring will be tempted to desert. Thus, members of the sex that invests little in each offspring (usually males) will compete for access to members of the high-investing sex (usually females). Maynard Smith (1978) viewed the problem in game theoretic terms; the optimal behaviour for one parent depends on what the other parent is doing. If the female can rely on the male to complete the care of a nest of young, then her optimal behaviour might be to abandon the nest and begin on a new offspring. The ESS solution depends on a variety of factors: the effectiveness of parental care by one or two parents; the chance that a deserting male will get to mate again; a female that exhausts her resources in laying eggs is less able to guard them; and whether a male can be confident that he was the father of the eggs he might help in rearing.
>
> In terms of plant reproduction, Bateman's principle has important implications for investment in male and female function in hermaphrodites and in the conditions under which evolution might favour a trend towards single-sex (dioecious) plants. The principle is mostly likely to apply in cases where seed set is resource-limited rather than pollen-limited, and where the relationship between pollen production and the number of seeds fathered is more or less linear (i.e. not rapidly saturating). These points are discussed in the text (Burd 1994b).

Randomness in pollinator behaviour can produce wide variation in reproductive success between different flowers, and selection may have caused plants to produce more flowers, or more ovules per flower than will usually be fertilized, in order to benefit from chance fluctuations that bring in large numbers of high-quality pollen grains (Burd 1994a). When seed production is pollen-limited, then increased investment in floral display can increase fitness via male and female function more or less equally. When seed production is resource-limited, then male function will benefit much more than female function from increases in floral display. Bell (1985) goes so far as to suggest that the bisexual flower is 'primarily a male organ, in the sense that the bulk of allocation to secondary floral structures is designed to procure the export of pollen rather than the fertilization of ovules'. He found that insects visit larger flowers more often, and when flower numbers are experimentally reduced the surviving flowers are visited at a lower rate, whereas removal of attractive structures had no impact on the number of seeds per fruit. In species with dimorphic flowers, male flowers are generally larger than female flowers, and male inflorescences contain more flowers. In dioecious species, male plants typically have more inflorescences than female individuals.

6.8.3 Population regulation

We must be careful to distinguish between effects of pollination on plant population dynamics and effects on plant evolution. It may be that recruitment of many plant species is not seed-limited in most years or in most of the plant's habitat (Crawley 1990a). The simple experiment of sowing extra seeds may not lead to increased plant recruitment. In this case, factors leading to relatively large changes in seed production may have no demographic consequences at all.

However, this is not to say that these same processes have no evolutionary consequences. So long as there is genetic variation for the traits in question, and plant recruitment is genotype-specific, then traits can increase in frequency in microsite-limited populations, just as they can in pollen-limited populations.

6.9 Monocarpy and polycarpy

Annual plants (therophytes) flower once then die, generally within a few months of germination. Some longer lived plants also flower once, then die; these are the monocarpic (semelparous) species. Most non-annual monocarpic species are relatively short-lived (they are sometimes called pauciennial ('few year') plants); familiar examples are thistles, like *Cirsium vulgare* (Asteraceae) and melilot *Melilotus officinalis* (Fabaceae), which form a rosette in their first growing season then flower and die during their second year (these are strict biennial plants). Other short-lived monocarpic plants are facultative perennials and may be capable of repeat flowering if they are prevented from ripening a full seed crop by herbivores or by mowing (e.g. ragwort *Senecio jacobaea* (Asteraceae) will typically flower and die in its second year, but it can survive flowering to form another rosette from basal buds in the third year and flower again in its fourth year, and so on). A few monocarpic plants are very long-lived; famous examples include the spectacular silverswords from

Hawaii *Argyroxiphium sandwicensis* (Asteraceae), many bamboos (Janzen 1976) and some tropical rain forest trees (see Chapter 9).

The great majority of perennial plants are capable of repeat flowering (polycarpy or iteroparity). The questions of how long they wait before first flowering, and how they manage the trade-off between reproductive effort and subsequent survival are discussed in Chapter 4. Here we are concerned with questions that have to do with sexual allocation; for example, should individuals invest equally in male and female function throughout their reproductive lives, or should they vary sexual invest-ment as they age?

In their analysis of risk and reproductive effort, Real and Ellner (1992) divided fitness into arithmetic mean and variance components. Because traits map non-linearly on to fitness components like offspring number or growth rate, the expected value of a fitness component is a function of both the arithmetic mean and the variance of the traits. Thus, the optimum is affected by the quality and variability of the habitat as well as by the curvature of the trait-fitness relationship. The model can predict the full spectrum from monocarpic to varying degrees of polycarpic investment, and from open- to closed-flower pollinations (chasmogamous versus cleistogamous pollination). Mono-carpy is especially favoured if offspring types exhibit negative covariance in their contribution to growth rate.

6.10 Pollination by wind

Wind pollination comes much closer to meeting the assumptions of panmixis (see above) than does insect pollination. Distances travelled are typically greater, and the likelihood of pollen mixing is enhanced. Distance travelled decreases as the density of the surrounding vegetation increases. Wind velocity and circulation pattern are important, and slow wind speed has a greater effect on the distance travelled by heavy compared with light pollen grains. Pollen flow can be restricted if the wind is not strongly directional or if the weather is unusually wet (Willson & Burley 1983). The fundamental trade-offs are between pollen number, pollen size and pollen dispersal distance (larger pollen grains are likely to travel less far, but large size might make for greater pollen tube competitiveness).

There is no expectation that pollen fitness would become saturated in wind-pollinated species (there is no practical limit to the amount of pollen that the wind can carry, and the wind (unlike some clumsy pollinating insects) does not cause pollen to be lost). On the other hand, it is likely that male fitness is a decelerating function of invest-ment in pollen, since dispersal distances are limited and stigmas within pollination range are likely to become saturated (see below). Wind pollination is strongly associated with monoecy and with protogyny (Gleaves 1973, Bertin & Newman 1993) and is more common in

tropical forest trees than has previously been reported (Bullock 1994).
Roubik (1993) argues that wind pollination in temperate forests is
significantly more efficient (cheaper) than animal pollination in neotropical moist forest, where the trees may invest as much as 3% of their net
primary production in pollinator rewards.

Wind pollination has been shown to be extremely effective in
dioecious *Staberoha banksii* (Restionaceae; rush-like monocots from
South Africa); the number of pollen grains per pistil varied between 0
and more than 100, but the average pollen load per pistil per plant
ranged from 4.4 to 28 (Honig *et al.* 1992). Seed set was very high
(average 93% of maximum) and showed no decline in isolated female
plants of the sort that might suggest pollen limitation. In their study of
the monoecious, wind-pollinated and highly inbred cocklebur *Xanth-
ium strumarium* (Asteraceae), Farris and Lechowicz (1990) used hierarchical multiple regression to formulate a path diagram to assess which
plant traits had significant impacts on reproductive success. Fruit production was strongly correlated with a plant's vegetative biomass, and
fruit size was negatively correlated with the total number of fruits
produced. Plants that flowered later produced more fruits, at least in
part because they grew to a larger final size than early emerging
individuals (in contrast to most studies which show that early establishment leads to larger final size because of intraspecific competition and
superior position in the size hierarchy; see Chapter 11). All other trait
effects were indirect, through their impact on plant size.

6.11 Pollination by animals

It is essential to adopt a spatially explicit approach in any theoretical
analysis of animal pollination. This is because: (i) distance of pollen
movement is extremely restricted; (ii) the spatial distribution of mature
plant genotypes is such that related individuals tend to occur in clumps,
so that inbreeding is much more likely than it would be in a panmictic
population; and (iii) pollinator behaviour often leads to assortative
mating (see below).

The central concept of animal pollination is the notion of *carry-over*;
this has to do with the proportion of the pollen picked up from the first
flower, which is deposited on the second and subsequent flowers. In
many cases, carry-over is so low that if pollination does not occur in the
first flower visited after leaving a particular male flower, then it is not
likely to occur at all. In other cases, successful pollination could occur in
the twentieth flower visited after leaving a given pollen donor. In
general, carry-over means that the average pollinator flight distance
from one flower to the next underestimates the distance of gene
movement; the greater the carry-over, the greater the underestimate
(Levin 1983). Likewise, carry-over tends to be increased when flowers
offer little reward, because pollinators spend little time on them. Carry-

over is typically low in plants like orchids that package their pollen grains into discrete pollinia (Box 6.4).

In considering the relationship between flowers and pollinators, it is important to remember that the plant and the pollinator have quite different objectives. A harried, underfed, yet constant pollinator is ideal for the plant. Best for the pollinator is precisely the opposite: large sources of food in close proximity to one another (Feinsinger 1983). The object of the exercise from the plant's perspective is to obtain the maximum number of visitations by pollinators, and to send them off to receptive stigmas on as many different plants as possible per unit resource invested in attraction and reward. The pollinator is interested in maximizing its foraging success for nectar or pollen; generally, it has no interest at all in the effectiveness of pollination (but see the counter examples of fig wasps and yucca moths in Box 6.5). The period of pollen production is typically short because of: (i) seasonality of pollinating animals; (ii) competition between male plants for access to unpollinated stigmas; and (iii) the opportunity cost of extended male investment; instead of exporting pollen, extra vegetative shoots could be produced instead, and these could be used to fuel the filling of seeds and the provisioning of fruits later in the season.

Bell (1985) experimentally reduced petal size in *Impatiens capensis* (Balsaminaceae) and found that even severe mutilation had no clear effect on seed set. There was, however, a big reduction in the pollen mass exported. This led him to speculate that the main purpose of floral display was for male sexual function. However, in *Campanula americana* Johnson *et al.* (1995) found the opposite. They removed 0, 50 and 100% of the petal lobes and measured significantly reduced seed set in the flowers with their petals reduced, but no reduction in pollen export (this was because a relatively ineffective bee pollinator *Halictus* quickly removed pollen from the reduced flowers). It is certainly not always true, therefore, that petals serve male function substantially more than female function.

6.11.1 Flowering phenology

The timing of reproductive events is important in determining the efficiency of animal pollination. If allocation to floral attraction and reward is made from rate-limited income (current photosynthesis) rather than from a pool of stored reserves, then investment in male function can impose an opportunity cost for late-season female investment. If resources are diverted from vegetative shoots whose net production could have fuelled later female investment in fruits and seeds, then one expects a more female-biased sex ratio than predicted by a model based on simultaneous division of a common pool of resources. The ratio becomes even more female biased if the fruiting tissue is photosynthetic (Burd & Head 1992).

Box 6.4 Pollination by pseudocopulation in *Ophrys.*

The bee orchids are a spectacularly attractive group of small (*c.* 10 cm), rosette-forming terrestrial orchids found in great diversity during late spring in open ground throughout Mediterranean Europe. Their attractiveness to humans derives from the similarity of their flowers to the body form and colouration of the females of their pollinating insect species (Plate 6, facing p. 366). The specificity of their floral morphology to particular pollinating insect species has allowed extraordinary adaptive radiation (more than 130 species), and this genus provides the best example of a case in which pollinating insects have been the main cause of speciation. Chromosome numbers are the same, and all the species tested so far can be crossed experimentally with one another to produce fertile hybrids, demonstrating that it is the reproductive isolation caused by the specificity of the pollinating insects that maintains genetic integrity under field conditions.

The genus can be divided into two on the basis of the behaviour of the male during pseudocopulation: (i) males align head-first on the labellum and the pollinia are attached to the insect's forehead (section *Euophrys*); (ii) males align tail-first and the pollinia are attached to the tip of the abdomen (section *Pseudophrys*). The orientation of the insect on the labellum is such an efficient isolating mechanism that two co-occurring species of *Ophrys* can share the same species of pollinator without cross-hybridizing (e.g. *O. iricolora* (abdominal pollination) and *O. transhyrcana* (head pollination) are both pollinated in Cyprus by the same large bee *Andrena morio*).

Apart from *O. apifera*, which is often self-pollinating, all *Ophrys* spp. attract insects by mimicking the scent, visual signals and tactile stimuli of their female mates, inducing the males to attempt to copulate with the flowers. Each species of *Ophrys* generally attracts a single species that is not attracted to other co-occurring *Ophrys* spp. Chemical analysis of the labellum shows that it contains up to 100 different scents which are similar, if not identical, to insect pheromones. Less specific scents attract the insects from further away, then close-range orientation and attraction is determined by visual and tactile cues. There are striking visual similarities between the hairy labella of the *O. omegaifera* group and the bodies of anthophorid bees, and between the black and white markings of *O. cretica* and the black and white abdomen of *Melecta tuberculata* (Anthophoridae). In orchids pollinated by *Andrena* spp., where the insects are visually similar to one another, visual attraction is less important and specificity of pollinators to individual orchid species is less pronounced. One pseudocopulation can be as brief as a few seconds or as long as 15 minutes. After a few

continued on p. 180

Box 6.4 *Continued.*

copulations with *Ophrys*, it appears that the male insects soon learn to avoid the flowers and seek out real females with which to mate; this makes the phenology of flowering in relation to emergence of naive male insects vitally important. In cases where two coexisting *Ophrys* spp. share the same pollinating insect, it is normal that their flowering seasons do not overlap (the later species does not begin to flower until the first species has finished).

6.11.2 Nectar reward

Male–male competitiveness can be improved by increasing attraction to pollinators by producing more nectar or a more imposing floral display. The downside of this strategy is that if the rewards are too high, then the pollinators are likely to stay on the plant for a long time and carry out lots of geitonogamous self-pollinations. Pollinators can develop preferences for different scent morphs (Galen *et al.* 1987) as well as different petal sizes and colours (Devlin & Ellstrand 1990a,b). Pollinators can also learn which flowers are currently offering superior nectar rewards (see below).

The within-plant control over pollinator behaviour is nicely illustrated by the mountain herb *Delphinium nelsoni* (Ranunculaceae). The spike-like inflorescence has buds at the top and maturing fruits at the bottom. Just above the maturing fruits are flowers with receptive stigmas but with pollen production switched off and, above them, the first flowers to open produce pollen but do not have receptive stigmas. Clearly, the ideal way for a bee to forage through the plant is to begin with the lower flowers, then progressively move up the inflorescence, during which time pollen from other plants is cleaned from the insect's body, so that when it reaches the pollen-shedding upper flowers it can carry away a full body-load. Pleasingly, this is what real bees actually do under field conditions. But why should the bees be so cooperative? The answer is that they would not be expected to be cooperative. Pyke (1978) discovered why the bees begin at the bottom of the inflorescence and forage upwards: the nectar reward per flower is higher in the lower female flowers. Charnov (1982b) makes the interesting point that it might be in the plant's (male) reproductive interest that there be adaptations in the lower (female) flowers to clean the bees of as much foreign pollen as possible, prior to their visiting the upper, pollen-dispensing flowers.

This strategy does not always work in precisely this way. For example, the blue (mature) flowers of *Anchusa strigosa* (Boraginaceae) produced nectar at higher rates and received more visits per unit time from

Box 6.5 Pollination by herbivorous insects: fig wasps and yucca moths.

Each of the 750 or so species of figs (*Ficus*, Moraceae) is pollinated by its own unique species of fig wasp (Hymenoptera; Chalcidoidea; Agaonidae). The relationship is one of the most intimate between any plant and insect. The fig depends absolutely on the wasp for pollination of its female flowers, and the wasp depends absolutely on the fig as a source of food for its larval stages. The female fig wasp pollinates the female fig flowers, then lays her eggs through the tissues of the style into the developing tissues of the ovule. She oviposits only into the short-styled flowers, and flowers with long styles are safe from herbivory. A tiny gall containing a single larval fig wasp forms inside the ovary. In due course, the next generation of fig wasps emerges inside the cavity of the fig (the synconium). Males emerge first and cut a hole in the side of the developing ovary, inseminating the female fig wasp before she emerges. After emergence, the mated female loads up her specialized baskets with pollen, then leaves the fig through an exit hole bored for her by the males (they have specialized mouthparts for this purpose). The tiny, wingless males die inside the fig, having never seen the light of day. (Note that the monoecious figs exhibit extreme protogyny, with the female flowers receptive several weeks before the male flowers shed their pollen; anthesis occurs as the new generation of wasps emerge from their galls in the female flowers.) The female then flies off into the forest and is attracted to a conspecific fig tree by volatiles given off by fig fruit at exactly the right stage of development for pollination (Ware & Compton 1994). She squeezes into the synconium through the ostiole, often losing her wings in the process (and sometimes her legs and antennae as well). She then goes about the business of pollinating the stigmas of the female fig flowers, and the cycle is completed. The details of the story are somewhat different for dioecious figs (Janzen 1979; Herre 1987), but the basis of the mutualism is the same.

The interaction between yucca moths *Tegeticula* spp. (Lepidoptera; Incurvariidae) and *Yucca* spp. (Agavaceae) is another obligate pollination–seed predation mutualism in which adult female yucca moths pollinate yuccas and yucca moth larvae feed on developing yucca seeds. Mated female moths arrive in a yucca flower and oviposit into young ovules. They then take pollen from their specialized baskets and push the pollen down the stigmatic tube, after which they collect fresh pollen then fly off to another inflorescence (James *et al.* 1993). Individual moths 'cheat' by ovipositing in yucca pistils without attempting to transfer pollen, and pollination often fails because moth populations

continued on p. 182

Box 6.5 *Continued.*

contain some pollinators and some non-pollinating cheaters (Powell 1992). Interspecific mutualisms have an inherent conflict of interest in that the fitness of the moth increases at the expense of the yucca, and current theory predicts that mutualistic interactions are only evolutionarily stable when both interacting species possess mechanisms to prevent excessive exploitation. Pellmyr and Huth (1994) discovered such a mechanism in yucca; there was a strong negative relationship between moth egg number and the probability of flower retention, and a strong positive relationship between the number of pollinations received and the probability of flower retention. Selective maturation of fruit with low egg loads and high pollen loads selects against moths that lay many eggs per flower or provide low-quality pollinations. Evidence from molecular phylogenies suggests that the yucca–yucca moth relationship has evolved independently more than once by colonization of a new host (Bogler *et al.* 1995).

bees compared with violet (young) flowers. Bees usually start at the bottom of the inflorescence and work their way upwards, but pre-foraging nectar volume was not correlated with flower height as it was in *Delphinium*. Bottom flowers were depleted more frequently than upper ones (Kadmon *et al.* 1991).

It is clear that the plant can manipulate pollinator behaviour by varying nectar reward levels; for both birds and bees, lower rewards mean less turning, longer moves and hence greater likely distance of pollen movement. Really high rewards might bring about a numerical response by the pollinators (e.g. honey-bees recruiting extra workers) and this might increase long-distance pollen flow. Departure rules used by solitary long-tongued bees in the genera *Anthophora* and *Eucera* were established by collecting nectar in microcapillary tubes from the flowers *Anchusa strigosa* in Israel. Nectar accumulation rates in experimentally emptied flowers indicated that time since a bee visit was a good estimate of nectar standing crop for these flowers. Nectar reward influenced both the probability of departure from a flower and the behaviour of the bee after departure (e.g. subsequent distance moved within the same plant). Kadmon and Shmida (1992) established a probabilistic departure rule (rather than a simple threshold rule) that depended on rewards from both the current and the previously visited flower. The two kinds of departure decisions (leave the neighbourhood of an individual flower or leave the whole plant) were influenced by the size of the reward; the smaller the nectar reward, the more likely the bees were to depart.

The tendency of bees to forage upwards and the degree of pollen carry-over mean that the frequency of selfing should increase from the bottom to the top of a multiflowered inflorescence; these should act as strong selective forces maintaining anti-selfing mechanisms in mass-flowering species and protandry in species with vertical, spike-like inflorescences (Barrett *et al.* 1994).

Low reproductive success through natural bee pollination can occur when the stigmas become clogged with heterospecific pollen from abundant, simultaneously flowering plants. This was the reason for poor seed set in the rare perennial *Gentiana cruiciata* pollinated by the bumblebee *Bombus pascuorum* in The Netherlands; its stigmas were clogged by pollen of *Rubus caesius* (Rosaceae) growing nearby (Petanidou *et al.* 1995).

6.11.3 Pollen reward

Rewarding pollinators with pollen rather than nectar is an option adopted in many bee-pollinated families. Pollen reward is no use in dioecious or monoecious species, however, because the pollinators would soon learn not to visit the pollen-free female flowers. Pollen-collecting bees may behave differently than nectar-collecting conspecifics in relation to reward levels (Zimmerman 1982). Various adaptations reduce the likelihood of pollen wastage: devices to confound pollen thieves (Faegri & van der Pijl 1979), or presentation of pollen in an extended series of small packets throughout the life of a flower rather than in a single large dose (e.g. through staggered anther dehiscence). Bumble-bees are capable of responding to variation in pollen availability (presumably by vision) before landing on the flowers of *Campanula rotundifolia*, and of avoiding flowers with low pollen availability (Cresswell & Robertson 1994).

6.11.4 Plant spatial pattern

Plant population density and plant spatial pattern can affect both the number of pollinators attracted to an individual plant, and the behaviour of the pollinators that are attracted. Using the self-incompatible annual crucifer *Brassica kaber* grown in fan-shaped arrays with different floral backgrounds, Kunin (1993) found that widely separated plants had significantly reduced seed set in all floral backgrounds as a result of pollen limitation, and that the reduction was greatest when the floral background comprised very similar flowers (*B. hirta*) so that the pollinators behaved as generalist (low-quality) pollinators rather than as specialists. As Rathcke (1983) has noted, self-incompatible plants growing at low densities suffer a double jeopardy: they are likely to suffer low visitation rates by pollinators, and to receive a high proportion of useless, foreign pollen when they do get visited.

It is important, however, not to jump too readily to the conclusion that these isolated, self-incompatible, pollen-limited plants are necessarily less fit than their conspecifics producing larger seed crops in high-density patches. After all, their offspring are dispersed into an environment where intraspecific competition (e.g. access to microsites) is likely to be very low, and where specialist seed and seedling-feeding herbivores are likely to be scarce. Of course, it may be that recruitment is difficult or impossible for these isolated plants, especially if satiation of the local seed-feeding animals is required before there is any recruitment at all (see Chapter 13). As ever, we need whole-generation data on fitness (seeding plant to seeding plant) rather than part-life-cycle data (like seed production per plant), which is always likely to be a poor surrogate in circumstances where density dependence is important.

6.12 Sexual investment by hermaphrodites

This section deals with the question of investment in male and female reproductive function in cosexual and monoecious hermaphrodites; the issues of adjusting the sex ratio in dioecious populations and of determining the degree of maleness and femaleness in species with labile sexuality are dealt with later. Let us assume that total investment is limited and the issue is simply what proportion of reproductive investment to allocate to each sex. We need to consider two cases: (i) simultaneous hermaphroditism (most cosexuals); (ii) sequential hermaphroditism (most monoecious species and some cosexuals). The allocation of limited resources between male and female function is expected to depend on the way that efficiency changes with changing investment (i.e. on the marginal efficiency; see Charnov 1982b, 1984).

Figure 6.4 shows a graph of female fitness against male fitness. If the plant is totally female, we place a point at 1 on the *y* axis and if the plant is totally male we place a point at 1 on the *x* axis. If male and female fitness are directly substitutable, then the fitness of a 50 : 50 hermaphrodite individual is exactly 1. Notice that the model allows for other ratios of male to female fitness (e.g. 1.5 units of female fitness and 0.5 units of male fitness lies on the same straight line as 1 : 1 investment). The direction along this line in which selection might drive the relative investment does not concern us here (see below).

However, the trade-off between male and female fitness might not be linear. For example, some investment (e.g. peduncls, sepals and petals) may be shared between the sexes in cosexual flowers. This produces a curve that is convex to the origin, and which lies everywhere above the linear relationship (Fig. 6.4a). Alternatively, the plant may gain in fitness by specializing in being male or female, so that once the predominant sex has been determined (e.g. by environmental conditions; see below) then any resources would be better spent in extra effort in that same sex, rather than in a small investment in the other sex; this gives the concave curve in Fig. 6.4a. Selection will favour hermaphrodites

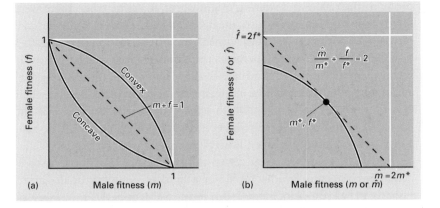

Fig. 6.4 The trade-off between male and female fitness. (a) On the straight line, hermaphrodites have exactly the same fitness, no matter what proportion they devote to male and female function. When costs are shared between the sexes, the curve is convex and hermaphrodites are favoured because their fitness is higher than any single-sexed plant. When specialization in male or female function increases fitness, then the curve is concave and natural selection could lead to dioecy. (b) Graphical demonstration that fitness is greater for hermaphrodites than males or females when the curve is convex. (From Charnov 1982b.)

when the curve is convex, and will tend towards the evolution of dioecy when the curve is concave (Charnov 1982b).

It is worth looking at one of the male-function trade-offs in some detail because it is informative of a whole class of design and behaviour constraints facing plant reproduction. Consider the question of advertising for pollinator services. For a given investment, the plant could produce a small number of flowers each day over a protracted period, or one massive burst of flowers on a single day. What are the costs and benefits of each approach? The benefit of a brief, mass flowering is that it would attract large numbers of pollinators. Furthermore, these pollinators might be attracted from further away, and therefore they might take pollen grains to a larger total number of females and those females are less likely to be close relatives. On the downside, however, the mass flowering by its very attractiveness might tend to keep the pollinators on the same plant, with the result that most of the pollinations, bought at such a high cost, would be geitonogamous selfings. This question turns back on another investment decision. Is the floral advertisement backed up by a large or a small reward? Suppose that the main pollinators are nectar feeders. If the nectar reward is too high, the insect will not travel to another plant before returning to its nest, so pollen (male) reproductive success will be negligible. However, if the reward is too low, the insect might learn to avoid flowers of this species. Clearly, there must be an optimal, intermediate nectar reward for a given kind of pollinator.

What about the strategy of producing a small number of flowers per day over a longer period? This has the advantage of being much less prone to the risk that flowering might occur at a time when there were few receptive stigmas. It also means that each flower could be well provisioned with nectar. On the other hand, the attraction of pollinators now becomes a problem, because the floral display will be relatively inconspicuous (so it will tend not to draw pollinators from a long way away) and pollinators will be reluctant to visit the plant because their rate of resource gathering will be low. Proportionately more of the flowers may end up being selfed, because each pollinator is likely to visit a high proportion of the open flowers on any visit to the plant.

Equal allocation of resources to the two sexual functions was predicted by Maynard Smith (1978) in a kind of energetic analogue to Fisher's equal sex ratio model (Box 6.6). Subsequently, a number of authors have pointed out why equal investment by hermaphrodites in the two sexual functions might *not* be expected: (i) when there is inbreeding and a high degree of self-fertilization (Charlesworth & Charlesworth 1981); (ii) if male and female gametes are dispersed over different distances; (iii) if fitness is a non-linear function of investment in either sex (Charnov 1979); (iv) there is local mate competition (Charnov 1980); or (v) the environment is temporally or spatially

---

Box 6.6 Sex ratio and sexual allocation in hermaphrodites.

Fisher (1930) argued that the 'sex ratio will so adjust itself, under the influence of Natural Selection, such that the total parental expenditure incurred in respect of children of each sex shall be equal; for if this were not so and the total expenditure incurred in producing males, for instance, were less than the total expenditure incurred in producing females, then since the total reproductive value of the males is equal to that of the females, it would follow that those parents, the innate tendencies of which caused them to produce males in excess, would, for the same expenditure, produce a greater amount of reproductive value; and in consequence would be the progenitors of a larger fraction of future generations than would parents having a congenital bias towards the production of females. Selection would thus raise the sex ratio until the expenditure upon males became equal to that on females.

Suppose that in a given population there are more females than males. Then a parent who produced only males would have more grandchildren than a typical member of the population. This means that the gene for producing males would increase in frequency. The same argument shows that if there were more males than females, then a mutant gene producing only females would increase in frequency. So in a population with an excess of either sex, selection would always favour the production of the other, rarer sex. The only stable sex ratio is 1 : 1.'

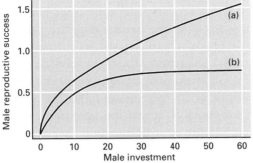

Fig. 6.5 Male reproductive success and male investment. The number of flowers and the size of their attraction (petals) and reward (nectar or pollen) are a good measure of male investment. Diminishing returns are expected, however, because of local pollen dispersal and relatedness to close neighbours (see text). It is likely that investment continues to increase for a wider range of investments for (a) wind-pollinated plants compared with (b) animal-pollinated species. The units are arbitrary.

variable (Lloyd & Bawa 1984). Any of these processes might cause a shift away from a 50 : 50 allocation pattern.

What is the shape of the function relating male fitness to male investment? As we have already seen, this is affected strongly by whether the flowers are pollinated by wind or by animals (Fig. 6.5). It is reasonably clear that male function (i) saturates more quickly for insect-pollinated flowers, and (ii) becomes asymptotic at a lower total investment for insect pollination compared with wind pollination. The two main causes of rapid saturation are: (i) limited pollen-flow distances; and (ii) inbreeding caused by spatial aggregation of related genotypes. In the insect-pollinated radish *Raphanus raphanistrum* (Brassicaceae), for example, diminishing returns in male function scale roughly with the square root of male investment (Young & Stanton 1990).

6.12.1 Measuring the costs of male and female function

How do you work out the costs of male and female function? What currency should be used (dry matter, carbon, nitrogen, phosphorus or protein)? What investments are shared, and which are unequivocally attributable to one sexual function or the other? Lloyd (1982) reasons as follows. Since most of the published data use dry matter, then this is the only practical measure of currency to adopt. Some functions are obviously shared between the sexes (support organs like pedicels and peduncles, protective tissues like bracts, bud scales and sepals, attraction signals like petals and rewards such as nectar) and though it is not obvious that they contribute equally to male and female function, there is no theoretical justification for splitting their costs other than 50 : 50. Pure male costs involve the stamens (filaments, anthers and the pollen contained within them) and pure female costs involve the pistil (stigma, style and ovary, plus all the investments in the endosperm, fruit and dispersal structures). There are some grey areas. For example, in plants that use pollen as a reward, the amount of pollen produced affects both male and female fitness. Of the six studies reviewed by Lloyd (1982) only one (the wind-pollinated *Microlaena polynoda* (Poaceae)) showed even roughly 50 : 50 investment; male investment in the other five cases ranged from as little as 4% to a high of 33%. The modal investment in male function was about 10%, and wind-pollinated species appeared to invest consistently more in male function than did animal-pollinated species. Similarly, data on pollen/ovule ratios in a diverse array of angiosperms support the distinction between wind-pollinated and insect-pollinated species; investment in male function was consistently greater in wind-pollinated taxa (Cruden & Jensen 1979, Damgaard & Abbott 1995). Schoen (1982) reported an interesting pattern in male investment from the herb *Gilia achilleifolia* (Polemoniaceae). Across six populations, the degree of selfing varied from 20 to 85% and there was a strong negative correlation between the degree of selfing and the amount of dry matter invested in male function.

6.12.2 Theory of male and female investment

The combined fitness of any parent plant *i* from male and female function is assumed to be simply additive:

$$w_i = m_i + p_i \qquad (6.2)$$

where w is total fitness, m is maternal and p is paternal fitness (Lloyd 1984). This can be combined with Bateman's rule (Box 6.3) that paternal fitness is limited by access to receptive stigmas. Male fitness will therefore be given by the maternal fitness of its mates' eggs that it has been able to fertilize in competition with other males, plus those of its own (selfed) seeds:

$$w_i = m_i + \sum_{j=1}^{K} m_j e_i c_i \qquad (6.3)$$

where K is the number of mated plants, m_j is their maternal fitness, e_i is the fraction of 'eligible' eggs and c_i is the 'competitive share' of these eggs obtained by the pollen of plant *i*. When there is partial self-fertilization, then some of the eggs on female plant *j* are not available for cross-pollination, so $e_i < 1$, but for the sake of simplicity assume that $e_i = 1$. Likewise, let us assume that all the other K plants that receive pollen have equal female fitness, so we can replace the summation by K equal contributions:

$$w_i = m_i + K m_j c_i \qquad (6.4)$$

Now we make the simplifying assumption that pollination is a pure lottery; every pollen grain has the same probability of fathering a given ovule, so that male success is directly proportional to the fraction of the total number of pollen grains that was produced by that particular male (panmixis, see above). Assume further that each plant has a similar total reproductive investment (this is a big assumption, given the phenotypic plasticity and the tremendous size variation of mature plants; see Chapter 11) and that the losses of pollen to predators, accidents, etc. are similar for all plants. Now the competitive share can be written as

$$c_i = \frac{a_i}{Ka_k} \tag{6.5}$$

where a_i is the fraction of reproductive resources given over to male function, and we assume that all the K other plants in the population are allocating the same fraction to male function a_k. The simplest assumption is that an increment of investment in either function confers a constant gain in fitness for all proportional allocations a_i and $1 - a_i$. This gives linear paternal and maternal fitness curves (see Fig. 6.6). This linearity assumption requires that: (i) the number of potential mates K is very large, so that the investment of an individual

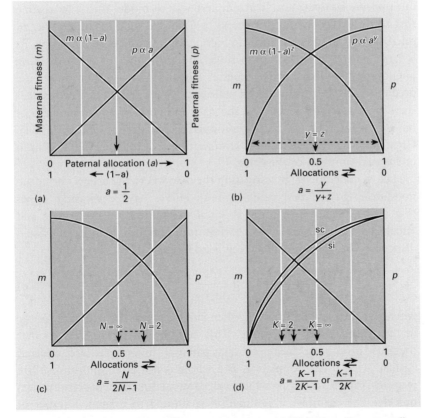

Fig. 6.6 Lloyd's model of fitness and sexual investment. (a) With linear fitness pay-offs for both sexual functions the model predicts 50 : 50 investment in both sexes. (b) With convex non-linear functions, there is a wide range of strategies and optimal investment favours the sex with the greatest exponent (y or z), i.e. the species whose fitness decelerates least quickly; (c) sib competition favours reduced female investment; and (d) self-incompatibility causes an increase in optimal male investment (see text for details).

male has very little effect on the probability that each of its pollen grains will fertilize an egg; and (ii) that there is no density dependence in seed performance, so that the success of a seed is independent of the number of seeds produced by that maternal parent. Exceptions to both these assumptions are discussed below.

It turns out that the ESS allocation strategy for this highly simplified model is neutrally stable (see Lloyd 1984 for details). If the whole population allocates its resources 50 : 50 to male and female function, then a mutant with a different allocation strategy would do equally well, no matter how extreme its allocation (e.g. a totally male plant would be just as successful as a total female; see also Shaw & Mohler 1953 for a model with neutrally stable sex ratios in dioecious populations).

Suppose now that the success of individual seeds is density dependent. In principle, an individual seed might do better in a large cohort of seeds because of predator satiation (see Chapter 13), or it might do better in a sparse seedling population where competition was less intense, or the probability of pathogen infection was lower (see Chapter 12). Assume that the density dependence is negative and works through intraspecific competition. The simplest way to model this is to assume that seeds are competing for access to a limited (and fixed) number of microsites, x, in which plants can recruit. If the number of parent plants is large and if levels of competition are the same for all competing phenotypes (so their relative fitnesses are all the same), then the model predicts a 50 : 50 investment in male and female function. The result is stable and does not depend on the sizes and costs of pollen or seeds or on the magnitude of total reproductive investment (Lloyd 1984). Greater efficiency leads to more competition, not to lower investment in one sexual function or the other.

Let us conclude by investigating the case where the fitness curves are non-linear. In the first case, let both curves have decreasing slopes (Fig. 6.6b). This might come about if the returns on both male and female investment were subject to diminishing returns because of limited dispersal. The optimum allocation is no longer 50 : 50, but rather

$$a = \frac{y}{y + z} \tag{6.6}$$

where y is the exponent of the pollen curve and z is the exponent of the maternal curve (see Fig. 6.6b). This equilibrium is an ESS, so that mutants producing more or less paternal investment are selected against, even if the population as a whole is not at equilibrium. More resources are invested in whichever parental function decelerates less quickly. For example, if pollination is limited to a small number of neighbouring plants, then male function will suffer diminishing returns as more and more competition on stigmas is with the male's own pollen (this is analogous to local mate competition; see below). Similarly, if seeds are subject to negative density dependence because of competition for a limited number of microsites in the vicinity of the female parent, then female reproductive function will be subject to diminishing returns; the more seeds produced per parent, the greater the competition for microsites with seeds from the same maternal parent. As Lloyd (1984) stresses, although both these mechanisms are highly stabilizing for the equilibrium allocation, neither of them is likely to cause a large departure from 50 : 50 investment.

Male and female function might interact with each other (e.g. increasing investment in pollen could increase both male and female function simultaneously by attracting more honey-bees and the bees might come from greater distances; alternatively, increasing investment in pollen might clog up the stigmas on the same plant, thereby reducing female fitness). In the first case, it is much more likely that increased investment in male function would improve female fitness than the reverse since, as we have seen, most investment in female function happens late in the season, long after pollen production has ceased. In the second case of negative interaction, not surprisingly the ESS allocation reduces investment in the reproductive function that causes the problem (pollen in our example).

Where there are biases away from 50 : 50 investment, they appear from field data to be consistently in the direction of increased allocation to female function. Yet, of the processes we have investigated here, we find either no major impact on proportional investment (local pollen competition, competition between seeds for limited microsites) or processes tending to favour a relative increase in male function (e.g. increased attraction to pollen-feeding pollinators). Lloyd (1984) argues that the reason for the prevalence of relatively high investment in female function in animal-pollinated plants is that, rather than being subject to diminishing returns (as we assumed above), male fitness is actually subject to a fixed upper limit (a strict asymptote rather than a decelerating power function), and that for many plant species this upper limit is reached with relatively low proportional investments (about 10%; see Table 6.2).

In a study of sexual resource allocation in the gynodioecious Mediterranean dwarf shrub *Thymus vulgaris* (Lamiaceae), Atlan *et al.* (1992)

Table 6.2 Estimates of paternal allocations in outcrossing cosexual angiosperms (after Lloyd 1984).

Species	Pollinating agent (wind or insects)	Paternal allocation (%)	Frequency of outcrossing (%)
*Microlaena polynoda**	W	51.2	?
Zizania palustris	W	16.0–33.3	*c.* 90
*Impatiens biflora**	I	9.9	? (Protandrous)
*Impatiens pallida**	I	9.6	? (Protandrous)
Gilia achilleifolia	I	9.3	Extrapolation to 100
Smyrnium olusatrum	I	4.0	*c.* 100

*Chasmogamous flowers only, in species with cleistogamous flowers as well.

found that female plants produced twice as many seeds as hermaphrodites. There was a clear trade-off between male and female function within the hermaphrodites, some of the plants functioning as virtual males, others as virtual females. The female advantage in seed production was due simply to reallocation of resources that were not used in male function. Sex allocation had a genetic component, and an interaction between nuclear and cytoplasmic genomes was implicated.

6.13 Agamospermy: seeds without sex

If you ask meteorologists what the weather is going to be like tomorrow, they are likely to tell you that tomorrow's weather will be like today's. Likewise, the most probable set of environmental conditions facing a plant in the next generation is the set of environmental conditions to which the parent plant was exposed. Thus, you might think, the genes that made the most successful plants in the present generation ought to be copied exactly, and these would maximize fitness of the offspring. Why go in for sex and run the risk of throwing away half of a successful genome? 'If it works, don't fix it'. This is exactly the strategy that apomictic plants have adopted. They get all the benefits of seed reproduction (dispersal and dormancy, the potential for risk-spreading in space and time), with none of the potential problems of inheriting useless paternal genes that might reduce the fitness of the progeny.

Some apomictic plants are extremely successful. Dandelions of the *Taraxacum officinale* (Asteraceae) aggregate and brambles of the *Rubus fruticosus* (Rosaceae) aggregate have spread to become weeds throughout the temperate world. The basic chromosome number found throughout *Taraxacum* is $n = 8$. Diploids ($2n = 16$) represent about 10% of all species; they are invariably sexual and typically self-sterile. It is possible that these plants occasionally demonstrate facultative agamospermy (Sorensen 1958). The bulk of species, however, are triploid ($2n = 24$) obligate agamosperms, producing progeny of identical genotype to the parent without sexual fusion, and with little or no recombination or segregation (Richards 1972). Probably all of the British brambles in the subgenus *Rubus*, except for the abundant *R. ulmifolius* which is a sexual diploid, are apomicts. Most of them are tetraploids ($2n = 28$) but triploids, pentaploids and hexaploids are also known. The curious feature of bramble sexual biology is that most of the apomictic species are *pseudogamous*, i.e. a diploid embryo forms without fertilization, but a stimulus from the male gamete is required, so pollination is necessary. Apomixis is often facultative in brambles so that new apomictic genotypes can arise by hybridization, giving rise to a constant stream of new apomictic taxa (Edees & Newton 1988).

Facultative apomixis (the production of asexual seeds if pollination fails) may help to account for the successful invasion of islands by some dioecious plants (you would expect that dioecious plants would make the worst kind of invaders, since both sexes need to be present simultaneously if a new population is to be established; see below). Thus *Pandanus tectorius* has been able to colonize many Pacific islands (Cox 1985) despite its dioecious habit because of the production of asexual seeds.

So if apomixis is such a good idea, then why don't more plants do it? The simplest explanation is that they can't; there may be phylogenetic constraints against the evolution of apomixis in most plant taxa. A more interesting possibility is that the assumptions of our model for the benefits of apomixis are wrong. We assumed environmental constancy; this is a reasonable short-term prediction, but it is clearly flawed since all environments are certain to change in the long term. Second, we assumed that recombination and outcrossing produced inferior genotypes. As we saw earlier, this assumption is also likely to be false in the long term. The persistence of apomicts may depend on occasional outbreeding. Alternatively, it is important to remember that populations tend to *move* as conditions change, rather than go extinct (Eldredge 1995). Thus, an apomictic species with a reasonably wide geographic range might persist through times of change in localized refugia, then expand its range again, if and when good conditions return.

6.14 Sex ratios and variable sex expression

It is important to distinguish sex ratio (the relative numbers of males and females in a population) and sex allocation (relative amounts of resources dedicated to male and female reproductive function; Frank 1990). We have already discussed sex allocation in hermaphrodites; here we discuss the sex ratios of dioecious plant populations, and the proportional representation of functional males and females in populations that exhibit labile sexuality. These are plants that you might think were dioecious on the basis of a single field visit, but which long-term study of cohorts of marked individuals shows to vary between the production of male and female flowers at different times.

The problems of sex allocation can be stated as follows (Charnov 1982b).

1 For a dioecious species, what is the equilibrium sex ratio maintained by natural selection (measured as the proportion of males amongst the offspring)?

2 For a sequential hermaphrodite, what is the equilibrium sex order (male first or female first) and what is the optimal time of sex change?

3 Under what conditions are the various degrees of hermaphroditism or dioecy evolutionarily stable? When, for example, does selection favour genes for protandry over dioecy? When is a mixture of sexual types stable?

4 When does selection favour the ability of an individual to alter its allocation to male versus female function in response to particular environmental or life-history situations?

6.14.1 Sex determination in plants

Humans are familiar with chromosomal sex determination (XX = female, XY = male), in which gametes from the father determine the sex of each offspring (X sperm makes a female, Y sperm makes a male). Thus, if there were controlling genes on the Y chromosome, then selection would lead to very male-biased sex ratios, because daughters do not contribute to a Y chromosome's fitness. With respect to the Y chromosome, fathers and daughters are unrelated. If there were controlling genes on the X chromosome, the picture is more complicated. If the gene is expressed in the mother, then it is like a haplodiploid system and, with inbreeding, mothers are more closely related to daughters than to sons. If the gene works in fathers, then selection favours the production of all daughters, since fathers never contribute an X chromosome to a son. In humans, however, the sex ratio at birth is as close to 50 : 50 as makes no difference. Fisher (1930) worked out why equality of sex ratio was the expectation in all but a few rather special circumstances (see Box 6.6).

It is those special circumstances that concern us here, because sex determination in plants is seldom caused by the possession of single sex chromosomes (although exceptions include some well-known dioecious plants like *Silene* (Caryophyllaceae) and *Cannabis*). In other dioecious plants, sex is determined by a single pair of alleles (Frankel & Galun 1977; Karlin & Lessard 1986). However, perhaps the commonest cause of femaleness in typically hermaphrodite species is *cytoplasmic male sterility*, a matrilineally inherited cytoplasmic gene that switches off male function. The possession of *restorer genes* can switch maleness back on again. For example, in a gynodioecious population (a mix of females and hermaphrodites) femaleness is often due to simple cytoplasmic inheritance while hermaphrodites carry a non-sterility cytoplasm and differ in their nuclear genotype at a single locus that does not affect the females (Maurice *et al.* 1993). This mechanism allows less restrictive evolution of dioecy and trioecy (males, females and hermaphrodites) than if sex determination were by nuclear genes alone (Maurice *et al.* 1994).

6.14.2 Labile sex expression and environmental conditions

Broadly speaking, conditions associated with low plant growth rates (shade, low nutrients, drought, etc.) tend to lead to increased investment in male reproductive function (Maekawa 1924; Freeman *et al.* 1980; Cruden & Lyon 1985). Experimentally imposed drought stress, for example, can shift the sex ratio of a spinach population from 66%

female under wet conditions to 37% in the dry environment (Freeman & Kalled in Charnov 1982b). The response was not due to differential mortality and is best explained as an example of labile sexual expression. Female *Arisaema triphyllum* (Araceae) were dug up from good growing conditions, had a substantial portion of their roots and foliage cut off, and then were replanted under poor growing conditions; 21 of 25 plants had become male by the following year. When male plants were removed from the field and cultured under ideal conditions, 31 of 32 plants turned into females by the following year (Schaffner 1922).

Allocation to female reproductive function in *Rumex acetosella* (Polygonaceae) declined across sites with increasing successional age, but male function did not decline (Escarre & Thompson 1991). In bisexual populations of maples, sex ratios alone are no use for inferring proximate mechanisms for biased sex ratios or sex-related differences in resource allocation, because sex is labile with age and interacts with habitat conditions (e.g. there are more male trees in dry years and in xeric habitats); large female trees could become bisexual, with male flowers in the upper branches and female flowers in the lower canopy (Sakai 1990).

Females of the dioecious understorey tree *Ocotea tenera* (Lauraceae) in montane forest in Costa Rica were much more likely to have a male tree as their nearest neighbour than expected by chance, given that the population sex ratio was 50 : 50. Plants monitored over 10 years changed their sex, and changes in the sex ratio of populations on experimental plots were due to labile sex expression by marked plants rather than to differential mortality. Wheelwright and Bruneau (1992) suggest that the non-random distribution of the sexes might be caused by labile sex expression modified by the presence of neighbouring trees (see also Trivers & Willard 1973; Lloyd & Bawa 1984; Charnov & Bull 1986).

It might appear from these examples that an attractive option for plants is to change their sex expression as they grow larger; small plants could express maleness (because this is less expensive) and larger plants could change to become female. The prediction would be that there should be a threshold size below which the plants should be male and above which they should be female. There is very little evidence, however, to support the notion of permanent, adaptive switches in sex expression. Where plants do alter their principal mode of sex expression, they typically change back again when conditions change. Thus after a heavy bout of female reproduction, some monoecious trees revert to maleness for a variable period before the next heavy seed crop is produced (e.g. *Quercus robur* (Fagaceae); Crawley & Long 1995).

6.14.3 Monoecy

Although monoecious plants make up only a small proportion of all plant species (about 5%), they can be the ecological dominants in many temperate and Arctic communities. Many of our most familiar plants, including angiosperms like oak, birch and beech, and gymnosperms like spruce and pine, have separate male and female flowers on the same individual. The majority of monoecious trees exhibit protogyny, so that on a given plant the stigmas are receptive before pollen is shed from the individual's own male flowers. In a minority of cases (e.g. *Quercus*), the male flowers shed their pollen before the stigmas on the female flowers are receptive (protandry). There is often pronounced spatial separation of male and female function within the canopy of monoecious trees; for example, whole branches of *Cedrus libani* (Pinaceae) can be densely covered by male cones, while lower branches of the same individual support a sparse population of large, female cones. Rather few monoecious trees are self-incompatible, but the separate phenology of male and female function means that self-pollination is an exception rather than the rule. A population of monoecious trees will typically contain individuals exhibiting a range of reproductive phenologies; some individuals will be consistently early in their dates of anthesis, others consistently late (Crawley & Akhteruzzaman 1988).

A monoecious annual cucurbit, *Lagenaria siceraria* shows strong protandry with an average 17 male flowers opening before the first female and a strongly male-biased flower ratio (20 male to 1 female). This gives rise to a much higher male investment in flowers and a higher female allocation to fruits (large gourds); female flowers had no measurable cost to the plant. Experimental removal of all flower buds of each sex allowed the resulting extra vegetative growth to be used as an estimate of the opportunity costs of that sexual function. The flowers on these regrowth branches were almost exclusively female; male and female functions are effectively separated in time. Pollinated plants stopped allocating resources to male flowers immediately. The overwhelming cost to the plant is in the production of the one or two large fruits. There is a great deal of variation within populations in the timing of first female flower production. Individuals that produce their female flowers early and are successfully pollinated will shut down male function early and therefore have relatively low investments in male function (Delesalle & Mooreside 1995).

6.14.4 Dioecy

Most of the theory of sexual selection and optimal reproductive behaviour has been developed for dioecious animal species. It is a moot point how much of this theory applies to plants because such a small proportion of plant species is dioecious. The dioecious habit is much more common amongst woody plants (10–25% of species), than herbs (1–5% of species). Some islands, like Hawaii, have unusually large numbers of dioecious species (Sakai *et al.* 1995). Amongst woody plants, there is a well-established correlation between dioecious habit and the possession of fleshy fruits (Cox 1985); dioecious species are

Table 6.3 Frequency of dioecious species in different life forms of angiosperms. (From Bawa 1980; see for original references.) Note that the UK figure is particularly high because the main dioecious tree family, Salicaceae, is unusually rich in species (*Salix* and *Populus*).

	Dioecious species (%)				
Life form	North Carolina	Barro Colorado Island	California	UK	Tropical wet forest
Trees	12	21	20–33	20–30	~ 25
Shrubs	14	11	0–23		
Vines	16	11	–		
Herbs	1	2	4–9		

more likely to exhibit animal dispersal of their fruits than wind dispersal in both dry deciduous forest and wet evergreen forest (Bawa 1980). Bawa and Givnish argue that this correlation results from natural selection imposed by frugivores that prefer large fruit displays, and hence increased female investment (i.e. that dioecy is a consequence of the fleshy fruited habit; Tables 6.3–6.6). Sex ratios in dioecious plants are characteristically male biased (often strongly so). This is consistent with what we know about the relationship between plant size and reproductive function (small plants tend to be male) and about the size structure of most populations (populations tend to consist of a large number of small plants and a small number of large individuals; see Chapter 11). These biased sex ratios might come about as a result of bias in the sexuality of the seed crop, or from differential mortality from an initially 50 : 50 seed population; the evidence strongly favours differential mortality as the usual cause. Female plants have higher resource requirements and reproduction imposes a greater drain on their resources; so, to the extent that survivorship is resource dependent, we

Table 6.4 Correlation between dioecy and pollination systems in a dry forest in Costa Rica. (From Bawa 1980; see for details of classification of pollination systems.)

	Percentage of tree species	
Pollination systems	Hermaphroditic/ monoecious (*n* = 94)	Dioecious (*n* = 28)
Medium-large bee*	25	1
Small bee or opportunistic†	26	80
Beetle	14	3
Fly	1	2
Wasp	3	2
Moth	19	9
Butterfly	1	0
Humming-bird	3	0
Bat	8	0
Wind	0	3

*Mostly Anthophoridae, some Xylocopids.
†Mostly Halictidae, Megachilidae and/or Meliponini (Apidae).

Table 6.5 Correlation between breeding systems and modes of dispersal for angiosperms. (From Bawa 1980.)

Locality/taxonomic group	Breeding system	Number of species	
		Animal dispersed	Wind dispersed
Tropical lowland dry deciduous forest (Palo Verde, Costa Rica)	Dioecious	30	3
	Hermaphroditic and monoecious	60	26
Tropical lowland wet evergreen forest (La Selva, Costa Rica)	Dioecious	66	0
	Hermaphroditic and monoecious	222	29
Meliaceae	Dioecious	16	0
	Hermaphroditic and monoecious	9	12

Table 6.6 Correlation between breeding system and dispersal syndrome in gymnosperms. (From Givnish 1980.)

Number of species that are:	Monoecious	Dioecious
Wind dispersed	339	18
Animal dispersed	45	402

would predict that the death rate of females would exceed the death rate of males. Note, however, that the sex ratio in *Silene latifolia* is female biased in a way suggesting genetic variation for progeny sex ratio, expressed before germination (i.e. the difference is not the result of differential mortality of males). Individuals of the two sexes are poly-morphic; males have many more flowers than females, and females show more pronounced symptoms of resource limitation than do males (Lyons *et al.* 1994).

One view of the evolution of dioecy is that it is an extreme mecha-nism for the avoidance of inbreeding; it prohibits selfing and hence leads to fitter, outbred progeny (Charlesworth & Charlesworth 1978; Thompson & Barrett 1981). It may be that the evolution of dioecy is easier than the evolution of self-sterility (Barrett 1996, and see above). Given sufficiently high levels of selfing and inbreeding depression, population genetic models demonstrate that single-sex function can evolve in hermaphrodite populations (Lloyd 1974; Charlesworth & Charlesworth 1979). The alternative view is that dioecy is an adaptation to particular kinds of ecological conditions that confer a selective advantage to sexual specialization in separate individuals. The middle ground is that sexual function in many hermaphrodites is so labile that what we see as dioecious species are just hermaphrodites expressing femaleness under one (typically favourable) set of growing conditions and maleness under another (typically unfavourable) set of conditions (see above). It is interesting to note that dioecy is rare or absent in plant groups that exhibit self-incompatibility, an observation that supports the avoidance of selfing as a prime cause for the evolution of dioecy (Bawa 1980; Freeman *et al.* 1980).

Bawa (1980) also notes that almost all animal-pollinated dioecious species in the tropics use insects, mainly small opportunistic bees, whereas temperate zone dioecy is associated with wind pollination. In the tropics, dioecy is also associated with characteristic patterns of fruit dispersal, with single or few-seeded fruits dispersed mainly by birds (see Chapter 9). For gymnosperms (all of which are wind pollinated) seed dispersal is also correlated with dioecy: 88% of monoecious species are wind dispersed while 96% of dioecious species are animal dispersed (Givnish 1980). Wind pollination is hypothesized as making the male gain curve fairly linear rather than strongly saturating, while animal dispersal means that a larger seed crop attracts a disproportionately large number of dispersers, causing the female gain curve to bend upwards (Bawa 1980; Givnish 1980; Charnov 1982b, 1984).

6.15 Population genetics and genetic neighbourhoods

Mating between plants takes place within restricted spatial subsets of the population (neighbourhoods). The assumption of panmixis (see above) was a mathematical convenience, but quantitative population geneticists have long realized the importance of spatially restricted gene movement (Wright 1952; Falconer 1989). The classic example of spatial structure within a plant population comes from the work of Epling *et al.* (1960) on flower colour polymorphism in the outcrossing annual *Limnanthes parryae*. In 260 quadrats placed every 3 m along a transect through the Mojave desert, the percentage of blue morphs varied from 1% to more than 80%, with white flowers making up the remainder. The colour morphs fell into well-marked patches, such that the average gene frequency correlation between adjacent quadrats was 0.66, but half of the correlation was lost by 100 m; these patches were not associated with underlying environmental heterogeneity. Most of the gene flow through pollen and seeds appeared to be over a few metres or less, so that drift-generated variation patterns were able to persist for many generations. Epling *et al.* (1960) estimated that the neighbourhood area for these *Limnanthes parryae* was about 10 m^2.

6.15.1 Minimum viable population (MVP)

Conservation biologists speculate about the smallest number of individuals that might reasonably be expected to form a persistent population. There have been many attempts to define MVP (Frankel 1974; Shaffer 1981; Nunney & Campbell 1993) but they all flounder on the rocks of prediction. Since we cannot predict next year's weather, let alone predict genotype-specific plant death rates a few years hence, how can we possibly predict the population size and genetic attributes of a population that will allow it to persist decades or centuries into the future? Some of the definitions attempt to put figures on both the probability of the population surviving and on the number of years it is expected to survive (e.g. MVP is the current size of a population that guarantees 95% probability of survival for 100 years; Ewens *et al.* 1987). In truth, all attempts to define MVP are bound to fail, because the persistence of a real population of plants is so context dependent and will often have little or nothing to do with the *number* of plants in the population (see Chapter 19). Similarly, the MVP will depend on the genetic composition of the individuals within it; if the population is to pass through a particular bottleneck in the close future, then N individuals containing the genotypes that can make it through the catastrophe will persist whereas a population of exactly the same size, but lacking these particular genotypes, is doomed to extinction. Nor is it clear at what spatial scale the concept of MVP is supposed to apply: is it a patch within a metapopulation or the entire metapopulation itself?

Assessments of the persistence of populations of different types can be investigated by population viability analysis (Boyce 1992). This analysis takes account of human impacts (habitat destruction, climate change, grazing, harvesting, etc.) and environmental uncertainties (demographic stochasticity, environmental stochasticity, natural catastrophes, genetic stochasticity; Shaffer 1981). Demographic stochasticity refers to the probability of all the plants of a given genotype dying or failing to reproduce by chance alone. Local extinction is even more likely if low-density effects are inversely density dependent (i.e. things get worse, not better, as population density declines); these are known as Allee effects (failure of pollination is a possible example). Most analyses show that demographic stochasticity is the least likely cause of local extinction in plant populations (Menges 1991; Lande 1993; Lande *et al.* 1994). Environmental stochasticity and natural catastrophes grade into one another; extremes of weather, volcanic eruption, flooding and outbreaks of pathogens or herbivores are highly likely to cause local extinction of small populations, and the risk of extinction will be a non-linear declining function of population size (Lande 1993). Genetic stochasticity involves changes in gene frequency due to small population effects like inbreeding, founder effects and random drift.

In his study of population viability analysis of the rare perennial herb *Pedicularis furbishiae* (Scrophulariaceae) on the St John River in Maine, Menges (1990) concluded that a metapopulation approach was vital because local patches of the plant were certain to succumb. If they were not subject to erosion or ice-scouring on unstable regions of the river bank, then they would be ousted by competition from shrubs on the more stable reaches.

6.15.2 Genetic drift

The genetics of scarcity involves taking a restricted and randomly selected subset of the population's genes into the next generation. Each individual leaves half of its genes behind at each fertilization, and if the population is sufficiently small, and the number of surviving offspring is low enough, then a substantial fraction of the gene pool can be lost at each breeding season. The magnitude of random drift is measured by the effective size of the population N_e. This is defined as the size of an ideal, panmictic population of diploid hermaphrodites whose genetic make-up is affected by random drift to the same degree as our study population. In such a population, selfing occurs at a rate of $1/N_e$ and if there is no selection, migration or mutation the prediction of certain consequences of genetic drift is relatively straightforward (e.g. loss of alleles and loss of heterozygosity; Falconer 1989). Drift tends to override selection in determining the fate of alleles when the selection coefficient $s < 1/N_e$, which means that more and more of the allelic variation behaves as if it were selectively neutral as population size declines (Kimura & Crow 1964).

6.15.3 Effective population size

What is the relationship between N_e and N? The answer has to do with population dynamics as well as genetics. Real plant populations tend to have N_e between about 10 and 20% of census population size, N, mainly as a result of neighbourhood effects (spatial non-randomness in mating coupled with limited gene-flow distances through pollen and seed dispersal). For populations fluctuating in size from generation to generation, it is traditional for population ecologists to use the geometric mean to characterize central tendency (this is the antilog of the average of log population size; standard deviation of log population size is the accepted way of measuring population stability; Williamson 1972). Population geneticists, on the other hand, use the harmonic mean of population size (this is the reciprocal of the average of the reciprocals), because loss of alleles is so strongly dependent on the *smallest* population in a given time series (the bottleneck population size). The smallest population size has a disproportionately great effect in determining the amount of genetic drift that occurs during a given period.

Because the effective population size is given by

$$N_e = \frac{4N_f N_m}{N_f + N_m} \tag{6.7}$$

then populations with functionally biased sex ratios can have very low effective population sizes. If all the pollen came from one large plant ($N_m = 1$), for example, then the effective population size could not exceed 4, no matter how many female plants were present. Fertility variation from plant to plant also influences the size of N_e. In most real plant populations, seed production is concentrated in a few large individuals (Heywood 1986), and the variance in seed production is often much greater than the mean (e.g. fecundity distribution is often well described by the negative binomial distribution; see Chapter 11). Likewise, paternity tends to be concentrated in just a few large individuals (Schoen & Stewart 1987).

Both of these processes have the result of further reducing N_e relative to N (Hedrick 1983) and hence have conservation implications, since controlling the variance in male and female reproductive performance, by making sure that as many plants as possible are both male and female parents (e.g. by hand pollinations), could increase N_e by a factor of two (Frankel *et al.* 1995). The presence of a seed bank can increase N_e by introducing combinations of genes from times long past, following soil disturbance and seedling recruitment.

6.15.4 Mutation

What is the minimum population size in which the loss of alleles through random genetic drift is likely to be counterbalanced by the gain in genetic variation through mutation? Franklin (1980) computed $N_e = 500$ to maintain additive genetic variance, and this figure has had a huge influence in conservation genetics (Frankel *et al.* 1995). Patch coalescence in a metapopulation can reduce genetic variability, even if each patch has *c.* 500 occupants (Gilpin 1991).

6.15.5 Selection

For traits under stabilizing selection, a population with $N_e > 500$ individuals can maintain nearly as much genetic variance in typical quantitative characters as an infinitely large population (Lande & Barrowclough 1987). In a metapopulation, genetic bottlenecks following colonization, cause an increase in the between patch variance in genetic make-up. The genetic consequences of this depend upon whether or not natural selection differs from path to patch.

6.15.6 Components of variance

Understanding the genetics of a metric character centres on the study of phenotypic variation; how much of the observed variation is genetic, how much is environmental, and how much is attributable to genotype-by-environment interaction? Phenotypic variation V_P is partitioned into three components.

$$V_P = V_G + V_E + V_{GE} \qquad (6.8)$$

The genetic component V_G is traditionally split into two parts: (i) additive genetic variance V_A and (ii) dominance deviation V_D, and these are used in the calculation of heritability. The first measure, known as heritability in the broad sense, is given by V_G/V_P and is of limited practical value; the second, known as heritability in the narrow sense (or just *heritability*, h^2) is given by:

$$h^2 = \frac{V_A}{V_P} \tag{6.9}$$

which determines the degree of resemblance between relatives and hence is of the greatest importance in plant breeding programmes (see Falconer 1989 for details).

6.16 Gene flow through migration

The movement of genes in pollen and seeds, and the movement of plant fragments capable of regeneration, can have profound effects on population dynamics and population genetics, especially when migration leads to the establishment of new populations. Gene migration introducing novelty can lead to the formation of hybrids and sometimes can disturb favourable gene complexes. Patches of plants are coupled through limited dispersal into genetic metapopulations, and this may retard population divergence; small populations are expected to receive gene flow at a higher rate than large populations (Ellstrand & Ellam 1993), and large populations are more likely to be the donors, even when patches are in close proximity. Migration of genes into established plant populations tends to have rather little impact on demography (plant numbers and size structures), but it has potentially great genetic impact, depending on the intensity of selection and the relative fitness of the migrant genes.

Migration may rapidly regenerate gene diversity in populations that have passed through bottlenecks (Levin 1988). The seed bank can be important in temporal gene migration, allowing the recycling of old genes, sometimes from long in the past. Some authors argue that pollen is more important in gene flow than seed dispersal (Levin & Kerster 1974; Beattie & Culver 1979; Meagher & Thompson 1987; Weiblen & Thomson 1995), but the jury is still out on this; the best-documented cases of ultra-long distance gene flow have all been by seed or vegetative fragments (e.g. the establishment of alien plants on new continents; Crawley 1997).

6.16.1 Gene flow through pollen

6.16.1.1 *Wind pollination*

Gene-flow distances in wind-pollinated species are likely to be much greater than with insect pollination, but it is unlikely that pollen of many species is regularly transported more than 10 km, and most pollinations occur at distances much less than this. Plant breeders working with wind-pollinated crops like maize or sugar beet traditionally use isolation distances of about 1 km, and they have money to lose if they make mistakes. As the pollen cloud moves away from a source plant, it becomes progressively more diffuse and the ratio of indigenous to extraneous pollen reaching a stigma is bound to increase rapidly with distance from the source of extraneous pollen (Levin 1983). In wind-pollinated maize, for example, about 50% of pollen originated from within 12 m of target female plants (Paterniani & Short 1974).

6.16.1.2 Insect pollination

Gene movement distances tend to be greater than flower-to-flower pollinator movement distances because of carry-over, and because pollen–pistil compatibility may be lower between near neighbours than between moderately spaced plants (see above). Also, we must bear in mind the occasional rare event (e.g. the bee that gets on a train), which draws attention to the fact that it is essentially impossible to study long-distance pollen-flow events, so flat and so extended is the tail of the leptokurtic distribution of gene-flow distances.

The distance of gene movement by insects is strongly dependent on plant population density, and pollen dispersal distances are much greater in sparse populations than in dense (Ellstrand 1992). For example, in *Liatris aspera* (Asteraceae) pollinating insects moved three times as far in a sparse population (1 plant m^{-2}) than in a dense population (11 plants m^{-2}). The result of this difference in pollinator behaviour was that the *neighbourhood size* was positively correlated with density, rising from 45 plants in the sparse population to 363 in the dense population, whereas the *neighbourhood area* was negatively correlated with density, declining from 45 m^2 in the sparse population to 33 m^2 in the dense population (Levin 1988). Using a genetic marker for hypocotyl colour, Motten and Antonovics (1992) studied outcrossing rates in the self-compatible annual weed *Datura stramonium* (Solanaceae). They found that population-wide outcrossing rates were surprisingly low for a plant with such showy entomophilous flowers, and ranged from 1.9% in an experimental population with a clumped spatial pattern to 8.5% in a population with a dispersed spatial arrangement.

One of the best ways to determine pollen-flow distances experimentally is to use transgenic plants as the male parents, and then to look for the transgene in seeds produced by maternal plants at different distances away from the transgenic source plants. These studies emphasize the extremely small distances travelled by most pollen grains under field conditions. These methods are not suitable, however, for detecting very rare but very long distance pollen movements, because the screening protocol is too expensive to handle the vast samples that would be required. In a study with the crucifer *Brassica napus* subsp. *oleifera* (oilseed rape), which is pollinated by honey-bees, the selectable marker was a dominant transgene conferring tolerance of the herbicide glufosinate. Seeds were harvested from maternal plants and screened with the herbicide both in the greenhouse and under field conditions to estimate the frequency of pollen dispersal at different distances. Herbicide-tolerant plants (those that survived the spraying) had their genotype confirmed by Southern blot analysis. At 1 m from the source transgenics, the frequency was estimated as 1.5%; at 3 m it was 0.4%. The frequency decreased sharply to 0.02% at 12 m and only one transgenic was detected at 47 m (0.00033%); no transgenics were found at distances greater than 47 m (Scheffler *et al.* 1993). In a parallel study with transgenic potatoes *Solanum tuberosum*, which are pollinated mainly by bumble-bees, the transgenes (*nptII*) conferred tolerance of the antibiotic kanamycin which could be used as a selectable marker. Where transgenic and non-transgenic potato plants were grown in alternate rows with their leaves touching, 24% of seedlings from non-transgenic maternal plants were kanamycin tolerant. Comparable seedlings from plants at up to 3 m distance had a tolerance frequency of only 2%, at 10 m the frequency was 0.017% and at 20 m no transgenic plants were found (McPartlan & Dale 1994). Much greater pollen flow distances have been recorded by Timmons *et al.* (1996) using emasculated rape plants.

6.16.2 Assortative and disassortative mating

This occurs when there is a tendency for individuals to mate with partners of the same phenotype as themselves at a rate significantly higher (assortative) or lower (disassortative) than would be expected on the basis of random mating. This could occur because of floral traits affecting pollinator behaviour, flowering phenology, spatial distribution or plant size. If the heritability is high, then progeny of assortative mating will be likely to inherit genes for the character. If the intensity is sufficiently high, this could in principle lead to sympatric speciation (Maynard Smith 1966 but see Futuyma & Mayer 1980). Many insect pollinators are known to fly more or less horizontally from flowers of one phenotype to flowers of similar phenotype on another individual, and to ignore flowers of a different size or colour, or flowers of the same phenotype at different heights, in between.

6.16.3 Venereal diseases of plants

The herb *Silene alba* (Caryophyllaceae) and the anther-smut fungus *Ustilago violacea* form a fascinating plant–pathogen interaction (see Chapter 13). The fungus has two life stages, vegetative and floral. Dispersal of infective propagules and successful infection of hosts is brought about by promiscuous pollinating insects. Infection causes both male and female plants to produce male flowers in which the anther sacs are filled with the spores of *Ustilago* rather than with pollen. This means that both male and female flowers are sterilized by the disease; the males produce no pollen and the females suffer abortion of all their ovaries. When spores are transmitted to healthy flowers (e.g. by pollinating insects), the diploid spore germinates and undergoes meiosis; when the meiotic products of opposite mating types conjugate, an infective stage can attack the healthy plant and invade its vegetative tissues (a systemic infection). The dynamics of infection have been likened to those of a sexually transmitted disease by Antonovics *et al.* (1995). In classic density-dependent transmission (Anderson & May 1979), the probability of an individual becoming infected is a function of the density of infectives, but with *Silene* it is the *proportion* of other *Silene* that are infected by *Ustilago* that is the important determinant of the likelihood of infection by pollinating insects (so-called frequency-dependent transmission). This means that when there is a limited number of plants visited per insect, transmission may *decline* with increasing plant density, a behaviour not noted with density-dependent transmission, but supported by field work on *Silene* populations of different sizes.

6.16.4 Gene flow through seed dispersal

As with pollen-flow distances, most seed dispersal is extremely local and highly clumped. Typical seed-dispersal distances are measured in centimetres or metres (e.g. *Echeveria* (Crassulaceae) 1.1 m, *Liatris* (Asteraceae) 2.5 m, *Phlox* (Polenoniaceae) 1.1 m, *Viola* 0.8–2.1 m, *Erythronium* (Liliaceae) 0.33 m; Levin 1988), although the oaks returned to Britain after the retreat of the ice age at an average rate of about 400 m per year thanks, presumably, to occasional long-distance dispersal of acorns by birds like jays (Bossema 1968). Rare, extremely long-distance dispersal occurs when seeds stow away on a ship in some foreign port, and find themselves deposited some months later on the other side of the world, perhaps 10 000 km away (Crawley 1977). For example, gene flow in an alien plant like the annual grass *Bromus tectorum*, which now occupies thousands of square kilometres of dry grassland in the north-west USA, was revolutionized by the five or six founder events that followed long-distance dispersal from its native home in Mediterranean Europe (Mack 1981; Novak *et al.* 1993).

If recruitment is seed-limited, and seed-dispersal profiles ('seed shadows') are strongly leptokurtic (Fig. 6.2), then occasional long-distance dispersal is likely to lead to the establishment of new population foci. New foci are preserved by heavy local seed rain and give rise, in turn, to other loci. The spread of new loci is not likely to follow a simple travelling wave (the 'ripples in a pool' model) because the dispersal distances are so haphazard (Willson 1993; Portnoy & Willson 1993). This leads to the prediction that spatial patterns of adult plants should be more clumped (e.g. show smaller values of k of the negative binomial) for seed-limited (potentially dominant) species compared with the typically less-competitive species whose recruitment is not seed-limited (see Crawley 1990a and Chapters 4, 8 & 15).

Secondary seed dispersal occurs when animals like ants or rodents pick up the dispersed seed and carry it away. In *Viola nuttallii*, whose seeds bear elaiosomes (ant bodies), most seeds are dispersed within 1 m of the parent and the modal dispersal distance is less than 20 cm. An important issue is whether the seed is more or less likely to be found and killed by granivores following dispersal (e.g. rodents and ants remove and consume seeds from bird and ungulate faeces). If secondary seed dispersal increases the probability of burial, then this can have a huge impact on fitness (see Chapter 13).

6.17 Sex on islands

Plants that evolved on remote islands differ in many ways from their continental cousins. One of the most conspicuous differences is in their pollinators. Very few of the specialist pollinating insects found in continental regions occur on remote islands, and this has had a major impact on floral evolution. This being the case, we might predict that pollinator service would be inferior on islands. Typical flowers of remote island plants are small, white or greenish and pollinated by generalist, promiscuous pollinators like flies, moths and short-tongued bees. Their continental relatives are more likely to have showy coloured flowers (blues and reds) and to be pollinated by specialists like long-tongued bees, humming-birds or butterflies, all of which would find the journey to a remote island difficult or impossible (Barrett 1996).

The altered pollination environment on islands imposes constraints on floral evolution by increasing the likelihood of self-pollination. Several plant taxa show smaller flower size with increasing isolation along offshore archipelagos (e.g. *Campanula* off Japan; Inoue *et al.* 1996). Likewise, tristyly is often reduced to distyly or monomorphic self-compatibility on islands (e.g. *Eichhornia paniculata* (Pontederiaceae); Barrett 1996). An outbreeding distylous ancestor might evolve to become a homostylous inbreeder under island conditions, and then subsequently re-evolve outbreeding through herkogamy (same ends, different means).

There are two evolutionary responses to unsatisfactory pollinator service: evolution of selfing or of dioecy. The response depends on the kinds of service involved, because pollination service can be unsatisfactory in several different ways. For example, if there is *insufficient* pollination, this leads to pollen-limited seed production. Alternatively, there may be plenty of pollinations but they are inferior, leading to inbreeding. The response to pollen limitation might be evolution towards self-compatibility and selfing. The response to inferior pollination might be adaptations favouring outcrossing. Two conspicuous features of island floras are that incompatibility systems are rare or absent, but dioecy is much more common than in continental floras. Perhaps on islands it is more difficult to evolve incompatibility systems than it is to evolve dioecy? Or perhaps it is the correlation between dioeciousness and fleshy fruitedness that is the key (relatively higher dispersal rate of dioecious plants; see above)?

6.18 Local mate competition

Suppose that a single seed lands on an uninhabited island, and suppose that the female has control over the sex of her offspring. How many male and female seeds should she produce? This question was first investigated by Hamilton (1967, 1979) in the context of parasitic wasps, which have the ability to control the sex of their offspring. Because of their haplodiploid sex-determining mechanism, fertilized eggs (diploids) become females while unfertilized eggs (haploids) will develop into males. The answer to the question is intuitively obvious: the female should only produce as many males as are necessary to fertilize all of her female offspring; if males are capable of mating with large numbers of females, then the ideal would be one male and all the remainder ($n - 1$) females (Hamilton 1979). Likewise, if the island was already populated and the sex ratio of the existing residents was strongly biased towards one sex or the other, then the new arrival would do best to put all its efforts into producing progeny of the rarer sex (see Box 6.6). Given that the ability of plants to determine the sex of their offspring is limited (see above) you might ask what relevance all this has for plant ecology.

The key to the problem is the expected intensity of reproductive competition between sibling plants. Plants are likely to be closely related to their neighbours when limited seed dispersal is coupled with high rates of inbreeding. Under these circumstances, there are sharply diminishing returns for investment in male reproductive effort but not in female effort (Fig. 6.5a), and selection should favour mechanisms that bias the sex investment in favour of females. This would happen if most pollen was likely to find its way to stigmas of the same genotype (e.g. if plant population density was very low, or if pollinators were very scarce). In general, we expect male investment to go down as the likelihood of local mate competition goes up. In circumstances where most of the available stigmas belong to the same genet (e.g. in extensive, spreading, clonal perennials), so that most pollinations are bound to be selfings, then theory predicts a reduction in allocation to male reproductive function (Charnov 1982b). The rationale is that there is no point in investing in pollen grains that are destined only to compete with one another on related stigmas.

Recall, however, that competition amongst sibling seedlings selects for increasing male function (see above). How important is sib competition as a selection pressure in plants? How is it affected by plant breeding system and dispersal mechanism? How important is it in the evolution of seed dormancy? Does it affect population-level genetic structure? Do some individuals benefit (kin selection)? These questions are reviewed by Cheplick (1992) who highlights the woeful lack of hard data on sib competition between plants under field conditions.

6.19 Mate choice in plants

Mate choice by females involves selecting certain potential pollen parents and rejecting others. This could happen in several different ways: (i) by selective attraction of faithful pollinating insects; (ii) selection of pollen grains on the stigma or during the course of pollen tube growth through the style; (iii) differential fertilization (rejection of inferior male gametes); (iv) delayed fertilization ('waiting for Mr Right'); or (v) differential (i.e. genotype-specific) abortion of fertilized ovules. All these processes are discussed in Willson and Burley's (1983) stimulating book on mate choice, which applies ideas from mainstream behavioural ecology to plants.

Tests of many of the hypotheses suggested by Willson and Burley have been beset by problems, and experiments supporting the notion of sexual selection in plants (either competition amongst males for access to matings, or overt preferences by females for particular males) are few and equivocal. Mate choice by females is not expected to be important in cases where seed set is pollen-limited rather than resource-limited (see Bateman's principle, Box 6.3), nor in cases where self-pollination is frequent (see above). Nevertheless, this is an exciting area deserving further study using genetic engineering to produce carefully controlled experimental genotypes.

6.20 Conflicts of interest

There are several places in the plant's life cycle where genetic conflicts of interest may arise, and theoretical approaches from sociobiology have thrown light on the evolution of many of the reproductive features exhibited by seed plants (Hamilton 1972; Trivers 1974; Charnov 1986; Haig & Westoby 1988). Consider the question of how much the maternal plant should invest in provisioning each of its seeds. What each seed wants for itself may be in conflict with the overall best interests of the parent; the seed would like the maximum possible amount of resources, but the parent is interested in producing the maximum number of potentially viable offspring. According to Trivers (1974) three factors determine the optimal allocation of any parental resource from the perspective of a particular offspring: (i) the benefit it receives from the resource; (ii) the cost to siblings of *not* receiving the resource; and (iii) the degree of relatedness between the sibs and the offspring in question. If the costs to the sib are more than twice the benefits to the seed in question, and the coefficient of relatedness (Hamilton 1964) between the two seeds is 0.5, then both the offspring and the parent would gain in fitness by giving the resources to the sib, rather than to our selfish seed. If the cost–benefit ratio is less than 1, then both parent and offspring would benefit from the resources being given to the selfish seed rather than to its sibling. Conflicts of interest between parents and offspring arise when the cost–benefit ratio varies between 1 and 2. These issues are discussed in detail by Godfray (1995), but the key points have to do with signalling of the seed's 'needs' to the parent and the ability of the parent to detect deceit on the part of its offspring.

An interesting twist is given to the debate by double fertilization, which means that the resources to sustain the seedling come from triploid endosperm tissues that have twice the genetic contribution from the maternal parent as from the pollen parent. It is argued that this allows the female greater control over the use of her resources and reduces the scope for male–female and seed–female competition (Haig & Westoby 1988; Uma Shaanker *et al.* 1988; Queller 1989; Dominguez 1995). An interesting point raised by Godfray (1995) is whether female investment in seeds should be regarded as female reproductive allocation (as we dealt with it above) or as parental care; the distinction has important implications for the way that we should view the evolution of seed-provisioning traits.

Ellner (1986) used kin-selection arguments to investigate conflicts about germination decisions that might arise between the seeds in multiseeded fruits that are dispersed as a single unit. Here, the optimal solution for the maternal plant could be a dormancy mechanism that causes germination of the seeds at different times (e.g. one per year over a series of years). This both reduces competition between simultaneously germinating seedlings and allows the group of seeds to sample a wide variety of different kinds of years, only a few of which might be suitable for recruitment. The conflict arises because each individual seed's fitness is likely to be maximized by germinating immediately. It is noteworthy that this habit of staggered germination is particularly well developed in multiseeded desert plants, where water availability is unpredictable from year to year, and where complete failure of a cohort of seedlings is commonplace (Gutterman 1992).

6.21 Case studies

The following three examples show plant sexuality can be studied under field conditions.

6.21.1 Paternity analysis

Seldom is the question addressed of how many seeds each male plant actually sires. In the dioecious *Chamaelirion luteum* (Liliaceae), Meagher (1986; Devlin *et al.* 1988) was surprised to find no relationship between male size and paternity, but there were alternating positive and negative

Neither female availability (localized sex ratios) nor local female size
distribution had any impact on male success.

Genealogical analysis is a powerful tool for the analysis of reproductive performance. It assigns paternity on the basis of genetic likelihood
criteria using an iterative procedure for fractional allocation of paternity
within a progeny pool (Smouse & Meagher 1994). Because of duplication of genotypes, 102 out of the total of 273 male plants in the study
population could be used (but they turned out to be an unbiased sample
of plants in relation to the ecological variables investigated in this
study). Different males made unequal contributions to the overall
progeny pool, with many males contributing little or nothing to the
next generation. There was no indication of pollen limitation of seed set
(cf. Gill 1989).

6.21.2 Male fitness and pollen flow

Devlin and Ellstrand (1990a,b) studied the hermaphrodite annual crucifer *Raphanus raphanistrum*. Female fertility was measured as seed yield
relative to the seed yield of the entire population (the proportion of
seeds contributed by the individual in question). Male fertility was far
more difficult to measure because of pollen distribution over stigmas,
incompatibility, pollen competition, seed abortion and fruit abortion.
Thus, pollen movement may not reflect the relative number of mature
seeds fathered by the plant in question. Paternity analysis was carried
out, using multilocus genetic markers, which allowed ranked likelihoods to be worked out for the probable paternity of any seed (Roeder
et al. 1989). The technique works well if there are relatively few
identical genotypes amongst the mature male plant population. There
was a 15-fold difference in male fertility (from 0.004 to 0.060; see
Fig. 6.7). This is the proportion of seeds pollinated by the plant in
question; if they all pollinated with equal efficiency, then fertility would
be 0.024 since there were 41 plants in the study, so the best father was
doing about 2.5 times as well as the average plant. Male plants with dark
purple flowers had higher fertility than those with light purple flowers
and these, in turn, had higher fertility than the white-flowered plants.
There was no evidence for assortative mating, however. Peak fitness
occured when neighbours of opposite gender peaked in synchrony; this
is likely if flowering phenology is attuned to local microclimate and
substrate, but this was not always obvious. Male fitness increased with
total flower production but with diminishing returns, presumably because there is a limited number of receptive stigmas within the pollen
dispersal range. Male and female function were poorly correlated with a
wide range of functional genders (0.03–0.91; virtual single sex reproductive function) but contrary to expectation, the variance in female

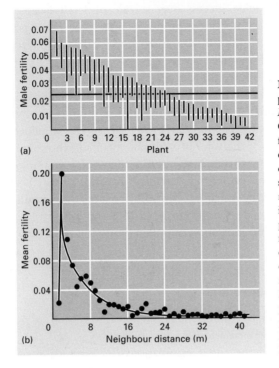

Fig. 6.7 Male reproductive performance in radish *Raphanus sativus*. (a) Confidence intervals (95%) for the mean male fertility of each of 41 plants; mean taken over 39 female parents, showing the great range in male performance by different individuals in the same population. (b) Mean male fertility as a function of distance to the female parent; only females within 8 m of the male are likely to produce high rewards in terms of paternity (selfing, the first point on the graph, produces low male fertility). (After Devlin & Ellstrand 1990a,b.)

function was greater than the variance in male function for the year in question. Female fitness varied 200-fold and total (male plus female fertility) varied 25-fold. Much of this variation, of course, is environmental and little of it may have had any genetic basis.

6.21.3 Selfing and inbreeding depression

Selfing has evolved repeatedly in outcrossing taxa, and theory predicts that an increase in the level of self-fertilization should occur in concert with changes in reproductive allocation and in the magnitude of inbreeding depression. A comparison of reproductive allocation and the fitness consequences of self-fertilization in two species of *Epilobium* (Onagraceae) was carried out by Parker *et al.* (1995). They found that selfing was very high in the weedy *E. ciliatum*, pollinator visits were few and electrophoresis revealed no genetic variation at 11 putative isoenzyme loci. In contrast, the patch-forming perennial *E.* (*Chamerion*) *angustifolium* had low rates of selfing, high pollinator visitation rates and an electrophoresis-estimated outcrossing rate of $t = 0.64$ (s.e. $= 0.08$). The pollen–ovule ratio was 10 times greater in *E. angustifolium* than *E. ciliatum* because of reduced pollen production by the selfing *E. ciliatum*. Unsurprisingly, there was less inbreeding depression in *E. ciliatum*. With *E. angustifolium* there was substantial variation from individual to individual, with some maternal parents showing strong inbreeding depression, but others exhibiting strong outbreeding depression. This genetic variability probably reflects differences in the history of inbreeding and in the long-distance migration of individuals from different populations (Parker *et al.* 1995).

6.22 Conclusions

As these case studies show, there is a vast amount of exciting work to be carried out on plant reproductive performance under field conditions. Techniques from molecular genetics are certain to revolutionize our understanding of what makes a successful male and female parent. It may turn out to be almost entirely phenotypic (e.g. big plants do better), or it may be the result of genotype-by-environment interactions. The replacement of mean-field models of population genetics by spatially explicit models is likely to have as great an impact on our understanding of the genetics of neighbourhoods as the switch from point models to coupled map lattices has had on our understanding of plant–herbivore interactions.

7: Seed Dormancy

Mark Rees

7.1 Introduction

Plant ecologists often focus their studies on properties of the established or regenerative phases of the life cycle. Some have even suggested that traits of these two phases are uncoupled (Grime 1979, 1988). In this chapter, I argue that such an assumption is theoretically and empirically untenable, and emphasize that an understanding of ecological and evolutionary processes requires an integrated approach to the whole life cycle.

I begin by explaining what constitutes a seed and what is dormancy. Next, seed germination behaviour is discussed in terms of its effects on population persistence, dynamics and coexistence. Finally, the evolution of dormancy and its relationship with other plant traits, such as adult longevity and seed size, is explored.

7.1.1 Types of seeds

Seeds are the structures that develop from fertilized ovules and each usually contains one embryo together with its supply of food. The food resources may be contained in a specially developed endosperm or in the cotyledons. Fertilization can be either (i) allogamous, where the fusing nuclei are from different individuals, or (ii) selfing, where the fusing nuclei are from the same individual (see Chapter 6). A detailed discussion of the various types of reproduction in plants and whether they should be considered as sexual or asexual is given in Mogie (1992). Generally, seeds are not dispersed in the form of ripened ovules but have various associated structures, so forming fruits such as caryopses, nuts, achenes, berries and the like. In this chapter the term 'seed' will be used in a very broad sense to indicate the fertilized ovule and the associated structures with which it is dispersed (Harper *et al.* 1970).

Seeds typically consist of a variety of genetically diverse tissues. The embryo is diploid and has equal contributions from the father (pollen parent) and the mother. The endosperm is typically triploid with genetic material contributed by the mother and father in the ratio $2 : 1$. Surrounding these is the seed coat, which is derived entirely from maternal tissues. Thus, a seed is parent and offspring combined in a single unit (Ellner 1986; Haig & Westoby 1988). This can have important evolutionary implications, particularly when the parent's and seed's interests are in conflict. For example, theoretical models demonstrate potential conflicts between offspring and parent concerning the level of seed provisioning, with the fitness of individual seeds being maximized at higher levels of provisioning than would maximize parental fitness (Haig & Westoby 1988). Conflicts may also arise as to the way that seeds should be dispersed (Ellner 1986; see Chapter 6 for details).

7.1.2 Definitions of dormancy

Dormancy is an inactive phase during which growth and development are deferred and respiration is greatly reduced. As Harper (1957) put it: 'Some seeds are born dormant, some achieve dormancy and some have dormancy thrust upon 'em'. Hence, we may identify three types of dormancy: innate, induced and enforced (Harper 1957, 1977). Innate dormancy prevents seeds from germinating while they are still attached to the parent plant. This may be due to: (i) immaturity of the embryo (e.g. ash *Fraxinus excelsior* (Oleaceae)); (ii) the presence of an impermeable seed coat (e.g. many of the Fabaceae) or a chemical inhibitor; or (iii) the seeds requiring a particular environmental cue (e.g. chilling, fluctuating temperatures). Innate dormancy is often interpreted as a mechanism that ensures germination during a favourable season for establishment.

In many species, the ability of a seed to germinate may be altered by the environment it experiences after ripening. This is known as induced or secondary dormancy, and is elegantly demonstrated by the experiments of Wesson and Wareing (1969a,b). They showed that burial induced a dependence on exposure to light in several species which, prior to burial, had no light requirement for germination. In other species, which had varying degrees of light sensitivity when fresh, germination following burial was completely dependent on exposure to light (Wesson & Wareing 1969b). They also conducted a series of detailed field experiments where a grass sward was disturbed at night to prevent seeds from being exposed to light. The disturbances were then allocated to one of three treatments: (i) left uncovered; (ii) covered with glass; or (iii) covered with asbestos. This allowed the effects of disturbance, sealing the plots and sealing the plots while not exposing seeds to the light to be differentiated. The numbers of seedlings observed in each of the treatments at three different depths is shown in Fig. 7.1. Clearly many of the buried seeds required exposure to light in order to germinate.

Enforced germination occurs when seeds are deprived of one or more essential requirements for germination (e.g. moisture, oxygen, light). No special physiological mechanisms are involved, and the seeds are simply unable to germinate. Before germinating, seeds often pass through different dormancy states as conditions change. For example, in *Persicaria maculath* (Polygonaceae) the seeds are shed in autumn and

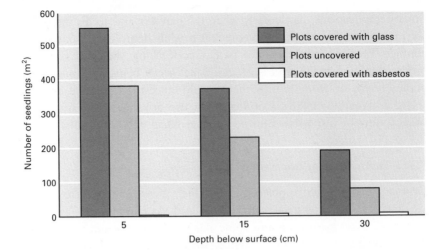

Fig. 7.1 Number of seedlings emerging in field plots at three depths. Note that in the plots covered with asbestos virtually no seedlings emerged; however in those plots where the seeds were exposed to the light many seedlings were recorded. Also covering a plot with glass increased the number of seedlings recorded relative to the uncovered plots. The effect could be the result of increased temperature and/or humidity levels. (Data from Wesson & Wareing 1969a.)

germination is prevented by an after-ripening requirement (innate dormancy). During winter an enforced dormancy occurs as a result of low temperatures. If the seeds fail to germinate during the following spring then an induced dormancy occurs that can only be broken by a second period of chilling (Staniforth & Cavers 1979).

7.2 Seeds and the environment

It should be clear by this stage that seeds have subtle responses to their environment. These responses are now briefly reviewed; more complete discussions can be found in Fenner (1992) and Leck *et al.* (1989).

7.2.1 Effects of light

Seeds are sensitive to several aspects of their light environment (its intensity, spectral composition and duration), and the effects of each depend, to a certain extent, on the others. The light environment experienced by a seed depends on how deeply it is buried in the soil, and whether or not there is a leaf canopy above it. Obviously, seeds that are buried experience very low light conditions. The presence of a leaf canopy reduces the intensity of light in all wavelengths relative to full daylight; the photosynthetically active part of the spectrum (400–700 nm) is particularly reduced relative to the near far-red (700–1000 nm). The ratio of red to far-red light in daylight is typically 1.2 whereas the green light that has passed through a leaf canopy has a ratio of about 0.2.

Short exposures (less than 1 hour) to red light (approximately 660 nm) tend to break dormancy whereas far-red light tends to impose dormancy (Flint & McAllister 1937). In some species long exposures (e.g greater than 8 hours) are required to break dormancy (Pons 1991), but in most uncultivated species long exposures to light with a low red : far-red ratio tend to inhibit germination (Fig. 7.2). Among the species studied by Gorski *et al.* (1977), 58% of the uncultivated species showed strong (80–100%) inhibition of germination by a leaf canopy, whereas only 10% of the cultivated species had a similar response, suggesting the importance of the response in natural populations (this is explored further in section 7.4).

Changes in the spectral composition of light are detected by the seed's phytochrome system. The various transformations of the pigment system are well understood, although less is known about the way that this mediates the control of germination (Bewley & Black 1982). In a fascinating series of experiments Cresswell and Grime (1981) demonstrated how a seed's light requirement for germination can be determined by the light-filtering properties of the maternal tissues that surround the seed during maturation. The chlorophyll content as the seed dries is crucial, because photoconversion of phytochrome occurs only when the seed is moist. If the tissues around the seed remain green during seed maturation a light requirement will be induced. This is because the phytochrome in the seed will be largely in the inactive (P_r) form, requiring a light stimulus for germination to occur. However, seeds which are exposed to unfiltered light before drying will have phytochrome in the active (P_{fr}) form and will be able to germinate under light or dark conditions (Fig. 7.3). This form of dormancy is

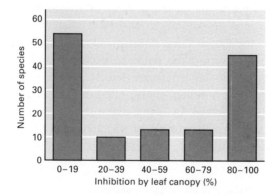

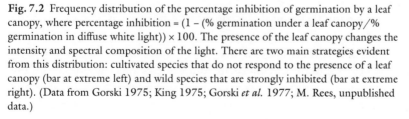

Fig. 7.2 Frequency distribution of the percentage inhibition of germination by a leaf canopy, where percentage inhibition = (1 − (% germination under a leaf canopy/% germination in diffuse white light)) × 100. The presence of the leaf canopy changes the intensity and spectral composition of the light. There are two main strategies evident from this distribution: cultivated species that do not respond to the presence of a leaf canopy (bar at extreme left) and wild species that are strongly inhibited (bar at extreme right). (Data from Gorski 1975; King 1975; Gorski *et al.* 1977; M. Rees, unpublished data.)

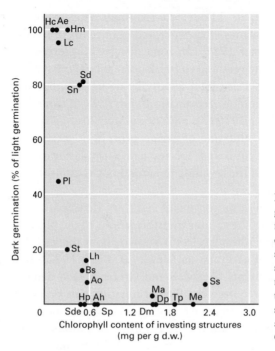

Fig. 7.3 Percentage germination in the dark as a function of the chlorophyll concentration of the tissues surrounding the maturing seed. Values for chlorophyll refer to the concentration at the mid-point of the range of seed moisture contents. (For abbreviations, see Cresswell & Grime 1981.)

mediated by the maternal tissues, but the same effect could be obtained if the light were filtered by neighbouring, possibly competing, plants. In this case, individuals whose seeds ripened in dense vegetation would have different germination characteristics from seeds that matured in the open (Fenner 1991; Platenkamp & Shaw 1993). The ratio of red to far-red light transmitted by leaf canopies also varies between species (Thompson & Harper 1988), which raises the intriguing possibility that seed germination may depend on the specific identity of neighbouring plants.

Changes in seed dormancy can also be mediated by the day-length experienced by the parent plant during seed maturation. This has been demonstrated by covering the fruits to exclude daylight (Gutterman 1977). The changes in day-length are detected in the vegetative parts of the plant and some product translocated to the seeds. In general, germination is promoted by short-day regimes (i.e. spring-like or autumn-like) and dormancy increases with day-length (e.g. midsummer conditions; Gutterman 1973, 1974; Heide *et al.* 1976).

7.2.2 Effects of the chemical environment

Soil is the natural chemical environment of most seeds, with its complex three-phase physical structure of solids, liquids and gases. The nature of the seed–soil contact and the process of water uptake are reviewed by Hadas (1982). Many chemicals in solution can affect germination, but inorganic ions, with the exception of nitrate and ammonium ions, tend not to have any specific effects on germination (Egley & Duke 1985).

Nitrate is the main inorganic ion known to stimulate germination (Lehmann 1909; Roberts & Smith 1977). The concentration of nitrate depends on the rate of nitrogen mineralization, which in turn depends on temperature, and the rate of nitrate consumption, which tends, of course, to be greatest during the growing season (see Chapter 3). So there is no general pattern of seasonal nitrate levels across different soils, and several other factors such as soil type and pH may also be important (Davy & Taylor 1974; Runge 1983). Seeds may obtain their nitrate from the parent plant during development, or directly from the soil (Goudey *et al.* 1988; Bouwmeester 1990). In several weeds of arable land there is a positive relationship between endogenous nitrate content and germination (Saini *et al.* 1985), and species with nitrate-dependent germination often require light. Hence, the effect of nitrate on germination may be to provide a gap-detection mechanism, which when combined with the light response, described above, may allow seeds considerable subtlety in their germination behaviour (Pons 1989).

High concentrations of inorganic ions may inhibit germination in a non-specific manner. For example, maximal germination of many halo-phyte species occurs at intermediate levels of salinity, and it has been suggested that this allows the seeds to determine whether rainfall has been sufficient for growth and establishment (Ungar *et al.* 1979; Ungar 1987).

The role of organic compounds in the inhibition of germination is still a matter of contention. Some authors, notably Rice (1983), suggest that organic inhibition of seed germination plays a major role in succession, but others express doubts (Williams & Hoogland 1982). The importance of organic compounds in stimulating germination in many species of parasitic plants is well established (Sunderland 1960; Logan & Stewart 1992). Parasitic plants obtain some, or all, of their nutrients from the host plant. Many different compounds released by host and non-host roots can stimulate germination of root-parasitic plants, and the stimulatory effects can vary between different parasite species (Sunderland 1960; Cook *et al.* 1972). Many of the compounds are active at extremely low concentrations (e.g. strigol is active at 10^{-9} mol m^{-3} in soil solution). Interestingly, strigol can also stimulate germination in dormant seeds of non-parasitic species, which suggests that the biochemical pathways that control germination may be similar in parasitic and non-parasitic plants.

7.2.3 Effects of temperature

There are three separate physiological processes that are affected by temperature in seeds (Roberts 1988). First, temperature and moisture content determine the rate of seed deterioration. Second, temperature affects the rate of dormancy loss in dry seeds and the pattern of dormancy change in moist seeds. Finally, temperature can determine the rate of germination in non-dormant seeds.

In hard-seeded species, the seeds are impermeable when initially dispersed, and there is often a specific structural weakness that causes the seed coat to rupture when heated up and cooled down. This allows the embryo to become moist, which in turn allows germination (Tran & Cavanagh 1984). Intense short-term heating effects that result from exposure to fire are also important in many hard-seeded species (e.g. *Acacia* spp. (Mimosaceae)). In chaparral and other Mediterranean shrub communities seeds of several species are retained in serotinous cones stored in the canopy. Fire is required for the opening of cones and both the timing and quantity of seed release are correlated with fire temperature (Bradstock & Myerscough 1981). In South African Proteaceae, however, serotiny is relatively weak, seeds are rarely retained for more than 5–6 years and fire is not essential for seed release (Bond 1985).

In many temperate woody species dormancy is broken by chilling, particularly in species that possess embryonic dormancy (Bewley & Black 1982), and the chilling requirement ensures that germination occurs in spring. In some aquatic grasses, whose seeds become incorporated into lake sediments, near-freezing water temperatures over winter reduce seed dormancy and increase the germination when the water temperature rises in spring (Probert & Longley 1989). Again this seed behaviour can be viewed as a mechanism that ensures germination in the appropriate season.

The range of temperature fluctuation that a seed experiences is also important in determining whether it germinates (Morinaga 1926; Warington 1936; Thompson & Grime 1983). In many wetland species, germination is stimulated by increasing temperature fluctuations and this appears to provide a mechanism whereby spring germination is promoted by increasing irradiance and a falling water table (Thompson *et al.* 1977). In some species the impact of temperature fluctuations depends on the light regime. Typically, complete darkness increases the temperature fluctuation required for germination to occur. In some species germination does not occur at all in complete darkness, irrespective of the temperature fluctuation (Fig. 7.4). Sensitivity to temperature fluctuations in darkness is especially characteristic of species from grassland, wetland and disturbed habitats and is thought to represent a depth-sensing mechanism.

Established plants provide an insulating blanket that affects not only average soil temperatures but also the range of temperature fluctuation. In an interesting experimental study, Rice (1985) explored the germination responses of two *Erodium* spp. (Geraniaceae) to changes in temperature regime. He exposed dry seeds of both species to alternating and constant temperatures and then moistened the seeds and assessed germination. Seeds from the alternating temperature regime germinated readily while those from the constant temperature regime did not

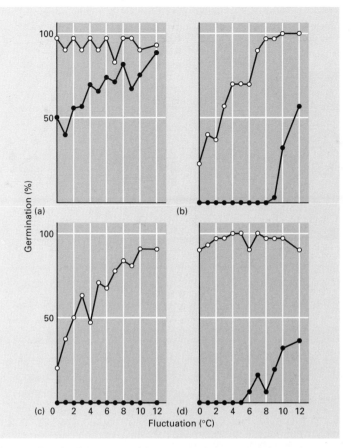

Fig. 7.4 Germination as a function of range of temperature fluctuation. Open circles indicate an 18-hour photoperiod, the closed circles total darkness. (a) *Holcus lanatus* (Poaceae), (b) *Chenopodium rubrum*, (c) *Juncus effusus* and (d) *Epilobium hirsutum* (Onagraceae). (After Thompson & Grime 1983.)

(Fig. 7.5a). Rice measured the pattern of diurnal temperature change in three types of microsite: (i) under litter from *Bromus mollis*, *Avena barbata* and *Vulpia* spp. (all Poaceae); (ii) bare soil; and (iii) under soil mounds created by pocket gophers (*Thomonomys bottae*). He found that the range of temperature fluctuations varied greatly between the microsites (Fig. 7.5b). By burying bags of seeds in each microsite he was able to assess germination rates; he also assessed plant performance by planting pregerminated seeds and estimating their seed production in each of the microsites. He found that germination was greatest in the bare ground microsites and least under the gopher mound, and that plants produced most seeds in the bare ground and least under the gopher mound (Fig. 7.5c,d). This experiment demonstrates clearly the adaptive nature of seed germination behaviour and helps to explain the spatial pattern of *Erodium* distribution in Californian grasslands (Rice 1985). Simply observing that *Erodium* spp. generally occur in gaps one might naively assume that this was the direct result of competitive

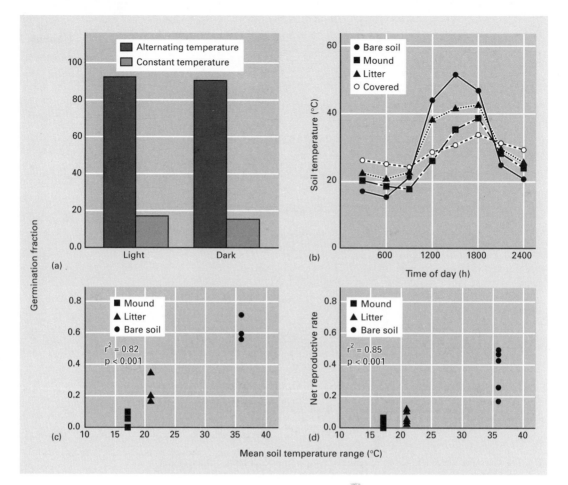

Fig. 7.5 (a) The percentage germination of *Erodium botrys* seeds that were previously stored at either constant temperature or alternating between 12-hour periods at 25 or 45 °C. (b) Pattern of diurnal fluctuation in soil temperatures in different germination microsites. 'Covered', bare soil shaded by a reflective cover. (c) Relationship between diurnal temperature range and germination response of *E. botrys*. (d) Relationship between diurnal temperature range and net reproductive rate, number of seeds produced per seedling planted (after Rice 1985).

exclusion but the real cause is the seed germination biology, which has evolved as a way of avoiding intense competition.

7.2.4 Other germination cues

Various forms of abrasion and chemical scarification that damage the seed coat often increase the rate of germination (Baskin & Baskin 1989). The detailed mechanisms that underlie this process are poorly understood and can be complex. For example, in *Witheringia solanacea* (Solanaceae) seeds are often dispersed by birds. Germination is promoted if the seeds are in the gut for short periods of time

(< 30 minutes), but longer periods of gut retention reduce germination success. The plant appears to produce a chemical laxative in the pulp that surrounds the seed, and in this way can manipulate the time spent in the bird's gut and so increase germination success (Murray *et al.* 1994). However, in other bird species increased retention time leads to increased germination (Barnea *et al.* 1991).

In fire-prone ecosystems it has been suggested that smoke may be an important cue that stimulates germination after a fire (Brown *et al.* 1993). Although germination is promoted by smoke in about 50% of fire-prone species tested, a recent study has shown that non-fire-prone species also show increased germination (Pierce *et al.* 1995). The adaptive significance of smoke-induced germination is therefore difficult to assess.

7.3 Seed banks

7.3.1 Temporal dynamics

The pattern of seedling emergence from the seed bank can be studied on several time scales. For example, we can study the emergence of seedlings within a year or we could add together the seedlings emerging over a period of a year in order to obtain an emergence trajectory for a cohort over several years. Patterns at one level determine what we observe at higher levels of organization but, as we shall see, the relationships are not always straightforward.

Many temperate plants exhibit peaks of germination in spring or autumn. However, even within these categories there is often considerable spread in the timing of germination and in some species seedlings can be observed in virtually every month (Fig. 7.6). Differences in the timing of emergence can have a profound effect on the outcome of competition both within and between species (see Chapter 11).

Many tropical species have no dormancy and their seeds remain viable for only a few months (Garwood 1989). Many species characteristic of mature, undisturbed forest, including economically important timber and fruit trees, have large, non-dormant, recalcitrant seeds (drying causes loss of viability; Chin & Roberts 1980). Species from a wide range of regeneration strategies (i.e. weeds, pioneers and species from mature forest) have germination restricted to particular times of the year, with little or no germination occurring during adverse periods (this is often the dry season in tropical environments; Vázquez-Yanes & Orozco-Segovia 1984). Some species show delayed germination where seedling emergence occurs asynchronously (often not immediately after seasonally adverse periods) and this germination strategy is most common in large-seeded species with hard or fibrous seed coats (Foster 1986).

Lonsdale (1989) noted that populations of established plants often

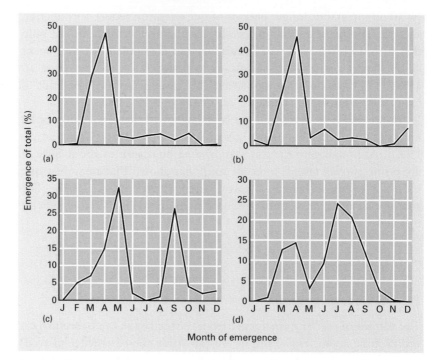

Fig. 7.6 Variation in the timing of seedling emergence within a year for (a) *Raphanus raphanistrum*, (b) *Sinapis arvensis*, (c) *Capsella bursa-pastoris* and (d) *Sisymbrium officinale* (all Brassicaceae). Some species have single peaks of germination, whereas others show complex patterns with peaks in spring and autumn. In all cases there is considerable variation in the timing of seedling emergence. (Data from Roberts 1964; Roberts & Boddrell 1983.)

show a wide range of survivorship curves (see Chapter 12) whereas, based on the studies of Roberts (Roberts 1964, 1979; Roberts & Neilson 1980; Roberts & Boddrell 1983; Roberts 1986), seed populations had traditionally been regarded as following log-linear (Deevey type II) decay curves (Harper 1977; Fenner 1985; Begon & Mortimer 1986; Watkinson 1986; Silvertown & Lovett Doust 1993). Lonsdale (1989) suggested that the most likely explanation for this apparent constancy of pattern was that the data were simply inadequate to reject a null hypothesis of exponential decay.

Seed survivorship curves are often studied indirectly. The pattern of seed-bank decay is inferred from the distribution of seedling emergence through time, although a number of studies have directly assessed the number of viable seeds present in successive years (e.g. Roberts 1962; Roberts & Dawkins 1967). In these studies the pattern of seedling emergence often closely mimics the decline in the number of viable seeds. The observed patterns of recruitment reflect two underlying processes: (i) germination, which may or may not result in successful establishment; and (ii) seed mortality. Germination is only partially observable, through those seedlings that successfully establish, while seed mortality is completely unobserved. This makes it difficult to determine the relative importance of germination and seed mortality in producing the observed patterns. Cook (1980) provides a number of plausible arguments to suggest that germination is the dominant process.

Recent reanalysis of the classic data collected by Roberts (Roberts 1964, 1979; Roberts & Neilson 1980; Roberts & Boddrell 1983; Roberts 1986) has demonstrated that seed-bank decay is, in fact, highly variable, with many species showing age-dependent patterns of recruitment rather than exponential decay (Fig. 7.7; Rees & Long 1993). These age-dependent recruitment patterns could be the result of one or a combination of the following processes: (i) age-dependent germination; (ii) age-dependent mortality; (iii) age-dependent establishment success of those seeds that germinate; (iv) changes in the abiotic or

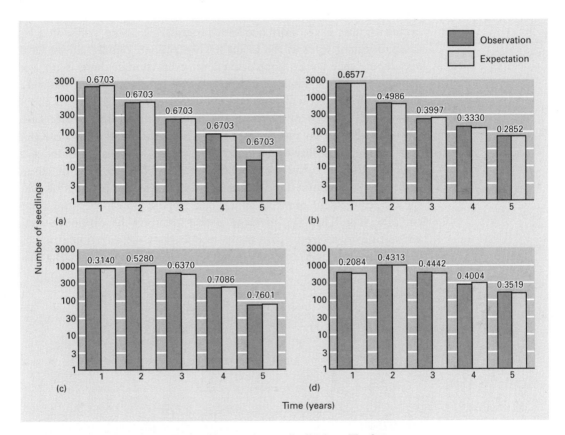

Fig. 7.7 Observed ■ and expected □ seedling recruitment distributions. The figures above the bars are the age-specific recruitment probabilities. In each case the expected values were calculated from a fitted statistical distribution. (a) *Geranium pusillum*, fitted distribution exponential $\chi_3^2 = 6.08$, $P > 0.10$; the age-specific probability of recruitment is constant. (b) *Rumex sanguineus* (Polygonaceae), fitted distribution compound exponential $\chi_2^2 = 2.80$, $P > 0.30$; the age-specific probability of recruitment declines with seed age. (c) *Veronica arvensis* (Scrophulariaceae), fitted distribution Weibull $\chi_2^2 = 1.81$, $P > 0.30$; the age-specific probability of recruitment increases with seed age. (d) *Vicia hirsuta* (Fabaceae), fitted distribution log logistic $\chi_2^2 = 3.64$, $P > 0.20$; the age-specific probability of recruitment initially increases then declines with seed age.

biotic environment; (v) variability between individuals in the probability of recruitment; (vi) density-dependent germination or mortality.

The effects of between-individual variability are subtle and require some explanation. Consider first a population consisting of two seed types, say large and small seeds. Small seeds are assumed to be more dormant than large ones, and the age-specific probability of recruitment is assumed to be constant. A constant age-specific probability of recruitment implies that a pure population of either seed type will decay according to a negative exponential. Now we ask what happens to a population made up of both large and small seeds. At first sight you might think that the mixed cohort would also show an exponential decay but at some average rate. Closer consideration shows that this is not the case and recruitment is in fact age dependent, with the age-specific probability of recruitment declining as seeds grow older. This occurs because the cohort becomes progressively more dominated by small, dormant seeds as the larger seeds germinate rapidly. So we have age dependence at the population level even though there is no age dependence at the level of individual seed behaviour (Rees & Long 1993). Things are even more complicated if different seed types show age-specific patterns of recruitment. Again consider the simple case where there are two types of seed, say large and small, but now the age-specific probability of germination increases as seeds get older as a result of the breakdown of the seed coat or the leaching of some germination inhibitor. Under these circumstances the patterns of age-specific recruitment observed in a mixed population can become quite complex (Fig. 7.8). The message is that we must be cautious when attempting to interpret patterns at one level of organization (the cohort) using information from lower levels (the individual seed). This

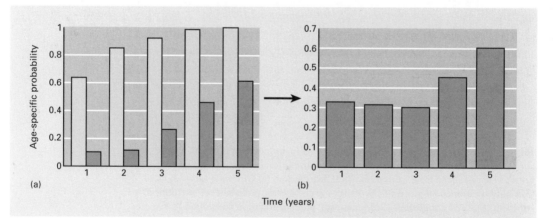

Fig. 7.8 A model population (a) comprising two types of seed each with a different increasing age-specific probability of germination. The composite population (b) has a complex aggregate age-specific pattern of recruitment, even though the pattern at the individual level is simple.

problem is considerably reduced by the use of explicit mathematical models that allow the consequences of individual behaviour on population phenomena to be explored (Hassell & May 1985; Rees & Long 1992, 1993; Rees 1993a).

Density-dependent germination can generate spurious patterns of age-dependent recruitment. This occurs because as seeds germinate the density is reduced and so older seeds experience lower densities. If the probability of germination is lower at high seed densities then the age-specific probability of recruitment will appear to increase as seeds get older. If germination is promoted at high seed densities then the opposite pattern will be observed (Rees & Brown 1991; Rees & Long 1993).

7.3.2 Physical structure

The spatial distribution of a seed bank reflects past patterns of land use, herbivory and the previous spatial structure of the established plant population; the ghost of communities past (Roberts 1981; Thompson 1992). The mean density of seeds varies dramatically between habitats; in arable soils seed densities may reach $70\,000-80\,000\,m^{-2}$, whereas forest soils may have *only* a few thousand seeds per square metre (Roberts 1981). Seed densities are generally greatest in the top few centimetres of soil and typically decline rapidly with depth (Chippendale & Milton 1934).

As with nearly all biological populations, seed banks are spatially aggregated (Thompson 1986; see Chapters 11 & 14). This aggregation can be brought about by many factors: (i) seeds may become trapped in cracks or depressions in the soil; (ii) caching by herbivores; or (iii) limited dispersal from the parent. In the field it is difficult to tease apart the relative importance of these processes; nevertheless spatial aggregation can have important population and community consequences (see Chapter 15).

7.4 Population persistence

Populations inhabit spatially and temporally variable environments and this obviously has important implications for population persistence. I now develop a simple model for an annual plant inhabiting an environment where the factors that limit population growth vary in both time and space. At first, we consider only spatial variation. The condition for population persistence can be simply stated: we require that on average each seed replaces itself with more than one seed when the population is small. If this is the case, then the population will increase when rare and eventually reach some density-dependent equilibrium (see Chapter 12). In contrast, if each seed replaces itself with less than one seed then population size will decrease and the population become extinct. By applying these intuitive criteria it is relatively easy to explore how changes in the environment alter the likelihood of population persistence. Before we do so, however, note two things. First, population persistence is not the same as population stability because a population can persist without necessarily being stable. Second, although the models are derived with annual plants in mind, the qualitative insights apply to perennials as well. More complex, stage-structured models must be constructed to predict persistence for perennials (Rees 1994) and the algebra is more involved, but they add nothing substantially new.

First, assume the habitat consists of a large number of microsites. Some fraction $(1 - E)$ of these contain perennial plants and any seeds from the annual that germinate at a site occupied by a perennial die before reproduction (Crawley & May 1987). Although this may seem unrealistic, it is often the case that recruitment from seed fails completely in the presence of perennial vegetation (Fenner 1985; Eriksson 1989; Crawley 1990a; Rees & Brown 1991; Rees & Long 1992). Now assume that some fraction d of the seeds die over winter and that in spring a fraction g seeds germinate. Those that germinate in empty microsites produce F seeds per microsite and those that germinate in microsites occupied by perennials produce none. Hence, the number of seeds next year, S_{t+1}, produced from the S_t seeds this year is

$$S_{t+1} = (1 - d)(1 - g)S_t + (1 - d)gEFS_t \qquad (7.1)$$

The first term describes the seeds that survive but do not germinate (the seed bank), while the second predicts the total number of seeds produced by those seeds that survive and germinate in empty microsites. Dividing S_{t+1} by S_t gives the finite rate of increase (λ) and for persistence we require $\lambda > 1$. Note that because we are only interested in the behaviour of the population when population size is small we have not included any density dependence in the model (see Chapter 12). Now, intuitively you might expect that the formation of a seed bank ($g < 1$) would promote the persistence of the annual. Intriguingly, this is not the case: the formation of a seed bank can in fact drive the population extinct (Rees & Long 1992). This result occurs because delaying germination does not increase the probability of a seed germinating in an empty microsite and so cannot increase the finite rate of increase. In fact, there is a cost to forming a seed bank because long-dormant seeds are more likely to die before germination, and this is why forming a seed bank makes persistence more difficult. The consideration of simple models immediately focuses attention on the entire life cycle, not just on one part of it (e.g. adult fecundity).

This simple model assumes that germination biology may be summarized by a single parameter (g). However, we have seen from experimental work that the probability of a seed germinating can depend on the presence of established plants (King 1975; Gorski *et al.* 1977; Rice 1985; Van Tooren & Pons 1988; Rees & Brown 1991). The main result from these studies is that the presence of established plants inhibits seed germination (see Fig. 7.2).

It is important, therefore, to modify the model to allow for different germination probabilities in empty microsites and in microsites occupied by perennials. If we assume that the probability of germinating in an unoccupied microsite (i.e. one that does not contain a perennial) is g_u, and the probability of germination in a microsite occupied by an established perennial plant is g_o, we obtain the following model:

$$S_{t+1} = (1-d)(1-\Omega)S_t + (1-d)g_u EFS_t \qquad (7.2)$$

where $\Omega = g_u E + g_o(1-E)$. Now for persistence we require

$$(1-d)[1 - g_o(1-E) + g_u E(F-1)] > 1 \qquad (7.3)$$

Comparison with the first model is simpler if we put $g = g_o(1-E) + g_u E$ so that the fraction of seeds that germinate in each time interval is equal (this ensures that the cost of forming a seed bank, as a result of seed mortality, is the same in both models). Persistence then becomes easier whenever $g_u > g_o$. In other words, inhibition of germination by perennial plants promotes persistence.

The data collected by Gorski *et al.* (1977) suggest that for the uncultivated species $g_o < 0.2g_u$ to a reasonable approximation, assuming that g_o and g_u can be estimated by the probability of germination under a leaf canopy and in diffuse light respectively. Therefore, the germination biology described above is likely to be important in promoting persistence in many plant populations. It is worth reiterating that it is not delaying germination *per se* that promotes persistence but rather the seed's germination response to the presence of established plants.

These simple models assume that all microsites not containing established perennials are equally suitable for growth. However, we would expect that the presence of perennial neighbours would reduce microsite quality by shading or nutrient or water uptake. Several studies have demonstrated that the relationship between plant fecundity (F_c) and the weight or number of neighbours is a non-linear, decreasing function (Pacala & Silander 1985; Goldberg 1987; McConnaughay & Bazzaz 1987; Miller & Werner 1987). The relationship is often well described by the simple hyperbolic function

$$F_c = \frac{F}{1 + \alpha i} \qquad (7.4)$$

or by an exponential function

$$F_c = Fe^{-\alpha i} \qquad (7.5)$$

where F is the fecundity of a plant with no neighbours, i is the number of perennial neighbours and α is a decay parameter. In order to determine the condition for persistence of an annual plant in such an environment, we must calculate its average fecundity. This involves a new complicating factor because average fecundity will depend on the spatial distribution of perennial plants. When fecundity declines as an exponential function of the number or weight of perennial neighbours we may approximate the average fecundity by the following expression:

$$\text{average fecundity} \approx F \exp(-\alpha\,\bar{\imath}) + \frac{F\alpha^2\sigma_i^2}{2}\exp(-\alpha\,\bar{\imath}) \qquad (7.6)$$

where $\bar{\imath}$ and σ_i^2 are the mean and variance in the number of perennial neighbours respectively. The first term on the right-hand side is the fecundity of a plant with the average number of perennial neighbours, and the second, positive term is proportional to the variance in the number of neighbours, demonstrating that spatial variance in microsite quality promotes persistence relative to the average environment. Thus, the spatial arrangement of perennial plants is important in determining persistence of annuals. If perennials are spatially aggregated (resulting in a large variance term) then this will promote persistence relative to a more even spatial distribution.

The results presented so far have assumed that the fraction of sites available for colonization is constant from year to year. However, many annuals live in successional environments where the fraction of sites available for colonization (E) varies through time. In the simplest successional environment virtually all microsites will be available for colonization after a large-scale disturbance ($E \approx 1$), whereas if there is no disturbance then virtually all sites will be occupied by perennial plants ($E \approx 0$) Consider the case where a constant proportion of the seeds germinate in any year. Earlier, in a constant environment, we showed that the formation of a seed bank made persistence more difficult. However, in a random environment this result no longer holds. If the probability of germination is high, the population rapidly declines in years when there is no large-scale disturbance, resulting in extinction. A lower germination rate results in slower decay between disturbances, resulting in persistence. However, the germination rate cannot be too low, otherwise most seeds will die before they have a chance to germinate and this would drive the population to extinction. Conversely, if the presence of perennials inhibits germination, then persistence is strongly promoted because seeds avoid germinating in the years between large-scale disturbances (Rees & Long 1992).

For simple models it is possible to calculate the critical probability of large-scale disturbance, and so the critical time between disturbances, required for a population to persist (Rees & Long 1992). This depends on average plant fecundity and on the probabilities of seed germination and mortality, all of which can easily be estimated from experimental data. Using experimentally determined parameters Rees and Long determined the effect of changes in (i) mollusc herbivory, (ii) the ability of seeds to detect perennial vegetation and (iii) seed mortality on the critical time between disturbances required for persistence. Experimental results indicated that mollusc herbivores reduced plant fecundity by approximately 30% (Rees & Brown 1992), which most ecologists would assume has a profound effect on the condition for persistence. Surprisingly, it has little effect on persistence compared to changes in: (i) the inhibitory effect of perennials of germination; and (ii) the probability of seed mortality. For example, reducing average fecundity from 930 seeds to 612, as a result of mollusc herbivory, reduced the average time between disturbances required for persistence from 20 to 19 years. In contrast, assuming that perennials have no effect on the probability of germination decreased the average time to 13 years and reducing the probability of seed mortality from 0.2 to 0.1 increased the average time to 38 years. This demonstrates the importance of understanding the processes that determine the rate of decay of the seed bank. Equally, these quantitative insights could only be obtained from the analysis of parameterized population models (see Crawley & Rees 1996).

7.5 Population dynamics and coexistence

Seed banks have usually been incorporated into models of population dynamics by assuming that seeds obtain no information on the quality of the environment, and that the probability of germination depends on seed age alone (MacDonald & Watkinson 1981; Pacala 1986b). The inclusion of seed dormancy affects two important aspects of population structure: (i) lifetime probability that a seed becomes a seedling; and (ii) the time-lag between reproduction and germination (Pacala 1986b). It is also assumed that seed age has no effects on subsequent plant performance. Under these circumstances seed dormancy is generally stabilizing (Fig. 7.9), although it can slow down the approach to equilibrium (MacDonald & Watkinson 1981; Pacala 1986b).

In model populations in a constant environment, species that differ only in their germination fractions cannot coexist; the species with the highest probability of germination will always exclude all the others. This result holds for a wide range of life histories (Rees 1994), but when there is temporal variation in the environment, and species have differential responses to the environment, then the presence of a seed bank can promote coexistence. The interaction between environmental variability and the formation of seed banks is illustrated in Figs 7.10 and 7.11. First consider the case where two species both form seed banks, and both are equivalent in all respects except that one species has a higher fecundity. Under these circumstances, in a constant environment, the species with the higher fecundity will always exclude the other (Fig. 7.10a). However, if we include variation in the probability of germination, say as a result of species-specific germination responses to fluctuations in the weather, then the species are able to coexist (Fig. 7.10b). In order for environmental variability to promote coexistence there must be a seed bank and species must respond differently to

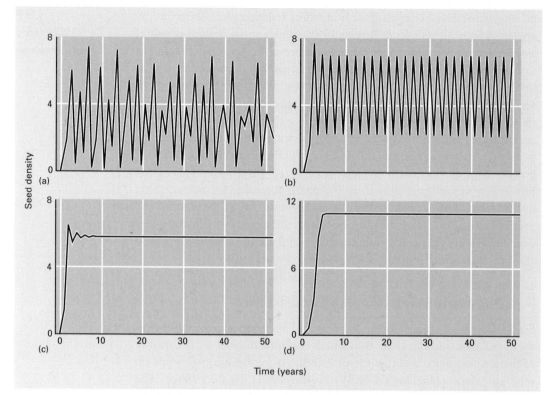

Fig. 7.9 Impact of forming a seed bank on the dynamics of an annual plant population. The model has been chosen so that when there is no seed bank the population has chaotic dynamics generated by density-dependent reproduction. Moving from (a) to (d) a higher proportion of seeds remain dormant and the dynamics change from chaos, when there is no seed bank, through population cycles to damped oscillations and finally a smooth approach to a stable point equilibrium.

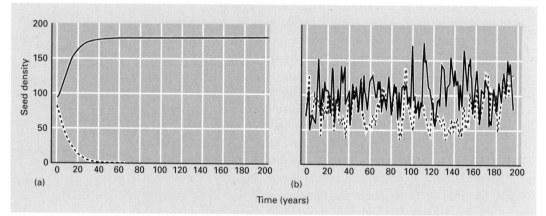

Fig. 7.10 Simulation of an annual plant competition model. (a) Both species form a seed bank and are competitively equivalent (all competition coefficients were set at 1), but one of the species (solid line) has a higher fecundity and as a result quickly excludes the other species (broken line). (b) As in (a) but with stochastic variation in the probability of germination, even though the average values of the parameters are the same as in (a); here the species coexist. Hence stochasticity (random fluctuations) in the probability of germination promotes coexistence.

the varying environment (this 'storage effect' is illustrated in Fig. 7.11).
When there is temporal variation in the competition coefficients (so the
identity of the competitive dominant varies from year to year) but no
seed bank, we see rapid exclusion of the inferior competitor despite the
environmental variability (Fig. 7.11a). In contrast, if we include a seed
bank we see long-term persistence (Fig. 7.11b). These two examples
illustrate how the combination of temporal variation in the environment and the existence of a seed bank can promote coexistence, even
though either process on its own has no such effect. The interaction
between temporal variation and plant life-history traits is complex and
can be highly counter-intuitive. For example, variation in the probability
of seed mortality promotes exclusion, whereas variation in germination
probability, as we have seen, promotes coexistence. Chesson (1986,
1988) provides a detailed discussion of these complexities (see also
Chapter 15).

7.6 Evolution of dormancy

Why is dormancy a problem? The general answer is that we would
normally expect individual plants to reproduce as soon as they could.
There are two main reasons for this: (i) in an increasing population,
early reproduction is favoured because of the multiplicative nature of
population growth (see Chapter 12); and (ii) individuals that forgo
reproduction may die before they are able to reproduce (see section
7.4). Both processes impose a cost of delayed reproduction (Bulmer
1985). Hence, from an evolutionary perspective, seed dormancy, which

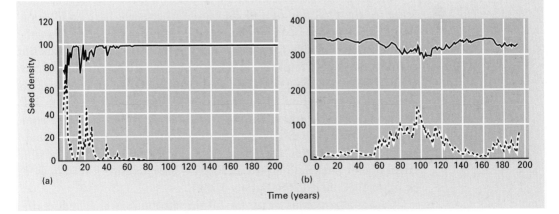

Fig. 7.11 Simulation of an annual plant competition model. (a) Assuming there is no
seed bank but there is temporal variation in the competition coefficients so that on
average one species is competitively dominant; this results in competitive exclusion. (b)
As in (a) but assuming there is a seed bank. The presence of a seed bank produces a
qualitative change in the outcome of the interaction from exclusion to long-term
persistence, by allowing the species that is on average competitively inferior to exploit
those years when it is competitively dominant (the 'storage effect').

is a form of delayed reproduction, presents a problem; and this has led evolutionary biologists to explore the conditions under which the evolution of seed dormancy might be favoured.

It is instructive to begin by looking carefully at the assumptions of models suggesting that there is a cost to remaining dormant. The models assume: (i) seeds have no information on the quality of the environment; (ii) competition occurs primarily between individuals that are not related rather than between siblings; and (iii) there is no temporal variation in the quality of the environment. As we have seen, allowing seeds to detect the quality of the environment will favour the evolution of dormancy even in a constant environment. Not surprisingly the evolutionary stable strategy (ESS) germination strategy in the model presented in section 7.4 is for complete germination in microsites that are unoccupied by perennials and complete dormancy in microsites that are occupied (de Jong *et al.* 1987). This strategy maximizes the number of progeny produced.

If competition occurs primarily between sibs, perhaps as a result of dispersal inside multiseeded fruits, then dormancy may be favoured in a temporally constant environment, even when seeds have no information on the quality of the environment (Ellner 1986). This is a result of a parent–offspring conflict, which the parent wins by using the seed coat to prevent the embryo from germinating. The conflict occurs because parents wish to reduce competition between offspring, whereas the individual offspring whose germination is delayed have lower inclusive fitness than those which germinate. Thus, selection on embryos favours complete germination, as in the models described above, whereas selection on the parent favours delayed germination for some of the offspring (seeChapter 6).

Allowing the environment to vary from one year to the next can also select for the evolution of dormancy even when seeds have no information on the quality of the environment. Consider the case where the conditions are suitable for growth and reproduction in some years but reproduction fails completely in other years. In such an environment an annual plant genotype with no seed dormancy would maximize its arithmetic average population growth rate (λ, introduced in section 7.4) but would become extinct the first time that reproduction failed completely. At the other extreme, a genotype that never germinated would also become extinct as a result of seed mortality. Hence, in a variable environment we would expect an intermediate germination strategy to be optimal (Cohen 1966; Bulmer 1984).

Theory also predicts that life-history attributes which reduce the impact of environmental variation on fitness will show patterns of negative covariation (Venable & Brown 1988; Rees 1993b, 1994). For example, species with efficient seed dispersal in space reduce the likelihood that all seeds will be exposed to unfavourable conditions in any one year, and so we would expect a negative relationship between the efficiency of seed dispersal and dormancy. It has also been suggested that a trade-off between spatial and temporal seed dispersal might arise as a result of physical and biochemical constraints (Lokesha *et al.* 1992). These authors argue that packing seeds with fats allows seed weight to be reduced while maintaining energy content because fats contain more energy per unit weight than proteins or carbohydrates. Hence in wind-dispersed species, where the efficiency of dispersal depends, in part, on seed weight because heavy seeds fall more rapidly than light ones, seeds should contain a higher proportion of fat than protein or carbohydrate. However, the use of fats has several disadvantages: (i) their synthesis is more energy demanding than the production of proteins or carbohydrates; and (ii) lipid autoxidation is thought to cause the disruption of several cell components resulting in loss of viability (Ponquett *et al.* 1992). Therefore, species with wind-dispersed seeds should have a high fat content and as a result the seeds are short-lived, whereas species that are not wind dispersed will have a lower fat content and so have greater seed longevity. There is some evidence that wind-dispersed seeds do indeed contain a higher proportion of fat than seeds that are passively dispersed or dispersed by animals (Lokesha *et al.* 1992). Likewise, species with large individual seeds are predicted to have reduced dormancy because their seedlings can draw on a larger food reserve, and hence establish in relatively unfavourable environments. This reduces the realized variance in environmental quality and this, in turn, selects against seed dormancy (Venable & Brown 1988). Recently these ideas were tested using modern comparative methods by Rees (1993b), who found that: (i) large-seeded species do have less dormant seeds; and (ii) that species with efficient seed dispersal in space also have less seed dormancy.

7.6.1 Relationships between regenerative and established plant traits

It has been suggested that because seeds and seedlings may respond to selective pressures rather differently from established plants, there will be an uncoupling of the traits of seeds and seedlings from those of adults (Grime 1979, 1988). If this view is correct, then there should be no correlation between seed/seedling traits and those of the adult. However, the parent plant is an important component of the environment for many seeds and seedlings, which suggests that adult traits should influence those of the juvenile (Rees 1993b, 1994). If this view is correct then we should expect to see correlations between traits of the regenerative and established plant stages.

Numerous studies have shown a positive correlation between seed size and adult longevity (or adult size; see Fig. 7.12; Salisbury 1942; Baker 1972; Silvertown 1981; Mazer 1989; Thompson & Rabinowitz 1989). These patterns clearly contradict the notion that adult and seed

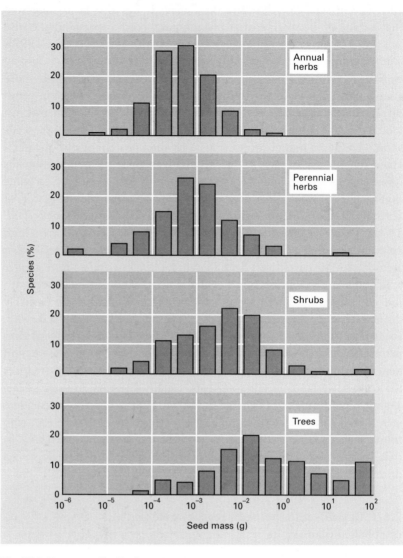

Fig. 7.12 Frequency distribution of seed mass for species of different growth form and longevity within the Californian flora. (Redrawn from Westoby *et al.* 1992.)

traits are uncoupled. The patterns could be generated by one or a combination of several different mechanisms.

1 Small plants simply cannot produce large seeds as a result of mechanical constraints.

2 Large long-lived plants are typical of closed habitats where competition between seedlings and established plants will be intense, and this favours species with large seeds.

3 When seeds of similar morphology are released from the same height small seeds travel further than large ones (Augspurger & Franson 1986; this prevents small plants having large seeds due to the difficulties of dispersal). Mechanisms 1 and 3 do not preclude large long-lived species from having small seeds and, of course, such species do occur (Fig. 7.12).

[237]
Chapter 7
Seed Dormancy

4 Annual species are under stronger selection for risk-spreading within clutches than perennial species because annuals can only reproduce once. So if we also assume that smaller seeds allow greater risk-spreading, this would generate a positive correlation between seed size and adult longevity (Leishman & Westoby 1994). One difficulty with this hypothesis is that producing small seeds might be a more risky strategy because each seed has a lower chance of establishment. If this were the case we would expect a negative relationship between seed size and adult longevity.

Distinguishing these hypotheses is likely to be difficult since many of the traits vary together (Leishman & Westoby 1994). For example, large plants are often long-lived simply because it takes a long time to become large. Also the traits are often measured with very different levels of accuracy. Leishman and Westoby (1994) found that seed size was positively correlated with both adult longevity and adult height, and both these traits are themselves positively correlated. Using multiple regression it is possible to look at the impact of adult longevity having first controlled for differences in height and vice versa. This analysis shows that there is a correlation between height and seed size having corrected for longevity, but no correlation with longevity having corrected for height. From this analysis it is tempting to suggest that the effect of adult longevity on seed size arises purely as a result of changes in plant height. However, longevity was measured on an extremely crude scale, because the authors scored their species as either annual or perennial, whereas height was measured on a continuous scale. Under such circumstances the lack of a correlation between adult longevity and seed size, having corrected for plant height, is perhaps not surprising. Also this study ignored the effects of phylogeny, which can qualitatively change the results obtained (see Harvey *et al.* 1995; Rees 1995). Differences between open and closed habitats are also confounded by changes in plant longevity and size. For example Salisbury (1942, 1974) showed, by comparing congeneric species, that species from open (early successional) habitats typically had lighter seeds than those from closed (later successional) habitats. However, in Salisbury's study, 63% of the species from open habitats were annuals compared with only 2% from the closed habitats. Clearly adult longevity, size and habitat will often be confounded and this will make interpretation of comparative analyses difficult.

Adult longevity and seed dormancy can be viewed as alternative ways of dealing with an uncertain environment (Rees 1993b, 1994), so we might expect short-lived species to have long-lived seeds, whereas long-lived species have short-lived seeds. This pattern appears to hold for a wide range of British plants ranging from arable weeds to perennial grasses, even after correction for the effects of seed size (Rees 1993b).

Again there is evidence of links between adult traits and those of the regenerative stages.

7.7 Conclusions

Understanding plant population and community dynamics requires a knowledge of seed ecology. Patterns of seed dispersal (see Chapter 9) and dormancy (this chapter) play an important role in determining how often individuals of different species interact and this in turn determines the strength of interactions between species (see Chapters 8 & 15). It should also be equally clear that in order to understand the evolution of plant life histories we need to consider not only the values a particular trait may take but also the life history in which the trait is embedded.

8: Mechanisms of Plant Competition

David Tilman

8.1 Introduction

The presence of a plant species in a locality, its abundance and the number of other plant species with which it coexists are influenced by numerous biological and physical processes. Interspecific interactions such as competition, herbivory, seed predation, mutualism, parasitism and disease may greatly affect plant dynamics and community structure (e.g Janzen 1970; Hubbell 1980; Fowler 1981; Tilman 1982; Schoener 1983, 1985; Crawley 1983; Berendse 1985; Coley *et al.* 1985; Brown *et al.* 1986; McNaughton 1986; Clay 1990; Huntly 1991). In addition, soil pH, temperature, rainfall and other physical factors, fire, trampling, burial, erosion, windfall, landslides and other disturbances, and seed or pathogen or herbivore dispersal and other spatial processes also greatly influence plant distribution, dynamics and diversity (e.g. Connell 1978; Sprugel & Bormann 1981; Goldberg 1985; Pickett & White 1985; Vitousek & Matson 1985; Whitney 1986; Clark 1989; Petraitis *et al.* 1989; Gilpin & Hanski 1991). Plant ecology is an exciting and intriguing area of study because of this multiplicity of interacting forces. This chapter focuses on one of these, competition among plant species. The basic mechanisms of plant competition are developed first, and then these mechanisms are integrated with some of the other potentially important forces to build more complete explanations for observed patterns.

8.2 Competition in natural plant communities

The ability of an individual of one species to inhibit the survival and/or growth of individuals of another species is called interspecific competition, i.e. competition between different species. Ecologists define interspecific competition as an interaction in which an increase in the population density or biomass of one species leads to a decrease in the population growth rate and the population density or biomass of another species. For plants, the strength of interspecific competition is measured in terms of the magnitude of growth suppression caused by each unit biomass of the neighbouring species. Interspecific competition can be a major force in natural plant communities. Numerous experimental studies have shown that the survival and growth of an

[239]

individual plant may be strongly influenced by competition with its neighbours (e.g. Connell 1983; Schoener 1983; Wilson & Keddy 1986a,b; Goldberg 1987; Gurevitch *et al.* 1990).

8.2.1 Competition in a grassland field

Wilson and Tilman (1991) studied competition in a Minnesota field that was dominated by native North American prairie species. To measure the strength of competition, they planted seedlings of three different grass species into existing vegetation (All in Fig. 8.1a) and determined how rapidly they grew during a growing season. They also inserted these seedlings into sites from which they had removed all neighbouring plants and excluded roots of other plants within 25-cm metal rings that had been driven into the soil (– All in Fig. 8.1a). At the end of the growing season, the seedlings of a native bunchgrass called little bluestem (*Schizachyrium scoparium*) weighed 8.5 g per plant without neighbours but weighed only 0.5 g per plant in the presence of neighbours (Fig. 8.1a). Thus, the neighbouring plants led to a 16-fold inhibition of growth of little bluestem. Similarly, the presence of neighbours caused an eightfold inhibition for Kentucky bluegrass (*Poa pratensis*) and a sixfold inhibition for quack grass (*Agropyron repens*). This inhibition indicates that all three of these grass species experience strong interspecific competition in this field.

Although this demonstrates strong interspecific competition, it does not specify the actual mechanisms whereby each species inhibits the other. For plants, at least, it seems unlikely that inhibition is a direct

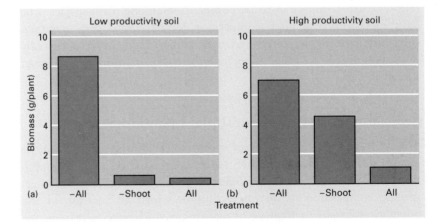

Fig. 8.1 (a) Biomass of little bluestem plants at the end of one growing season for seedlings planted into a prairie field. Both roots and shoots of all plants were removed from some plots (– All, just shoots were held back to prevent shading in some plots (– Shoot) or all existing plants were left intact (All) for some plots. (Data from Wilson & Tilman 1991.) (b) Biomass of little bluestem plants in similar plots in the same field, except these plots were much more productive because they had been fertilized for the preceding 5 years.

result of changes in plant density or biomass. Rather, changes in the biomass or population density of one species are likely to affect the availabilities of various resources, which then influence the growth and possibly the survival of the other species. Thus, the most likely mode of competition among plants is competition for resources.

8.2.2 Limiting resources

All plants require light, water and the same set of approximately 15 inorganic nutrients (N, C, K, P, Ca, Mg, S, etc.; see Chapter 3). The growth rate of a plant population, which is often best measured in terms of the specific rate of biomass change (i.e. as $1/B \, dB/Bdt$, where B is plant biomass per unit area), depends on the concentrations of these resources in that habitat. It is often the case that the growth rate of a species in a habitat is determined by the one resource in lowest availability relative to the plant's requirements for all resources. This is called the limiting resource for that plant species in that habitat. In general, the specific rate of biomass change of a species is an increasing, but saturating, function of the environmental concentration of its limiting resource. The growth of an individual plant would be decreased by the presence of neighbouring plants if these plants consumed and thus reduced the environmental concentration of its limiting resource. This is the basic mechanism of resource competition.

What may have caused the observed interspecific competition among grass species at Cedar Creek Natural History Area? A variety of lines of evidence suggest that these plants were competing with each other for a single limiting resource, soil nitrogen. Nutrient addition experiments that have been performed in this field have shown that addition of inorganic nitrogen (as ammonium nitrate fertilizer) led to significant increases in plant biomass, but that addition of P, K, Ca, Mg, S, water and trace metals had no effect on plant biomass (Tilman 1987, 1988, 1990b). Measurements of the concentration of soil ammonium and nitrate in the Wilson and Tilman competition plots described above found that these concentrations were very low in the plots that had the normal complement of neighbouring plants living in this field, but that concentrations were much higher in plots from which all neighbouring plants had been removed. These higher concentrations may explain the higher growth rates of seedlings in these plots.

8.2.3 Competition for nitrogen and light

Moreover, Wilson and Tilman's competition experiment included an additional treatment. Seedlings of all three grass species were also planted into plots that contained the roots of neighbouring plants, but the shoots (the above-ground parts of the plants) were kept from shading a seedling by clear nylon mesh that held them back from the seedling (– Shoot in Fig. 8.1a). At the end of the growing season, there was no detectable difference in the biomass of the seedlings between those grown in the presence of both the roots and shoots of the neighbouring plants versus those grown just in the presence of the roots. Both treatments greatly inhibited seedling growth compared with plots from which all neighbours had been removed. This indicates that the three grass species were inhibited by the roots of their neighbours and unaffected by the shoots of the neighbours on this nitrogen-poor soil. This further suggests that these plants were competing for a soil resource but were not competing for light.

Identical treatments were performed in other replicated plots in this field that had been fertilized with ammonium nitrate for the preceding 5 years. These fertilized plots had much greater above-ground plant biomass and had markedly different species composition and species diversity than the unfertilized plots. In the fertilized plots, the higher plant biomass led to much greater interception of light, and thus to low intensity of light at the soil surface. Comparisons of seedling biomass at the end of the growing season in the various treatments in these highly productive plots showed that plants competed as strongly in productive plots as they had in unproductive plots (Wilson & Tilman 1991). However, there was a shift in the mechanism of competition. In the productive plots, there was a highly significant effect of shading by neighbours. Unshaded plants grew significantly faster than those experiencing the shoots and roots of neighbours, but not as rapidly as those without neighbours (Fig. 8.1b). Thus, in the productive plots, the three grass species were competing for both light and nitrogen, whereas they were competing mainly for nitrogen in the unproductive plots.

8.3 A single limiting resource

For the grassland habitat discussed above, the major limiting resource was nitrogen. Let us now consider a simple, general theoretical framework that should apply to and predict the outcome of competition for a limiting soil resource (O'Brien 1974; Tilman 1976, 1977, 1982; Hsu *et al.* 1977; Armstrong & McGehee 1980). One piece of information needed is the dependence of plant growth rate on resource availability. Figure 8.2 shows hypothetical resource-dependent growth curves for two species, species A and B. Note that the specific growth rate (dB/Bdt, which is also called the relative growth rate, or RGR) increases with resource concentration, but that each plant species has a maximal specific growth rate that is approached at high resource concentrations. Each plant population also experiences various sources of loss or mortality. For instance, herbivores or pathogens may consume leaves, roots or stems, seed may be consumed, physical processes such as wind or frost may damage plants, and various factors may lead to plant death. All such sources of loss reduce the effective growth rate of a plant population. These losses can be expressed in terms of their effect on the specific rate of change of plant biomass, i.e. in the same dB/Bdt currency as used for plant growth (Hubbell & Werner 1979). This then leads to a simple relationship:

$$\text{rate of biomass change} = \text{growth} - \text{loss}$$

or

$$dB/Bdt = f(R) - m \qquad (8.1)$$

where $f(R)$ is a function describing the resource dependence of the specific growth rate of a plant (as shown by the curves in Fig. 8.2) and m is the total mortality and loss rate, again expressed on a specific biomass basis.

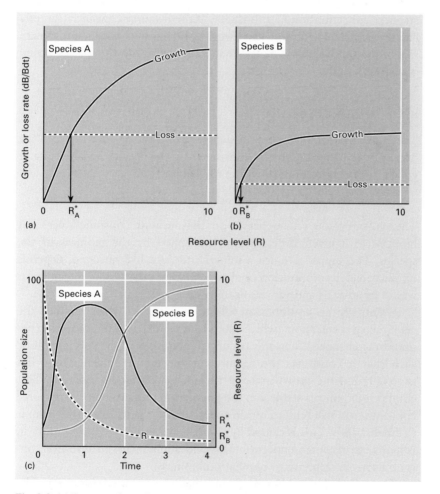

Fig. 8.2 (a) Resource-dependent growth and loss for species A. The resource concentration at which growth equals loss is R_A^*. (b) Resource-dependent growth and loss for species B and its R_B^*. (c) Dynamics of resource competition between species A and B. Note that species B, which has the lower R^*, displaces species A and drives the resource concentration down to R_B^*.

8.3.1 The R* concept ('R star')

When $dB/Bdt = 0$, the biomass of the species is no longer changing, and is thus at equilibrium. The long-term outcome of resource-dependent growth is thus determined by setting $dB/Bdt = 0$. When $dB/Bdt = 0$, $f(R)$ must be equal to m, i.e. the resource-dependent growth of the species must equal its loss rate. The implication of this is immediately grasped by inspection of Fig. 8.2. Note that $dB/Bdt = 0$ at the resource concentration at which the line labeled m intersects with the resource-dependent growth curve. The resource concentration at which this occurs is called $R*$ (Fig. 8.2).

$R*$ represents the environmental resource concentration required for the growth rate of a species to exactly balance its various sources of loss. A species will only be able to survive in a habitat if the habitat has a resource concentration of at least $R*$. If the resource concentration of a habitat were held at a value less that $R*$, the resource-dependent growth rate of the species would be less than its loss rate and the species would go extinct locally.

8.3.2 Resource dynamics

The concentration of the resource depends on the balance between the rate at which the resource is consumed by the species and the rate at which it is being resupplied within the habitat, i.e.

rate of resource change = supply rate – consumption rate

The consumption rate depends on the amount consumed per unit biomass in a given time interval multiplied by the biomass of the species. The supply rate, for a mineral nutrient like nitrogen, depends on microbial decomposition of organic matter, excretion of nitrogenous wastes by animals and various fixation processes.

Eventually an equilibrium will be reached at which neither the resource concentration (which would be at $R*$) nor the biomass (per unit area) change. When this occurs, the biomass of the species will be such that it consumes the resource at the rate at which it is being supplied, and the growth rate of the species equals its loss rate. This equilibrium point is stable, i.e. the population biomass and the resource concentration will tend to return to this point if perturbed away from it (Tilman 1980). This is caused by adjustments in population biomass in response to resource concentrations and adjustments in resource concentrations in response to population biomass.

There are many different forms that can be used for $f(R)$. The simplest form is the Michaelis–Menten equation, which states that

$$f(R) = rR/(R+k) \qquad (8.2)$$

where r is the maximal specific growth rate of the species (also called its

RGR$_{max}$), R is the resource concentration and k is the half-saturation constant for growth (the resource concentration at which the species grows at half of its maximal growth rate, i.e. at $r/2$). This form of the model, when substituted into Equation 8.1, leads to

$$R* = mk/(r - m) \tag{8.3}$$

Other, more complicated models have been proposed because the Michaelis–Menten formulation seems to exclude many of the interesting features of terrestrial plants, such as their production of leaves, roots, stems and seeds, the differences in the longevity of these structures and differences in nutrient conservation abilities (Tilman 1990a). Interestingly, even the most complex models lead to a rather simple conclusion. At equilibrium, all such models predict that a species will have a unique $R*$ value, and that this value will be determined by all the plant traits included in the model.

Thus, when a single species lives in a habitat, it reduces the concentration of its limiting resource down to its $R*$. $R*$ is the concentration of resource that a species requires to survive in a habitat. The $R*$ of a species is a summary variable that incorporates the effects of all types of nutrient and tissue loss, herbivory and mortality, as well as the effects of all traits that influence growth, including the nutrient dependence of photosynthesis and respiration, the plant pattern of allocation to root, leaf, stem and seed, and the nutrient conservation ability of the species.

$R*$ is a critical variable because it is the direct measure of the effect of one plant species on a potential competitor. Plants compete through their effects on environmental resource levels, and $R*$ quantifies these effects. It is the level to which each species can reduce the concentration of the limiting resource. There are two qualitatively different ways to determine the $R*$ of a species for a limiting resource. The simplest way is to observe it directly by allowing a single species to grow, in monoculture, in an environment in which a particular resource is known to be the only limiting resource. Once the biomass of the monoculture reaches an equilibrium, the level to which it has reduced the concentration of the limiting resource is the $R*$ of that species. The other way is to construct a realistic model of processes controlling growth and loss, determine the values of the parameters for a particular plant species and predict the $R*$ from the underlying model.

8.3.3 Competition for a limiting resource

If several different plants compete for the same single limiting resource, the single species with the lowest $R*$ should displace all other species from the habitat (Tilman 1976, 1977; Hsu *et al.* 1977). This displacement occurs because the species with the lowest $R*$ can continue increasing in abundance and reducing the resource concentration down to its $R*$. However, once the resource concentration falls below the $R*$

values of other species, they are unable to survive in the habitat because their resource-dependent growth rates are held to a lower rate than their losses.

This is illustrated in Fig. 8.2b,c. Note that species A and B have different resource-dependent growth curves and different loss rates. Species B has a lower $R*$ than species A. As the two species grow in the same habitat, they will compete for the limiting resource. Species B starts to displace species A when it first reduces the resource concentration below R_A*. As equilibrium is approached, the resource concentration approaches R_B*, the biomass of species B stops changing and the biomass of species A approaches 0.

8.3.4 Tests of the $R*$ hypothesis

There have been numerous tests of resource competition theory, which has proven to be surprisingly robust. The first was an experimental study of competition between two species of freshwater algae, both diatoms (Tilman 1976). *Asterionella formosa* (Fragilariaceae) had a lower $R*$ for phosphate than did *Cyclotella meneghiniana* (Bacillariophyceae). When they were grown together with phosphate as the only limiting nutrient, *Asterionella* displaced *Cyclotella*. *Cyclotella* had a lower $R*$ for silicate (which diatoms require to make their frustule, which functions like a cell wall), and *Cyclotella* displaced *Asterionella* when they were silicate limited. Hansen and Hubbell (1980) showed, for several different cases, that the bacterial species with the lower $R*$ for tryptophan, the limiting resource, competitively displaced the other species. Rothhaupt (1988) showed that $R*$ successfully predicted the outcome of interactions among two zooplankton species that were consuming two different algal species. Tilman *et al.* (1981) found that the temperature dependence of $R*$ could be used to predict the outcome of competition among diatoms for silicate along a temperature gradient. Other cases, reviewed in Tilman (1982), also show that $R*$ can correctly predict the long-term outcome of competition among aquatic organisms for a single limiting resource.

However, there were questions about the applicability of resource competition theory to more complex organisms, such as terrestrial vascular plants. Indeed, the complexity of terrestrial plants may make it quite difficult to predict the $R*$ of a species based on its underlying physiology, morphology and life history. However, does this mean that measured $R*$ values would be unable to predict the outcome of resource competition?

We studied nitrogen competition among five grass species (Tilman 1990b; Tilman & Wedin 1991a,b; Wedin & Tilman 1993). Each of the five species was grown in replicated, long-term monocultures in garden plots with soils prepared such that nitrogen was the only limiting resource. By the third growing season, the above- and below-ground living biomass of the five species has equilibrated, and the species differed significantly in levels to which they had reduced the concentration of nitrogen (nitrate and ammonium) in the soil solution (Fig. 8.3). Little bluestem (*Schizachyrium scoparium*) and big bluestem (*Andropogon gerardi*) had the lowest R^* values for nitrogen of the five grass species. They also had the greatest allocation to root and the greatest belowground living biomass of the five species. In contrast, *Agrostis scabra* had the lowest allocation to root and the greatest R^* of these species, and *Poa pratensis* and *Agropyron repens* had intermediate R^* values. Thus species that allocated a greater proportion of their biomass to root drove soil ammonium and nitrate concentrations down to lower levels.

Various pairs of these species were also planted together in replicated garden plots in which soil nitrogen was the only limiting resource. Three qualitatively different starting conditions were used: seed competing against seed, seed competing against established plants, and established plants competing by invading into other established plants (Wedin & Tilman 1993). The long-term outcome of competition after 5 years was identical for all three types of experiments. In all cases, if a species had a significantly lower R^* than another species, it displaced that other species. The results of competition between *Agrostis* and *Schizachyrium*, which differed greatly in their R^* values, can be compared to that between *Agropyron* and *Poa*, which have quite similar R^* values (Fig. 8.4a,b). *Agrostis*, which has an R^* for nitrogen that is 5.5 times greater than the R^* of *Schizachyrium*, was rapidly displaced by *Schizachyrium* from low nitrogen soils (Tilman & Wedin 1991b). In contrast, the R^* of *Agropyron* was only 1.05 times that of *Poa*, and competitive displacement of *Agropyron* by *Poa* is much slower. Indeed, for the various pairs of species tested, the rate of competitive displacement was more rapid the greater the difference in the R^* values, as is predicted by theory. This indicates that resource competition theory can correctly predict the outcome of competition between terrestrial plants.

Fig. 8.3 The concentration to which a species reduces its limiting resource in monoculture at equilibrium is its R^*. When growing in monoculture on soils in which nitrogen was the only limiting resource, five grass species differed significantly in the level to which they could reduce the soil solution concentration of ammonium plus nitrate. Species with greater allocation to root had lower R^* values for soil nitrogen. (Data from Tilman & Wedin 1991a.)

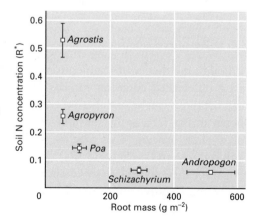

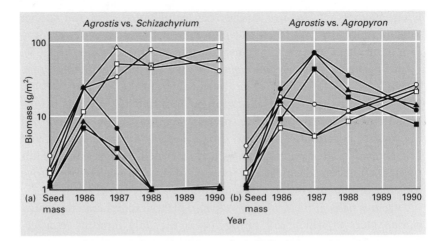

Fig. 8.4 (a) The dynamics of competition for limiting nitrogen between *Agrostis* (solid symbols) and *Schizachyrium* (open symbols), for seeds planted into long-term garden plots. Each point is a mean of three replicates. Each line represents a different initial ratio of seeds of the two species. *Schizachyrium* has a lower R^* for nitrogen and, as predicted, is the superior competitor. (Data from Tilman & Wedin 1991b.) (b) A similar case of competition, except between *Agrostis* (solid symbols) and *Agropyron* (open symbols). These species have more similar R^* values. *Agropyron*, which has the lower R^*, is the predicted winner, and did come to dominate these plots.

8.4 Competition for two resources

In order to predict the outcome of plant competition for two or more resources, it is necessary to know the dependence of the growth of each species on the availability of *all* limiting resources. Plants require many different resources, most of which are essential for plant survival, i.e. plants cannot survive without them. Some species may be limited simultaneously by two or more resources and it is possible for different species to be limited by different resources in the same habitat. However, different forms of an essential resource, such as the ammonia and nitrate forms of nitrogen, can be substituted for each other. This chapter only considers competition for limiting, essential resources. A more complete development of theory can be found in Tilman (1982).

8.4.1 Resource isoclines

Although the theory of competition for two resources is best expressed using differential equation models (Tilman 1977, 1985, 1990a), its major features can be understood graphically (Tilman, 1980). The curve shown in Fig. 8.5a is the resource-dependent zero net growth isocline for a plant species. This isocline shows the concentrations of two limiting resources (R_1 and R_2) for which the resource-dependent growth of this species exactly balances all its losses (i.e. $dB/dt = 0$). For any of the resource concentrations on this curve, the population biomass of this species will be constant. The two 'ends' of the curve give, in

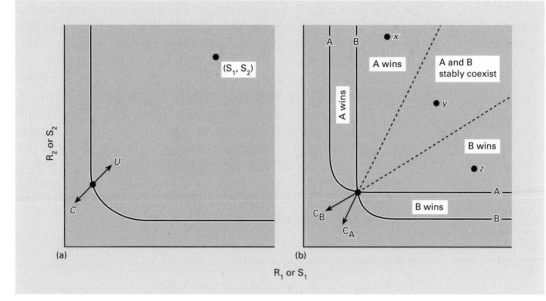

Fig. 8.5 (a) The curve is the resource-dependent zero net growth isocline of a plant species. The point labelled (S_1, S_2) is the resource supply point, U is the resource supply vector and C is the consumption vector. The dot on the isocline is the equilibrium point associated with this supply point. See Tilman (1982) for additional details. (b) Curves labelled A and B are isoclines for species A and B. These, and the consumption vectors, define regions in which one species wins in competition, the two stably coexist or the other wins. See Tilman (1982) for further details.

their limits, the R^* values of this species for R_1 and R_2. The shape of this curve means that these two resources are 'interactive-essential' for this species. This occurs if a plant can adjust its morphology or physiology in response to concentrations of two essential resources so as to acquire them in the proportion in which they are required. Plants that are not so plastic, such as single-celled algae, have an isocline that is closely approximated by a right-angled corner.

Population biomass will decrease $(dB/dt < 0)$ if resource availabilities fall inside the isocline, and it will increase if an environment has resource availabilities that fall outside the isocline. On the horizontal portion of the isocline, far from the origin, the species is limited by R_2, and changes in R_1 do not affect its growth and vice versa. To see this, consider the extreme right-hand end of the isocline. At this point, addition of R_2 would lead to increased growth and its removal would lead to decreased growth, but there would be almost no effect on growth from changes in R_1, which would just lead to motion along the isocline.

8.4.2 Resource consumption vectors

Equilibrium will occur when resource consumption equals resource supply, and reproduction equals mortality. Reproduction equals mortality for any point on the growth isocline. The actual point on the isocline that will be the equilibrium point in a given habitat is determined by the rates of resource consumption and supply. Optimal foraging theory (Rapport 1971) predicts that a plant should consume nutritionally essential resources in the proportion in which the plant is equally limited by them. This is represented by a consumption vector, C (Fig. 8.5a). The length of the consumption vector is the amount of the two resources consumed by this species. The slope of this consumption vector gives the ratio in which this species requires R_2 and R_1. This consumption vector can be empirically determined, or it can be approximated by assuming that it is parallel to a line from the origin through the 'corner' of the isocline. The slope of such a line should approximate the ratio in which the plant requires these resources.

8.4.3 Resource supply vectors

Let S_1 and S_2 be the maximal amounts of all forms of resources 1 and 2 in a given habitat. The point (S_1, S_2) is called the resource supply point of that habitat. Each habitat is considered to have a particular resource supply point. A simple model of resource supply assumes that rate of supply of a resource should be proportional to the amount of the resource that is not already in the available form. This would give

$$\mathrm{d}R_j \mathrm{d}t = a(S_j - R_j) \tag{8.4}$$

where a is a rate constant and j refers to resource j. This equation defines a resource supply vector, U, that always points toward the supply point. Thus, the resource supply point (S_1, S_2) of Fig. 8.5a leads to the equilibrium point shown with a dot. At the equilibrium point, the population density of the species is such that its total rate of resource consumption of each resource equals the total rate of supply of each resource. This occurs because, at this point, the two vectors are equal in length and opposite in direction. Because this occurs at a point on the zero net growth isocline, the plant's reproductive rate equals its mortality rate.

8.4.4 Coexistence and displacement

To predict the outcome of competition between two or more species for two limiting resources, it is only necessary to superimpose their isoclines and consumption vectors. Consider the two isoclines shown in Fig. 8.5b. These isoclines cross because species A is a better competitor for R_1 and species B is a better competitor for R_2. The point at which these isoclines cross is a two-species equilibrium point, i.e. a point of potential coexistence of these two species. At this point, the reproductive rates of both species A and B equal their mortality rates. However, these species will only be able to coexist in habitats with certain resource supply points (Fig. 8.5b). If habitats have low supply rates of R_1 and high supply rates of R_2, such as at supply point x, both species will be limited by R_1. Species A, which is the superior competitor for R_1, will reduce the level of R_1 down to a point on its isocline at which there is insufficient R_1 for the survival of species B. Thus, species A will competitively displace species B from habitats with low $S_1 : S_2$ ratios (Fig. 8.5b). Comparably, species B is a superior competitor for R_2 (has a lower $R*$ for R_2) and it will displace species A from habitats in which both species are limited by R_2. Such habitats have high $S_1 : S_2$ ratios.

The two species can coexist in intermediate habitats in which each species is relatively more limited by a different resource, species A by R_2 and species B by R_1. For habitats with supply points within the region defined by the consumption rates of these species at the two-species equilibrium point (Fig. 8.5b), the consumption of the two species will eventually reduce resource levels down to the equilibrium point. Thus, the resource requirements of these two species define habitats in which one species is dominant, both stably coexist or the other is dominant. Along a resource ratio gradient, such as from supply point x to supply point y to supply point z, there is a smooth transition from dominance by species A, to coexistence of A and B, to dominance by species B. This resource ratio gradient is a gradient from low $S_1 : S_2$ ratios to high $S_1 : S_2$ ratios. Such separation along the $S_1 : S_2$ gradient only occurs if the species are differentiated in their requirements for R_1 and R_2, with the superior competitor for one resource being the inferior competitor for the other resource.

8.4.5 Experimental tests

Several experimental studies of algal competition for limiting phosphate and silicate have shown that this simple theory can predict the outcome of interspecific competition, including stable coexistence (Tilman 1976, 1977, 1982; Tilman *et al.* 1982; Carney *et al.* 1988). Additionally, the distributional patterns of algal species along natural resource ratio gradients in Lake Michigan are consistent with their requirements for the limiting resources and the outcome of laboratory competition experiments among the species (Tilman 1982). There have not yet been studies of the applicability of this theory to competition for two or more resources by terrestrial plants.

8.5 Multispecies communities

As developed so far, resource competition theory seems to predict the outcome of resource competition experiments, but the predicted outcomes seem discordant with nature. If a single resource is limiting to all plants, as seems to be the case for nitrogen in the old fields and prairie of Cedar Creek Natural History Area, Minnesota, this theory predicts that a single species should eventually dominate these habitats. Two species could coexist if there were two limiting resources, and so on, but there could never be more species than there were limiting resources in the habitat. However, despite the strong experimental evidence that nitrogen is the only limiting resource, except in a drought year when water becomes limiting, Cedar Creek fields contain from 70 to more than 200 different species of vascular plants. Observations over the past decade show that these species are persisting in these fields, and that species that were driven to low density because of the drought are reinvading and becoming re-established. All of this suggests that 70–200 species are stably coexisting in a habitat in which there is only one demonstrably limiting resource. How could this occur?

Clearly, any explanation for the existence of such biodiversity must be more complex than the simple resource competition theory presented above. Of the many processes that can, in theory, explain the stable coexistence of multispecies communities (reviewed in Tilman & Pacala 1993), there are four distinct explanations that incorporate resource competition among plants. Each of these explanations starts with the simplifying assumptions of the basic model of resource competition: (i) that organisms live in a spatially homogeneous habitat, (ii) that all interactions go to equilibrium, (iii) that the entire food web can be approximated by just the link between consumer species and their resources, and (iv) that all organisms within a habitat experience exactly the same resource concentrations at the same time. Each of these theories accepts three of the four simplifying assumptions above, but modifies one of these to make it more realistic. Thus, each of these explanations is one logical step more realistic than the simple model of resource competition discussed previously.

8.5.1 Spatially discrete individuals

The models of resource competition presented above assumed that all individuals of all species experienced exactly the same resource concentrations at any given instant. This mathematically convenient assumption is not biologically realistic. Terrestrial plants occur as discrete individuals. Each individual consumes resources in its immediate neighbourhood. Thus, an individual mainly competes with its immediate neighbours (Pacala 1986a,b; Goldberg 1987; Pacala & Silander 1990). An individual plant with a low $R*$ for a limiting resource would reduce the concentration of that resource in its immediate vicinity down to its $R*$, but this would not affect the growth of a plant species with a higher $R*$ that did not live in its neighbourhood.

Models can be formulated to deal with plants as spatially discrete individuals that only influence resource concentrations in their immediate vicinity, but such models are not yet analytically tractable. The essential, though highly abstracted, features of resource competition among spatially discrete individual plants can be captured in 'metapopulation' models, such as those of Levins and Culver (1971), Horn and MacArthur (1972), Hastings (1980) or Shmida and Ellner (1984). Their models have shown that spatial subdivision can allow the stable coexistence of two competitors even though one is a superior competitor in any given habitat. The inferior competitor could coexist, but only if it were a superior disperser or had a lower loss rate than the superior competitor.

These models do not explicitly include resources, but rather rank species from being the best competitor (i.e. species with the lowest $R*$ for a limiting resource) to the worst competitor (species with the highest $R*$). A habitat is considered as being divided into a series of patches, each the size of an individual plant. The abundance of a species is represented as p, the proportion of patches that are occupied by that species. Species disperse randomly among all patches. If a superior competitor invades a patch occupied by an inferior competitor, the superior competitor displaces the inferior competitor from that patch. These ecological assumptions, as applied to a single species, are expressed in the following equation, which was first proposed by Levins (1979):

$$\frac{\mathrm{d}p}{\mathrm{d}t} = cp(1-p) - mp \tag{8.5}$$

where m is the mortality (local extinction) rate and c is the colonization rate. Propagules disperse at random among all sites. The propagule production rate by the occupied sites, cp, is multiplied by the proportion of unoccupied sites. $1-p$, to give the rate of production of newly occupied sites. A site becomes vacant when the individual occupying that site dies. The local mortality rate, m, is multiplied by p, the proportion of occupied sites, to give the rate at which occupied sites become empty.

Building on Hastings (1980), Tilman (1993) extended this work to show that it could explain, in theory, the stable coexistence of a potentially unlimited number of species competing for a single limiting resource. For this work, the equation describing the dynamics of site occupancy by the ith species, for a series of n species ranked from the best competitor (species 1) to the worst competitor (species n), is:

$$\frac{\mathrm{d}p_i}{\mathrm{d}t} = c_i p_i \left(1 - \sum_{j=1}^{i} p_j\right) - m_i p_i - \left(\sum_{j=1}^{i-1} c_j p_j p_i\right) \tag{8.6}$$

There is one such equation for each species. The dynamics of each species depend on colonization (the first term), on mortality (the term $-m_i p_i$), and on competitive displacement (the last term). A species is only affected by species that are superior competitors, i.e. have a lower rank (lower $R*$).

It is possible for any number of competing species to stably coexist, at equilibrium, with this model if there are the appropriate interspecific trade-offs among them (Tilman 1993). Specifically, coexistence requires that an inferior competitor be a superior colonist and/or have greater longevity than its superior competitor. Moreover, there is an analytical limit to similarity of competitively adjacent species. This limit to similarity means that it is not sufficient for an inferior competitor to be a better colonist, for example, than its superior competitor, but rather that it must be better by a specified amount that is greater the greater the abundance of its superior competitor (Tilman 1993).

Coexistence occurs because the superior competitor is unable to occupy all sites within a given habitat. Even in the absence of competitors, there is a portion of the habitat that is always unoccupied by the best competitor. These open sites can allow the long-term persistence of another species if it has sufficiently great dispersal and/or longevity to survive in the open sites, despite the competitive displacement that comes when its superior competitors invade and despite the lost reproduction that comes from its propagules landing in sites occupied by its superiors. However, this second species also has a finite dispersal ability and experiences some mortality. It thus cannot occupy the full remainder of the habitat. The sites it leaves open can allow the persistence of an inferior competitor, if it has sufficiently greater colonization and/or longevity. This process may continue without limit. No matter how many species there are, they cannot occupy all potential sites within an infinite environment as long as each species has finite dispersal ability and all experience some mortality (Tilman 1993; see also Chapters 14 & 15).

Although there have been no direct tests of the applicability of this hypothesis to Cedar Creek grasslands, it is supported by much circumstantial evidence. Monocultures of *Schizachyrium*, which were planted at high densities, were rarely invaded by other plant species, whereas monocultures of species with higher $R*$ values for nitrogen were frequently invaded and had to be constantly weeded. There was also a marked interspecific trade-off between the nitrogen competitive abilities of the five grass species we studied and their colonization abilities. *Schizachyrium* and *Andropogon* allocate less than 1% of their biomass to seed, have seed with a low viability and require 11–17 years to invade most abandoned fields. In contrast, *Agrostis* and *Agropyron*, which have much higher $R*$ values, have much greater allocation to seed, also reproduce via rhizomes and invade most abandoned fields within 1 or 2 years. A comparable trade-off occurs during succession (Gleeson & Tilman 1990). Because plant species that allocated more to root were better nitrogen competitors in our garden experiments (Tilman & Wedin 1991b), this allocation pattern suggests that there is a strong trade-off between competitive ability and colonization ability. If this is so, the theory discussed above may explain the stable coexistence of numerous species in these grassland fields.

8.5.2 Spatial heterogeneity

It has long been suggested that spatial heterogeneity within a habitat might account for the coexistence of numerous competitors. This hypothesis modifies the assumption of spatial homogeneity, but retains the other three simplifying assumptions made by the original theory. The essential idea is that each species is a superior competitor for a particular suite or intensity of environmental factors, and that point-to-point variability in these factors could allow numerous species to coexist.

This hypothesis cannot explain the coexistence of numerous species living in a habitat in which there is a single limiting resource and no other limiting factors. In such a case, point-to-point heterogeneity in the supply rate of the single limiting resource could not allow more than one species to persist, at equilibrium. This occurs because the species with the lowest $R*$ for this resource would reduce resource concentration to its $R*$ at all points in the habitat. That species would be more abundant in areas with higher rates of resource supply and less abundant in areas with lower rates of supply, but it would be the only species at equilibrium.

There are, however, other ways that spatial heterogeneity could influence species diversity. For instance, the growth (and thus $R*$) of species could be influenced by a physical factor such as temperature or pH, and this physical factor may vary spatially in intensity. If species were differentiated in their abilities to compete for a single limiting resource at different intensities of this physical factor, such heterogeneity could allow a potentially unlimited number of species to coexist. A situation with five competitors is illustrated in Fig. 8.6. Species A is the superior resource competitor at 10 °C, species B at 11 °C, species C at 12 °C, species D at 13 °C and species E at 14 °C. In any homogeneous habitat, in which plants living in all localities experience exactly the

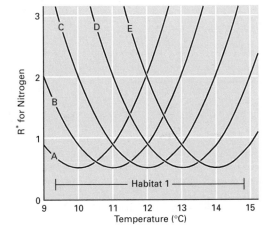

Fig. 8.6 Nitrogen competition along a temperature gradient. The five species are differentiated in the temperature at which they have their greatest competitive ability (i.e. lowest $R*$). A single species would win in nitrogen competition in a homogeneous habitat, but all five species could coexist in the range of temperatures of Habitat 1.

same temperature, only a single species will win. However, if there were spatial heterogeneity in temperature, such as might occur along an elevational gradient or with respect to aspect and microtopography, it is possible for several or all of these species to coexist stably as long as propagules arrive in the proper locations. Thus, the interspecific differentiation and trade-off illustrated in Fig. 8.6 could allow a large, potentially unlimited, number of species to coexist in a habitat with spatial heterogeneity in a physical factor such as temperature or pH.

Alternatively, if there are two limiting essential resources, such as nitrogen and phosphate or nitrogen and light, spatial heterogeneity could allow numerous species to coexist. Let us consider a case in which five different species compete for two essential resources. Again, let us assume that these species are differentiated such that each species is a superior competitor for a particular ratio of the limiting resources, and that each species consumes the resources in the ratio in which it is equally limited by them. This gives regions in which various pairs of species can coexist (Fig. 8.7a) and predicts that the species should, at equilibrium, be separated along an $S_1 : S_2$ resource ratio gradient (Fig. 8.7b). The point at which each species reaches its greatest abundance along the resource ratio gradient is determined by its requirements for the limiting resources. Thus, species A is dominant at low $S_1 : S_2$ ratios because it is the best competitor for R_1 but the worst

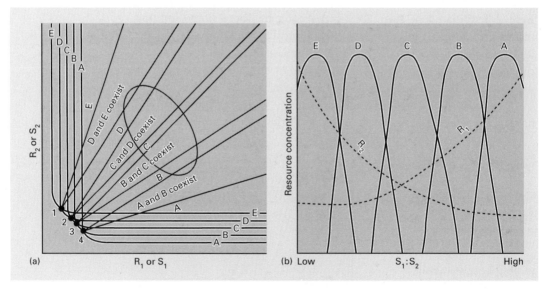

Fig. 8.7 (a) Zero net growth isoclines for five species and their consumption vectors (not shown) create regions in which various pairs of species can stably coexist or a single species is dominant. The five species have an interspecific trade-off in their competitive abilities for R_1 versus R_2. If habitats have spatial heterogeneity in resource supply rates, such as indicated by the circle, more species could potentially coexist than were limiting resources. (b) The interspecific trade-off above leads to separation of these five species along a gradient in the relative rates of supply of these two resources.

competitor for R_2. At all points along the gradient, R_1 and R_2 are important limiting resources for some of the species.

[257]
Chapter 8
Competition

Spatial heterogeneity in S_1 and S_2 might occur along elevational gradients or in response to heterogeneity in substrates and parent material from which soils form or from excretion of wastes by herbivores (e.g Lehman 1982). Such heterogeneity is represented not by a single resource supply point, but by a cloud of such supply points, with each point representing the average resource supply rates experienced by an individual plant. There is much spatial heterogeneity in the levels of nutrients in both soils and lakes (e.g. Burgess & Webster 1980; Lehman 1982; Yost *et al*. 1982). Such variance can be illustrated graphically by showing all of the different supply points experienced by the individual plants in a heterogeneous habitat. Let the circles in Fig. 8.7a include 99% of the point-to-point spatial heterogeneity in the (S_1, S_2) as experienced by individual plants in a habitat. Such spatial heterogeneity can allow many more species to coexist than there are limiting resources. All that is required for a species to exist in a habitat is that there be some site with a suitable (S_1, S_2), i.e. that the circle overlaps its region of existence, and that its propagules arrive in that site. For instance, five species could stably coexist in a habitat with the heterogeneity represented by the circle in Fig. 8.7a because there are sites in which each species can survive.

8.5.3 Resource fluctuations and non-equilibrium conditions

It is also possible that resource concentrations and populations do not go to equilibrium but fluctuate. Armstrong and McGehee (1976a,b, 1980) showed that such resource fluctuations could allow an unlimited number of species to stably persist in a spatially homogeneous habitat in which all individuals of all species experienced the same resource concentrations at any given time. Fluctuations allow species to persist because of non-linearities in the resource-dependent growth curves of the consumer species.

Levins (1979) provided a clear example of this. Consider two species, both of which are limited by the same resource. Species A has a linear resource-dependent growth curve, whereas species B has a non-linear curve (Fig. 8.8a). Both species experience the same mortality rate, m, giving an R_A^* that is less than R_B^*. Thus, in a non-fluctuating environment, species A should competitively displace species B. How would resource fluctuations around a mean of $\bar{R} = R_A^*$ affect the long-term average growth rates of species A and B?

Because species A has a linear resource-dependent growth curve, fluctuations below $\bar{R}$ lead to decreases in growth rate that are exactly balanced by the increases in growth caused by fluctuations above the mean. Thus, the long-term average growth rate of species A is independent of the variance in resource availability (Fig. 8.8b). In contrast, the

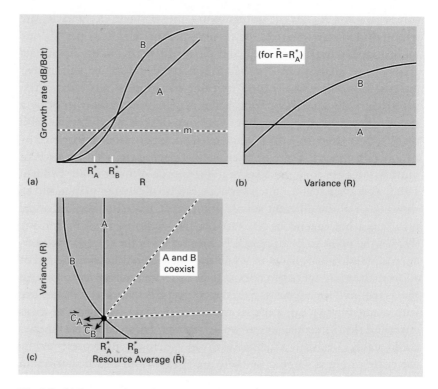

Fig. 8.8 (a) Resource-dependent growth and loss (*m*) curves for species A and B. The *R** values of these species predict that species A should displace species B from an unvarying environment. (b) Dependence of long-term average growth rate on the magnitude of temporal variance in resource concentration for a case in which the average concentration is held at R_A^*. (c) Zero net growth isoclines for species A and B in relation to long-term average resource availability and its temporal variance. The two isoclines cross because species B has a higher *R** but can exploit resource pulses, and thus requires less resource to survive in a fluctuating habitat compared with species A.

sigmoid shape of the resource-dependent growth curve of species B means that fluctuations below the mean of $\bar{R}$ lead to small decreases in growth, whereas comparable fluctuations above the mean lead to greater increases in growth. Thus, the long-term average growth rate of species B increases as the magnitude of the fluctuations, as measured by the variance around $\bar{R}$, increases (Fig. 8.8b). This means that the variance in R is functioning as if it were a resource for species B.

The responses of these two species can be used to create zero net growth isoclines for the case in which a single fluctuating resource is decomposed into two components, its mean and its variance (Fig. 8.8c). The isocline for species A is a straight line because it is unaffected by variance. The curved isocline of species B shows that species B has a higher R* for the resource in an unvarying habitat at equilibrium (variance = 0), but that its requirements are lower as the magnitude of resource fluctuations increases (Fig. 8.8c). The point at which the two curves cross represents coexistence. This point can be stable if species B 'consumes' the variance more than does species A. This is likely because species B is a specialized consumer of resource pulses and thus could quickly decrease resource concentration when it is above the mean. Thus, a species with opportunistic foraging can coexist with a resource competitor that has a lower R*, if resource concentration fluctuates. The relative abundances of the superior competitor and the opportunist should depend on the magnitude of the resource fluctuations. With small fluctuations, the superior competitor would displace the opportunist. They would coexist with intermediate fluctuations, and the opportunist would displace the superior competitor in a highly fluctuating environment.

Sommer (1984, 1985) showed that pulses of phosphate and/or silicate could allow the long-term persistence of many more algal species than there were limiting resources, whereas the number persisting in habitats at equilibrium did not exceed the number of limiting resources. Grover (1988, 1989, 1990, 1991) obtained similar results, and demonstrated the underlying physiological bases for such coexistence. It seems highly likely that desert annual plants are exploiters of pulses of rainfall, and that this allows numerous species of annuals, each specialized on a different pattern or timing or magnitude of rainfall, to persist with long-lived, deep-rooted perennials that are better water competitors in general. Many perennial plants differ in the season of their growth and may differ in the timing of their exploitation of soil resources (McKane *et al.* 1990), which may be an important factor allowing them to persist in their natural habitats.

8.5.4 Multiple trophic levels

Hutchinson (1959) suggested that trophic complexity might allow numerous competitors to coexist. This was supported by Paine's (1966, 1969) work on intertidal invertebrates, which showed that removal of starfish, the top predator, caused the loss of many rocky intertidal species. Levin *et al.* (1977) explicitly modelled multitrophic-level food webs and tested their model in laboratory food webs containing phage viruses, their bacterial prey, and the sugar that limited the growth of the bacteria. Their model predicted that the number of distinct viral strains could not exceed the number of distinct bacterial strains, but that the number of bacterial strains could be greater than the number of limiting sugars if some bacterial strains were more limited by viral attack and others more by the sugar. Their experimental food webs confirmed this prediction. They found that two bacterial strains could coexist on a single limiting substrate, but only if, as predicted, the bacterial strain that was more susceptible to viral attack was also the better competitor for the substrate.

This theory has been generalized to any food web (Tilman 1982). At equilibrium in a spatially homogeneous habitat in which all organisms simultaneously experience the same resource concentrations, the number of predator species cannot exceed the number of their prey species, and the number of prey species cannot exceed the sum of the number of their limiting resources plus the number of predator species. Thus, a single limiting resource can support a single prey (consumer) species, which can support a single predator. However, if there is a prey species that is either a better resource competitor but more susceptible to predation, or a poorer competitor but more resistant to predation, it may be able to invade into and persist in this food web. If this happens, there are now two prey species, which could allow the invasion of another predator species. However, the addition of another predator species could allow a further prey species to invade. This process could continue indefinitely if all species have the appropriate trade-offs. Thus, it is conceptually possible for a potentially unlimited number of consumer (prey) species to coexist in a homogeneous non-spatial habitat at equilibrium in which there is a single limiting resource if the habitat also has the right suite of predators (Tilman 1982).

Thus, a grassland field, such as an old field or stand of prairie at Cedar Creek Natural History Area, could conceivably contain over 100 plant species, all competing for a single limiting resource (nitrogen), if the plant species were appropriately differentiated with respect to competitive ability versus susceptibility to herbivory. At the present time, there are few data with which to evaluate this idea (see Chapter 13).

8.6 Conclusions

There is strong evidence that both terrestrial and aquatic plants compete for limiting resources in natural, managed and agricultural habitats. Competition both within and among species is thus one of the major forces determining the distribution and abundance of plant species and the biodiversity of plant communities. The simple theory of resource competition presented in this chapter captures the essence of this interaction.

If two plants compete for a limiting resource, each influences the other through its effect on the environmental concentration of the limiting resource. For spatially homogeneous habitats that have reached equilibrium, the effect of each species on the other is measured by the $R*$ of each species. A single variable, $R*$, which is directly measurable by growing a plant in monoculture, integrates the effects of the morphology, physiology and life history of a plant species on its resource competitive ability. The available experimental evidence suggests that $R*$ can correctly predict the long-term outcome of nutrient competition among terrestrial plants and among various aquatic organisms. It is a useful first step in understanding the mechanisms of competition.

However, natural habitats are unlikely to be spatially homogeneous, individual plants are spatially discrete, resource levels often oscillate because of external environmental forces and plants live in complex food webs in which they are influenced not only by resource competition but also by various herbivores, pathogens, parasites and mutualists. All such factors directly influence plant dynamics and thus can modify the effects of plant competition. The four hybrid models discussed above all emphasize the need to consider the joint effects of interspecific resource competition and other constraints on plant fitness. Each of these is presented as a distinct model for mathematical convenience. Clearly, plants living in any natural habitat may experience all these constraints simultaneously. One of the major challenges facing plant ecology is the synthesis of such constraints into a more general theory that can help us better understand the forces that control the dynamics and diversity of plant communities.

9: Ecology of Pollination and Seed Dispersal

Henry F. Howe and Lynn C. Westley

9.1 Introduction

It is easy for us to understand how animals like ourselves compete for mates, care for young and enhance our chances of contributing to future generations. It is difficult for us to 'think like a plant'. To understand plants, we must shift our frame of reference to that of a sedentary and mindless organism that cannot choose where it lives, know its mate or mating success, or assist its offspring for more than a minute fraction of its potentially long life. Rules of life are very different for plants compared with animals capable of decision and movement.

9.2 Challenges of a sedentary existence

Consider the reproductive challenges faced by organisms as different as a squirrel and an oak. A squirrel actively searches for food and for a place to live, while the oak must grow where an acorn happens to land or where a squirrel buries it. A squirrel looks for a mate, while an oak casts millions of pollen grains to the wind and snares the pollen of neighbours on its stigmas. A squirrel nourishes and protects a small litter until the youngsters have a reasonable chance of survival; perhaps 1 in 10 newborn squirrels survive to reproduce. Acorns are doomed unless a squirrel or jay removes them from beneath the oak *and* forgets to eat them; perhaps 1 in 100 000 or 1 in 1 000 000 acorns survives to reproduce. If a squirrel fails to breed in its second year, only a chance or two remain before its genetic inheritance is lost forever. An oak may have centuries in which to leave its offspring, perhaps in dramatic episodes of seedling success after decades of failure.

This comparison underscores both the constraints and the potentials of reproduction among plants. Often high fecundity and long life give plants a hedge against disastrous losses of vulnerable seeds and seedlings, while offering opportunities that are not available to most animals. For example, an extraordinarily successful mammal might raise several healthy youngsters in a lifetime, but a single tropical fig, producing 5 000 000 seeds each year, could leave tens of thousands of vigorous offspring if it happened to grow at the edge of a mudslide where conditions favoured its rapidly growing saplings (Garwood *et al.* 1979). Even an oak, with a much lower fecundity than a fig, could leave dozens

[262]

of adult offspring if it lived in the right place at the right time. All the parental care that a squirrel can muster could result in only a few offspring. The extravagance that allows a plant to hedge against huge losses of seeds and seedlings carries with it enormous potential for parental fitness (Salisbury 1942).

In short, the rhythms of plant reproduction tax our animate sensibilities, but raise fascinating questions. How do plants mate and produce offspring against apparently enormous odds? Why do some plants rely on wind or water for transport of pollen or seeds, while others consort with self-interested animals that eat nectar, pollen or seeds? How do conditions alter conflicts between self-interested mutualists? This chapter cannot do justice to all these issues, which should be explored in depth elsewhere (references in Fleming & Estrada 1993; Lloyd & Barrett 1996; Howe & Westley 1988). Our aim here is to bring the dynamics of pollination and seed dissemination into bold relief, and to suggest routes towards answers to many questions that plant reproduction poses.

9.3 Adaptive trends

Flowers disperse pollen, while fruits are organs of dispersal. Parallels between pollen and seed dispersal are inexact because pollen and seeds have different targets (Table 9.1). The target for pollen from one flower is the receptive stigma of another, while the target for a fruit is a place where a seedling can establish. Most pollen grains fail because they never find a target. Most seeds and seedlings die after they reach a potentially suitable place.

Strictly speaking, no animals are adapted for pollination or seed dispersal. Animals exploit flowers and fruits. They are adapted for feeding on pollen, nectar, seeds or fleshy portions of fruits or, in odd cases like orchid bees (Euglossinae), for collecting chemical fragrances used in courtship rituals. Some animals possess morphological, behavioural or physiological traits that help them to feed, and coincidently increase or decrease chances of effective pollination or seed dissemination. A discussion of pollination or seed dispersal from the plant perspective concerns plant reproductive processes. A relevant discussion of animal traits concerns foraging and its coincidental role in plant

Table 9.1 Contrasts between pollen and seed dispersal, from the plant perspective.

	Pollen dispersal	Seed dispersal
Target	Stigma of conspecific flower	Any site suitable for establishment
Animal motivation to target	Collect fragrances, nectar or pollen	None: seeds discarded when convenient
Advantage to visitor constancy	High: promotes correct transfer	Usually low

reproduction, rather than animal reproduction *per se* (Howe 1984, Crawley & Krebs 1992).

9.3.1 Flowers and pollinators

In the best of all plant worlds, an individual would disperse all of its pollen grains to receptive stigmas, and would capture genetically superior pollen on each of its own stigmas. A myriad adaptations in flower form and function constitute various evolutionary means of achieving this theoretical maximum. In the best of all animal worlds, a pollinator uses pollen or other floral rewards to further its own fitness, defined as its genetic contribution to future generations (Hamilton 1964). Its adaptations 'for pollination' are really adaptations for finding and eating either the male gametes (pollen) or collecting other bribes that the plant may offer.

The ultimate goal of angiosperm pollination is fertilization. Pollination occurs when: (i) a pollen grain lands on a conspecific stigma; (ii) the tube cell extends a pollen tube to the nucellus; (iii) one generative cell fertilizes the egg cell to form a diploid zygote; and (iv) the second generative cell unites with the polar nuclei to form a triploid endosperm nucleus. A mature fertilized ovule is a seed, consisting of a diploid zygote or embryo, a triploid endosperm invested with nutrients from the maternal parent and diploid cells derived from cells surrounding the original nucellus (see Raven *et al.* 1992). This fundamental process is often assisted by coloured petals, sepals, bracts, nectaries or other attractants that promote effective pollen transfer.

Through long-term evolutionary adjustments, pollinators have forged some order out of the vast array of potential adaptations for pollen transfer (Table 9.2). Suites of flower characters, such as colour, shape, size and reward, are broadly adapted to taxonomic associations of pollinating agents, forming pollination syndromes that illustrate convergent evolution of characters, often among unrelated plant taxa, to distinctive pollination environments. Similarly, pollinator characteristics match the challenges of distinctive arrays of flower characters.

Common pollination adaptations may be ancient or recent in origin. The first angiosperms, like contemporary *Magnolia* or nutmegs *Myristica*, were probably pollinated by beetles that ate pollen but nonetheless carried some pollen from one flower to another more reliably than wind or water (Kevan & Baker 1983; Irvine & Armstrong 1990). Even today, pollen-eating beetles show little modification for flower visitation beyond an occasional elongation of the thorax. Other common pollination adaptations are more recent. Explosive anthers in 'buzz-pollinated' *Solanum* spp. (Buchmann 1983) and fragrant but nectarless orchids like *Cypripedium* (Dressler 1981) epitomize specializations among very different plants for more recently evolved pollinators, such as bees. Bees make such specialization possible because they recognize and forage for distinctive flower shapes and colour patterns, and collect quantities of pollen on bristles or in basket-like corbiculae (Barth 1985). Flower-visiting vertebrates are also relatively recent in the fossil record. Conspicuous shaving brushes and tough, tubular bat-flowers found in the canopies of many tropical forests (e.g. *Pseudobombax* and *Ochroma* (both Bombacaccae); Fleming 1988) and red bird-flowers near the ground in the some of the same forests (*Heliconia*; Carpenter 1983; Dobkin 1984) probably evolved during the last tens of millions of years. Some plant

Table 9.2 Pollination syndromes.

Agent	Anthesis	Colour	Odour	Flower shape	Nectar
Insects the primary agents of pollination					
Beetles	Day and night	Usually dull	Fruity or aminoid	Flat or bowl-shaped; radial symmetry	Undistinguished if present
Carrion and dung flies	Day and night	Purple-brown or greenish	Decaying protein	Flat or deep; radial symmetry; often traps	If present, rich in amino acids
Syrphid and bee flies	Day and night	Variable	Variable	Moderately deep; usually radial symmetry	Hexose-rich
Bees	Day and night or diurnal	Variable but not pure red	Usually sweet	Flat to broad tube; bilateral or radial symmetry; may be closed	Sucrose-rich for long-tongued bees; hexose-rich for short-tongued bees
Hawkmoths	Crepuscular or nocturnal	White pale or green	Sweet	Deep, often with spur; usually radial symmetry	Ample and sucrose-rich
Settling moths	Day and night or diurnal	Variable but not pure red	Sweet	Flat or moderately deep; bilateral or radial symmetry	Sucrose-rich
Butterflies	Day and night or diurnal	Variable; pink very common	Sweet	Upright; radial symmetry; deep or with spur	Variable, often sucrose-rich
Vertebrates the primary agents of pollination					
Bats	Night	Drab, pale, often green	Musty	Flat 'shaving brush' or deep tube; radial symmetry; much pollen; often upright, hanging outside foliage, or borne on branch or trunk	Ample and hexose-rich
Birds	Day	Vivid, often red	None	Tubular, sometimes curved; radial or bilateral symmetry, robust corolla; often hanging	Ample and sucrose-rich
Primarily abiotic pollination					
Wind	Day or night	Drab, green	None	Small; sepals and petals absent or much reduced; large stigmata; much pollen; often catkins	None or vestigial
Water	Variable	Variable	None	Minute; sepals and petals absent or much reduced; entire male flower may be released	None

lineages include taxa that have faced and responded to a wide variety of selection pressures from pollinators. Grant and Grant (1965), for instance, found that different species of the genus *Phlox* (Polemoniaceae) were adapted to attract bees, flies, butterflies, moths, beetles, birds or even bats. A modern forest or meadow is a living museum of many threads of evolutionary descent.

Abiotic gamete dissemination is the ancestral condition in vascular plants (Sporne 1965). Flagellated gametes of primitive mosses and ferns swim through a film of water to reach the egg, a mechanism that does not survive in flowering plants. In contrast, water pollination in flowering plants involves a highly specialized loss of flower parts from insect-pollinated ancestors. In the waterweeds, *Vallisneria* (Hydrocharitaceae), for instance, the entire male flower is released into the water where pollen can float (not swim) to female flowers (Proctor & Yeo 1972). Wind pollination (anemophily) is common and perhaps ancestral in gymnosperms, but derived from animal-pollinated forms in angiosperms. Common wind-pollinated grasses and sedges have greatly reduced floral parts rather than a primitive architecture. Among dicots, familiar wind-pollinated trees and shrubs of temperate Europe and North America often have insect-pollinated relatives in the tropics.

Much of the exciting work in pollination biology explores situations in which pollination syndromes do not seem to fit expected patterns. Sometimes, exigency forces insects to visit the 'wrong' flowers. In other cases, however, flexibility is adaptive. Flowers of scarlet gilia (*Ipomopsis aggregata* (Polemoniaceae)) are pollinated by humming-birds early in the season and hawkmoths later, allowing the plant population to take advantage of seasonal changes in pollinator abundances (Paige & Whitham 1985). In a somewhat different example, white morphs of *Succisa pratensis* (Dipsacaceae) attract bees while purple morphs attract butterflies, thereby allowing a single plant population to avoid competition for the same, sometimes scarce, pollinator populations (Kay 1982). Each species appears to conform to at least two syndromes, but for quite different adaptive reasons.

Less exotic variation in flower or plant characteristics may reduce the predictiveness of pollination biology. For instance, ratios of sucrose to hexose sugar, amino acid content and lipid composition in nectar vary from individual to individual, flower to flower and often with time of day, though values cluster around modes associated with different syndromes (Simpson & Neff 1983). Similarly, flower number, which varies widely among individuals within a species, affects the attractiveness of a plant to pollinators. In giant *Lobelia* (Campanulaceae) of high African mountains, flower number influences the chance that pollinators move from one flower to another of the same individual plant or move on to other individuals (Burd 1994). In wild radishes *Raphanus sativus* (Brassicaceae), the success with which plants contribute genes to the next generation through pollen (paternal function) increases with flower production, but shows a diminishing marginal gain that varies from year to year due to variation in the size of a limited pollinator assemblage (Devlin *et al.* 1992). Inherent imprecision must be enormous in mass-flowering tropical trees, which bear so many flowers that tree size and proximity to neighbours rather than characteristics of individual flowers have overwhelming effects on pollination success (Frankie & Haber 1983). Such intrinsically variable consequences of flower characteristics do not fit the spirit of fixed syndromes (see Table 9.2). Labile syndromes have the added advantage that they may hedge against the failure of any particular pollinator, or they may simply represent the limits of adaptive specialization to a world that is uncertain in different ways and degrees.

Finally, we should remember that syndromes are often defined on the basis of the way that *we* perceive the flowers, not as animals see them. Birds see red but do not smell; bats smell but hardly see at all; many bees see ultraviolet light reflected by flower petals that is invisible to humans (Kevan 1983). Bees may visit flowers of the wrong apparent colour simply because we do not see the colours that actually attract them. Similarly, humans cannot experience the landscape of smells and radar images that guide a nectarivorous bat through the darkness of a tropical forest.

9.3.2 Fruits and frugivores

In the best of all possible plant worlds, an individual plant orchestrates the efficient dispersal of its embryos away from insect or pathogen infestations and competition, and provisions them with enough lipid, protein and starch to ensure a start in life. In the best of all frugivore worlds, an animal gorges itself on a balanced diet from fruits of a single plant, digesting fruit pulp but discarding inedible seeds that take up space in the gut.

Plants meet the challenges of dispersal and provisioning by relying on animals or physical forces to scatter seeds (Table 9.3). In general, animal characteristics facilitate digestion rather than carrying seeds, so the process is far from precise from the plant's vantage (Janzen 1983; Howe 1984; Herrera 1985, 1995). Reduced gizzards in fruit-eating birds and shortened guts in fruit-eating birds and mammals clearly speed seed passage rather than retain seeds for wide dispersal. Moreover, modifications for fruit eating usually lead to only modest specialization on different fruit types. African birds, ground mammals and monkeys, for instance, sort fruits of a forest by size and to some extent by fruit morphology, but much overlap in diet occurs between very different guilds of animals (Gautier-Hion *et al.* 1985). Perhaps the most potent mechanisms for fruit specialization will turn out to be physiological (Karasov & Levey 1990; Levey & Grajal 1991). Differences in digestive enzymes and absorption in the intestine may allow some frugivorous birds and mammals to eat foods that competitors cannot.

In the real world, dispersal and provisioning conflict (Howe 1984,

1993a,b). Heavy wind-dispersed seeds fall closer to parents than light ones (Augspurger 1986), and within some tree populations fleshy fruits containing large seeds are less likely to be eaten and carried away by birds than fruits with lighter seeds (Howe & Vande Kerckhove 1980, 1981). Large seed size should therefore be a disadvantage because undispersed seeds do not find vacant sites or because insects, rodents, competition or disease kills higher proportions of seeds or seedlings in high densities near parents than further away (Clark & Clark 1984; Schupp 1992; Howe 1993b). Yet the larger the seed, the more vigorous

Table 9.3 Dispersal syndromes.

Agent	Colour	Odour	Form	Reward
Primarily self-dispersed				
Gravity	Various	None	Undistinguished	None
Explosive dehiscence	Various	None	Explosive capsules or pods	None
Bristle contraction	Various	None	Hygroscopic bristles in varying humidity	None
Primarily abiotic dispersal				
Water	Various, usually green or brown	None	Hairs, slime, small size, or corky tissue resists sinking or imparts low specific gravity	None
Wind	Various, usually green or brown	None	Minute size, wings, plumes, or balloons impart high surface to volume ratio	None
Primarily vertebrate dispersal				
Hoarding mammals	Brown	Weak or aromatic	Tough thick-walled nuts; indehiscent	Seed itself
Hoarding birds	Green or brown	None	Rounded wingless seeds or nuts	Seed itself
Arboreal frugivorous mammals	Brown, green, white, orange, yellow	Aromatic	Often arillate seeds or drupes; often compound; often dehiscent	Aril or pulp rich in protein, sugar, or starch
Bats	Green, white, or pale yellow	Aromatic or musty	Various; often pendant	Pulp rich in lipid or starch
Terrestrial frugivorous mammals	Often green or brown	None	Tough, indehiscent often > 50 mm long	Pulp rich in lipid
Highly frugivorous birds	Black, blue, red, green or purple	None	Large arillate seeds or drupes; often dehiscent; seeds > 10 mm long	Pulp rich in lipid or protein
Any frugivorous birds	Black, blue, red, orange or white	None	Small or medium-sized arillate seeds, berries or drupes; seeds < 10 mm long	Various; often only sugar or starch
Animal fur or feathers	Undistinguished	None	Barbs, hooks, or sticky hairs	None
Primarily insect dispersal				
Ants	Undistinguished	None to humans	Elaiosome attached to seed coat	Oil of starch body with chemical attractant

the seedling. There will be no such thing as the 'perfect seed' from the vantage of *both* seed dispersal and seedling vigour, simply because a seed cannot be both large and small at the same time.

Most fruits are adapted for dispersal by animals (Howe 1986). Regular seed dispersal by ants is common among understorey forest herbs and in deserts (Davidson & Morton 1981; Handel *et al.* 1981), but most seed dispersal by animals is by vertebrates that eat fruits and pass or regurgitate seeds intact, hide seeds in caches or passively carry seeds stuck to their fur or feathers. Dispersal agents include odd cases of fish (Goulding 1980) and frogs or lizards (Da Silva *et al.* 1989; Fialho 1990), but birds and mammals are the most common vertebrate dispersal agents.

Seeds regularly hoarded by rodents or birds often have no obvious adaptations for dispersal other than a thickened seed coat and a rounded form (Vander Wall 1990). For instance, pinon pines *Pinus edulis* in the western USA lure seed-eating jays *Gymnorhinus cyanocephalus* and nut-crackers *Nucifraga columbiana* to open cones that display large, round, wingless seeds. Seeds are carried up to 20 km by the birds, buried in small caches and used as food reserves. Those that are forgotten in the ground, or whose owner dies, may survive. Some tropical nuts have a pulp that partially satiates a rodent, and perhaps encourages burial rather than consumption (Bradford & Smith 1977; Tang 1989). Such fruits may be adapted to huge frugivorous mammals long extinct (Janzen & Martin 1982; cf. Howe 1985) or they may represent an offering by which plants bribe a seed-eating animal into using something other than the seed itself, much as flowers offer nectar in addition to pollen.

Most spectacular are fruits adapted for consumption by frugivorous birds and mammals (van der Pijl 1972). Beyond the familiar blue, black and red berries of temperate shrubs, tropical fruits include many brilliant and often multicoloured forms (Snow 1981; Wheelwright & Janson 1985). Some like *Swartzia prouacensis* (Fabaceae) even dangle on long threads as an enticement to fruit-eating bats. Whatever the particular evolutionary route for such elaborate adaptations, their existence highlights the importance of dispersal to plants.

As with abiotic pollination, abiotic dispersal is a derived condition in angiosperms (Burrows 1986; Murray 1986). Water is probably an ancient mode of seed dispersal, but the corky husk, watertight shell and massive endosperm of ocean-going coconuts *Cocos nucifera* (Arecaceae) show contemporary specialization for long exposure to salt water. Similarly, winged samaras are secondarily derived in both ancient (e.g. Magnoliaceae) and much more recent (e.g. Fabaceae) families. Nearly universal wind dispersal in the recent and rapidly evolving Asteraceae is accomplished with highly specialized plumed seeds (e.g. the familiar dandelion 'clock' of *Taraxacum officinale*). Abiotic dispersal has evolved many times in the plant kingdom.

The classification of dispersal syndromes suffers from many of the same shortcomings as pollination syndromes. Plants are more creative of structure and ornament than biologists are of categories, and much of the time frugivores eat what they please. This is not surprising because specializations for eating fruits do not greatly restrict the *kinds* of fruits that can be eaten. Frugivorous birds often have wide gapes, short intestines and lack a grinding gizzard (Moermond & Denslow 1985; Wheelwright 1985). Frugivorous monkeys also have shorter guts than leaf-eating relatives (Hladik 1967). Such adaptations for processing large quantities of fruits reduce the time that seeds are in the gut, and therefore increase the chances that seeds will be discarded unharmed.

Recently, physiologists have suggested that taxonomic differences in sugar-digesting enzymes could restrict diet choice (Martinez del Rio *et al.* 1992; Martinez del Rio & Restrepo 1993). For instance, waxwings produce the digestive enzyme sucrase and easily eat fruits high in sucrose, but starlings do not have the enzyme and cannot digest such fruits easily. Physiological constraints limit frugivores to fruits with similar digestive properties, whether those plants are related or not.

As with pollination syndromes, fruit dispersal syndromes offer insight into the history of selection on a plant taxon. For instance, most Lauraceae produce unprotected green, blue or purple-black drupes, which are eaten by large birds; the cosmopolitan genus *Beilschmiedia* (Lauraceae) attracts toucans in tropical America, huge flightless cassowaries in Australia and paradise birds in New Guinea (Crome 1975; Beehler 1983; Stocker & Irvine 1983). A more diversified evolutionary history is illustrated by the Fabaceae whose 12 000 species show fruit adaptations for consumption and dispersal by birds, bats, fish, ungulates or dissemination by wind, water, adhesion to fur or feathers, ballistics or simple gravity (van der Pijl 1972). Syndromes are not always accurate predictors of particular ecological interactions (Schupp 1993), but they do suggest long-term conditions under which plant taxa evolved and to which animals adjust.

Finally, syndromes may give clues to the ecology of a community. For instance, most fruits in the rain forest of Barro Colorado Island, Panama, appear to be adapted for dispersal by birds, bats or hoarding rodents (Leigh 1982). In a climatically comparable forest in Amazonian Peru where there are 11 species of primates, the birds, bats and rodents play lesser roles and a disproportionate share of fruits are yellow 'monkey fruits' (Janson 1983; Terborgh 1986). Syndromes are not precise enough for fail-safe inferences about interactions between *particular* animals and plants, but the representation of syndromes in a community does offer a general insight into the way in which a forest works.

9.3.3 Coevolution or co-occurrence?

Coevolution may in theory represent reciprocal adjustments of one species to another, or general adjustments of higher plant taxa to animal taxa and vice versa (J.N. Thompson 1994). The first would result in species by species coevolution, the second a more diffuse reciprocal evolution of higher taxa. Alternatively, relationships between animals and plants may reflect transitory ecological circumstance rather than a history of evolutionary adjustment.

Species by species coevolution occurs in figs and fig wasps. Each of 700 *Ficus* spp. (Moraceae) attracts a unique species of pollinating wasp that raises its brood on fig ovules within the flowering structure (syconium), and carries pollen to other conspecific fig plants (Wiebes 1979; Bronstein & McKey 1989). The mutualism is basically symbiotic; wasp and fig live in close proximity, with each entirely dependent for reproduction on its partner (see Box 6.5). Such specificity, however, is very unusual.

Reciprocal, symmetrical coevolution is rare in non-symbiotic flower–pollinator or fruit–disperser systems where each animal has a choice of foods and each plant is exploited by a range of pollinators or dispersal agents (Howe 1984; Thompson 1989). Coevolution of fruiting plants and dispersal agents, when it does occur, is often asymmetrical. In an examination of phylogenetic records, for instance, Herrera (1985) found that angiosperm taxa survived about 30 times longer than did their probable animal dispersal agents. This suggests that dispersal agents are more likely to adjust to existing fruit environments than plants are to the disperser environment.

9.4 Reproductive imperatives of success and failure

A cardinal lesson from field biology is that the reproductive efforts of plants almost always fail. If an oak and each of its 100 young were to produce 100 reproductive offspring, we would be blessed with 10^{19} oaks in 10 generations (there are only 10^{22} stars in the known universe)! Given that an oak may produce hundreds of thousands of acorns in a lifetime, and that many trees are a great deal more fecund than oaks, it is obvious that almost all pollen grains and ovules are doomed to failure. Plant breeding and dispersal systems circumvent this almost certain failure through evolutionary strategies that promote fitness through both pollen (male function) and ovules (female function).

9.4.1 Pollen success and failure

A key stage in the sexual reproduction of a plant is successful fertilization through pollination. A pollen grain may fail to disperse to stigmas, may fail in competition with other pollen grains after reaching a stigma, may be rejected by the maternal plant, or may produce a fertilized ovule that exhibits genetically inferior characteristics (Willson & Burley 1983). Each aspect of failure or success has ecological and evolutionary causes and consequences.

Some pollen fails to leave the male parent; this happens when pollinator activity is low or when wet or calm weather precludes wind pollination. Most pollen ends up near the parent, with a long tail of low and diminishing pollen density away from the parent (Levin & Kerster 1974). Pollen carried by wind usually settles on foliage of neighbouring plants; the chance that a pollen grain will find an appropriate stigma declines dramatically with distance, especially for widely dispersed plants or plants in a heterogeneous habitat (Regal 1977). For pollen carried by animals, the chance of ending up on the right stigma depends on the behaviour of the pollinator and on the stickiness of pollen on animals that visit flowers (Lertzman & Gass 1983). Pollen grains that are easily brushed off an insect will be left on neighbouring flowers, while those that stick tight may be carried to stigmas much further away. The longer a pollen grain stays on a pollinator, the more likely it is to be deposited on a stigma of a plant other than its parent (a benefit), but equally the more likely it is to be delivered to the wrong flower species, brushed off on foliage or eaten by a bee, beetle or bat (a cost).

Plant breeding systems give an indication of how many pollen grains, on average, it takes for pollen to fertilize available ovules of a species (Cruden 1977). Wind-pollinated species may produce 1 000 000 pollen grains for each ovule, while obligately self-fertilized species may produce as few as three pollen grains per ovule. Species that differ in opportunities for outcrossing produce predictably different modes in the number of pollen grains for each ovule (Table 9.4) Natural selection also favours variation of pollen–ovule ratios within species as conditions favouring pollen donation versus selfing occur (Bertin 1988; Ashman & Stanton 1991). For example, populations of the tropical bean *Caesalpinia pulcherrima* (Fabaceae) in Mexican highlands, where butterflies are scarce, have 88% bisexual flowers and 12% male, while the same species in lowlands, where pollinators are common, have 89% male flowers and only 11% are bisexual (Cruden 1976). Evidently an abundance of butterflies makes maleness profitable.

Once the pollen grain reaches a receptive stigma, it must germinate and grow to the ovule. The race to the ovary can be competitive. Mulcahy *et al.* (1983) found that it takes over an hour for a pollen tube to reach a *Geranium maculatum* ovary, and that this places a premium on speed of pollen germination and order of deposition. If visits are few and far between, the first pollen grains deposited have an enormous

Table 9.4 Pollen–ovule ratios of plants with different breeding systems. (From Cruden 1977.)

Breeding system	Number of species	Pollen–ovule ratio (mean ± SE)
Cleistogamy (flower does not open)	6	5 ± 1
Obligate selfing (flower open)	7	28 ± 3
Facultative selfing	20	168 ± 22
Facultative outcrossing	38	797 ± 88
Obligate outcrossing	25	5858 ± 936

advantage over those that arrive later. As the rate of deposition increases with increasing pollinator activity, so the advantages of early deposition diminish and differences in pollen vigour become important.

Causes of variation in male fitness are often difficult to assess from field observations, but they can be teased out with experimental manipulation (Schlessman 1988). In the radish *Raphanus raphanistrum* (Brassicaceae), a population may contain individuals with genotypes for both yellow and white flowers, but butterflies and bees prefer the yellow morphs. The genetic marker for yellow petal colour indicates that yellow plants are far more successful *as fathers* in donating pollen, even though yellow and white morphs have equivalent 100% success *as mothers* (Fig. 9.1). Maureen Stanton and her colleagues (1991) have since used electrophoretic analysis of more numerous genetic markers in

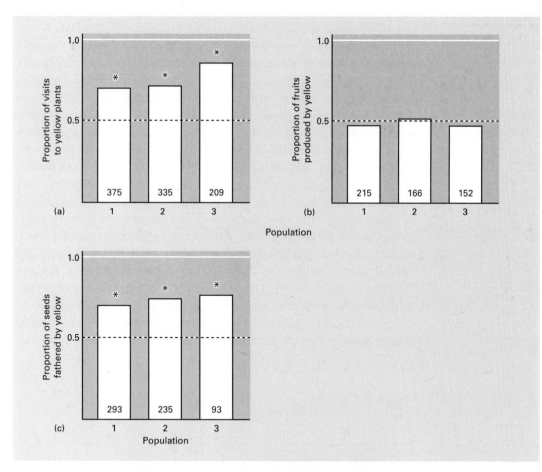

Fig. 9.1 Performance of yellow and white morphs of the radish (*Raphanus raphanistrum*). Pollinators favour yellow morphs. All flowers produce seeds, so neither yellow nor white flowers are pollen-limited (a). Pollinator preference for yellow does influence the extent to which yellow-flowered plants donate pollen (b). Stars indicate significantly higher proportions of seeds fathered by yellow compared with white flowers. (From Stanton *et al.* 1986.)

the related *Raphanus sativus* to demonstrate how different pollinators can influence whether male or female function has the advantage in sexual competition.

Pollen efficiency does not stop with seed set; the resulting seed must be viable and competitive. In an American pink, *Sidalcea oregana* subsp. *spicata* (Malvaceae), pollen source makes a difference to the size and flower production of resulting offspring (Fig. 9.2; Ashman 1992). Plants derived from outcrossing were clearly larger and bore more flowers than those from sib matings or selfed flowers; likewise, sib crosses produced larger offspring than self-fertilizations. In another study, large pollen loads in the cucumber *Cucurbita pepo* enhanced seedling vigour, even though small loads fertilized all the ovules (Winsor *et al.* 1987). A follow-up genetic study succeeded in showing a small but significant heritable effect of large pollen loads for some traits in this cucumber (Schlichting *et al.* 1990). In *Sidalcea* and *Cucurbita*, some genetic contributions are clearly more advantageous than others.

Breeding systems exist because they maximize reproduction of individuals with different allocations to maternal or paternal function under quite different circumstances. At least in theory, inbreeding increases fitness of parents in stable environments and decreases it in changing or unpredictable environments (Williams 1975). Recent evidence partly supports this assertion in the cucumber *Ecballium elaterium*, which exists in dioecious or monoecious populations around the Mediterranean. One expects and finds greater allelic diversity in dioecious than in monoecious populations (Costich & Meagher 1992). Further, one expects and finds that the dioecious populations naturally predominate in 'unpredictable' dry sites, the monoecious populations in more mesic sites (Costich & Galan 1988). An intriguing discovery is that transplanted dioecious plants outperform monoecious plants in xeric and mesic habitats, raising the question of why all populations are not dioecious (Costich 1995). The answer may lie in a physiology of gender differentiation of the two sexes in dry sites that favours transmission of alleles for dioecy. Alternatively, high seedling survival may favour female function in all mesic sites, including mesic patches in dry

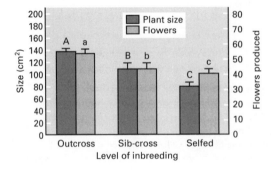

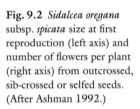

Fig. 9.2 *Sidalcea oregana* subsp. *spicata* size at first reproduction (left axis) and number of flowers per plant (right axis) from outcrossed, sib-crossed or selfed seeds. (After Ashman 1992.)

habitats. Such a demography would favour gender differentiation in dry places, monoecy in mesic sites.

Finally, sexual reproduction through pollination and seed dispersal exists in the context of plant life cycles that include other means of procreation. In many or even most plants that live more than 1 year, vegetative reproduction through runners, rhizomes, bulblets or tubers generate asexual stems (ramets) of the same genetic individual (genet) (Harper 1977). Trade-offs between sexual and asexual reproduction are sometimes straightforward. For instance, the Jerusalem artichoke (*Helianthus tuberosus* (Asteraceae)) is a 'pseudoannual' in which the above-ground shoot dies every autumn, but leaves several to several hundred genetically identical underground offspring as tubers to sprout the next spring. Westley (1993) has shown that individuals from which flowers are removed invest more in asexual tubers than individuals allowed to bear flowers and seeds. Some clones reallocate resources diverted from sexual reproduction to large tubers, while other clones reallocate to a profusion of small tubers. The ability to reallocate resources from sexual to asexual reproduction is clearly adaptive in a world in which herbivores and weather often frustrate sexual reproduction. Inherited ability to reallocate to large or numerous asexual offspring probably reflects as yet unexamined conditions experienced by different populations. For instance, populations experiencing predictably stiff competition from other plants may invest in highly competitive ramets from large tubers, while those more likely to experience high random mortality may invest in numerous small tubers.

9.4.2 Fertilized, unfertilized and aborted ovules

Ovules often fail to yield seeds (Young & Young 1992). Fertilized ovules may fail to develop because: (i) herbivores eat them; (ii) damaged fruits are aborted; or (iii) limited resources cause fruit abortion. If pollinators are limiting or wind-pollinated plants are widely dispersed, some ovules may never be fertilized.

Selective abortion of fruits may optimize the number of healthy seeds produced, or may not occur if pollinators are scarce. A maternal plant often has the theoretical opportunity to select among pollen sources or balance number of healthy seeds against nutrient or light limitations (Stephenson 1981). On the other hand, selective abortion is unlikely if most ovules are not fertilized. For instance, the American aroid jack-in-the-pulpit *Arisaema triphyllum* is pollinated by fungus gnats, which are often in scarce supply. Bierzychudek (1982) found that in naturally pollinated aroids seed number bore no relation to plant size, yet in hand pollinations seed number was strongly correlated with plant size. In this system, seed production appeared to be pollen-limited.

An ideal study of pollination and seed dispersal might integrate ecological factors that influence fertilization, fruit maturation, dispersal and establishment. An ongoing study of a large tropical herb in Mexico, *Calathea ovadensis* (Marantaceae), provides a case history and a promising analytical framework.

Calathea is visited by several pollinators. Schemske and Horvitz (1984, 1988) have shown that butterflies and moths account for most of its pollinator visits, but bees account for almost all of the successfully tripped flower mechanisms. Two bee species, *Bombus medius* and *Rhathymus* sp., are far more efficient at transferring pollen than other insects, but they are so uncommon (i.e. have so few visits) that they are much less important to the plant than the abundant but less efficient bee *Euglossa heterosticta*. The picture is complicated by the fact that a moth larva *Eurybia elvina* secretes food for ants that would otherwise protect flowers, while the larva eats flowers and fruits. Ants can increase fruit production, but the presence of *Eurybia* decreases fruit production whether or not ants are present.

A path analysis (Fig. 9.3) shows the effect of each factor on flower production, number of initiated fruits and the number of mature fruits. In the year shown, *Euglossa* and *Rhathymus* bees have strong positive effects on the numbers of fruits initiated, *Exaerete* has a weak positive effect and a *Eulaema* bee species has a negative effect. Much of the variation remains unexplained (U_n), and year-to-year variations undoubtedly would alter the coefficients for each pollinator. In a year other than the one illustrated, for instance, *Rhathymus* was absent.

This path analysis of *Calathea* offers important insights. It suggests

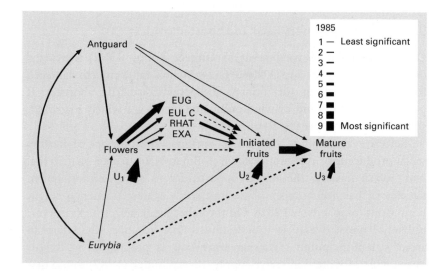

Fig. 9.3 Path diagram for contribution of different pollinators, antguards and *Eurybia* on fruit production of *Calathea ovadensis* during 1 year. Widths of solid arrows indicate relative importance of significant effects; broken lines indicate non-significant effects. Unmeasured factors (U_1, U_2, U_3) are indicated. EUG, *Euglossa*; EUL C, *Eulaema cingulata*; RHAT, *Rhathymus* sp.; EXA, *Exaerete smaragdina*. (From Schemske & Horvitz 1988.)

that pollinator limitation exists in *Calathea* despite heavy visitation, and
shows that the degree of pollinator limitation changes with the composition of the pollinator assemblage. If existence of different efficiencies
of pollinators is a precondition for coevolution, path analyses from
different years quantify ecological constraints on coevolutionary opportunities and show how those constraints change from year to year.

9.4.3 Dispersed and undispersed seeds

Seed dispersal is a reproductive bottleneck for many plant species. For
reproduction to result in recruitment of the next generation, seeds must
find uninhabited sites suitable for growth; this usually requires dispersal
away from the immediate vicinity of the parent plant, where competition is intense and the risk of death from pathogens or insect herbivores
is disproportionately high (Howe & Smallwood 1982) (Fig. 9.4). For
wind-dispersed species, seed-dispersal distance is determined by terminal
velocity, which in turn is modulated by height of the plant, mass of the
seed, surface area presented as a plume or flat wing, and shape (Burrows
1986; Augspurger 1986, 1989). Where neighbouring trees obstruct
wind dispersal, seeds of the tropical 30-m tree *Platypodium elegans*
(Fabaceae) may travel no further than those of the 1-m milkweeds
Asclepias or dogbanes *Apocynum* (Fig. 9.4; Augspurger 1983). In open

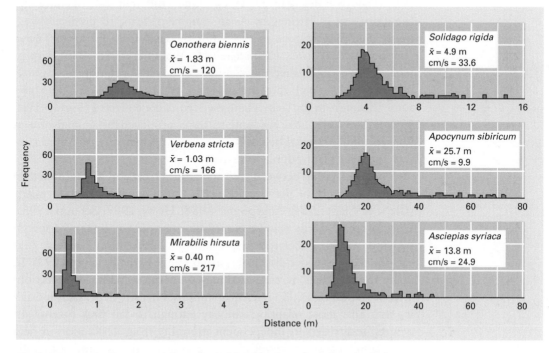

Fig. 9.4 Propagule dispersion of six species in 10–15 km hour⁻¹ winds at the Caylor
Prairie in Iowa. Mean dispersal distances and terminal velocities are given for each
species. (After Platt & Weis 1977.)

habitats at the edge of landslides, some *Platypodium* seeds might travel much further but most seeds still land close to the parent.

Animal hoarding patterns, foraging behaviour and breeding habits can all determine patterns of seed dispersal (Fogden 1972; Levey 1986; Sorensen 1986; Vander Wall 1990), and digestive systems of different animal species may or may not enhance germination (Lieberman *et al.* 1979). Sometimes, as in the central American solitaire *Myadestes melanops*, extension of the time that seeds are in the gut by only a few minutes may decrease seed viability (Murray *et al.* 1994). In other cases, animal guts have no effect on germination.

Animal behaviour is usually important because it results in either scattered or clumped seed distributions, which influence seedling distributions in a variety of ways (Howe 1989; Masaki *et al.* 1994). Animals may leave seeds in distributions not unlike those produced by wind dispersal. *Sciurus* squirrels sometimes bury seeds 100 m from walnut trees but they leave most nuts within 20–40 m (Stapanian & Smith 1978), resulting in seed distributions not notably different in shape or maximum likely dispersal distance from a milkweed or a *Platypodium*. Other instances of animal dispersal of seeds may lead to very different patterns of seedling emergence from those that wind could produce. For instance, a North American chipmunk *Tamias amoenus*, a small squirrel that buries several to many seeds in a cache, is responsible for clumps of bitterbrush *Purshia tridentata* (Rosaceae) seedlings of 2–102 individuals (Vander Wall 1994). On quite a different scale of clumping, fastidious use of latrines by Asian rhinoceros leads to predictably extreme clumping of tens of thousands of seeds from favourite *Trewia* (Euphorbiaceae) fruits (Dinnerstein & Wemmer 1988). Not surprisingly, *Trewia* saplings recruit in clumps of dozens or hundreds. Animals may carry seeds much further than wind. Jays that hoard pine nuts bury some 20 km from the trees where the seeds were collected (Vander Wall 1990). Thus seed and seedling patterns produced by animals are as idiosyncratic as animal behaviour itself, with potentially profound consequences for plant population structure.

The relationship between seed and adult dispersion is determined by successive events in the lives of seeds, seedlings and juvenile plants (Clark & Clark 1984; Augspurger 1988; Howe 1989; Schupp 1993). Once dropped, seeds may be moved during a second bout of dispersal by wind, water, ants or rodents, resulting in either mortality (e.g. a rat eats the seed) or wider dispersal (Janzen 1982; Byrne & Levey 1993). In many tropical trees, intense seedling competition, pathogen infection, seed predation or seedling herbivory imposes density-dependent mortality that kills most offspring unlucky enough to fall near parents. In deserts, severe density-independent mortality from desiccation or heat may reduce density without changing eventual distributions of adults (i.e. germinating seeds are far enough apart that the adults produced do not compete; Ellner & Shmida 1981). Similar processes in uneven habitats may relegate recruits to protected sites. For instance, saguaro cactus *Carnegia* (Cactaceae) seeds and seedlings in the Sonoran desert of North America suffer mortality from density-independent heat and desiccation in exposed sun, leaving survivors in rocky shaded refuges or under cover of nurse plants (Turner *et al.* 1966; Steenbergh & Lowe 1969, 1977). Adult plant dispersion is influenced by seed dispersion, but because of many intervening events adult location may not reflect initial seed distributions.

In many plants, both density-dependent and density-independent mortality mediate the transition from seed to adult distributions. For the toucan-dispersed neotropical nutmeg *Virola nobilis* (Myrsiticaceae) (formerly *V. surinamensis*), disproportionately high seed and early seedling mortality from weevil infestations kills virtually all offspring near parents, leading to populations that are loosely clumped at low densities (Fig. 9.5). This density-dependent mortality lowers density and increases dispersion of ensuing adults. Seed, seedling and yearling mortality caused by other herbivores is at least as heavy, but is random with regard to distance from parent trees and affects only population density not dispersion (Howe 1993b; see Terborgh *et al.* 1993 for other examples). While density-independent rodent herbivory does not affect dispersion except to reduce density, density-independent drought mortality increases plant aggregation by favouring groups of recruits in wet ravines or near treefall gaps where the young plants receive enough light to grow extensive root systems before the dry season begins (Howe 1990).

Such data on seed and seedling mortality show how different dispersal agents play different roles for *Virola nobilis* (Howe 1993b). On Barro Colorado Island, an average of half of each crop of 300 to as many

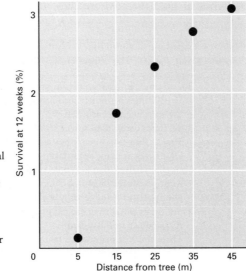

Fig. 9.5 Seed and seedling survival 3 months after fruit drop in the toucan-dispersed tree *Virola nobilis* in Panama. A strong advantage to dispersal occurs because *Conotrachelus* weevils lay eggs on and kill a disproportionate number near parent trees. (Data from Howe 1993b.)

as 30 000 fruits per tree are taken by six species of birds, by a frugivorous monkey *Ateles geoffroyi* that knocks down more than it eats, and in productive years by the nocturnal mammal *Potos flavus*, which eats fruits left by the more common diurnal visitors. The assemblage changes little from year to year despite a fourfold annual variation in median fruit crops. Birds take away > 90% of all fruits removed by animals. The toucan *Ramphastos swainsonii* eats 35% of those taken; over half of these are regurgitated more than 40–50 m from parent trees. Other birds take fewer seeds or drop them near the parent tree or both, making the toucan far more efficient from the plant's perspective than other potentially important dispersal agents (Table 9.5). If both toucans and guans were hunted out of a forest, *Virola* recruitment would decline precipitously and the tree population might drift to local extinction.

There is every reason to expect that the interplay of variation in biotic and abiotic factors will strongly influence dispersal processes and their consequences. A recently demonstrated positive effect of dispersal, and of chance encounters with light gaps for the *Calathea* system discussed earlier, again illustrates that biotic and abiotic factors may interact in important ways (Horvitz & Schemske 1994). For some colonists of forest clearings, like pin cherry *Prunus pensylvanica* (Rosaceae) in the eastern USA, seed germination may occur when forests are cut 50 years or more after the seeds were shed (Marks 1974); dormant seeds wait for forest openings to come to them. Other species wait for openings as robust shade-tolerant seedlings rather than seeds (Martinez-Ramos & Soto-Castro 1993). Nearly 80% of the large, mammal-scattered seeds of *Gustavia superba* (Lecythidaceae) survive seed and seedling stages through the first 6 months of life in the same Panamanian forest where *Virola nobilis* seed and seedling mortality is 2000 times higher (Sork 1985, 1987). Some species benefit from being moved to special sites. For instance, birds disperse mistletoe seeds to branches of host trees where they must be deposited to survive (Reid

Table 9.5 Chestnut-mandibled toucans as dispersal agents of *Virola nobilis* on Barro Colorado Island, Panama. (Data from Howe 1993b.)

30–60 times better than rufous motmots and slaty-tailed trogons
15–30 times better per seed handled
Twice as many seeds taken
8–16 times better than spider monkeys
Two to four times better per seed handled
Four times as many seeds taken
Four times better than keel-billed toucans
As good per seed handled
Four times as many seeds taken
Three times better than crested guans
As good per seed handled
Three times as many seeds taken

1986, 1989). Nevertheless, the seedlings still suffer serious density-dependent and density-independent mortality (Davidar 1983; Larson 1991). These examples highlight the importance of seed dispersal for a few species of plants. They only begin to suggest the variety of ecological factors affecting the success or failure of seed dispersal.

9.5 Adjusting to physical and biological reality

9.5.1 Physical environment

Physical realities restrict plant options. Cold or dry weather may preclude flowering or fruiting directly, or influence availability of pollinators or dispersal agents. Plant reproduction is generally timed to make best use of seasonally available insects, humming-birds or dispersal agents, or to take advantage of conditions when wind pollination or dispersal is likely to be most effective. In temperate forests of eastern North America, for example, the vast majority of tree species are wind pollinated (Regal 1982). They usually flower before the spring leaf flush, which would interfere with brisk air currents, and before large populations of insect pollinators emerge. Insects and birds pollinate the later-flowering herbs and shrubs closer to the ground, where it is too sheltered for effective wind pollination. Trees that flower later, when insects are more abundant and air-flow more obstructed, tend to be animal pollinated and dispersed.

Climate also constrains pollination in the tropics (Regal 1982). Most lowland tropical trees and shrubs are pollinated by insects, birds or bats. Flowering peaks often coincide with the dry seasons that favour insect foraging. In montane tropical forests, cool temperatures mean that small insects are less effective pollinators, and nectar-eating birds, large-bodied bees and large hawkmoths that are capable of thermoregulation are the principal pollinators.

Seasonality also influences seed success. In cold, arid and seasonally dry climates, seeds are often dormant for several months in the tropics, or one to many years in temperate habitats, until conditions are right for germination. Climate sets the conditions for both fruit production and seed and seedling demography.

Seasonality influences seeding ecology, even in tropical forests. In the seasonal forest of Barro Colorado Island, an 8-month wet season (with 2300 mm of rain) is followed by a 4-month dry season with little rain (only 100 mm). The fruits of most tree species are shed either at the end of the wet season or at the start of the next wet season (Fig. 9.6). Garwood (1983) found that most seeds germinate at the beginning of the wet season directly after they are shed or after brief dormancy at the end of the dry season. Both tactics allow seedlings to establish with minimal drought stress before they suffer severe desiccation during the next dry season.

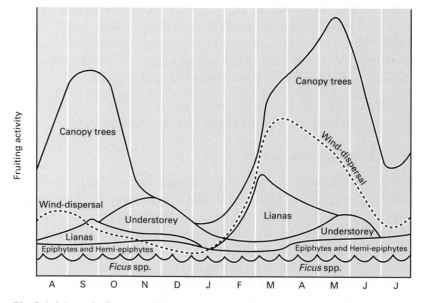

Fig. 9.6 Schematic diagram of the seasonality of fruiting activity on Barro Colorado Island, Panama, showing the proportion of the forest vegetation in fruit. (After Foster 1982.)

9.5.2 Adjusting to neighbours

Climatic realities of seasonal cold or drought force periods of intense interaction among species in a community. During each period of flowering or fruiting, members of each plant species may compete with or facilitate the reproduction of others of their own species, or other species that might share animal mutualists or enemies.

An example of forced flowering synchrony is the shrub *Hybanthus prunifolius* (Violaceae), which bursts into flower in the understorey of Barro Colorado Island within a week of a rare dry season rain (Augspurger 1980, 1981). The primary pollinator, the solitary bee *Melipona interrupta*, is attracted away from other flower sources by huge displays of *Hybanthus*. Augspurger found that scattered individual shrubs that were stimulated to flower early by experimental watering, while the bulk of the plants were still waiting for rain, set much less seed (58%, compared with 86% for plants flowering in synchrony with the rest of the population). Moreover, fruits that did develop on artificially watered plants experienced 11% predispersal mortality from caterpillars of flower-eating microlepidopterans, while fruits on plants developing in synchrony with others experienced 5% mortality from caterpillars. Evidently, the pollinator did not pay attention to infrequent early stragglers, whereas flower-eating caterpillars were attracted to early flowering plants but were satiated by mass flowering of the synchronized plants.

In other species, flowering asynchrony may be more advantageous than synchrony (Bawa 1983; Frankie & Haber 1983). Massive flower or fruit displays of huge canopy trees in tropical forests so tax the pollinator and disperser assemblages that asynchrony is the only way of ensuring visits by the appropriate animals. For instance, when *Virola nobilis* trees reach peak fruit production at the same time as neighbours, dispersal agents are satiated and the percentage of seeds removed from individual trees declines (Howe 1983; Manasse & Howe 1983). In this case the disadvantage of synchrony does not outweigh the advantage of shedding fruits in time for rapid germination and seedling growth before the dry season sets in. On the other hand, in North America and Europe shrubs and vines ripen berries or drupes late in the growing season and retain them well into the winter (Thompson & Willson 1979; Stiles 1980; Jordano & Herrera 1981; Herrera 1995), thereby extending dispersal several months after fruits ripen. In these cases asynchrony enhances the movement of seeds, all of which wait until spring or later to germinate.

Definitive tests of competition between plants for the services of animal mutualists involve manipulative experiments. Demonstration that two plant species compete for pollinators may be as straightforward as in the case of *Delphinium nelsonii* (Ranunculaceae) and *Ipomopsis aggregata* (Polemoniaceae) in the Rocky Mountains of North America, where exclusion of humming-birds from one of the species led to increased seed set in the other (Waser 1978). Alternatively, competition for pollinators may be subtle, as in the case of the spring flower *Claytonia virginica* (Portulacaceae). Competing chickweed *Stellaria pubera* (Caryophyllaceae) flowers draw pollinators away and capture *Claytonia* pollen, thereby reducing the effective population size (Campbell 1985). Under competition from *Stellaria*, far fewer *Claytonia* actually breed than the same number of individuals would without *Stellaria* present.

9.6 Conclusions

A remarkable feature of the ecology of pollination and seed dispersal is how much remains to be explored, despite rapid progress in understanding particular systems and processes. In many parts of the world, basic natural history must be done before processes can be understood. Yet where basic information is available, experimental manipulation or sophisticated analytical techniques such as path analysis can tease apart ecological causes and effects far more reliably than was possible even a few years ago. The ecology of pollination and seed dispersal offer unparalleled opportunities for students of nature with any number of aptitudes and interests.

10: Plant Chemistry and Herbivory, or Why the World is Green

Susan E. Hartley and Clive G. Jones

10.1 Why is the world green?

Herbivores consume, on average, 10–20% of the annual net primary production in terrestrial ecosystems (see Chapter 13). Yet many herbivores, particularly insects, have high reproductive potential and can periodically reach high densities, causing defoliation, reduced plant growth and reproduction, and increased plant mortality (Strong *et al.* 1984) particularly in agricultural and forestry monocultures (Barbosa & Schultz 1987). Outbreaks are, however, generally rare and unusual events in natural ecosystems. The terrestrial world is 'green' (Hairston *et al.* 1960) and tends to stay that way.

Why is the terrestrial world green? It does not have to be that way. The aquatic world is not green: in many aquatic ecosystems, pelagic and benthic herbivores remove 80% or more of the available primary production (Hay & Fenical 1988; Cyr & Pace 1993; Falkowski 1995; Fig. 10.1). In fact, we might say that in aquatic systems there are outbreaks of plants rather than herbivores!

Understanding why terrestrial plants remain largely uneaten in the face of herbivores, from both an ecological and evolutionary perspective, is arguably the fundamental question in plant–herbivore interactions. However, the answer can also tell us a lot about general ecological processes, such as resource–consumer interactions, tritrophic interactions and evolution. Given that plants, insect herbivores and their natural enemies comprise most (80%) of the macrospecies diversity on the planet (Strong *et al.* 1984) and a substantial fraction of the biomass, understanding why the world is green ought to provide some insight into how many organisms interact.

What answers do we have as to why the world is green? Several have been proposed (Box 10.1). Hairston *et al.* (1960) suggested that predators, parasites and disease, keep herbivores rare. An alternative view is that plants are responsible for this state of affairs (McNeill & Southwood 1978; Janzen 1988; White 1993), or that herbivores are between the devil (natural enemies) and the deep blue sea (poor food) (Lawton & McNeill 1979). We subscribe to the view that the world is green because plants are fundamentally poor as food for herbivores. Plants have many characteristics that herbivores find very hard to deal with, at least most of the time. We focus on the idea that the chemical

[284]

Fig. 10.1 How much of the world's green do herbivores consume? Frequency distributions of the proportion of annual net primary productivity removed by herbivores in (a) aquatic algae (phytoplankton, $n = 17$, and reef periphyton, $n = 8$); (b) submerged ($n = 5$) and emergent ($n = 14$) vascular plants; (c) terrestrial plants ($n = 67$). Primary productivity and herbivory data for aquatic and terrestrial rooted plants are limited to the above-ground portion of the plants. Comparison of herbivory on algae and rooted plants therefore assumes that the proportion of above-ground primary productivity grazed is representative of the whole plant, an assumption supported by the limited data available from terrestrial systems (above ground: median, 18%, $n = 69$; below ground: median, 13%, $n = 14$). Arrows indicate median values (aquatic algae, 79%; aquatic macrophytes, 30%; terrestrial plants, 18%). (From Cyr & Pace 1993.)

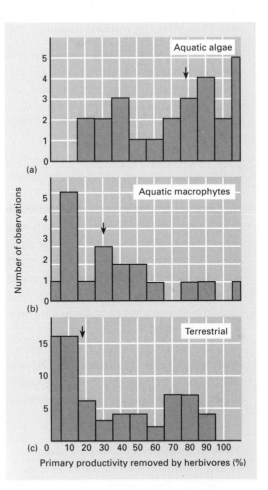

composition of plants makes them poor food. Plants have to survive in a world in which resources are often limited, competition is strong, and stress and damage from many abiotic and biotic sources are frequent. As a consequence, plants are endowed with chemical, physical and physiological functional attributes that make them inherently unsuitable food for herbivores.

Plants are poor food because they mostly contain 'cruddy' ingredients and because the occasional morsels of good food are hard to find amongst the bad. We define 'cruddy' ingredients as those that are of low nutritional quality and/or prevent herbivores from readily obtaining nutrients. Examples of 'cruddy' plant ingredients are tough fibres, high carbohydrate and cellulose content, low nitrogen and water content and toxins and/or digestibility-reducing substances. Finding suitable food amongst all the unsuitable material is also difficult for herbivores because plant characteristics are intrinsically heterogeneous in space and time, and because plant characteristics vary continuously in response to the environment.

Box 10.1 Why the world is green.

'The small fraction of insect orders that have adapted to feed on green plants is remarkable because plants are the most obvious and readily available source of food in terrestrial communities . . . life on higher plants is a formidable evolutionary hurdle that most groups of insects have conspicuously failed to overcome.' (Strong *et al.* 1984, p. 15)

'Cases of obvious depletion of green plants by herbivores are exceptions to the general picture, in which the plants are abundant and largely intact.

There are temporary exceptions to the general lack of depletion of green plants by herbivores. This occurs when herbivores are protected either by man or natural events, and it indicates that the herbivores are able to deplete the vegetation whenever they become numerous enough, as in the cases of the Kaibab deer herd, rodent plagues, and many insect outbreaks. It therefore follows that the usual condition is for populations of herbivores not to be limited by their food supply.' (Hairston *et al.* 1960, pp. 421–422, 424)

'Natural food is often far from ideal. The conclusion is that the foliage of seed plants is, even for those taxa that have evolved the ability to live on it, often only marginally adequate nutritionally.' (Southwood 1973, p. 10)

'Plant-feeding insects live in a world dominated on the one hand by their natural enemies and on the other by a sea of food that, at best, is often nutritionally inadequate and at worst is simply poisonous.' (Lawton & McNeill 1979, p. 223)

'A lack of nitrogen looms largest of all for herbivores . . . not only is their food loaded with excess carbon, much of the limited nitrogen that it does contain is not readily assimilated. Small wonder 90% of plant material ends up with decomposers not herbivores.' (White 1993, p. 13)

'Why don't all the herbivores eat up all the green world? Because most of the green world is inedible to any given species of herbivore. Why don't the herbivores that can readily consume a species of plant eat their host into oblivion? Because the carnivory and climate regimes stop them.' (Janzen 1988, p. 905)

We suggest that poor food quality of plants makes it possible for some herbivore populations to be controlled by parasites and predators, at least some of the time. Low food quality mediates these effects by constraining herbivore reproductive output (Lawton & McNeill 1979; Strong *et al.* 1984; Hunter & Price 1992). In some systems proximal bottom-up (plant) control seems to predominate (Price *et al.* 1990; Price 1992); in others, particularly biocontrol, proximal top-down control (natural enemies) is easy to demonstrate (Price 1987; Greathead & Greathead 1992). The current consensus is that a combination of top-down and bottom-up proximal forces structure communities, and attention has shifted to identifying the *relative* importance of these forces under different circumstances (Crawley 1983; Hunter & Price 1992; Hunter *et al.* 1992; Matson & Hunter 1992; Price 1992; Power 1992; Belovsky & Joern 1995; Hunter 1996; Karban 1996).

Our approach to answering why the world is green is phytocentric (Coleman & Jones 1991) rather than herbivore-centric, so we do not specifically address herbivore specialization, adaptation or population dynamics, which are covered in Chapter 13. We also focus on plant interactions with insect herbivores because we know this area best, and because it constitutes a large percentage of research in this field. We emphasize plant chemistry because we believe that this is the most important 'currency' in plant–insect herbivore interactions. We attempt to take both an ecological and an evolutionary perspective, looking for general patterns, where possible. We aim to point out what we know, or think we know, and what we do not know, but would like to know. This means we take a constructively critical look at the theories that attempt to explain ecological and evolutionary patterns. This is not a depressing enterprise, quite the reverse, but until we have a good idea of the limitations of existing knowledge we cannot clearly identify gaps and move to fill them. Consequently, we conclude by suggesting how ecologists and evolutionary biologists might develop more integrated theoretical frameworks for understanding chemical interactions between plants and herbivores, and how these can help us answer the question 'Why is the world green?'

10.2 Plants are poor food: they have 'cruddy' ingredients

The world is not actually green, it just looks that way to us. Most plant biomass is not eaten by herbivores, because all that is green is not edible (Sinclair 1975; Crawley 1983, 1989). For functional reasons, plants have a different elemental stoichiometry from animals, e.g. a much lower nitrogen content, so they are poor-quality food for herbivores. There are many different ways for plants to be 'cruddy' (Box 10.2). Many hypotheses addressing herbivore performance or abundance tend to focus on a single aspect of 'cruddiness' (e.g. low nitrogen levels), explaining outcomes based on that variable alone (e.g. White 1993). The problem with this approach is that the 'poor food' variables tend to

---

Box 10.2 'Cruddy' ingredients: some examples.

Plants contain low amounts of nitrogen and protein (McNeill & Southwood 1978; Mattson 1980; White 1984, 1993).

Plants have tough leaves; plants are lignified and fibrous, rich in cellulose and carbohydrate (Feeny 1976; Lowman & Box 1983; Nichols-Orians & Schultz 1989; Vicari & Bazely 1993).

Some plant tissues generally have a lower water content than animals (Scriber 1977, 1979; Scriber & Slansky 1981).

Plants contain toxins, repellents, growth inhibitors and digestibility-reducing compounds that can be present all the time, i.e. constitutive secondary metabolites (Conn 1981; Rosenthal & Berenbaum 1992), or produced following damage, i.e. induced secondary metabolites (Tallamy & Raupp 1991; Karban & Baldwin 1997).

---

be correlated with each other. Low nitrogen content is often correlated with high lignin content, for example (Waring *et al.* 1985). It is difficult to tease apart the effects of these different variables (Reese & Beck 1978; Mattson & Scriber 1987; Feeny 1992). A second problem is that the effects of chemicals are context dependent. The structure and physiology of the plant, the way the insect feeds and the conditions in the gut of the insect can all influence the effects of secondary metabolites (Martin & Martin 1984; Martin *et al.* 1987; Feeny 1990; Bernays 1991).

It is not always necessary to invoke herbivore selection for these 'cruddiness' traits. Many of them are a primary consequence of the way that plants work (Coleman & Jones 1991). For example, the primary reason that terrestrial higher plants have high fibre and lignin content is that they need to stand upright, overshadow their neighbours and transport resources. Similarly, they often have low nitrogen levels because nitrogen is a scarce resource in many soils. Many secondary compounds may well be exceptions to this principle: herbivores and pathogens might be an important selective influence, at least some of the time (Feeny 1992; Marquis 1992; but see Futuyma 1983a,b; Jermy 1984, 1993; Jones & Firn 1991; see Chapter 5).

How limited are herbivores by plant 'cruddiness'? Surprisingly, we do not have a general quantitative answer to this question, not least because there are very few synoptic field estimates of herbivore consumption, performance and abundance in the absence versus presence of natural enemies. However we can get some insights from the fact that herbivore growth rates are often higher on artificial diets than on host plants, although the benefits of high nitrogen-containing artificial diets decline with larval age (Joseph *et al.* 1993). Furthermore, the assimilation and growth efficiencies of insect herbivores are much lower than those of predatory insects (Southwood 1973). Lastly, insect herbivores show remarkable precision and discrimination in selecting food, both within and between plants (White 1970; Eastop 1973; Kidd *et al.* 1985; Kimmerer & Potter 1987; Price 1989), which presumably indicates that not all is suitable.

10.2.1 Nitrogen limitation of herbivores

10.2.1.1 The arguments

The stoichiometry (i.e. elemental proportions) of nitrogen in plant–herbivore interactions is problematical: insect herbivores contain nearly 10 times more nitrogen than the plants they eat (or much more than this if they eat wood, and a bit less if they eat nitrogen-rich seeds) (Fig. 10.2; Mattson 1980). Even if we assumed that insects could always achieve a high nitrogen utilization efficiency, say a generous 50%, herbivores would still have to consume 20 times their final biomass to grow to adults, and even more to provide nitrogen in eggs of their offspring. We know the nitrogen utilization efficiency of relatively few insect herbivores (Scriber & Feeny 1979; Abrahamson & Weis 1986; Slansky & Rodriguez 1987), and have complete nitrogen budgets for even fewer (e.g. Montgomery 1982). Nevertheless, 50% nitrogen utili-

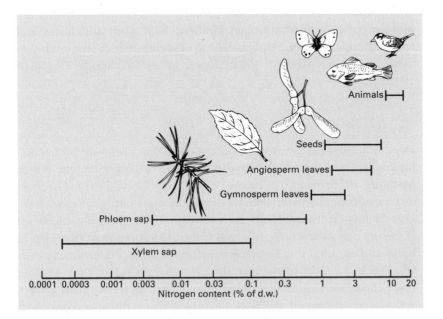

Fig. 10.2 Comparing plant and animal nitrogen. Ranges of nutrient nitrogen concentration in different plant and animal tissues. (From Strong *et al.* 1984.)

zation efficiency may be achieved only rarely. We also know very little about the variation in, and constraints on, nitrogen stoichiometry for insect herbivores in terrestrial ecosystems, compared to what is known for aquatic systems (e.g. Sterner 1995); and unlike these systems we have more or less ignored other elements such as phosphorus.

There are many ways that nitrogen can be limiting for herbivores, even if the total nitrogen content of plants was adequate. It is the combination of all these ways that makes plants such a poor nitrogen source (Mattson 1980; White 1993). For example, the total nitrogen content of plants is low, but the concentration of specific forms of nitrogen (e.g. soluble or extractable protein and particular amino acids) may be even lower. Also, nitrogen is a very heterogeneous resource in plants in both space and time (e.g. as leaves age their nitrogen content decreases; Feeny 1970; Mattson 1980; Raupp & Denno 1983). Furthermore, plant nitrogen is diluted by carbohydrates, cellulose, fibre and lignin. Herbivores have to eat a lot of carbon to obtain enough nitrogen, and this problem can be exacerbated in the presence of toxins, which may limit consumption. In addition, the nitrogen in plants is often complexed, or becomes complexed on eating, with refractile materials such as tannins (Feeny 1970 but see Bernays 1981; Mole & Waterman 1985). This makes it difficult for herbivores to extract and digest nitrogen, and this problem is compounded by leaf toughness, fibre and lignin. Finally, not just the concentration but the flux of nitrogen (i.e. concentration per visit time) can be important. Xylem and phloem feeders use resources of very low nitrogen content but high flux rates, and their throughput of sap is correspondingly high (Brodbeck & Strong 1987). These insects have perhaps come closest to solving the problem of low plant nitrogen content, high plant toughness and toxins, although perhaps only by virtue of their gut flora (the endosymbiotic microbes; Mattson 1980; Jones 1984; Bernays & Chapman 1995).

10.2.1.2 The evidence

Does increasing plant nitrogen content mean that plants are better food for herbivores? Many studies have investigated this question (e.g. Fox & Macaulay 1977; Slansky & Feeny 1977; Ohmart *et al.* 1985). The most direct evidence that herbivores are limited by plant nitrogen comes from experiments fertilizing plants with nitrogen and measuring effects on herbivore performance. Waring and Cobb (1992) reported that 67% of these studies found that herbivores performed better on fertilized plants (Fig. 10.3). In most of these studies, however, it is not clear whether these effects were due to increased plant tissue, increased plant growth and vigour, decreased defences, increased tissue nitrogen concentrations, or some combination of the above. Many studies also found that whilst some herbivore performance parameters (e.g. growth rate and

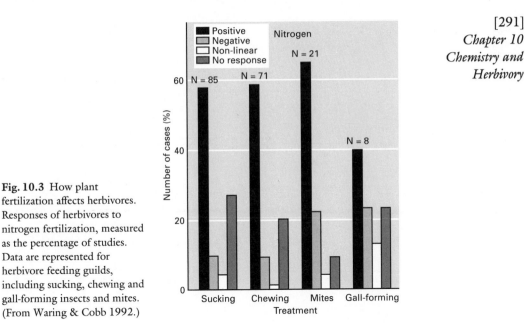

Fig. 10.3 How plant fertilization affects herbivores. Responses of herbivores to nitrogen fertilization, measured as the percentage of studies. Data are represented for herbivore feeding guilds, including sucking, chewing and gall-forming insects and mites. (From Waring & Cobb 1992.)

fecundity) are often increased on fertilized plants, others (e.g. survivorship) are not, so the overall benefits of increased nitrogen are still somewhat unclear (e.g. Stark 1965; Auerbach & Strong 1981; McCullough & Kulman 1991; Hartley & Lawton 1992; Ohgushi 1992).

Although there are many studies demonstrating an increase in at least one or more aspects of insect survival, development, growth rate or fecundity on fertilized plants (e.g. Fig. 10.4), few studies have determined whether or not there are any subsequent effects on herbivore population dynamics. Studies that causally relate changes in plant chemistry to long-term fluctuations in insect abundance are very rare (some ecologists would say there is not a single one!). From the plant's point of view, there is an even bigger omission: we do not know how increases in herbivore performance on fertilized plants feed back to plant growth. For example, do fertilized plants suffer more or less damage than unfertilized ones (Hunter & Schultz 1995)? Do fertilized plants tolerate increased herbivory better than unfertilized plants with lower herbivore levels? If we fertilize plants, do they perform better or worse overall? Fertilized plants may be preferred by herbivores, they may even receive more damage, but if they grow more, does it matter? We have relatively few experimental studies that address these issues and they produce conflicting results (Onuf *et al.* 1977; Gyphis & Puttick 1989; Dudt & Shure 1994; Hartvigsen *et al.* 1995). So, somewhat surprisingly, we do not know if a fertilized world would be more or less 'green'.

One 'natural' experiment where plant fertilization actually has affected plant populations via changes in herbivore performance is the case of nutrient inputs from pollution. In Holland, nitrogen inputs to

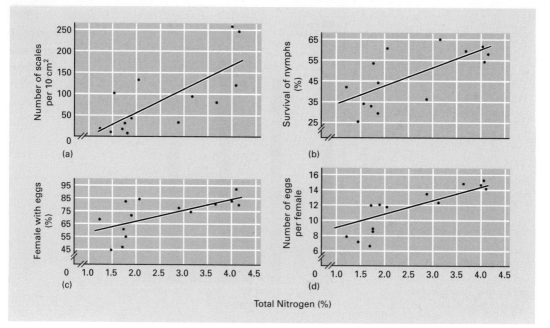

Fig. 10.4 How plant nitrogen affects herbivores. Relationships between the total nitrogen concentration of the young foliage of 14 host species during peak colonization by nymphs of *Fiofinia externa* and the density of these colonists that survived to maturity (a, $r = 0.74**$), their survival (b, $r = 0.07**$), their development rate (measured as percentage of females with eggs; c, $r = 0.64**$), and their fecundity (d, $r = 0.74**$). *, $P < 0.05$; **, $P \leqslant 0.01$ by linear regression analysis. (From McClure 1983, with permission.)

heathlands from wet and dry atmospheric deposition are associated with increased populations of the heather beetle and increased heather defoliation, which opens up the heather canopy and leads to the eventual replacement of heather by grasses (Brunsting & Heil 1985). In this case, however, even though fertilization dramatically increases insect density and damage, the main consequence is a change in plant community composition, rather than a reduction in the amount of 'green' in the world (see Chapters 14 & 15).

It is difficult to discover the mechanisms whereby fertilizing plants affect herbivore performance, because fertilization has so many effects on plants. For example, nitrogen fertilization usually decreases fibre and lignin content and hence toughness but it also tends to decrease concentrations of some secondary metabolites, particularly phenolics and tannins (Bryant *et al.* 1987; Hartley *et al.* 1995; Fig. 10.5). In addition, the effects of changes in plant quality are difficult to separate from changes in plant quantity because plants usually grow faster when fertilized. The 'plant vigour hypothesis' (Price 1991) proposes that herbivores perform best on faster growing plants, because plant growth rate is usually correlated with high nitrogen. However, faster growth is correlated with many other factors as well as nitrogen; it usually implies

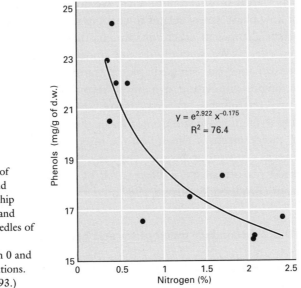

Fig. 10.5 Correlation of secondary chemistry and nitrogen. The relationship between total phenols and nitrogen in the new needles of 3-year-old Douglas fir seedlings fertilized with 0 and 200 ppm nitrogen solutions. (From Joseph *et al.* 1993.)

$$y = e^{2.922} x^{-0.175}$$
$$R^2 = 76.4$$

less lignification or greater turnover rate of leaves, for example. The most vigorously growing plants are not necessarily those with the highest nitrogen content, because rapid growth tends to dilute nitrogen concentrations. Also, senescing plant tissue often contains high levels of soluble nitrogen and some sucking insects track this nitrogen (Jones 1983; Coleman & Jones 1988).

There have been a number of attempts to explain insect outbreaks on the basis of increased nitrogen under some conditions. Outbreaks often correlate to climatic factors, such as drought, that are thought to 'stress' plants and increase their soluble nitrogen content. This, in turn, leads to improved performance of insects (Mattson & Haack 1987; Major 1990; White 1993). Some of the operational problems with the 'plant stress hypothesis' have been discussed elsewhere (Larsson 1989; Watt 1994; Fig. 10.6) including defining what stress is or is not, and the non-linear responses of herbivores to stressed plants. We note here at least two

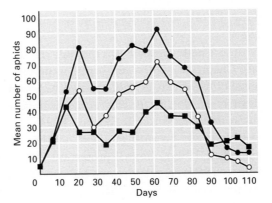

Fig. 10.6 Contingent effects of plant stress. The mean number of aphids per tree ($\pm$SE) on control (○), intermittently stressed (●), and continuously stressed (■) *Picea sitchensis*. (From Major 1990, with permission.)

major areas of uncertainty. First, we do not have a clear understanding of how plant 'stress' relates to changes in the nitrogen content of plants versus changes in other constituents that might 'cancel out' any nitrogen effects. We certainly do not understand how improved herbivore performance on 'stressed' plants translates into changes in plant or herbivore population dynamics. Secondly, even if we do establish a causal relationship between climate, 'stress', nitrogen and insect population dynamics, we still do not know the additional contributions made by direct effects of climate on the insect (Wellings & Dixon 1987). At present we cannot be sure that any 'stress'-induced improvement in insect performance is a direct result of increasing nitrogen levels in the plant (Brodbeck & Strong 1987).

10.2.2 Secondary metabolites and herbivores

10.2.2.1 The general argument and evidence

The world may remain green because plants are poisonous, contain repellent or deterrent chemicals, and compounds that inhibit digestion (Ehrlich & Raven 1964; Janzen 1979; Schultz 1988). Plants are 'coloured' with secondary compounds (Janzen 1978), natural products that do not appear to be universally necessary for primary biochemical activities affecting plant growth, development and reproduction. These secondary metabolites usually have restricted taxonomic distributions (Fraenkel 1959; Conn 1981; Haslam 1986) and express the chemical individuality of the plant species (Bu'Lock 1980; Haslam 1988). There are many tens of thousands of these chemicals (Waterman 1992) and some are known to have deleterious effects on herbivores. Compounds may be toxic (Berenbaum 1978; Isman & Duffey 1982), repellent and deterrent (Bernays & Chapman 1977; Jones & Klocke 1987; Bernays 1990), growth inhibitory (Feeny 1969; Klocke *et al.* 1986), and can reduce fecundity (Berenbaum & Feeny 1981), slow development (Roehrig & Capinera 1983; Bergelson *et al.* 1986), reduce digestibility (Klocke & Chan 1982; Felton & Duffey 1991a), or have combinations of several effects (Bernays *et al.* 1989; Feeny 1992; Bennet & Wallsgrove 1994), sometimes acting synergistically (Berenbaum & Zanger 1988).

Insect herbivores are not the only 'targets' of secondary metabolites (see Chapter 5). There are many examples of 'functions' of secondary metabolites which do not involve herbivores (Seigler & Price 1976; Beart *et al.* 1985). Briefly, some examples include ultraviolet (UV) absorption (Stapleton & Walbot 1994), growth regulation (Kefeli & Dashek 1984; Abrahamson *et al.* 1991), signalling (Baas 1989; Silverman *et al.* 1995), thermal protection (Monson *et al.* 1994; Sharkey & Singass 1995), regulators of gene expression (Peters & Verma 1990), attraction of pollinators (Pichersky *et al.* 1994), allelopathy (Rice 1984) and control of nutrient cycling (Northup *et al.* 1995). Of course, these compounds also have well-established roles in defence against bacterial, viral and fungal attack (Kosuge 1969; Hahlbrock *et al.* 1981; Dixon 1986; Hain *et al.* 1993 but see Knogge *et al.* 1987), which may perhaps have more serious consequences for plants than insect herbivory.

10.2.2.2 *Specific arguments and evidence*

What patterns do we see in the amounts, types, distributions and activities of secondary metabolites in plant–herbivore interactions? Many of the relationships we discuss are still somewhat unclear and/or speculative, and broad, confirmed generalizations are hard to make. Nevertheless, such generalizations are worth attempting. Much time and energy has gone into trying to understand these patterns, and they provide an important context for the theories that we explore later, which have been developed to explain the patterns.

Quantity: the effects of the amount of secondary metabolites. When comparisons are made within a single plant species against a particular herbivore, there is often a reasonably good correlation between an increase in the amount of one or more secondary compounds and a decrease in herbivore performance. For example, there are often higher levels of secondary compounds in resistant compared to susceptible plant genotypes (Liu *et al.* 1992), although other plant constituents may also differ in concentration. It is also often possible to show clear dose–response relationships *in vitro*, such as in artificial diets, and in many (Hunter *et al.* 1994) but by no means all cases the concentrations found in the plant are sufficient to account for these effects (Feeny 1969; Jones & Firn 1979; Cates & Zou 1990).

There are, however, many idiosyncrasies in the relationship between the concentrations of secondary compounds and their effects on herbivores. It is often surprisingly hard to correlate secondary compound levels and insect performance, particularly in the field. It is even more difficult to demonstrate a causal relationship, in other words to attribute effects to secondary compounds alone, in the face of confounding effects, such as other plant constituents which are often correlated with the secondary compounds. Furthermore, effects are far from universal: one insect's 'meat' is another insect's 'poison' (e.g. Bernays & Woodhead 1982), so the relationships often depend on which herbivores and which chemicals are under consideration. Many insects clearly have many biochemical adaptations, such as detoxification or sequestration mechanisms (e.g. Brattsten 1992; Rowell-Rahier & Pasteels 1992) to deal with secondary compounds. Sophisticated behavioural adaptations such as petiole cutting or trenching of leaves (Carroll & Hoffman 1980; Edwards & Wanjura 1989; Doussard & Denno 1994) are certainly indicative of the efforts some herbivores take to make plants edible.

There are some clear and useful patterns between the concentrations of secondary metabolites, plant growth rates and habitat. Slow-growing plants from nutrient-poor habitats tend to have higher amounts of some secondary compounds than fast-growing plants from nutrient-rich habitats, and can probably be considered to be better defended (McKey 1979; Coley *et al.* 1985). Of course, lignin and toughness are negatively correlated with growth rate, while nitrogen is positively correlated with growth rate, which makes it difficult to say which factors are causal. It may well be reasonable to assume that oak trees, with high levels of phenolic compounds, are better defended than grasses, which appear to have lower phenolic levels. On the other hand, it is difficult to argue that oaks are better or less well defended than, say, umbellifers, which have a diversity of different types of secondary metabolites including furano-coumarins (Jones & Lawton 1991). If plants have very different types of compounds, the confounding effects of this diversity make legitimate comparisons problematical.

Diversity: the effects of the number of secondary compounds. There is a staggering, fascinating and baffling diversity of secondary metabolites. Why, for example, do plants have over 14 000 different terpenes (Gershenzon & Croteau 1992)? Later, we discuss several ideas that have emerged as to why this diversity has arisen and how it is maintained. At this point we ask: what is the relationship between the absolute diversity of secondary chemical structures in plants and the degree of defence? Does having more compounds mean that a plant is better defended?

If a plant has a greater variety of secondary metabolites we might expect a greater likelihood that at least some of them will be biologically active, and that a greater diversity of targets (e.g. different insects) could be affected (Jones & Firn 1991). There is also an increased chance of synergistic effects, i.e. two compounds might have more of an adverse effect on herbivores than either compound on its own (Hay *et al.* 1994). The interaction of myristicin with furanocoumarins in parsnip and their effects on the parsnip webworm illustrate this point (Berenbaum & Neal 1985; Berenbaum & Zangerl 1988). It also seems reasonable to suppose that herbivores might find it harder to adapt to plants containing many secondary compounds than to those containing only a few (Gould 1983; Prestridge 1996), although the evidence here is ambiguous (Gould 1988a,b).

For all this, however, we do not really understand the relationships between the diversity of compounds, the degree of defence and the ecological or evolutionary consequences. For example, the number of insect herbivore species on British umbellifers does not appear to correlate to the diversity of biosynthetically distinct classes of secondary metabolites they contain (Jones & Lawton 1991). There are at least two problems in this area. First, we do not really know how many different chemicals a given plant species contains (Jones & Firn 1991). Second, we do not have a good functional definition of diversity: how do we measure chemical diversity in ways that mean something to herbivores?

Quality: the effects of different types of secondary compounds. The diversity of secondary metabolites inevitably means that there are many different types of compounds (Mann 1978; Seigler 1981; Waterman 1992). How many types of compounds does a plant really need to be well defended? Does having many different types make a plant better defended? Which plants have which type of compounds and why? Establishing clear general patterns here is problematic. One of the biggest difficulties is that we have no clear functional basis for distinguishing different types of secondary compounds. We can define them in terms of chemical structure (see Chapter 5), but the relationships between structure and function are largely unknown. Both biosynthetic origin and taxonomic relatedness of the plants can be used, but neither of these characteristics has a clear relationship to function. If we have non-functional characterizations of types of compounds, it is perhaps not surprising that it is very difficult to predict the degree of plant defence. Asking if plants with types '1', '2' and '3' are better defended than plants with compounds of types '4', '5' and '6' is like comparing the flavours of apples and oranges. In most cases we cannot say how similar or different plants with different types of compounds are from a herbivore's point of view. This makes it rather hard to relate herbivore host shifts to chemical differences between hosts for example (Futuyma 1983a,b). Even when we can be reasonably certain that a plant species really does have many different types of secondary compounds, it does not necessarily seem to have fewer herbivores than other plant species that appear to have fewer types of secondary compounds. Bracken fern *Pteriduim aquilinum* (Dennstaedtiaçeae) is a good example of a plant with a large diversity of different types of secondary compounds (at least 12 biosynthetically distinct classes). However, bracken has about as many herbivore species as would be expected for plants with a similar distribution, abundance and architecture (Lawton 1976, 1982; Jones 1983). Generalizations are also difficult because plant chemistry is not the sole determinant of herbivore distribution and abundance (e.g. Tahvanainen 1983; see Strong *et al.* 1984).

Context dependency of effects. Effects of secondary metabolites are often modified by the prevailing conditions in both the cells of the plant and the tissues (gut, haemolymph, etc.) of the insect herbivore. For example, many plant secondary compounds are stored in bound or conjugated form and are compartmentalized or separated from the enzymes necessary for their activation (Stafford 1981; Duffey & Felton 1989). It is only when the plant tissue is damaged by the herbivore that the compounds become active. The activity of some compounds is dependent on the redox state of the environment. Activity of phenolics, for example, may be dependent upon oxidation by enzymes or chemical oxidants found in leaves and in digestive tracts of herbivores (Appel 1993). Furthermore, alkaline pH and the presence of surfactants in insect guts can reduce the precipitation of proteins by tannins (Martin & Martin 1984; Martin *et al.* 1987; Felton & Duffey 1991b). In addition, herbivores have a range of detoxification mechanisms in the gut or haemolymph which allow them to cope with plant secondary metabolites (Krieger *et al.* 1971; Brattsten 1992).

10.2.2.3 Theories

Theories about broad, general patterns in plant secondary metabolism and its role in defence against herbivores can be grouped into three somewhat overlapping domains that ask questions about the quantity, diversity and quality of secondary chemicals respectively.

1 Why do some individuals have more or less secondary metabolites than others (e.g. what controls the allocation to phenolic compounds within plants)?

2 What determines the number of secondary metabolites found in plants (e.g. is an arms race going on)?

3 Why do different plants have different types of secondary compounds (e.g. why do some plants contain tannins while others contain alkaloids)?

All the questions can be asked within phenotypic/ecological and genotypic/evolutionary contexts (Table 10.1). We consider these theories in the historical order that they appeared in the literature. The

Table 10.1 Domains of conceptual frameworks.

Question	Context	
	Phenotypic/ecological	Genotypic/evolutionary
Quantity (amount)	Carbon/nutrient balance (Bryant *et al.* 1983) Growth/differentiation balance (Herms & Mattson 1992)	Resource availability (Coley *et al.* 1985) Apparency (Feeny 1976; Rhoades & Cates 1976) Growth/differentiation balance
Diversity (number)	Common chemistry Biochemical barrier Diverse defence Enemy escape (Jones & Lawton 1991)	*Raison d'être* (Fraenkel 1959) Coevolution (Ehrlich & Raven 1964) Sequential evolution (Jermy 1976, 1993) Screening hypothesis (Jones & Firn 1991)
Quality (type)	Carbon/nutrient balance	Resource availability Apparency

> **Box 10.3** *Raison d'être* of secondary plant substances.
>
> 'Plants also contain a vast array of what have been called "secondary" plant substances. These may be conveniently grouped as glucosides, saponins, tannins, alkaloids, essential oils, organic acids, and others, many thousands of which have been described in the literature. Their occurrence is sporadic but may be specific for families, subfamilies, and genera and sometimes even for species or subspecies. Their role in the metabolism of plants has never been satisfactorily explained, but in view of their sporadic occurrence and of the differences in their chemical constitution, it is almost inconceivable that they play a function in the basic metabolism of plants. For the same reasons, it is also highly improbable that they are of nutritional importance for the insects in the same sense as the "primary" substances are, namely that they are metabolized and utilized in tissue synthesis.
>
> It is suggested that the food specificity of insects is based solely on the presence or absence of these odd compounds in plants, which serve as repellents to insects (and other animals) in general and as attractants to those few which feed on each plant species. The immense variety and number of compounds concerned thus corresponds to the equally immense variety of specific nutritional relationships between insects and plant hosts. The compounds concerned need not play any role in the basic metabolism of either the plant or the insect, since they serve merely as trigger substances which induce, or prevent, uptake of the true nutrients.' (Fraenkel 1959, p. 129)

'essence' of the theories is presented in boxes, while the text provides a commentary. We have not attempted to cover every theory on plant secondary metabolism nor every aspect of the theories that we do cover.

The raison d'être *of secondary plant substances.* Fraenkel's 1959 paper put chemical ecology on the map. This seminal work (Box 10.3) was the first to articulate clearly the idea that plant secondary metabolites evolved as defences against insect herbivores, and that insects were therefore important selective agents on plants. Prior to his paper, plant secondary metabolites were more or less ignored by ecologists and evolutionary biologists, and were considered to be waste products by many chemists. Fraenkel's central idea still lies at the heart of most subsequent theories on the evolution of plant secondary metabolites, although many would now broaden the scope to include defence against microbial pathogens and perhaps plant competitors (i.e. allelopathy).

Coevolution. In a classic and hugely influential paper, Ehrlich and Raven (1964) proposed the idea of reciprocal, step-wise coevolution (Box 10.4). This is essentially a continuous evolutionary 'arms race' between

Box 10.4 Coevolution.

'A systematic evaluation of the kinds of plants fed upon by the larvae of certain subgroups of butterflies leads unambiguously to the conclusion that secondary plant substances play the leading role in determining patterns of utilization.

Angiosperms have, through occasional mutations and recombination, produced a series of chemical compounds not directly related to their basic metabolic pathways but not inimical to normal growth and development. Some of these compounds, by chance, serve to reduce or destroy the palatability of the plant in which they are produced. Such a plant, protected from the attacks of phytophagous animals, would in a sense have entered a new adaptive zone. Evolutionary radiation of the plants might follow, and eventually what began as a chance mutation or recombination might characterize an entire family or group of related families. Phytophagous insects, however, can evolve in response to physiological obstacles. If a recombinant or mutation appeared in a population of insects that enabled individuals to feed on some previously protected plant group, selection could carry the line into a new adaptive zone. Here it would be free to diversify largely in the absence of competition from other phytophagous animals.' (Ehrlich & Raven 1964, pp. 601–602)

' "Coevolution" may be usefully defined as an evolutionary change in a trait of the individuals in one population in response to a trait of the individuals of a second population, followed by an evolutionary response by the second population to the change in the first. "Diffuse coevolution" occurs when either or both populations in the above definition are represented by an array of populations that generate a selective presure as a group.' (Janzen 1980, p. 611)

'The working model for interpreting coevolutionary interactions has itself evolved over the past 10 years. The original "stepwise" model included two species, each evolving primarily in response to selective pressures exerted by the other; this model implied a perpetual "arms race" in which each species developed more specialized adaptations to the properties of the other. With increased understanding of ecological and evolutionary patterns, an alternative model of "diffuse coevolution" has been invoked. Diffuse coevolution implies that many species, on the same or different trophic levels, may simultaneously exert selective pressures on one another and be affected by changes in other component members. However, no appropriate metaphor has emerged that succinctly describes diffuse coevolution, and conveys the notion of

continued on p. 301

plants evolving new secondary metabolites and herbivores evolving adaptations to these compounds. The requirement for step-wise reci-procity has proved difficult to demonstrate (Bernays & Graham 1988; Courtney 1988; Spencer 1988; Berenbaum & Zangerl 1992; also see Jermy 1984, 1993) and the original ideas of coevolution have them-selves evolved into 'diffuse coevolution' or 'multispecies coevolution' (Janzen 1980; Fox 1981, 1988; Gould 1988b; Rausher 1988). Thomp-son (1988, 1994) provides excellent reviews of the ideas and the evidence.

Coevolution is an intrinsically appealing concept that has driven much of the research in plant–herbivore interactions. Evidence does exist for different parts of the scenario. We have good evidence that secondary chemicals can affect herbivores, that herbivores can adapt and that defences are heritable. Evidence that insect herbivores have affected the evolution of plant chemical composition is less compelling, largely because of uncertainties about the effects of these consumers on plant fitness or plant populations (Crawley 1989). It has long been assumed that herbivores do reduce the fitness of individual plants (Coley 1993). However there are still relatively few studies showing this (e.g. Rausher & Feeny 1980; Marquis 1984; Crawley 1985; Schultz 1988; Bergelson & Crawley 1992; Gomez & Zamora 1994; Schierenbeck *et al.* 1994). On the other hand, outbreaks do occur that clearly affect individual plants and even plant species, and herbivores clearly can change the structure and dominance hierarchies in plant communities (see Chapter 13). No doubt the debate over selection by herbivores will continue.

Overall, much of the evidence for coevolution is case-specific and subject to alternative explanation (see later). In particular, a conclusive demonstration of *all* the necessary components in any one system has so far eluded researchers. No doubt modern molecular techniques will help resolve some of the issues. It is important to recognize that even if we demonstrate coevolution in one system or five systems, this will not resolve the issue of whether or not coevolution has been the most prevalent force determining the patterns we see today. 'Insufficient evidence is available to delineate the generality and strength of selection exerted by herbivores for resistance traits in plants' (Marquis 1992). It is also worth asking how 'diffuse' can the relationships be and yet still be considered coevolution? For example, should we include all of the herbivores, all the microorganisms and even plant competitors interacting with a plant? What about the soil microorganisms with which plants compete for nitrogen? What about the selective influences caused by abiotic factors for which there can be no feedback?

Sequential evolution. While this theory (Box 10.5) places primary emphasis on evolution and adaptation of insect herbivores, rather than focusing on plant defence *per se*, it has been an important lone voice proclaiming a legitimate alternative to coevolution. Plants evolve characteristics and herbivores adapt, tracking plant changes but not markedly influencing them. One can consider this theory a null model that should hold if convincing evidence for coevolution is not forthcoming. With increasing recognition of some of the problems with coevolution theory (see earlier), this theory is finally getting some of the attention it deserves, particularly from entomologists.

Tibor Jermy's argument is essentially a recognition of the fact that plants loom large in the lives of insect herbivores, but not the other way round, in contrast to the ideas of coevolution. From the plant perspective we can argue that they have to deal with many problems (resources, competition for resources with other plants and soil microbes, belowground communities of consumers/mutualists, abiotic stress and dam-

---

Box 10.5 Sequential evolution.

'(1) most phytophagous insects have very low population densities compared to the biomass of their host-plants, therefore, they can hardly be important selection factors for the plant; (2) insect–host-plant interactions are not necessarily antagonistic: mono- and oligophagous insects, if their number is fairly high, may ideally regulate the abundance of their host-plants (mutual advantage); consequently, (3) resistance to insects is not a general necessity in plants and it cannot explain the presence of secondary plant substances; (4) parallel evolutionary lines of plants and insects which should result from co-evolutionary interactions are rare, while many closely related insects feed on botanically very distant plant taxa – a relationship which cannot be related to co-evolution.

Therefore, the theory of *sequential evolution* is proposed: the evolution of flowering plants propelled by selection factors (e.g. climate, soil, plant–plant interactions, etc.), which are much more potent than insect attacks, creates the biochemically diversified trophic base for the evolution of phytophagous insects, while the latter do not appreciably influence the evolution of plants.' (Jermy 1976, p. 109)

age), as well as with a diversity of consumers. Not all consumers are herbivores; plant pathogens may, in general, have a bigger impact than insects (see Chapter 13). Many of the attributes of plants that are invoked as adaptations against insect herbivores in coevolutionary scenarios may be general responses to stress and damage that happen to affect insects (Jones & Coleman 1991). Similarly, adaptations by insects are a necessary requirement of making a living on a marginal food that is constantly changing as plants grow and develop and respond to their environment. This is not to say that insects do not have an effect on plants; together with other forces, they no doubt influence plant attributes. It is just that herbivores may not have exclusive right to the origin and maintenance of plant chemical defences. Such a conceptual model is more inclusive and less idealized than coevolution: the plant is seen as central to a matrix of interactions with the abiotic and biotic environment.

Apparency. This theory (Box 10.6), simultaneously developed by Feeny (1976) and Rhoades and Cates (1976), was one of the first attempts to explain the evolutionary origin of the observed patterns in the distribution of different types of secondary compounds across plant species. The theory suggested that long-lived plants ('apparent' to herbivores) tend to have a high content of quantitative, digestibility-reducing defences such as tannins and resins, whilst herbaceous 'unapparent' plants tend to be defended by qualitative or toxic defensive compounds, such as alkaloids or glucosinolates. 'Apparent' plants are likely to be found by herbivores and therefore need to allocate resources to defences that are particularly effective against specialist herbivores that are adapted to toxins (i.e. these plants rely on chemicals that reduce food quality). In contrast, ephemeral plants are unapparent and less likely to be found by their specialist enemies, so they can rely on 'cheaper' toxic chemicals that will be most effective against non-adapted generalist herbivores that may accidentally find them.

Although this theory seems to be currently out of favour, its central notion of relating the types of plant defence to the likelihood of discovery by herbivores is important. Grubb (1992) recently re-examined this set of ideas in a more-or-less chemical-free context and found much that was reasonable. There are, however, a number of conceptual and operational problems that have become, dare we say it, apparent. It certainly appears that trees and woody shrubs have a preponderance of phenolics, tannins, resins and lignin, relative to more ephemeral species, although this is by no means universal (Barbosa & Krischick 1987). The explanation for these patterns is herbivore rather than plant centred in this theory (Swain 1977, 1979). For example, structural requirements (e.g. lignification) are clearly correlated to plant longevity but are not included in the theory. The theory also takes little account of taxonomy. Plants can have any sort of defence according to

Box 10.6 Apparency.

'Most crucifers are ephemeral plants, characteristic of early stages of community succession, and are likely to be relatively "hard to find" by their adapted enemies, against which they seem to rely primarily on escape in time and space. Selection in such species seems to have favored fast growth to early maturity, high reproductive output, and ability to disperse and colonize new areas rapidly. Such plants might not be able to "afford" the metabolic cost of high concentrations of defensive compounds, since this would be likely to reduce metabolic allocation for growth and reproduction and, perhaps also, competitive ability against other plants of both their own and other species. Oaks, by contrast, are climax forest dominants and are "bound to be found" by their enemies in ecological time; predation pressure on undefended plants would probably be severe. In such plants allocation of the energy and nutrients required for quantitative defense is more likely to bring commensurate increases in fitness.

The susceptibility of an individual plant to discovery by its enemies may be influenced not only by its size, growth form and persistence, but also by the relative abundance of its species within the overall community. To denote the interaction of abundance, persistence and other plant characteristics which influence likelihood referred to of discovery, I now prefer to describe "bound to be found" plants by the more convenient term "apparent", meaning "visible, plainly seen, conspicuous, palpable, obvious". Plants which are "hard to find" by their enemies will be referred to as "unapparent", the antonym of apparent. The vulnerability of an individual to discovery by its enemies may then be referred to as its "apparency", meaning "the quality of being apparent; visibility". Since animals, fungi and pathogens may use means other than vision to locate their host-plants, I shall consider apparency to mean "susceptibility to discovery" by whatever means enemies may employ.' (Feeny 1976, p.5)

'Assuming that plant defenses are costly to the time and energy budget of plants, the observed distribution of toxic and digestibility-reducing defensive systems, both between leaves of different stages of maturity and between plant species, can be explained in terms of greater invest-ment in chemical defense for predictable plants and tissues than for ephemeral plants and tissues. Since escape, particularly escape from specialist herbivores, is high for ephemeral plants and ephemeral leaf tissues (usually young leaves), they are defended by a cheap, divergent, toxic chemical defense affording some protection against generalist herbivores. Escape is low for predictable plants and predictable plant

continued on p. 305

Box 10.6 *Continued*

tissues (usually mature leaves) which thus utilize a more costly convergent digestibility-reducing chemical defense, effective against both specialist and generalist herbivores. Predictable plants utilize toxins in
their ephemeral leaf tissues and generalized digestibility-reducing systems, particularly tannins, in their predictable leaf tissues. Ephemeral
plants utilize toxins in their ephemeral tissues and are postulated to
utilize specific digestive enzyme inhibitors, in their mature leaves.'
(Rhoades & Cates 1976, p. 205)

the idealization of apparency, yet physiology, phylogeny and evolutionary history surely place constraints on what is possible.

A more fundamental, and as yet unresolved, conceptual problem
resides in the distinction between 'qualitative' and 'quantitative' secondary metabolites. There is, as yet, no clear toxicological or pharmacologi-
cal basis for this distinction. We do not know the relationships between
digestibility, toxicity and chemical structure, so we do not know how to
divide chemicals into 'qualitative' and 'quantitative' types, or even
whether this is possible. All chemicals that have biological activity act in
a dose-dependent manner, not just 'quantitative' defences. Furthermore, there is a real diversity of modes of action of chemicals that
preclude such a simple dichotomous key (Jones & Firn 1991).

The main operational problem with the theory is a set of clear
criteria for defining and measuring what apparency is, an obvious
prerequisite for effectively testing the theory (Fox 1981; Crawley 1983).
What does an apparent plant look like? What attributes of plants are
important in defining apparency to herbivores and over what scale in
time and space? We know that many plants recruit a herbivore community very quickly (years not centuries) (Strong *et al.* 1984), and attempts
to demonstrate experimentally that ephemeral, rare or quickly growing
plant species suffer less herbivory have not always been successful
(Futuyma & Wasserman 1980; Coley 1983).

Despite these difficulties, apparency theory has made important
contributions to the debate on the selection pressures influencing types
of plant defences and it has stimulated development of other theories
that have emphasized plant growth and physiology, rather than discovery by herbivores. It is clear that some of the central issues of apparency
theory have not been resolved, but this does *not* mean that the entire
edifice should be discarded, which is the impression one might get from
reading comments in some of the literature.

Carbon/nutrient balance (CNB). This important physiological trade-off-
based hypothesis (Box 10.7) has much experimental support (e.g.

Box 10.7 Carbon/nutrient balance hypothesis.

'The evolutionary response of plants to herbivory is constrained by the availability of resources in the environment. Woody plants adapted to low-resource environments have intrinsically slow growth rates that limit their capacity to grow rapidly beyond the reach of most browsing mammals. Their low capacity to acquire resources limits their potential for compensatory growth which would otherwise enable them to replace tissue destroyed by browsing. Plants adapted to low-resource environments have responded to browsing by evolving strong constitutive defenses with relatively low ontogenetic plasticity. Because nutrients are often more limiting than light in boreal forests, slowly growing boreal forest trees utilize carbon-based rather than nitrogen-based defenses.

We suggest that the response of all plants to reduced nutrient availability is qualitatively similar.

Growth is the process most strongly affected by nutrient stress. The decline in growth with nutrient stress is generally greater than the decline in photosynthesis so that carbohydrates and carbon-based secondary metabolites such as phenols accumulate. Under conditions of nutrient limitation carbon is relatively "cheap", and the nutrients in leaves are difficult to replace, for the reasons outlined in the previous sections. Therefore it is not surprising to see carbon-based secondary defenses rise under conditions of nutrient limitation, whereas nitrogen-based defenses decline.

The plant phenotypic response to carbon stress due to insufficient light is essentially the converse of that described above. Photosynthesis and carbohydrate concentrations decline. Growth rate is reduced more severely than is nutrient absorption, so that tissue nutrient concentrations accumulate above levels necessary to support growth. Under such circumstances one finds a reduction in carbon-based defense such as terpenes and phenolics and in some species, particularly herbaceous plants, accumulation of alkaloids, cyanogenic glycosides and other nitrogen-based defense compounds, because under such circumstances nutrients are cheap relative to carbon. Similarly, following fertilization or the nutrient release that accompanies fire, tissue nutrient concentrations increase and growth is stimulated more strongly than photosynthesis, so that concentrations of carbohydrates and carbon-based defenses decline and nitrogen-based defenses increase.' (Bryant *et al.* 1983, pp. 357, 363–364)

Gershenzon 1984; Waring *et al.* 1985; Larsson *et al.* 1986; Bryant *et al.* 1987, 1989; Price *et al.* 1989). It is based on the idea that the availabilities of carbon and nitrogen in the environment phenotypically determine the amount and kind of chemicals that a plant allocates to defence versus growth. If soil nutrients (particularly nitrogen) are limiting relative to carbon (light), growth is limited to a greater extent than is photosynthesis. Consequently, carbon is 'in excess' relative to nitrogen, and so carbon-based secondary metabolites such as phenolics, terpenoids and tannins accumulate. If nutrient availability is increased (e.g. by fertilizing) or carbon supply is reduced (e.g. by shading), allocation to carbon-based compounds declines. Allocation to nitrogen-based defences (e.g. alkaloids) is essentially the reverse. High relative carbon availability decreases nitrogen allocation, while high relative nitrogen availability increases nitrogen allocation.

The theory has had considerable success in predicting phenotypic responses of plant-based secondary metabolites to changes in nutrient or light availability. Carbon-based defences have received more attention than nitrogen-based defences. There are, however, some problems with this framework (Mole 1994; Berenbaum 1995). The theory does not distinguish between different carbon pools: free phenolics, tannins and lignin are all part of the same biosynthetic pathway, but free phenolics and tannins are considered defensive, while lignin is primarily structural. Consequently it is not a strict requirement of the theory that carbon accumulation under nutrient limitation should necessarily result in increased defence. One consequence of this non-mechanistic, black-box aspect of the theory (all carbon is lumped together) is that we do not know why the theory sometimes fails to predict outcomes (McCanny *et al.* 1990; Herms & Mattson 1992). For example, there are deviations from predictions under extreme or novel environmental conditions. In extremely low nutrient environments plants tend to be unresponsive (Iason & Hester 1993; Hartley *et al.* 1995) or show species-specific responses (Chapin 1980; Chapin & Shaver 1985), and under elevated CO_2 the framework has been rather unsuccessful (Ayres 1993; Lambers 1993; Lincoln 1993; Lincoln *et al.* 1993). Furthermore, certain classes of compounds seem less responsive than others to the carbon/nutrient status of the plant. Products of the shikimate pathway (e.g. phenolics and tannins) tend to behave as predicted by the CNB hypothesis rather more often than terpenes (Björkman *et al.* 1991; McCullough & Kulman 1991; Reichardt *et al.* 1991; Muzika & Pregitzer 1992). However, some chemical ecologists feel this hypothesis is so imprecise about the mechanisms involved that it cannot make reliable predictions about the physiological responses of plants to variation in their environment. Particularly elegant work measuring whole-plant allocation patterns by Baldwin and colleagues (Baldwin *et al.* 1993; Ohnmeiss & Baldwin 1994) has come closest to a comprehensive and detailed testing of the predictions of the CNB hypothesis. Their results for the production of nicotine under nitrogen-limited growth did not match the predictions of the theory.

While the idealizations and simplifications of this (and any) theory have heuristic value, the 'carbon-based' versus 'nitrogen-based' defence

---

Box 10.8 Resource availability hypothesis.

'Resource availability in the environment is proposed as the major determinant of both the amount and type of plant defense. When resources are limited, plants with inherently slow growth are favored over those with fast growth rates; slow rates in turn favor large investments in antiherbivore defenses. Leaf lifetime, also determined by resource availability, affects the relative advantages of defenses with different turnover rates. Relative limitation of different resources also constrains the types of defences.' (Coley *et al.* 1985, p. 895)

---

dichotomy is probably overly simplistic. 'Nitrogen-based' defences obviously contain carbon and many 'carbon-based' compounds are derived from amino acids. In reality we have not yet ascertained what is limiting to plants and how this relates to the controls over secondary metabolism. Focus on these regulatory controls and resource limitation could do much to enhance the predictability of this theory.

Resource availability. This theory (Box 10.8) is essentially a genotypic (between species) cost–benefit or trade-off model based around growth rate and leaf lifetime. Slow-growing species have an optimal defence level that is higher than that of more rapidly growing species (Coley *et al.* 1985; Coley 1988; Fig. 10.7). The theory does not invoke specific physiological trade-offs between growth and defence (in contrast to the growth/differentiation balance hypothesis; see later), but it does invoke

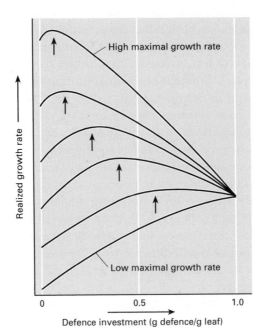

Fig. 10.7 Trade-off between growth and defence. Effect of defence investment on realized growth. Each curve represents a plant species with a different maximum inherent growth rate. Levels of defence that maximize realized growth are indicated by an arrow. (From Coley *et al.* 1985.)

selection trade-offs: it is not cost-effective, in evolutionary terms, for plant species to attempt both at once. The theory is sometimes confused with the CNB hypothesis, which has a phenotypic focus addressing within-species responses to the local environment. In contrast, the resource availability hypothesis is primarily concerned with different plant strategies across species. The theory has subsequently been extended to a within-species framework (Bazzaz *et al.* 1987). The link with plant strategies has allowed Coley (1987) to combine resource availability and apparency into a single general model of plant defence that follows Grime's (1979) triangular classification of plant strategies.

The resource availability hypothesis has much experimental support (Coley 1988; Bryant *et al.* 1989; Jing & Coley 1990; Shure & Wilson 1993; Fig. 10.8; but see Myers 1987; McCanny *et al.* 1990), although research comparing defence production in resource-rich and resource-poor environments has produced conflicting results (reviewed in Basey

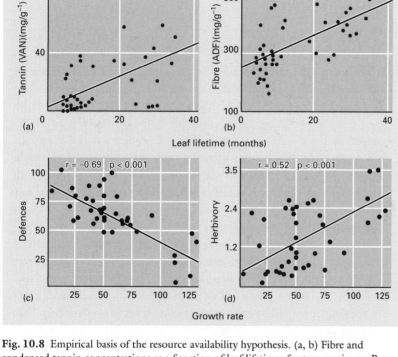

Fig. 10.8 Empirical basis of the resource availability hypothesis. (a, b) Fibre and condensed tannin concentrations as a function of leaf lifetimes for tree species on Barro Colorado Island, Panama ($n = 41$). Fibre content was measured using the acid detergent method and tannin (VAN) content using the vanillin assay for catechin-based tannins. (c, d) Herbivory and defence as a function of height growth (cm year^{-1}) for tree species on Barro Colorado Island, Panama ($n = 41$). Defence is a linear combination of defences, including measures of fibre, tannin, toughness and pubescence (units are arbitrary). Herbivory is expressed as a transformation of measured rates of leaf damage (units are % per day) and is calculated as the natural log of the mean annual rate for each species. (From Coley 1987.)

& Jenkins 1993). Like the CNB hypothesis, it is not a mechanistic theory. One area in particular deserves clarification. The theory proposes 'mobile' defences in short-lived leaves and 'immobile' defences in long-lived leaves. It is not exactly clear what is meant by immobile and mobile compounds. We do not have any good evidence that so-called mobile compounds turn over in a way that allows the plant to 'recover' the investment during the lifetime of a leaf (Briggs & Schultz 1990; Baldwin & Ohnmeiss 1994; Baldwin *et al.* 1994; but see Coley 1993; Shure & Wilson 1993). On the other hand, turnover during the lifetime of a leaf may not be necessary since many of the resources are often reabsorbed at senescence, prior to abscission. This aspect of 'turnover' has received no attention, but may help to resolve this conflict over requirements of the mass-balance component of the theory.

Chemical diversity in ecological time. The hypotheses described in Box 10.9 formalized both new and previously postulated relationships between extant patterns of plant chemical diversity and insect species richness. The framework was erected to analyse quantitatively a set of

---

Box 10.9 Chemical diversity in ecological time.

'It is widely expected that plants with unusual or diverse secondary chemistry should have reduced insect species richness. The mechanisms are defined by two hypotheses. These are named and explicitly developed here for the first time, although both are implicit in earlier studies. In the *diverse defence hypothesis*, we predict a negative correlation between insect species richness and plant chemical diversity, because several allelochemicals should be more difficult to circumvent than one or two. Similarly, the *biochemical barrier hypothesis* predicts low insect diversity on biochemically unusual plants, because rates of insect colonization should be reduced over time by novel allelochemicals. Alternatively, and contrary to general beliefs, diverse or unusual chemistry might enhance species richness, with mechanisms defined by two additional hypotheses. The *common chemistry hypothesis*, first formulated by Dethier (1941), predicts that insect species richness and plant chemical diversity will be positively correlated, because plants with high chemical diversity are more likely to share at least some classes of compounds with other plant species, facilitating host shifts by preadapted insects. Moreover, because many natural enemies of phytophagous insects use specific cues derived from the plant to locate prey, a new *enemy escape hypothesis* predicts enhanced diversity on biochemically unusual plants, because colonizing phytophages experience reduced levels of parasitism and/or predation again facilitating host shifts.' (Jones & Lawton 1991, pp. 768–769)

ecological patterns (and evolutionary implications) in chemical and insect databases compiled from literature on the British umbellifers (Jones & Lawton 1991). Surprisingly, this study found no correlations between chemical diversity, at the level of biosynthetically distinct classes, and insect species richness on these plants. However, a single test is hardly definitive. It is clear that 'mining' the literature has obvious problems, such as an absence of data on the absence of chemicals, and biases due to intensity of biochemical studies on particular plants. Nevertheless, it ought to be possible to gain some synoptic insights from the vast amounts of published data on natural product distribution, diversity, toxicity and insect host plant records. Such data remain more or less an unexplored resource in the field of chemical ecology. Quantitative analyses of broad patterns would help to balance the tendency of most coevolutionary studies to be species specific. After all, why do we collect all these data if we don't then use them as fully as possible?

Screening hypotheses of chemical diversity. The theory (Box 10.10) offers an explanation for evolution of plant chemical diversity that arises as a consequence of formal, toxicological constraint. The central argument hinges on the low individual probability that any chemical will have high, specific biological activity against targets at naturally occurring concentrations. This probability is inherently low because such activity is dependent on tight stereospecific fit of chemical to receptor, or chemical to chemical. In essence, this constraint on activity means that most compounds are inactive, and many compounds must be made and 'screened' against targets for any activity to arise. Compounds retained by plants beget further diversity, thereby increasing the overall probability of any activity. Plants that lose too much diversity, by eliminating

---

Box 10.10 Screening hypothesis of chemical diversity.

'A common-sense evolutionary scenario predicts that well-defended plants should have a moderate diversity of secondary compounds with high biological activity. We contend that plants actually contain a very high diversity of mostly inactive secondary compounds. These patterns result because compounds arising via mutation have an inherently low probability of possessing any biological activity. Only those plants that make a lot of compounds will be well defended because only high diversity confers a reasonable probability of producing active compounds. Inactive compounds are retained, not eliminated, because they increase the probability of producing new active compounds. Plants should therefore have predictable metabolic traits maximizing secondary chemical diversity while minimizing cost.' (Jones & Firn 1991, p. 97)

inactive compounds, have a reduced probability of producing new, active chemical variants in the face of target adaptation. Plants should therefore retain chemicals, most of which are inactive, and should have evolved mechanisms for maximizing diversity whilst minimizing its costs.

The theory thus derives from a very different premise from that of coevolution, and can be considered a null model that parallels, from the plant side, the sequential evolution model for insects of Jermy (1976). The theory is relatively new, and the specific predictions that relate to mechanisms of diversification at minimal cost have yet to be tested.

Growth/differentiation balance (GDB). This framework (Box 10.11) examines the dilemma of plants: to grow and outcompete their neighbours, whilst at the same time defending themselves against herbivores and pathogens. It is an optimality model, based around the physiological constraints of a trade-off between growth and differentiation. The hypothesis addresses this issue at a range of levels, from the cell (rapidly dividing cells tend to have low secondary compound levels) to the

---

Box 10.11 Growth/differentiation balance hypothesis.

'Physiological and ecological constraints play key roles in the evolution of plant growth patterns, especially in relation to defenses against herbivores. Phenotypic and life history theories are unified within the growth-differentiation balance (GDB) framework, forming an integrated system of theories explaining and predicting patterns of plant defense and competitive interactions in ecological and evolutionary time.

Plant activity at the cellular level can be classified as growth (cell division and enlargement) or differentiation (chemical and morphological changes leading to cell maturation and specialization). The GDB hypothesis of plant defense is premised upon a physiological trade-off between growth and differentiation processes. The trade-off between growth and defense exists because secondary metabolism and structural reinforcement are physiologically constrained in dividing and enlarging cells, and because they divert resources from the production of new leaf area. Hence the dilemma of plants. They must grow fast enough to compete, yet maintain the defenses necessary to survive in the presence of pathogens and herbivores.

The physiological trade-off between growth and differentiation processes interacts with herbivory and plant–plant competition to manifest itself as a genetic trade-off between growth and defense in the evolution of plant life history strategies.' (Herms & Mattson 1992, p. 283)

species level (rapidly growing species tend to have low levels of secondary metabolites, which is an equivalent trade-off to that of the resource availability hypothesis; see above). GDB was developed partly in response to the fact that some experimental results are inconsistent with CNB. GDB is essentially similar to resource-based allocation optimality modes such as the CNB and resource availability hypotheses (Tuomi 1992; Mole 1994), but it adopts a more comprehensive approach.

A potential problem with GDB is that it can be difficult, and perhaps inappropriate, to try and squeeze all plant physiological processes into a growth/differentiation trade-off. There is not necessarily a clear positive relationship between differentiation and defence or a clear negative one between growth and defence (Vrieling & van Wijk 1994). The current evidence for trade-offs is rather weak, and is usually measured at a different organizational scale from those of the predictions of GDB (Mole 1994). We need more physiological studies in order to critically evaluate these trade-off models.

A more general point has recently been raised in response to GDB and many of the trade-off-based frameworks: to what extent should allocation to defence be considered to be driven by 'source' (i.e. plant resource) or 'sink' (i.e. herbivory) factors? Are abiotic factors the major selective forces determining whether plants invest highly in secondary plant compounds or is it herbivory (Baas 1989)? GDB and CNB predominantly assume the former, while apparency predominantly assumes the latter. This question is still the subject of lively debate (Herms & Mattson 1994; Lederau *et al.* 1994; Lederau 1995).

10.2.3 Last thoughts on secondary metabolism and how green the world is

The rich, diverse and continuously developing body of theory on secondary metabolism is a clear reflection of the vigour in this field. While there is, and will no doubt continue to be, substantial debate on the causes and consequences of interactions between plant secondary metabolites and herbivores, few would dispute that these compounds play an important role. We may not be sure exactly how important secondary metabolism is compared to low nitrogen, other 'cruddy ingredients', heterogeneity or natural enemies, but uncertainties about relative importance should not be construed as implying no importance. There is little doubt that plant secondary metabolites play key roles in keeping the world green; it is just that we have some way to go in figuring out exactly how and why.

10.3 Plants are poor food: they are unpredictable

Plants are not just poor-quality resources for herbivores, they are unpredictable resources. Plants are very heterogeneous in space and time

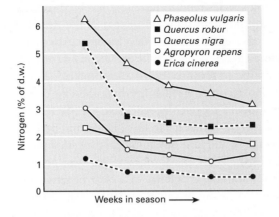

Fig. 10.9 Temporal heterogeneity of plant nitrogen. Change in levels of leaf nitrogen of five different plant species through a temperate region growing season. The time span involved was divided into four sectors, since the actual times are different in each case. (From Bernays & Chapman 1995.)

(Denno & McClure 1983), both qualitatively and quantitatively, and they can suddenly change in quality in response to abiotic and biotic stress and damage.

10.3.1 Intrinsic heterogeneity

Plants are intrinsically heterogeneous because that is how they work. In other words, much heterogeneity is a direct result of primary plant functions and processes of growth, development, reproduction and senescence. This heterogeneity has consequences for herbivores, irrespective of the particular chemicals involved. There is a great deal of evidence showing intrinsic heterogeneity in plant characteristics is important to herbivores (Cates 1975; Schultz *et al.* 1982; Denno & McClure 1983; Jones 1983; Baldwin *et al.* 1987; Marquis 1992; Figs 10.9–10.12; Box 10.12).

There is variation in plant characteristics at every spatial and temporal scale. Plants are moving targets and this can cause problems for herbivores (Jones 1983). What is the relationship between heterogeneity and defence? Are the most heterogeneous plants the most well defended? At what scale is heterogeneity relevant to insect herbivores? Do plants escape from herbivore attack in space and time by being heterogeneous? We do not really know. Many studies have related variation in plant characteristics to variation in herbivore damage or numbers (see Marquis 1992 for review; Hunter 1996), but there are only a few experiments directly addressing the effects of plant heterogeneity on herbivore performance. Several studies show that variability in bud burst affects herbivores (e.g. Crawley & Akhteruzzaman 1988; Quiring 1993; Mopper & Simberloff 1995). Asynchrony of hatching or development with bud burst can result in severe mortality to some folivores, such as winter moth feeding on apple (Holliday 1977) or gypsy moth on oak (Hunter 1993). However, date of bud burst had no effect on winter moth feeding on Sitka spruce or heather (Watt & McFarlane 1991; Kerslake & Martley 1997). Heterogeneity tends to make resources more finely grained and short-lived, which can increase

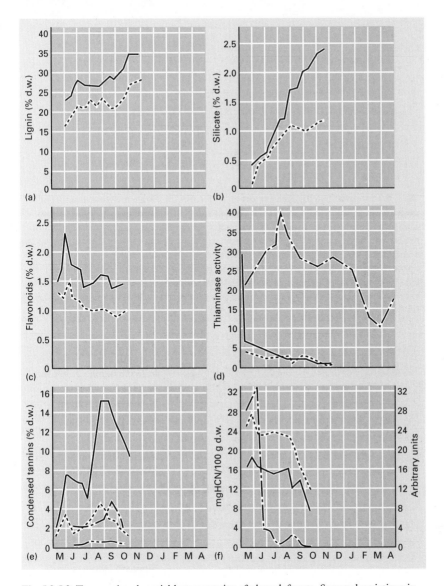

Fig. 10.10 Temporal and spatial heterogeneity of plant defences. Seasonal variations in allelochemicals of bracken fern. (a) Lignin. Frond (——), rachis (- - -) (b) Silicate. Frond (——), rachis (- - -). (c) Flavonoids. Frond: open (——); frond: shade (- - -). (d) Thiaminase. As micrograms of thiamine destroyed per minute per gram dry weight. Frond (——), rachis (- - -), rhizome (– – –). (e) Condensed tannins. Frond (- - -), rachis (– – – –). Frond: open (——); frond: shade (- - -). (f) Cyanogenesis (populational). As micrograms of HCN per gram dry weight. Frond (– – –). Cyanogenesis in an individual pinna. As arbitrary units. Pinna: open (——); pinna: shade (- - -). (From Jones 1983.)

the likelihood of competition between herbivores and increase the risk of 'missing the boat' (Jones 1983). Gall-formers, which require very specific young stages of plant tissue for oviposition sites, are a good example because few suitable oviposition sites on a plant may exist and timing is crucial (Craig *et al.* 1986, 1990). We actually have very few

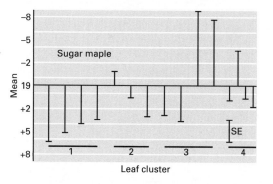

Fig. 10.11 Spatial heterogeneity in defence of leaves on a branch. Value in terms of digestibility of individual leaves on individual sprays of sugar maple (*Acer saccharum*). Horizontal axes are mean values (as percentage tannic acid equivalents) of all leaves on the branch. Each vertical bar represents one leaf, expressed in units of deviation from the branch mean. Upward-deviating bars indicate more-digestible, higher-quality leaves; downward deviations represent less-digestible leaves. In this case, branch terminus is at left. This indicates that the sugar maple leaves are all even aged and grow in clusters. One standard error of the branch mean shown lower right. HEME, hemoglobin assay. (From Schultz 1983.)

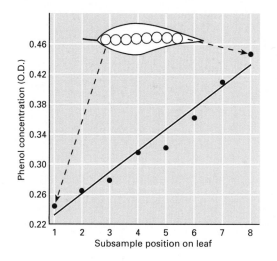

Fig. 10.12 Spatial heterogeneity in defence of a leaf. Subsamples of leaf tissue collected on or along the midrib of the leaf blade, where galls are formed, reveal that the concentration of phenolics (O.D., optical density) is lowest at the base of the leaf, where stem mothers prefer to form galls ($r = 0.980$, $P < 001$, $n = 8$). The data presented here are from a single representative leaf. Leaves were collected at the time of aphid settling. (From Whitham 1983.)

clear cases of a reduction in herbivore population density and a corresponding decrease in plant damage as a direct result of plant heterogeneity (Morrow & Fox 1980; Senn *et al.* 1992; Linhart & Thompson 1995; Hunter 1996). There is some evidence that damage dispersed within a plant is less detrimental to the plant than damage that is concentrated in one area (Mauricio *et al.* 1993).

The study of plant–insect interactions is becoming increasingly spatially explicit so it seems appropriate that ecologists give greater emphasis to the role of spatial heterogeneity in plant chemistry in the study of insect population dynamics. This topic has recently been reviewed by Hunter (1996) who concludes 'there is no such thing as a herbivore population immune to the effects of variation in plant quality,

Box 10.12 Intrinsic heterogeneity: some examples.

In time

Diurnal fluctuations (Haukioja *et al.* 1978)

Leaf age and stage of development (Raupp & Denno 1983; Meyer &
Montgomery 1987)

Seasonal variation (Dement & Mooney 1974; Schultz *et al.* 1982;
Potter & Kimmerer 1986; Ayres & McLean 1987)

Plant maturity (Kearsley & Whitham 1989; Bowers & Stamp 1993)

In space

Within individual leaves (Zucker 1982; Kimmerer & Potter 1987)

Between leaves within shoots (Schultz 1983; Jones *et al.* 1993)

Between branches on the same trees (Whitham 1983; Tuomi *et al.*
1991)

Whole plants (Crawley & Akhteruzzaman; Senn *et al.* 1992; Soumela
& Ayres 1994; Hahnimäki *et al.* 1995; Hemming & Lindroth
1995)

even if its dynamics are dominated by predation ... I would argue that
spatial variation in plant quality is a near ubiquitous feature of insect–
plant systems'. It has also been suggested (Belovsky & Joern 1995) that
spatial variation in plant quality may have sufficient impact on insect
population dynamics to cause an insect species to be endemic in one
area and epidemic in others.

Indirect effects of plant heterogeneity are mediated by increasing the
time spent searching for good quality food since this can lead to
increased predation or parasitism (Heinrich & Collins 1983; Schultz
1983; Courtney 1988), but increased death risk has proved difficult to
demonstrate in the field (Fowler & MacGarvin 1986; Damman 1987;
Bergelson & Lawton 1988; but see Häggstrom & Larsson 1995). Even
if it does occur we do not know if it then results in less damage to
plants. So overall, we know a lot more about the intrinsic heterogeneity
of plant characteristics than we know about the consequences of the
heterogeneity to either plants or herbivores.

10.3.2 Extrinsic heterogeneity

Because they cannot run away plants respond to almost everything.
These responses to the environment affect their quality as food for
herbivores (Woodhead 1981; Waterman & Mole 1989; Tallamy &
Raupp 1991; Box 10.13). We know that this heterogeneity has effects
on herbivore behavioural performance (e.g. Nichols-Orians 1991;
Sagers 1992), but the overall impact on herbivore populations and

subsequent plant performance remains far from clear (Rhoades 1983; Cates & Zou 1990). It is particularly difficult to know if reduced herbivore performance translates into better plant performance amidst other effects of the environment (Crawley 1989; Karban 1993a,b).

10.3.2.1 Responses to abiotic stress

The effects of single abiotic stresses (e.g. Box 10.13) on plant chemistry and herbivores have received considerable attention (see reviews by English-Leob 1990; Jones & Coleman 1991; Waring & Cobb 1992), although we are only now beginning to draw general conclusions. We know a lot less about the effects of multiple stresses, despite the fact that plants live in a complex and continuously changing world. How do different environmental variables interact to change plant food quality? For example, what is the sum effect of simultaneous changes in ozone, nitrogen and drought? We have not carried out enough multifactor studies, particularly related to the complex changes expected with human-accelerated environmental changes (but see Rao *et al.* 1995). We do not really know what the balance sheet looks like: a factor may improve some aspects of plant quality and decrease others simulta-

---

Box 10.13 Extrinsic plant heterogeneity: some examples.

Light: increases levels of carbohydrates and carbon-based secondary compounds (Larsson *et al.* 1986; Mole *et al.* 1988)

Nitrogen: increases levels of protein; decreases levels of some secondary compounds (Gershenzon 1984; Mihaliak *et al.* 1989; Estiarte *et al.* 1994; Hartley *et al.* 1995)

Drought: may increase nitrogen levels; may decrease secondary compound levels (Rhoades 1983; White 1984; Mattson & Haack 1987; Major 1990)

Ozone and other pollutants: may alter amino acid content and carbon-based defences (Rowland *et al.* 1988; Jackson 1991; Jones *et al.* 1994)

Temperature: may alter plant quality; alters phenology (Wellings & Dixon 1987; Benedict & Hatfield 1988)

Pathogens: can increase secondary compound levels (Bazzalo *et al.* 1985; Dixon 1986; Campbell & Ellis 1992)

Herbivores: can increase secondary compound levels (Tallamy & Raupp 1991; Karban & Baldwin 1997)

neously. This makes it extremely difficult to predict the overall effect on herbivores. One of the most basic problems is that the impacts of stress on plant and herbivore are highly variable and contingent. They depend on the nature of the perturbation, the state of the plant, the plant's sensitivity to the perturbation, and the herbivore's sensitivity to the plant's responses (Larsson 1989; Watt 1990; Jones & Coleman 1991). Multiply that complexity by several simultaneous stress and damage factors and it is not surprising that it is difficult to make predictions. We really need to develop theories that can deal explicitly with the high level of contingency.

10.3.2.2 Responses to biotic damage

Are the effects of biotic damage to plants and the subsequent responses of herbivores any less contingent and more predictable? It is well established that wounding by herbivores, as well as other biotic and abiotic factors, causes chemical changes in plants (Box 10.13, Fig. 10.13). These so-called induced defences can then have adverse effects on herbivores. We have lots of information on the effects of damage on plant quality (Karban & Myers 1989; Tallamy & Raupp 1991; Karban & Baldwin 1996). Damage-induced chemical changes can take place over different temporal scales – delayed versus rapid induction (Hartley & Lawton 1991; Neuvonen & Haukioja 1991; Bryant *et al.* 1993), and different spatial scales – systemic versus local induction (Ryan 1974; Hartley &

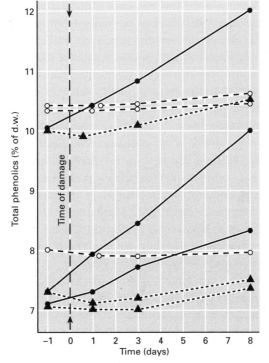

Fig. 10.13 Induction of chemical defences by herbivores. Changes in total phenolics in leaves of six birch trees after damage to the leaves of the three experimental trees produced by allowing *Apochemia pilosaria* caterpillars to eat approximately 15% of leaf-area of damaged leaves. —●—, grazed leaves; ··▲··, ungrazed leaves on the grazed trees; ––○––, control (undamaged) trees. (From Lawton 1987.)

Firn 1989; Jones *et al.* 1993; Simms & Vision 1995). A diversity of different compounds are involved, and over 100 plant species have been shown to be inducible (Karban & Baldwin 1997) although many are not (see Herms & Mattson 1992). There has also been extensive research into the effectiveness of these induced changes against herbivores and other organisms, accompanied by much debate (Fowler & Lawton 1985). Effects on insect herbivores are often variable, and while some species prefer and perform better on damaged leaves (Hartley & Lawton 1987; Myers 1987; Hartley 1988), there are many examples where feeding and performance decline (Coleman & Jones 1991). There is some good evidence for induced chemical changes being effective against pathogens (Kuc 1982; Simms & Vision 1995). Resistance to fungi can be related to induction of phenolics, for example (Moesta & Griesbach 1982; Carver *et al.* 1991), and damage-induced increases in phenolic content of some plants can convert susceptible to resistant strains (Carrasco *et al.* 1978).

Despite this wealth of information and mechanistic detail (Ryan *et al.* 1986; Baldwin 1989), we still have some way to go to establish broad, general patterns. Are there consistent patterns across plant species in terms of which compounds are induced and when, and how long induction lasts? Are there general mechanisms of induction (e.g. the involvement of signalling by compounds like salicylic acid and methyl jasmonate; Bennett & Wallsgrove 1994)? Can we apply what we know about, say, the induction of phenolics in birch to the induction of other sorts of compounds in other plant species? At the moment the answer to these questions is a tentative 'no'. How much specificity is there in the plant response (Mullick 1977; Baldwin 1991; Hartley & Lawton 1991; Karban & Niiho 1995; Silverman *et al.* 1995)? What is the nature of the differences in induction responses to insect versus mechanical versus pathogen damage (Rhoades 1985)? Can we make an evolutionary distinction between active and passive induction (Tuomi *et al.* 1984; Bryant *et al.* 1993; Hunter & Schultz 1995)? One area in particular which is proving interesting, if complex, is the idea of cost. We often assume that defence has a cost, so induction is defence on the cheap. There is some evidence supporting this idea of cost (Baldwin *et al.* 1990; Han & Lincoln 1994; Fineblum & Rausher 1995) but trade-offs between inducible defences and, say, plant competitive ability may be hard to demonstrate (Karban 1993a,b; Adler & Karban 1994; Sagers & Coley 1995). The evolutionary aspects of these costs are well reviewed by Simms (1992) and a recent survey of the literature on costs of resistance in plants demonstrated some intriguing general patterns (Bergelson & Purrington 1996).

One fundamental question that is still unresolved is whether or not induced changes really are effective defences against herbivores. Despite an accumulating body of evidence in support of this idea, there are still some reasons for doubt (Williams & Myers 1984; Roland & Myers

1987). In particular, there are relatively few studies showing any effects of these induced chemical changes on insect population dynamics (Fowler & Lawton 1985; Karban & Myers 1989; Karban 1993a; Harrison 1995). Consequently, we are not really sure if observed effects on feeding preference (e.g. Hartley & Lawton 1987) really do translate into density effects in the real world. Even if we were confident that a general reduction in insect population density would result from induced defences, we still need to show that if plants induce, they get less

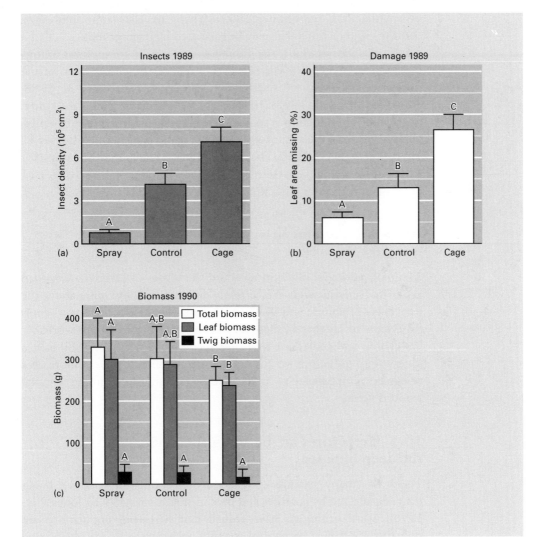

Fig. 10.14 Herbivore damage and its effect on subsequent plant growth. Treatment effects on insect numbers, leaf damage and biomass production. Letters indicate treatment differences at $P < 0.05$. (a) Total insect numbers summed over the 1989 season. (b) Leaf damage in 1989. (c) Biomass production in 1990. Different letters above the bars indicate significant differences among treatments at $P < 0.05$ and among treatments with a biomass category at $P < 0.05$ for (c). Error bars represent one standard error of the mean. (From Marquis & Whelan 1994.)

subsequent damage (Edwards & Wratten 1983; Edwards *et al.* 1992 but see Rausher *et al.* 1993). The benefit in terms of fitness ought to be enough to cover the costs of the induction. Few studies have attempted to address this issue (Silkstone 1987; Fig. 10.14), but induced responses are known to confer resistance to fungal infection in cotton (Karban *et al.* 1987). Similar effects have been demonstrated between leaf-miners and folivores in birch (Hartley & Lawton 1987, 1991; Valledares & Hartley 1992), early-season and late-season herbivores on oak (West 1985; Faeth 1986, 1988; Hunter & Willmer 1989; Hunter & West 1990; but effects vary with guild, see Hunter 1992). There are also some interesting tritrophic interactions that may be mediated by damage-induced chemical changes. For example, the induction of phenolic compounds following damage by gypsy moth larvae may reduce the effectiveness of the nuclear polyhedrosis virus attacking gypsy moth (Keating *et al.* 1988, 1990; Schultz *et al.* 1992). In this case the plant may suffer the consequences of its induction response. In contrast, plant damage by herbivores may release cues that attract predators and parasites (Turlings *et al.* 1990; Vet & Dicke 1992; Dicke *et al.* 1993). Plants may be calling for help, but we are not sure (Firn & Jones 1995), and it is also possible that other plant responses hinder parasitoids (Price 1988).

10.3.3 Last thoughts on unpredictability and how green the world is

All in all, we know much more about the causes of heterogeneity than about its consequences. We certainly do need to know more about the net effects of abiotic and biotic stress and damage. At present it seems reasonable to say that all these factors create an unpredictable world for herbivores, and contribute to its green colour. However, we still have a long way to go before we can say that heterogeneity *per se* is more, less or equal in importance to 'cruddy' ingredients in determining why the world is green.

10.4 Herbivores are between the devil and the deep blue sea

There are many examples of herbivore population density being markedly affected by parasites and predators (Crawley 1992a; Jones *et al.* 1996). Some ecologists have argued that herbivore populations are primarily controlled by natural enemies, rather than by the plant or by competition between herbivores (see Chapter 13), and strong regulation of herbivore populations by natural enemies may only be possible if plants are of such low nutritional quality that herbivore birth rates are low (Schultz 1992). Control of herbivore populations most likely results from a combination of bottom-up and top-down forces (Lawton & McNeill 1979; Faeth 1985, 1987; Hunter & Price 1992; Hunter

---

Box 10.14 Suggested future hot topics.

How much heterogeneity in plants is driven by intrinsic forces (i.e. being a plant), compared to extrinsic ones (biotic and abiotic)? What are the consequences of heterogeneity to herbivores? Can heterogeneity *per se* be a defence?

How does plant quality and quantity interact with natural enemies? When does competition between herbivores become important? We could try manipulating food quality in the presence and absence of natural enemies and look at effects on both herbivores and plants.

What is the relative importance of idiosyncratic characteristics such as unusual secondary metabolites compared to more general ones like nitrogen content? How do we get answers to this question?

What is the link between mechanisms like damage-induced chemical changes or fertilization and subsequent impacts on the plant? Let's stop studying just the herbivores. How many different ways to be 'cruddy' do plants really need? What is the relative importance to the plant of different chemical defence mechanisms?

How different are plant interactions with herbivores from interactions with pathogens, both mechanistically and in terms of the bottom line on consumer and plant?

Can we deal with the contingencies that seem inherent in the complex interactions among abiotic environment, plant and herbivores? How can we deal with simultaneous multiple stress, damage and feedback to plants?

How does chemical diversity arise and how is it maintained? Can we get a good answer to Fraenkel's original postulate?

1996; Karban 1996). There are, as yet, few studies that allow us to evaluate the relative importance of these two controls, but some experimental work has recently begun to address this issue (Karban 1989; Kato 1994; Ohsaki & Sato 1994; Bustamante *et al.* 1995).

Competition among insect herbivores is now thought to be more common than previously thought (Denno *et al.* 1995), and given that plants are marginal and unpredictable food we might expect to see competition, particularly among the immobile herbivores that are more likely to be resource-limited. Gall-forming and leaf-mining insects have been shown to compete in some cases (Bultman & Faeth 1986; Craig *et al.* 1986, 1990; Fritz *et al.* 1987; Faeth 1988; Valledares & Hartley 1994), and plant-mediated interactions between spatially and temporally separated herbivores have also been demonstrated (Faeth 1986; Harrison & Karban 1986; Karban 1986; Masters & Brown 1992). We also now realize that plants can directly influence a herbivore's risk of mortality, via direct or indirect effects on natural enemy effectiveness (Schultz 1983, 1988, 1992).

There is now increasing interest in plant resources as a fundamental structuring force in insect herbivore population dynamics (Ohgushi 1992; Price 1992) and while few would dispute the importance of natural enemies or plant characteristics, we still have much work to do before we can say what factors are the most important, or when, where and how these factors interact (Box 10.14).

10.5 Conclusions

The world is green because plants are the way they are, and herbivores have to live with this. They usually find it hard to live with, so herbivore numbers are usually low and are kept even lower by natural enemies. Herbivores are indeed 'between the devil and the deep blue sea' (Lawton & McNeill 1979). They rarely escape from this dual control, and we do not know exactly what allows them to escape. So the world stays green. All that looks green is not edible. Plants are poor food because they contain 'cruddy' ingredients and because the occasional morsels of good food are hard to find amongst the bad, and the plants keep changing. We are still not sure why plants have such a huge array of secondary chemicals nor how this came to be. We do know that these chemicals are important in keeping the world green, but we are not sure exactly *how* important.

11: The Structure of Plant Populations

Michael J. Hutchings

11.1 Introduction

The structures that we can identify in populations of plants result from the action of biotic and abiotic forces to which their members, and in some cases their ancestors, have been exposed in the past. The forces experienced by the ancestors of a population can clearly affect its genetical structure (although this may not be perceptible), but the spatial structure of plants in a population is also a legacy of the spatial arrangement of parent plants and of the interactions that have taken place between plants in the past. Plant interactions together with abiotic factors also mould the 'performance structure' of populations. This can be viewed as an expression of the opportunities for growth realized by each member of the population in the course of its development. Finally, the age structure of a population reflects both past opportunities for recruitment and the mortality risks to which each recruit has subsequently been exposed. These four aspects of structure–performance, spatial arrangement, age and genetical structure–are the subject of this chapter.

All of these aspects of population structure are interrelated, so that change in one generates changes in the others. The growth of the individuals in a population, for example, sets in motion a chain reaction of modifications to population structure. While it is convenient to discuss each aspect of population structure in isolation, it is important to remember the close links between them. I begin with a description of plant performance, the most conspicuous aspect of population structure, and follow this with discussions of spacing, age and genetical structure. Finally, the influence of abiotic environmental variation on population structure is considered.

11.2 Performance structure in plant populations

11.2.1 Plant weights

The sizes of individuals in plant populations are far from uniform. This is true even in most monocultures, including cereal crops and forest plantations, in which plants are even-aged. Since many aspects of performance are correlated with size, they too vary from plant to plant.

While seedling populations usually have a symmetrical (but not necessarily normal) weight distribution (Obeid *et al.* 1967; Harper *et al.* 1970; Rabinowitz 1979), several factors combine to produce an asymmetric, positively skewed distribution of adult plant weights (see Box 11.1). The factors responsible are considered below. The population therefore develops a marked 'size hierarchy', with a small number of large plants, which account for most of the population's biomass (Weiner & Solbrig 1984), and many small ones (Fig. 11.1). Three important points to note are that: (i) all plants in the population do not

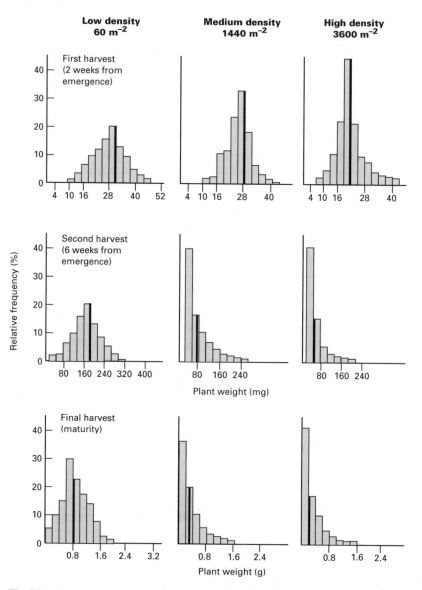

Fig. 11.1 Frequency distributions of plant weights in populations of flax, *Linum usitatissimum*, sown at three densities. Mean plant weight is shown by the black bars. (From data in Obeid *et al.* 1967.)

Box 11.1 Measuring the size inequality of plants.

The simplest measures used to characterize plant size distributions are mean, variance, skew and kurtosis. The mean is simply the average plant size. Variance describes the variability in size (calculated as the average of the *squared* residuals; a residual is the difference between an individual plant's size and the mean plant size). Skew indicates whether the distribution has long 'tails' to the left (skew < 0) or to the right (skew > 0), or whether the distribution is symmetrically bell-shaped (skew $= 0$). Skew is calculated on the basis of the *cubes* of the residuals. Kurtosis measures the degree to which the distribution is more pointy (leptokurtic) or more flat-topped (platykurtic) than a normal distribution. Kurtosis is based on the *fourth powers* of the residuals, with negative values indicating platykurtosis and positive values showing leptokurtosis. Normal distributions have kurtosis of 0.

If y is the weight of an individual plant, and there are n different plants in a sample, the parameters are calculated as follows:

Mean $\quad \bar{y} = \Sigma y / n$

Variance $\quad s^2 = (\Sigma y^2 - (\Sigma y)^2 / n) / (n - 1)$

Skew $\quad g_1 = (n\Sigma y^3 - 3\Sigma y \Sigma y^2 + 2(\Sigma y)^3 / n) / (s^3(n - 1)(n - 2))$

Kurtosis $\quad g_2 = ((n + 1)(n\Sigma y^4 - 4\Sigma y \Sigma y^3 + (6(\Sigma y)^2 \Sigma y^2 / n) - 3(\Sigma y)^4 / n^2) / (s^4(n - 1)(n - 2)(n - 3))) - (3(n - 1)^2 / (n - 2)(n - 3))$

where $\quad \Sigma y \quad$ is the sum of all the plant weights

$\quad\quad\quad \Sigma y^2 \quad$ is the sum of all the squared plant weights

$\quad\quad\quad \Sigma y^3 \quad$ is the sum of all the cubed plant weights

$\quad\quad\quad \Sigma y^4 \quad$ is the sum of all the fourthed plant weights

$\quad\quad\quad (\Sigma y)^2 \quad$ is the grand total of the plant weights squared

$\quad\quad\quad (\Sigma y)^3 \quad$ is the grand total of the plant weights cubed

$\quad\quad\quad (\Sigma y)^4 \quad$ is the grand total of the plant weights fourthed

$\quad\quad\quad s \quad$ is the standard deviation; the square root of s^2

For worked examples, see Sokal and Rohlf (1981, pp. 114–117).

Gini coefficient

When the emphasis is to be placed on size inequality rather than on asymmetry (skewness) in describing the size hierarchy, then the Gini coefficient is an appropriate measure (Weiner & Solbrig 1984). For n plants of mean size $\bar{y}$, this is given by:

$$G = \frac{\sum_{i=1}^{n} \sum_{j=1}^{n} [y_i - y_j]}{2n^2\bar{y}}$$

continued on p. 328

Box 11.1 *Continued*

When all the plants are the same size, $G = 0$, while in a population of infinite size where all but one of the plants are infinitely small, $G = 1$.

Coefficient of variation

Size inequality is often calculated using the coefficient of variation (CV), which is closely correlated with the value of the Gini coefficient. For plants of mean size $\bar{y}$, with a standard deviation of s, CV is given as:

$$CV = (s/\bar{y}) \times 100\%$$

The CV is independent of the units in which the mean and standard deviation are expressed.

Since data on plant dry weights often follow a log-normal distribution (Box 11.2), it is usually better to carry out statistical analyses on log weights rather than on weights. This has the added advantage of making the calculation of relative growth rates more straightforward (Box 11.3).

directly influence the performance of all other plants, although they may do so indirectly; (ii) the strength of interaction between plants is not necessarily reflected by their respective ranks in the size hierarchy; and (iii) the closest, 'first-order' neighbours of a plant have by far the greatest impact upon its performance. This is true almost irrespective of rank in the size hierarchy.

Many factors promote variation in the sizes of plants in populations. First, seed size is rarely constant within a species (Hendrix & Sun 1989; see section 11.2.2.3), and seedling size at a given age is often correlated with the size of the seed from which it grew (Black 1957; Fenner 1983; Stanton 1984; Crawley & Nachapong 1985). Second, relative growth rate (RGR) is genetically determined. Third, the time of germination relative to that of neighbours is a major determinant of the future growth of plants. In monocultures of *Dactylis glomerata* (Poaceae), for example, the growth of the earliest germinating plants was hardly affected by competition, whereas those germinating 10 days later had achieved negligible weight increase after 35 days of growth (Ross & Harper 1972; Fig. 11.2a). In addition, the size of the unoccupied space in which plants germinated had a significant effect on their subsequent growth, although position within the space had little effect (Fig. 11.2b,c). Lastly, several studies show positive correlations between plant performance and distance to close neighbours (Pielou 1961; Yeaton & Cody 1976; Liddle *et al.* 1982). Distance, size, species and spatial arrangement of neighbours together can account for much of the

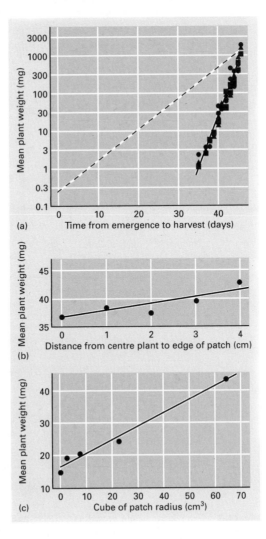

(a)

(b)

(c)

Fig. 11.2 Gap size and seedling establishment. (a) Mean weight of groups of plants emerging at different times in three populations. The dotted line indicates the weights of plants growing for different lengths of time in the absence of competition. (b) Response of individual plant weight to varying the distance of the centre plant from the perimeter of patches of a given size, showing the rather slight effect of distance. (c) Response of individual plant weight to varying sizes of patches, showing a much more pronounced effect. (From Ross & Harper 1972.)

variation in individual plant weight (Mack & Harper 1977). In the words of Ross and Harper (1972), 'an individual's potential to capture resources is dictated by the number and proximity of neighbours already capturing resources' from the same resource pool.

Cannell *et al.* (1984) conducted a detailed analysis of the effects of close neighbours on growth rates in monocultures of Sitka spruce (*Picea sitchensis* Pinaceae) and lodgepole pine (*Pinus contorta*) planted in hexagonal lattices. In a hexagonal lattice, each tree has six first-order neighbours, all the same distance away from it. Every tree in the populations was classified as having either 0, 1, 2, 3, 4, 5 or 6 first-order neighbours taller than itself. Trees taller than all of their neighbours had high relative height growth rates (RHGRs), while those with many taller neighbours had significantly lower RHGRs (Fig. 11.3a,b). Thus, the smallest, most suppressed trees experienced the most severe competition. Many trees died during the experiment; there was an inverse

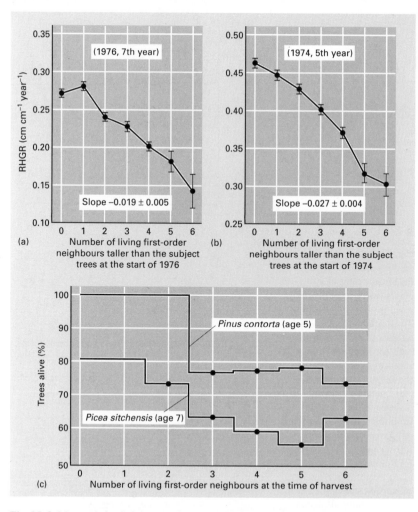

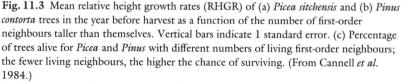

Fig. 11.3 Mean relative height growth rates (RHGR) of (a) *Picea sitchensis* and (b) *Pinus contorta* trees in the year before harvest as a function of the number of first-order neighbours taller than themselves. Vertical bars indicate 1 standard error. (c) Percentage of trees alive for *Picea* and *Pinus* with different numbers of living first-order neighbours; the fewer living neighbours, the higher the chance of surviving. (From Cannell *et al.* 1984.)

relationship between the probability of a tree surviving and the number of its first-order neighbours that survived (Fig. 11.3c). In other words, trees were more likely to survive if their neighbours died. This study clearly illustrates the way in which competition from neighbours can affect plant size, survival and the spatial structure of the population.

Herbivory could also influence the size hierarchy. Weiner (1993) has remarked that the variety of forms of herbivory, and of plant communities in which it occurs, may make generalizations about its effects upon population structure impossible. For example, if herbivores preferentially attack, but do not kill, the small plants in a population, they might increase size variation, whereas herbivory on the large plants would reduce it (Crawley 1983; Weiner 1988). The effects would be different if the attacked plants were killed and thus removed from the size hierarchy. Few studies have shown correlations between probability of herbivore attack and plant size. Amongst these, large plants of winter wheat suffered more rabbit grazing than small plants (Crawley & Weiner 1991), whereas herbivores damaged smaller plants more in *Impatiens pallida* (Balsaminaceae) populations (Thomas & Weiner 1989). Most experiments have shown that herbivory increases variation in plant sizes (Windle & Franz 1979; Cottam 1986; Gange & Brown 1989). Recent information shows that the growth and size achieved by plants following simulated herbivory depends greatly on the parts of the plant removed (Price & Hutchings 1992). Thus, differences in the sites and patterns of herbivory on different plants may also frustrate attempts to generalize about its effects on population structure.

Although inter-plant competition promotes variation in plant weights (Koyama & Kira 1956), a size hierarchy (Box 11.2) will develop even in its complete absence. Populations with very low density, in which few if any plants are competing, develop size inequality because seed size, RGR and time of germination all vary, even in the absence of competition. At higher densities, competition commences as a symmetric or two-sided interaction between plants (Weiner & Thomas 1986), but as time passes it becomes strongly asymmetric (one-sided), such that although competing plants of any size reduce each other's growth, larger plants have a far more adverse effect on the growth of a smaller plant than vice versa. Because asymmetric competition increases variation in growth rates between dominant and suppressed plants, its onset promotes size inequality, and this effect is heightened as time passes (see Fig. 11.1). Note that the size of the largest plants in crowded populations is not usually as great as that of plants growing in isolation (see for example Weiner *et al.* 1990), implying that *complete* asymmetry (i.e. the larger competitor suppresses growth of the smaller, but the smaller plant has no adverse effect on the larger), does not occur. Morphology affects the severity of competitive interactions between plants, with species that cast greater shade provoking more strongly asymmetric interactions (Ellison & Rabinowitz 1989; Geber 1989).

It has been hypothesized that the change through time from symmetric to asymmetric competition is caused by a change in the resources for which plants compete. While plant roots may compete for soil-based resources from an early stage in growth, competition for light cannot begin until substantial leaf canopies have been produced. Whereas overtopping a competitor blocks light effectively (competition is highly asymmetric, since the large plant prevents its competitor from obtaining light whereas the reverse is not true), access to edaphic resources cannot be gained by one plant and denied to another so conclusively. Several models have clarified the role of symmetric and asymmetric competition in the development of size hierarchies (Huston 1986; Miller & Weiner

Box 11.2 Plant size distributions.

Assume that we begin with a population of seeds with a normal distribution of sizes, and that they germinate to produce a population of young seedlings with a normal distribution of shoot dry weights. At any subsequent time, the shape of the frequency distribution of plant sizes reflects the interplay between size-specific growth and death rates and such factors as plant competition and size-specific attack by natural enemies. The point to be made here is that strongly skewed distributions can arise without competition, so the existence of a size hierarchy is not *evidence* of competition. There could be a small number of large plants and a large number of small ones, even in the absence of competition (Hara 1984).

The degree to which variance, skew and kurtosis of a plant size distribution change through time depends on the relationship between growth rate and size (specifically on the relationship between the increment of a size measure and the same size measure):

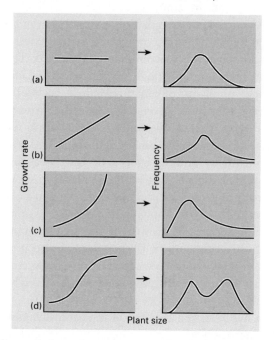

(a) Regardless of plant size, when growth rate is constant, mean size increases without any change in variance. (b) When growth rate increases linearly with size, the variance increases with size, but skew is unaltered. (c) When growth rate increases faster than linearly with size, the variance and skew increase with time, resulting in a log-normal distribution of plant weights. Many data from even-aged monocultures exhibit this pattern. (d) When growth rate is a sigmoid function of size, bimodality can develop, and variance, skew and kurtosis all change through time. (See Westoby 1982 and Hara 1984 for details.)

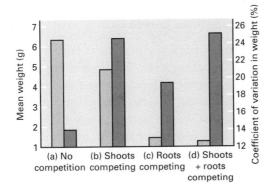

Fig. 11.4 An experiment on the vine *Ipomoea tricolor* (Convolvulaceae), in which plants were grown (a) with no competition, (b) with above-ground competition, (c) with below-ground competition, (d) with both above- and below-ground competition. Mean weight (lighter bars) was significantly different ($P < 0.01$) for all comparisons between treatments except (c) and (d). The coefficient of variation in weight (darker bars) was significantly different for the comparison of treatments (a) and (b) ($P < 0.05$), and for comparisons of treatments (a) and (d) ($P < 0.01$). (From Weiner 1986.)

1989; Bonan 1991), but an elegant experiment by Weiner (1986) is especially informative. He adopts the premise that competition for light is asymmetric while competition for edaphic resources is relatively symmetric. Plants of the twining vine morning glory (*Ipomoea tricolor*, Convolvulaceae) were grown in arrangements allowing either no competition, above- or below-ground competition alone, or both above- and below-ground competition (Fig. 11.4). Below-ground competition reduced mean vine weight far more than above-ground competition, and nearly as much as when there was both above- and below-ground competition. Size inequality, however, was much greater when the vines competed above ground compared with below ground. Weiner points out that when there is both above- and below-ground competition, mean weight is reduced primarily by competition for edaphic resources, whereas size inequality is primarily generated by competition for light. The asymmetry of the interaction when both shoots and roots compete is thus determined by competition for a resource that is not the one limiting population growth (Weiner 1990).

Many of the plants in competing populations have RGRs (see Box 11.3) close to or even below zero. For example, Thomas and Weiner (1989) plotted the relationship between weight at the start of a period of growth in a natural stand of *Impatiens pallida* and the change in plant mass during this period (Fig 11.5). Small plants had a zero or even negative change in mass, but the change in weight of the large plants was directly correlated with their weight at the start of the period. Mortality caused by competition was virtually confined to the small plants with zero or negative growth (see also White & Harper 1970; Ford 1975), so that the weight frequency distribution of the population was being progressively truncated at its left-hand (low weight) end.

Box 11.3 Expressing plant growth rate.

The simplest way to understand plant growth rates is to plot a graph of log shoot dry weight against time. The slope of the graph changes as the plant ages. The steeper the slope, the faster the plant is growing. At maturity (or under severe competition) plants may stop growing (slope = 0) or even lose weight (in which case the slope is negative). The usual means of expressing the rate of growth is simply to determine the slope of the graph at any point in time (i.e. change in *log* weight per unit time). This is called the relative growth rate (RGR) of the plant at that time (or at that size).

$$RGR = \frac{\log_e W_2 - \log_e W_1}{t_2 - t_1}$$

where W_2 is the shoot dry weight at time 2 (t_2) and W_1 is the weight at time 1 (t_1). To find the maximum relative growth rate we find the slope of the steepest part of the curve; this is called RGR_{max}. Note that the slope one measures is influenced by the length of time over which the measurement is made. The longer the period, the *lower* the estimate of RGR_{max}:

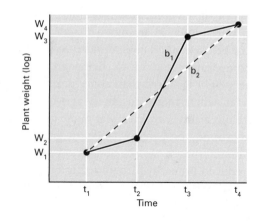

where the slope b_2 calculated over the longer period is shallower than the slope b_1 calculated over a single time period.

It is important to note the distinction between growth rate and relative growth rate. Growth rate (*sensu* Box 11.1) is the rate of change in plant weight, dW/dt (with dimensions g time^{-1}). Relative growth rate is the rate of change in log plant weight, $d(\log W)/dt$ (with dimensions time^{-1}).

Other important measures used in the quantitative analysis of plant growth (with their dimensions in parentheses) are the following:

continued on p. 335

> **Box 11.3** *Continued*
>
> | | |
> |---|---|
> | **ULR** (unit leaf rate) | Mean rate of interest of total plant dry weight per unit leaf area $(g\,m^{-2}\,time^{-1})$ |
> | **LAR** (leaf area ratio) | Ratio of leaf area to total dry weight $(m^2\,g^{-1})$ |
> | **RGR** (Relative growth rate) = ULR × LAR $(time^{-1})$ | |
> | **SLA** (specific leaf area) | Leaf area per unit leaf dry weight $(m^2\,g^{-1})$. ULR and SLA tend to be negatively correlated (see Chapter 1) |
> | **LAI** (leaf area index) | Ratio of the area of leaf (measured on one surface only) to the area of ground beneath (dimensionless) |
> | **LAD** (leaf area duration) | Area beneath a graph of LAI against time computed over the whole growing season (time) |
> | **RWR** (root weight ratio) | Ratio of total root dry weight to total plant dry weight (roots plus shoots) (dimensionless) |
>
> More detailed definitions can be found in Evans (1972).

However, rapid growth of the dominant plants in the population would ensure that as time passed the range of sizes of plants in the population continued to increase. Mohler *et al.* (1978) illustrated this in the fir tree *Abies balsamea* (Pinaceae). As time passed, plant density declined from 11.5 m^{-2} at 3 years old to 0.2 m^{-2} at 59 years (Fig. 11.6). Peak death rates occurred when the trees were between 19 and 35 years old.

Almost all of the earlier literature used skewness (Box 11.1) of the

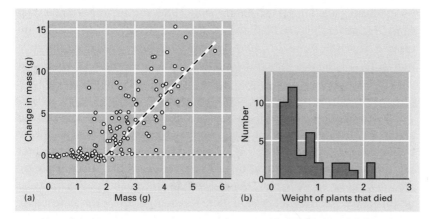

Fig. 11.5 (a) Change in shoot mass versus mass 8 weeks earlier for surviving plants in a population of *Impatiens pallida* and (b) the distribution of initial masses for the plants that died.

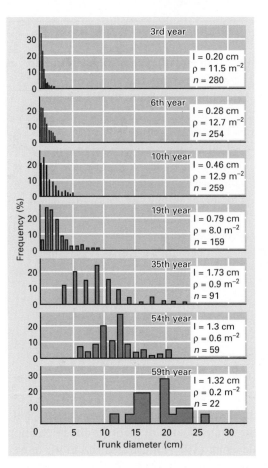

Fig. 11.6 Frequency distributions of trunk diameter for *Abies balsamea* stands arranged in age sequence. The interval of trunk diameter between successive bars (I) in the histograms was determined by dividing the range of observed diameters into 12 equal intervals. ρ indicates stand density and n shows the number of trees in the sample. (From Mohler *et al.* 1978.)

weight frequency distribution as a measure of the size hierarchy in plant populations. This statistic measures the asymmetry of the distribution. Nearly all published values for weight frequency skewness were positive, increased through time and were higher by a given date in denser populations (see Turner & Rabinowitz 1983 for an important exception). However, Figure 11.6 shows that the degree of asymmetry decreased through time in *Abies balsamea*. When competition causes mortality, it becomes difficult to predict the way in which skewness will change through time (see Begon 1984). Weiner and Solbrig (1984) also point out that measuring the asymmetry of the distribution of weights yields little biological insight, and that statistics providing information about the *inequality* of plant sizes, such as the Gini coefficient and the coefficient of variation (Box 11.1), would be more useful for interpreting the behaviour of populations of competing plants. These statistics are now used more often than skewness to quantify variability in performance (see Bendel *et al.* 1989; Knox *et al.* 1989 for assessments of the merits of these measures). Figure 11.6 clearly shows that size inequality of surviving *Abies balsamea* trees increased through time, at least until 35 years of age.

Relative position in the size hierarchy of a plant and its immediate neighbours clearly has dramatic effects upon future growth and survivorship. It is also a major determinant of lifetime reproductive success, as shown by long-term field studies on *Astrocaryum mexicanum* (Palmaceae), a dominant understorey palm of tropical rain forests in southeastern Mexico (Sarukhán *et al.* 1984). The probability of surviving from the second to the fifth year was strongly correlated with the size of the plant when it was 2 years old (Fig. 11.7a), and both the frequency of flowering (Fig. 11.7b) and mean annual fecundity (Fig. 11.7c) were much higher for plants from larger size classes.

Most studies of plant size structure have involved monocultures, but Ogden's (1970) study of a mixed annual herb community is an exception. Each of the constituent species populations had positively skewed frequency distributions of shoot weight. Other studies have involved

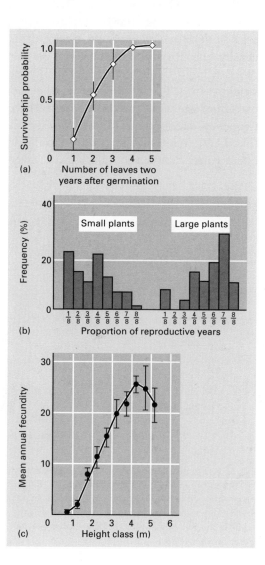

Fig. 11.7 Population structure and plant performance in the palm *Astrocaryum mexicanum*. (a) Probability of survival increases with seedling size. (b) Most large plants reproduce nearly every year, but most small plants reproduce rather infrequently. (c) Taller plants have significantly greater mean annual fecundity. (From Sarukhán *et al.* 1984.)

two species. Butcher (1983) investigated size hierarchies in monocultures and two-species mixtures of wild oats (*Avena fatua*), leafless pea (*Pisum sativum* var. Filby) and conventional pea (*P. sativum* var. Birte). In monoculture, wild oat developed the greatest size inequality as measured by the Gini coefficient (Box 11.1), followed by conventional pea and leafless pea. In two-species mixtures, the degree of inequality was related to the ability of one of the species to outcompete the other. In a mixture of wild oat and conventional pea, the pea considerably outgrew the oat, resulting in a virtually complete separation of their weight frequency distributions; the mixture displayed far greater inequality than either of its components in monoculture. With wild oat and leafless pea, however, the two weight frequency distributions did not separate, and the size inequality of the whole population was similar to those from the individual monocultures (Fig. 11.8).

Weiner (1985) also investigated size hierarchies in monocultures and mixtures, using strawberry clover (*Trifolium incarnatum*) and ryegrass (*Lolium multiflorum*). In monocultures at low fertility and high density, the clover population had a lower mean weight per plant than ryegrass, but a far greater size inequality, because of a greater range of weights. In a 50 : 50 mixture at the same density, the mean and maximum size of ryegrass plants far exceeded that of clover, and the size frequency distributions of the two species were markedly separated (Fig. 11.9). While size inequality for all the plants in the mixture was intermediate between its values in the two monocultures, the compo-

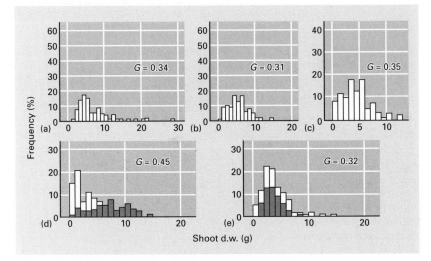

Fig. 11.8 The distribution of shoot dry weight in (a) monocultures of conventional pea, var. Birte; (b) monocultures of leafless pea, var. Filby; (c) monocultures of wild oat, *Avena fatua*; (d) a 50 : 50 mixture of Birte and *Avena*; (e) a 50 : 50 mixture of Filby and *Avena*. All populations were sown at the same density. The contribution made by the pea plants in mixture is shown by the dark-shaded area. G = value of the Gini coefficient. (After Butcher 1983.)

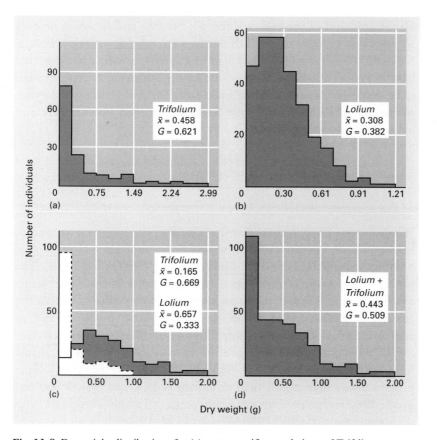

Fig. 11.9 Dry weight distributions for (a) monospecific populations of *Trifolium incarnatum*; (b) monospecific populations of *Lolium multiflorum*; (c) mixtures of *Trifolium* (white area) and *Lolium* (solid area) with their weight distributions shown separately; (d) the mixture as a whole, both species taken together. *x̄*, mean weight; G, Gini coefficient. (From Weiner 1985.)

nent populations showed *opposite* changes in their inequality. The suppressed clover population had higher inequality than in monoculture, while the ryegrass showed reduced inequality. Under these experimental conditions, ryegrass is clearly a better competitor than clover. In comparison with the monocultures, the majority of clover plants in mixtures suffered because weak intraspecific competition from other clover plants was replaced by strong interspecific competition from ryegrass. In contrast, rygrass benefited in mixtures, since strong intraspecific competition was alleviated by replacing ryegrass individuals by less competitive clover plants.

In mixed populations, the development of widely differing size distributions for the constituent species has dramatic consequences for subsequent performance and survival. Although the behaviour of the individual species is poorly understood, the mixtures themselves often conform broadly to the patterns expected from a knowledge of the behaviour of monocultures. Bazzaz and Harper (1976) found that

50 : 50 mixtures of the two crucifers *Sinapis alba* and *Lepidium sativum* accumulated biomass and self-thinned in accordance with the expectations of $-3/2$ thinning, and had size frequency distributions that began close to normality and became highly positively skewed through time. However, *Sinapis* suffered little mortality and accumulated much of the mixture's biomass, while *Lepidium* thinned considerably and made little growth (see section 12.7).

11.2.2 Other aspects of performance

11.2.2.1 Height distribution

Whereas population weight distributions become positively skewed as time passes, the distribution of heights either remains symmetrical or becomes negatively skewed (Koyama & Kira 1956; Ogden 1970). The reason is that many of the short plants die because they receive inadequate light and because surviving suppressed plants display etiolation (greater internode extension in the direction of the light) and so approach the height of more competitive plants. These processes reduce height inequality, and the tendency to etiolate is greater in shade-intolerant species (Morgan & Smith 1979; Hara *et al.* 1991). Suppressed plants branch less than dominant plants, and branch and leaf production is confined more to the tops of stems compared with dominant plants (Geber 1989; Weiner *et al.* 1990). The allometry of growth also differs, with the stem height of etiolated plants increasing at a faster rate than stem weight. As a consequence, the mechanical strength of the stems and trunks of these plants is low, and they often die through lodging or being blown over.

11.2.2.2 Productivity

Studies on plant productivity often pay close attention to the structure of the population and to the dynamics of turnover of different plant parts. The combination of the sampling methods from traditional energetics with the detailed census techniques of plant demography has allowed much greater precision in the estimation of primary productivity (Flower-Ellis & Persson 1980; Callaghan & Collins 1981), and may revolutionize the study of the impact of herbivores on the demography of plant modules (see Chapter 13).

11.2.2.3 Seed sizes

Compared with other aspects of performance, seed weight within a species has often been regarded as remarkably constant under a wide range of conditions. This view was based on *mean* weights calculated from bulk samples of many seeds. More recent detailed studies, however, have revealed wide variation in individual seed masses. In the umbellifer *Lomatium grayi*, for instance, individual seed weights showed a 16-fold variation (Thompson 1984); while this may be exceptional (but see Black 1957, 1959), Thompson quotes several examples of seed masses varying from twofold to sixfold within species. Seed size is rarely constant for a species because of the influence of environmental, genetic and developmental factors such as seed number, timing of inflorescence production, location of different inflorescences on the plant and location of seeds and number of seeds produced. In the oak *Quercus robur* there are more than fivefold differences in mean acorn weight from one individual to another, and those differences (which presumably have a genetic basis) are consistent from year to year (Crawley & Long 1995). In many cases the frequency distribution of seed weight approximates at least to symmetry, if not to a normal distribution (Schaal 1980; Pitelka *et al.* 1983). Where more than one seed morph is produced, the weight distribution may have a more complex form (Harper *et al.* 1970). When more seeds are initiated by a plant than it can support to maturity, many inviable seeds may be produced, resulting in a positively skewed distribution of seed mass (Gross & Werner 1983; Thompson 1984). This can occur when environmental conditions deteriorate between seed initiation and maturation (Fuller *et al.* 1983; see Chapter 13).

Variation in individual seed weights has rarely been considered in studies of population structure, but the strength of correlations between seed weight and the size and competitive ability of the resulting seedling (Black 1957; Crawley & Nachapong 1985) suggests that we should pay closer attention to seed weight variation in future.

11.2.2.4 *Fitness*

The ultimate measure of a plant's performance is the number of its offspring that reproduce in future generations. However, in field studies involving many plants it is normally not possible to tell which parent produced which emerging seedling. In such cases, the most practical estimate of offspring contributed to future generations may simply be a direct count of the seed production of individual plants. Generally, seed production is correlated with the size of the producing plant (Solbrig 1981). In many species, particularly monocarpic perennials, the probability of flowering is also size dependent, so that plants must exceed a critical threshold size before flowering (e.g. Kachi & Hirose 1983; Klinkhamer *et al.* 1987). Since weights in plant populations are positively skewed and seed output is correlated with plant size, the frequency distribution of seed production per plant is also very positively skewed. Thus, most of the seeds are produced by a very small proportion of the population, and many individuals produce no seeds at all (Mack & Pyke 1983; Scheiner 1987). The implications of this for population genetics are discussed in Chapter 6.

11.3 Spatial structure of plant populations

The terms 'regular' and 'aggregated' describe spatial patterns that are respectively more uniform and more clumped than random distributions. Methods for analysing spatial pattern are described in Chapter 14.

11.3.1 Spatial structure of seed and seedling populations

The spatial distribution of the seeds and seedlings of a species is determined by the interplay between the following factors.

11.3.1.1 *The distribution of seed-producing parents and of the seed rain*

In many species, flowers are spatially aggregated and thus both seed dispersal and germination are also often aggregated (Rabinowitz & Rapp 1980). Seeds of some species (e.g. thistles, dandelion, fireweed) disperse in clumps, and many other passively dispersed seeds are deposited at high local densities. If all dispersed seeds were equally likely to germinate, dense clumps of seedlings would be produced, resulting in intense competition between the seedlings. While intense competition between seedlings certainly does occur in some cases, all seeds are not equally likely to germinate in the field. Germination requires that a seed is both in the correct physiological state and in a suitable microsite (see below). The accumulation of a bank of viable seeds in most soils testifies to the fact that many seeds remain dormant in the soil after dispersal, perhaps for a very long time (see Chapter 7).

The pattern of seed dispersal is influenced by wind and water movements, and by the abundance and activity of animal-dispersal agents (see Chapter 9). Many seeds exhibit secondary dispersal after reaching the ground, as a result of factors such as rain splash, frost heaving and animal movements (Watkinson 1978; Liddle *et al.* 1982). Established vegetation intercepts many dispersing seeds, and acts as a potent mortality factor by preventing them reaching the ground (Symonides 1988).

11.3.1.2 *The behaviour of seed- and seedling-feeding herbivores*

Some seed-feeding animals modify the spatial pattern of dispersed seeds by caching or scatter-hoarding, and these activities modify the spatial pattern of seedlings that germinate. In the case of caching, scattered seeds are collected into clumps. If there is a large crop of seeds and many caches are made, some may escape rediscovery and survive to produce seedlings (Bossema 1968; Abbott & Quink 1970; Smith 1975). In the case of scatter-hoarding, animals redistribute seeds in a more regular pattern, which may later increase the probability of individual seeds escaping predation and reduce the intensity of competition between seedlings (Smith & Follmer 1972).

11.3.1.3 The spatial distribution of suitable germination sites

In a given area of habitat there is a finite number of microsites (termed 'safe-sites' by Harper 1961) that fulfil all the conditions (ample water and oxygen, the correct light and temperature conditions, etc.) required for germination of any species. Once all of the microsites are occupied by seeds of this species, addition of more seeds can result in reduced overall rates of germination (Palmblad 1968). When safe-site abundance limits germination in this way, seedling distribution may simply reflect the distribution of safe-sites; such control of recruitment patterns occurs in ragwort *Senecio jacobaea* (Asteraceae) (Crawley & Nachapong 1985) and in the crucifer *Cardamine pratensis* (Duggan 1985).

Germinating seeds may themselves modify the microenvironment in ways that either promote or inhibit germination of conspecific seeds (Linhart 1976; Waite & Hutchings 1978). Such density-dependent effects are extremely localized, perhaps requiring physical contact between the seeds. Some advantages of density-dependent promotion of germination can be envisaged (e.g. better moisture retention, stabilization of the ambient microenvironment and easier soil penetration by roots when many seedlings germinate together; Linhart 1976). However, the resultant intense clumping of seedlings will probably lead to density-dependent mortality through intraspecific competition and an increased risk of herbivore or pathogen attack. Indeed, some species with positive density-dependent germination show identical establishment rates when the same number of seeds are sown in clumps and in isolation (Waite & Hutchings 1979). Species with positive density-dependent germination may thus merely be recipients of an external stimulus that confers no consistent selective advantage.

The major advantage of germination that responds *negatively* to seed density is that the production of dense, strongly competing clumps of seedlings would be avoided. This might allow a reservoir of ungerminated seeds to accumulate, which would confer future colonizing potential, and buffer the population against the risk of the growing plants undergoing wholesale mortality (see Chapter 7). Such inhibition of germination is caused by the release of allelopathic chemicals from germinating seeds, growing plants or decaying plant remains. An example is seen in reedmace (*Typha latifolia*), which produces a chemical that inhibits germination and growth of *T. latifolia* seeds and seedlings (McNaughton 1968).

Established plants of many species suppress seedlings in their immediate vicinity by casting deep shade, competing vigorously for water and nutrients in the upper layers of the soil, and producing inhibitory chemicals. They may also support herbivores that consume establishing seedlings or pathogens that kill seedlings outright; in such cases, the probability of a seedling producing an established plant is much lower in the vicinity of an adult conspecific compared with other places (see

Chapter 7). Successful establishment of many species from seed, particularly pioneers, requires gaps in the vegetation cover (Fenner 1978; Swaine & Whitmore 1988; Fisher *et al.* 1991). The growth of pioneer species is far more suppressed by vegetation than that of species typical of closed habitats (Fenner 1978). Canopy gaps are caused by the death of large plants (e.g. a tree fall in a forest) or by small-scale animal disturbances such as footprints, dung piles, digs or scrapes. Large-scale disturbances through fire, flood or landslide and the various disruptive activities of humans can provide extensive areas suitable for the establishment of many species from seed. Not all species require gaps for germination, however. For example, many woodland species are stimulated to germinate beneath leaf canopies that bias the spectral composition of transmitted light, decreasing its red : far-red ratio in comparison with that of direct sunlight.

11.3.2 Spatial structure of populations of established plants

The spatial distribution of mature plants reflects the spatial pattern of recruitment (see section 11.3.1) and the modification of this pattern by mortality factors, which differ in intensity from place to place. Where density-dependent mortality is strong, the distribution of surviving adult plants is less aggregated than that of seedlings. In contrast, when abiotic mortality factors are important, mortality may be negatively density dependent, because most plant deaths are expected to occur at the edge of the population distribution, where conditions are least favourable and where recruitment is likely to have been lower and more patchy. Such inverse density-dependent mortality would produce greater aggregation in the adult population than in the seedlings. In this section I describe how biotic factors, especially plant competition, influence the spatial distribution of mature plants.

Strictly random spatial patterns are rare in plant populations. Regular distributions are also uncommon, and many of the reported cases are disputed; these distributions are discussed later in this section. Most plant populations display some degree of spatial aggregation. Among the many factors responsible, the following are important.

1 Most dispersed seeds land close to the parent plant. The probability of mortality caused by seed predators, seedling herbivores or fungal pathogens may also be highest nearest the parent plant, and may fall as distance increases (this is the 'distance hypothesis'; Janzen 1970). An alternative model suggests that the mortality risk of seeds and seedlings due to herbivores and pathogens is a function of seed density rather than distance from parent plants (this is the 'density hypothesis'; Connell 1979). By plotting seed density and mortality rate against distance from the parent plant, a predictive graph of recruitment against distance can be generated. Janzen (1970) argued that recruitment should peak some distance away from the parent because mortality beneath the parent is so

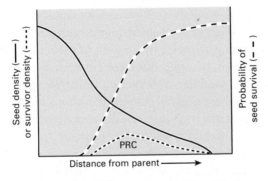

Fig. 11.10 Janzen's (1970) original 'distance hypothesis' model of tree recruitment. Seed density (solid line) is assumed to fall off with distance from the parent. Probability of seed survival (dashed line) is assumed to increase with distance from the parent as a result of reduced risk from natural enemies. Consequently, the population recruitment curve (PRC) peaks at some distance from the parent tree. Peak probability of seed survival is always much less than 1.0. See section 15.2.4 for a modern interpretation.

high (Fig. 11.10), and that this may help to explain why individual trees of a given species are often widely spaced out in tropical rain forest. However, Hubbell (1980) suggested that when the axes of seed density and mortality rate are scaled realistically, recruitment might peak beneath the parent plant because of the high seed density deposited there. The fact that, at least in tropical dry forest, many tree species show *clumped* distributions of seedlings beneath parents (Hubbell 1979) supports this view, as do simple mathematical models (Geritz *et al.* 1984). Experimental comparisons of the 'distance' and 'density' hypotheses (Wilson & Janzen 1972; Connell 1979; O'Dowd & Hay 1980) do not give clear grounds for preferring one over the other; in some cases mortality of seeds and seedlings depends on their density and is virtually independent of distance to parent plants, while in others distance appears to be more important than density. However, predictions of offspring recruitment distances generally ignore the fact that the probability of surviving at different distances from parent plants may change with time. If this is true, the position of the recruitment curve would alter as time passes. For example, in the tropical tree *Platypodium elegans* (Fabaceae), the distances between surviving offspring and parent trees increased with time, showing density- or distance-dependent mortality (Augspurger 1983). Median distance from parents to 3-month-old plants was greater than parent to seedling distance; many seedlings near parents were killed by fungal pathogens. Subsequent survival of seedlings depended on the availability of light gaps on the forest floor (see Chapter 15). Finally, older saplings were further from the parents than were 1-year-old plants. Thus, for *P. elegans*, the spaced-out adult plants appear to support Janzen's distance model, whereas the distribution of the younger recruits is better explained by Hubbell's prediction of clumped seedlings.

2 Germination, as we have seen, may be positively density dependent, and suitable safe-sites for germination may be clumped due to the

activities of digging animals, soil microtopographic features, etc. On a larger scale, a heterogeneous habitat might represent a mosaic of suitable and unsuitable patches for the growth of particular species.

3 While there may be large distances between adjacent plants of a clonal species, the modular construction units (shoots, ramets, tillers) of individual clones are generally clumped within a habitat because they are produced vegetatively. The mean distance at which a parent clone establishes a new module is selected in response to several potentially conflicting considerations, including:

(a) minimizing competition between connected modules of the same clone (the habitat patch from which they extract resources should not overlap extensively);

(b) the investment and metabolic costs of differing lengths of stolons (typically annual structures) or rhizomes (typically perennial structures) on which new modules are produced (see Eriksson & Jerling 1990);

(c) the ability of daughter modules to gain resources directly from the parent (Caraco & Kelly 1991; see Chapter 3);

(d) the need to counter invasion by, or avoid competition from, other plants;

(e) the patchiness of the habitat – as patch size alters, so the length of stolons or rhizomes that will maximize the probability of escaping from an unfavourable patch or remaining within a favourable patch will change (Hutchings & Bradbury 1986; Sutherland & Stillman 1988).

Clonal growth forms have been described using the relative terms 'phalanx' (highly branched and slowly expanding clones with closely packed modules) and 'guerilla' (less branched, invasive clones with loosely packed modules; see Chapter 4). Morphology is far from fixed, however. It responds to local resource availability, enabling the pattern of placement of modules in the habitat to react to local conditions (Hutchings & de Kroon, 199X). Many species are able, through such adjustments in their architecture, to exploit patches of habitat with abundant resources efficiently, and rapidly vacate patches with scarce resources. In some clonal species, module placement appears to follow simple rules that might allow rather precise predictions about growth and form (see Chapter 4). In some situations, however, basic growth rules may be so disrupted by edaphic variability (e.g. the presence of stones, patchily distributed nutrients), by neighbours or by herbivores that the rules cannot be worked out from the plant's final form (Cook 1988).

Whenever plants are irregularly spaced, differential demands are made on the resources supplied by the habitat. Where local demand for resources causes competition, plants will suffer reduced performance, and some may die. Such competition-induced mortality is likely to follow a density-dependent pattern, so that local density differences would decrease and aggregated or random distributions would become more regular through time. Kenkel (1988), for example, showed that a jack pine (*Pinus banksiana*) population with a random distribution at establishment developed a highly regular distribution through self-thinning. From maps of the distributions of the trees, Kenkel *et al.* (1989) then tesselated the habitat, a technique that allocates a 'territory' to each plant, consisting of all the space closer to it than to any other plant. Mean territory area was significantly greater for surviving trees than for trees that had died, and the size of surviving trees was positively correlated with their territory size. The smallest surviving trees had small near-neighbours, implying that density-dependent mortality is strongest when local competitive effects are greatest.

Regular distributions of plants have often been reported in desert perennial communities. It has been suggested (Woodell *et al.* 1969) that competition for water has converted initially aggregated or random spatial patterns into more regular patterns through spatially-density-dependent mortality. Competition in such communities produces positive correlations between the distances separating nearest neighbour plants and the sum of their sizes (Yeaton & Cody 1976; Yeaton *et al.* 1977; Phillips & McMahon 1981). Desert perennials need large, laterally spreading root systems to obtain enough water to ensure survival between the rare falls of rain. Although above-ground parts might form less than 10% ground cover, the root systems of neighbouring plants often abut. Established plants are often extremely old and recruitment is rare. Chapin *et al.* (1989) also found regular spacing in populations of competing alder shrubs in Alaskan tundra. They suggest that regular spacing is found only in resource-poor habitats such as deserts and tundra because there is strong competition for some limited resource and because the communities contain only one or a few dominant species, individuals of which compete for this resource at a given rooting depth or canopy height. This proposal is worth more extensive assessment.

Not all desert perennial communities exhibit regular spacing or competition for water. For example, populations of the cactus *Copiapoa cinerea* were randomly distributed or clumped (Gulmon *et al.* 1979), and water use by established plants was far below that available annually. Establishment of young plants was limited by their low water-storage capacity and the long intervals between successive rain storms, rather than by competition. Clearly, even in relatively simple communities, generalizations about spatial structure are difficult.

Finally, whereas random distribution of sexes in dioecious species (those with separate male and female individuals) should optimize dispersal of pollen to female plants and minimize seed predation by promoting more even dispersal, sexes are segregated in many dioecious species. For example, Meagher (1980) has shown that plants of the lily *Chamaelirium luteum* were more likely to have nearest neighbours of the *same* sex, and that males were in higher proportion where density was greater (see Chapter 6). The sex ratio of adult *Populus tremuloides* (Salicaceae) is altitude dependent, with females predominating at lower elevations and males at higher (Grant & Mitton 1979). Several dioecious species exhibit segregation of sexes along moisture gradients (Freeman *et al.* 1976 and Shea *et al.* 1993) give further examples. In most cases female plants occupy 'better' conditions than males, probably because their reproductive function is more severely resource-limited (Willson & Burley 1983). Males are often larger, probably because they allocate less energy to reproduction, leaving more energy for growth. This may lead to competitive superiority over females and, eventually, to a male-biased sex ratio (Lloyd & Webb 1977). Dioecious species reported to show sexual segregation are almost exclusively wind pollinated; insect-pollinated species appear to lack sexual segregation (Shea *et al.* 1993), presumably because of the reduced probability of cross-pollination if the two sexes were to grow in different patches of habitat.

11.4 Age structure in plant populations

11.4.1 The seed bank: dispersal in time

Detailed investigations of temporal changes in the seed banks of soils enabled Thompson and Grime (1979) to distinguish species with transient seed banks, in which all seeds germinate or die within a year of dispersal, and species with persistent seed banks, in which a fraction of the dispersed seed survives for more than a year in a dormant condition in the soil (see Chapter 7 for a consideration of seed dormancy).

The rate at which seeds emerge from dormancy, and length of time they can remain viable, are subject to different kinds of selection in different environments. Short-term viability and rapid germination of all seeds may lead to a population explosion if germination coincides with favourable conditions, or to extinction if conditions become hostile after germination. Conversely, periodic germination of a few seeds from a population with long-term viability limits the immediate capacity for population growth, but may ensure that at least some seeds germinate during favourable conditions. It also lessens the probability of mortality caused by intraspecific competition between seedlings (see section 11.3.1). Protracted germination of a cohort of seeds has been called 'dispersal in time' (Levins 1969). In many species which germinate in this way, dormant seeds greatly outnumber growing plants, just as dormant buds and meristems in clonal plants may outnumber growing parts. Because of the practical problems of accurately recording the fates of seeds in species with long-term dormancy, investigations of age structure in plant populations tend to ignore the individuals in the seed bank (but see Smith 1983; Kalisz 1991; Rees & Long 1992).

11.4.2 Age structure of the growing plants in populations

There are few short-cut methods for determining age structure in populations of growing plants. Attempts to correlate size or other easily measured aspects of performance with age tend always to give inaccurate results, since size variation increases rapidly as plants age and this variation is increased by habitat heterogeneity and competition. Three approaches can be used to determine population age structure (Hutchings 1985). First, if the population can be sacrificed, some type of annually produced morphological marker (e.g. tree rings or bud scars) can be counted. Second, with large trees in non-tropical environments it may be possible to make non-destructive ring-counts using a core-borer. Third, individual plants can be recorded as they enter the population (as 0 year olds), and uniquely tagged or mapped so that their survival can be followed (e.g. Hutchings 1987). This method is laborious and time-consuming, and requires repeated censuses, perhaps over many years, but it provides the most valuable data for demographic analyses (Hubbell & Foster 1986c). Age structures for species of differing taxonomic status and ecology are shown in Figs 11.11–11.13.

Law's (1981) study of a colonizing population of the grass *Poa annua* enabled him to determine age structures directly, because the arrival of each plant, and the time of its death, were recorded in monthly censuses. Despite the fact that *P. annua* is a colonist of open habitats,

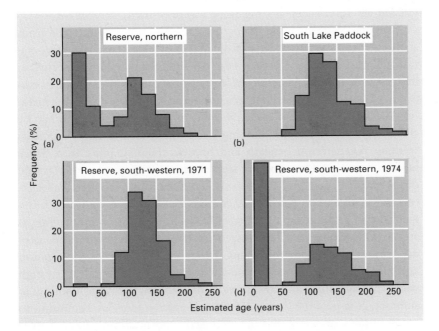

Fig. 11.11 Age structures of *Acacia burkittii* populations: (a) sheep excluded since 1920; (b) sheep grazed continuously; (c) fenced against sheep but rabbit-grazed until 1969; (d) the same site, 3 years later, following rabbit eradication in 1969. (After Crisp & Lange 1976.)

recruitment occurred throughout the study, and age structure changed through time (see Fig. 12.19). While the population soon after establishment inevitably contained only young plants, the proportion of young plants decreased sharply through time. This was because the first colonists had high survivorship throughout the study, whereas the mortality rates increased steadily as time of colonization was delayed.

While opportunities for recruitment by colonizing species like *P. annua* tend to decrease as a site becomes vegetated, site clearance is not always necessary to encourage plant recruitment. Sometimes a change in management that alters the risks attending seedling establishment is sufficient, as found in a study of the arid zone shrub *Acacia burkittii* (Fabaceae) (Crisp & Lange 1976). Plant ages were estimated by regression using a combined measure involving height and canopy diameter. Age frequency histograms (Fig. 11.11) show that while some populations have shown little recruitment over the last century, others have made substantial gains in the last 25 years. In 1920, an area was fenced to prevent sheep grazing; the effect on recruitment is readily seen by examining Fig. 11.11a,b. Despite fencing against sheep, rabbits prevented recruitment where they were abundant. The effect of rabbit eradication in 1969 is shown in huge recruitment by 1974 (Fig. 11.11c,d). The age-structure data obtained in this study provided a sound basis for a management strategy to conserve *A. burkittii*. Without this, Crisp and Lange predicted virtual extinction of the species within a century in areas where sheep grazing persisted, and greatly reduced populations due to rabbit grazing wherever sheep grazing alone was prevented.

Dynamic patterns in age structure can sometimes be inferred from analysis of community-wide spatial patterns of senescence and regeneration (Watt 1947; see Chapter 14). An interesting example occurs in high-altitude balsam fir (*Abies balsamea* (Pinaceae)) forests in the northeastern USA, where a striking phenomenon known as 'wave regeneration' occurs (Sprugel 1976). Large areas of green canopy are broken by crescent-shaped bands of standing dead trees, with areas of vigorous seedling regeneration beneath and beyond them, grading into dense sapling stands, and into mature trees about 100 m away on the far side of the wave. Figure 11.12 shows the age structure of the trees along a transect across a complete wave. Sprugel believes that the waves originate when a small group of old trees blows down or succumbs to fungal attack. This exposes a small crescent of trees, and the wave propagates downwind, away from this focus. Increased wind speeds, desiccation and increased deposition of rime ice at the exposed edge lead to high winter foliage loss, and production loss through foliage cooling in summer. It is possible that at 60 years of age the trees are beginning to senesce in any case, and that the increased exposure simply hastens their death.

Populations of late successional species often have little recent recruitment and their age structures are dominated by older plants.

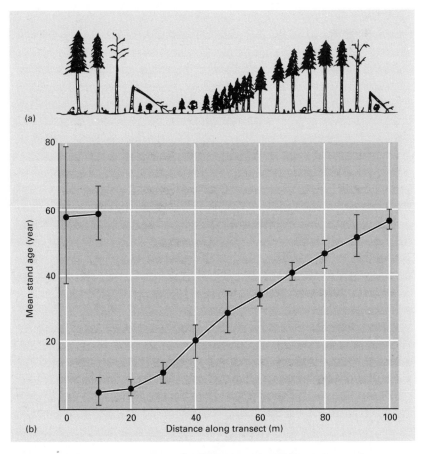

Fig. 11.12 (a) Wave regeneration in high-altitude balsam fir *Abies balsamea*. (b) Mean stand age plotted against distance along a transect across an average 'wave'. (After Sprugel 1976.)

Knowles and Grant (1983) distinguished climax, colonizing and fugitive species of coniferous trees in the Colorado Rocky Mountains on the basis of their age structures. Climax species, such as Engelmann spruce (*Picea engelmannii*) and ponderosa pine (*Pinus ponderosa*), live a very long time (up to 500 and 300 years respectively) and have distinct points of inflection in their cumulative age distributions (at 220 and 120 years respectively; Fig. 11.13a). Knowles and Grant believe that this indicates a transition to a comparatively stable condition for a substantial fraction of the population. The colonizing (early successional) lodgepole pine (*Pinus contorta*) has a much lower life expectancy (about 100 years) and no inflection in its age structure (despite a single datum point for a tree aged 230 years). A fugitive species, limber pine (*Pinus flexilis*), has an intermediate type of age structure. Note the similarity between the *size* distributions of ponderosa pine and Engelmann spruce (Fig. 11.13b), despite the wide differences in their age distributions. Knowles and Grant suggest that this similarity in size distribution reflects

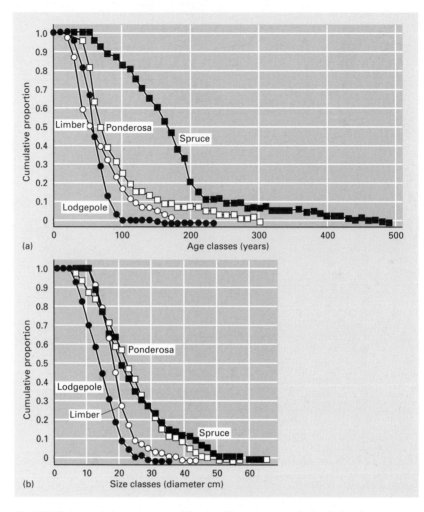

Fig. 11.13 Age and size structures of four coniferous tree species in the Rocky Mountains, Colorado, USA. (a) Cumulative age distributions for all species and all sites. (b) Cumulative size class distributions for all species at all sites. All pairwise comparisons of curves are significantly different by the Kolmogorov–Smirnov two-sample goodness-of-fit procedure, except for the comparison between spruce and ponderosa pine in (b). (From Knowles & Grant 1983.)

the fact that both are climax dominants, and argue that community-level interactions, which are virtually independent of age, determine size structure.

The two hypotheses that (i) an inflection in the cumulative age distribution is characteristic of climax species in mature stands of vegetation and (ii) species characteristic of the climax have similar size distributions, have been strongly disputed by Parker and Peet (1984). These authors present data that contradict the hypotheses, and they argue persuasively that the vegetation studied by Knowles and Grant has suffered considerable disturbance in the past and is far from successional climax. When site history is imperfectly understood, age structures of populations that have developed over a long period are clearly extremely difficult to interpret in detail.

Population age structures can be used to infer the survivorship through time of cohorts of plants. Survivorship curves generated from such data often contain substantial error, however, since they are produced from observations made on a single date, and the age structure observed must be assumed to be the outcome of constant annual rates of recruitment and age-specific mortality. As we have seen, such assumptions are often unjustified, as for example when grazing pressure changes (Fig. 11.11). The use of age structures to generate static life tables (also known as time-specific or vertical life tables) and age-specific survivorship curves is discussed by Begon and Mortimer (1986, pp. 13–17). Accurate documentation of the survivorship curve of a single-aged cohort requires regular censusing from recruitment of the cohort until death of its last member. Depletion curves, which plot the survival through time of multi-aged populations, also require periodic censuses of the population (see Chapter 12). Both survivorship and depletion curves can be used to calculate half-lives, i.e. the time taken for half of the cohort (or original population) to die (Harper 1967). Characteristically, half-lives are longer for multi-aged populations than for cohorts because multi-aged populations contain a proportion of established plants when observations begin, whereas all the plants in a cohort are observed from the time of their recruitment as seedlings, when they usually experience a death rate that is much higher than average.

Brief mention should be made of the use of 'age-states' to describe plant populations (Sharitz & McCormick 1973; Rabotnov 1978; Gatsuk *et al.* 1980). Here, the stage of development that plants have reached (e.g. juvenile, vegetative, reproductive, senescent) is recorded, rather than their ages. The same developmental stage may be reached by plants of very different ages, and the attainment of a particular state may not preclude a plant from returning to a state that it has held previously (e.g. reverting from reproductive to vegetative; see the contrast between age-based Leslie matrices and stage-based Lefkovitch matrices in Chapter 12). The value of determining the age-state structure of a population is that the spectrum of developmental states may be a much better indicator of its condition than its age structure. For example, Mehrhoff (1989) has shown that plants in stable populations of the endangered orchid *Isotria medeoloides* either flowered every year or alternated between flowering and non-flowering, whereas plants in declining populations were increasingly non-flowering or dormant and subterranean. Gatsuk *et al.* (1980) recommend determination of both developmental stages and calendar ages whenever possible.

11.4.3 Age structure of populations of modules

Up to this point, age structures have been discussed only with respect to genets. However, valuable information can also be obtained by considering the age structure of plant parts such as shoots, tillers and leaves (see also Chapter 12). For example, Noble *et al.* (1979) analysed the age structure of shoot populations of the sand sedge *Carex arenaria* in two dune systems. At each site plots were either given NPK fertilizer or left as controls; fertilizer addition produced an increased stable density of shoots with higher birth and death rates, so that the age structure of the population came to be dominated by younger shoots. This study emphasizes that if attention had not been paid to the detailed turnover of shoots, and if only their densities at the beginning and end of the experiment had been recorded, then the dramatic impact of fertilizers on the shoot dynamics of the sedge would have been overlooked.

Abiotic conditions markedly affected shoot age structures in the sand sedge. Turkington's (1983) study of cloned genets of *Trifolium repens* shows that the local *biotic* environment (specifically the identity of neighbouring plant species) can also influence the age structure of plant parts. The presence of different neighbour species caused a single cloned genet of *T. repens* to exhibit widely different leaf birth and death rates, and to accumulate different leaf population sizes with differing age structures.

11.5 Genetic structure of plant populations

Seeds of many species have a high probability of dying before germination, but whether mortality during the seed phase alters the genetic structure of plant populations randomly or directionally is largely unknown. Cohorts of buried seed of three genetic lines of oilseed rape *Brassica napus* (Brassicaceae) showed significant differences in seed death rate at some sites but not at others (Crawley *et al.* 1993). In the growing phase, the probabilities of plants surviving and reproducing are partially genotype-dependent (e.g. Krannitz *et al.* 1991) but, as stated elsewhere, also strongly dependent on the quality of the microsite in which the seed germinates, and its neighbourhood. In most cases a very limited proportion of the genetic diversity found in the seedling population survives to reproduce. Although the 'cost' of plant mortality could in theory be offset to some extent by greater speed and precision of genetic adaptation (Antonovics 1978), mortality at the seedling stage does not necessarily spare those plants that possess the genotypes which would confer maximum survival and fecundity under the prevailing conditions.

Directional changes in genetic structure have been documented in some growing plant populations that are undergoing mortality. For example, Bazzaz *et al.* (1982) established populations of *Phlox drummondii* from equal proportions of 10 cultivars that could be recognized by the colour and markings of their petals. Mortality rates differed between cultivars, and changed through time, so that as the population aged the proportions of each cultivar among the surviving plants changed significantly. Natural populations of *Echium plantagineum* (Boraginaceae) also show genotype-specific thinning. Burdon *et al.* (1983) used electrophoretic techniques to demonstrate that certain alleles altered in their frequency in the surviving population with the passage of time; plants heterozygous at particular genetic loci had characteristically higher survival rates than plants homozygous at the same loci.

Several authors have compared the genetic structure of populations of species with limited geographical range and small numbers of individuals with that of widespread species with abundant individuals (Loveless & Hamrick 1984; Karron 1987; Karron *et al.* 1988). In many but not all cases, restricted species have fewer polymorphic loci, and fewer alleles per polymorphic locus, than widespread congeners. There could be several causes for this, including founder effects, genetic drift, strong directional selection producing genetic uniformity in a small range of habitat types and, in some but by no means all cases (Karron 1987), high levels of inbreeding followed by selection against homozygous individuals (Karron *et al.* 1988).

Techniques such as DNA fingerprinting enable both the genetic variation in populations and its spatial structure to be analysed. For example, Nybom and Schaal (1990) sampled plants of the sexual species *Rubus occidentalis* (Rosaceae) and of the purportedly apomictic *R. pensilvanicus* (Rosaceae) from different locations in a site where they occurred together; 20 samples of the sexual species yielded 15 different genotypes. Samples with the same genotype were found no more than 4 m apart. The population therefore consisted of many small plants, each originating from a sexually produced seed with a recombinant genotype. Twenty samples of the presumed apomict yielded five genotypes. This is not wholly unexpected, because even apomictic *Rubus* spp. set some fertilized seed and this introduces some genetic variation into their populations. However, samples up to 500 m apart shared the same genotype, a fact that could not be explained by vegetative spread. Instead, it demonstrates that a high proportion of the seeds from which plants of *R. pensilvanicus* were established had been derived apomictically, because they exhibited no genetic variation.

Genetic variability in plant populations is sometimes far higher than anticipated. For example, Jefferies and Gottlieb (1983) predicted that *Puccinellia ×phryganodes* (Poaceae), an apparently sterile triploid with extensive clonal growth, would have very little genetic variation, either within or between populations. However, electrophoretic analysis revealed that most assayed material was genetically unique, leading to the hypothesis that although fertile seed has never been recorded, it is occasionally produced (flowering is uncommon, but small amounts of viable pollen are produced). Alternatively, somatic mutations within the clones may have increased genetic variability in the population (see Gill 1986).

The prediction of low genetic variation in populations of clonal species that rarely recruit from sexually produced seed sounds reasonable. Even if such populations established by recruitment from many seeds, the fittest clone might be expected gradually to eliminate all others, resulting in genetically monotonous vegetation. However, comprehensive reviews of genotypic diversity in clonal species in which recruitment from seed is extremely rare showed that populations of most such species are multiclonal (Silander 1985; Ellstrand & Roose 1987). They are not dominated by single genotypes, nor are their constituent clones equally abundant. In addition, few genotypes occur in more than one population, making each population of such clonal species genetically unique. The only exceptions are cases where the sampled 'populations' were separated by no more than a few metres, so providing ample opportunity for sharing of genotypes. These results call into question the common view of clonality as an evolutionary dead-end.

A full analysis of the reasons for maintenance of genetic diversity in populations of clonal species is presented by Silander (1985). A major factor is that few clonal species are *completely* unable to produce sexual progeny, so that recruitment from seed may occur, albeit extremely rarely. The few species shown by Ellstrand and Roose (1987) to have uniclonal populations were *obligately* clonal with extremely limited ranges. The persistence of individual genotypes, as favoured by the clonal habit, together with rare recruitment of sexually derived progeny, can sustain genetic diversity in clonal populations (Carter & Robinson 1993), as models predict (Soane & Watkinson 1979). For example, *Populus tremuloides*, the most widespread tree in North America, is strongly clonal. Its populations contain very large clones of great age, but exhibit similar levels of genetic diversity to many non-clonal forest trees. Jelinski and Cheliak (1992) attribute this to rare seed recruitment events during the last postglacial, despite the rarity of sexual reproduction under present climatic conditions. While populations of *P. tremuloides* are genetically differentiated, even at a small geographical scale, greater differentiation has probably been retarded by long-distance dispersal of both pollen and seeds (Jelinski & Cheliak 1992). A similar situation is found on a smaller scale in the clonal grass *Setaria incrassata* (Carter & Robinson 1993).

11.6 Abiotic influences on population structure

As we have seen, each plant in a population experiences a biotically unique microenvironment that exerts a great influence on its performance and its likelihood of survival; the sizes, distances, spatial arrangement and specific identity of its neighbours differ from those of other plants. Abiotic conditions also vary throughout the habitat, perhaps even over distances as small as seeds themselves, and this variation also affects performance. For example, to a seedling, the two sides of a shallow hoof-print may differ as dramatically in microclimate (radiant cooling, frost incidence, cold air drainage) as the two sides of a steep mountain valley several kilometres apart (Wellington & Trimble 1984).

That local environmental conditions have a major impact on individual plants in natural populations is suggested by Gottlieb's (1977) study of the annual herb *Stephanomeria exigua* (Asteraceae). Using electrophoretic techniques, no significant genetic differences were found between large and small plants, thus apparently vindicating, at least for this species, the ecologist's tendency to ignore genotype when studying differences in performance between individual plants in populations. Unfortunately, there is no guarantee that the genetic loci studied by Gottlieb have any control over plant size, while others, more related to size, might have differed between large and small plants. Gottlieb proposed that plant size and fecundity were primarily controlled by local environmental factors, and unlikely to be the basis of directed evolutionary changes.

The extent to which such small-scale abiotic heterogeneity influences population structure and individual fitness was investigated by Hartgerink and Bazzaz (1984) in a greenhouse experiment on the annual herb *Abutilon theophrasti* (Malvaceae). Heterogeneity was created at three densities in experimental plots by either adding nutrients, sand or stones, or by compacting the soil in seedling-sized patches. Single seedlings were grown in each of the patch types. Seedling populations were also grown in the same spatial arrangement in plots in which every effort was made to eliminate microhabitat heterogeneity. Several aspects of performance were measured for the seedling and adult plants in each treatment. The effects of density and patch type on seedling masses are shown in Fig. 11.14. Both density and the interaction of patch type with presence or absence of heterogeneity had a significant effect on plant mass; the pattern of response to different types of patch in the heterogeneous treatment was significantly different from that of plants in the same locations but in homogeneous environments. Seed production varied significantly with the type of patch in which the plants were grown. Much of the variance (47–62% depending on density) in seedling height after 2 days of growth could be explained by patch type differences, but as growth proceeded the level of explanation due to patch type fell, reaching 20–25% at the final harvest. One-third of the variance in individual mass at final harvest was attributable to differences in patch type. Overall, at final harvest, 58–76% of the variance in plant mass, depending on density, could be explained by a multiple regression in which most of the explained variance was due to patch type and seedling size early in growth. In contrast to the experiments cited in section 11.2.1, individual plant sizes were *not* correlated with the sizes of their immediate neighbours (see also Fowler 1984).

The conclusions drawn from this study are highly relevant to a discussion of population structure. Hartgerink and Bazzaz (1984) pro-

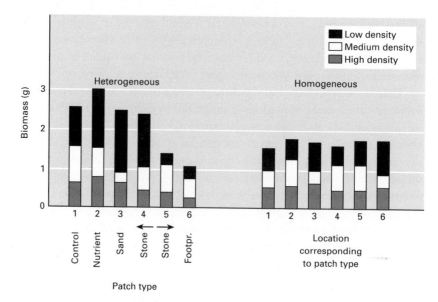

Fig. 11.14 Soil surface heterogeneity and plant performance. Mean above-ground biomass of plants on each patch type on a heterogeneous substrate, or in the corresponding locations in the container on a homogeneous substrate, at three plant population densities: (a) 130 m^{-2}; (b) 410 m^{-2}; (c) 770 m^{-2}. In patch type 2 extra nutrients were added to the soil, whereas in patch type 3 sand was mixed with the soil. Patch type 4 indicates a ceramic tile placed above the seed; type 5 indicates a tile below the seed. Patch type 6 is a 'footprint' created by compacting the soil. (From Hartgerink & Bazzaz 1984.)

pose that small-scale stochastic events that are unpredictable in time and space (including, for example, the availability of seedling-sized patches differing in suitability for plant growth) may strongly influence the position a plant achieves in the size hierarchy. Those components of plant genotypes which might determine fitness under controlled conditions may have their influence completely nullified in the field as a result of such stochastic events. Consequently, as Gottlieb also suggests, selection against less favourable genotypes may not take place in any predictable fashion. This, ultimately, will contribute to the maintenance of genetic diversity in populations. Environmental heterogeneity at the scale of seeds and seedlings can exert an overwhelming effect on individual fitness, and can also largely control variance in population parameters, as Hartgerink and Bazzaz (1984) demonstrate. As Harper (1981) and Watkinson (Chapter 12) emphasize, the *variance* in population attributes is more relevant to evolution than mean values. I hope that this chapter has demonstrated that too much emphasis on the attributes of the 'mean plant' leads us towards almost meaningless and biologically misleading abstraction, and prevents us from focusing on the variation in attributes, which is of far greater ecological and evolutionary importance.

12: Plant Population Dynamics

Andrew R. Watkinson

12.1 Introduction

In studying the dynamics of populations we are concerned with the ways that various biological and physical factors interact to bring about changes in plant numbers through time and space. By quantifying births and deaths, immigrations and emigrations, we confront such questions as why some species are rare and others common, and what processes are responsible for fluctuation in their numbers. Related to these topics are questions about why the age and size distributions of individuals vary from one population to another (see Chapter 11), and what factors determine the genetic structure of plant populations (see Chapter 6). While these structural aspects of populations are objects of study in their own right, it should be remembered that the same processes that determine the dynamics of populations are also largely responsible for determining their size, age and genetic structures.

12.2 Population flux

Observations on natural populations show that when conditions are suitable and resources are freely available (as in the early stages of secondary succession), all populations have the potential to increase exponentially. The exponential growth of a number of tree species, as reflected in the pollen record, is clearly seen at the end of the Pleistocene as trees reinvaded habitats at higher latitudes (Bennett 1983). With time, however, the rate of population growth slows and eventually a maximum population size is reached. For example, in the Botanic Gardens in Liverpool, a derelict piece of land was laid bare in late December. Law (1981) found that the first colonists of weedy grass *Poa annua* appeared in the following April (Fig. 12.1a). Most of these colonists survived and produced large numbers of seeds that germinated from August to September, leading to a massive increase in recruitment and a corresponding increase in population size. Following this initial very rapid increase in population size, the growth rate of the population declined. The growth rate is density-dependent (the rate of increase declines as the density of the population increases), because seedling recruits to the population at high densities germinate in dense aggregations around the parent plant where they have to compete for a limited

[359]

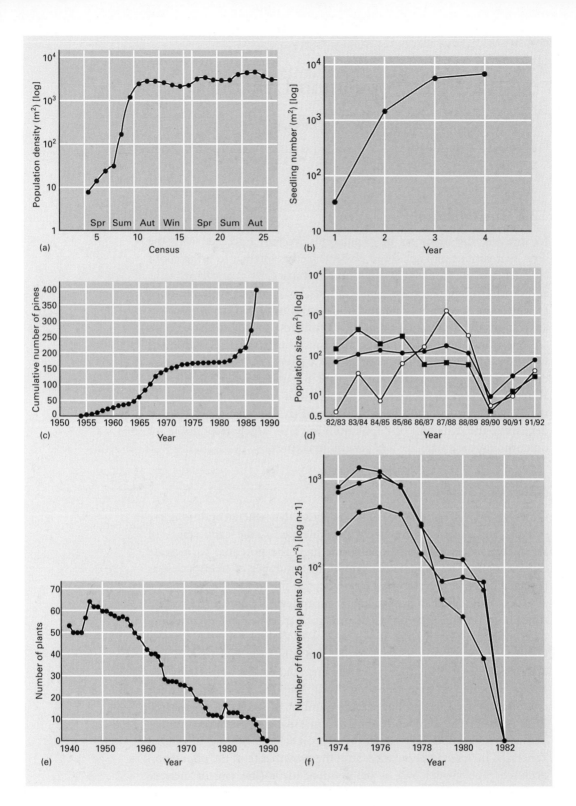

supply of resources. As a consequence they grow much more slowly, produce fewer seeds and are more likely to die at an early age.

In a different environment, an agricultural field cropped to winter wheat and subject to no weed control, the colonization by the annual grass *Bromus sterilis* shows a similar pattern (Fig. 12.1b). In this case, however, the time taken for the population to reach its maximum density was longer, because *Bromus sterilis* has discrete generations and produces only one seed crop per year (Watkinson *et al.* 1993). A rather different pattern of invasion is seen by the pine, *Pinus radiata*, invading a eucalypt dry sclerophyll forest (Burdon & Chilvers 1994). Following an initial period of invasion in which the source of propagules was from an adjacent plantation, the population appeared to reach a plateau (Fig. 12.1c). However, the population began to increase again once the early colonizing plants reached reproductive maturity and started to produce seeds, suggesting that the population was temporarily recruitment-limited.

That population growth is *regulated* and limited by the availability of some resource can be seen clearly in populations of both *Poa annua* and *Bromus sterilis*. Observations on these populations, however, were carried out over a relatively short period. A much longer study on a number of annual plants in a desert community in Arizona, USA (Venable *et al.* 1993) shows considerable variation in the dynamics of three species over a 10-year period (Fig. 12.1d). There is however no typical pattern of plant population dynamics, rather a continuum from virtual stasis to violent year-to-year fluctuations, with the variation dependent on habitat and life-history differences (Crawley 1990a). In a successional environment, for example, most populations are doomed to extinction (Fig. 12.1e,f) (Watkinson 1990; Tamm 1991).

Populations of course vary not only in time but also in space. Examination of the fossil tree pollen record in North America has allowed the migration of hemlock, *Tsuga canadensis* (Pinaceae), to be recorded over the Holocene (Davis 1988; Davis *et al.* 1994). Figure 12.2 shows isochrones representing the approximate position of the frontier of the continuous species population at 2000-year intervals from 12 000 years ago to the present. The rate of spread of this and other boreal and temperate trees into deglaciated regions averaged 100–400 m year^{-1} (Huntley & Birks 1983; Davis 1988). On a more recent time scale the invasion of the alien shrub *Mimosa pigra* (Fabaceae)

Fig. 12.1 (*Opposite.*) Changes in the population density of (a) *Poa annua* colonizing derelict land (Law 1981); (b) *Bromus sterilis* invading a field of winter wheat (Watkinson *et al.* 1993); (c) *Pinus radiata* invading eucalypt forest (Burdon & Chilvers 1994); (d) three species of desert annual in the sonoran desert: *Erioplyllum lanosum* (Asteraceae) (●), *Evax multicaulis* (Asteraceae) (■) and *Pectocarya recurvata* (Boraginaceae) (○) (Venable *et al.* 1993); (e) the orchid *Dactylorhiza sambucina* on a successional site (Tamm 1991); (f) three populations of the annual grass *Vulpia fasciculata* on sand dunes (Watkinson 1990).

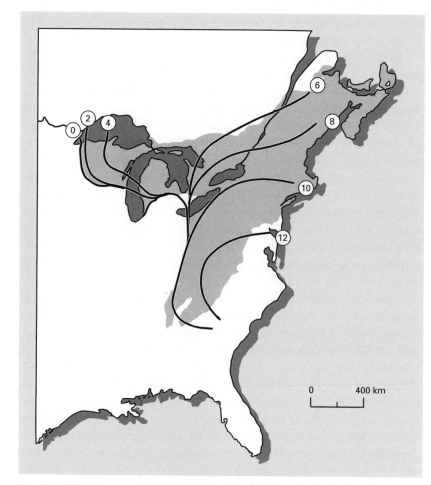

Fig. 12.2 Migration map for hemlock (*Tsuga canadensis*) during the Holocene. Isochrones represent the approximate position of the frontier of the population from 12 000 years ago (12) in 2000 year intervals, to the present (0). Shaded area shows present distribution. (From Davis *et al.* 1994.)

into the wetlands of northern Australia has shown that within a flood-plain the rate of spread averages 76 m year^{-1}, giving a doubling time for the area of a population of just over 1 year (Lonsdale 1993). In contrast, the number of infestations in the Northern Territory as a whole has shown a doubling time of almost 7 years, indicating periodic quantum leaps in dispersal (Fig. 12.3). On a much smaller scale, Symonides (1983, 1988) provides data on both the spatial and temporal flux of the annual *Erophila verna* (Brassicaceae) in areas of 2×2 m divided into 10×10 cm quadrats. Despite the pioneering work of Watt (1947) it is only relatively recently that plant population biologists have turned their attention towards exploring both the temporal and spatial dynam-ics of populations. Even within a population, however, the simple monitoring of population size at periodic intervals is insufficient if one is to understand the dynamics of populations. Even in those popula-

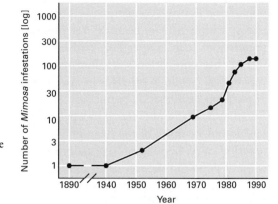

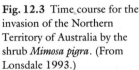

Fig. 12.3 Time course for the invasion of the Northern Territory of Australia by the shrub *Mimosa pigra*. (From Lonsdale 1993.)

tions where the number of individuals remains relatively constant from one census to the next, there may be a rapid flux of births and deaths and a considerable turnover of individuals. This underlying flux can only be revealed if the fate of mapped or tagged individual plants or their component modules is followed in detail. In a study of the clonal perennial *Hieracium pilosella* (Asteraceae) in which the fates of individual rosettes were followed, Bishop *et al.* (1978) found that the total number of rosettes in a grazed population varied little over 4.5 years, but that this concealed a rapid flux of rosettes, approximately equivalent to a 25% turnover per year. Very few seedlings established during this period, and most new rosettes resulted from clonal growth. The one major change in population size occurred in the summer of 1976, coincident with the worst drought in England at that time for 200 years.

12.3 Population regulation

There are only four factors that determine the numbers of individuals in a population: the numbers of births (B), deaths (D), immigrants (I) and emigrants (E). The number of individuals in a population at time $t + 1$ is related to those at time t by the equation

$$N_{t+1} = N_t + B - D + I - E \qquad (12.1)$$

The change in population size from one census to the next can be obtained by subtracting N_t from both sides of the equation to give

$$\Delta N = B - D + I - E \qquad (12.2)$$

Clearly if the sum of the terms on the right-hand side of the equation is positive the population will increase in size, whereas if ΔN is negative the population will decline. If the numbers of births, deaths, immigrants and emigrants do not change with density, then a population will increase to infinity if ΔN is positive, or decline to extinction if ΔN is negative. These two kinds of population are said to be unregulated.

No factor that alters the growth rate of a population, independent of population density, can be invoked to explain why population densities persist within narrow bounds. Yet observations on a large number of populations (e.g. Harper 1977; Silvertown & Lovett Doust 1993) have shown that population size often remains more or less constant from year to year, while in others it has been shown that the growth rate of a population decreases as population size increases. Clearly the assumption that B, D, I and E are all independent of plant density is untenable. Unless one or more of the demographic variables in Equation 12.1 is a function of N, population size cannot be regulated, and the population will not persist in the face of an uncertain and fluctuating environment.

Whilst the density-dependent control of one or all of the variables in Equation 12.1 is necessary to explain population regulation, it is the *interaction* between density-dependent and density-independent processes that determines the actual size of a population. Consider a population in which the birth rate (i.e. seed production) is density-dependent, while the death rate is density-independent but varies with the physical conditions. The rates of immigration and emigration are assumed to be equal. In Fig. 12.4a there are three equilibrium populations (n_1, n_2, n_3) corresponding to different death rates, which reflect the physical conditions in the three environments. In real populations, the actual death rates will inevitably vary (Fig. 12.4b) and the population will fluctuate round the equilibrium. Whilst it is naive to expect all populations to be at equilibrium, Fig. 12.4 nevertheless illustrates clearly how population size will vary depending on the interaction between physical and biotic conditions. The relative importance of different density-dependent and density-independent processes will vary considerably from population to population. For example, in Fig. 12.4c a major change in the density-dependent birth rate results in only a minor change in the equilibrium population size, whereas in Fig. 12.4d the same shift in birth rate produces a major change in population size. The difference between the two populations is in the *slope* of the density-dependent mortality curve; the shape of the density-dependence curve is vitally important in affecting both population size and stability.

12.4 The individual and the population

Population biologists are interested in why populations are as large as they are, and why they vary in size through both space and time. Yet, what is a population? For a geneticist a population is generally considered to be a group of individuals genetically connected through parenthood or mating (see Chapter 6), but for an ecologist a population is often defined simply as the total number of individuals of a single species in an area circumscribed for the purpose of study, often a quadrat or a series of quadrats. The result of such an approach is a tendency to ignore the role of plant dispersal in population dynamics. All populations, though, have a patchy distribution at some scale with local population turnover. Watt (1947) did much to emphasize the idea

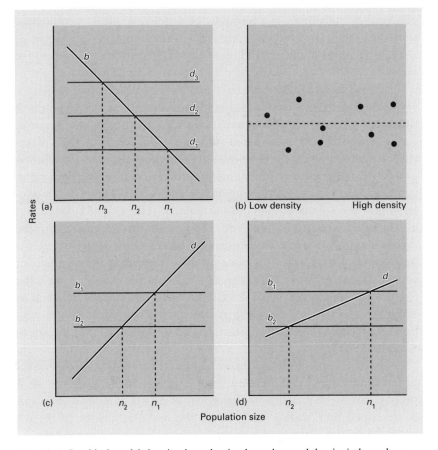

Fig. 12.4 Graphical model showing how density-dependent and density-independent processes interact to determine population size. (a) Birth rate, *b*, is density-dependent. Three different levels of density-independent mortality (d_1, d_2 and d_3) lead to three different average population sizes (n_1, n_2 and n_3). The higher the death rate, the lower the equilibrium population size. (b) In reality the average death rates will vary and the population will fluctuate around the equilibrium. The existence of fluctuation does *not* mean that the populations are unregulated. (c,d) How the same change in density-independent birth rate can lead to very different changes in population size when the *form* of the density-dependent death rate varies between populations (the shallower the slope, the greater the change in equilibrium population density).

of 'shifting mosaic' population dynamics in his classic paper on pattern and process in the plant community, but the incorporation of a spatial component into plant population studies is generally a recent phenomenon. The metapopulation concept of population structure, in which metapopulations are viewed as sets of populations persisting in a balance between local extinction and colonization, has provided a considerable impetus to studies of the spatial and temporal dynamics of populations (Hanski 1991). Few species, however, fit the classic metapopulation model, and Harrison (1994) argues for a much broader and vaguer view of metapopulations as sets of spatially distributed populations, among which dispersal and turnover are possible but do not necessarily occur. The important point to emphasize here is that, because of the spatial structure of individuals at a hierarchy of scales (e.g. local, metapopulation, geographical), a full understanding of the population dynamics of a species requires an understanding of how they operate across a range of scales.

It is also necessary to consider exactly what an individual is. This is discussed in detail in Chapter 11, but it is necessary to make two points at this stage: (i) we should recognize that individual plants may occur in various life states, for example as seeds or vegetative plants (for plants such as desert annuals, the large majority of individuals may occur below ground in the seed bank, only emerging occasionally above ground as vegetative plants to produce more seeds); and (ii) plant growth is modular, unlike the unitary growth of most animals (Harper 1981; Jackson *et al.* 1985; Watkinson & White 1986). This means that for most plants the zygote develops into a modular organism where a basic structural unit is iterated (see Chapter 11). In some modular organisms the products of growth fragment into physiologically independent parts, as with the fronds of *Lemna minor* or the rosettes of *Hieracium pilosella*, whereas in others, like an oak tree or clonal pasture grass, the module may remain part of a more or less closely integrated physiological whole (Pitelka & Ashmun 1985).

Populations of modular organisms are composed of N individual genets (individuals that develop from zygotes), each made up of η subpopulations of modules. The number of genets in a population can be described by Equation 12.1, while the modular growth of the ith genet can be described by the equation

$$\eta_{i,t+1} = \eta_{i,t} + B' - D' \tag{12.3}$$

where B' is the number of module births and D' is the total number of module deaths (Noble *et al.* 1979; Harper 1981). Equation 12.3 allows the growth of individual plants to be studied at a demographic level; the net result is that individual genets, even of the same age, may show considerable variation in size depending on the birth and death rate of modules (see Chapter 11). The total number of modules (n) in a population can be calculated from the equation

$$n = \sum_{j=1}^{N} \eta_j \tag{12.4}$$

Few studies have attempted to measure the flux in the number of genets *and* the modular flux within individual genets. Rather, most studies have been concerned with the flux of either genets (annuals, facultative biennials, trees) or the total number of modules (herbaceous perennials) in a population:

$$n_{t+1} = n_t + b - d + i - e \tag{12.5}$$

where b, d, i and e define the numbers of births, deaths, immigrants and emigrants of modules irrespective of the genet which produced them, or whether they arise from reproduction or clonal growth.

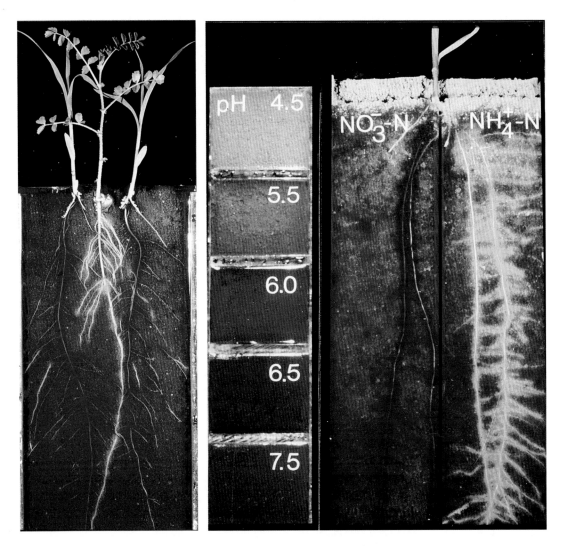

Plate 1 Both the source of nitrogen taken up by the plant and the species of plant involved affect the rhizosphere pH. The right-hand column shows a split-root experiment with maize (*Zea mays*) grown in soil infiltrated with agar containing a pH indicator, in which plants were fertilised either with nitrate (left) or ammonium (right). The nitrate-fertilized plants have a more alkaline rhizosphere due to OH^- extrusion; the ammonium-fed plants have a more acid rhizosphere due to H^+ loss. The central column gives the pH calibration. The left-hand column shows two *Sorghum bicolor* plants and a central chickpea (*Cicer arietinum*), fed with N as nitrate. The sorghum produces an alkaline rhizosphere, but even under these conditions chickpea acidifies the soil because it is N-fixing and so metabolizing ammonium rather than utilizing soil nitrate. (Reproduced with permission for Marschner *et al.* 1986.)

[*facing page 366*]

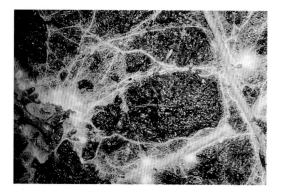

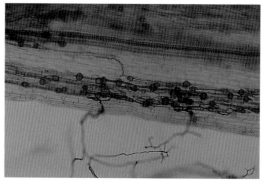

Plate 2 The characteristic dichotomously branching and stunted root tips of an ectomycorrhizal root of *Pinus contorta* surrounded by the hyphae of the fungus *Suillus bovinus*. (Photograph courtesy of Professor R. Finlay).

Plate 3 A root of bluebell *Hyacinthoides non-scripta* stained to show colonization by an arbuscular mycorrhizal fungus (Glomus sp.). The external mycelium is clearly visible, along with the entry points that connect the external and internal mycelia. The most prominent internal structures are the vesicles, found in most of these fungi. At this scale, the characteristic arbuscules are not identifiable.

Plate 4 Branching and plant shape. Trees grown in groups only produce branches on their outer sides.

Plate 5 Topiary caused by herbivores: hawthorn, *Crataegus monogyna* browsed by deer. The unbrowsed upper part of the canopy is in full flower, while the lower browsed part produced no flowers at all.

Plate 6 The species-rich orchid genus *Ophrys* consists of out-crossing species pollinated through pseudo-copulation with specialist wasps and solitary bees. Different *Ophrys* species growing in the same habitat use different insect species as their pollinators.

Plate 8 Repeated browsing of this oak seedling by rabbits has led to repeated regrowth; some seedling-sized plants may be as much as twenty years old.

Plate 7 Herbivores and grass flowering. The fenced community on the right has produced copious flowers and is straw-coloured by midsummer; the closely grazed community on the left has been prevented from flowering and is still bright green in midsummer.

Plate 9 Herbivores and plant community structure: the mature oak/birch woodland was caused by the myxoma epidemic of the mid 1950s. Rabbits still exert a profound impact on the ground flora of this woodland; the absence of plant cover is not due to shade, but to current rabbit grazing, as indicated by the growth of *Holcus mollis* inside the rabbit enclosure.

Plate 10 Weed biocontrol. Before: the waterway is smothered beneath a solid blanket of floating fern, *Salvinia molesta*. Beetles were introduced into the cage in the centre.

Plate 11 Weed biocontrol. After: the beetle *Cyrtobagous salvinae* has completely and permanently controlled the weed. (Both photographs by Peter Room.)

Plate 12 The Park Grass Experiment, Rothamsted. Foreground: treated with P and K; species-rich mosaic of herbs and legumes; middle distance: received N, P and K until 1989 but only P and K since then and is now dominated by yellow *Ranunculus acris*; background: still gets N, P and K and is dominated by the white umbellifer *Anthriscus sylvestris*.

Plate 14 Alien plants assemble into curious but structured plant communities. Here, the shrub *Buddleja davidii* from China grows with the herb *Conyza sumatrensis* (Asteraceae) from South America on urban waste ground in London; there are 25 alien species and only five native plant species on this site.

Plate 13 Biodiversity in Hawaii. Natural disturbance by lava flows leaves isolated island refugia of moist forest which serve as sources of organisms which recolonize the lava during primary succession.

12.5 The fates of individuals

In order to understand what factors determine the abundance of plants, it is necessary to understand how the demographic parameters in Equations 12.1, 12.3 and 12.5 are affected by the age and size of plants, and by population density, competitors, herbivores, pathogens, weather, soil conditions and various hazards like fire or burial by sand. Monitoring the fate of individuals from birth provides us with some insight into these processes, but the amount of useful information gained from such studies is limited, without additional experimentation, as the factors causing death and affecting fecundity and dispersal are rarely revealed by the simple monitoring of natural populations.

Ideally we would monitor the fates of individuals from the time of zygote formation, since it is at this point that the life of a genet begins, but this is an extremely difficult stage at which to begin counting (Harper 1977). Most studies follow the fates of genets from the time of seed maturation or (much less satisfactory) from seedling emergence. An exception is provided by the study of De Steven (1982) who monitored the fates of witch-hazel (*Hamamelis virginiana*) fruits from initial fruit set to seed maturation and dispersal. Only 14–16% of the fruits set in each of 2 years survived to maturation and dispersal. Physiological abortion resulted in the death of approximately 60% of the fruits. Other major causes of mortality included damage by caterpillars, infestation by a host-specific seed weevil and consumption by squirrels. De Steven speculated that much of the physiological abortion was the result of foliage-feeding caterpillars reducing leaf area, and thus limiting the internal resources for fruit maturation.

Factors other than resource limitation may, however, influence the percentage of fertilized ovules developing into seeds. Wiens (1984) reported that the seed–ovule ratio is about 85% in many annuals and approximately 50% in perennials, while Wiens *et al.* (1987) reported that pre-emergent reproductive success averages approximately 90% in inbreeding species and 22% in outbreeding species. They argue that genetic load and developmental selection provide the best explanation for reduced fruit and seed set in outcrossing species generally (see Chapter 6).

12.5.1 Fates of seeds

Following seed maturation, most plants stand still and wait to have their seeds dispersed by animals, wind or water (see Chapter 9). Undoubtedly some seeds are transported long distances from the parent plant, but the large majority of seeds are deposited close to the parent (see Chapter 7). The density of seeds then declines sharply with distance. The most widely used models to describe the relationship between the number of seeds (y) and the distance from the source (x) for the part of the dispersal curve distal to the mode are the inverse power law ($y = ax^{-m}$)

and the negative exponential model ($y = ae^{-mx}$) where a and m are fitted parameters (Okubo & Levin 1989); for the large majority of species Willson (1993) concluded that the negative exponential model provided an adequate fit to data. Harper (1977) argued that the steeper the dispersal curve, the more likely that a species would spread as an advancing front, rather than as isolated individuals over a greater distance. Examples of plant invasions into new areas are provided by Mack (1981), Lonsdale (1993) and Perrins *et al.* (1993).

Predispersal seed predation accounts for the death of many seeds (Crawley 1992b), but in cases where the seed feeders are also the principal agents of dispersal, it is often difficult to determine what fraction are killed and what fraction are dispersed. In general, the fates of seeds are difficult to follow once they reach the ground and are incorporated in the surface litter or soil. Occasionally it is possible to follow the fates of individually labelled seeds (Watkinson 1978a,b), but more often it has been found necessary to sow or bury replicated samples of viable seeds into small areas in which natural seed dispersal is prevented. Replicate samples are then retrieved at intervals and the number of emerged seedlings and buried seeds counted. The seed population in enforced dormancy (see Chapter 7) can be estimated by counting the number of seedlings that emerge from seeds maintained under favourable conditions. Those seeds that do not germinate (the innate and induced fraction of dormant seeds) can then be tested for viability using tetrazolium chloride (Moore 1985). The fates of the seeds of two species, determined using these methods, are shown in Fig. 12.5. For the winter annual *Vulpia fasciculata* (Poaceae) (Watkinson 1978a) a small proportion of the seeds at the time of seed dissemination were dormant, but after a short period of after-ripening most of the seeds had lost their innate dormancy and were capable of germination. More than 99.5% of the seeds germinated in the year they were produced, and few seeds were eaten or died. In contrast, approximately 30% of the seeds of *Ranunculus repens* remained dormant after 14 months and approximately 50% were lost to seed predators (Sarukhán 1974). Postdispersal seed predation by rodents, birds and ants is an important source of mortality and in experimental studies losses of 100% are frequently recorded (Crawley 1988).

The examples illustrated in Fig. 12.5 provide a particularly detailed picture of the fates of seeds. Other studies have shown how the dormancy characteristics of seeds may vary through time and how the longevity of seeds differs between species and habitats (Leck *et al.* 1989; Murdoch & Ellis 1992; Thompson 1992). As a consequence of the different patterns of germination and dormancy in different habitats, densities in the seed bank are typically lowest in primary forest and highest in highly disturbed habitats such as arable fields, where seed densities may range from just a few hundred to over 100 000 m^{-2}. In arable and horticultural fields the size of the seed bank for individual

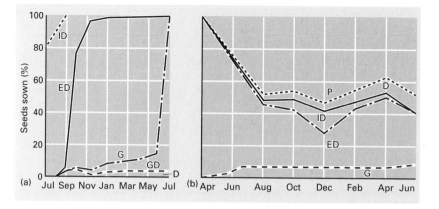

Fig. 12.5 Fates of seeds of (a) *Vulpia fasciculata* (Watkinson 1978a) and (b) *Ranunculus repens* (Sarukhán 1974). G, seeds observed to germinate; GD, individuals that germinated and died; D, fraction of seeds that died without germinating; P, seeds lost to predation (this category is included in D for *Vulpia*); ED, seeds in enforced dormancy; ID, seeds in innate or induced dormancy. Components always add to 100%.

species may be correlated with the size of the weed population above ground (Debacke 1988; Li 1995), and consequently there is considerable interest in the persistence of the seed bank in such habitats. On the basis of a range of studies on the survival of weed seeds on agricultural land (e.g. Roberts 1964, 1986), it has frequently been suggested that the number of viable seeds declines exponentially with time:

$$N_t = N_0 e^{-gt} \tag{12.6}$$

where N_t is the number of survivors at time t, N_0 is the initial seed density, and g is a constant that expresses the decay rate of the population. The implication is that the rate of loss of seeds from the seed bank is constant. Clearly the data demonstrate variation in the rates of seed survival between species (Fig. 12.6) and also dependence on conditions, but a more detailed analysis of the data by Rees and Long (1993) shows that weed seed banks do not in general decay according to a negative exponential pattern, but rather show a wide range of patterns that are difficult to detect on semilog plots where the logarithm of seed number is plotted against time (see Chapter 7).

12.5.2 Fates of individuals classified according to age and stage

The survivorship of seedlings is very difficult to measure accurately, especially early in the life cycle, unless frequent observations are made. Even then, those seedlings that die before emergence will not be recorded. For example, estimates of seedling mortality in populations of the annual *Vulpia fasciculata*, obtained by mapping permanent plots at 2–3 weeks intervals, ranged from 1 to 10% (Watkinson & Harper 1978), whereas estimates from an experiment in which seeds were radioactively

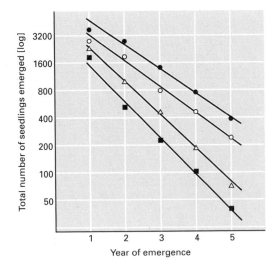

Fig. 12.6 Number of
seedlings emerging in
successive years from a
horticultural soil where seed
immigration was prevented.
The decline in seedling
emergence reflects a similar
decline in the size of the seed
bank. ●, *Thlaspi arvense*; △,
Stellaria media; ○, *Capsella
bursa-pastoris*; ■, *Senecio
vulgaris*. Combined data from
experiments that began in
1953, 1954 and 1955. (From
Roberts 1964.)

labelled varied from 12 to 28% (Watkinson 1978a). Observations on a range of tree seedlings (e.g. *Anaxagorea crassipetala* (Annonaceae); Fig. 12.7) show that the risk of death typically declines continuously from birth (Li *et al.* 1996) during the early stages of life, presumably because the small tree seedlings are particularly vulnerable to such mortality factors as pathogens, shading, tree falls and herbivore activity.

The subsequent fate of plants depends on their age and size. Data on the fates of plants followed through time indicate that, following the initial stages of establishment, the death rate is remarkably constant, at least for many perennial herbs (Harper 1977). There is often no indication of good or bad years for survival, despite considerable variation in the weather from year to year. This point is illustrated (Fig. 12.8) for five cohorts of the man orchid, *Aceras anthropophorum* (Wells 1981). The data also indicate that herbaceous perennials may be extremely long-lived; Tamm (1972), for example, found that the half-life of a population of *Primula veris* was 50 years. Where the fates of herbaceous perennials have been recorded at more regular intervals, it has generally been found

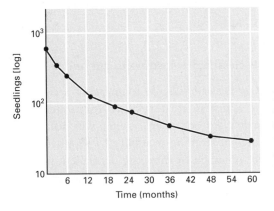

Fig. 12.7 Observed
survivorship curve of a
seedling cohort of the
understorey tree *Anaxagorea
crassipetala* in undisturbed
tropical wet forest in Costa
Rica. (From Li *et al.* 1996.)
Note the log scale.

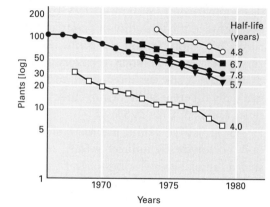

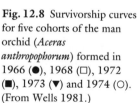

Fig. 12.8 Survivorship curves for five cohorts of the man orchid (*Aceras anthropophorum*) formed in 1966 (●), 1968 (□), 1972 (■), 1973 (▼) and 1974 (○). (From Wells 1981.)

that there is a marked seasonal rhythm to mortality (e.g. Sarukhán & Harper 1973). Under temperate conditions in the UK, the risk of death is generally greatest in spring and early summer when the growth of plants is most rapid, and rather low during the winter. The major cause of death in such cases appears to be interference from actively growing neighbours, rather than from harsh weather conditions directly.

Except for short-lived plants, however, it is difficult to follow the survival of cohorts of individual genets through their entire lifespan. Figure 12.9 illustrates the survivorship curves for 10 species of winter annual from various habitats from seed maturation to flowering. The curves exhibit a wide range of shapes, from the extreme Deevey type I curve of *Vulpia fasciculata* to the type III curve of *Spergula vernalis*

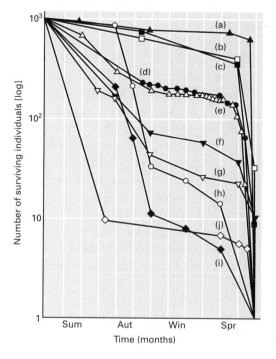

Fig. 12.9 Survivorship curves for natural populations of 10 winter annuals from seed production to maturity. The average number of seeds per plant follows the species name. (a) *Vulpia fasciculata*, 2; (b) *Avena barbata*, 4; (c) *Avena fatua*, 4; (d) *Cerastium atrovirens*, 7; (e) *Phlox drummondii*, 23; (f) *Bromus mollis*, 47; (g) *Bromus rubens*, 76; (h) *Sedum smallii*, 114; (i) *Minuartia uniflora*, 305; (j) *Spergula vernalis*, 100–414. Note how the shape of the survivorship curve changes as mean fecundity increases. (From Watkinson 1981b.)

(Caryophyllaceae) (Watkinson 1981b). Clearly there is a broad correlation between the fecundity of plants and the shape of their survivorship curves; the more fecund the plant, the more concave its survivorship curve. However, a single survivorship curve cannot be expected to characterize a species; different cohorts of the same species may show markedly different survivorship curves in different places or at different times (Mack & Pyke 1983).

For long-lived plants, information on age-specific survival can often only be sought by indirect means, (e.g. some morphological marker such as annual growth rings or leaf and bud scars), from which the age structure of the population can be determined. This allows an estimate to be made of the probability of survival from one age class to the next, assuming that the age-specific survival rates have remained the same from year to year and that recruitment into the population is constant from year to year. However, the number of cases where these two assumptions hold must be very small, and consequently there are considerable hazards in the interpretation of static age structures (Johnson *et al.* 1994). While animal population biologists have traditionally classified life histories in relation to age, the difficulty of obtaining data on age-specific fecundity and survival has led to plant population biologists classifying life cycles by stage rather than age, but this is only one of the reasons.

The existence of seed dormancy before plants enter a phase of active growth means that seedling plants may vary enormously in age. In such cases it is the stage of the plant rather than its age that provides the best description of fate. Also, because of the modular nature of plant growth, age and size are only loosely correlated, and two individuals of the same age may vary considerably in size. Moreover, in plants, it has been shown that size or stage rather than age is the major factor influencing survival and fecundity (Fig. 12.10). Small plants are typically much more vulnerable to death than adult plants (Sarukhán *et al.* 1984; Fowler 1986). The probability of flowering, especially in monocarpic perennials, is also critically dependent on plant size (Gross 1981; Kachi & Hirose 1985) and, once reproduction begins, reproduction is strongly dependent on plant size (Watkinson & White 1986). Moreover, the classification of plants by stage/size rather than age allows a population to be classified and then monitored through time to calculate growth, survival and fecundity in relation to the initial stage/size classification of the population. This allows population models to be developed for even long-lived trees (see Chapter 15).

There can be little doubt that both stage/size and age are important determinants of survival and reproduction for many higher plants. Ideally it would always be preferable to monitor the fate of individual plants according to both their age and their stage/size (Law 1983), but this has seldom been attempted (see Werner & Caswell 1977; van Groenendael & Slim 1988). Such large samples and such detailed

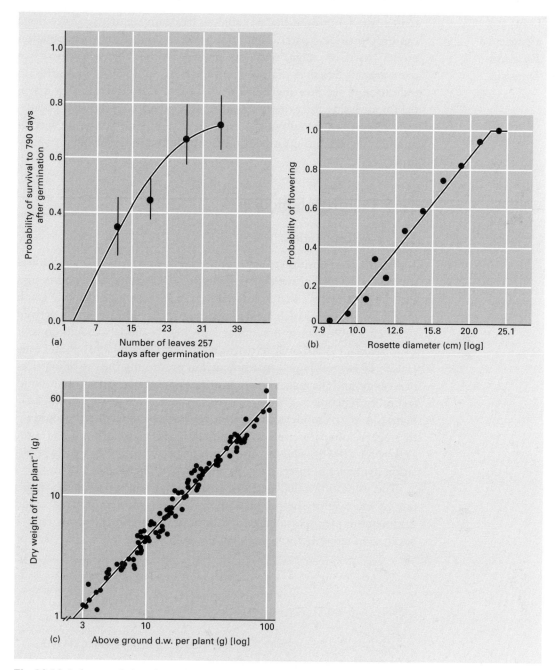

Fig. 12.10 Influence of plant size on (a) probability of survival of seedlings of the desert shrub *Simmondsia chinensis* (Sarukhán *et al.* 1984); (b) probability of flowering of the monocarpic perennial *Oenothera glazioviana* (Kachi & Hirose 1985); and (c) seed production of the iteroparous perennial *Rumex crispus* (Weaver & Cavers 1980).

monitoring are required to calculate all the necessary transition probabilities between the various age and stage/size classes that the labour involved is daunting. Most studies have analysed population behaviour in terms of either age or, much more frequently, stage/size. It must be remembered, however, that any analysis based on age or stage/size alone can only provide a partial description of the population. A population model based on stage/size may often provide a better predictor of demographic fate than one based on age (Caswell 1989), but it gives no indication of the flux of genets through time. This information is essential for analysing the genetic structure of populations. Similarly, population analyses of herbaceous perennials that monitor only the flux of modules provide no information on the number of genets in a population.

12.6 Population models

12.6.1 Matrix models

We are now in a position to ask how the various patterns of fecundity and survival affect the growth of populations. The data on fecundity and survival can be conveniently summarized in the form of a life-cycle graph (Fig. 12.11), from which a corresponding matrix model can be derived (Caswell 1989). Those unfamiliar with matrix models should consult Begon *et al.* (1995b) or Silvertown and Lovett Doust (1993). P.H. Leslie (1945, 1948) was largely instrumental in promoting the use of matrices in ecology, and the matrix that summarizes the age-specific mortality and fecundity schedules of a population is often referred to as the Leslie matrix. A matrix model that summarizes survial and fecundity according to stage is often referred to as a Lefkovitch matrix (Box 12.1) after Lefkovitch (1965), who first used such matrices to investigate the population dynamics of insect pests classified according to stage.

The value of λ calculated from a matrix is referred to as the 'finite rate of increase of the population', and is the most important single parameter in plant population dynamics. It represents the actual rate of population growth for a population characterized by a given set of

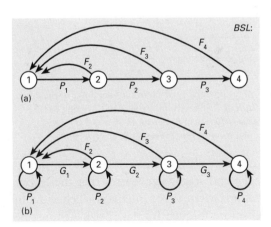

Fig. 12.11 Examples of life-cycle graphs for populations classified by (a) age and (b) size or stage. Four age or size/stage classes are shown in each example. (a) F_x is the fecundity of individuals of age x and P_x is the probability of surviving from one age to the next. (b) F_x is the fecundity of individuals of size/stage x, P_x is the probability of surviving from one size/stage to the next and G_x is the probability of surviving and moving into the next size or stage class. (From Caswell 1989.)

Box 12.1 The stage-classified matrix model

The projection matrix corresponding to the life-cycle in Fig. 12.11b is a square matrix defined by the expression

$$M = \begin{bmatrix} P_1 & F_2 & F_3 & F_4 \\ G_1 & P_2 & 0 & 0 \\ 0 & G_2 & P_3 & 0 \\ 0 & 0 & G_3 & P_4 \end{bmatrix} \tag{12.1.1}$$

P_i defines the probability that an individual will survive and remain in the same size or stage class, G_i is the probability that an individual will survive and grow to the next size or stage class, and F_i is the rate of reproduction for an individual in size or stage class i. We assume that the plants are hermaphrodite; matrix procedures for dioecious species are described by Caswell (1989). While matrices of the form outlined above have been widely used, the exact form of the matrix depends on the biology of the species, the classification used to define population structure and the time interval used in population projection. For example, depending on the time interval, it is possible for an individual to remain in a particular stage, or move up one or more stage classes. All of the elements below the leading diagonal relate to the probability of growth to a larger size or stage class while all those above relate to fecundity, clonal growth or shrinkage (Hughes 1984).

Corresponding to the projection matrix is a population vector that describes the stage structure of the population. For the population illustrated in Fig. 12.11b there are four stage classes, and the number of individuals in each stage class can be written in the form of a column vector

$$n = \begin{bmatrix} n_1 \\ n_2 \\ n_3 \\ n_4 \end{bmatrix} \tag{12.1.2}$$

This vector summarizes the stage distribution of the population at a given time and $\sum_{i=1}^{4} n_i$ gives the total size of the population. In order to calculate the state of the population after one time interval it is necessary only to multiply the column vector by the matrix.

$$\begin{bmatrix} P_1 & F_2 & F_3 & F_4 \\ G_1 & P_2 & 0 & 0 \\ 0 & G_2 & P_3 & 0 \\ 0 & 0 & G_3 & P_4 \end{bmatrix} \times \begin{bmatrix} n_{1,t} \\ n_{2,t} \\ n_{3,t} \\ n_{4,t} \end{bmatrix} = \begin{bmatrix} P_1 n_{1,t} + F_2 n_{2,t} + F_3 n_{3,t} + F_4 n_{4,t} \\ G_1 n_{1,t} + P_2 n_{2,t} \\ G_2 n_{2,t} + P_3 n_{3,t} \\ G_3 n_{3,t} + P_4 n_{4,t} \end{bmatrix}$$

$$= \begin{bmatrix} n_{1,t+1} \\ n_{2,t+1} \\ n_{3,t+1} \\ n_{4,t+1} \end{bmatrix} \tag{12.1.3}$$

continue on page 376

Box 12.1 *Continued*

If the matrices are represented by symbols then:

$$Mn_t = n_{t+1} \qquad (12.1.4)$$

and

$$n_t = M^t n_0 \qquad (12.1.5)$$

After repeated multiplication we find that oscillations in the relative numbers of individuals in each stage class dampen down, and that the *proportion* of individuals in each stage class remains the same. This 'stable stage distribution' occurs in a population that is either *increasing or decreasing exponentially* (depending on whether recruitment exceeds mortality), and is a characteristic of the transition matrix. It arises whatever the initial stage distribution, n_0. Note also that after repeated multiplication, each stage class is increasing at a constant rate, and that multiplying the population vector by the transition matrix is equivalent to multiplying it by a single number λ called the 'finite rate of increase' of the population.

$$Mn_t = \lambda n_t \qquad (12.1.6)$$

An example of a transition matrix for a tree, the nikau palm (*Rhopalostylis sapida* (Arecaceae)) in New Zealand, is provided by Enright and Watson (1992). The transition matrix was calculated from life-table data collected over 7 years for plants classified into eight height classes

$$
\begin{bmatrix}
0.334 & 0 & 0 & 0 & 113 & 273 & 455 & 353 \\
0.002 & 0.785 & 0 & 0 & 0 & 0 & 0 & 0 \\
0 & 0.045 & 0.922 & 0 & 0 & 0 & 0 & 0 \\
0 & 0 & 0.013 & 0.976 & 0 & 0 & 0 & 0 \\
0 & 0 & 0 & 0.014 & 0.935 & 0 & 0 & 0 \\
0 & 0 & 0 & 0 & 0.060 & 0.956 & 0 & 0 \\
0 & 0 & 0 & 0 & 0 & 0.041 & 0.982 & 0 \\
0 & 0 & 0 & 0 & 0 & 0 & 0.013 & 0.985
\end{bmatrix} \qquad (12.1.7)
$$

This stage-classified matrix can be used to predict the numbers in each height class at any subsequent census (as long as the conditions for the model hold). It may also be used to calculate the population's finite rate of increase and stable size distribution. The estimate of λ from this matrix is 1.004, indicating a relatively stable population size or one that is only just increasing. The stable stage distribution is [98.24, 0.89, 0.47, 0.20, 0.04, 0.05, 0.07, 0.04], listed in percentages.

survival and fecundity schedules. The value of λ determines whether the population will increase ($\lambda > 1$), decrease ($\lambda < 1$) or remain constant ($\lambda = 1$) in size. Estimates of λ for 66 species of herb, shrub and tree provided by Silvertown *et al.* (1993a,b) range from 0.50 in the case of *Ranunculus repens* to 11.82 for *Digitalis purpurea* (Scrophulariaceae). Not surprisingly, given the short duration of most studies, many of the estimates for tree species were very close to unity; for example, the average value of λ for *Astrocaryum mexicanum* (Arecaceae) calculated from six plots in a tropical rain forest in Veracruz, Mexico was 1.0046 (Piñero *et al.* 1984). This indicates a relatively stable population size with a doubling time of approximately 150 years.

The sensitivity of the finite rate of increase to changes in the life-history parameters allows an investigation into which of the elements in the life table play a key role in determining the numbers in a population through elasticity analysis (de Kroon *et al.* 1986; Caswell 1989). A comparison of the relative importance of the recruitment of seeds, recruitment of seedlings, clonal growth, retrogression, survival from one year to the next in the same stage class (stasis) and progression to later stage classes (Silvertown *et al.* 1993) showed that seedling recruitment was more important in herbs than in woody plants; retrogression occurred only in herbs, particularly those with a tuber; stasis occurred in nearly all species but especially woody plants; and progression was more important than fecundity in nearly all species. Grouping the elements in the matrix according to their contribution to the finite rate of population increase, through either fecundity (F), survival (L) or growth (G), allowed an ordination of the species in G–L–F space (Fig. 12.12); the species clearly segregate according to their life history and habitat. For example, semelparous herbs fall along the F–G axis whilst woody plants of forests lie near the L = 1 vertex and woody plants of open habitats lie towards the centre of the triangle. Obviously, the finite rate of population increase λ is a function of all of the elements in the matrix and the importance of each will depend on the values of the others, but this comparative study nevertheless clearly demonstrates variation in the importance of fecundity, survival and growth in the demography of plants with contrasting life cycles. Moreover, Silvertown and Franco (1993) have shown that succession and a variety of environmental factors such as grazing and fire alter the relative importance of fecundity, survival and growth to the demography of species' populations.

12.6.2 Difference equations

For species that breed at discrete intervals and have non-overlapping generations (e.g. many annuals), it is possible to relate the number of individuals at time $t + 1$ to those at time t by a difference equation of the

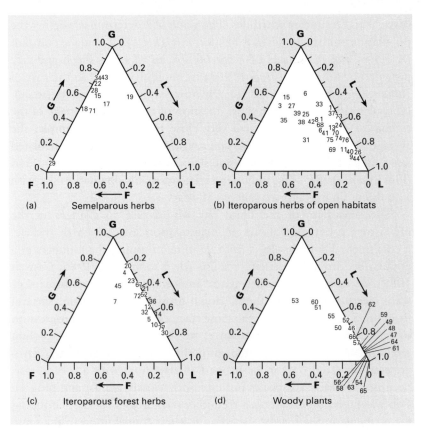

Fig. 12.12 Distribution of 77 specis in G–L–F space: G = growth, L = survival, F = fecundity (a) semelparous herbs, (b) iteroparous herbs of open habitats, (c) iteroparous forest herbs, (d) woody plants. (From Silvertown & Franco 1993.) See text for details.

form:

$$N_{t+1} = \lambda N_t \qquad (12.7)$$

where λ is again the finite rate of increase of the population and is a measure of the balance between survival and reproduction. If N is the number of reproducing plants then λ takes into account both the number of seeds produced per plant and the survival of those offspring to maturity. The inclusion of a seed bank into difference equation models is considered in Chapter 7. For *Phlox drummondii* (Polemoniaceae), Leverich and Levin (1979) estimated λ as 2.42 (i.e. the population will multiply by a factor of approximately two and a half per year). Over a number of years population growth is described by the equation

$$N_t = N_0 \lambda^t \qquad (12.8)$$

or, in this case

$$N_t = N_0 \, 2.42^t$$

It is plain to see that this equation leads to a population which, like the matrix models described in the previous section, grows exponentially for ever!

There are two important limitations in all the models discussed so far. First, population size cannot increase indefinitely, and some or all of the parameters in the life table must be density-dependent. This is the subject of the next section. Second, the models assume that survival and reproduction are constant from year to year, and from place to place within a year, and are not affected by changing physical conditions. Detailed study of the winter annual grass *Bromus tectorum* by Mack and Pyke (1983) has shown convincingly that the survival of seedlings varies greatly both between years and between different cohorts of the same generation, depending on year-to-year variation in the environment.

12.7 Density-dependence

It was emphasized in section 12.5 that in order to understand what determines the number of individuals in a population, it is not sufficient just to record survivorship and fecundity. It is also necessary to find out how the various elements in the life table depend on the number and sizes of individuals in the population. Ideally, this should be carried out by experimental manipulation of natural populations to densities above and below their existing levels. If this is not practicable, it may be possible to measure plant performance at a range of naturally occurring densities. However, interpretation of this type of study requires caution because differences in plant performance due to density will be confounded by microenvironmental differences between sites (Antonovics & Levin 1980).

Examples of density-dependence have been recorded in all aspects of plant demography, including seed dispersal (Baker & O'Dowd 1982), seed predation (Watkinson *et al.* 1989), germination (Inouye 1980), seedling establishment (Hett 1971), survival and growth during the vegetative phase of the life cycle (Symonides 1983; Condit *et al.* 1994), flower formation (Bishop & Davy 1984), flowering phenology and outcrossing (Schmitt *et al.* 1987) and the number of seeds set per plant (Watkinson 1985b); see Fig. 12.13. Variations in plant performance with the level of crowding may result from factors both internal and external to the population. For example, an increase in population density may lead to a larger number of individuals competing for limited resources compared with plants at low densities. As a consequence fewer resources will be available to individual plants, fewer modules will be iterated and individual plant size and fecundity will be lower. At extremely high densities the death rate of modules on some plants may exceed the birth rate, and whole genets will die, the number increasing with density. Thus intraspecific competition for limiting resources could lead to both reduced survivorship and fecundity. The level of crowding in a population can also affect the level of herbivore

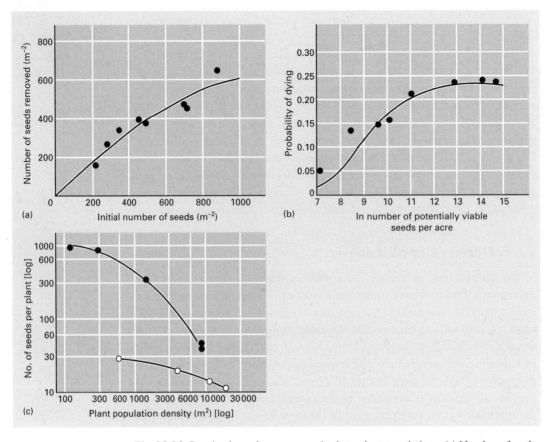

Fig. 12.13 Density-dependent processes in three plant populations. (a) Number of seeds removed by ants in relation to the initial density of seeds of the annual grass *Sorghum intrans* (Watkinson *et al.* 1989). (b) Mortality in a population of sugar maple establishing from seed; 1 acre = 0.4 ha (Hett 1971). (c) Influence of flowering plant density on seed production per plant by *Salicornia europaea* on a low (●) and high (○) marsh (Watkinson & Davy 1985).

and pathogen activity. For example, Augspurger and Kelly (1984) found that mortality due to damping-off disease increased with seedling density in populations of a tropical tree, *Platypodium elegans* (Fabaceae), while Burdon *et al.* (1992) have demonstrated density-dependent mortality in 5–10-year-old stands of *Pinus sylvestris* caused by the snow blight pathogen *Phacidium infestans*. Similarly, Watkinson *et al.* (1989) have shown that the predation of seeds of *Sorghum intrans* (Poaceae) by ants increases with seed density.

Particular attention has focused on the consequences of intraspecific competition for density-dependent growth and survival in plant monocultures and mixtures (Firbank & Watkinson 1990; Silvertown & Lovett Doust 1993) as the more closely plants are crowded together the more they interfere with each other's growth for the same essential resources. As a consequence individual plants tend to be smaller in high density populations and to have a greater risk of death. The relationship between individual plant performance (w) and the density of plants (N) in even-aged populations at any one time can typically be described by an equation of the form

$$w = w_{\mathrm{m}}(1 + aN)^{-b} \qquad (12.9)$$

where the parameter w_{m} can be interpreted as the mean performance of an isolated plant, and a and b are parameters that describe the strength of the density-dependent feedback response (Watkinson 1980, 1985b). Although Equation 12.9 refers to the mean performance (e.g. dry weight) of plants, this is not to imply that there is an approximately equal partitioning of resources between individuals; rather a hierarchy of exploitation develops (see Chapter 11). Such inequalities develop more quickly in high-density populations, and as the individuals continue to grow the capacity of some individuals to absorb competition by plastic responses is exceeded. Once this point has been reached the plastic response of plants may be augmented by self-thinning and density-dependent mortality. The time course of self-thinning can typically be approximated by an equation of the form

$$w = cN^{-k} \qquad (12.10)$$

or

$$\log w = \log c - k \log N$$

(Yoda et al. 1963) where c is a constant that varies from species to species and k has a value of approximately $3/2$ for a wide range of species (White 1985). There has recently been considerable debate over the generality of the self-thinning rule or $-3/2$ power law (see Weller 1987; Zeide 1987; Lonsdale 1990), but what is clear is that the self-thinning rule can be used to describe the time trajectory of a population undergoing density-dependent mortality, and that more generally it defines the boundary line for those combinations of weight and density that are possible in plant populations. For a wide variety of species (White 1985) the value of $\log_{10} c$ lies between 3.5 and 5.0 (where weight is measured in grams per plant and N in plants per square metre). This narrow range holds true over an impressive seven orders of magnitude of plant density and eleven orders of magnitude of mean biomass per plant. All combinations of weight and density are potentially possible to the left of the lines in Fig. 12.14 but none are possible to the right.

12.8 Population dynamics

12.8.1 Annual plants

The population dynamics of annual plants with discrete generations can be described by an equation of the form

$$N_{t+1} = \lambda N_t f(N_t) \qquad (12.11)$$

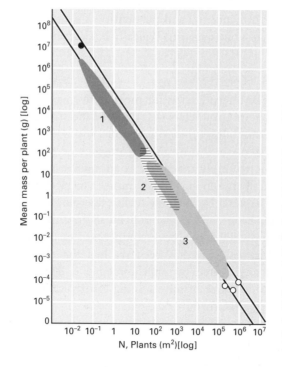

Mean mass per plant (g) [log]

N, Plants (m²)[log]

Fig. 12.14 Relationship between mean plant mass (w) and density (N) in single-species populations of plants. Each shaded area encloses the limits of many self-thinning lines or points of maximum mass and density. 1, Trees; 2, clonal perennial herbs; 3, annual and perennial herbs grown at high density in experimental systems. ○, *Lemna* spp.; ●, *Sequoia sempervirens*. The boundary lines represent $w = 10^4 N^{-3/2}$ and $w = 10^5 N^{-3/2}$ and are not regression lines. (From White 1985.)

where N is the number of individuals, t is time, λ is the finite rate of population increase (see Equation 12.7) and $f(N_t)$ is a density-dependent feedback term. How much can this equation tell us about the population dynamics of plants in the field? Assuming that $f(N_t)$ can be approximated by Equation 12.9 together with a term for self-thinning, estimates of the parameter values have been made for a range of annual species. In the case of *Salicornia europaea* (Chenopodiaceae) colonizing mud flats on salt marshes (Watkinson & Davy 1985), it has been shown that populations can be expected to reach an equilibrium density within 3 years if density-independent mortality is low. Population regulation occurs through the density-dependent control of fecundity; as density increases intraspecific competition results in a decline in individual fecundity. Comparison of the theoretical growth curves for *Salicornia* in England with the observed colonization curve given by Joenje (1978) for the Lauwerszeepolder in The Netherlands indicates that the model mimics the growth of natural populations rather closely (Fig. 12.15).

For populations of annuals plants that may persist over a number of years, models of the form outlined in Equation 12.11, in which fecundity and survival vary only with intraspecific competition, have been used to predict approximately and explain the densities that occur in the field. For *Salicornia europaea* (Jefferies *et al.* 1981), *Sorghum intrans* (Watkinson *et al.* 1989) and *Vulpia fasciculata* (Watkinson & Harper 1978), it has been shown that population size is essentially determined by the interaction between the density-dependent control of fecundity and the level of density-independent mortality (Fig. 12.16). It

Fig. 12.15 Population changes calculated from Equation 12.11 showing the effects of varying the probability of surviving density-independent mortality, p, on the rate of colonization and equilibrium population size in *Salicornia europaea*. The actual colonization curve for *Salicornia europaea* on the Lauwerszeepolder, Netherlands, is shown as a dashed line. If it is assumed that p declines from 0.35 to 0.04 after two generations, then the model closely mimics the growth of the natural population. These predicted values are very close to those actually observed in the field. (From Watkinson & Davy 1985.)

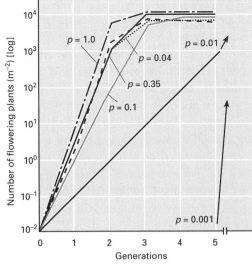

is variation in the level of density-independent mortality over space and time that, in particular, results in variations in the abundance of all three species.

There is the possibility that fluctuations in abundance may result from intrinsic factors to the plant population, as when high values of the finite rate of population increase, λ, are coupled with over compensating density-dependence. Variation in these intrinsic processes provides the potential for a wide range of dynamic processes including exponential and oscillatory damping towards a stable equilibrium point, stable limit cycles and chaotic behaviour (Hassell *et al.* 1976; May & Oster 1976). It has been argued, however, that plant populations are in general unlikely to display such dynamics, as a consequence of the asymmetric nature of intraspecific competition in plant populations, self-thinning, reproductive thresholds and the presence of a seed bank (Watkinson

Fig. 12.16 Mean observed (●) population densities of *Sorghum intrans* in the wet–dry tropics of northern Australia over a 7-year period compared with the predicted population densities (○) derived from a stochastic model in which fecundity was negatively density-dependent and in which all sources of mortality except self-thinning were density-independent. The 95% confidence limits for population size are also given. (From Watkinson *et al.* 1989.)

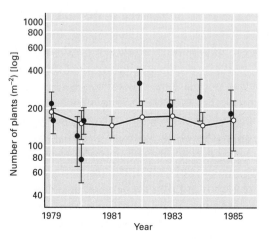

1980; Pacala 1986; Rees & Crawley 1989). While there has been some debate over the likelihood of plant populations showing limit cycles or chaotic dynamics (Rees & Crawley 1991; Silvertown 1991; Cousens 1995), it is only for *Erophila verna* (Brassicaceae) that two-point cycles have been demonstrated in the field and then only over a short period and within a few small quadrats (Symonides *et al.* 1986).

Fluctuations in the relative abundance of annual plants in the annual grasslands of northern California depend strongly on weather (Pitt & Heady 1978). Similarly Venable *et al.* (1993) have shown that year-to-year fluctuations of annual plants in the Sonoran Desert can be attributed to environmental variation through time. However, different species responded differently to the temporal variation in the environment, a factor that has implications for coexistence in these communities of annual plants (Venable *et al.* 1993; Pake & Venable 1995). Rees *et al.* (1996) studied the population dynamics of four species of winter annuals – *Erophila verna*, *Cerastium semidecandrum* (Caryophyllaceae), *Myosotis ramosissima* (Boraginaceae) and *Valerianella locusta* subsp. *dunensis* (Valerianellaceae) – over a 10-year period in sand dunes at Holkham on the north coast of Norfolk in England. Data were gathered from 500 permanent 10×10 cm quadrats on two replicate transects. Populations fluctuated in parallel on the two transects (Fig. 12.17) highlighting the importance of year-to-year weather variation as a cause of ups and downs in population density. Detailed statistical analysis demonstrated that population growth was density-dependent for all species, and that intraspecific competition was much more important than interspecific competition in all years. Had it not been for the within-year spatial density-dependence, population densities would have been roughly two-fold higher than those observed (range from 1.5-fold for *Myosotis* to 3.0-fold for *Cerastium*). All eight populations were characterized by stabilizing density-dependence. There was no tendency towards cyclic or chaotic dynamics in any case; note that two-point cycles had been suggested for *Erophila verna* in Poland (Symonides *et al.* 1986).

The role of dispersal in the population dynamics of plants is clearly seen in colonizing environments, but its role in the patch dynamics of plant populations is only just beginning to be explored. One of the most striking examples of the role of dispersal in the maintenance of annual plant populations is described by Keddy (1981), who found that the abundance of *Cakile edentula* (Brassicaceae) along a sand dune gradient in Nova Scotia was typically greatest in the middle of a dune ridge and lower towards the seaward and landward ends. Along this gradient the levels of fecundity and mortality were such that the populations could be maintained only at the seaward end of the gradient. On the middle and landward sites the level of mortality exceeded reproductive output, so the populations would not have been able to persist in isolation. A computer simulation of the populations (Watkinson 1985a) shows

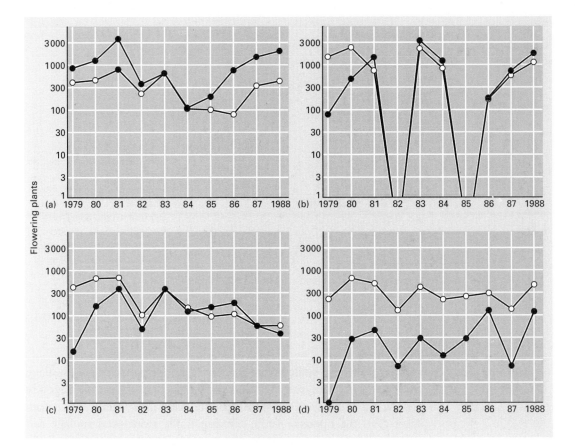

Fig. 12.17 Population trajectories in two areas (O, ●) for (a) *Erophila verna*,
(b) *Cerastium semidecandrum*, (c) *Myosotis ramosissima* and (d) *Valerianella locusta*. (From
Rees *et al.* 1996.)

that the plants at the middle and landward ends of the gradients exist
only because of immigration from the seaward end. There is a high
annual dispersal of seeds (> 50%) landward from the highly fecund
plants on the beach (Keddy 1982). This example (Fig. 12.18) neatly
illustrates the complex interactions between various density-dependent
and density-independent processes that affect abundance and performance of plants.

In a very different environment, the verge of the M25 motorway
round London, Crawley and Brown (1995) found that the yellow drifts
of oilseed rape (*Brassica napus* subsp. *oleifera*) seen each spring concealed complex dynamics from year to year. There was a substantial
turnover in patch occupancy with approximately 50% of the patches
occupied in one year being extinct the next. In the absence of soil
disturbance, rapid secondary succession led to the local extinction of
populations, but given sufficient soil disturbance rape population density was seed-limited. Patches of rape have the potential to establish
from a seed bank, seeds dispersed from nearby fields or feral popula-

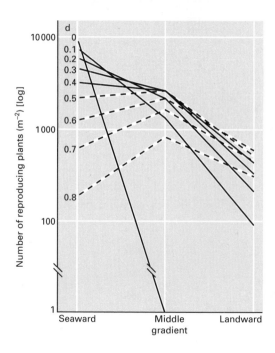

Fig. 12.18 Results of a computer simulation based on Equation 12.11 to demonstrate the effect of seed dispersal, *d*, along a sand dune gradient on the population size of the annual crucifer *Cakile edentula*. Only a net landward movement of seeds in excess of 0.5 (dashed lines) produces a pattern of abundance similar to that observed in the field. (From Watkinson 1985a.)

tions, or from seed spilled from lorries. The importance of seed spillage could be seen from the fact that the verge next to the carriageway carrying traffic to the rape-seed crushing plant had significantly more plants than the opposite verge carrying traffic away; seed spillage can cause a twofold to fivefold increase in population density (see Fig. 4.8).

12.8.2 Perennial plants

Our understanding of how density affects plants of different ages and sizes is unfortunately limited. Yet this type of information is essential if we are to understand the dynamics of populations structured by size or stage. There are, nevertheless, studies of plants with a range of different types of life history that provide insights into the dynamics of structured populations.

Small plants presumably perceive the level of crowding to be much higher than their larger neighbours. Indeed Law (1975) found that the survival of seedlings to young adults in populations of *Poa annua* was density-dependent whilst the survival of older plants was not. In contrast, the age-specific fecundity of three adult categories (young, medium, old) of plants was density-dependent in each case, although the medium plants were the most fecund. An appropriate matrix model to describe the population dynamics of *Poa annua* (see Begon *et al.* 1995b) categorized into four age groups of approximately 8 weeks' duration (seedling, young adult, medium adult, old adult) together with a seed bank is described in Box 12.2. Iteration of the model with appropriate parameter values gives an indication of how a population

Box 12.2 Dynamics of a model population of *Poa annua*

Law (1975) modified the basic Leslie matrix to describe the population dynamics of *Poa annua* by making seed production and seedling survival density-dependent, and by allowing for the existence of a seed bank

$$\begin{bmatrix} p_s & 0 & b_1(N) & b_2(N) & b_3(N) \\ g_s & 0 & 0 & 0 & 0 \\ 0 & p_0(N) & 0 & 0 & 0 \\ 0 & 0 & p_1 & 0 & 0 \\ 0 & 0 & 0 & p_2 & 0 \end{bmatrix} \qquad (12.2.1)$$

where p_s is the probability of a seed surviving in the seed bank, g_s is the probability of germination, $p_0(N)$ is the probability of seedlings surviving to become young adults, p_1 is the probability of young adults becoming mature adults and p_2 is the probability of mature adults becoming old adults; $b_1(N)$, $b_2(N)$ and $b_3(N)$ are the number of seeds produced by young, medium and old adults respectively. The parameters that are functions of (N) are assumed to be density-dependent. In each case, the functions are assumed to decrease monotonically with increasing density. The dynamics of this model are depicted in Fig. 12.19.

establishing with 100 seeds grows towards an equilibrium population density with a stable age structure after 18 time periods (Fig. 12.19). Compare this stimulated colonization curve with the actual colonization curve (Fig. 12.1a) observed by Law (1981).

Studies on populations growing under a range of conditions are particularly informative and allow us to tease apart the various factors determining abundance and dynamics. For the monocarpic perennial

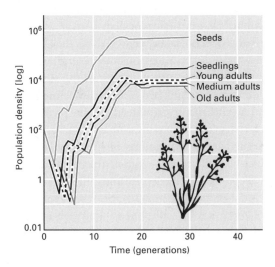

Fig. 12.19 Results of a simulation of the density-regulated growth of five age-classes of *Poa annua*, based on the density-dependent matrix model given in Box 12.2. (From Law 1975.)

Cirsium vulgare (Asteraceae), matrix analysis has been used to examine the factors determining abundance and dynamics in a range of experimental populations involving sheep grazing (Gillman *et al.* 1993; Bullock *et al.* 1994). Heavy grazing resulted in higher seedling emergence and survival of seedlings and rosettes, resulting in higher finite rates of population increase, λ. Density-dependence was shown to operate primarily through the effects of rosette density on the number of small rosettes produced per flowering plant. A density-dependent matrix model incorporating these demographic elements explained over 80% of the between-treatment variation in mean observed number of *Cirsium vulgare*. An alternative modelling approach, using a safe-site model, to explore the effects of density-dependence during germination showed that the abundance of *Cirsium vulgare* depended critically on the density of gaps (Silvertown & Smith 1989); a threshold density of gaps was discovered below which thistle populations went extinct, and above which populations grew geometrically.

For the woodland perennial herb *Viola fimbriatula* seedling survival is related to the local density of adults, while adult survival and fecundity are size dependent; seeds may remain viable in the soil for many years (Solbrig *et al.* 1988). A matrix population model that incorporates all of these elements predicts that the population will show stable oscillations around an equilibrium density, induced by density-dependence and the time lag between the death of adults and their replacement by recruits. Unfortunately the period of observation in the field (5 years) was too short to validate this prediction. Long-term simulations have also been carried out for another woodland herb, the bulb-forming *Narcissus pseudonarcissus* (Liliaceae) (Barkham & Hance 1982). A spatially explicit cellular automaton model incorporating density-dependent clonal growth was used in this case and showed that the measured rates of reproduction and mortality were sufficient to stabilize *Narcissus pseudonarcissus* populations at realistic bulb densities. Moreover it was possible to show that the management practice of coppicing in these woodlands may allow populations to persist that might otherwise be expected to decline to extinction in unmanaged woods. These simulations were based on the number of bulbs or ramets. Varying seedling recruitment in the simulation model predicts that the number of ramets per genet will decline as the ratio of the probability of an adult arising from seed to the probability of an adult arising from clonal growth increases. Watkinson and Powell (1993) have similarly argued from simulations of the population dynamics of the stoloniferous herb *Ranunculus repens* that the ratio of seedling to ramet recruits will be critical in determining the number of genets in populations of clonal plants (Fig. 12.20), but that even low ratios of seedling to ramet recruits may be sufficient to maintain quite high genet densities relative to the total numbers of ramets in populations. Whether seedling recruitment occurs within patches of established genets is clearly crucial

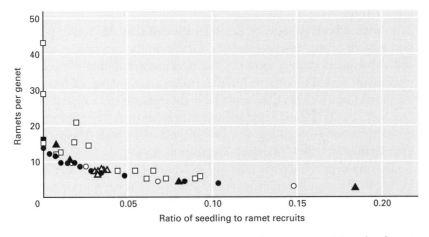

Fig. 12.20 Results of a computer simulation showing the influence of the ratio of seedling to ramet recruits in populations of clonal plants on the number of ramets per genet (or conversely the ratio of genets to the total number of ramets in a population). The different symbols represent the results of different simulations in which seedling numbers, population size, density-dependent mortality and clonal growth were varied. (From Watkinson & Powell 1993.)

here (Eriksson 1993). For many clonal plants (60%), seedling recruitment has not been observed within adult patches (Eriksson 1989), and the genet density can be expected to reflect the initial colonizing cohort and its survivorship.

Scale is clearly critical in understanding the population dynamics of plants. Observations on established populations of many clonal plants will clearly fail to provide any information on the dynamics of seedling populations and consequently of the dynamics of the population as a whole. On a much larger scale, Menges (1990) has shown that the persistence of the endangered perennial herb *Pedicularis furbishiae* (Scrophulariaceae) along river banks of a 200-km stretch of the St John River in Maine depends critically on the probabilities of local population extinction and re-establishment in its riverine habitat. Each individual population of the species is virtually certain to die out as a result of varying hydrology.

Within forests the recruitment of new individuals into tree populations is controlled by the appearance of gaps and consequently the dynamics and abundance of tree species depends critically on an understanding of the dynamics of gap formation (see Brokaw 1982, 1985). For example, Hubbell and Foster (1986c) have shown that the abundance of shade-intolerant species is correlated with the percentage of open canopy on Barro Colorado Island, Panama. In a study of the understorey palm *Astrocaryum mexicanum* in the tropical forests of Veracruz, Mexico, it was found that survivorship, growth and fecundity varied between gaps and the mature forest (Martínez-Ramos *et al.* 1988), such that the finite rate of population increase was higher in the mature forest than in young gaps. Incorporating a matrix model of the palm with a forest growth cycle model showed that the overall population growth of the palm depended critically on the rate of gap opening and subsequent canopy closure. The data indicated that *Astrocaryum mexicanum* would be able to increase under a wide range of forest gap conditions but was likely to grow to highest densities in closed forest. One source of population regulation has been shown to be the increased mortality amongst seedling and juvenile plants in high-density stands (Sarukhán *et al.* 1984; Martínez-Ramos *et al.* 1988).

Density-dependence has rarely been reported in tropical forests and its role in the population dynamics of trees remains open to debate (see Chapter 15; Hubbell *et al.* 1990). Evidence for the role of density-dependence in the dynamics of tropical trees comes from studies of three species in the neotropics: *Astrocaryum mexicanum*, *Cecropia obtusifolia* and *Trichilia tuberculata* (Meliaceae). Larger estimates of the finite rate of population increase for *Astrocaryum mexicanum* were obtained for populations growing in low-density plots (Piñero *et al.* 1984), while Hubbell *et al.* (1990) reported that the population growth rate of a population of *Trichilia tuberculata* decreased monotonically from low- to high-density stands as a result of density effects on the survivorship of small plants. Moreover, the model developed for *Trichilia tuberculata* predicted population densities at equilibrium and size class distributions similar to those found in the field. Density-dependent effects on the fecundity and survival of the pioneer species *Cecropia obtusifolia* incorporated into a matrix model were similarly shown to produce population densities at equilibrium similar to those found in the field (Alvarez-Buylla 1994). It is important to note that the effects of density were strongly dependent on the type of model used. A model that included only density-dependent effects was found to underestimate the effects of density on populations by a factor of $1/13$ compared with models that also included the density-independent environmental changes caused by patch dynamics.

12.9 Interactions in mixtures of species

So far we have considered the population dynamics of single species, almost as though they occurred in an ecological vacuum. Natural communities, however, contain an assemblage of species that may interact in a great many ways. The number is almost endless but if the interactions are classified on the basis of the effect that the population of one species has on another then the number is greatly reduced. Essentially the population of one species may cause an increase (+), decrease (−) or have no effect (0) on another species, resulting in five different types of two-way interaction (Williamson 1972), the most important of which are − + (plant–herbivore or plant–pathogen), − − (competition) and + + (mutualism). The impact of interspecific competition and mutualism on the population dynamics of plants are discussed below. The dynamics of plant–herbivore and plant–pathogen interactions are discussed in Chapter 13. Commensalism (+0) and amensalism (–0) have received far less consideration than the other three types of interaction and are perhaps better considered as highly asymmetric forms of mutualism and competition respectively.

A note of caution is necessary in applying the above definitions. For example, if the removal of one species results in an increase in the abundance of another, it might be tempting to assume that it is competition which determines the relative abundance of species. It is possible, however, that the removal of one species from a community leads to a reduction in the abundance of shared herbivores, and that it is the reduction of herbivore feeding, rather than reduced competition, which leads to the observed increase in the abundance of other species in the community, i.e. apparent competition (see Chapter 16).

12.9.1 Interspecific competition

Where interspecific competition occurs for a resource that is in limited supply, the above definition leads us to expect that there will be a reciprocal reduction in the fitness of individuals of both species. Nevertheless it is possible that the individuals of one species may be very much more affected than those of the other; competition in this case is said to be one-sided or asymmetrical. Many studies have focused on the growth and survival of only one of the species in a mixture, so it is impossible to define the nature of the interactions between them. Moreover, relatively few studies have investigated how the survival and fecundity of plants are affected by the presence of a second species. Thus despite the vast literature from both experimental and field studies on how plants interact in mixtures to affect yield or final biomass, our understanding of how interspecific competition affects the abundance and dynamics of plant populations remains poor.

Much of our knowledge of how interactions between plants may affect abundance comes from perturbation experiments in which species have been selectively removed from natural communities (see Schoener 1983; Goldberg & Barton 1992). The removal of *Spartina patens* (Poaceae) from a high marsh in North Carolina by Silander and Antonovics (1982), for example, produced a large and significant increase in the abundance of *Fimbristylis spadiceae* (Cyperaceae) alone among six potential competitors (Fig. 12.21). Similarly the removal of *Fimbristylis spadiceae* produced a large increase in the percentage cover of *Spartina patens*, indicating reciprocal, symmetric competition. In contrast, the removal of *Spartina patens* on the low marsh had much less effect on the abundance of *Fimbristylis spadiceae*, and no effect on the dominant *Spartina alterniflora*. Four of the subdominant species (including *Spartina patens*) increased by almost equal amounts when *Spartina alterniflora* was

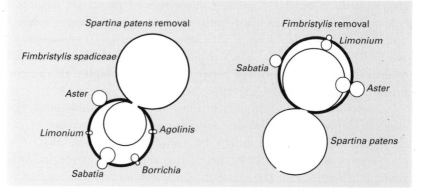

Fig. 12.21 Perturbation response diagrams for *Spartina patens* and *Fimbristylis spadiceae* removal treatments on a high marsh in North Carolina. The heavy circle represents the relative abundance (percentage cover) of the species removed, before its removal. The areas of the outer circles represent the abundances of the species in the controls. The areas of the inner circles within each of the heavy circles represent the increase in abundance of species following a removal. (From Silander & Antonovics 1982.)

removed. The interaction between *Spartina alterniflora* and *Spartina patens* therefore appears to be highly asymmetric.

While many field experiments have demonstrated competition in the field by showing that the fitness of individuals or total population size differs between treatments that differ in absolute abundance of neighbours, few provide any other information apart from whether or not competition affects components of fitness for a single species at a single site and time (Goldberg & Barton 1992). Moreover, many of the manipulation experiments involve the complete removal of the most abundant species, presumably on the basis that the larger the perturbation the greater the chance of observing a response. Manipulation experiments involving more than one species are particularly few and yet such information is vital if we are to understand the competitive structure of guilds of species and the dynamics of competing species. In a review of nine studies involving the manipulation of at least four single-species manipulations within a community, Law and Watkinson (1989) found that most species showed no response to manipulation, whilst among the pairwise interactions that were not 00, most were usually asymmetric of the form – 0 or + 0; – – interactions, usually taken to indicate competition, were conspicuous by their absence. How do we reconcile these observations with the fact that if we sufficiently reduce the number of neighbours around a target plant, a large increment in its growth will be observed? On the one hand the evidence seems to point towards weak interactions between pairs of species and on the other to competition being a major force determining community structure and dynamics. The resolution of the paradox of a strongly interactive but weakly connected community can perhaps be found in the non-linearity of the per capita rate of increase of species to density

(Law & Watkinson 1989; Law *et al.* 1993); the per capita rate of species increase decreases rapidly with increasing densities at low densities, but decreases slowly at high densities. The consequence of this non-linearity is that the dynamics of a species' population, when in the range of densities typical of the community, may depend only weakly on the densities of the species present. The dynamics of the community are then driven more by forces external to the community than by density-dependent effects within the community.

The idea of a competition–colonization trade-off is consistent with weak interactions between species. Large-seeded species tend to be rare because they produce fewer seeds. Large-seeded species have a competitive advantage over smaller-seeded species, but these strong interactions between individuals of different species do not translate into strong population-level effects for two reasons: (i) the large-seeded species are rarer and so they interact with the smaller-seeded species infrequently (Rees 1995); and (ii) the species occur in intraspecifically aggregated patches, and this further lowers the frequency of contact between individuals of different species (Rees *et al.* 1996; see Chapter 15).

Goldberg and Werner (1983) have in addition argued that competitive interactions are often likely to be non-species-specific and that there will be a large degree of equivalence in the effect of species of similar growth form on the ability of any particular species to establish within a community. Thus Peart (1982) found that grass seedling establishment and growth in monocultures of different grass species was dependent more on the biomass in the patches than on which species dominated the patch. Similarly, both Werner (1977) and Gross (1980) found that it was the abundance and biomass of particular growth forms (e.g. grasses, perennial dicots, shrubs, etc.) they removed, rather than the identity of individual species, that determined seedling success in populations of *Dipsacus sylvestris* and *Verbascum thapsus* (Scrophulariaceae) respectively.

Unravelling the role of competition in the dynamics of weed populations in crops poses less of a problem than in the species-rich communities dominated by perennials, described above. Numerous competition experiments have been carried out using a range of experimental designs (Firbank & Watkinson 1990; Silvertown & Lovett Doust 1993), but they essentially fall into two groups (Law & Watkinson 1989). Firstly, there are those where the competing individuals are counted over a number of generations at regular intervals. Secondly, there are those where the species are grown together with various combinations of densities, so that the dynamics of the competing species can then be inferred indirectly. For plants, the most comprehensive sets of data on competitive ability come from experiments in which wheat has been grown over a range of densities with a variety of weeds, including *Agrostemma githago* (Caryophyllaceae) (Watkinson 1981a; Firbank & Watkinson 1986), *Bromus sterilis* (Firbank *et al.* 1985) and *Elytrigia* (= *Agropyron*) *repens*

(Poaceae) (Firbank & Motimer 1985). The estimates of the competition coefficients, giving a measure of the equivalence between species, range from 0.41 to 1.5 for the effect of the wheat on the weed and from 0.06 to 1.63 for the effect of the weed on the wheat. Obviously the value of the competition coefficient can be expected to vary with the conditions and in particular on the relative emergence time of crop and weed. Such data have been used to explore the potential dynamics of single species of weeds growing with a crop (e.g. Doyle *et al.* 1986; Firbank & Watkinson 1986; Ballaré *et al.* 1987; Gonzalez-Andujar & Ferandez-Quintanilla 1991) and the effects of various control measures on weed abundance. For example, Doyle *et al.* (1986) have explored the effects of a variety of management options on the population dynamics of *Alopecurus myosuroides* (Poaceae) growing in winter wheat and have shown that both ploughing and herbicides are necessary to maintain low levels of the weed and ensure small cereal losses. The influence of competition from winter wheat and herbicides on the net rate of population increase and equilibrium population size of *Avena fatua* (Poaceae) is clearly shown in Fig. 12.22 (Mortimer 1987).

Weed–crop studies have typically focused on the interaction between the crop and a single weed species, despite the fact that most weed–crop communities are multispecies communities. In a study of three weeds (*Bromus sterilis*, *Galium aparine* (Rubiaceae), *Papaver rhoeas*) growing with winter wheat over 4 years, Watkinson *et al.* (1993) found that *Bromus sterilis* dominated the dynamics of the weed community. For all three species it was possible to obtain estimates of the finite rate of population increase and also the strength of the density-dependent feedback response. It was also possible to quantify the competitive effect of *Bromus sterilis* on the other two species and to show how the abundance of *Galium aparine* and *Papaver rhoeas* in mixture was dependent on the abundance of *Bromus sterilis*. The abundance of *Bromus sterilis* was, however, largely a function of its high finite rate of population increase ($\lambda = 54$) and its ability to swamp the other species rather than its competitiveness. Similarly, the high seed output per plant of *Avena barbata* relative to *Avena fatua* allows the two species potentially to coexist, despite the competitive advantage of *Avena fatua* over *Avena barbata* in

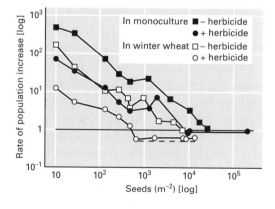

Fig. 12.22 Influence of competition from winter wheat and herbicide application on the rate of population increase of the weed *Avena fatua* over a range of densities. (From Mortimer 1987.)

terms of depressing seed yield (Marshall & Jain 1969; Firbank & Watkinson 1985).

Evidence for the importance of competition in determining plant abundance also comes from studies of succession (Tilman 1988), spatial environmental gradients (Keddy 1990; Kadmon 1995) and changing environments through time. In a study of heathlands in The Netherlands, Berendse and Elberse (1990) and their colleagues have shown that increasing atmospheric deposition of nitrogen has caused an increase in the abundance of the deciduous perennial grass *Molinia caerulea* and a decrease in the formerly dominant evergreen shrub *Erica tetralix*; experimental studies indicate that the relative nutrient requirement of *Molinia* is higher than that of *Erica* and that while *Erica* is able to maintain itself as the dominant under nutrient poor conditions, *Molinia* is competitively dominant at higher nutrient levels.

Although there is a huge literature on competition (Keddy 1989; Grace & Tilman 1990), much of it is concerned with competition intensity and community organization, and the role of competition in the dynamics of single species remains largely unexplored. Clearly there are real problems in quantifying the role of competition in diverse communities if competition is treated as a phenomenological density-dependent process (Law & Watkinson 1989; Silander & Pacala 1990). Consequently others have urged that a potentially much more predictive approach is to study the mechanisms of interspecific competition (Tilman 1990), but there are problems here too (Silander & Pacala 1990). An attempt to reconcile these two approaches is provided by Pacala and Tilman (1994) (see Chapter 15).

12.9.2 Mutualism

A wide range of mutualistic interactions have been recorded involving plants, including the well-known ant–acacia interaction and numerous pollination mutualisms. There is also a widespread mutualistic association that goes largely unseen, between the roots of plants and certain microorganisms, namely arbuscular mycorrhizal fungi, ericoid mycorrhizas and ectomycorrhizas, together with nitrogen-fixing associations of *Rhizobium* with legumes and *Frankia* with certain woody plants.

In comparison with our understanding of the role of competition, herbivory and even pathogens in the population dynamics of plants, our understanding of the role of mutualisms is still in its infancy (Law 1988). There is not even a sound theoretical framework for the treatment of mutualistic interactions. One of the reasons for this is that the Lotka–Volterra models that have formed the basis for theoretical studies of competition and predator–prey interactions lead to both populations undergoing unbounded exponential growth in what May (1981) refers to as an 'orgy of mutual benefaction'. To counter this, most models of mutualisms involve the density of the host plant species being limited by factors other than the mutualistic partner (e.g. mycorrhiza, pollinators), whereas the equilibrium density of the latter is directly proportional to the density of the former (Law 1988). In contrast, Crawley (1986a) includes mutualism in his model as a negative term, arguing that the full capacity for population growth by the plant, λ, can only be exhibited when the individual has its full complement of obligate mutualists. When these mutualists are in limited supply the actual rate of increase is reduced. There remain, however, real problems in making mathematical models that capture the essence of mutualistic interactions.

The lack of a theoretical framework is perhaps the least of our problems. Demonstrating empirically that mutualistic interactions play a part in the population dynamics of plants is a major problem in itself. It has been suggested that a great many plant invasions probably fail for want of specialized pollinators or mycorrhiza (Crawley 1986a) and the difficulties sometimes experienced in growing host populations without effective mycorrhizal endophytes suggest that mycorrhizal fungi have important effects on host dynamics (Law 1988). Bond (1995) also discusses the possibility of plant extinction due to the probability of mutualisms failing. However, there is very little information in the literature concerning the effects of mutualists on the dynamics of their host plant populations, and virtually none on the dynamics of mutualistic associations.

Perhaps the most spectacular mutualistic interactions involve *Homo sapiens* and crop plants (Begon *et al.* 1996a). The increase in the numbers of individuals of cereal crops and the areas that these crops occupy results directly from the development of agriculture and largely parallels increases in the human population. Failure of these crops would have a dramatic impact on the human population.

More often studies of mutualisms focus on one of the partners in the mutualistic association, typically the host plant. There is considerable functional diversity in the mutualisms between ants and plants; some plants house ants and others feed them, often with important consequences for plant-eating insects (Strong *et al.* 1984) and consequently plant growth and survival. For example, Janzen (1966) has shown in the relationship between the swollen thorn acacias and acacia ants (*Pseudomyrmex* spp.) that the removal of ants from trees reduces both survival and growth. Numerous studies on animal-pollinated species have demonstrated the complex interactions involved, where the flowers offer nectar or pollen or both as a reward to their visitors. A number of studies have demonstrated pollen limitation on plant reproduction, often by artifically pollinating flowers, but whilst the number of seeds may be increased it is frequently not clear to what extent pollinator activity is limiting seed set in the field and what impact this has on population dynamics. For example, Primack and Hall (1990) have shown that where pollination is augmented in a population of the pink lady's slipper orchid, *Cypripedium acaule*, that future growth and flowering is reduced; there is a cost of reproduction and the consequences for population abundance are unclear. *Cypripedium acaule* is primarily pollinated by bumble-bees, as is *Lathyrus vernus* (Fabaceae). Ehrlén and Eriksson (1995), in a study of the latter species, have shown that, whilst pollen addition increased fruit set, a matrix analysis of population growth demonstrated that the projected population growth of pollen-supplemented individuals ($\lambda = 1.019$) did not exceed that of controls ($\lambda = 1.031$), indicating a demographic cost of reproduction large enough to offset the increased seed production. Clearly further studies of this type are required if we are to quantify the effects of pollinators on population dynamics and further our understanding of the evolution of mutualistic interactions.

Perhaps the most widespread mutualism involving higher plants is that with mycorrhizal fungi and yet these have been little studied from a population perspective (Allen 1991). Mycorrhizal fungi require carbon from the host plant to survive, whilst the benefits of mycorrhizal fungi to the host plant include the facilitation of phosphorus uptake, the uptake of other nutrients such as nitrogen and copper, the improvement of water relations and protection against root pathogens (Newsham *et al.* 1995a). The lack of data on this mutualistic interaction is in part undoubtedly due to the technical problems involved in studying the dynamics of an interaction that occurs below ground and the difficulty of experimentally manipulating the system. The effects of arbuscular mycorrhiza (AM) on the growth of host plants in pot culture experiments using sterilized soil have been intensively studied, showing that phosphorus uptake per unit root length is typically enhanced by colonization. This enhancement leads to a greater growth rate in phosphorus-deficient soils. A striking example of the effect of mycorrhizal fungi in phosphorus-deficient soils is provided by the greenhouse study of Hartnett *et al.* (1993) in which the prairie grass *Andropogon gerardii* was grown at a range of densities to look at the effects of AM fungi and phosphorus availability on intraspecific competition. Under field conditions, however, the evidence for increased growth is less convincing (Fitter 1990).

Two basic approaches have been used to examine the effect of mycorrhizal fungi in the field on plant growth. The first, involving the attempted inoculation of plants, has produced contradictory results (McGonigle 1988). In some cases the increase in the growth of the host plants has been considerable, whereas in others the gains have been small or negligible (Law 1988). Colonization of plants in control plots by naturally occurring inoculum is one explanation of these results. An alternative approach to the use of inoculation is to remove mycorrhizal fungi from a proportion of the population by fumigation, sterilization or fungicides. Again the results have been contradictory (e.g. Carey *et al.* 1992), but it can be argued that fungicides in particular offer the only currently available method of directly comparing plants growing with and without mycorrhiza in the field (West *et al.* 1993). Inevitably this approach also results in comparisons being made of plants growing in environments differing in terms of other fungi and possibly also bacteria and animals that are affected by the fungicide.

In a study of the role that AM fungi play in the population dynamics of the winter annual grass *Vulpia ciliata* subsp. *ambigua* growing on sandy soils in the east of England, Newsham *et al.* (1995b) have shown that the benefit of the fungus to the plant is unrelated to plant nutrition. Rather, it would appear that the benefit lies in protecting the plant from the deleterious effects of root pathogenic fungi (Fig. 12.23). The benefit, in terms of seed production, has been estimated at over 30% (Carey *et al.* 1992; Newsham *et al.* 1995b), and this is likely to translate into an effect on population dynamics, as the abundance of *Vulpia ciliata* is directly related to fecundity (Carey *et al.* 1995).

That mycorrhizal fungi have an impact on the abundance of plants can be seen from an analysis of the effects of the fungicide benomyl on the communities in which *Vulpia ciliata* occurs (Newsham *et al.* 1995c). Perhaps not surprisingly the long-term application of the fungicide resulted in the elimination of the lichens (predominantly *Cladonia rangiformis*) in the community. This resulted in a large increase in one of the moss species, *Ceratodon purpureus*, implying that the lichen was competitively dominant to the moss. There were also changes in the abundance of several higher plant species that could be related directly to mycorrhizal colonization. Of the angiosperm species, two showed significant declines (*Erodium cicutarium* (Geraniaceae) and *Crepis capillaris* (Asteraceae)) and two significant increases (*Rumex acetosella* (Poly-

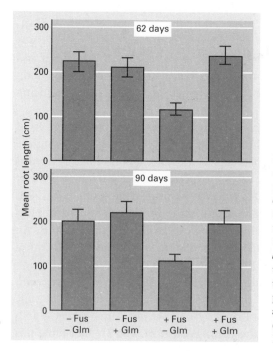

Fig. 12.23 Effects of factorial combinations of the pathogenic fungus *Fusarium oxysporum* (Fus) and the AM fungus *Glomus* sp. (Glm) on the mean root length (±standard error) of *Vulpia ciliata* plants sampled from the field 62 and 90 days after transplantation. Only the plants infected with *Fusarium* alone show reduced growth. (From Newsham *et al.* 1995b.)

gonaceae) and *Arenaria serpyllifolia* (Caryophyllaceae)). Strikingly the first two are mycorrhizal and the latter two non-mycorrhizal, and further analysis revealed that the percentage change in frequency of the species that did form mycorrhizas at the site was a simple function of the percentage reduction in mycorrhizal colonization brought about by fungicide treatment (Fig. 12.24). Other studies (Koide *et al.* 1988; Gange *et al.* 1990, 1993) have similarly concluded from field-based fungicide experiments that AM fungi may modify population abundance. In particular, Gange *et al.* (1990) found that the fungicide iprodione depressed AM fungal colonization and resulted in reduced cover abundance of four mycorrhizal species. A number of studies have shown that AM fungi can influence the outcome of competition between species (Fitter 1977; Hartnett *et al.* 1993). Interestingly the results from the study by Newsham *et al.* (1995c) reveal that two groups of symbiotic fungi are important regulators of plant community structure: the lichen *Cladonia rangiformis* was a keystone organism in the community and appeared to be able to suppress the moss *Ceratodon purpureus*, and AM fungi appeared to play a significant role in determining the interspecific competitive abilities of higher plant species. Therefore it can be expected that other symbioses, in addition to those formed by the AM fungi and lichens, may be important determinants of structure and function in a range of communities. The role of such

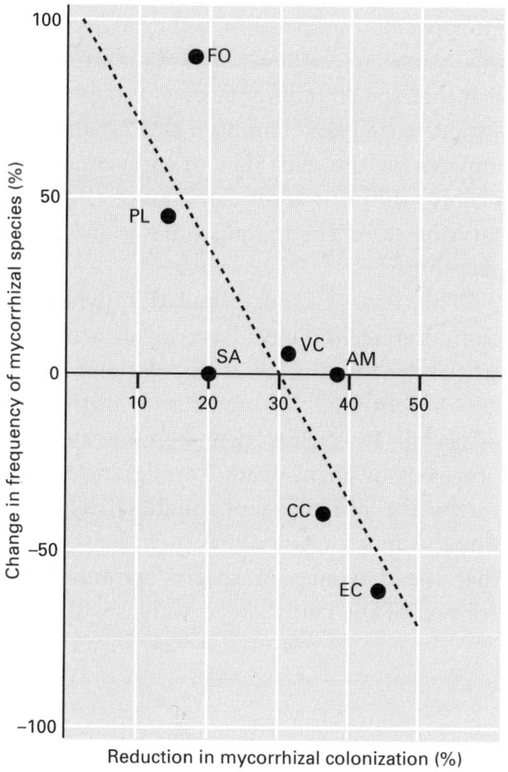

Fig. 12.24 Relationship between the change in frequency in the vegetation of mycorrhizal species in a long-term fungicide experiment and the reduction in colonization of roots by AM fungi brought about by the fungicide benomyl. EC, *Erodium cicutarium*; CC, *Crepis capillaris*; AM, *Achillea millefolium*; VC, *Vulpia ciliata*; FO, *Festuca ovina*; PL, *Plantago lanceolata*; SA, *Sedum acre*. (From Newsham *et al.* 1995c.)

symbioses in determining plant population and community dynamics remains a largely unexplored area of research.

12.10 Concluding remarks

Since the publication of John Harper's book *Population Biology of Plants* in 1977, there has been an explosion of studies on plant populations. Many of the earlier studies were concerned primarily with monitoring the fates of individuals within quadrats where a species was already relatively abundant. This has led to a wealth of information that can now be used in a comparative demography of plants (Silvertown *et al.* 1993). However, there are still relatively few studies that have looked at what determines the abundance and dynamics of populations.

Plant populations are regulated by density-dependent processes, but our understanding of the impact of density-dependent processes on dynamics remains poor, except in the case of annual plants. Most of the evidence to date, however, indicates that the dynamics of plant populations are relatively tame as a result of intrinsic density-dependent processes; there is virtually no evidence for cyclic dynamics and none for chaotic dynamics. Causes of changes in abundance remain poorly documented, although they may frequently be driven by changes in, for example, weather, the abundance of herbivores and succession.

Crawley (1990a) suggested that the relative importance of various processes in plant population dynamics could be ranked as follows: interspecific competition > herbivory > intraspecific competition for microsites > seed limitation. It is ironic that despite the general view that interspecific competition is a major driving force in determining the structure of plant communities we understand so little about how it impacts on the dynamics of individual species. What role mutualisms play is again not at all clear, given the few studies that have addressed this question. The initial studies indicate that they may have a considerable impact.

The theory of metapopulation dynamics, together with advances in computer technology, have given a major impetus to studies on the spatial dynamics of plant populations. Most population studies have so far concentrated on monitoring births and deaths or on monitoring dispersal. It is clear that we now need to integrate studies on the processes of birth, death, immigration and emigration within the context of the availability of suitable patches of habitat for species' populations. A major challenge is now the scaling up of population studies so that the dynamics of species' populations can be viewed within the context of the landscape as well as within the quadrat.

13: Plant–Herbivore Dynamics

Michael J. Crawley

13.1 Introduction

Most plants are inedible to most herbivores, yet all plants are attacked by one or more herbivore species. So why don't the herbivores that *can* eat a plant increase in abundance to the point at which the plants are devastated? There are two main schools of thought on this. The first argues that predators, parasites and diseases keep the herbivores at such low densities that they do no perceptible damage to the plants (see Crawley 1992a for details). The second argues that all that is green is not edible (at least not all of the time) either because of a shortage of vital nutrients like nitrogen, or through the presence of toxins and digestibility-reducing substances (see Chapters 5 & 10). Likewise, there are two extreme views about the role of herbivores in ecosystems; we can call these the 'top-down' and 'bottom-up' views of ecosystem function. The top-down school believes that herbivores as a group are kept scarce by a suite of effective natural enemies. The bottom-up school sees ecosystem function as driven by the rate of resource supply to the plants, with herbivores acting as little more than a vehicle for channelling plant productivity into top predators. Neither of these two models sees herbivores as playing a key role in community dynamics. At this point it is worth recalling Albert Einstein's famous remark: 'A model should be as simple as possible. But no simpler'. As we shall see, both top-down and bottom-up models are just too simple. Neither of them even hints at the variety of impacts that we know herbivores can have on the distribution and abundance of plants.

13.2 Herbivores and plant performance

The impact of herbivory on plant performance depends on its timing (phenology), location (the tissue attacked and how old it is), intensity (how much is eaten) and frequency (how often the plants are attacked). In the following sections I consider the effects of a wide range of natural enemies of plants, from pathogens and nematodes through molluscs and insects to vertebrate herbivores like geese, mice and elephants. Each phase of the plant's life cycle is considered in turn (seedling establishment, plant growth, seed production) before we investigate the role of herbivory in plant population dynamics. I have concentrated on work

[401]

published after 1992; the earlier literature is accessible through reviews by Crawley (1983), Louda *et al.* (1989) and Fritz and Simms (1992).

13.2.1 Seedling growth and survival

Abiotic factors often account for entire cohorts of seedlings. Severe winter cold can lift seedlings out of the ground (frost-heave) and drought can kill 100% of seedlings for several years in a row. Seedling survival may be extremely low unless environmental phenology is just right; for example, Brewer and Platt (1994) manipulated the fire season in a pine sandhill community in Florida and found that successful reproduction of *Pityopsis graminifolia* (Asteraceae) from 1991 cohorts following 1992 fires occurred only in May-burned plots.

13.2.1.1 Pathogens

For the purposes of this chapter, pathogens are defined as viral, bacterial and fungal enemies of plants, along with more unusual groups like the mycoplasma-like organisms (MLOs) and the flagellate protozoa. A good introduction to the biology of these groups, with illustrations of their impacts on plant growth and form, is Agrios (1988). Fungal damping-off and root rot diseases can cause mass mortality of seedlings, especially under humid conditions and when seedling densities are very high. The survival of seedlings of the tropical forest tree *Platypodium elegans* (Fabaceae) on Barro Colorado island was strongly affected by damping-off fungus, which accounted for 64–95% of deaths during the first 3 months after emergence; seedling deaths from pathogen attack were more likely for seedings close to the parent plant (Augspurger 1983a,b).

13.2.1.2 Nematodes

Invasion of the roots of seedling *Trifolium repens* (Fabaceae) by two nematode species *Heterodera trifolii* and *Meloidogyne hapla* during the first week following germination caused such severe stunting that plants from infested soil were only 6% of the size of plants in chloroform-fumigated soil (Sarathchandra *et al.* 1995).

13.2.1.3 Molluscs

These animals are generally seen as the principal herbivores of seedlings in many temperate habitats; farmers and gardeners would certainly subscribe to this view. In an experiment where molluscs were excluded from plots into which six common grassland species has been sown in artificially created gaps in autumn, there were significantly more seed-lings at the beginning of the following spring for four of the species (*Agrostis capillaris* (Poaceae), *Senecio jacobaea* (Asteraceae), *Stellaria graminea* (Caryophyllaceae) and *Taraxacum officinale* (Asteraceae)) but not for two others (*Trifolium repens* and *Ranunculus acris* (Ranunculaceae); Hanley *et al.* 1995). In some mollusc-exclusion experiments the ultimate fate of seedlings is unclear, and it is not possible to know whether mature plant density would have been any higher on the mollusc-free plots. For instance, in a study on feral oilseed rape carried out at 12 different sites throughout Britain, Crawley *et al.* (1993) found higher seedling densities following mollusc exclusion, but in no case did these lead to increased plant densities at flowering time at any of the sites, in any of the 3 years that the experiment was repeated, because subsequent mortality (mainly through interspecific competition from perennial grasses) eliminated most of the young rape plants on the mollusc-free plots.

13.2.1.4 Insects

The Janzen–Connell hypothesis (see Chapter 15) predicts that high tree-species diversity in tropical forests results from distance-dependent or density-dependent mortality of tree seedlings caused, at least in part, by the incessant rain of specialist insect herbivores from the canopy of large parent trees overhead. While many tests of the hypothesis have been equivocal (Clark & Clark 1984; Condit *et al.* 1992; Forget 1992; Schupp 1992; McKee 1995; Shibata & Nakashizuka 1995), insect herbivory is often amongst the most potent factors causing seedling mortality of forest tree species. In a field study of oak *Quercus emoryi* (Fagaceae) recruitment in southern Arizona, McPherson (1993) found that insect herbivory was the leading cause of seedling mortality. Similarly, in subtropical humid forests of north-east India, seedling mortality from insect feeding was generally important but its impact was greater for early than for late successional tree species and in open stands compared with dense forest, where pathogens and spiders webs were the main cause of seedling death (Khan & Tripathi 1991).

13.2.1.5 Rodents

In mesic grasslands in Silwood Park (an area of acid sands in Berkshire in south-east England), arthropods played only a minor role in seedling mortality compared with molluscs and rodents. In spring, seedling herbivory was infrequent and neither rodents nor molluscs encountered more than about 10% of seedlings. In autumn, however, both rodent and mollusc numbers were much higher and molluscs and rodents showed similar rates of exploitation (*c.* 30% of seedlings were encountered). With increasing seedling size, mollusc exploitation declined but rodent exploitation increased; molluscs were likely to take small chunks out of many seedlings, whereas rodents were likely to remove the entire seedling. Legume seedlings were less likely to suffer herbivore damage than were grass seedlings (Hulme 1994). Woodmice *Apodemus sylvaticus* can be serious seed and seedling pests of crops like sugar beet, where they move systematically along the rows, removing each plant in turn.

13.2.1.6 Birds

Birds, too, can be serious predators of crop seedlings, especially during late winter (woodpigeon *Columba palumbus*, partridges *Perdix perdix* and *Alectoris rufa*, skylark *Alauda arvensis* and rook *Corvus frugilegus*; Murton 1971).

13.2.1.7 Ungulates and lagomorphs

Large vertebrate herbivores can be devastating to seedling survival. In seed-addition experiments in upland and lowland sites in Colorado, Milchunas *et al.* (1992) found that large herbivores had a bigger impact on seedling recruitment than did the removal of perennial plants; very few seedlings emerged on long-term grazed plots and none at all survived to the end of the growing season on the currently grazed treatments. In South African karoo, grazing by sheep reduced the size, but not the survival, of seedlings of perennial plant species (Milton 1995). However, by opening up the plant canopy, and disturbing the soil with their digs and scrapes, large herbivores can create opportunities for seedling recruitment, especially for unpalatable plant species. For example, recruitment of the thistle *Cirsium vulgare* (Asteraceae) increased as grazing pressure was increased in experimental sheep paddocks because both seedling emergence and seedling survival were enhanced (Bullock *et al.* 1994).

13.2.1.8 Litter

Plant litter can have a major impact on seedling establishment and subsequent performance through shading, crushing, allelopathy and by isolating the seedling roots from the mineral substrate, so increasing their risk of death through desiccation (Bergelson 1990; Bosy & Reader 1995; Milton 1995). Facelli (1994) studied the effects of insect herbivory (insecticide application), herb competition (removal) and oak leaf litter (addition) on seedling establishment by the Chinese alien tree *Ailanthus altissima* (Simarubaceae) in New Jersey. Insect exclusion increased seedling emergence and reduced seedling mortality, the more so in the presence of litter. Herb competition reduced the growth of woody seedlings, but the addition of litter had a stronger negative effect on herb growth and hence improved the growth of woody seedlings. Thus, litter had a mixture of positive and negative effects on tree seedlings; litter addition reduced competition from herbs, but exacerbated the negative effects of insect herbivory on seedling survival (Facelli 1994).

13.2.2 Shoot growth

This section deals with the most familiar and most conspicuous kinds of herbivores; the species that we see when we inspect leaves and stems in the field. However, many of the less familiar and more inconspicuous enemies of plants (viruses, bacterial vascular wilts and MLOs) also have major impacts on shoot growth. Through growth (cell division and enlargement) and differentiation (cell maturation and specialization) plants must accumulate shoot biomass fast enough to compete in the struggle for light, but they must retain sufficient defences to survive in the presence of pathogens and herbivores (Herms & Mattson 1992; see Chapters 5 & 10). The most obvious effect of defoliation is felt through a reduction in leaf area; less photosynthetic surface means a lower rate of net photosynthesis. While this is bound to be true when the leaf area index (LAI) is low, it is not necessarily true when LAI is high, because of self-shading of leaves. If the curve of net photosynthesis versus LAI has a hump, because at high leaf areas the shaded leaves use up more in respiration than they produce in gross photosynthesis, then it is possible that herbivory can increase the rate of carbon fixation by reducing self-shading (Fig. 13.1).

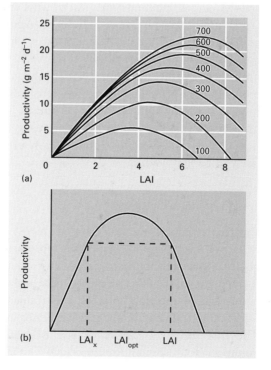

Fig. 13.1 Net productivity is an n-shaped function of leaf area index (LAI) in subterranean clover. (a) Optimum LAI increases with light intensity (cal cm^{-2} day^{-1}) since light penetrates deeper into the canopy, and a higher fraction of the leaves remains above the compensation point (Black 1964). (b) At LAI, any reduction in leaf area from herbivory that leaves a residual area greater than LAI$_x$ will cause an increase in productivity because of reduced self-shading of leaves. Productivity is maximized when the leaf area is held at LAI$_{opt}$.

The impact of a given level of defoliation depends critically on the age of the leaves removed (old leaves are generally less valuable than young ones) and on the spatial distribution of damage, both within and between leaves. For example, simulated 10% leaf area loss that was early (3 months before flowering) and aggregated (all leaves taken from a single branch) had a much bigger impact on total seed production in the forest shrub *Piper arieianum* (Piperaceae) than the removal of a similar leaf area that was late (at flowering) or diffuse (distributed throughout the canopy; Marquis 1992). The precise location of insect feeding within the leaf blade, and the spatial dispersal of damage across different leaves, can have major effects on the impact of a given level of insect feeding on plant performance. Because the tip of many dicot leaves matures and ceases expansion well before the base, removal of tissue from the base of an expanding leaf will have much bigger consequences than removal from the tip (Coleman & Leonard 1995), and dispersed damage can have much less impact than the same amount of aggregated damage, particularly if feeding is aggregated on the youngest, most valuable leaves (Edwards *et al.* 1992; Mauricio *et al.* 1993).

13.2.2.1 Viruses

The impact of virus infection on shoot growth can vary between symptomless and complete devastation. The typical signs of virus infection are mosaics and ring spots on leaves, but stunt, dwarfness, leaf-roll, yellows, streak, pox, enation (outgrowths from otherwise smooth surface), tumours or pitting can all occur (Agrios 1988). The viruses may be transmitted by seed, through insects with piercing mouthparts, by nematodes and occasionally by fungi. The numbers of virus taxa in a multispecies plant community can be high (> 50) and presumably many more taxa are present than can be detected by current screening techniques. Understanding the role of viruses in altering interspecific and intraspecific plant competition represents an enormous challenge to field ecologists, especially since the host range of individual viruses is likely to be highly vagile.

13.2.2.2 Fungal pathogens

In their review, Jarosz and Davelos (1995) showed that fungal pathogens can alter plant survivorship, reproduction, growth, competitive ability and susceptibility to herbivores. They emphasize that pathogen incidence is highly variable across space and time, and that because host plants can be multiply infected it is unlikely that there is widespread selection in favour of less aggressive strains in natural plant populations. The rust fungus *Uromyces rumicis* is found on the leaves of both common dock species *Rumex crispus* and *Rumex obtusifolius* (Polygonaceae) where it causes reduced leaf area, shoot and root weight. On its own, the rust caused greater damage to *Rumex crispus* than did the beetle *Gastrophysa viridula*, but early attack by the beetle appeared to reduce the probability of subsequent attack by the rust (Hatcher *et al.* 1994).

13.2.2.3 Nematodes

Infestation of *Trifolium repens* by the stem nematode *Ditylenchus dipsaci* greatly reduced establishment from seed and reduced shoot growth at first harvest to about 30% of clover yield on aldicarb-treated plots where nematode numbers were greatly reduced (Cook *et al.* 1992); competitive ability and nitrogen fixation by clover were also reduced by *Ditylenchus dipsaci* infection (Mercer 1994). The flower gall nematode *Anguina amsinckiae* on fiddleneck *Amsinckia intermedia* (Boraginaceae), growing in competition with wheat, caused a 25% reduction in shoot biomass, but this had no impact on wheat yield (Pantone 1995).

13.2.2.4 Molluscs

Meristem damage by molluscs was the single most important herbivore impact on the population dynamics of the perennial herb *Lathyrus vernus* (Fabaceae) during a 4-year study in sweden. Mollusc feeding reduced survival, growth and fruit production, and individuals suffering meristem or leaf damage in one year were more likely to suffer the same kind of damage during the following year (Ehrlen 1995).

13.2.2.5 Insects

The rates of defoliation attributable to insect herbivores are traditionally considered to lie within the range of 5–15% of leaf area per year (Landsburg & Ohmart 1989). Critics have argued, however, that these percentages represent a serious underestimate of defoliation rates because they fail to account for leaf turnover during the growing season (i.e. for the appearance and disappearance of leaves between one sampling period and the next). It is certainly true that far too many estimates of defoliation have been based on a single sample of leaves, taken toward the end of the growing period, once insect feeding was assumed to have ceased. Such studies have provided no estimates of the numbers of leaves that were consumed completely, leaving no trace, or of the effect of partial damage on the rate of premature abscission of leaves. In a detailed study of herbivory and leaf turnover in water lilies, for example, it was discovered that feeding by adult and larval *Pyrrhalta* beetles reduced average leaf longevity from 45 to 17 days. This degree of leaf turnover would have led to severe underestimation if peak leaf standing crop been used to estimate production (Wallace & O'Hop 1985).

Studies involving careful, frequent measurements on large samples of individually marked leaves have been rare. In work on cohorts of several hundred individually marked birch leaves *Betula pendula* (Betulaceae) from each of six trees, Zakaria (1989) measured every leaf on every day. He found that 11% of leaves produced between April and mid-July fell prematurely or were consumed entirely, so that no trace of their existence remained by the date at which a single damage sample would have been taken. Between mid-July and August a further 6% were lost. Premature losses of leaves varied from tree to tree between 0 and 35%. Of the total annual leaf production, 13–35% of leaves were produced after the first spring flush was complete and would not have been detected if a single cohort of leaves had been tagged in spring. He found no evidence that the probability of insect damage on a given day was influenced by the history of damage of a given leaf.

In a comparison of the impact of three species of insect herbivores on goldenrod *Solidago altissima* (Asteraceae), Meyer (1993) found, somewhat surprisingly, that the xylem-sucking spittlebug *Philaenus spumarius* was the most damaging, the leaf-chewing beetle *Trirhabda* sp. was intermediate and the phloem-sucking aphid *Uroleucon caligatum* was least damaging. After 3 weeks of feeding, the spittlebug and the beetle had reduced total leaf mass, total leaf area and root mass, but the impact of the spittlebug was five to six times greater. Spittlebug feeding also reduced the mass of apical buds, stem mass and the number of lateral stems. These three insect herbivores had no impact on proportional biomass allocation to leaves, stems or roots nor did they affect the net assimilation rate. The mechanism of herbivore impact appeared to be through a reduction in leaf area per unit leaf mass, rather than via an alteration in plant physiology (Meyer 1993). Similarly, Karban (1980) found that the xylem-feeding cicada *Magicicada septendecim* caused a 30% reduction in ring width in the wood of *Quercus ilicifolia* (Fagaceae) without causing any reduction in acorn production. In other cases, phloem-sucking insects have been shown to have a major impact on shoot growth. For example, even relatively low levels of feeding by sycamore aphid *Drepanosiphon platanoides* can, if sustained, have a significant effect on the growth of young trees of *Acer pseudoplatanus* (Aceraceae) (Dixon 1971a,b). Shoot feeding by insects tends to slow or stop root growth and may initiate root die-back; the general rule seems to be 'stress the shoot, reduce the root' (and vice versa; see Crawley 1983 for examples). Thus, defoliation by the chrysomelid beetle *Gastrophysa viridula* of dock *Rumex crispus* growing on shingle spits in fast-flowing rivers led to reduced root growth with the result that the plants were more likely to be killed by being washed away during storms (Whittaker 1982).

13.2.2.6 *Grazing birds*

Arctic ecosystems are particularly sensitive to grazing because of their low overall net primary productivity. Greater snow geese *Chen caeru-*

lescens atlantica reduced above-ground biomass of *Eriophorum scheuchzeri* (Cyperaceae) and *Dupontia fisheri* (Poaceae), taking 65–113% of cumulative net above-ground primary production (NAPP) of *Eriophorum* and 30–78% of *Dupontia*. Nitrogen content of plants was higher after grazing than in ungrazed plants, and although grazed plants were able to grow new foliage, goose grazing did not enhance NAPP (Gauthier *et al.* 1995; cf. Hik & Jefferies 1990). Geese in temperate grasslands in winter are often regarded as serious pasture pests. In coastal Norfolk, for example, brent geese *Branta bernicla bernicla* feed on marine algal beds when they first arrive, moving inland to salt marsh, grass, then arable fields as each habitat is successively exploited, before returning to feed on the marsh in spring, where they build up their fat reserves for the return flight to the breeding grounds. Various options have been considered for management, including culling the birds, paying compensation to farmers or setting up alternative feeding areas (Vickery *et al.* 1994). The expected increase in the brent goose population will mean that algal resources are depleted more quickly, so that the birds arrive on farmland earlier in the winter, thus increasing the levels of conflict between brent geese and agriculture (Vickery *et al.* 1995).

13.2.2.7 Grazing ungulates

The effects of selective defoliation by large, mobile polyphagous ungulates like sheep, cattle and goats on the relative abundance of pasture plants has been studied since the 1930s (Jones 1933; Milton 1940). This early work demonstrated that by managing the timing, duration and intensity of grazing, the pasture manager could control both the standing crop and the botanical composition of the sward. Grazing-sensitive or highly preferred species declined in abundance ('decreasers'), while grazing-tolerant or unpalatable species became more abundant ('increasers'). Generally, the decreasers were the high-quality pasture plants and the increasers were less valuable (lower digestibility and nutrient content). For an animal of a given body mass, bite size depends on the biomass, height and bulk density of the sward. Size of teeth and the morphology of mouthparts are important determinants of the degree of selectivity exhibited, so cattle are less able to graze individual plant parts (e.g. leaves versus stems) as selectively as sheep, goats or horses, and lambs graze more selectively than calves (Matches 1992). Under open rangeland conditions, domestic herbivores are less mobile than many wild ungulates and this impedes large-scale selectivity. Likewise, domestic animals are typically kept at less variable and usually much higher densities, and this reduces the potential for small-scale selectivity (Skarpe 1991). In Serengeti, the grazing ungulates that migrate are specialists on the earlier growth stages of grass which tend to be transient, while the species that are resident tend to specialize on later growth stages which are more enduring (Murray & Brown 1993). Grazing affects both the amount and the quality of forage available at any given season. In general, midsummer biomass is lower on grazed than on ungrazed swards, but total annual productivity is higher. Current year defoliation tends to have positive effects on forage nitrogen concentration and digestibility, and optimizing the quantity and year-to-year stability of digestible forage yield may best be achieved with light grazing rather than with heavy or no grazing (Milchunas *et al.* 1995).

On a methodological note, McNaughton (1992) highlighted the problems of interpretation that can arise when experiments on the impact of defoliation are carried out using isolated plants growing in separated pots in greenhouses. He grew grass plants inside an opaque cardboard collar that could be raised, as the plants grew taller, to simulate canopy closure. The plants inside collars grew to twice the height, produced fewer, larger tillers, produced relatively more stem and sheath, and shifted from a carbohydrate-dominated economy to a more nitrogen-balanced economy compared with plants grown in isolated plots without simulated canopy closure.

13.2.2.8 Browsing ungulates

The moose *Alces alces* of Isle Royale in Lake Superior, Michigan have had a substantial impact on tree growth. Exclosures set up in the late 1940s currently show significantly higher tree biomass (230 versus 150 t ha^{-1}), lower shrub biomass (1.9 versus 3.1 t ha^{-1}) and much lower herb biomass (0.2 versus 0.8 t ha^{-1}) than in browsed plots outside the fence. Moose browsing prevented saplings of preferred species from growing into the tree canopy, resulting in a forest with fewer canopy trees and a well-developed understorey of shrubs and herbs (McInnes *et al.* 1992). Birch trees *Betula pendula* stripped of their long-shoot leaves in June or July to simulate summer moose browsing, showed reduced shoot growth in the year following damage; both height growth and leaf biomass produced on short shoots were reduced (Bergstrom & Danell 1995). Vertebrate browsers (moose and deer) significantly preferred taller plants of pin cherry *Prunus pensylvanica* (Rosaceae) in the first two seasons after forest clearcutting, but despite the preference of browsers for large plants there was still a clear net growth advantage for plants of large initial size when the effects of competition, browsing and regrowth were combined (Shabel & Peart 1994).

13.2.3 Root growth

Studies of root herbivory have suffered severely from the 'out of sight, out of mind' syndrome. It is clear, however, that below-ground herbivores often have more impact on plant dynamics than their above-ground counterparts (e.g. the root-feeding flea beetle *Longitarsus jacobaeae* compared with the shoot-feeding cinnabar moth *Tyria jacobaeae* on ragwort *Senecio jacobaea*; see below). Fine roots, tap roots, rhizomes, tubers and bulbs all support a range of specialist herbivores and pathogens. The effects of root-pruning by subterranean herbivores on shoot physiology are demonstrated by a field experiment that involved a single, early-season experimental severing of lateral roots and rhizomes of *Cardamine cordifolia* (Brassicaceae). Root cutting caused greater transient leaf water deficits at midday, elevated nitrate-nitrogen concentrations and soluble carbohydrates in leaves, but had no impact on most primary nutrients. In the longer term, root pruning led to increased herbivory by leaf-chewing and leaf-mining insects, but caused no increase in sap-feeding insect herbivores (Louda & Collinge 1992 and see Oesterheld 1992).

13.2.3.1 Pathogens

The effect of vesicular-arbuscular mycorrhiza (VAM) on the fecundity of *Vulpia ciliata* subsp. *ambuigua* (Poaceae) was investigated by applying soil fungicides at two sites in eastern England. Newsham *et al.* (1994) concluded that asymptomatic root pathogenic fungi were important determinants of fitness in this grass, and that the main benefit supplied by the VAM was apparently in protecting the root system from attack by soil pathogenic fungi, rather than by increasing the rate of phosphorus uptake. Mycorrhizal fungi interact with other soil organisms in the root, the rhizosphere and the bulk soil; some of these interactions are inhibitory, some stimulatory, some competitive, others mutualistic. Grazing on mycorrhizae (e.g. by collembola) can reduce the efficiency of the root system in foraging for water and minerals. Exudation of carbohydrates from roots into soil can occur at substantial rates (Kozlowski 1992), but Griffiths and Robinson (1992) argue that rhizosphere bacteria do not use plant-derived carbon to mineralize soils organic nitrogen to any great extent and that root-induced nitrogen mineralization is relatively unimportant.

13.2.3.2 Nematodes

The effects of root-feeding nematodes on the growth and physiology of parasitized plants act through altered mineral nutrition, water uptake, photosynthesis, respiration and biosynthesis. It appears that nematodes are attracted by root exudates, and that in response to penetration and migration of nematodes in roots there can be disruption of symbiotic nitrogen fixation and enzyme regulation (Fitter & Garbaye 1994; Mateille 1994). Potato cyst nematodes *Globodera pallida* and *Globodera rostochiensis* are amongst the most serious pests of commercial potato crops (Mercer 1994). Potato root growth was reduced within a day when juvenile nematodes were inoculated directly onto root tips, but the degree of root growth reduction varied greatly from genotype to genotype of the host plant (Arnitzen *et al.* 1994). Root dry weight of the grass *Stenotaphrum secundtaum* inoculated with the sting nematode *Belonolaimus longicaudatus* stopped increasing after 84 days while uninfested roots continued to grow, so that after 210 days they had more than double the root dry mass (Giblin-Davis *et al.* 1992). After 120 days, the root-lesion nematode *Pratylenchus penetrans* had reduced the shoot dry weight of kiwi fruit *Actinidia chinensis* (Actinidiaceae) by 15% and the root dry weight by 25% compared with nematode-free plants (Vrain 1993).

Penetration of the roots of *Narcissus tazetta* (Amaryllidaceae) by *Aphelenchoides subnitens* caused yellowing of the foliage accompanied by root rot; bulbs of infected plants were smaller, their foliage became necrotic prematurely, and secondary infection by *Fusarium* penetrated the bulb and caused it to rot (Mor & Spiegel 1993). Nematodes can interact with fungi to exacerbate impacts on plants; for example growth of lettuce *Lactuca sativa* (Asteraceae) was reduced by damping-off fungus *Pythium tracheiphilum* and was reduced further by the nematodes *Meloidogyne hapla* and *Pratylenchus penetrans*. The fungus, however, had a negative effect on the populations of both nematodes in roots (Gracia *et al.* 1991).

13.2.3.3 Insects

The interaction between root herbivory, plant competition and nutrient supply rate was investigated using pot-grown seedlings of *Centaurea maculosa* (Asteraceae). The root herbivores were caterpillars of a moth *Agapeta zoegana* and grubs of a weevil *Cyphocleonus achates*. The plants were grown with or without competition by grass, and with or without experimental nitrogen addition. Competition with grass affected plant growth more than herbivory, but nitrogen addition increased the plant's capacity to compensate for root herbivory. While the moth had little impact, the weevil was capable of causing a 63% reduction in shoot biomass in low nitrogen and 30% reduction under high nitrogen. Root herbviory caused an increased in biomass allocated to root, leading to an increased root/shoot ratio (Fig. 13.2; Muller-Scharer 1991; Steinger & Muller-Scharer 1992). The weevil *Ceuthorhyncus cruciger* was the most important root herbivore of the biennial herb *Cynoglossum officinale* (Boraginaceae) growing in coastal sand dunes in The Netherlands, infesting 22–61% of flowering plants and 0–6% of non-flowering rosette plants over a 3-year period. Weevil attack was concentrated on large, flowering individuals and caused significant reductions in total seed production and seed mass per unit shoot biomass (Prins *et al.* 1992). The main soil-dwelling root herbivores of temperate grasslands are leatherjackets (Diptera, Tipulidae) and chafers (Coleoptera, Scarabeidae) many of which are serious pests of pasture, ornamental grasslands and golf courses. Root-damaged saplings of ash *Fraxinus excelsior* (Oleaceae) suffered significantly more damage from the leaf-feeding weevil *Stereonychus fraxini* in choice tests, apparently as a result of reduced leaf

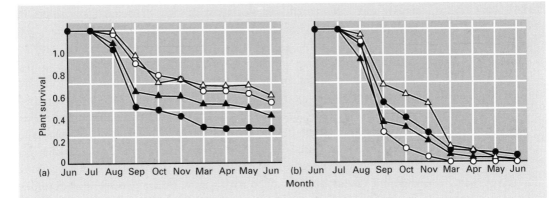

Fig. 13.2 Plant survivorship and root herbivory in *Centaurea maculosa*. (a) Without competition from grass. (b) With competition from grass. Open symbols, low plant density; solid symbols, high plant density. Triangles, no herbivores; circles, with root herbivores. (After Muller-Scharer 1991.)

toughness; Foggo *et al.* (1994) suggest that root damage by agricultural machinery to hedgerow ash trees makes the trees more susceptible to insect herbivores. Insects that feed on the underground parts of rosette-forming herbaceous biennials and perennials can cause a significant increase in plant death rate, and hold considerable potential for biological weed control (see below).

13.2.3.4 Fossorial vertebrate herbivores

Digging and tunnelling herbivores are particularly common in arid and semi-arid regions. For example, the nocturnal crested porcupine *Hystrix cristata* is the principal herbivore of the bulbs of geophytes like *Pancraticum sickenbergeri* (Amaryllidaceae) in the Negev Desert of Israel (Kamenetsky & Gutterman 1994). One of the main effects of digging and tunnelling herbivores comes through their long-term impact on soil structure and microtopography, and the indirect consequences of these changes for plant community structure. For instance, mean plant biomass is often reduced and plant species richness increased in the vicinity of gopher mounds and prairie dog towns (for details and examples, see Huntly & Reichman 1994).

13.2.4 Plant shape

Herbivores affect plant shape directly by pruning shoots and indirectly by altering the patterns of apical dominance. When meristems are removed or damaged by herbivores, the production of auxin is interrupted and this releases previously suppressed buds in non-adjacent, orthostichous axils with direct vascular connections to the bud. Release may occur both up and down the stem (but generally not in unlinked axils), and leads to the development of new branches (new tillers in grasses). Repeated herbivore attack can create highly branched, extremely bushy plants (much as gardeners nip out the apical bud to encourage branching in their petunias). Apical dominance and the suppression of lateral branching is presumably an adaptation for light competition, since in the race to capture a light gap, non-branched individuals are more likely to succeed than individuals that divert scarce resources away from height growth into the production of basal and lateral branches. Care is needed in attributing specific causes to changes in plant shape, because apparently similar deformities can have quite different causes; some may be pathogenic, others genetic. For example, 'witches broom' is a phrase used to describe bushy shoot growths found in the crowns of certain trees, but witches brooms can be caused by somatic bud mutations, fungi or MLOs in different cases. Likewise, fasciation (the formation of flattened, strap-like shoots) can be viral, bacterial, fungal or somatic in origin.

13.2.4.1 Viruses

Several viruses cause massive alteration of plant shape. One of the most serious pests of citrus trees is tristeza virus, which can cause total canopy die-back. Other tree species are prone to viral diseases like bunchy top, swollen shoot, canker or fasciation (Agrios 1988).

13.2.4.2 Mycoplasma-like organisms (phytoplasmas)

MLOs are essentially bacteria without coats (Agrios 1988), most of which were formerly thought to be viruses. Like viruses, MLOs tend to be transmitted from plant to plant by herbivorous insects with piercing mouthparts like aphids and plant-hoppers (Hemiptera) or by soil nematodes. They cause witches broom in a wide range of woody species including walnut, paulownia, elm, alder and ash trees, and yellows disease in monocot crops like rice and herbaceous ornamentals like aster. Other MLO diseases include white leaf disease of sugar cane and clover proliferation. All MLO infections tend to be highly aggregated within the tissues of diseased individuals so that detection can be highly irregular (Chen *et al.* 1992; Gundersen *et al.* 1994). Detection and identification of these organisms has improved dramatically in recent years, through the use of DNA probes, polymerase chain reaction (PCR) and immunofluorescence microscopy with specific monoclonal antibodies (Nakashima & Murata 1993; Griffiths *et al.* 1994).

13.2.4.3 Fungi

Taphrina betulina causes the familar witches broom on birch *Betula pubescens*. The fungus causes the development of large numbers of axillary shoots from a single infected bud; these shoots grow vigorously during the first season but are dead by the second or third year after infection (Jump & Woodward 1994). The fungus *Crinipellis perniciosa* causes witches brooms in cooca and is a serious economic pest in South America (Griffith & Hedger 1994).

13.2.4.4 *Bacteria*

Rhodococcus fascians induces fasciation (leafy galls) on several dicots and monocots. Virulence is due to at least three loci on a linear plasmid encoding a cytokinin synthase gene (Crespi *et al.* 1994).

13.2.4.5 *Insects*

The modular demography of plant growth reveals several of the subtleties of herbivore impact on plants. Some of the most exquisite modifications of plant shape are caused by galling insects of the wasp family Cynipidae (Hymenoptera) on various species of *Quercus* (Fig. 13.3). It is not yet clear how the genetic plans for the gall are executed (e.g. whether the female insect injects a DNA template along with egg), but the fact that the same individual oak tree can produce 20 uniquely distinctive gall morphologies for 20 different wasp species suggests that the process involves more than just a generalized hormonal response by the plant to the presence of the feeding larva (Cornell 1983). Galling of apical buds of *Salix exigua* (Salicaceae) by the midge *Rabdophaga* sp. (Diptera) caused stunting of shoots and significant reduction in future growth and reproductive potential of galled shoots; for example galling of lateral vegetative buds caused the loss of 19–23 buds from the next generation (Declerk-Floate & Price 1994). The arroyo willow *Salix lasiolepis* is attacked by a guild of gall-forming sawflies in the genera *Euura*, *Pontania* and *Phyllocolpa*. Male clones of the willow tend to support higher gall densities than female clones, and differences between the sexes in leaf phenology appear to be more important in determining galling rates than differences in leaf chemistry. While female clones are investing in seed production, the male clones continue to produce the rapidly growing vegetative shoots favoured by the gall formers (Boecklen *et al.* 1990). The galling aphid *Pemphigus betae* attacks the leaves of the cottonwood tree *Populus angustifolia* (Salicaceae); galls formed on larger leaves, and galls formed in more distal positions within a leaf produce greater numbers of offspring. Colonizing stem mothers discriminate and select only the best microhabitats within the leaf for gall formation, trading off the benefits of galling on a larger leaf with the costs of sharing the leaf with other galls (multiple galls tend only to be found on the largest leaves; Whitham 1980). Long-term study of the tephritid fly *Eurosta solidaginis*, which induces stem galls on the perennial herb *Solidago altissima*, suggests that natural

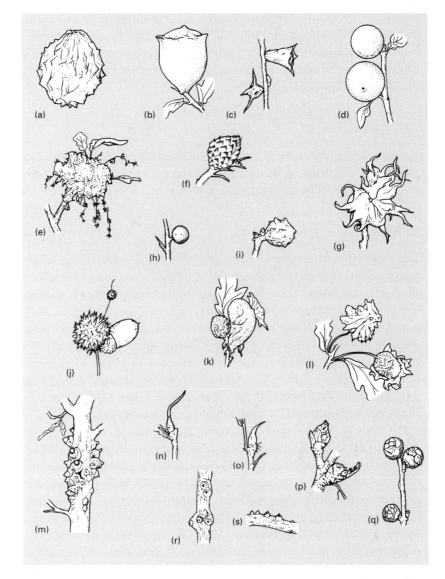

Fig. 13.3 Cynipid galls on oak. The wide variation in gall morphology is produced by 19 species of the same insect genus (*Andricus*) attacking a single host-plant species (*Quercus robur*). (a) *A. hungaricus*, (b) *A. quercustozae*, (c) *A. polycerus*, (d) *A. kollari*, (e) *A. quercusramuli*, (f) *A. fecundator*, (g) *A. coriarius*, (h) *A. gallaetinctoriae*, (i) *A. tinctoriusnostrus*, (j) *A. sekendorffi*, (k) *A. dentimitratus*, (l) *A. quercuscalicis*, (m) *A. testaceipes*, (n) *A. aries*, (o) *A. solitarius*, (p) *A. inflator*, (q) *A. lignicola*, (r) *A. rhyzomae*, (s) *A. quercuscorticis*.

selection on gall size is imposed by parasitoids (host larvae do better in larger galls) and bird predation (large galls are more likely to be attacked; Weis *et al.* 1992). Herbivores that cause branching can have profound effects on subsequent herbivore infestation. Both the aphid *Uroleucon tissoti* and the spittlebug *Philaenus spumarius* were more abundant on branched than unbranched individuals of goldenrod *Solidago altissima* (Pilson 1992).

Other effects of insect attack on plant shape derive mainly from bud-feeding species that cause lateral branching and leaf-feeding species that induce the production of regrowth shoots from epicormic buds. Examples are provided by work on coniferous trees (Whitham & Mopper 1985) and broad-leaved trees (Carne 1969).

13.2.4.6 Grazing mammals

Ungulate grazing imposes intense selection on the growth form of grasses; prostrate grasses are rapidly replaced by upright grasses following the erection of fences to exclude herbivores from Serengeti grasslands (Belsky 1992). Early-season grazing by musk ox *Ovibos moschatus* on *Oxytropis viscida* (Fabaceae) leads to the creation of more tightly clustered growth tips, and greener, more viscid leaves; the regrowth plants are less likely to be grazed again in the year following herbivory (Mulder & Harmsen 1995). Rabbit grazing on the stoloniferous herb *Hieracium pilosella* (Asteraceae) caused a significant increase in the rate of turnover of rosettes. By clipping off the flower heads, rabbits cause premature death of the rosette, but rapid proliferation of new rosettes meant that rosette density was not reduced by grazing, even though mean rosette age and reproductive success were both considerably reduced (rosette half-life declined from 2.3 years inside the rabbit exclosure to 1.4 years in the presence of rabbits; Bishop & Davy 1984).

13.2.4.7 Browsing mammals

Perhaps the most familiar example of browsing and plant shape comes from giraffe-browsed *Acacia* (Fabaceae) trees in East Africa (Plate 5, facing p. 366, colour); browsing by gardeners with shears produces the bizarre shapes of topiary in yew *Taxus baccata*, box *Buxus sempervirens* and privet *Ligustrum ovalifolium* (Oleaceae). The impact of feeding by moose *Alces alces* on their woody food plants varied with the phenology and intensity of damage; branching of annual shoots increased after winter browsing but decreased after summer leaf-stripping, leaf size increased after winter browsing, but summer feeding produced variable results (no effect or increases), and shoot size increased after moderate or high levels of winter browsing (Danell *et al.* 1994, Edenius *et al.* 1995). The quality of regrowth foliage and twigs produced in response to winter herbivory is different (usually lower) in quality as food for browsers in the following year. This can introduce a time-lag into the plant–herbivore interaction and may be responsible for the production of multi-year cycles in herbivore abundance (see below).

13.2.5 Flowering

The switch that changes a vegetative meristem into a flowering meristem may be triggered by day-length or temperature, and can be suppressed by increased nutrient supply rate or herbivory (Crawley 1983). The switch is vitally important, since once a meristem has become floral its days are numbered, and the module of which it is a part will wither and die, when the seeds ripen. Plants that have only one or a few shoot meristems are typically killed by flowering (these are the monocarpic (semelparous) species), and such species are especially susceptible to herbivores which attack and kill flower buds (see below). Both the amount and timing of flowering are strongly affected by weather, directly affected by temperature or drought and indirectly affected via weather events that influence the pollination rate (see Chapter 6); this is because non-pollinated flowers typically stay open much longer than pollinated flowers, which soon shed their petals and begin the process of fruit maturation.

13.2.5.1 *Fungi*

Anther smut *Ustilago violacea* has been intensively studied in the dioecious *Silene alba* and *Silene dioica* (Caryophyllaceae) (see Shykoff & Bucheli 1995 for references). The fungus is systemic and causes both male and female plants to produce anthers; instead of pollen, however, the anthers contain spores of the fungus so that pollinating insects carry the disease from plant to plant. Infected female plants produce no seeds at all and infected male plants produce no pollen. Therefore, flowering in infected plants is important for the dispersal of the fungus, and all the factors influencing pollinator attraction and reward (see Chapter 6), like nectar production, the mass and duration of floral display, are now subverted to determining the fitness of the pathogen rather than the host plant (Alexander & Antonovics 1988). Antonovics *et al.* (1995) liken the process to the spread of veneral diseases in animal populations, where the infection dynamics depend on the loading of the fungal spores onto the 'pollinating' insect and the unloading of the fungal spores into the flowers of uninfected individuals; force of infection depends on the searching efficiency and vagility of the insects. There is high turnover of infection in patches of the plant, and floral infection does not always lead to systemic infection and sterilization (Antonovics 1992). Large patches are more likely to be infected than small patches, but within an infected patch the proportion of plants infected declines as population size increases (inverse density dependence; see Section 6.16.3).

13.2.5.2 *Molluscs*

The application of slug pellets to grassland vegetation caused an increase in flowering by several species including *Agrostis capillaris*, *Stellaria graminea* and *Taraxacum officinale* but a reduction in flowering (presumably as a result of increased interspecific competition) by *Senecio jacobaea* and *Ranunculus acris* (Hanley *et al.* 1995). It is not clear whether the slugs and snails reduced flowering by removing young flowering shoots, by pruning flower buds from expanded flowering shoots, or by preventing the initiation of flowering stems as a result of defoliation.

13.2.5.3 Insects

English-Loeb and Karban (1992) studied 20 similarly sized clones of *Erigeron glaucus* (Asteraceae), a long-lived iteroparous herb growing at Bodega Bay in northern California. Clones differed in the proportion of flowers produced close to the modal flowering date (floral synchrony) and in the number and density of flower heads (floral display). Both of these traits affected seed-head herbivory and reproductive success. Capitula maturing during autumn escaped attack by the tephritid fly *Tephritis ovatipennis* and clones producing dense floral displays were favoured both in terms of reduced herbivory and increased seed production. During a 6-year study (3 years with controlled herbivory, then 3 years post-herbivory follow-up), the effect of the thrips *Apterothrips apteris* on *Erigeron glaucus* was felt mainly through reduced pollination, because capitula with ray florets damaged by thrips were less likely to be visited by pollinating insects. *Erigeron glaucus* plants with plume moth caterpillars *Platyptilia williamsii* produced more rosettes than those without moths because consumption of apical buds released axillary buds, but this difference in plant architecture did not result in increased flower production. Both spittlebugs *Philaenus spumarius* and plume moths caused significantly reduced seed production. One year after the herbivory treatment ended, plants that had been attacked by spittlebugs still suffered reduced flower production, but by the second year no long-term effects of herbivory could be detected (Karban & Strauss 1993).

Defoliation of lupine bushes *Lupinus arboreus* (Fabaceae) by outbreak populations of tussock moths *Orygia vetusta* can be intense and may last for more than 10 years (Harrison & Maron 1995). The immediate effects of defoliation are the production of new leaves that are much smaller and lower in water content than the new leaves of undamaged bushes, but flowering and seed produciton are eventually reduced by up to 80% on heavily damaged bushes. After 1 or 2 years of rapid growth, however, surviving bushes were not affected in terms of height, basal stem diameter or canopy volume compared with their unattacked counterparts; there was a tendency for juvenile bushes to produce more seeds, the more heavily attacked they had been the previous year. Harrison and Maron (1995) argue that the ability of the host plant to recover its biomass helps to explain the sustained nature of the insect outbreaks.

13.2.5.4 Birds

Many birds are attracted to feed on the plump flower buds of fruit trees (Rosaceae) in early spring, and bullfinches *Pyrrhula pyrrhula* can be serious pests of orchards (Matthews & Flegg 1981). An interesting indirect form of defoliation in trees is caused by insectivorous birds like tits *Parus* spp. searching for cynipid galls in the closed buds of *Quercus cerris* (Hails & Crawley 1991); destruction of buds during winter can delay spring leafing following refoliation from epicormic buds (an example of 'invisible' herbivory).

13.2.5.5 Mammals

One of the most conspicuous effects of livestock grazing is that mesic grasslands remain green through the summer while ungrazed meadows turn to straw following flowering of the grasses (Plate 7, facing p. 366). Defoliation does not affect flowering of all species equally, and although regular mowing of a lawn stops the grass from flowering it actively encourages flowering by daisy *Bellis perennis* (Asteraceae), whose prostrate rosette leaves escape mower blades set at the regulation height of 12 mm. Browsing of the shrub *Pteronia pallens* (Asteraceae) by sheep in karoo rangeland in South Africa during floral development caused an 85% reduction in flowering that led to an eventual reduction of 40% in viable seed production per capitulum (Milton 1995).

13.2.6 Fruiting and fruit dispersal

The function of ripe fleshy fruits is to get the seeds inside them moved to suitable recruitment microsites. The herbivores that eat the fruits have different concerns, of course (see Chapter 9); they are deterred by infestation of the fruits by insects (*Crataegus monogyna* (Rosaceae); Courtney & Manzur 1985) or by bacterial rots and fungi (Janzen 1977). Sometimes the seed itself is the reward and the plant buys the services of a dispersal agent using viable seeds as the currency. One of the best-documented examples of a predatory seed disperser concerns the jay *Garalus garalus*, which can disperse the acorns of English oak *Quercus robur* sufficiently quickly that the edge of the oak's geographic range expanded northwards at the end of the last ice age at an average of about 400 m per year (Bossema 1979). The best examples of seed-predatory specialist pollinators are the fig wasps and yucca moths described in Box 6.5.

13.2.6.1 Ants

Seed-feeding ants are important for the dispersal of many plant species in deserts and semi-arid habitats. The floras of countries like South Africa and Australia are especially rich in species whose fruits bear special 'ant bodies' known as elaiosomes; back at the nest, these oil-rich tissues are stripped off and the seed is cast out on the midden. There is a debate as to whether or not this directed dispersal leads to the seeds being deposited in a more nutrient-rich microsite; Rice and Westoby

(1986) did not find that the soil of the midden was any more nutrient rich than the surrounding. It could be that the real benefit may derive from the seed being placed in a position where the desert crust has been broken by the ants.

[421]
Chapter 13
Plant–Herbivore
Dynamics

(1986) did not find that the soil of the midden was any more nutrient rich than the surrounding. It could be that the real benefit may derive from the seed being placed in a position where the desert crust has been broken by the ants.

13.2.6.2 Rodents

There is no doubt that small mammals are important postdispersal seed predators, but many of them are also influential seed dispersers. By storing seeds in buried caches they protect the seeds from other surface-feeding granivores and, if the rodent dies or forgets the location of the cache, the seedlings may germinate with greater probability than un-cached seeds. A different strategy, known as scatter-hoarding, involves the herbivore burying the seeds in a spaced-out (more regular) pattern (Stapanian & Smith 1984). Woodmice *Apodemus sylvaticus* gathering acorns of *Quercus robur* in Silwood Park took healthy acorns but rejected and left behind those acorns that contained grubs of the weevil *Curculio glandium* (Crawley & Long 1995); it is not clear whether weevil infestation increased or decreased the probability of an acorn producing an oak sapling.

13.2.6.3 Primates

In Kibale Forest, Uganda, chimpanzees *Pan troglodytes* are important fruit dispersers; seeds removed from their dung germinated more rapidly, and were more likely to germinate, than seeds that had not passed through chimp guts. A single chimp dropping might contain many large seeds (e.g. 30 or more *Mimusops bagshawei* (Sapotaceae) seeds, each 15 mm diameter) in addition to hundreds of small seeds (e.g. *Ficus* spp. (Moraceae); Wrangham *et al.* 1994). Primate fruit feeding provides differing degrees of benefit to the plant in different places. Monkeys (*C. pogonias* and *C. wolfi*) in Gabon were mainly fruit pulp-eaters but in Zaire the same species were seed-eaters, aril-eaters or leaf-eaters, with the results that the Gabon primates were mainly seed dispersers while the Zaire monkeys were principally seed predators. Gautier-Hion *et al.* (1985) and Gautier-Hion & Maisels (1994) speculate that the low availability of fleshy fruited species in Zaire results from poor soil conditions. Leighton (1993) assessed the criteria used by orangutans *Pongo pygmaeus* in selecting fruits in Bornean lowland forest. The principal determinant was fruit patch availability, and an analysis of selectivity for 52 chemically unprotected 'primate-fruit' pulp species showed that the most important traits were large crop size (ripe fruits per patch), high pulp weight per fruit and high pulp mass per swallowed unit of pulp plus seed (fruit handling time), percentage digestible carbohydrate and percentage phenolic compounds in the pulp. Comparative studies suggest that fruit traits such as colour, size and protection have evolved as covarying complexes (dispersal syndromes) in response to selection by frugivorous primates, but careful comparison of floras like New Guinea, which lacks primates and other diurnal mammalian frugivores, with floras where primates are common shows no difference in the frequency of traits associated with dispersal by diurnal animals (Fischer & Chapman 1993).

13.2.6.4 Birds

The likelihood that a plant's seeds will be dispersed by fruit-eating birds depends on the size, shape, colour, ripeness and astringency of the fruits and on whether the fruits are damaged by insect herbivores or fungi (Traveset *et al.* 1995). Birds like emerald toucanets, keel-billed toucans, resplendent quetzals and three-wattled bellbirds removed 46% of the single-seeded fruits of *Ocotea tenera* (Lauraceae) in Monteverde Forest, Costa Rica (Wheelwright 1993). The remainder of the fruits were destroyed by insects (25%) or vertebrate pulp feeders (4%), or aborted after remaining ripe but uneaten on the tree for as long as 100 days. Birds preferred plants with larger-than-average fruits and selected larger fruits within a given tree, apparently because net pulp mass increased with fruit diameter. Fruits produced earlier in the season were more likely to be removed and were removed more quickly than late-ripening fruits (Wheelwright 1993).

French and Westoby (1992) studied fruit dispersal by birds in 16 sites around Sydney, Australia. Most fruit removal was by diurnal birds rather than by nocturnal mammals, and neither bird diversity nor abundance differed with the soil fertility of a site. It appears that plants with vertebrate-dispersed fruits were favoured on fertile sites, rather than there being a greater probability of fruit removal on fertile soils. In Britain this would certainly be the case, since the main plant family with bird-dispersed fruits is Rosaceae, and these plants are more or less restricted to high-fertility sites.

13.2.6.5 Ungulates

Janzen's (1984) hypothesis of 'foliage as fruit' to explain the dispersal of grass seeds in the guts of large ungulates was tested by Quinn *et al.* (1994). They found that passage of buffalograss *Buchloe dactyloides* (Poaceae) seeds through cattle guts had a positive effect on both germination and seedling growth following gut retention of 1–5 days and that only about 15% of seeds were destroyed by chewing. The germinability of seeds of *Biserrula pelecinus* (Fabaceae) increased almost fourfold following passage through the gut of cattle, with the result that adult plant density was almost threefold higher in places where dung-pats had been deposited 3 years earlier than in the surrounding, non-dunged pasture (Malo & Suarez 1995).

13.2.7 Seed production

The modular construction of plants ensures that seed production increases monotonically with shoot dry weight (Samson & Werk 1986); most of the evidence suggests that the relationship between size and fecundity is linear, at least for annual and short-lived perennial plant species (Rees & Crawley 1989). While many long-lived perennial plants do not begin to reproduce until they have reached a threshold size, there appears from the limited evidence to be a direct proportionality between shoot mass and the total number of seeds produced per year once the plants have reached that size. In long-lived, iteroparous plant species like trees, we have little information on what determines long-term average fecundity, but it is abundantly clear that seed crop varies enormously from year to year and from tree to tree within a year (Sork *et al.* 1993; Crawley & Long 1995). Some of the plant-to-plant variation is caused by weather during the growing season (especially by frosts and rainfall), by the recent history of fruiting (e.g. seed production tends to be low for a year or two after the production of a peak seed crop), and by pollination failure (e.g. bad weather for insect flight or wind dispersal), but some appears to be genetic (Crawley & Long 1995).

An important general question is whether the effect of herbivory on plant fecundity is related to plant productivity, e.g. does herbivory have a bigger proportional impact in reducing seed production in unproductive environments?

13.2.7.1 Viruses

Studies on virus impacts on wild plant populations are few and far between. Yahara and Oyama (1993) investigated the incidence of the geminivirus, tobacco leaf curl, and its impact on mortality, growth and reproduction of its host plant *Eupatorium chinense* (Asteraceae). The virus was an important, but not the sole, cause of plant mortality (half the initial cohort of plants died in the first year of the study and of these 82% were infested by the virus). Virus-infected plants had significantly lower growth rates and prdouced fewer seeds, but virus infection had no influence on the probability of flowering. Yahara and Oyama (1993) point out that these viruses had a higher quantitative impact than many plant pathogens or insect herbivores.

13.2.7.2 Fungi

Both white rust *Albugo candida* and downy mildew *Peronospera parasitica* reduce the reproductive output of infected *Capsella bursa-pastoris* (Brassicaceae) (Alexander & Burdon 1984). Grasses have relatively few intrinsic toxins, relying more on growth habit to survive defoliation and on endophytic fungal toxins as chemical defences. Endophytic fungus reduces the probability of the grass being eaten by vertebrate herbivores, insects, nematodes and some plant diseases. Endophytes also make the plants more drought tolerant, may increase their rate of tillering and stimulate the production of chitinase by the grass (Latch 1994; Joost 1995).

13.2.7.3 Bacteria

Several poisoning diseases of livestock are caused by bacterial infection (e.g. floodplain staggers is caused by *Clavibacter toxicus* in the seed-heads of *Agrostis avenacea* (Poaceae) and annual ryegrass toxicity is caused by the same bacterium in *Lolium rigidum*. The bacterium appears to be carried into the grass by a seed-galling nematode *Anguina funesta* (McKay *et al.* 1993).

13.2.7.4 Insects

Insects affect seed production by leaf-feeding, root-feeding and stem-boring, but perhaps the most pronounced effects are caused by sucking insects (phloem-feeders like aphids and xylem-feeders like cicadas or spittlebugs; see Crawley 1985). The chrysomelid beetle *Chrysomela speciosissima* removed 30% of leaf area of *Senecio ovatus* in the first half of the growing season (half of the loss due directly to beetle feeding, the other half to indirect associated causes), and although the difference in leaf area disappeared during the second half of the season as a result of compensatory growth, the grazed plots produced 36.5% fewer seeds at the end of August (Pysek 1992). The desert geophyte *Asphodelus ramosus* (Liliaceae) is attacked by a univoltine mirid bug *Capsodes infuscatus* that feeds on inflorescence meristems, flowers and fruits, completely stripping the flowering spike in years of peak abundance and destroying more than 95% of the population's expected seed production (Ayal 1994). The numerical response of the bug and the irregular flowering pattern of the plant are such that dense bug populations tend to crash through mass starvation in years immediately after peak flowering. As a result, the bug has a limited impact on plant dynamics, since subsequent large flower crops will satiate the relatively low number of bugs. Few studies have addressed the lifetime consequences of herbivore attack for perennial plants but Doak (1992) studied the long-lived clonal herb *Epilobium latifolium* (Onagraceae) in south-central Alaska. The plant grows as clumps of interconnected shoots and suffers chronic attack by its principal herbivore, the moth *Mompha albapalpella*, with little reduction in growth from low-intensity attacks, even if these are quite frequent. High-intensity attacks, however, even at very low frequencies, suppress growth and greatly reduce seed production. Knowing the average level of herbivore attack will be insufficient to predict the impacts of herbivory, particularly since intense herbivore attack in-

creases the inequality in seed production between clumps. While defoliating insects had little impact on acorn production by oaks *Quercus
robur* (defoliation by insects never exceeded 15% over a 17-year period),
the exclusion of sucking insects by insecticide application over a period
of 7 years led to consistently increased acorn production (Crawley 1985;
Akhteruzzaman 1991). It appears that sexual reproduction may be more
sensitive to insect damage than asexual reproduction or girth increment
in perennial plants (see also Myer & Root 1993). Insects are also major
agents of predispersal seed mortality, attacking whole fruits (e.g. orange
tip butterfly *Anthocharis cardamines* feeding on the crucifer *Cardamine
pratensis*; Duggan 1985) or eating individual seeds within fruits (e.g. seed
weevils *Apion ulicis* on gorse *Ulex europaeus* (Fabaceae); see below).

13.2.7.5 Crustacea

Some of the most unusual seed predators are mangrove crabs *Goniopsis
cruentata* and *Ucides cordatus* which eat the seeds of various mangrove
species in Belize, taking between 18% and 60% of the seed crop of
different tree species (McKee 1995).

13.2.7.6 Vertebrates

Seed production is reduced by vertebrate herbivores in many different
ways. Grazing ungulates stop flowering and can curtail seed production
altogether. Squirrels eat the immature male and female cones of pine
trees and consume the seeds once they have ripened; they can account for
more than 50% of the potential seed crop (Snyder 1993; Allred *et al.*
1994). Cattle consumed 98% of the inflorescences of *Yucca elata* (Agavaceae) but reduced recruitment could not be attributed to this massive
reduction in seed production, since recruitment also declined in populations that had not been grazed at flowering time. It turns out that the
population decline was caused by cattle grazing on the small caudices,
both ramets and genets (Kerley *et al.* 1993).

13.2.8 Seed predation

It is useful to deal with predispersal and postdispersal seed predators
separately. Perhaps the most important reason is that the cast of
characters amongst the herbivores is quite different. Most of the species
involved in predispersal seed predation are small, sedentary, specialist
feeders belonging mainly to the insect orders Diptera, Lepidoptera,
Coleoptera and Hymenoptera. In contrast, the postdispersal seed predators tend to be larger, more mobile, generalist herbivores like ants,
rodents and granivorous birds. Also, the spatial distributions of the
seeds and the implications of this for predator foraging are quite
distinct. Predispersal seed predators exploit a spatially and temporally

aggregated resource, and are able to use searching cues based on a conspicuous parent plant. Postdispersal seed predators usually have no such cues, and must search for inconspicuous items scattered (or buried), often at low density, against a cryptic background. Rates of seed predation are notoriously variable; between individual plants in a given year, the rate of predispersal seed predation can vary between 0 and 100%, and for the same individual plant the rate of seed predation might be 0% in one year and 100% in the next (Crawley 1992b).

As perceived by seed-feeding animals, the annual seed crop is a highly variable and unpredictable resource. This is partly because variations in the size of the seed crop tend not to be caused by the activity of seed-feeders themselves, at least in the short term. In the case of annual plants, this is because the impact of seed losses in one year is often buffered by recruitment from the bank of dormant seeds in the soil (see Chapter 7), or by the immigration of wind-borne propagules from elsewhere (see Chapter 12). In the case of long-lived, iteroparous plants, there tends to be an asymmetric relationship between the plant and its seed predators, in which the numbers of the seed predators are affected by seed production, but seed production is more or less independent of seed predation.

However, a major determinant of the risk of death for many seeds is burial; death rates are substantially lower for buried seed than for seed exposed on the surface (Crawley *et al.* 1993; Crawley & Long 1995). For buried seed, the probability of predation increases rapidly with seed size. Once buried, small seeds are relatively secure from small mammal predation, but rodents will dig up larger seeds from considerable depths (Hulme 1996).

We have already considered the role of seed predators as partial mutualists when they disperse seeds to potential recruitment microsites and enhance seed germination (see Section 13.2.6). However, deposition of seeds in dung is not the end of the story. Many animals act as secondary seed dispersers, removing viable seeds from dung piles and killing some or all of them. For example, the ants *Pheidole* spp. harvest seeds from the dung of frugivores and cache them in their nests inside partially decomposed twigs. Levey and Byrne (1993) consider that these ants are simultaneously antagonistic and mutualistic, killing most seeds but significantly benefiting some. They found that surviving seeds cast out onto the ants' refuse piles grew faster and survived better than seedlings in other microsites under light levels typical of small clearings in the forest. Their work challenges several notions: (i) small seeds are largely protected from predation because of their size; (ii) postdispersal seed harvesting is equivalent to seed predation; (iii) competition between seedlings within frugivore droppings is common; and (iv) small seeds accumulate over time through the formation of a seed bank. None of these was the case for *Miconia nervosa* (Melastomataceae) at La Selva in Costa Rica (Levey & Byrne 1993).

13.2.9 Mast fruiting and predator satiation

Salisbury (1942) was the first to point out the potential fitness benefits of mast fruiting in trees when he explained that seedlings of beech *Fagus sylvatica*

only persist in Britain in 'mast' years when the number of the progeny is so large that after the depredations of field mice and other enemies there still remains a residue that survive, whereas in the intervening seasons between the 'mast years' the seedlings are entirely destroyed by their natural enemies. It is, indeed, not improbable that a pronounced fluctuation in numbers of offspring may have a definite survival value, since an intermittent high reproductive capacity, owing to the lag in the increase of predators and parasites, might well increase the abundance of a species whereas a reproductive capacity maintained at the same high level would have little or no effect, since the plant enemies would likewise be maintained at a high level (Salisbury 1942, p. 2).

The theory that periodic synchronous seed production in long-lived plants is an adaptation which allows satiation of seed-feeding animals, and hence increases the probability of seedling recruitment following years of peak seed production, is now widely accepted (Janzen 1971, 1978; Waller 1979, 1993; Silvertown 1980; Crawley 1990a; Fenner 1991), although alternative hypotheses have been advanced such as enhanced pollination success (Norton & Kelly 1988; Smith *et al.* 1990) or enhanced rates of seed dispersal (Ballardie & Whelan 1986; Davidar & Morton 1986; Christensen & Whitham 1991). The rationale for the evolution of the masting habit is based on five putative selective advantages (Smith *et al.* 1990):

1 populations of seed predators are reduced during years of small seed crops, and hence are unable to exploit a high percentage of the seeds during mast years;
2 larger mast crops cause seeds to be dispersed over greater distances;
3 the use of weather cues for the timing of mast crops may also provide optimal conditions for reproductive growth;
4 weather cues may predict optimum future conditions for seed germination and seedling growth; and
5 concentrations of pollen production in mast years increases the probability of pollination for wind-pollinated species.

Taking these conditions together leads to the prediction that masting species should be long-lived, wind-pollinated plants with large, edible seeds that would otherwise suffer high rates of seed predation. While many masting species fit comfortably into this classification, there is a sufficiently large number of exceptions to caution against complacency (e.g. dipterocarps, Appanah 1993; fleshy fruited New Zealand tree species, Ogden 1985; cycads, Ballardie & Whelan 1986).

As yet, there have been few long-term studies which have documented that predator satiation occurs in the field, and that the probability of a seed producing a seedling is significantly higher after years of peak seed produciton (notable exceptions are studies on beech *Fagus sylvatica* (Jensen 1985; Nilsson & Wastljung 1987) and ash *Fraxinus excelsior* (Tapper 1992)). Most studies of seed production by trees do not report how fluctuations in seed production map to changes in tree numbers; the implicit assumption is that tree recruitment is seed-limited, but this is seldom if ever demonstrated. Recruitment might not be seed-limited because of microsite limitation (a shortage of suitable places for recruitment) or predator limitation (too many seed-feeding or seedling-feeding animals). For *Quercus robur* in south-east England it appears that seed density, microsites and predators all affect oak recruitment, but to differing degrees in different places (Fig. 13.4). At one site, recruitment was seed-limited so that seedling density was correlated with acorn production over the full range of acorn densities. This was because vertebrate herbivores were scarce as a result of high numbers of cats and foxes, and there was a continuous supply of freshly disturbed ground in which squirrels and jays could bury acorns. In a second site, oak recruitment was herbivore-limited; high rabbit densities ensured that most acorns were eaten before they could be buried, and such oak seedlings as did appear were repeatedly browsed (Plate 8). Sowing extra oak seed did not lead to increased recruitment. In a third site, oak recruitment was microsite-limited; burying extra acorns inside rabbit exclosures in the shade beneath closed-canopy oak woodland did not lead to sapling recruitment. It is clear that acorn-feeding insects on their own did not prevent oak regeneration, despite the fact that they killed 30–80% of the acorn crop, because there was some recruitment at the first site even after low acorn years. Invertebrate seed-feeders are important, however, in that they can determine which of the peak years

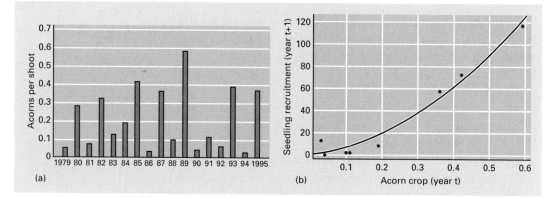

(a) (b)

Fig. 13.4 Acorn production and seedling recruitment in *Quercus robur*. (a) Pattern of alternate-bearing over 17 years in Silwood Park. (b) Seed-limited recruitment at one of three sites (Sunningdale). In Silwood Park, recruitment was not seed-limited; it was herbivore-limited in rabbit-grazed grasslands, and microsite-limited under a closed oak canopy. (From data in Crawley & Long 1995.)

produce sufficient sound acorns to satiate the vertebrate herbivores (e.g. 1989) and which do not (e.g. 1993; Crawley & Long 1995).

13.2.10 Mature plant death rate

It is relatively unusual for herbivores to kill their host plants directly (exceptions are discussed later), and most herbivores behave more like parasites than predators of plants. Typically, plants suffering high rates of herbivore damage become less competitive and are eventually killed by competition from other plants (e.g. through shading or water depletion). An important general question is whether herbivory has a more important effect on plant death rates for slow- or fast-growing plants.

13.2.10.1 Mycoplasma-like organisms

Palm lethal yellowing disease of various palm tree species in Florida, USA, is caused by an MLO (Harrison *et al.* 1994), and yellowing and decline symptoms in various fruit trees in the genus *Prunus* (Rosaceae) in Germany were attributed to a homogeneous group of MLOs by the use of Southern blot analysis (Ahrens *et al.* 1993). Decline of two alder species *Alnus glutinosa* and *Alnus cordata* (Betulaceae) in southern Italy that showed symptoms of yellowing, sparse foliage, premature autumn coloration, sprouting, deliquescent branching, phloem necrosis, dieback and witches brooms was attributed to an MLO by Marcone *et al.* (1994). In other cases, it appears that insect injury rather than MLO infection was the likely cause of declining vigour and die-back (e.g. ash yellows in *Fraxinus velutina* (Oleaceae) in Zion Nation Park, Utah; Sinclair *et al.* 1994).

13.2.10.2 Flagellate protozoa

These organisms have only recently been recognized as pathogens of plants. The best-known case involves the protozoan responsible for a lethal heartrot of coconut palms; the symptoms are similar to attack by certain MLOs, but the cause of death is distinctive (Agrios 1988).

13.2.10.3 Fungi

The classic examples of large-scale mortality of mature plants come from introduced species of fungal pathogens eradicating abundant, widespread, native trees. In the USA, the best-known case is chestnut blight. The fungus *Endothia parasitica* was introduced from eastern Asia and rapidly eliminated mature chestnut trees *Castanea dentata* (Fagaceae) from a huge area of eastern deciduous forest, transmitted from tree to tree by woodpeckers. The basal area occupied by chestnut declined from about 30% in 1950 to virtually nothing in 1970. The demise of the chestnuts allowed an increase in formerly suppressed trees like *Quercus prinus, Acer rubrum, Liriodendron tulipifera* (Magnoliaceae) and *Carya* spp. (Juglandaceae) through competitor release (Day & Monk 1974). In Britain, virtually all of the hedgerow and woodland elm trees have been eliminated since 1960 by Dutch elm disease *Ceratocystis ulmi*, a phloem-blocking fungus transmitted from tree to tree by scolytid bark beetles. Elms now persist as juvenile hedgerow suckers, and it is unclear if the disease will recur if and when these plants become large enough to attract the bark beetles. Dry sclerophyll forest in Victoria, Australia, was devastated by an epidemic of the root rot *Phytophthora cinnamomi*, a pathogen with a wide host range encompassing at least 48 different plant families. After 30 years, the density of the formerly dominant *Eucalyptus* sp. (Myrtaceae) was still 50% lower than on disease-free plots. The surviving trees showed vigorous crown growth, but no seedlings were present (Weste & Ashton 1994). The rust fungus *Melampsora lini* is a pathogen of the herbaceous perennial flax *Linum marginale* (Linaceae) whose main effect is to increase plant mortality during the winter following infection. Disease had no consistent impact on flowering levels, mainly because infection in the field typically occurred after flowering had begun (Jarosz & Burdon 1992). There are various resistance phenotypes within the *Linum marginale* population, and during a major epidemic of the rust in summer 1989 there was a marked change in the resistance structure of the population, in which the dominance of three resistance phenotypes was lost. It is not clear whether this genetic diversity for resistance results from metapopulation dynamics (i.e. from the combined effects of the presence of a pathotype that is virulent on all host phenotypes (at least in some locations), coupled with patchy recruitment patterns of the host plant and low pathogen transmission frequencies from patch to patch) or from genetic linkage between resistance genes and other traits that are under more intense selection (Burdon & Thompson 1995). Snow blight *Phacidium infestans* on *Pinus sylvestris* caused density-dependent mortality (as related to original stand density) in 12 out of 26 sites investigated by Burdon *et al.* (1992). Plants that had been killed by the disease up to 2 years previously contributed to inoculum production. In two young pine populations (5–10 years) the pathogen was also a significant cause of plant morality, in some cases leading to thinning within dense patches (reduced aggregation) but not in others (Burdon *et al.* 1994).

13.2.10.4 *Nematodes*

Die-out of the dune-stabilizing grass *Ammophila breviligulata* (Poaceae) on the mid-Atlantic coast of the USA has been observed for the last 15 years, and pathogenic nematodes have been identified as the probable causal agents. By creating conditions conducive to vigorous plant growth, Seliskar (1995) hypothesized that the plants could better withstand the stress of nematode attack. Addition of NPK fertilizer resulted in increased growth and spread of plants that had been introduced to a site where the grass had been dead for only 1 or 2 years, and application of NPK to moribund sites not only rescued the plants but also increased their growth, vigour and spread, thereby preventing further loss of plant cover and improving dune stabilization (Seliskar 1995). The European equivalent, marram grass *Ammophila arenaria* (Poaceae) remains vigorous under regular burial conditions on the seaward slope of sand dunes, but begins to degenerate as soon as sand accumulation slows down. De Rooi-Van der Goes *et al.* (1995) hypothesized that upward growth of plants following burial allowed them to escape, temporarily, from harmful soil organisms. To test this, they buried shoots in sterilized and non-sterilized soil. Burial in both kinds of sand resulted in stem elongation and increased biomass of root and shoot, but numbers of shoots were significantly higher only after burial in sterilized soil, consistent with the idea that upward growth through nematode-free sand benefits the plants, at least temporarily, by providing them with enemy-free space (De Rooi-Van der Goes *et al.* 1995). The pine wood nematode *Bursaphelenchus xylophilus* attacked and killed 260 000 mature pine trees in a single location in Japan (Numata 1989), but in most plant species deaths attributable to nematode feeding occur at the seedling stage.

13.2.10.5 Insects

Outbreaks of certain forest Lepidoptera can cause mass mortality of trees, but this generally requires repeated defoliation over a period of 3–5 years. Often, the insect population crashes during the second or third year of an outbreak, and the trees recover without suffering any substantial mortality (Stalter & Serrao 1983). Severe mortality sometimes results from bark beetle attacks on mature coniferous trees (Berryman *et al.* 1985; Romme *et al.* 1986), especially in the kind of large, uniform-aged stands that are caused by extensive fires or blow-downs. Attack by a few individual bark beetles cannot overcome the resinosis defence of a healthy tree, but mass attack (enhanced by the aggregating pheromones given off by egg-laying female beetles) can overcome the defences of healthy trees and set in train a positive feedback that ends only when all the trees have been killed (Berryman 1991).

Sucking insects have been recorded as increasing mature plant death rate although they seldom cause mass mortality. Waloff and Richards (1977) carried out a classic study on insect exclusion from broom *Cytisus scoparius* (Fabaceae) in Silwood Park, where over an 11-year period insecticide spraying significantly increased the longevity of the bushes and dramatically increased their fecundity.

On the coast of central California, large stands of bush lupine *Lupinus arboreus* suffer periodic patchy die-off in which thousands of plants die while mature plants nearby live on. Leaf damage ranges from nil to moderate in instances of die-off, and it appears that root damage by the ghost swift moth *Hepialus californicus* is the principal cause of death. The caterpillars hollow out the rootstock and are liable to kill plants in their second or third flowering seasons (older plants are better able to withstand root damage). It is noteworthy that attacked plants are able to set seed before dying and the population recovers by recruitment of seedlings (Strong *et al.* 1995).

Some plant species have extremely few meristems, and this makes them extremely vulnerable to meristem-feeding herbivores. For example, coconut palm *Cocos nucifera* (Arecaceae) has a single meristem and the plant is killed if this meristem is attacked by rhinoceros beetle *Oryctes rhinoceros* or rats (Mohan & Pillai 1993).

Insect herbivory often interacts with plant competition to affect plant death rates. For example, when levels of plant competition experienced by the arid-land shrub *Gutierrezia microcephala* (Asteraceae) were reduced experimentally by removing neighbouring individuals, the death rate of plants decreased from 47 to 22% on plots exposed to intense herbivory by the grasshopper *Hesperotettix viridis*. Ungrazed plants, protected with cages, showed no response to reduced competition; their death rate was 8% whether or not their neighbours were removed (Parker & Salzman 1985).

The flea beetle *Longitarsus jacobaeae* is capable of killing vegetative rosettes of ragwort *Senecio jacobaea* and has been employed as a successful biocontrol agent in western USA (McEvoy & Rudd 1993). In experimental plots, flea beetle attack reduced ragwort rosette density by 95% and flower production on post-rosette plants by 39% compared with control plots. It also reduced the ability of flowering plants to compensate for defoliation and defloration by cinnabar moths *Tyria jacobaeae*. On its own, cinnabar moth reduced capitulum production by 77% but in combination with flea beetle capitulum production was reduced by 98% and no viable seeds were produced (James *et al.* 1992). Although the cinnabar moth is usually a hopeless failure as a biocontrol agent of ragwort (Crawley 1989c) because its feeding is restricted to a brief period during midsummer and the plant possesses such substantial powers of late-season regrowth (Islam & Crawley 1983), cinnabar moth introduction did lead to successful biocontrol of ragwort in eastern Canada (Harris *et al.* 1978). This was because severely defoliated plants were killed by the early autumn frosts that are characteristic of this part of Canada, before the plants had a chance to recover from defoliation.

Monocarpic (semelparous) plants like ragwort may suffer *lower* death rates when herbivory reduces the size of the flower crop produced or prevents flowering altogether. Ragwort invests much of its stored reserves in seed production during the second summer of life but if the plant is stripped of its flower heads by cinnabar moth *Tyria jacobaeae* this investment is greatly reduced, so that following herbivory the plant is much more likely to perennate. Rootstocks of deflorated plants are much more likely to survive the winter than plants of a similar initial size that produced large seed crops (Gillman & Crawley 1990; see Fig. 4.4). This effect can sometimes produced bizarre results. In an experiment involving hand-removal of cinnabar moth caterpillars from replicated plots in Silwood Park that were fenced against rabbits, the host plant went extinct within 3 years on plots from which the insect herbivores were removed, but survived on the plots exposed to herbivory. The reason for this was that the ragwort plants on the caterpillar-free plots produced large seed crops and died during the following winter. However, because the plots were fenced against rabbits, the grassland was dense and tall, and ragwort seeds did not produce any seedlings. The plants on the plots with cinnabar moth produced very few seeds or seedlings, but many of the adult plants survived and flowered throughout the study period because four successive years of defloration kept them alive.

Ragwort control is sometimes attempted by hand-pulling of the flowering stems. This removes the tap root but leaves a ring of five to ten broken lateral roots in the soil, each of which is capable of sprouting a new vegetative bud that will form a rosette in the following year. The result of this human herbivory, therefore, is to increase ragwort density five to tenfold, rather than to reduce it, as intended.

13.2.10.6 Mammals

Ring-barking and bark-stripping by squirrels, rabbits and hares can kill large trees in their prime (Gill 1992). During the long-grass phase of plantation forestry (the period between planting and canopy closure) rodent populations are at their highest, and ring-barking of young forest trees by voles can be a serious problem. In natural boreal forests, bark-stripping by voles is an important cause of mortality in both broadleaf trees (Danell *et al.* 1991; Ericson *et al.* 1992) and coniferous trees (Bucyanayandi *et al.* 1992; Hansson 1994). Elephants have been recorded as killing as many as 96% of the mature trees in a *Terminalia glaucescens* (Combretaceae) woodland in tropical Africa (Laws *et al.* 1975). An experiment involving simulated winter browsing by moose *Alces alces* on *Pinus sylvestris* had two clipping intensities. With light clipping, mortality was confined to the slow-growing pines, but severe clipping caused tree deaths across the whole range of growth rates. If light clipping was to kill a tree, then death tended to occur within 1 year, but severe browsing could cause tree death 2 years or more after the event (Edenius 1993).

13.3 Herbivores and plant vigour

Current theory predicts that the outcome of herbivory on plant performance depends on plant productivity (Coley *et al.* 1985). Slow-growing plants are thought to be less able to compensate for biomass losses than fast-growing plants and are likely, therefore, to be more susceptible to herbivory if attacked (see Chapter 10). Plants with higher herbivore-free growth rates might be expected to suffer higher rates of damage and, in general, we might expect to observe a positive correlation between palatability and competitive ability (see below). Although the costs of resistance to herbivores are notoriously difficult to measure, it is clear from recent work using genetic engineering with carefully controlled genotypes that these costs can be non-trivial (Bergelson & Purrington 1996), so that resistant phenotypes would not be expected to prosper in the absence of herbivores (Simms 1992). Nevertheless, there is often substantial variation for resistance traits within present-day populations, so it is clear that selection does not rapidly eliminate suboptimal genotypes, perhaps because the costs of resistance are generally rather low relative to other fitness-affecting traits (Bergelson & Purrington 1996).

Plants that have slow growth rates by reason of their age or the soil conditions in their particular microhabitat might be more prone to insect attack or may suffer greater loss of performance with each individual insect they support.

13.3.1 Herbivory and plant productivity

It is plausible that the probability of attack by herbivores, the amount of feeding per herbivore per attack, and the ability of the plant to compensate for herbivory are all affected by plant productivity. The probability of attack may be lower for plants in unproductive habitats, but that the consequences of attack, should it occur, are more serious. Likewise, plants in productive environments may be more likely to be attacked, but are better able to compensate for herbivory if it should occur. At the ecosystem level, Fretwell and Oksanen (Oksanen *et al.* 1981, Oksanen 1983, 1988) argued that the importance of herbivory varies systematically with primary productivity; herbivory is hypothesized as having little impact in very unproductive systems, moderate impact in systems with low to intermediate productivity and low impact in more productive ecosystems (Fig. 13.5). The argument goes that productive environments are capable of supporting natural enemy populations that are sufficiently abundant to keep herbivores scarce, but below a threshold level of productivity the impact of natural enemies becomes trivial, and the herbivores in these systems are food-limited, with a consequently large (but perhaps fluctuating) impact on plant abundance. The evidence, however, is somewhat mixed. Cebrian and Duarte (1994) reviewed 56 published accounts relating plant growth rate to the percentage of photosynthetic biomass consumed daily by herbivores (Fig. 13.6). Herbivory increased with plant turnover rate, and there was a tendency of fast-growing plants to support a lower biomass of photosynthetic tissue than slow-growing ones. These authors found that

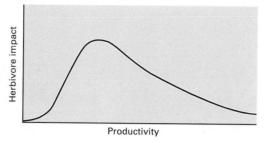

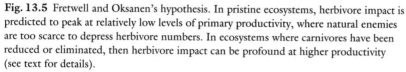

Fig. 13.5 Fretwell and Oksanen's hypothesis. In pristine ecosystems, herbivore impact is predicted to peak at relatively low levels of primary productivity, where natural enemies are too scarce to depress herbivore numbers. In ecosystems where carnivores have been reduced or eliminated, then herbivore impact can be profound at higher productivity (see text for details).

herbivory was independent of ecosystem primary production, and concluded that herbivory was more likely to be an important mechanism depressing plant biomass in fast-growing plant communities (Cebrian & Duarte 1994).

In Arctic ecosystems, the graminoids exhibit rapid regrowth but the slow-growing, highly defended shrubs regrow much more slowly (Mulder & Harmsen 1995). In southern Spain, the response of *Viola cazorlensis* (Violaceae) to herbivory varied with substrate, and plants growing on soil suffered a greater loss of reproductive output than plants growing on cliffs or bare rocks (Herrera 1993). Edenius (1993) found that herbivory had a bigger impact on the fecundity of fast-growing trees compared with slow-growing trees; the proportion of pine trees bearing cones increased with growth rate for control plants but not for experimentally clipped plants, so that herbivory affected plant performance differently across a gradient of plant productivity. Defoliation of seedings of *Pinus contorta* by cutworms of *Actebia fennica* (Lepidoptera) caused *c.* 40% mortality, but amongst the survivors recovery of height growth was completed by the second year on good sites. However, reduced height growth was still evident in the third year on poor sites (Maher & Shepherd 1992). A cautionary note is sounded by a comparison of regrowth in six semi-arid shrubs. Wandera *et al.* (1992) found that the species with the highest inherent growth rate, *Artemisia tridentata* (Asteraceae), was actually the least able to compensate for removal of 90% of the previous years growth (all the severely clipped plants died). Amongst the other five species, there was no correlation between compensatory growth ability and inherent growth rate. Common garden experiments comparing the responses of four species of Brassicaceae to early attack by the flea beetle *Phyllotreta cruciferae* showed that *Sinapis arvensis* seedlings were most tolerant of feeding damage and showed rapid compensatory growth, while *Brassica napus* was intolerant of damage at all growth stages; the other two species were intermediate. The level of tolerance was species specific and growth stage specific, and was *not* related to species' herbivore-free growth rates

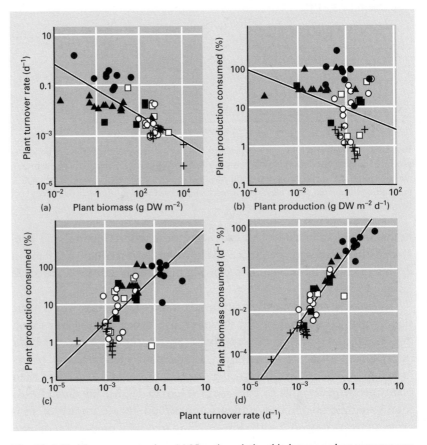

Fig. 13.6 Herbivore consumption. (a) Negative relationship between plant turnover rate and plant biomass. (b) Slight negative correlation between the percentage of primary production consumed by herbivores and the rate of plant production (plant production can exceed 100% when there is overcompensation; see text). (c) Positive correlation between herbivore consumption as percentage production eaten and plant turnover rate. (d) The much tighter positive correlation between the rate of plant consumption (percentage per day) and plant turnover rate. (After Cebrian & Duarte 1994.)

(Brandt & Lamb 1994). Fernandes and Price (1992) found that population sizes for most plant-galling taxa were significantly larger in xeric habitats compared with mesic habitats and they suggest that this pattern was caused by higher mortality of gall makers from parasitoids and fungal pathogens in the mesic habitats.

There are two competing hypotheses about plant stress and herbivore performance: the plant vigour hypothesis holds that herbivores feed preferentially on vigorous plants or vigorous modules within plants, while the plant stress hypothesis proposes that stressed plants are beneficial to plant herbivores (their defences are lower or the nutritional quality of their tissues (e.g. soluble nitrogen) is higher). Herbivore damage can elicit various kinds of response in the host plant, ranging from deterioration of food quality (inducible resistance) to improvement in food quality (inducible amelioration); these issues are discussed by Larson (1989), Tuomi *et al.* (1994), Brown and Weis (1995) and in Chapter 10.

13.3.2 Plant stress hypothesis

The main proponent of the stress hypothesis is White (1993). He argues that nitrogen is usually the nutrient which limits the reproductive performance of herbivores, and that plant stresses like drought tend to increase nitrogen avaiablility within plants (e.g. soluble amino acids in phloem). Likewise, there have been numerous reports of proliferation of insect herbivores on plants stressed by air pollution (Dohmen *et al.* 1984), soil acidity or impeded drainage (Bink 1986; Leather & Barbour 1987), shade (Maiorana 1981; Lincoln & Mooney 1984), soil nutrient deficiency (Larsson *et al.* 1985), previous defoliation (see Chapter 10), or previous low plant growth rate due to intense plant competition (see below). Recent field studies using controlled release of air pollutants (SO_2) have pointed to extra yield losses attributable to enhanced feeding by cereal and conifer aphids (Riemer & Whittaker 1989).

13.3.3 Plant vigour hypothesis

Price (1991) found evidence in favour of the vigour hypothesis in his survey of galling insects, listing as evidence the facts that: (i) female galling insects select relatively large plant modules on which to lay their eggs, and larvae survive better on larger modules; (ii) plant species from nutrient-rich environments are heavily utilized compared with species from low-resource habitats; (iii) the forestry literature indicates that most attacks by insect herbivores occur on young and open grown trees. Price suggests that the vigour hypothesis works best with intimately associated herbivores like endophytic gallers and shoot borers. It may be that there are general problems for the plant in defending rapidly growing modules against these herbivores. Complications involve the selection of vigorous modules on plants in drier sites over those in adjacent wetter areas, and early induction and subsequent increase in resistance with age (see Chapter 10).

13.3.4 Herbivore–plant–herbivore interactions

Different herbivore species living on the same plant have the potential to influence one another indirectly by modifying the architecture, physiology, biochemistry or phenology of the host plant. For example, herbivore species feeding late in the season may be influenced by changes to the plant caused by early-season herbivores (Faeth 1992). Masters and Brown (1996) studied the interaction between a root-chewing larva *Phyllopertha horticola* (Coleoptera) and a leaf-mining fly *Chromatamyia syngensiae* feeding simultaneously on the annual herb *Sonchus oleraceus*

(Asteraceae). They found that below-ground insect herbivory resulted in a host-plant-mediated benefit to the shoot-feeding insect (increased pupal weight of the leaf miner) while above-ground herbivory had an indirect negative effect on the root-feeder (reduced mean relative growth rate of the chafer grubs). A few cases have been reported where leaf-feeding herbivores have led to enhanced densities of root-feeding insects (Roberts & Morton 1985; Seastedt *et al.* 1988). These studies involved vertebrate herbivores in grassland and reported a humped relationship between foliar feeding and the density of root herbivore populations. Defoliation, as usual, caused reduced root mass (see above), and the enhanced populations of root herbivores observed at intermediate levels of defoliation were probably due to an increase in food quality sufficient to offset the reduction in food quantity.

13.4 Plant compensation

Plant responses to herbivore damage vary enormously. The net effect of single or repeated grazing events on the cumulative growth of plants can be zero, negative or positive depending on the availability of leaf area, meristems, stored nutrients, soil resources and the frequency and intensity of defoliation (Alward & Joern 1993; Noy Meir 1993; Trenbath 1993). Clearly, the potential for regrowth and compensation depends critically on the timing of herbivore attack and, in general, the earlier the attack, the greater the possibility for regrowth. Late-season attack leaves no time for regrowth and may also cause grazed plants to enter the unfavourable season in a more vulnerable condition (e.g. more likely to succumb to frost or drought). Box 13.1 explains the main processes involved in plant compensation; reviews of compensation in response to vertebrate and invertebrate herbivory are provided by McNaughton (1983) and Trumble *et al.* (1993) respectively.

Perennial plants can accumulate carbohydrates during periods of excess production and deplete their reserves when rate of utilization exceeds the rate of production. Stored carbohydrates play an important role in metabolism, growth, defence, cold hardiness and postponement of plant mortality (see Kozlowski 1992).

The availability of buds and latent meristems is important and their sensitivity of activation is expected to be related to the likely frequency of herbivory. If plants are grazed only once before the herbivores move on, then selection is likely to favour high bud sensitivity, but repeated damage is likely to select for a more restrained pattern of bud activation. A model of bud dormancy by Tuomi *et al.* (1994) shows that plants with bud dormancy never have higher seed production than ungrazed plants that have no dormant buds (see p. 444). Predictions about when (and if) compensation is to be expected centre on the population dynamics, food limitation and degree of polyphagy of the principal herbivores (see below).

Box 13.1 Mechanisms of plant compensation.

1 Increased light intensity for surviving leaf area.

2 Increase in the rate of carbon fixation at a given light intensity.

3 Improved water and nutrient availability to the surviving leaf tissue.

4 Source–sink relations (Kozlowski 1992, King *et al.* 1967).

5 Delayed senescence (plus rejuvenation) of leaves.

6 Increased duration of the growing period.

7 Redistribution of photosynthate to production of new leaves and away from roots, flowers, fruits or storage.

8 Reduced rates of floral abortion.

9 Production of new shoots from dormant buds or newly produced epicormic buds.

10 Ungrazeable reserve (e.g. storage in roots or woody stems).

11 Importance of not regrowing while the herbivore is still around.

12 Seed bank.

See Crawley (1983) for details.

Grazed and ungrazed plants may have equal absolute growth rates (full compensation) so long as the relative growth rate (RGR) of grazed plants increases exponentially with grazing intensity (the proportion of biomass removed). Below-ground RGR tends to be reduced by defoliation of more than about 20% of shoot biomass, even when there is complete compensation of shoot biomass (Oesterheld 1992). Basic allometric relations like root/shoot ratio and leaf area/leaf weight tend to be restored after a time-lag of 40 days or so, and are independent of defoliation intensity. The term *herbivore optimization hypothesis* (Fig. 13.7) is somewhat unfortunate, carrying as it does the implication that the herbivores are striving to optimize plant productivity (shades of the 'prudent predator' argument; see May & Watts 1992). Over-compensation by grasses was most frequently observed in the soil environment where each species was naturally most abundant (Alward & Joern 1993).

13.4.1 Reduced rates of fruit and seed abortion

Plant species often produce many more flowers than they could ever turn into ripened fruits packed with proteins and carbohydrate. This allows substantial scope for compensating for predispersal seed predation through the differential abortion of damaged fruits prior to seed fill (Stephenson 1981; Crawley 1983). This kind of compensation is only likely to occur if fruit production is not pollinator-limited (see Chapter 6), and is not expected to be important in habitats or in years when pollinators are scarce (Nilsson & Wastljung 1987; Horvitz & Schemske 1988; Gill 1989, Lehtila & Syrjanen 1995b).

13.4.2 Grasses

It is commonly found that grazed tillers have higher relative growth rates than ungrazed tillers, and this can result in full compensation for tissue lost to defoliation. However, compensation ability may be reduced and tiller mortality increased if there was repeated defoliation of tillers in the previous year (Vinton & Hartnett 1992). In cut rather than grazed swards, above-ground production is often maximized by frequent mowing at intermediate cutting heights, but this is at the expense of storage of reserves in roots that would otherwise be used for subsequent flower and seed production. Replacement of lost leaf area means that forage quality is improved (digestibility and nitrogen concentration are typically higher in the regrowth) but frequent defoliation limits root growth and this can limit subsequent water uptake capaciy (Turner *et al.* 1993).

In an experiment with a short grass *Eustachys paspaloides* and a mid-height grass *Themeda triandras*, McNaughton (1992) found complete compensation for defoliation in total yield, yield of live tissues, crown mass and total leaf number, but neither species compensated in residual leaf mass, total leaf area, leaf area per unit biomass or stem mass. The greater defoliation tolerance of the short grass was due to its much greater tillering, greater nitrogen accumulation, greater allocation to leaf blade and sheath, less allocation to crown and root, higher concentration of foliage per unit canopy volume, lower rate of foliage death and the ability to sustain a considerably higher leaf area per unit of plant mass (McNaughton 1992). Grass species from different kinds of grasslands often show different responses to herbivory. For example, a comparison of tallgrass prairie and montane grassland species showed that few of the prairie but most of the montane grass species compensated for herbivory (Wallace & Macko 1993). Long-term observations indicate that *Schizachyrium scoparium* declines and *Paspalum plicatulum* increases in response to herbivory, partly because *Schizachyrium* tillers are selected by herbivores in preference to those of *Paspalum* regardless of season or stocking rate. In experiments under controlled conditions, however, *Schizachyrium* tillers compensated for artificial herbivory while *Paspalum* tillers responded negatively to defoliation and failed to compensate. Thus, it appears that herbivore selectivity rather than defoliation tolerance determines the long-term outcome of competition under field conditions (Brown & Stuth 1993). This highlights the potential for misleading extrapolation from experiments on single species' responses to defoliation. Species A might look better adapted to grazing than species B under controlled conditions, but under field conditions the herbivores might prefer species A over species B to such an extent that species A declines despite its superior regrowth abilities (as in the example of *Agrostis capillaris* and *Nardus stricta* (Poaceae) in sheep-grazed upland pastures in Britain; see Crawley 1983, pp. 301–302).

13.4.3 Trees

Compensation by trees often depends on whether the herbivores are removing leaves (defoliation) or twigs (browsing). In a study with *Betula pendula*, Hjalten *et al.* (1993) compared responses to browsing and defoliation at three levels of plant competition (three stand densities). Defoliated plants showed reduced growth irrespective of the level of intraspecific competition. Browsed plants growing in medium- and high-density plots were unaffected by treatment, while topped plants in the low-density plots showed enhanced growth. The outcome of stimulated herbivory on the leading shoot of *Pinus sylvestris* has been shown to depend on timing, extent and whether the damaged part was a source (needle) or a sink (bud). Damage to buds generally had a positive effect on growth while the response to needle removal was variable. The growth of new shoots situated above the damage was reduced, especially if damage occurred late in the season; growth of new shoots below the defoliated shoot was increased (Honkanen *et al.* 1994). Regenerating oaks *Quercus robur* in mesic grassland are repeatedly browsed back to ground level by rabbits *Oryctolagus cuniculus*, but resprouting from epicormic buds at, or just below, soil level allowed them to produce new shoots. On excavation, seeding-like plants with 1-year-old shoots proved on the basis of ring-counts to be as much as 20 years in age (Crawley & Long 1995; see Plate 8, facing p. 366).

13.4.4 Shrubs

There are often major differences in browse tolerance between coexisting shrub species. For example, both *Artemisia tridentata* (Asteraceae) and *Purshia tridenta* (Rosaceae) increased the frequency of new long shoots in response to simulated browsing, but relative growth rate was not correlated with herbivory tolerance. In fact, the faster-growing species, *Artemisia*, was killed by severe simulated browsing (Bilbrough & Richards 1993). The consequences of five harvesting intensities in clipping experiments on ramets of bilberry *Vaccinium myrtillus* (Ericaceae) were monitored for 5 years. New ramets emerged rapidly from dormant buds at the base of the removed ramets so that 70–97% of ramet density in the ungrazed controls was achieved by final harvest. Biomass recovery was much slower, however, and only 11–64% recovery was noted after 5 years. This suggests that bilberry is not capable of full recovery from the kind of severe damage that might be inflicted during peak years of the 3–4 year rodent population cycle. Extreme grazing pressure on *Vaccinium* is rare in these boreal ecosystems, however, through the presence of alternative foods and abundant natural enemies (Tolvanen *et al.* 1994). *Salix exigua* compensated for galling herbivory by insects through the release of newly formed lateral buds close to galls within the shoots, but the majority of lateral shoots produced in response to galling had abscised by the following growing season (Declerck-Floate & Price 1994). For *Indigofera spinosa* (Fabaceae) the compensatory response was positively associated with residual biomass, and highest cumulative regrowth biomass was produced by plants that were browsed during the dormant season (Oba 1994).

13.4.5 Herbs

In *Sanicula arctopoides* (Apiaceae) grazed by deer *Odocoileus hemionus*, full reproductive compensation can occur when umbels bearing up to 33% of the plant's flowers are removed. Artificial removal of developing fruits 20 days later, however, resulted in a 42% decrease in seed production. Removal of secondary umbels resulted in decreased seed abortion rates in later developing umbels, while removal of tertiary umbels resulted in decreased seed abortion rates in both earlier and later-formed umbels (Lowenberg 1994; and see Hendrix 1979). Defoliation reduced seed production by hemiparasitic *Melampyrum pratense* and *Melampyrum sylvaticum* (Scrophulariaceae). Removal of unripe fruits early in the season caused enhanced seed production late in the season, compensating totally (*M. pratense*), or partially (*M. sylvaticum*), as a result of reduced floral abortion; apparently, flower production substantially outstripped the ability of plants to provision all their seeds. In *M. sylvaticum*, clipping the main stem in May led to regrowth branching and to the production of more flowers, but stem clipping in June led to reduced seed abortion rather than increased branching (Lehtila & Syrjanen 1995b).

The common assumption that plants are better able to compensate for herbivore damage at higher levels of soil fertility is called into question by work on *Solidago altissima* by Meyer and Root (1993); they found that total seed production was reduced by herbivory but there was a strong interaction between insect impact and soil fertility so that, for all three insects investigated, total seed production was reduced only at the *high* level of soil fertility.

All of these mechanisms are limited in the extent to which they can help the plant to recover from defoliation; there will inevitably come a point where the negative effects of grazing (loss of leaf area, loss of nutrients) outstrip the positive responses of regrowth and compensation; increased grazing beyond this point will inevitably lead to reduced productivity and eventually to the replacement of the grazed plants by other, more grazing-tolerant or less palatable species. This break-point will occur at lower grazing intensities in low productivity, semi-arid communities than in well-watered, fertilized pastures; indeed productivity may be a monotonically declining function of grazing intensity in the most resource-poor plant communities (De Angelis & Huston 1993).

13.5 Herbivores and plant fitness

The fact that plants are capable of compensating for moderate levels of defoliation, coupled with the observation that total primary productivity can be greater under moderate grazing intensities than when the same plant species are not grazed at all, has misled some people into the belief that grazing can increase the Darwinian fitness of plants. There is simply no compelling evidence to support this belief. The most celebrated case of alleged overcompensation involves scarlet gilia *Ipomopsis aggregata* (Polemoniaceae) described by Paige and Whitham (1987). They reported that plants browsed by elk produced 2.4 times as many fruits as unbrowsed plants. Their study was based on only 20 clipped and 20 unclipped plants, and much more comprehensive studies, repeated under field conditions in more than 20 different locations throughout the Rocky Mountain states, have failed to detect a single case of overcompensation (Bergelson & Crawley 1992a,b; Bergelson *et al.* 1996). The most likely cause of Paige and Whitham's result is that the large plants were allocated to the grazed treatment and the small individuals to the unclipped controls (compare the sizes of the rootstocks of the two kinds of plants in their Fig. 1; and see Crawley 1993a, pp. 154–159). It is no surprise that large clipped plants can produce more fruits than small unclipped ones. In order that the Darwinian fitness of a plant be increased by herbivory, all that is required is that herbivory is the lesser of two evils. If the plant suffers less by being partially eaten than it would if it were not eaten, then that partial consumption has increased its fitness (some hypothetical examples are given by Crawley 1987). The fact of the matter is that examples of overcompensation have not been reported from well-designed, properly replicated experiments. Belsky *et al.* (1993) and Crawley (1993b) review the arguments for and against the notion that overcompensation could be an adaptation by plants to predictable herbivory and find no evidence to support this view. It is clear that there is a continuum of normal plant regrowth patterns, and that increased plant growth rates, higher total biomass and increased seed production are just one end of a spectrum of responses. Plants experience injury from a wide variety of sources other than herbivory including fire, wind, freezing, heat and trampling and it is argued that rapid regrowth may have been selected for any one (or all) of these. Until we obtain convincing data to the contrary, it is prudent to hold to conventional wisdom: herbivory is bad for the plants that get eaten and good for the ones that don't.

13.6 Overgrazing

Overgrazing is a term used to describe managed plant–herbivore systems, where the population of domestic livestock has been maintained at high densities for a protracted period, typically by a combination of supplemental feeding during the unfavourable season, reduction of predator numbers, improved veterinary care, development of extra watering holes (often from deep-drilled wells) and provision of mineral licks (McNaughton 1993). Overgrazed systems are characterized by reduced animal productivity, soil erosion, reduced water catchment efficiency, elimination of productive palatable pasture plants and increase in the abundance of toxic herbs and browse-resistant, unpalatable woody plants. Overgrazing is typical of semi-arid rangelands, upland pastures and other extensive pastoral systems, particularly where human population is high.

13.7 Herbivores and plant genetics

Variability and rarity are key concepts here (see Box 13.2). Pathogens are assumed to maintain high genetic diversity and to reinforce the need for sexual reproduction by continually reducing the fitness of genotypes that become dominant in the population (see Chapter 6). But does outcrossing provide an escape from larger herbivores in a similar way? Strauss and Karban (1994) compared the success of thrips *Apterothrips apteris* on plant progeny produced by selfing or outcrossing of the thrips' home clone. When the home clone had low infestation rates, all outcrossed progeny had, on average, higher infestation than selfed progeny. In contrast, outcrossed progeny of clones characterized by high thrips infestations had the same or lower infestation when the pollen parent was a low-infection clone. Thus the advantages of outcrossing (where they occurred) were caused by the alleles contributed to the progeny and not by progeny variability or rarity as such (Strauss & Karban 1994).

A study of 16 populations of rust fungus *Puccinia chondrillina* in Turkey in the native range of its host plant, skeleton weed *Chondrilla juncea* (Asteraceae), uncovered 48 different multilocus isozyme phenotypes, most of which were confined to a single locality. In 13 of the sites, the commonest host clone was always infected by the rust, and in 10 of these was the only clone infected by the disease. This supports the notion of a threshold host-plant density for persistence of each pathotype (Anderson & May 1991) and suggests that the pathogen may be imposing negative frequency-dependent selection on these *Chondrilla juncea* populations (Chaboudez & Burdon 1995). Natural infestations of mildew *Erysiphe cichoracearum* were allowed to develop on 32 stands of 16 cloned shoots of *Solidago altissima*. Infestation differed between years, indicating genetic variation amongst the mildew strains, but small-scale plant genetic diversity had a significant influence on pathogen levels which affect plant performance and ultimately fitness. It is possible that the pathogen plays a role in the maintenance of lateral clonal growth that leads to mixing of genotypes and the formation of polyclonal patches within plant populations (Schmid 1994).

In the USA, Jaindl *et al.* (1994) tested the hypothesis that 110 years

Box 13.2 Gene-for-gene coevolution.

The gene-for-gene hypothesis was originally formulated by Flor (1956) who wrote 'for each gene that conditions reaction in the host there is a corresponding gene that conditions pathogenicity in the parasite'. The accepted current meaning is that for each gene determining resistance in the host there is a corresponding gene for avirulence in the parasite with which it specifically interacts (Kerr 1987).

Suppose that R is a dominant gene conferring resistance to a pathogen and r is a recessive host gene conferring susceptibility to the pathogen, and V is a dominant pathogen gene conferring avirulence and v is a recessive pathogen gene conferring virulence. Of the four combinations of host and pathogen alleles (hosts R or r, and pathogens V or v) only one combination (host R, pathogen V) represents host resistance (a so-called 'incompatible reaction'). The other three combinations of alleles (Rv, rV and rv) all mean that the pathogen develops on the host (a compatible reaction). Sequential release of new crop varieties carrying different resistance genes is a a form of evolutionary arms race; this process has resulted in a series of now classic examples of microevolution as pathogens have responded to these new selective forces by acquiring the corresponding genes for virulence (Thompson & Burdon 1992). Gene-for-gene coevolution has been documented in various fungal pathogens including rusts, smuts, powdery mildews and downy mildews, and in bacteria and viruses. It may be the genetic uniformity of plants in agricultural crops and their homozygosity for resistance that means that gene-for-gene evolution is untypically common in these extensive uniform environments, and much less common in heterogeneous natural systems where the number of resistance loci is greater and there may be synergistic interactions among them.

Examples of gene-for-gene coevolution from insect herbivores are less clear-cut; 20 genes for resistance to hessian fly are known in wheat, and there are 11 biotypes of the fly that differ in their ability to attack wheat varieties with these various resistance genes (Weller *et al.* 1991). It is possible, however, that the appearance of gene-for-gene dynamics in this system is an artefact of the experimental protocols employed. It is likely that resistance is polygenically determined in most plant–insect interactions, and given the metapopulation structure of most plant–herbivore demes it seems likely that individual populations are strongly influenced by genetic drift, extinction and gene flow from between-patch migration rather than pure gene-for-gene interactions (Thompson & Burdon 1992). Costs of resistance and virulence, and time delays in the onset of frequency-dependent selection are important in determining the dynamics of gene frequency in mathematical models of these more realistic, metapopulation processes (Barrett 1988; Frank 1991).

of exposure of rangeland vegetation to livestock grazing might have caused the evolution of genotypes that were more tolerant of defoliation. Common garden experiments in Oregon demonstrated that grazing history had no consistent effect on the response of *Festuca idahoensis* (Poaceae) to defoliation. However, there were genetic differences in plant growth form; individuals from protected populations had greater height and relative growth rate than those form grazed areas, even under defoliation. It appears, therefore, that selection by grazing has been for a more prostrate growth form, rather than for altered phenology or increased compensation ability. Similarly, 10 genotypes of *Sporobolus kentrophyllus* (Poaceae) from an intensively grazed site in the Serengeti were grown under glass in Syracuse, New York, by Hartvigsen and McNaughton (1995). They found that there was polymorphism ranging from short, rapidly growing genotypes adapted to intense grazing conditions, to tall, slow-growing, grazer-susceptible genotypes that were superior competitors for light in the absence of herbivory.

13.8 Herbivores and atmospheric CO_2

Plants often respond to elevated atmospheric CO_2 levels by showing increased rates of carbon assimilation and growth, with reduced tissue nitrogen concentrations and altered patterns of carbon allocation, compared with leaves of the same age and developmental stage grown in ambient CO_2 concentrations (Mooney & Koch 1994). Another common response is an acceleration in developmental rate, so that tissues are at a more advanced developmental stage by a given calendar age in a CO_2-enriched atmosphere. Since nitrogen concentration in leaves declines with age from seedling to mature plant leaves, the observed CO_2-induced reduction in plant nitrogen concentration may not be due to physiological changes in plant nitrogen-use efficiency, but rather a size-dependent phenomenon resulting from accelerated plant growth (Coleman *et al.* 1993).

Elevated CO_2 alters the plant's quality as food for herbivores; concentrations of water, nitrogen and secondary compounds in the leaves are changed along with toughness, starch and fibre content. Anticipated rises in CO_2 levels could alter the dynamics of plant–herbivore interactions because herbivore consumption, growth and fitness are all likely to be affected by lower food quality (Lincoln *et al.* 1993).

Single-factor experiments on CO_2 enrichment in controlled environments have often failed to indicate the outcome of field experiments at larger spatial and temporal scales. This is sometimes due to ignorance of the relevant environmental factors, unforeseen interactions, unacceptable short-cuts (e.g. adding a pulse of fertilizer to simulate increased rates of mineralization), ignorance of scaling-up processes (Woodward 1992; Mooney & Koch 1994), or failure to select the right model systems for laboratory study so that an untested plant species turns out to be the dominant influence under field conditions.

13.9 Herbivores and plant population dynamics

It is one thing to show that herbivores affect plant performance. It is an entirely different matter to demonstrate that herbivory affects plant population dynamics. As we have seen, attack by defoliating, sucking, stem-mining, root-feeding and gall-forming species can delay seed ripening, reduce seed production and individual seed weights, reduce the rates of shoot and root growth, increase the susceptibility of plants to disease, and reduce the competitive ability of plants relative to their unattacked neighbours. This tells us virtualy nothing, however, about the importance of herbivores in community dynamics, chiefly because we have so little information about what regulates plant populations in the wild. Suppose, for example, that plant recruitment is not seed-limited—then insects that reduce seed production will have no impact on plant population dynamics.

The aim of this section is to understand how feeding by herbivorous animals affects average plant population size, population stability and the nature of short-term transient dynamics following disturbance. Because intuition is such a poor guide to questions like these, we need to construct simple theoretical models of dynamics that encompass the plants' entire life cycle and which incorporate all of the important density-dependent and density-independent processes. The structure of the model must reflect the sequence in which these various processes operate as well as describing the interactions between them.

13.9.1 Herbivory and plant competition

As we have seen, it is more likely that herbivory will weaken a plant to the point at which it succumbs to competition from another plant than that herbivory kills the plant directly. There is plenty of evidence that differential rates of herbivore attack can relax or even reverse the competitive relationships between plant species in natural communities (Crawley 1989b). For example, the grass aphid *Holcaphis holci* reduced the rate of tillering of its host plant *Holcus mollis* (Poaceae), and this reduction led to increases in the abundance of the low-growing herb *Galium saxatile* (Rubiaceae) in years when, or in places where, the aphid was abundant. In low-aphid years, the grass advanced at the expense of the herb (Crawley 1983). A native honeysuckle *Lonicera sempervirens* (Caprifoliaceae) and an alien species *Lonicera japonica* had different patterns of biomass allocation over nine harvest dates and three levels of herbivory (none, insect herbivory, insect and mammal herbivory). In the absence of herbivory, the native plant had a higher growth rate and accumulated higher biomass than the alien, but the native suffered higher rates of herbivory and the alien showed greater powers of compensation with higher allocation to leaves and stems (Schierenbeck *et al.* 1994). The grassland cactus *Opuntia fragilis* showed higher net growth in an open, unshaded treatment not because of competitor release, but because the grass cover enhanced damage by the cactus moth borer *Melitara dentata* (Burger & Louda 1994). Early-season grasshoppers preferred grasses like *Schizachyrium scoparium* and *Poa pratensis* and favoured the growth of forbs like *Solidago* spp. (Aster-aceae), whereas late-season grasshoppers preferred *Solidago* spp. and favoured the growth of grasses (Ritchie & Tilman 1992).

Root herbivory by the moth *Agapeta zoegana* had a much greater impact on the abundance of spotted knapweed *Centaurea maculosa* when the herb was under competition with grass; rosette survivial, shoot numbers and fecundity all declined as herbivore numbers increased (Muller-Scharer 1991).

Fungal parasitism and insect herbivory may interact to affect a plant's competitive ability. For certain grass species, infection by seed-borne *Acremonium* fungal endophytes may be entirely beneficial. A 3-year field study by Clay (1993) found infection levels approaching 100% and controlled greenhouse study showed that infected plants generally outcompeted non-infected plants in mixtures. Grasses infected with fungal endophytes (E+ individuals) were less damaged by her-bivory than uninfected (E−) individuals. The E+ plants were less preferred by caterpillars of *Spodoptera frugiperda* and less nutritious if they were consumed. Generally, the infected E+ plants were competi-tively superior in the presence of herbivores, and equally competitive, or not markedly less competitive, in the absence of herbivory (Clay *et al.* 1993).

When the most abundant genotype (type A) of the alien weed *Chondrilla juncea* was reduced following release of a single strain of the rust fungus *Puccinia chondrillina* in a biocontrol project in south-eastern Australia in 1971, two previously uncommon genotypes of the weed increased in abundance, suggesting that their previous rarity had been caused by competition with genotype A (Cullen & Groves 1977; Burdon *et al.* 1984).

It is worth noting that the competitive ability of plants measured in the greenhouse or common garden has often been considerably greater than competitive ability measured in the field. For example, compensa-tion for simulated deer browsing in two dune annuals was less in the field than in the greenhouse (Gedge & Maun 1994), and seedlings of *Quercus emoryi* were much less tolerant of clipping or defoliation in the field compared with the growth chamber (McPherson 1993).

13.9.2 Herbivores and plant demography

The general model framework for studying the dynamics of the interac-tion between two species in different trophic levels was established more than 70 years ago (Lotka 1925; Volterra 1926). The first equation describes the rate of change in plant abundance, dV/dt (where V is plant (vegetation) biomass). The equation has two components: one for

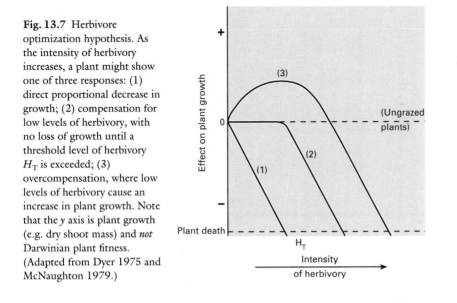

Fig. 13.7 Herbivore optimization hypothesis. As the intensity of herbivory increases, a plant might show one of three responses: (1) direct proportional decrease in growth; (2) compensation for low levels of herbivory, with no loss of growth until a threshold level of herbivory H_T is exceeded; (3) overcompensation, where low levels of herbivory cause an increase in plant growth. Note that the y axis is plant growth (e.g. dry shoot mass) and *not* Darwinian plant fitness. (Adapted from Dyer 1975 and McNaughton 1979.)

plant growth ($f(V,N)$) and one for plant losses through herbivore feeding ($g(V,N)$). The second equation describes the rate of change in herbivore numbers, dN/dt. Its two terms describe herbivore reproduction ($v(N,V)$), which depends on the herbivores' success at capturing plant biomass, and herbivore mortality ($w(N,V)$). The coupled differential equations look like this:

$$\frac{dV}{dt} = f(V,N) - g(V,N)$$

$$\frac{dN}{dt} = v(N,V) - w(N,V) \tag{13.1}$$

A variety of specific functional forms for f, g, v and w are presented in Table 13.1 where their effects on plant and herbivore dynamics are described. This continuous-time formulation is more appropriate for plant–herbivore interactions than the discrete-time formulations typically used in work on insect dynamics (see Crawley 1992a for details) because: (i) herbivory is more like parasitism and less like predation; and (ii) most plants are long-lived compared to the herbivores that feed on them. Discrete-time plant–herbivore models would be appropriate for specialist, univoltine insects attacking a self-replacing (i.e. non-successional) population of annual plants; examples of discrete-time models of plant dynamics are presented in Chapter 7. Equations 13.1 are mean-field (single-point) models; spatial models of plant–herbivore interactions are introduced later, and discussed in detail in Chapter 15.

13.9.3 Generalists and specialists

A thorough understanding of plant–herbivore interactions requires an

Table 13.1 Plant–herbivore population dynamics. The functional forms of f, g, v and w in Equation 13.1 and the numerical values of their various parameters determine the dynamic behaviour of the plant–herbivore system. A range of different functional forms have been used in the literature and each has different effects on dynamics.

		Specific form	Dynamics	Comments
Plant growth	$f(V,N)$	aV	Unbounded exponential plant growth	Totally unrealistic
		$aV(K-V)/K$	Logistic plant growth to a herbivore-free equilibrium, K	Simple, but extremely useful
		$a\exp(-eN/V)V(K-V)/K$	Impaired plant regrowth. High herbivore pressure (N/V) leads to an exponential reduction in plant growth rate, a	When parameter e is large this can produce persistent stable cycles in herbivore abundance
Herbivore functional response	$g(V,N)$	bVN	Linear unlimited functional response (the Lotka–Voterra assumption)	Unrealistic at high plant availability because it assumes unlimited appetite for each herbivore)
		$bN\,V/(m+V)$	Type II functional response	Allows for limited gut capacity and/or limited feeding time at high plant availability; destabilizing since herbivore feeding is inversely density dependent
		$bN\,V^2/(n+V^2)$	Sigmoid type III functional response, modelling switching by the herbivore at low plant availability	Stabilizing since it allows that herbivore feeding is directly density dependent when plants are scarce
Herbivore numerical response	$v(N,V)$	cNV	Lotka–Volterra assumes no limitation on functional response	Unrealistically simple
		cV	Ratio dependent numerical response	Assumes that the amount of plant available per herbivore V/N is the main determinant of herbivore reproductive success
		$cN(1-\exp(pV))$	Saturating functional response limits maximum herbivore population increase	Useful and realistic
		$c(NV)^z$	Starvation tolerance $(z>1)$ or starvation sensitivity $(z<1)$	Starvation tolerance causes over exploitation of the plants and is highly destabilizing
Herbivore death rate	$w(N,V)$	dN	Density-independent herbivore death rate	A reasonable null hypothesis
		dN^q	Density-dependent herbivore deaths (e.g territorially or Allee effects)	Highly stabilising when $q>1$ (e.g. disease of the herbivores) and destabilizing when $q<1$ (e.g. reduced herbivore breeding success at low densities)

appreciation of the importance of the degree of polyphagy exhibited by the herbivores. At one extreme we might have a strictly monophagous insect herbivore whose feeding is restricted to a single plant species. At the opposite end of the continuum we might have a polyphagous mammalian herbivore like the goat, which will eat virtually anything (but not quite everything). Broadly we can think of the specialists as small, sessile species (most insect herbivores) and the generalists as large, mobile and polyphagous (many ungulates). Of course there are exceptions: thus large polyphagous insect herbivores like locusts are honorary vertebrates, and sessile specialist vertebrates like koala bears are honorary insects.

To see the importance of the degree of polyphagy it is necessary only to consider one simple extreme case. Suppose we model the interaction of a plant and its specialist herbivore by the following equations:

$$\frac{dV}{dt} = aV\frac{(K-V)}{K} - bVN$$

$$\frac{dN}{dt} = cVN - dN \qquad (13.2)$$

In the absence of the herbivore, the plant increases to its herbivore-free equilibrium biomass, K. The herbivore is strictly food-limited and exhibits a type I functional response (see Table 13.1 for explanation). What does this model predict about steady-state plant abundance? At equilibrium, neither plant nor herbivore changes in abundance, so both equations are equal to zero; the positive and negative terms in each equation are equal. From the herbivore equation, therefore, we can cancel the Ns and divide both sides by c to discover that the equilibrium plant abundance $V*$ is:

$$V* = \frac{d}{c} \qquad (13.3)$$

This is a surprising result. It says that at equilibrium, the abundance of the plants has nothing whatever to do with either of the plants' parameters: their growth rate a or their herbivore-free equilibrium abundance, K. Plant abundance is determined solely by attributes of the herbivores. Plant abundance goes up as the herbivore death rate d goes up, and down as the herbivore numerical response c goes up. This counter-intuitive result arises because of the structure of the model and its implicit assumption about the degree of polyphagy. Because the herbivore is strictly food-limited and unaffected by any other form of density dependence, the faster the plant grows, the faster the herbivore eats it up. Thus, increasing plant growth rate makes the herbivores more abundant, but has no impact on equilibrium plant biomass. It is important to note that this does *not* mean that plant growth rate does not affect average plant biomass in a fluctuating environment. If transient dynamics (after disturbance, say) are a feature of the system, then

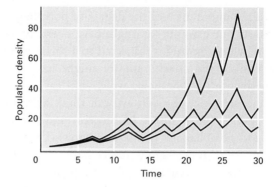

Fig. 13.8 Time-averaged population density and intrinsic rate of increase. Although equilibrium plant abundance is not determined by growth rate in a logistic model, the long-term average plant abundance in an environment where periodic catastrophes reduce plant numbers is positively correlated with intrinsic rate of increase; lower curve $r = 0.01$, mid $r = 0.02$, upper $r = 0.03$. All three populations have the same equilibrium population size (carrying capacity, $K = 100$).

plants with a higher growth rate will have a higher time-average biomass than plants with a slower growth rate (Fig. 13.8).

Now let's look at the simplest model for a generalist herbivore. Because the animal is a generalist, we can assume that its numbers are not determined by the abundance of the plant in question. This means that it might impose a constant rate of herbivory, g, independent of plant biomass. In this case, we don't need a herbivore equation, so the plant equation looks like this:

$$\frac{dV}{dt} = aV\frac{(K - V)}{K} - g \tag{13.4}$$

The equation is easily solved to find equilibrium plant biomass; because the solution is quadratic (it involves V^2) we get:

$$V* = \frac{aK \pm \sqrt{a^2K^2 - 4agK}}{2a} \tag{13.5}$$

Now, in contrast to the case with specialist herbivores, the equilibrium plant biomass *does* depend on plant growth rate a; the faster the plant grows, the higher its equilibrium biomass (see Fig. 13.9b).

The important point is that the *degree of polyphagy* is a vitally important parameter in understanding any given plant–herbivore interaction. Failure to include the degree of polyphagy in the model meant that in one case plant abundance was not affected by plant growth rate, while in the other plant growth rate had a pronounced effect on equilibrium plant abundance.

13.9.4 Plant growth

When both plants and herbivores are scarce it is reasonable to assume

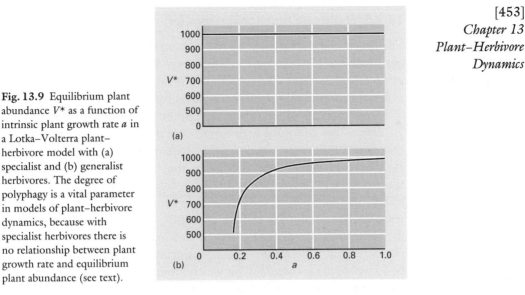

Fig. 13.9 Equilibrium plant abundance V^* as a function of intrinsic plant growth rate a in a Lotka–Volterra plant–herbivore model with (a) specialist and (b) generalist herbivores. The degree of polyphagy is a vital parameter in models of plant–herbivore dynamics, because with specialist herbivores there is no relationship between plant growth rate and equilibrium plant abundance (see text).

that plant biomass will increase exponentially at a per-unit-biomass rate a. If herbivores are abundant, plant growth may be reduced directly (e.g. by reducing leaf area) so that V is lower and hence dV/dt is reduced. Alternatively, high herbivore pressure may influence the rate at which the plant is capable of regrowing, so that both a and V are reduced. This has important consequences for dynamics, and impaired regrowth ability is capable of generating stable limit cycle behaviour (Table 13.1). Alternatively, plant growth may be reduced by less than the amount of biomass removed by herbivores because of plant compensation (see Section 13.4). If this is modelled by allowing that the net effect of herbivore feeding is less than the integral of their instantaneous rates of removal then we obtain a curious result: compensation does not benefit the plant at all, but simply causes the herbivore to become more abundant. This highlights the importance of modelling the phenology of herbivore feeding correctly. The plant should wait before beginning compensatory regrowth until it is likely that the herbivores have moved on (in the case of migratory ungulates) or pupated (in the case of univoltine insects). Thus, we should not expect to see compensatory plant growth in a symmetrically coupled, continuous-time plant–herbivore system involving food-limited specialist herbivores (like that modelled by Equation 13.2). In such a case, the most likely strategies that we might expect the plant to adopt will be escape in space (e.g. the production of copious wind-dispersed seeds) and/or an attempt to attract parasitoids or predators to reduce herbivore numbers (e.g. production of volatile attractants or the provisioning of extra-floral nectaries to reward predators like ants).

13.9.5 Herbivore functional responses

The functional response shows the amount of plant material consumed per herbivore per unit time as a function of V, the availability of that plant material. Thus, the instantaneous rate of consumption of plant material is N (the number of herbivores) times the functional response. The various forms of the functional response are shown in Fig. 13.10.

1 *Linear.* For generalist herbivores, some plant species are so attractive that they will be eaten whenever they are encountered (these are known as 'ice-cream plants' and they obey the 'zero-one rule' of foraging theory; see Crawley & Krebs 1992 for details). When such plants are scarce, the herbivores will show a linear, non-saturating functional response (they 'can't get enough of them').

2 *Saturating.* Most of the plants in an ecosystem will not be ice-cream plants, even for the most polyphagous generalist herbivores. There will be a level of plant abundance at which limited gut capacity, limited bite rate or limited egg load (for insect herbivores) imposes an upper level on the rate at which the herbivores can depress plant abundance. Thus, as plant abundance increases, the proportion of plant biomass eaten by herbivores will decline (Spalinger & Hobbs 1992). This is called *inverse*

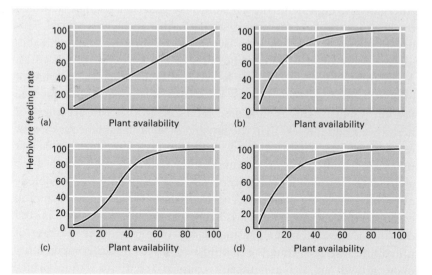

Fig. 13.10 Herbivore functional responses show the rate of feeding of an individual herbivore as a function of the amount of food available. (a) Type I functional response is a simple mathematical assumption, but it is unrealistic because it assumes unlimited herbivore appetite when food is abundant. (b) Type II functional response saturates at high food availabilities because of limited gut capacity or limited handling time. (c) Type III functional response is sigmoid, showing stabilizing density dependence at low food availabilities; it might come about as a result of switching by the herbivores or if there is a low-density refuge for the plants (d) Ratio-dependent functional responses have plant availability per herbivore on the x axis, rather than absolute plant availability.

density dependence and its influence on population dynamics is almost always destabilizing (see Crawley 1983, pp. 111–125).

3 *Sigmoid*. Some generalist herbivores exhibit a foraging behaviour known as switching (Crawley & Krebs 1992). This means that they do not feed from scarce plant species, perhaps because they have not developed a 'search image' for them. The plant species is only added to the diet after it increases beyond a threshold level of abundance (Table 13.1). This has a highly stabilizing impact on plant–herbivore dynamics because it provides a refuge from herbivory for the plant when it is scarce; the percentage of plant consumed increases with plant density and this introduces stabilizing, direct density dependence into the interaction. At high plant availabilities, however, the functional response is almost certain to saturate, so feeding becomes inversely density dependent (as with type II functional responses, Fig. 13.10b).

4 *Ratio dependence*. If the functional response works mainly in terms of the amount of plant food available *per herbivore* rather than the absolute density of plant food per unit area (as it might, for example, with a long-lived, strong-flying insect herbivore), then it may be more appropriate to model the functional responses in a ratio-dependent manner. There has been a heated debate about the merits and demerits of ratio-dependent models (Abrams 1994; Berryman *et al.* 1995) but the controversy matters little, as a simple example will show. Suppose that we take the herbivore equation from Equation 13.2 and assume a ratio-dependent numerical response coupled with density-independent herbivore mortality:

$$\frac{dN}{dt} = cN\left(\frac{V}{N}\right) - dN \qquad (13.6)$$

We can cancel the Ns and arrive at an equation for the equilibrium abundance of plants V^*.

$$V^* = \frac{d}{c}N^* \qquad (13.7)$$

Now suppose, instead, that we had assumed a linear functional response, but that herbivore death rates were directly density dependent, with the exponent $q = 2$ for simplicity. Then the herbivore equation becomes:

$$\frac{dN}{dt} = cNV - dN^2 \qquad (13.8)$$

At equilibrium the positive and negative terms are equal, so we can divide through by N. You will see that we get exactly the same equation for the equilibrium plant abundance $V^* - dN^*/c$ as when we assumed that there was a ratio-dependent functional response and density-independent herbivore mortality. The moral is as important as it is simple: *you cannot infer the mechanism by observing the dynamics*. A ratio-dependent functional response is indistinguishable in its effect on plant abundance from direct density dependence acting on the herbivore death rate. Our principal interest is always likely to be with the mechanisms underlying the dynamics, but there will always be several quite different model structures that are capable of generating the same pattern of dynamics. We want to know which of the many possible structures matches the true causes of the dynamics as closely as possible, and for that we need a knowledge of the mechanisms. This can only come from a combination of field observation and experiment, refined repeatedly in the light of new modelling.

13.9.6 Herbivore numerical responses

The numerical response predicts the way that a given amount of herbivore feeding is translated into surviving herbivore offspring. In many cases, it may be perfectly reasonable to assume a direct proportionality between the amount that is eaten by the herbivores in one time period and the number of herbivore births that there will be in the next time period. For many insect herbivores, for example, fecundity is a linear function of the female's body mass on emergence, and this is determined by her feeding success as a larva (Table 13.1). For generalist herbivores, it is usual that ice-cream plants have rather little impact on the numerical response; when generalist herbivores are food-limited, it is usually the abundance of the relatively abundant, relatively less preferred plant that determines their numbers (Crawley 1983). Complications arise when limits are placed on herbivore reproduction by processes other than feeding success. For example, the number of safe breeding sites may place an upper limit on the number of breeding pairs and hence on the total number of herbivore offspring, irrespective of the amount of food available. Alternatively, the breeding success of herbivores may be low when herbivore density is low because of an increased requirement for anti-predator vigilance by each female (this is an example of an 'Allee effect' where herbivore performance declines as population density gets lower; see Crawley 1992a). Even without these limitations, there is bound to be a maximum to the rate at which each female herbivore can produce female offspring. This means that when plants are very abundant there is a maximum to the net multiplication rate of herbivore population between one time period and the next. For herbivores with discrete generations (e.g. univoltine insects, and vertebrates with a single, brief breeding season each year) this creates the potential for herbivore satiation. For example, the cinnabar moth can increase by a maximum of roughly 20-fold from one year to the next, but ragwort, its host plant, can increase in biomass by a factor of 200–2000 from one year to the next (Crawley & Gillman 1989). Recruitment of the plant is determined (at least in mesic grasslands) by good weather for seedling recruitment in autumn, coupled with the right kind of soil disturbance at just the right time of year, with the result that cohorts of first-year, rosette plants vary enormously in abundance. If biomass of flowering plants is 2000-fold higher in one year than in the last, then the cinnabar moth caterpillars in the second year will be satiated. Moth numbers might increase by 20-fold from the previous year, but caterpillar feeding will cause only a minor reduction in seed production. The periodic production of mass seed crops in the years when the herbivore is satiated can replenish the seed bank for many years to come (a storage effect; see Chapter 14).

13.9.7 Herbivore density dependence

Herbivore populations are affected by many factors in addition to their food supply. The density-independent death rate d changes from year to year with winter weather, catastrophic storms, fires, and so on. In addition, the herbivores may be subject to density-dependent mortality over and above that which could result from exploitation competition for plant food (e.g. attack by predators or parasites). Alternatively, the herbivores might be territorial, like grouse *Lagopus lagopus scoticus* (Hudson 1992) or they may be prone to epidemic diseases (e.g. rabbits and myxomatosis or African ungulates and rinderpest; Dobson & Crawley 1994). These kinds of direct density dependence will have a stabilizing influence on the plant–herbivore interaction under most circumstances, because it prevents damaging overexploitation of the plants that might occur if herbivores were to reach very high densities.

13.9.8 Granivory: the dynamics of seed predation

Estimates of the magnitude of seed predation are common in agricultural folklore. Farmers, for example, sow four times as many seeds as they expect to see as mature plants:

> *One for the rook,*
> *One for the crow,*
> *One for the pigeon,*
> *And one to grow.*

This estimate of 75% as the postdispersal seed predation rate is remarkably close to the global average of a large number of field studies (Crawley 1992b). Seed predation is different from other kinds of herbivory in several ways. Even the most fervent advocates of the unimportance of competition amongst herbivores (such as Hairston 1989) have to admit that granivores are different. Indeed, some of the most convincing demonstrations of interspecific competition for resources have come from studies on granivorous desert rodents and ants (as we shall see below). Amongst the reasons that granivory differs from other kinds of herbivory are: (i) the food quality is relatively high (e.g. protein nitrogen levels are higher in seeds than in many other plant tissues); (ii) the food is parcelled into discrete packets that are often too small to allow the full development of an insect herbivore in a single seed; (iii) the seeds are available on the plant for only a brief period; and (iv) the production of seeds in any one year is typically less predictable than the production of other plant resources like new leaves.

We can summarize the impact of granivory on plant dynamics by reference to the recruitment curves in Fig. 13.11. Generally, plant recruitment curves have three zones: (i) an upper asymptote, when seed input is high; (ii) an increasing phase, where more seeds means more recruits; and (iii) a lower zone where there is no seedling recruitment at all. The relative locations of the break-points between these three zones of the graph will differ from species to species, and from habitat to habitat for any given species.

Zone I. When seed densities are high (Fig. 13.11b) then there may be intense competition for access to suitable microsites. Under

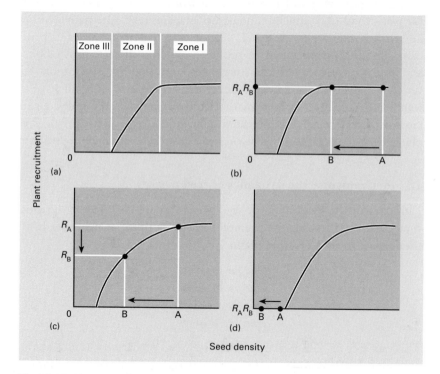

Fig. 13.11 Limitation of plant recruitment. (a) Recruitment curve plots the number of established plants versus the seed density from which the cohort was recruited. The curve can have three zones: Zone I, microsite (competition)-limited recruitment; Zone II, seed-limited recruitment; Zone III, predator (granivore)-limited recruitment. The effects of herbivores on plant recruitment depend on the prevailing zone. (b) Zone I: since recruitment is not seed-limited, large reductions in seed production as a result of herbivory (from A down to B) have no impact on plant recruitment ($R_B = R_A$). (c) Zone II: recruitment is seed-limited, so herbivory leads to reduced recruitment ($R_B < R_A$). (d) Zone III: recruitment is not seed-limited because seed production is below the threshold necessary for satiation of granivores, so herbivory has no impact on plant recruitment ($R_B = R_A = 0$).

these circumstances, large reductions in seed density resulting from seed predation may cause no measurable reduction in the number of plants that recruit to the juvenile population. Thus, when recruitment is microsite-limited, seed predation is unlikely to have any impact on mature plant density. For example, out of 20 species of native grassland plants sown into undisturbed mesic grassland at a rate of 1000 extra seeds per square metre, only two species (*Lotus corniculatus* (Fabaceae) and *Rumex acetosella* (Polygonaceae)) were seed-limited and showed enhanced recruitment (Johnston 1992).

Zone II. When seed densities are low relative to the availability of suitable microsites but sufficiently high to satiate the postdispersal granivores, then reduced fecundity will inevitably reduce plant recruitment (Fig. 13.11c). This is because at low seed densities there is no density dependence acting on seedling performance, and hence no opportunity for compensatory reductions in other mortality factors. Whether or not this reduced seedling recruitment leads to reduced densities of adult plants depends on the spacing of recruitment microsites and on the competitive interactions of the adult plants (see Chapters 12 & 15).

Zone III. In this zone there is no recruitment because of postdispersal seed predation and reducing seed production would have no impact on recruitment. Increasing seed production, however, can have sudden and dramatic effects on plant recruitment as the threshold density of seed for satiation of the granivores is passed (Fig. 13.11d).

If there is a generalization to be made, it appears that seed predation rates are more often inversely density dependent than directly density dependent. That is to say, the percentage seed loss to granivores tends to be lower in large seed crops than in small (see Section 13.2.9).

13.10 Case studies

Most studies of plant–herbivore dynamics are so short term that it is impossible to discover anything meaningful about the impact of herbivory on plant distribution or abundance. Also, a great many of the studies have been spoiled by serious shortcomings of experimental design, particularly by pseudoreplication (Hurlbert 1984; Crawley 1993a, pp. 45–61). In the worst cases, there are just two plots; one grazed and one control. In many more cases, plants within plots have been used as replicates, instead of taking the plot means as the response variable, and using the block-by-treatment interaction as the error term.

13.10.1 Keystone herbivores: the kangaroo rats of southern Arizona

Two long-term studies involving the exclusion of seed-feeding kangaroo rats *Dipodomys* spp. have been carried out by Jim Brown and his colleagues, the first in the Sonoran Desert (Davidson *et al.* 1984; Brown & Munger 1985) and the second in the Chihuahuan Desert of south-eastern Arizona (Brown & Heske 1990; Samson *et al.* 1992; Heske *et al.* 1994). In the Chihuahuan Desert, three classes of annual plants (winter, summer and biseason species) were monitored following rodent and ant removal. Over a 12-year period, rodent removal led to increased dominance of the winter annuals by initially rare, but competitively superior, large-seeded annuals. These plants eventually suppressed the small-seeded annuals. Removing granivorous ants caused an increase in small-seeded species, but had no impact on the larger-seeded species. Tall grasses like the annual *Aristida adscensionis* and the perennial *Eragrostis lehmanniana* also increased on the rodent-exclusion plots (Fig. 13.12). Species richness of summer annual dicots was increased on the plots where kangaroo rats were present, although their decline on the rodent-exclusion plots might be caused by increased grass competition rather than as a direct result of granivory. In the Sonoran Desert, ants and rodents interacted principally through exploitation competition of populations of winter annuals. Increasing dominance of large-seeded annuals on rodent-removal plots eventually led ant populations to decline. So asymmetric is the plant competition that in neither desert did differential predation by ants on small-seeded species lead to an increase in large-seeded annuals. In both deserts, however, the seed-feeding rodents were keystone species as a result of their selective predation of large-seeded species, which reduced their competitive superiority and maintained the diversity of winter annuals. This study is an excellent example of a case where the keystone species are invisible to the casual visitor to these deserts because of their small size and nocturnal habits. It was only by long-term experimental exclosure of the kangaroo rats that their pivotal role in ecosystem function was discovered.

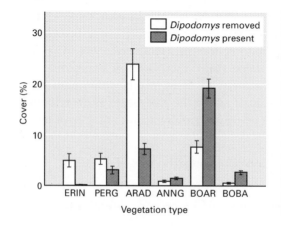

Fig. 13.12 Kangaroo rats as keystone herbivores. Long-term exclusion of seed-feeding rodents caused a marked change in plant community structure and function (see text for details). ERIN, *Eragrostis lehmanniana*; PERG, tall perennial grasses; ARAD, *Aristida adscensionis*; ANNG, tall annual grasses; BOAR, *Bouteloua aristidoides*; BOBA, *B. barbata* (all Poaceae). (After Brown & Heske 1990.)

13.10.2 Exclusion experiments using fences against large vertebrate herbivores

Exclusion of sheep from hill pasture in Snowdonia, Wales led to relatively rapid changes in the character of the vegetation over about 7 years. Low-growing plants such as *Danthonia decumbens* (Poaceae), *Juncus squarrosus* (Juncaceae), *Trifolium repens* and *Cirsium palustre* declined consistently while previously suppressed shrubs of *Calluna vulgaris* and *Erica cinerea* (both Ericaceae) grew into full-sized bushes that flourished for 20 and 12 years respectively, before degenerating. In the absence of sheep, voles became the dominant herbivores and caused large year-to-year variation in herbage biomass as their populations fluctuated. Species richness changed very little, and there were few cases of invasion by new species; an exception was the rowan tree *Sorbus aucuparia* (Rosaceae) whose bird-dispersed seeds were deposited next to the fence posts of the sheep exclosure, which were used as perches by the birds (Hill *et al.* 1992). Exclusion of rabbits for 50 years from creosote bush *Larrea tridentata* (Zygophyllaceae) in New Mexico led to a 30-fold increase in the grass *Sporobolus contractus* and lesser but significant, non-successional changes in several of the smaller shrubs (Gibbens *et al.* 1993). Erection of fences in Serengeti grasslands led to increased cover by tall rhizomatous perennial grasses, and reduced cover by annual species and short, typically stoloniferous, perennial species. Species diversity declined in two of three fenced areas from which herbivores were excluded (Belsky 1992). The impact of rabbit-exclusion from oak woodland is graphically illustrated in Plate 9, facing p. 366.

13.10.3 Cyclic herbivore populations

Several herbivore populations exhibit persistent, multiyear cycles of between 3 and 10 years in period (Crawley 1983). Some of these cycles are almost certainly predator–prey cycles (e.g. the 4-year cycles of vole numbers at high latitudes in Fennoscandia appear to be driven by the specialist predator, the least weasel *Mustella nivalis*; Hanski *et al.* 1991). Others look much more like plant–herbivore interactions (e.g. the 10-year cycle of the larch budmoth *Zeiraphera diniana* in the Engadine valley in Switzerland (Baltensweiler *et al.* 1977) and the 3–4 year cycle of the Norway lemming) where the cycles result from time-lags in herbivore population recovery caused by induced chemical changes resulting in reduced food quality for the herbivores (increased fibre in the case of larch budmoth; induced protease inhibitors in the case of the lemmings; Seldal *et al.* 1994). Other herbivore cycles might be plant–herbivore, predator–prey or multitrophic-level interactions (e.g. the famous 10-year cycles of snowshoe hare in Arctic Canada). Some of the best-known insect herbivore populations show different patterns of population dynamics in different places. In The Netherlands, the cinnabar moth *Tyria jacobaeae* has shown regular 5-year cycles over a 20-year period in coastal sand dunes (Fig. 13.13a–d; van der Meijden & van der Veen 1996). Over a somewhat shorter period in mesic grasslands in Silwood Park, the dynamics show no hint of regular periodicity

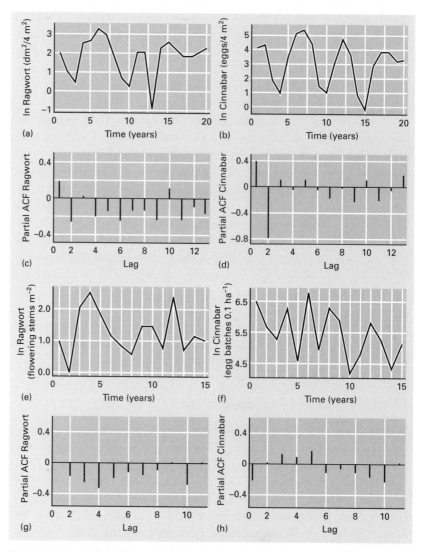

Fig. 13.13 Dynamics of the ragwort–cinnabar moth system in two contrasting locations in Dutch dunes and English grassland. In The Netherlands, the ragwort population (a) fluctuates over a wider range than ragwort in Silwood Park (e). Time-series analysis (partial autocorrelation plots) of ragwort populations shows no tendency towards cyclic behaviour either in The Netherlands (c) or Silwood Park (g). Cinnabar moth populations in The Netherlands (b) shows regular 5-year cycles in abundance, but there is no periodicity in the Silwood Park cinnabar moth populations (f). Time-series analysis of The Netherlands cinnabar moth (d) shows a highly significant negative term at lag 2, responsible for generating the cycles while there is no delayed density-dependent signal in the Silwood Park data (h). All four populations show very pronounced direct density dependence, and are all quite tightly regulated. (Data (a–d) from van der Meijden & van der Veen 1996 and (e–h) from M.J. Crawley, unpublished results.)

plant (ragwort *Senecio jacobaeae*) are uncoupled from the dynamics of the herbivore; the ragwort population is higher and fluctuates significantly more widely in the open sand dune community than in the closed grassland. The interaction is asymmetric; the insect's dynamics are food-limited, but the plant's dynamics are not herbivore-driven (fluctuations in ragwort abundance are caused principally by variation in recruitment, and this is determined chiefly by rainfall and soil disturbance; Crawley & Gillman 1989; van der Meijden & van der Veen 1996).

13.10.4 Weed biocontrol

Weed biocontrol projects have provided some of the best field data on plant–herbivore dynamics. Releases are carried out on a spatial scale that could never be managed in experiments by individual ecologists. Unfortunately, what these projects gain in scale, they often lose in rigour. Very few releases are either replicated or randomized. Sometimes releases are made with minimal prerelease study and virtually no follow-up. While there are some consistent successes (e.g. floating fern *Salvinia molesta* controlled by the weevil *Cyrtobagous salviniae*; Room & Thomas 1985; Plates 10 and 11, facing p. 366) and some consistent failures (nutsedge *Cyperus longus*; Holm *et al.* 1977), most of the systems are marked by a substantial degree of unpredictability. Table 13.2 shows the attributes of weed, control agent and release protocol that are most consistently associated with success or failure, but these must all be viewed with a good deal of circumspection. Current weed biocontrol practice is aimed at improving the scientific value of these release experiments by developing protocols that will allow detailed prerelease and follow-up studies and, where possible, by setting up control plots in addition to replicated release sites, where weed dynamics can be monitored (e.g. the campaign against the European wetland weed *Lythrum salicaria* (Lythraceae) in North America; Blossey 1995).

Some of the best work on biological weed control has been carried out in South Africa. The invasive shrub *Sesbania punicea* (Fabaceae) forms extensive thickets in river beds and valley bottoms, and three insect agents have been successfully introduced against it. The apionid *Trichapion lativentre* destroys nearly all of the flower buds and reduces seed set by an average of 98%. Despite this reduction of its food supply a second species, the curculionid *Rhyssomatus marginatus*, which feeds on seeds in ripening pods, has become established. Between then, the two introduced insects have almost completely arrested the reproductive performance of *Sesbania punicea* (Hoffmann & Moran 1992). A stem-boring weevil *Neodiplogrammus quadrivittatus* can cause the mature plant to die back, but on its own this just releases mass *Sesbania* recruitment from the seed bank. At most sites where all three agents are present, dense thickets of the weed have all but disappeared (Hoffmann

Table 13.2 Successes and failures of weed biocontrol showing the attributes of the weeds, the control agents and the release protocols associated with each outcome. The database is dominated by repeated attempts to control particular plant species: weeds that have been successfully controlled elsewhere in the past (e.g. *Opuntia* (Cactaceae) and *Lantana* (Verbenaceae) spp.), and weeds that are unsuitable for control by other means (e.g. plants infesting marginal grazing land where cost precludes any other means of control). This taxonomic bias can be seen from the fact that of 627 cases documented by Crawley (1985c), 152 involved *Lantana camara* and 117 involved various *Opuntia* spp.

	Successes	Failures
Plants	Genetic uniformity	Genetic polymorphism
	Water weeds	Terrestrial weeds
	Semi-arid rangeland weeds	Forest weeds
	Biennials	Annuals
	Shrubs	Trees
	No seed bank	Large seed bank
	Low powers of regrowth	High compensation ability
	Limited seed dispersal	Wide seed dispersal
Biocontrol agents	High *r*	Low *r*
	Small body size	Large body size
	Sedentary	Mobile
	Common as a native	Rare as a native
	Widespread as a native	Local as a native
	Sucking insects	Chewing insects
	Root-feeders	Shoot-feeders
	Stem-borers	Leaf-feeders
Protocol	Painstaking exploration for agents	Little exploration effort
	Good climatic matching of agent to release environment	Poor climatic matching
	Many release sites	Few release sites
	Repeated releases	Single releases
	Released in several years	Released just once
	Patience	Impatience

1990). *Acacia longifolia* is an excellent example of the danger of having preconceived notions about the attributes of ideal biocontrol agents. When the pteromalid gall wasp *Trichilogaster acaciaelongifoliae* was released against *Acacia longifolia* the situation looked hopeless; the wasp was uncommon in its native environment in Australia, it formed stem galls, and the adults were large and sluggish (see Table 13.2). Despite these apparent shortcomings the insect has been a spectacular success and seed production has been all but curtailed. It increased to such abundance that branches broke off the *Acacia* trees under the sheer weight of galls hanging from them (Dennill *et al.* 1993).

The relevance of weed biocontrol data to an understanding of plant–herbivore dynamics in native vegetation is open to question. There are several concerns:

1 the weeds are generally alien plants growing in plant communities that are often quite different from those in which they evolved;

2 the insects are also alien, imported especially for the purposes of control;

3 the insects have been freed from their native natural enemies by careful screening prior to release;

4 the range of genetic variability in both plant and insect populations may be lower than in native communities as a result of the small size of the initial introductions; and

5 the habitats in which biocontrol is practised are often highly disturbed (e.g. overgrazed semi-arid rangelands).

13.10.5 Exclusion experiments involving insect herbivores and chemical pesticides

Cantlon's (1969) demonstration of a dramatic increase in the abundance of the hemiparasitic woodland herb *Melampyrum lineare* (Scrophulariaceae) following the eradication of the katydid *Atalanticus testaceous* using insecticides was for many years the only report suggesting that a native insect population might maintain a natural plant population at low density. After shrubs of *Haplopappus squarrosus* and *Haplopappus venetus* (Asteraceae) were sprayed with insecticide, seed production was increased in both species, but plant recruitment was only increased in *H. squarrosus* (Louda 1982). Evidently, recruitment was seed-limited in *H. squarrosus* but not in *H. venetus*, so insect herbivory had no impact on the population dynamics of *H. venetus* despite it causing substantially reduced fecundity (Louda 1983). One of the few insect exclusion studies to encompass the whole life cycle, from mature plants in one generation to mature plants in the next, was carried out in localized blow-outs in Sandhills Prairie, Nebraska by Louda and Potvin (1995). They excluded capitulum-feeding insects of the monocarpic thistle *Cirsium canescens* (Asteraceae) in some of the blow-outs but not in others, and monitored seedling and flowering plant densities in the neighbourhood of individual sprayed and unsprayed thistles. Not only were there higher densities of seedlings around plants on sprayed sites, but these led to higher mature flowering plant densities in the next generation, indicating that recruitment was seed-limited rather than microsite-limited. Studies on the exclusion of insects (mainly frit flies of the genus *Oscinella* or hessian fly *Mayetiola destructor*) from grasslands have shown that by reducing the vigour of the pasture grasses on which they feed, the flies allowed the ingress of less competitive, weedy species, so that the quality of sown pasture declined less rapidly when insects were excluded by insecticide application (Thom *et al.* 1992; Standell & Clements 1994). Note, however, that when all the plants in a community are sprayed, it is impossible to tease apart the direct negative effects of herbivory from the indirect negative effects of competitor release of a plant's neighbours. Thus, if all plant species suffer more or less equally from herbivory, then blanket removal may show no effect at all, because all species may benefit equally from herbivore removal; we are more likely to detect effects of herbivory if herbivore species are excluded singly, or if individual plant species are freed from attack by their herbivore faunas.

13.11 Herbivores and plant diversity

It is difficult to generalize about the impact of herbivory on plant diversity because so few detailed long-term studies have been carried out, and the results of those which have are inconsistent. Several studies have shown increased plant species richness under herbivory (Belsky 1992; Pandey & Singh 1992; Montalvo *et al.* 1993), a few have shown reduced species richness (Milton 1940) and several have shown no effect of herbivory on species richness at all (Crawley 1989b; Hill *et al.* 1992; Bach 1994). Given the complexity of the factors affecting species richness, even in a simple experimental community like Park Grass (see Chapter 14), it would be naive to expect that the effect of herbivory on plant diversity would be simple. However, if there *is* an effect of herbivory on plant diversity, then the most likely mechanism is through a correlation between competitive ability and palatability amongst plant species. A positive correlation between palatability and competitive ability could lead to increased species richness in the presence of herbivores, and a negative correlation would lead to reduced diversity, through the rapid elimination of plants that were both preferred and uncompetitive. Even if the herbivores exhibited no preference, or all plants were attacked solely by monophagous herbivores, then changes in plant diversity could still result from herbivory, if plant species differed in their 'grazing tolerance' (i.e. in the degree to which a given amount of herbivore feeding led to reduced plant performance).

Some of the most persuasive data come from studies on marine systems (Menge 1976; Lubchenco 1978; Hay 1981), which show that diversity is generally increased when herbivores prefer the dominant plant species, and decreased when plants other than the competitive dominant are preferred. Grazing intensity is also important (Harper 1977): at low intensities, diversity might be low because of competitive exclusion by the dominant plant; diversity peaks at intermediate grazing intensities when the dominant is suppressed but other species are not substantially affected; diversity may be low at the highest grazing intensities if there is only a small pool of grazing-tolerant (or avoided) species.

If herbivores are to have an effect on plant community structure, then their influence is most likely to be felt during recruitment because, as we have seen: (i) herbivores rarely have a profound effect on mature plant survivorship; (ii) the effect of herbivory on plant fecundity is often marginal as a result of compensation; and (iii) recruitment from seed is often sufficiently rare in perennial plant communities that herbivores which reduce seed production might not reduce the rate of plant recruitment from seed (see Fig. 13.11). We should direct our experimental energies towards selective removal of herbivores during plant recruitment. If herbivores are important in plant community structure and dynamics, then this kind of experiment is more likely to demonstrate it.

13.11.1 Selective herbivory and the identity of the dominant plant species

It is sometimes said that selective herbivory on the dominant plant species is sufficient to promote plant species richness. This is wrong, as can be seen immediately from Fig. 13.14a. Suppose that two plant species compete for access to a gap in the canopy and that there can only be one winner (the gaps are the size of one mature plant). In such a case, other things being equal (like initial plant size and germination time), the individual with the faster relative growth rate will overtop the other plant and capture the light gap in a pure contest competition. In the absence of herbivores the winner is species A. Now suppose that a generalist herbivore attacks both species A and B but that it prefers species A so that, although the growth rate of both species is reduced by herbivory, species A suffers to a greater degree than species B. Figure 13.14b shows what happens in the long run; now speices B has a higher relative growth rate than species A, so species B will capture the

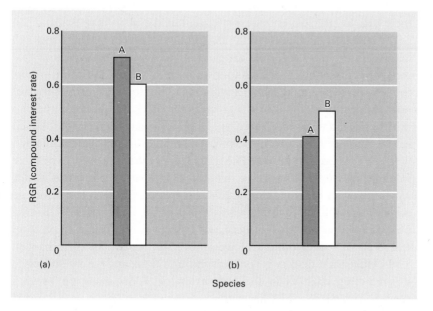

Fig. 13.14 Selective herbivory and the identity of the dominant plant. In the absence of herbivores (a) plant A has a higher relative growth rate and would eventually exclude species B. In the presence of selective herbivory (b), the growth rate of both plants is depressed, but A suffers more than B. Now the relative growth rate of B is greater than A, so B would form a monoculture and A would be excluded. Selective herbivory on its own does not increase plant species richness.

light gap. Selective herbivory has changed the identity of the dominant plant but has had no impact at all on plant species richness. Herbivory has changed a monoculture of species A into a monoculture of species B. Selective herbivory is *not* a sufficient condition for plant coexistence.

13.11.2 Selective herbivory and plant species richness

The theory underpinning the determination of plant species richness is based on an extremely simple principle: all persistent species must be able to increase when rare. This is the *invasion criterion* and it is a fundamental part of all models of plant species coexistence (see Chapter 14 & 19). There are essentially two ways that the invasion criterion can be fulfilled. First, there may be explicit frequency dependence, such that plant death rate declines as abundance decreases relative to the abundance of other plant species. This might come about because of a reduction in the rate of herbivory as generalist herbivores switched to more common foods, or as specialist herbivores were reduced in population density. Alternatively, frequency dependence might arise because there are parts of the individual plant (e.g. underground storage organs like rhizomes, bulbs or woody roots), or parts of the plant population, that are not exploited by the herbivores (e.g. plants on inaccessible cliffs or which grow in places where there are high densities of parasites or predators of the herbivore). In either case, the risk of death from herbivory declines as the abundance of the plant goes down. Second, there may be no frequency dependence, but the plant may escape when rare for purely stochastic reasons. Suppose that the plants and the herbivores are not uniformly distributed across the landscape, but that the spatial distribution of each follows some statistical distribution (e.g. the negative binomial; see Chapter 14). So long as the degree of non-uniformity is sufficiently high and the correlation between the two distributions is imperfect, there will be places where the plant finds itself in a temporary refuge from herbivory. To investigate this idea further, we need a theoretical model. The framework that has proved to be most valuable in studies of species richness is known as the *lottery model*, which owes its development largely to Chesson (1985). The structure differs from the Lotka–Volterra formulation of Equation 13.1 in that we assume that all space is occupied by one plant species or another (there is no bare ground and we ignore the process of disturbance). Thus, total plant density is constant through time, and we are interested in predicting changes in the proportion of plants made up by each species. The model looks like this:

$$\frac{dV_i}{dt} = -d_i V_i + \sum_{j=1}^{k} d_j V_j \left[\frac{q_i F_i V_i}{\Sigma q_j F_j V_j} \right] \qquad (13.9)$$

and is nothing like as daunting as it seems at first. The first term on the right-hand side says that in the absence of recruitment from seed our plant species i would decline exponentially in abundance at a species-specific rate d_i (a very reasonable assumption). Now because we assume that all space is filled by plants, the only space open for recruitment is the space vacated by the deaths of mature individuals. This is given by adding up the total number of plants that died from all k species, $\sum_{j=1}^{k} d_j V_j$. The final part of the model (in brackets) predicts the share of this open space that will be occupied by recruits of our species i. In the simplest case (all $q_j = 1$), this is given by the fraction of all seeds $\sum F_j V_j$ that were produced by species i. This is why it is called a lottery model; seeds are like lottery tickets, and your chance of winning is the number of tickets you have, $F_i V_i$, divided by the total number of tickets in the lottery. As it stands, this form of the model will not allow coexistence. The species that has the highest per-capita fecundity F_j will outcompete all the others and form a monoculture (Fig. 13.14). As explained already, species richness can only be preserved if all species have the ability to increase when rare. This could happen because of explicit frequency dependence (e.g. the impact of herbivory q_i might depend on the relative abundance of species i as in Fig. 13.15) or because of the spatial variance in co-occurrence between the plants and their herbivores. If herbivore populations are patchily distributed over their host plants, then by chance alone the inferior plant competitor may find itself herbivore free in a site where the superior competitor is subject to attack by its parasites. If the variance is sufficiently high and the population

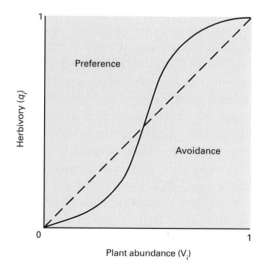

Fig. 13.15 Frequency-dependent herbivory q_i of the kind that might allow increased plant species richness. Above the 45° line (dotted) the herbivore exhibits preference (its diet contains a higher proportion of plant species i than presented in the environment) while below the 45° line the herbivore exhibits avoidance (its diet contains a lower proportion than the plant represents in the environment). The solid line shows frequency-dependent herbivory, involving a switch from preference for the plant when it is common, to avoidance when the plant is rare. This switching behaviour provides the plant with a refuge, allowing it to increase when rare.

densities of the two herbivores are not perfectly correlated, then the inferior competitor survives and plant diversity is enhanced. In this case, coexistence may occur in circumstances where traditional experimental protocols would fail completely to detect any frequency dependence (Pacala & Crawley 1992).

It is interesting to compare these findings with those of Harper (1969) in his influential essay on herbivores and plant diversity. He wrote:

It is easy to envisage a model in which plant species that are each mutually exclusive of the others because of wholly overlapping niches are demoted from dominance by specific predatory action: each demotion permitting a stepwise increase in the number of species present. The predators need (i) to be species specific in their food requirements, (ii) to transfer the regulation of plant density from dependence on the resources needed for plant growth to regulation by the behaviour of the predator, (iii) that the herbivore population shall be regulated by the availability of its specific host plant and not by its parasites and hyper-predators.

As to point (i), it is important to note that the plant species must suffer differential consequences of herbivory; this will often come about as a result of preferential feeding by the herbivores, but it need not. It could equally well result from differential grazing tolerance by the plants, in which the same amount of feeding has differential effects on plant performance (see Crawley 1983). As to point (ii), in a trivial sense it is true that plant competition is less important in models that include herbivores because, in the absence of herbivores, plant competition is the *only* regulating factor. Herbivory often appears to change the nature of plant competition (e.g. ungrazed stands are tall and typically structured by competition for light), but it is clear that individual plants benefit just as much from the removal of their neighbours in a grazed sward as they do in an in ungrazed one. With regard to point (iii), herbivores need not be food-limited in order to affect plant diversity. They are likely to have a greater impact if they are food-limited, but even the low levels of damage typically reported for invertebrate herbivores (5–15% leaf area loss) can change the identity of the dominant plant species. Thus, while none of Harper's points has proved to be wrong, they are all markedly less restrictive than he implied. It is now clear that a scarce, enemy-limited, polyphagous herbivore could increase plant diversity (Crawley & Pacala 1992; Pacala & Crawley 1992).

13.12 Herbivores and plant succession

The impact of herbivores on succession depends on the seral stage and on the processes currently responsible for limiting the rate of transition to the next stage. If succession is seed-limited, then herbivores that reduce seed production or dispersal by the next dominant species will delay their establishment and hence slow down the process of succession (some deciduous forest successions appear to be seed-limited; see Chapter 15). If, on the other hand, successional change is prevented by impenetrable stands of perennial vegetation (site pre-emption or 'inhibition' *sensu* Connell & Slatyer 1977), then herbivores that break up this perennial cover will relieve the inhibition and hence accelerate the succession, allowing the establishment of a cohort of later successional species. Finally, the current dominant species may be required to alter soil or microclimatic conditions before succeeding potential dominants can become established (this is 'facilitation' *sensu* Connell & Slatyer 1977). In this case, herbivory will slow down the course of succession by reducing the rate at which environmental amelioration occurs. To the extent that primary (soil-forming) successions tend to be driven by processes of facilitation (increasing soil nitrogen, declining light intensity), while secondary successions on established soils tend to be dominated by seed limitation or inhibition, then we would predict that herbivory would *delay* primary successions and seed-limited secondary successions and *accelerate* secondary successions limited by inhibition. Much of range-management thinking is based on the simple paradigm that grazing and succession are roughly equal and opposite processes: increase the rate of grazing and the plant community will move from woodland towards grassland; reduce the rate of grazing and the plant community will undergo succession towards woodland. The resilience and elasticity implict in this simplistic view can lead to overgrazing and habitat degradation, or to damaging range expansion by unpalatable (and usually valueless) woody plants. There are plenty of examples where alien vertebrate herbivores (rabbits in England, goats in Hawaii) stop grasslands undergoing succession to forest by preventing the establishment of trees. There are fewer examples where native herbivores have been documented as changing the rate (or altering the outcome) of succession. Davidson (1993) reviewed these studies and concluded that herbivory tends to accelerate the pace of plant community succession from grasses, shrubs and pioneer trees to persistent (climax) trees, but that herbivory typically retards succession from earlier seres to trees (as in old fields). She suggests that there is differentially high herbivory on intermediate seral species (grasses, shrubs and pioneer trees) whose main defence against herbivory is rapid compensatory growth in favourable resource environments.

13.12.1 Primary succession

Sand dunes on the coast of The Netherlands are built around the growth of marram grass *Ammophila arenaria* on the seaward face, followed in sequence by *Festuca rubra* (Poaceae), *Carex arenaria* (Cyperaceae) and *Elymus athericus* (Poaceae) on the landward side of the dune. What causes the marram grass to become prone to invasion by the fescue, and the fescue by the sedge and the sedge by the couch grass? Traditional explanations include resource depletion, alteration of the physical environment, accumulation of metabolic toxins and the actions of species-specific herbivores (see Chapter 8). Greenhouse experiments by van der Putten *et al.* (1993) demonstrate the role of specific soil-borne pathogens and nematodes. Seedings of *Ammophila* were grown in gamma-sterilized and unsterilized soil from the root zones of each of the four successional zones (*Ammophila, Festuca, Carex* and *Elymus*). Only in beach sand did sterilization not improve biomass production significantly, demonstrating the presence of soil organisms harmful to *Ammophila* in all but the fresh beach sand. In the second experiment the other three graminoids were grown reciprocally in sterilized and unsterilized soils from each of the four successional zones. Biomass production of each species was reduced by soil-borne diseases of their own and of their successors but not of their predecessors, suggesting that reduction of inhibition by soil herbivores was an important process determining the rate of transition between the four seral stages (van der Putten *et al.* 1993).

In resource-limited early successional habitats in Arctic environments, foraging by generalist vertebrate herbivores may accelerate or delay plant succession. They may increase the rate of nutrient turnover, particularly nitrogen, thereby sustaining net primary production of preferred forage plants. Or they may move from spot to spot, exploiting nutrient pulses in space and time in environments where high turnover rates cannot be sustained. In this case the long-term effects of herbivory on vegetation appear to be minimal. The top-down influence of humans can alter herbivore numbers and create a trophic cascade leading to the widespread destruction of plant communities (e.g. Arctic plant communities destroyed by increased numbers of greater snow geese; Jefferies *et al.* 1994; and see Bryant & Kuropat 1980). A specialist flea beetle *Altica subplicata* influences growth and survival of sand-dune willow *Salix cordata* and can affect the course of primary succession on Alaskan sand bars. A 3-year study using cages (rather than insecticides) to exclude the beetles showed a two-fold increase in height and diameter growth of *Salix* on ungrazed plants. Most of the plants that died during the study period were in plots with beetles present. Insect herbivory also had a significant impact on neighbouring plant species with greater increases in population density of both monocots and dicots on plots with beetle feeding. The most abundant herbaceous dicots, however, showed their greatest increases in plots with beetles absent. Despite these pronounced effects of *Altica* on reduced *Salix* growth, there were no differences after 3 years in plant species richness, species diversity or evenness in plots with and without beetles (Bach 1994).

13.12.2 Secondary succession

Reducing foliar herbivory in the early years of an old field secondary succession in Silwood Park by excluding sucking insects (mainly auchenorrhynchan Hemiptera) with chemical pesticides resulted in increased grass growth, and plant species richness was reduced as the grasses outcompeted the herbs. Reducing subterranean herbivory from chewing scarabeid beetles and tipulid flies using soil insecticides prolonged the persistence of annual herbs, increased perennial herb colonization and hence promoted plant species richness (Brown & Gange 1992). Deer, woodchuck, rabbits and small rodents were excluded in various combinations from old fields in Virginia, 1–4 years after disturbance. Plant species richness and evenness were greater in partial exclosures (open to woodchuck, rabbits and rodents but closed to deer) than in total exclosures or open control plots. Different species showed differing responses to herbivory: some increased with increasing herbivore feeding, some decreased and other species peaked at intermediate levels of herbivore access. *Aster pilosus* (Asteraceae) tended to dominate the total exclosure plots and *Celastrus scandens* (Celastraceae) to dominate the grazed controls. The main effects of herbivory were through alterations in the outcome of interspecific competition, rather than by direct effects on plant mortality (Bowers 1993). Pathogen outbreaks can cause changes to vegetation that have long-term consequences, and both animal and plant pathogens can be involved (Dobson & Crawley 1994). For example, there was mass regeneration of oaks *Quercus robur* in Silwood Park following the myxomatosis epidemic of the mid 1950s when rabbit numbers were reduced to very low levels and kept there for more than two decades. Most of the naturally regenerated oak woodlands within Silwood Park owe their existence to myxomatosis. Today, erection of a rabbit-proof fence in grassland leads rapidly to the establishment of an oak canopy from seedling recruitment and from the regrowth of suppressed seedlings (canopy closure occurs 10–20 years after fencing); matched plots outside the rabbit fence show no recruitment at all (Crawley 1990b; and see Pigott 1983). Invasion of dry sclerophyll forest in Victoria, Australia, by the polyphagous fungal pathogen *Phytophthora cinnamomi* led to massive die-back of the dominant *Eucalyptus* trees and understorey shrubs that initiated a protracted and continuing secondary succession. After 30 years, the density of *Eucalyptus* trees was still 50% lower than that on undiseased plots. Regeneration of the understorey had occurred in three stages. First, disease-resistance sedges and then seedlings of resistant opportunists rapidly colonized the ground left vacant by the destruction of the dominant understorey species *Xanthorrhoea australis* (Xanthorrheoaceae). Second, moderately susceptible species like *Banksia marginata* and *Grevillea steiglitziana* (both Proteaceae) resprouted from old stumps, and prostrate legumes increased in ground cover. In the third stage, some highly susceptible species including *Xanthorrhoea australis* began to regenerate from seed, but other species like *Isopogon ceratophyylus* (Proteaceae) had yet to show any signs of regeneration (Weste & Ashton 1994).

13.13 Summary

This chapter addresses the question of how feeding by herbivorous animals affects the distribution and abundance of plants. Because of the strongly non-linear relationship between plant recruitment and the level of herbivory, (Fig. 13.11) there is no necessary relationship between herbivore impacts on plant performance and their consequences on population and community dynamics. If herbivores *do* affect plant community dynamics, it is likely that their effects come about as a result of altering the outcome of an extreme contest competition. Herbivory can change the identity of the plant species that attains dominance, even when rates of defoliation are low and signs of herbivore damage on adult plants are scarce. There are strong and persistent asymmetries in herbivore impacts. For example, it appears that plants have more impact on the distribution and abundance of specialist herbivores than specialist herbivores have on the distribution and abundance of plants, and that generalist herbivores have more impact on the distribution and abundance of their preferred food plants than vice versa. Large mobile polyphagous herbivores (like ungulates) tend to have greater impact on plant populations than do small, sessile specialist herbivores (like leaf-feeding or sap-sucking insects). Ecosystem-level herbivore effects are more likely to be observed in low-productivity and open communities where plant recruitment is typically seed-limited than in more productive, closed communities where recruitment tends to be microsite-limited. There is still no clear consensus about the importance of herbivores in plant ecology, presumably because their impact differs so much from biome to biome and from community to community within each biome. An excellent flavour of the debate can be obtained by reading the exchange between Nelson Hairston and Andrew Sih in the 1991 *Bulletin of the Ecological Society of America* (Vol. 72, pp. 171–178). In order to resolve the debate we need to do lots more well-designed, properly replicated, long-term field studies, in which individual herbivore species are excluded in combination with seed-sowing and soil-disturbance treatments. Only then will we know the relative importance of seed limitation, microsite limitation and herbivory in plant recruitment.

14: The Structure of Plant Communities

Michael J. Crawley

14.1 Introduction

The stature, colour and texture of plants give landscape its unique character. As Darwin wrote, a 'traveller should be a botanist, for in all views plants form the chief embellishment'. The cast of the vegetation's features (its physiognomy) is determined by the size of the dominant plants (whether they are trees or bushes, herbs or mosses), by their spacing (whether they form continuous cover or are widely spaced-out) and by their seasonal prospect (whether the plants are deciduous or evergreen, and whether they undergo striking seasonal colour changes).

The first generation of plant ecologists (Warming 1909; Raunkiaer 1934) dedicated themselves to understanding why certain structures of vegetation are restricted to certain combinations of climate and soil. While these problems are far from resolved, modern plant community ecologists are occupied by questions involving species richness (why do so many (or so few) species of plants grow here?), species abundance (why is a single species dominant in one place, but many species codominant in another?), and patterns of spatial and temporal change (what determines the observed gradients in species composition we see as we climb up a mountain side, and what factors influence the succession of species we observe after disturbance?). The purpose of this chapter is to introduce the various structural attributes of plant communities, and Chapters 15 and 16 will consider the dynamics of how these patterns come about.

14.2 Definition of plant community

The plant 'community' is an abstraction of exactly the same kind as the 'population'; a community simply consists of all the plants occupying an area which an ecologist has circumscribed for the purposes of study. This definition draws attention immediately to the two key issues involved in studying plant communities: (i) how large should the area be and (ii) where, precisely, should the sample-area be put? These apparently trivial questions have fuelled the most heated controversy ever since the study of quantitative plant ecology began, towards the end of the nineteenth century (Goodall 1952). Failure to standardize the size of study areas and failure to agree on whether the areas should

[475]

be subjectively positioned in the most 'clearly typical' parts of the vegetation cover, or more objectively stationed by some form of randomization, have led to a great divergence of experimental approaches, and meant that clear comparisons between the findings of different studies are difficult or even impossible. Superimposed on these practical difficulties was a fundamental difference of opinion about the nature of the plant community itself. The two most polarized positions in this debate are represented by the views of the American ecologists Frederic Clements and H.A. Gleason. Clements believed that the plant community was a closely integrated system with numerous emergent properties, analogous to a 'super-organism'. In contrast, Gleason saw plant communities as random assemblages of adapted species exhibiting none of the properties of integrated organisms like homeostasis, repair and predictable development alleged by Clements.

14.2.1 Clements' view of community structure

Clements (1916, 1928) studied plant communities throughout North America and was struck by the vast extent of the rather uniform vegetation types he found. He called these 'climax communities', and proposed that the nature of the climax was determined principally by climate. Clements felt that the developmental study of vegetation rested on the assumption that the climax formation was an organic entity that arose, grew, matured and died. Each climax was seen as being able to reproduce itself, 'repeating with essential fidelity the stages of its development', so that the life history of a community was a complex, but definite and predictable process, comparable with the life history of an individual plant.

Clements recognized three major classes of climax vegetation in North America, grasslands, scrub and forests, and he subdivided each climax into a number of 'formations'. The grassland climax was divided into true grasslands (dominated by *Stipa* and *Bouteloua*) and sedgelands (*Carex* and *Poa*); the scrub climaxes into sagebrush (*Atriplex* (Chenopodiacae) and *Artemisia* (Asteraceae)), desert scrub (*Larrea* (Zygophyllaceae) and *Franseria* (Asteraceae)) and chaparral (*Quercus* (Fagaceae) and *Ceanothus* (Rhamnaceae)); the forest climaxes into woodland (*Pinus* and *Juniperus*), montane forest (*Pinus* and *Pseudotsuga*), coast forest (*Thuja* and *Tsuga*), subalpine forest (*Abies* and *Picea*), boreal forest (*Picea* and *Larix*), lake forest (*Pinus* and *Tsuga*), deciduous forest (*Quercus* and *Fagus*), isthmian forest and insular forest. Each formation was then further subdivided into 'associations' (today, these are simply called communities).

Clements has been accused of espousing a static plant community in which the climax was a fixed and spatially invariable organism, where individual plants were replaced faithfully by recruits of their own species. In fact, Clements was adamant that 'the most stable association is never in complete equilibrium, nor is it free from disturbed areas in which secondary succession is evident. Even when the final community seems most homogeneous and its factors uniform, quantitative study . . . reveals a swing of population and a variation in the controlling factors'.

Clements' view of succession was of a relatively orderly, predictable approach to a dynamic equilibrium. He identified six stages to the process: (i) nudation, (ii) migration, (iii) establishment, (iv) competition, (v) reaction and (vi) stabilization. Each of these might be 'successive or interacting', but the early processes tend to be successive and the later ones are more likely to be interacting (i.e. the frequency, strength and complexity of ecological interactions tend to increase through succession). Thus, in the course of development from bare substrate, through lichens and mosses to the final trees, there would be a series of recognizable (though ephemeral) communities before the climax was achieved (he called these 'seres'). He also distinguished clearly between primary successions, which were essentially soil-forming processes (and where there was no seed bank, and no reserve of vegetative propagules in the substrate at the outset), and secondary successions where the soil initially contained the propagules of many species characteristic of different stages of succession. Primary successions are dominated by the immigration of species from other areas, and occur on lava flows, dunes, rocks and in lakes, and tend to be associated with the accumulation of nutrients and organic matter in the soil. Secondary successions occur in fallow fields, drained areas, clear-cut forests, in the aftermath of fires, and so on. They involve only slight changes in soil (often nutrient *depletions*) and are less dependent on immigration of species.

14.2.2 Gleason's view of community structure

Gleason's name is associated with the 'individualistic concept' of plant community structure (Gleason, 1917, 1926, 1927). He freely admitted the *existence* of plant associations: 'we can walk over them, we can measure their extent, we can describe their structure in terms of their component species, we can correlate them with environment, we can frequently discover their past history and make inferences about their future'. In his view, however, recognizable plant communities owed their visible expression simply to the juxtaposition of individual plants, of the same or of different species, which may or may not interact directly with one another. The structure of the plant community was the result of continuously acting causes, chiefly 'migration and environmental selection', which operate independently in each place, and have 'no relation to the process on any other area; nor are they related to the vegetation of any other area, except as the latter may serve as a source of migrants or control the environment of the former'.

Where Gleason differed most from Clements was in his belief that the community 'is not an organism, but merely a coincidence'. He saw every species of plant as a law unto itself, and denied the emergent properties attributed to plant communities by Clements. He stressed the heterogeneity of community structure caused by accidents of seed dispersal, minor variations in environment, differences in the abundance of parent plants that could act as sources of seed, and the brevity of periods between disturbances.

Another fundamental difference from Clements' view was Gleason's denial of the determinism and directedness of succession. Gleason (following Cooper 1926) emphasized the random elements of seed immigration and seedling establishment, highlighted the different durations of similar successional stages in different places, and pointed out that different initial conditions and different histories of disturbance could lead to different end-points. (Clements stressed convergence to a single climatic climax.) Gleason also believed that disturbance was so frequent, and so patchily distributed in space, that most ecological communities exhibit transient dynamics and should not be regarded as equilibrium assemblages.

Gleason's views are best summarized by one of his own examples. Assume a series of artificial ponds has been created in farmland.

Annually the surrounding fields have been ineffectively planted with seeds of *Typha* and other wind-distributed hydrophytes, and in some of the new pools *Typha* seeds germinate at once. Water-loving birds bring various species to the other pools. Various sorts of accidents conspire to the planting of all of them. The result is that certain pools soon have a vegetation of *Typha latifolia*, others of *Typha angustifolia*, others of *Scirpus validus*; plants of *Iris versicolor* appear in one, and *Saggitaria* in another, of *Alisma* in a third, of *Juncus effusus* in a fourth. Only the chances of seed dispersal have determined the allocation of species to different pools, but in the course of three or four years each pool has a different appearance, although the environment, aside from the reaction of the various species, is precisely the same for each. Are we dealing here with several different associations, or with a single association, or with merely the embryonic stages of some future association? Under our view, these become merely academic questions, and any answer which may be suggested is equally academic (Gleason 1927).

14.2.3 The modern synthesis

The modern synthesis is very close to Gleason's view of community structures and dynamics (see Chapters 15 & 16, and the studies of recruitment in plant populations in Chapters 11 & 12). The issue is not whether there are identifiable (if vague) kinds of communities (no one seriously disputes this); the question is, to what extent do *biological* interactions (between one plant and another, between plants and their herbivores, or between herbivores and their natural enemies) influence community structure, compared with limitations imposed by the physical environment (abiotic conditions like soil, weather and exposure)?

Despite the severe criticisms levelled against most of the underlying assumptions of the climax concept (Cain 1947; Egler 1947; Mason 1947), the term 'climax community' is still quiet widely used. There may, indeed, be a place for a word to describe those communities that have been left alone long enough to pass through several generations of the dominant plants. Many ecologists, however, feel that the word 'climax' is so steeped in religious and ethical prejudices (continuous improvement, directed progress towards an ultimate goal, etc.), not to mention Freudian imagery, that its use is probably best avoided.

The legacy of Clements' 'super-organism' is to be found in Tansley's (1935) definition of the ecosystem: 'the whole system . . . including not only the organism-complex, but also the whole complex of physical factors. These ecosystems . . . form one category of the multitudinous physical systems of the universe, which range from the universe as a whole down to the atom'. It also provides the philosophy behind 'whole-systems ecology', as epitomized by the International Biological Programme of 1964–1974. This 'systems-thinking' has underpinned a good deal of the conservation ethic (Hardin 1968) and has inspired some terrible poetry (e.g. 'thou canst not stir a flower without troubling of a star'; Thompson 1918).

A denial of this holistic approach does not imply that ecological interactions are simple; far from it. There is nothing in the reductionist approach that precludes the possibility of plant communities displaying emergent properties. Indeed, there is a marriage to be achieved between the realistic aspects of the climax notion (context-specific interactions between species, multispecies effects, etc.) and the experimental approach of the individualists (Levins & Lewontin 1980). For a modern approach to the description of plant communities, the reader is referred to the five volumes of *British Plant Communities* (Rodwell 1991–1997), which describes the National Vegetation Classification (NVC) and provides an exemplary mix of history, biology, geology and human influence in the determination of botanical composition.

14.3 The niche concept

The niche is a multidimensional description of a species' resource needs, habitat requirements and environmental tolerances (Hutchinson 1957). For every vital attribute of a species' ecology it should be possible (at least in principle) to draw an axis to describe the range of possible values for the attribute, and then to plot the performance of the species at different positions along the axis. For example, a plant may not survive at very low or very high levels of soil moisture, but prospers at intermediate levels (Fig. 14.1).

The niche concept serves as a shorthand summary of a species' complex suite of ecological attributes, including its abiotic tolerances, its maximum relative growth rate, its phenology, its susceptibilities to

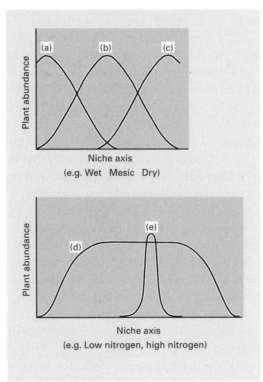

Fig. 14.1 The niche. Idealized relationship between plant abundance (cover, population density or biomass) and distance along a niche axis, of the kind that might be obtained from field transects. (a) A species with its peak abundance close to the minimum measured value. (b) Classic bell-shaped niche response curve showing suboptimal, optimal and superoptimal regions of niche space. (c) A species with its peak abundance close to the maximum measured value. (d) A species with a 'broad niche' growing abundantly over a wide range of niche conditions. (e) A species with a 'narrow niche' (a 'specialist') found only under a restricted range of conditions. It is important to distinguish between those broad-niched species that consist of rather uniform but highly plastic individuals and those comprising genetically polymorphic individuals, each having a rather narrow niche. Similarly, it is important to distinguish between plants where a narrow niche is caused by narrow tolerances of environmental factors and those with broader fundamental niches that are restricted to a narrow realized niche as a result of competition with other species.

various enemies and its relative competitive abilities with other plant species. Plant ecologists, however, have been rather slow to take this originally zoological definition of niche to their hearts. For whilst it is easy to see how resources might be divided up between animal species specializing on diets of different types (e.g. feeding on seeds of different sizes), it is much less obvious how plant resources like light, water or nitrogen could be apportioned between species. Habitat variables like soil moisture, pH or heavy metal concentration make obvious (and easily measurable) niche axes, and species attributes like temperature toler-ances, germination requirements, root depth, flowering phenology or disease resistance are amenable to straightforward interpretations. However, niche specialization in relation to plant resource partitioning is much less readily incorporated in Hutchinson's original model. In Chapter 8 Tilman presents an elegant solution to this problem by focusing on the performance of plants at different *ratios* of essential resources.

As every gardener knows, many plants can grow perfectly well under most soil conditions if they are freed from competition with the native vegetation by careful weeding. Of course the native vegetation, and the competitive milieu that it creates, is itself a prime component of the environment. The important point is not whether a species can grow in a particular soil, but whether it can hold its own on that soil in competition with the other members of the local flora (both as a seedling and as an established plant). Competitive ability, therefore, is not a species attribute, but depends on both the environmental conditions and the other plant species involved.

The distinction between what a plant could do on its own and what it actually does in the presence of other plants and natural enemies is formalized in the distinction between a species' *fundamental* and *realized* niches (Hutchinson 1957). One of the earliest experiments on realized plant niche was carried out by Ellenberg (1953). He grew five grass species in monoculture and monitored their performance in soils with different depths of water table. He found that all the species grown on their own did best at the same water table. He then grew the species in competition with each other at different water tables. Not one of the species did best at the same depth of water table in the presence of competitors as it had when growing on its own. This important result illustrates that the realized niches of these grasses were distinctly different from their fundamental niches. In general, the 'physiological optimum' of a species is not necessarily the same as its 'ecological optimum' (the position on the niche axis where the species achieves maximum abundance under field conditions; see Chapter 4). Ellenberg's rule states that plants are not found in the field under the conditions that greenhouse experiments would suggest are optimal for plant growth. Rather, species tend to reach peak abundance under field conditions in places where their competitors are less aggressive and their natural enemies less abundant. Thus, the realized niche is narrower than the fundamental niche, and the peak of the realized niche seldom occurs at the same position on the niche axis as the peak of the fundamental niche.

Niche breadth is the range of values along an axis at which the species can persist. Broad-niched (generalist) species can grow almost anywhere along the axis, while narrow-niched (specialist) species are restricted to a narrow band of values (Fig. 14.1). Species may be broad-niched in relation to some axes (e.g. soil pH) but narrow-niched in relation to others (e.g. germination temperature requirements). Whether a wide niche breadth derives from a high degree of phenotypic plasticity among rather uniform genotypes or from a high degree of genetic polymorphism in populations is an important question.

There are two main ways in which the niche concept is employed. The first is in classical 'autecological' studies like those reported in the *Biological Flora of the British Isles* (published in the *Journal of Ecology*). Individual accounts aim to explain the full range of biotic and abiotic conditions that influence the distribution, survival and reproduction of a particular species. A second use is in plant community ('synecological') studies, where certain niche parameters are fundamental to understanding community attributes like species richness and to explanations of such dynamic processes as successional change (Whittaker 1956). For instance, on a given length of niche axis, a community may contain a small number of broad-niched specialists or a large number of narrow-niched specialists. Species richness could be increased by: (i) increasing the length of the niche axis (so that more niches could be 'tacked on at the ends'); (ii) reducing the width of the existing niches; or (iii) increasing the degree of overlap tolerable between adjacent niches (so that more species could be packed into a fixed length of niche axis). Perhaps the most valuable aspect of the niche concept is that it requires us to distinguish clearly between the range of *conditions* under which a species can survive and the range of *resources* that it is capable of exploiting. These topics are expanded in Chapter 8.

14.4 Species richness

For every combination of soil, climate, altitude, slope and aspect there will be one species that grows better than any other, so that it produces more seeds or occupies more space by vegetative spread. In a spatially uniform and temporally constant world this single species would come to dominate the community to the exclusion of all others. This is known as the *competitive exclusion principle* or Gause's principle (see Box 14.1), and it leads us to the expectation that species richness should be uniformly low.

The reason that most communities contain so many species of plants is that the competitive exclusion principle simply doesn't work. Its logic is impeccable, but its assumptions are generally wrong. Environments are neither spatially uniform nor temporally constant; they are seasonal, spatially patchy, periodically disturbed and their constituent plant populations are subject to fluctuating competition from other species and variable levels of impact from herbivores, pathogens, pollinators and dispersal agents.

Explanations of species richness hinge on whether or not the community is in equilibrium. There are two extreme schools of thought on this, and the truth probably lies somewhere in between. The stochastic school believe that most communities exist in a state of non-equilibrium, where competitive exclusion is prevented by periodic population reductions and environmental fluctuations. The equilibrium school believe that coexistence is possible, even in uniform environments, if certain criteria are met (for example, the Lotka–Volterra

---

Box 14.1 Competitive exclusion principle.

The principle states that one species (the best competitor) will, given long enough, oust all others. This leads to the prediction that each separate community should contain just one species of each growth form. The logic is as follows. Two species must, by definition, be different. No matter how slight the difference, individuals of one species will produce more seeds (or occupy more microsites by vegetative spread) than individuals of the other. Since the total number of microsites is assumed to be limited, these differences, no matter how tiny, will eventually lead to the dominance of one species and the extinction of the other. For example, suppose 1000 seeds of two annual species (A and B) are sown each year; we start with an equal mixture of 500 A and 500 B. The seeds are harvested at the end of each growing season, mixed thoroughly and 1000 randomly selected seeds are replanted. This is repeated year after year. Even though species A has only a small advantage of species B, the second species is virtually extinct after 100 generations (Fig. 14.2).

More formally, let λ_A be the net reproductive rate of species A and λ_B of species B; we define $\lambda_A > \lambda_B$ by some arbitrarily small amount. At planting there are N_A seeds of species A and N_B of species B. Each species produces S_A and S_B seeds respectively, where $S_A = N_A\lambda_A$ and $S_B = N_B\lambda_B$ to give a total of $T = S_A + S_B$ seeds. The number of seeds of species A planted in the next generation is therefore:

$$N_{A(t+1)} = 1000\,(S_A/T)$$

and

$$N_{B(t+1)} = 1000 - N_{A(t+1)}$$

In the limit, N_A tends to 1000 and N_B tends to zero; this is the competitive exclusion principle.

Experimental tests have confirmed the model's major prediction. For example, Harlan and Martini (1938) sowed 11 varieties of barley in equal proportions, then gathered random samples of seed for resowing. Within 4–11 years, the less suitable varieties were completely eliminated at any given site.

---

competition criterion requires that intraspecific competition is more important than interspecific competition; see Begon *et al.* 1996).

Our task, in attempting to understand why there are so many species of plants, is to find what conditions are necessary and sufficient to invalidate the competitive exclusion principle. In doing so, we shall see why certain habitats are rich in species and others are relatively poor. I deal with each process in turn, but bear in mind that they are not mutually exclusive.

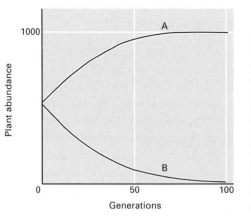

Fig. 14.2 Competitive exclusion. In a uniform environment one species will eventually exclude all others, no matter how small the differences in their reproductive output. The net reproductive rate of species A is only 5% greater than that of species B ($\lambda_A = 2.1$; $\lambda_B = 2.0$). By generation 100 species B is virtually extinct. The population is microsite-limited to a total of 1000 individuals in each generation.

14.4.1 Spatial heterogeneity

This hypothesis assumes that the competitive exclusion principle *does* operate, but that apparently uniform areas actually consist of many, subtly different microhabitats, in each of which one species excludes all others (see Chapter 8). This hypothesis is attractive, because spatial heterogeneity is so conspicuous in all real habitats (Ricklefs 1977; Grubb *et al.* 1982). However, as a model of species richness it is flawed, because by explaining everything, it explains nothing. It simply says that there must be as many different microhabitats (regeneration niches) as there are species in the community. Thus unless we know exactly what constitutes a suitable microhabitat for each species, and can measure the frequency with which different kinds of microhabitats occur, we cannot predict species richness or species' abundance distributions. A less stringent requirement is that the ranking of species' competitive abilities is different in different patches, so that each species is most competitive under certain *combinations* of conditions, but there is plenty of scope for stochastic efforts (a species finds itself, by chance alone, in a place where the superior competitors happen to be absent during the current generation).

14.4.2 Temporal variation

This model, first proposed by Chesson (1985), is a non-equilibrium explanation. The key is that all species must exhibit the ability to increase when rare (the *invasion criterion*; see Chapter 19). Populations decline during runs of 'bad years' but these declines are more than compensated by rapid increase during the occasional 'good years' for recruitment. Is the mere oscillation between good years and bad years sufficient for coexistence? The answer is no. The species that would be the inferior competitor under non-fluctuating conditions must do disproportionately well in its occasional good years, and these years must be bad for the superior competitor. Chesson calls this process *subadditivity* (Fig. 14.3). Also, the inferior competitor under non-fluctuating

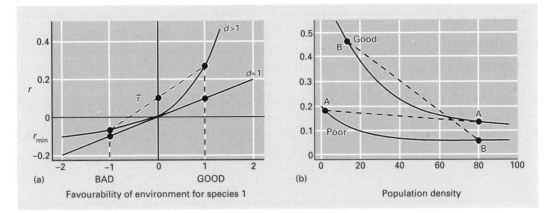

Fig. 14.3 Chesson's model for plant species richness in temporally fluctuating environments, showing the rate of increase, r, as a function of environmental conditions. The better the environment, the higher the value of r; in poor environments the population declines (negative values of r). (a) If, as the environment fluctuates from good ($+1$) to bad (-1), the species response is linear ($d = 1$) then fluctuation will not allow persistence because the average r value is zero (the plant will not be able to increase when rare). Persistence by this mechanism requires that the graph is accelerating ($d > 1$) and that the curve is asymptotic for very poor environments (r_{min}); this latter requirement is called the storage effect. (b) Subadditivity. The value of r declines as competition becomes more intense, but subadditivity means that the effect of competition on r is less in poor environments than in good environments. In the rare years that are good for the species that would be the inferior competitor at equilibrium (B), it must do disproportionately well (see text for details).

conditions must have sufficiently high survivorship (either as long-lived adults or as dormant seeds in the soil) to survive at least as long as the longest likely interval between two of its good years. Cheeson calls this the *storage effect*. Thus, coexistence by means of temporal variability is much more complex than coexistence by spatial heterogeneity, because it requires that three conditions are met simultaneously: there must be (i) niche differences between the species (their good and bad years must be different), (ii) sub-additivity, and (iii) storage (see Chesson 1985).

14.4.3 Competitive ability/dispersal trade-off

This model is based on migration between patches in a heterogeneous environment. Following Hutchinson's (1951) initial lead, a series of papers in the 1970s showed that coexistence of quite similar species is possible when, for example, the dispersal ability of the inferior competitor is sufficiently great that its rate of successful migration between patches is higher than the rate of extinction of populations within patches. The inferior competitor must be able to stay one jump ahead of the superior species (Levins & Culver 1971; Horn & MacArthur 1972; Levin 1974; Slatkin 1974). The limit to species richness is reached when migration rates become so great that the environment is no longer treated as patchy. It is patchiness of the environment rather than the number of different kinds of patch that allows coexistence to occur (Levin 1974).

Several different models have shown that there is no theoretical limit to the number of coexisting species, so long as the ranking of competitive ability is perfectly negatively correlated with the ranking of dispersal ability (i.e. the best competitor is the worst disperser and the best disperser is the worst competitor; see Skellam 1951; Crawley & May 1987; Nowak & May 1974; Tilman 1994; Crawley & Ree 1996). If this process is important, then the dominant species should be seed-limited, and the ruderal species should not be seed-limited.

14.4.4 Niche separation and resource partitioning

This is an equilibrium explanation of species richness, and assumes that the niches of coexisting plant species are sufficiently different that competitive exclusion does not occur. Species-rich communities may be composed of: (i) species with narrower niches; (ii) species with more broadly overlapping niches; (iii) habitats providing 'longer' niche axes; or (iv) a combination of these.

Just how different two species must be in order to allow them to coexist is a question of the greatest theoretical and practical interest. There is a substantial literature on the topic of the 'limiting similarity of species' (May & MacArthur 1972; Abrams 1976; Roughgarden 1976) but the main findings of the theoretical work are flawed by the limitations of the simple models on which they are based. For example, generalizations that are sound for one-dimensional niche separation break down for multidimensional niches. We do not yet have any simple generalizations about the limiting similarity of many dimensional niches. However, if plant communities *are* equilibrium assemblages structured by competition, then there will be rules for limiting similarity, and we shall need to work them out (Sugihara 1984).

Where the niche structure of complex plant communities has been investigated in some detail (e.g. limestone grasslands in northern England; Grime & Curtis 1976; Al-Mufti *et al.* 1977; Grime *et al.* 1981; Sydes & Grime 1984), niche differences have been found in germination behaviour, root depth, temperature thresholds, grazing tolerance, phenology (early growing versus late growing, early flowering versus late flowering) and many other factors. Whether or not these differences are necessary and sufficient to explain the maintenance of species richness remains to be discussed. It appears, however, that niche specificity is greater under more extreme environmental conditions. For example, plants of the extremely species-rich communities that occur on the ancient, nutrient-poor soils of Australia and South Africa show a much higher degree of soil specificity than species from more nutrient-rich communities (Shimda & Whittaker 1984). However, there is very little information to suggest that resource partitioning (in its original, zoological sense) is a significant process contributing to the maintenance of plant species diversity, at least in grasslands (Braakhekke 1980; but see Chapter 8).

14.4.5 Herbivory and the palatability/competitive ability trade-off

The effects of selective herbivory on plant species richness as discussed in detail in Chapter 13 and the trade-off between competitive ability and palatability are described in Chapter 4. Briefly, selective herbivory is a necessary but not a sufficient condition for increased plant species richness. Unless there is frequency dependence in herbivore attack or there is sufficient spatial randomness that the rare palatable species always has a refuge from herbivory, then competitive exclusion will not be prevented. Frequency-independent, selective herbivory could change the identity of the dominant plant species if it was the palatable plant species that was dominant in the absence of herbivory (see Chapter 13).

14.4.6 Disturbance

Disturbance of the soil surface and/or destruction of established plants may provide recruitment microsites, which allow the community to be invaded by new species, leading to increased species richness. At very high rates of disturbances, however, the pool of adapted species is small, so species richness would be low (e.g. on mobile sand or on frequently burned heathland). At very low rates of disturbance, competitive exclusion would occur, and species richness would again be low. Thus, species richness should be greatest at moderate levels of disturbance, because dominance is prevented, and the pool of potential colonists is relatively large (e.g. Armesto & Pickett 1985). This has been termed the *intermediate disturbance hypothesis* (Grime 1973; Connell 1979) and is another non-equilibrium explanation of species richness (and see Huston 1979; Underwood & Denley 1984).

14.4.7 Refuges

There is almost certainly some truth in *all* the foregoing models. We can be confident that species richness is enhanced whenever the inferior competitors have some kind of refuge in which their numbers can be maintained. The refuge acts as a source of propagules that are dispersed to those parts of the habitat where, in equilibrium, the species would be excluded.

Refuges may be provided by one or more of the following.
1 Tolerance of extreme conditions where no other species can survive (Proctor & Woodell 1975).
2 Spatial aggregation by the superior competitor which leaves a refuge of 'competitor-free space' (Atkinson & Shorrocks 1981; Pacala & Roughgarden 1982; Hanski 1983).

3 Existence of a long-lived seed bank (Grime 1979).

4 Herbivore resistance of one form or another (e.g. genetic polymorphism in chemical defences; see Chapters 5 & 10).

5 Seasonal effects whereby the inferior competitor grows much earlier (or much later) in the season than the superior competitors (Al-Mufti *et al.* 1977).

6 Long life and protracted reproduction of individual plants (Kelly 1985).

7 Good years for recruitment of one species may be bad years for another, and vice versa (Chapin & Shaver 1985).

8 Other aspects of niche differentiation (Grubb *et al.* 1982).

In any event, if communities really are equilibrium assemblages, then the maintenance of high species richness requires that the rare species obtain an advantage over common ones by: (i) the common species suffering density-dependent reductions in survival or fecundity (see Box 14.2); (ii) the rare species obtaining some frequency-dependent advantage in recruitment (see Chapters 12 & 15); or (iii) frequency-independent spatial stochasticity providing an ephemeral (but, over the long run, reliable) refuge from the superior competitor.

What we need is a theory of species richness that comes up with an answer like 20, rather than one (competitive excusion) or infinity (colonization/competition trade-off) as in present models.

14.4.8 Alpha, beta and gamma diversity

Whittaker (1975) distinguished three different kinds of species diversity that he called alpha, beta and gamma diversity. Alpha diversity refers to the number of species *within* a sample area (e.g. a single 20 × 50 m quadrat). Beta diversity describes the difference in species composition *between* two adjacent sample areas along a transect (e.g. beta diversity is low when the overlap between the species composition of two quadrats is high, and is highest when the samples have no species in common at all; see Fig. 14.4). Gamma diversity describes regional differences in species composition (e.g. the difference in botanical composition between comparable habitats on two adjacent mountain ranges; see Chapter 19). It is easy to see why different plants should be found on different substrates, and understanding beta diversity presents rather few problems. It is extremely difficult, on the other hand, to present a simple explanation of high alpha diversity (see Sections 14.4.1–14.4.7).

Areas with the very highest levels of species richness are those where there are many species per quadrat, but no two quadrats on a transect are alike in their species composition (i.e. those with high alpha, beta and gamma diversities; see Chapter 19). In regions of the lowest species richness there are few species in any sample, and the vegetation is spatially monotonous (low alpha and beta diversities).

Box 14.2 Density dependence.

What is density dependence?

The adjectival phrase 'density dependent' is applied to rates of birth, death, migration, growth and development. So, for example, when we say that a particular death rate is density dependent, we mean that the percentage of plants dying from this cause increases as population density rises. Thus, if 80% of the seedlings die when there are 10 m^{-2} but 95% die when there are 100 m^{-2}, then seedling death rate is density dependent.

Why does density dependence matter?

Populations without any density-dependent processes are doomed to extinction. Try the experiment of tossing a coin to generate random changes in population size. Start with, say, 64 plants and *double* population size when you throw a head, and *half* it when you throw a tail. If the population drops below 1 it is deemed to have gone extinct. Repeat the experiment several times. When population sizes are plotted against time, it will become clear that in very few cases do numbers remain close to the starting value of 64; they either go extinct or increase without bounds. This experiment has a very important hidden assumption. Since heads and tails are equally likely, then halfing and doubling are equally likely, so, on average, the population should stay at its starting value. We have to ask, however, what ecological process could ensure that births were exactly matched by deaths on average over many generations. The answer is that no density-independent process could achieve such a result, and the heads/tails model of population regulation is therefore fatally flawed. Even *with* the built-in assumption of balanced births and deaths, the populations still drift to extinction or increase to levels where competition (the archetypal density-dependent process) would occur. Density dependence matters, therefore, because without it populations could not persist.

What kinds of density dependence are there?

Most examples of density dependence are negative in their action. When we say, for example, that seed production is density dependent, we mean *negatively* density dependent (i.e. each plant produces *fewer* seeds at high population density than at low). In some cases, we may find positive density dependence (so-called 'Allee effects') where the number of seeds per plant *increases* with population density (though only at low densities). This might occur if pollination rates were

continued on page 490

Box 14.2 *Continued.*

extremely low when plants were widely spaced. Thus, 'density dependent' means negatively density dependent, and we use the explicit
phrase 'positively density dependent' to describe Allee effects. The
logistic model of population growth incorporates a very specific, simple
kind of density dependence that is linear, continuous and acts immediately (without a time-lag). More realistic models may need to include
density dependence that is strongly non-linear (it may not act at all
until very high densities are reached), intermittent (only acting in
certain seasons) or time-lagged (as when growth in one season influences seed production in the following year). These different kinds of
density dependence have very different implications for population
dynamics (see Chapter 12).

Key factors and regulating factors

A fundamental (and widely misunderstood) distinction is between 'key
factors' and 'regulating factors'. Key factors are those that are principally responsible for *fluctuations* in plant abundance. They tend to be
abiotic factors (like rainfall or fire), and cause substantial year-to-year
changes in population density. However, key factors are usually density
independent and, as we have seen, density-independent factors *cannot
regulate* population density.

Regulating factors are, by definition, density dependent. They may
not cause large amounts of mortality (or severe reductions in fecundity); all that is required is that they increase in intensity as plant
population density rises. Because regulating factors need not cause
massive mortality or dramatic reductions in growth, they are usually
difficult to detect and can easily be overlooked. For example, in a
population of desert annuals, the key factor influencing year-to-year
fluctuations in abundance might be early summer rainfall, whereas the
regulating factor might be density-dependent predation of seeds by
small mammals.

Detecting density dependence

The most widespread misconception about density dependence is seen
in the tendency to equate density dependence with intraspecific competition. This is encapsulated in the view that if the mature plants are
spaced so far apart that they could never compete with one another,
then density dependence cannot be operating. This overlooks two vital,
though elementary forms of density dependence. First, when plant
recruitment is microsite-limited (e.g. there is only a small number of

continued on page 491

Box 14.2 *Continued*.

suitably sized gaps in the plant canopy), then the rate of seed establishment will be density dependent. The larger the number of seeds produced, the lower the percentage of seeds that will become established as seedlings. Second, if the plants are subject to attack by mobile, polyphagous herbivores, then density-dependent seed predation can result when the animals aggregate in areas of relatively high plant density. In short, the fact that the mature plants are small or widely spaced out *does not* mean that the plant population is not regulated, nor that density-dependent processes are not operating.

Since density dependence is most likely to occur during the seed and seedling stages, it is almost always going to be difficult to detect, since these stages are the most difficult and laborious to study. Furthermore, density dependence is likely to vary spatially, within each generation. Thus, detailed life-table studies need to be performed in several places each year. Different rates of birth, death, growth and development will be detected in different places and some of these differences may be density dependent. When demographic rates are *averaged* over several patches, in order to produce a single figure for each generation, it may be impossible to detect the regulating factors.

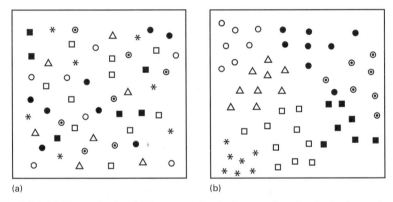

(a)   (b)

Fig. 14.4 (a) Fine-grained and (b) coarse-grained mixtures of species. In the fine-grained mixture, each individual is equally likely to have neighbours belonging to any of the other species, in proportion to their relative abundances. In the coarse-grained mixture, each individual is likely to have neighbours of its own species, despite the fact that species richness and relative abundance are the same in both cases. In (b), intraspecific competition is more important than interspecific competition.

14.4.9 Species richness in the Park Grass Experiment: a case study

One of the great difficulties that confronts anyone setting out on a study of plant species richness in a natural plant community is the lack of two key pieces of background information: (i) the *underlying heterogeneity* of the site (patchiness in soil conditions, microclimate, etc.); and (ii) the *history* of the site in terms of disturbance and episodes of plant recruitment. Ideally, we should like to work with experimentally manipulated communities that had developed from initially homogeneous conditions. Likewise, we should like the experiment to have been running for sufficiently long that the transient dynamics, which inevitably follow the imposition of novel experimental treatments, have had time to damp away. Such conditions are rare indeed, but the Park Grass Experiment at Rothamsted in south-east England fulfils both of them admirably Plate 12, facing p. 366. The experiment was started in 1856, 3 years before Darwin published *Origin of Species*. It was originally intended as a demonstration of the impact of different fertilizers on the yield of hay, but the botanical composition of the plots changed so rapidly that after only 6 years Gilbert and Lawes, the founders of the experiment, were writing that it looked more like an experiment on different seed mixtures than a fertilizer trial.

In 1995 there were 89 plots each of 200 m^2 or so, varying in species richness from more than 40 on the unfertilized control plots to virtual monocultures of *Holcus lanatus* or *Anthoxanthum odoratum* (both Poaceae) on unlimited plots that receive high rates of nitrogen (> 100 kg ha^{-1} year^{-1}) in the form of acidifying ammonium sulphate. There are two big patterns in the species richness data. First, species richness declines as biomass increases (Fig. 14.5c). This reflects a fundamental asymmetry, in that large plants are able to survive in phenotypically dwarfed forms on the low-biomass plots but small, low-growing plants have no chance of surviving in the dense shade cast on the high-biomass plots. Also, individual plants are larger on the high-biomass plots, with the consequence that there are fewer individuals per unit area, so it is a statistical inevitability that the number of plant species per unit area will be lower. Second, there are fewer plant species on the most acid plots (Fig. 14.5d). This probably reflects a 'pool-size effect' in that there are rather few plant species in this area of Hertfordshire adapted to soils of pH < 4 (toxic aluminium is soluble at this high level of soil acidity, earthworm activity is greatly reduced and symbionts like *Rhizobium* bacteria in legume nodules are unable to fix nitrogen). There is more to the story than this, and other factors have lesser, but significant, effects on plant species richness: the amount and type of nitrogen, and the application of phosphorus or potassium. The note-

Fig. 14.5 (*Opposite*.) Park Grass Experiment, Rothamsted. Times series of biomass measured as hay yield from two harvests (June and October) for two unlimed, fully fertilized plots: (a) Plot 9 has been acidified by the application of ammonium sulphate; (b) Plot 14 receives sodium nitrate and has not been aciified. Note that Plot 14 has a higher mean yield and that yield fluctuates much less than on Plot 9 (Silvertown *et al.* 1994). (c) Relationship between plant species richness and biomass, showing data from the plots that do not receive nitrogen fertilizer. (d) Relationship between plant species richness and soil pH for unlimed plots (unpublished data of Crawley & Silvertown).

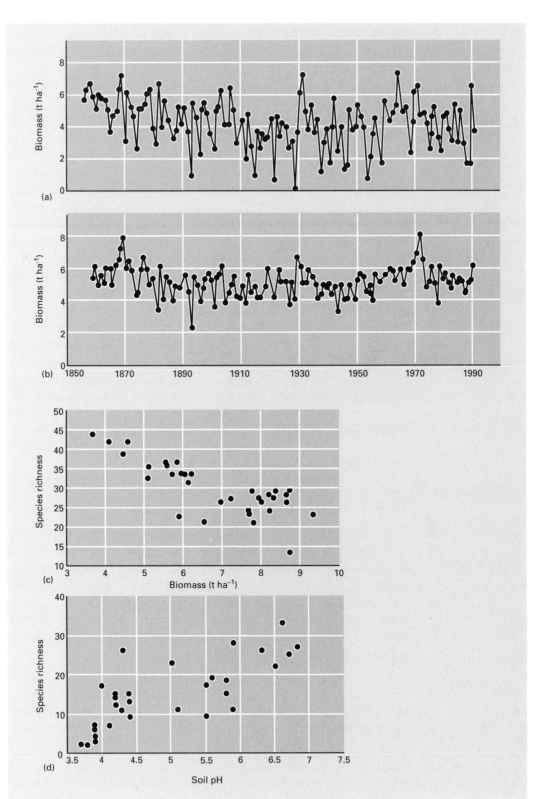

worthy feature of the Park Grass data is that *all* experimental nutrient inputs lead to reduced species richness (nitrogen has the biggest negative effect). Peak species richness is preserved only in the unfertilized control plots that have been limed to maintain their medieval soil pH (estimated to have been about pH 6.2). All the experimental inputs caused reduced species richness either through increased biomass, reduced soil pH, or as a result of more subtle changes that altered the outcome of interspecific plant competition.

In addition to their impact on species richness, nutrient additions affected plant diversity through a reduction in evenness. The limed control plots support a fine-grained mixture of small individual plants, and coarse-grainedness increases with fertility and soil acidity. Low evenness is most pronounced on the unlimed plots, where 140 years of increasingly acid rain (e.g. wet and dry deposition of sulphates) have led to increased dominance by a small number of grass species, and reduced relative abundance of many of the forb and legume species. Biomass variability on the Park Grass plots is closely related to soil pH and species richness (Dodd *et al.* 1995; Silvertown *et al.* 1994). Figure 14.5a,b contrasts two unlimed plots that both receive complete nutrients (N, P, K, Na, Mg) but obtain their nitrogen in different forms; Plot 9 receives acidifying ammonium sulphate while Plot 14 receives non-acidifying sodium nitrate. Plot 9 has a soil pH of 3.5 and supports only three plant species, while Plot 14 has a pH of 5.8 and supports 22 plant species. Biomass fluctuates much more on the species-poor Plot 9, probably because of its dominance by a shallow-rooted, drought-sensitive grass (*Anthoxanthum odoratum*).

14.5 Evenness and relative abundance

Two communities with exactly the same number of species, S, may differ from one another in many other ways, and it is important that we recognize how this may influence studies of species richness. The most important differences that affect the interpretation of species richness are: (i) the relative abundance of the different species; (ii) the degree of aggregation in the spatial distribution of each species; and (iii) the degree to which the spatial distributions of the different species are correlated with one another (see Fig. 14.6).

Fig. 14.6 (*Opposite.*) Species–area curves for communities of different structures.
(a) Fine-grained mixture with high equitability; the species–area curve follows the classic Arrhenius asymptotic form. (b) With the same species richness, but pronounced dominance, the curve rises more or less linearly as the rare species are added. (c) Coarse-grained mixtures produce stepped species–area curves. (d) When the patch sizes are very unequal, the shape of the curve depends on where the first quadrat is placed: (i) starting in a small patch; (ii) starting in a large patch and encountering another large patch at the first increase in quadrat size.

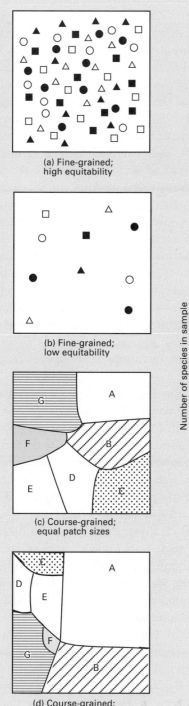

(a) Fine-grained;
high equitability

(b) Fine-grained;
low equitability

(c) Course-grained;
equal patch sizes

(d) Course-grained;
unequal patch sizes

Number of species in sample

Area sampled (A)

14.5.1 Species–area effects

Since the earliest studies of plant community structure it has been clear that the size of the quadrats employed was a vital consideration. The notion of 'minimal area' originated in floristic studies (Moravec 1973) and referred to the quadrat size that was likely to include some asymptotic percentage of the 'total flora' of the community (90% was a figure commonly employed; see Goodall 1952). Two thoroughly erroneous ideas are incorporated in this concept of minimal area: (i) that there is a fixed total flora for a community; and (ii) that there is a unique relationship between species richness and area. First, there is the tacit assumption that plant communities are made up from species that are essentially 'community faithful'. This is clearly not the case, and increasing quadrat size will almost always lead to increased species richness. Since the total cannot be known, then nor can some arbitrary fraction (say 90%) of the total be known. Second, the concept of minimal area ignores the important effects that the spatial patterns of species can have on the measurement of species richness based on frequency counts (see Box 14.3). We cannot use frequency data to estimate species richness unless we know both the degree of aggregation of each species and the extent of correlation between species' aggregation patterns (whether they tend to clump in the same places, or clumps of one species tend to preclude clumps of the other).

Plots of log species richness versus log area sampled ('species-area curves') are often presented as summaries of vegetation survey work, and both the slope and the intercept of such graphs have been imbued with great ecological significance (Goodall 1952; Connor & McCoy 1979; Grieg-Smith 1983). If species abundances can be described by the canonical log normal distribution of N individuals distributed between S species, and if N is directly proportional to the size of the sample area, A, then in the standard species–area relationship $S = cA^z$ the power z is roughly equal to 0.25 (May 1975). There is nothing mysterious, therefore, in the fact that most log–log plots of species richness versus area have slopes of about $1/4$ (Sugihara 1981). Williams (1995) advocates the use of the extreme value function (EVF) where $S = 1 - \exp(-\exp(c + zA))$ rather than the power function, but the interpretation of species–area curves is bound to be fraught with difficulties. For example, counts using political boundaries often give strange species–area curves, for the simple reason that large administrative areas tend to be large *because* they are barren (Crawley 1986; see Chapter 19).

Let us consider what shapes we would expect species–area curves to take under a variety of simple assumptions. First, let us assume that all the species are equally abundant and that they are independently randomly distributed. We call this a 'fine-grained' mixture of species (Fig. 14.4). In such a case, the number of species in a sample (S) increases with quadrat area (A, the *fraction* of the total area sampled)

Box 14.3 Quadrat size and plant frequency.

Data on plant frequency are extremely easy to gather. We simply put down a quadrat and score every plant species as either present or absent. No attempt is made to estimate plant density. The percentage of quadrats in which a species is found is called the frequency of that species. Common species will have frequency approaching 100%, and rare ones may have frequencies of 5% or less. Species with large individuals will register higher frequency scores than species at the same population density but with smaller individuals. It is important to understand how the measure of frequency depends on the size of the quadrats used in sampling. At one extreme, very large quadrats will contain the vast majority of species present in the community, no matter where the quadrat is placed. Therefore, with big quadrats, most species have frequency values approaching 100%. At the other extreme, a point quadrat (a long, metal pin with a sharp point) will only touch *one* species in a single canopy layer. The percentage of touches may give a good estimate of plant cover for the common species, but unless sampling effort is high only a small proportion of the plant species present in the community will be detected.

It is common practice to estimate species richness by counting the plant species in quadrats of a fixed size. The hope is that the quadrat is sufficiently large that most of the plant species will be found within it. The size of the quadrat that will contain 'most of the species' has been called the 'minimal area'. The lower the average plant population density, the larger this quadrat will need to be. This is not the only consideration, however, because the spatial pattern of the plants also influences the likelihood of their turning up in quadrats of a given size.

In the following graph we plot the probability of finding a plant in a quadrat for three different species. Each species had the *same* average population density. They differ only in the degree to which the individuals are aggregated in their spatial distributions.

For a quadrat of area A we are virtually certain to find a species whose individuals are distributed at random (i.e. A is a reasonable 'minimal area' in this case). If the individuals are densely clumped, however, we only have a 30% chance of finding a species in quadrats of this size, despite the fact that its mean density is the same. Indeed, at this level of aggregation, we would need a quadrat of size 100 (10 times the scale of this graph) to have better than a 50 : 50 chance of obtaining the aggregated plant species in a randomly placed quadrat. To have better than a 75% chance of finding it, the quadrat would have to have an area of 100 000 units.

The calculations for obtaining the graph are as follows. Assume we can describe the spatial pattern of the plant by a negative binomial

continued on page 498

Box 14.3 *Continued.*

distribution (Elliot 1977). The parameter k measures the degree of clumping; small values (like 0.1) mean tight clumping, while large values (like 10) mean the distribution of plants is random. In order to mark the species as 'present' we require one or more plants in the quadrat. The probability of getting one or more individuals is calculated most simply by finding the probability of getting *no* individuals (the so-called 'zero-term'), and taking this probability away from 1. The zero-term of the negative binomial distribution P_0 is:

$$P_0 = \left(\frac{k}{\mu + k}\right)^k$$

where μ is the mean number of plants in a quadrat of size A. Thus, the probability of 'presence' is $1 - P_0$. The graph was drawn for three values of k (10, 1 and 0.1) by working out $1 - P_0$ as the area of the quadrat was increased. Increasing the area increases the mean number of plants per quadrat (μ), and we write $\mu = NA$ where N is the population density of plants per unit area (of course, this is *not* affected by quadrat size or sampling method). For simplicity, N has been set to 1.0.

Because frequency depends so critically on quadrat size and on the size of individual plants and their spatial arrangement, the interpretation of plant frequency data requires considerable care (see Goddall 1952; Grieg-Smith 1983).

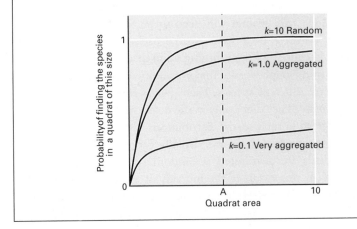

according to the equation:

$$S = S_{\max}(1 - (1 - A)^N) \tag{14.1}$$

where $S_{\max}$ is the maximum species richness (the total species pool) and N is the number of individuals of each species in the entire area (Fig. 14.6a; Arrhenius 1921). If the species are not equally abundant we must specify a statistical distribution to describe the relative abundances of the different species; it is traditional to use a truncated log normal distribution for this purpose. Now S depends on A in a manner influenced by the variance and skew of the distribution. In communities where species have more or less equal abundances (i.e. when 'equitability' is high and the N values are all roughly equal), the result is the same as in Fig. 14.6a. When the communities demonstrate strong dominance, however, the tail of the distribution is extremely long, with the consequence that species richness increases rather uniformly with quadrat size over an extended range of areas (Fig. 14.6b). The appropriate equation in this case is:

$$S = \sum_{i=1}^{S_{max}} (1 - (1 - A)^{N_i}) \qquad (14.2)$$

where the N_i values are distributed according to (say) a log normal distribution. If several equally abundant species are distributed in a 'coarse-grained' manner (i.e. in dense, non-overlapping aggregates), then the species–area curve will increase in a more or less linear, step-wise manner; species are simply added as more patches are included within a single quadrat (Fig. 14.6c). Where the patches are of very different sizes, the model is a little more complex, and the slope of the curve depends on whether the survey begins in a small or a large patch (Fig. 14.6d). In short, unless we know a great deal about the spatial distributions and relative abundances of the species, we can infer little from the species–area curve.

Similarly, on a much larger scale, comparisons of island or continental floras are confounded by processes affecting within- and between-area differences in species richness. For example, two regional floras with identical species richness will have very different species–area curves if one flora has high beta diversity and the other low. There are four basic configurations for such regional species–area curves:

1 alpha and beta diversity are both low (e.g. uniform, species-poor areas like central Australian spinifex);
2 alpha and beta diversity are both high (e.g. patchy, species-rich communities like tropical rain forests);
3 alpha diversity is high but beta diversity low (e.g. uniform but species-rich areas of goat-grazed Mediterranean maquis; and
4 alpha diversity is low but beta diversity high (e.g. some of the Cape heathlands of South Africa).

The slope of these species–area plots is proportional to the beta diversity, while the intercept reflects alpha diversity. An intriguing, but as yet unanswered, question concerns the ecological processes responsible for the apparent convergence in the number of species in fixed-area quadrats (e.g. 50×20 m) in comparable communities on different continents (Rice & Westoby 1983); i.e. what processes regulate alpha diversity?

14.5.2 Biogeography

The study of biogeography was revolutionized by the publication of MacArthur and Wilson's *equilibrium theory* in 1967. Botanists had known for years that distant islands supported fewer plant species than islands close to continents and also that, for a given degree of isolation, large islands supported more plant species than small ones. The genius of MacArthur and Wilson was in proposing a single, simple hypothesis to explain both these phenomena.

MacArthur and Wilson saw the number of species on an island not as a fixed, immutable quantity, but as a dynamic balance between gains and losses – as an equilibrium between immigrations and extinctions. They argued that the rate of immigration will be higher on near islands than on far, because of their proximity to continental 'sources' of potential immigrants. Further, they argued that the rate of extinction will be greater on small islands because population sizes will be smaller and the intensity of competition likely to be more intense.

The clever bit was to see that both immigration and extinction rates could be plotted against the number of species on the x-axis, using separate sets of curves for large islands or for small (Fig. 14.7a), and for near islands or for distant (Fig. 14.7b). The immigration curve falls because as more species become established fewer immigrants will belong to new species. The curve is likely to be concave, because the more rapidly dispersing species are established first and this causes a

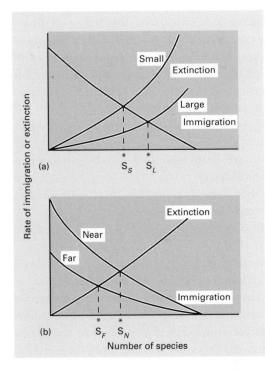

Fig. 14.7 MacArthur and Wilson's theory of island biogeography. (a) Effects of island size: larger islands (L) have higher equilibrium species richness (S^*) because the extinction rate is lower. (b) Effects of isolation: near islands (N) have higher equilibrium species richness because the immigration rate is higher. (From MacArthur & Wilson 1967.)

rapid drop in the overall immigration rate. The extinction curve rises with the number of species because the more species present the more there are to become extinct, and the more likely any one species is to become extinct because of reduced population size. The immigration and extinction curves intersect at the equilibrium species richness, where species gains are exactly balanced by losses. The equilibrium is high on large, near islands and low on small, distant ones.

A great many factors influence species richness on islands in addition to those considered by the equilibrium theory. For example, large islands usually have a greater variety of habitats than small ones, and most large islands have a wider range of altitudes than small ones (i.e. their beta diversity is higher). Species may also tend to become extinct faster on small islands because disturbances are more frequent or more devastating there (McGuiness 1984). Furthermore, the floras of many islands contain ancient, relic species. For instance, the island of Crete in the eastern Mediterranean contains about 10% endemic species, many of which were once distributed over much larger areas of southern Europe. These now persist in only a few rocky gorges where they have survived the changing climates of several ice ages, and escaped the attentions of the ubiquitous goats (Turrill 1929). In this case, the species survive on the island because of the *lower* rate of extinction there. The rapid spread of alien plants through modern plant communities (Crawley 1997) forcibly reminds us that species which grow together did not necessarily evolve together. The invasion of communities by preadapted species that evolved elsewhere may account for a considerable proportion of modern floras, compared with the number of species that actually coevolved with their current coinhabitants.

The value of the equilibrium theory is that it provides a baseline for explaining interesting patterns; the species–area curves that result can be used when comparing plant communities to highlight biologically interesting differences.

14.5.3 Species abundance distributions

For all the wealth of descriptive data on vegetation structure, there is still much to be learned about the factors controlling floristic composition, 'one of the most fascinating as well as urgent problems confronting the plant ecologist' (Watt 1971). The question of the determination of relative abundance is extremely complex, of course, for until we know what regulates the abundance of *single* species, it is unlikely that we shall be able to understand the determination of the relative abundance of all species.

Two communities composed of the same plant species can differ greatly in structure depending on their relative abundance distributions. Communities in which the species are all more or less equal in abundance are said to exhibit high 'evenness' (or high 'equitability'). Communities with one or two very abundant species and many rare ones are said to exhibit pronounced dominance (Fig. 14.8). The standard techniques for displaying relative abundance data are either to: (i) plot log abundance (measured as frequency, cover, biomass or primary productivity) ranked from the most to the least abundant species (Fig. 14.8a,b,c); or (ii) group the species into abundance 'octaves' (1, 2, 4, 8, 16, 32, 64, etc. individuals per species) and to plot the frequency of species in each class (Fig. 14.8d). The resulting histogram often approximates a truncated log normal distribution (Preston 1962; see May 1981 for details). There are many ways of characterizing the evenness of species abundance distributions (essentially the 'flatness' of the curves in Fig. 14.8a,b), but perhaps the best measure of dominance is simply the proportion of total abundance made up by the single most abundant species ($p_1 = n_1/T$, where n_1 is the abundance of the commonest species and T is the total abundance of all species; May 1981).

Since it only takes one species to fluctuate in order to alter relative abundance within the entire community, relative abundance is only as constant as the community's least constant species. Where relative abundance *is* observed to be constant, this may be due to low rates of turnover (e.g. individual plants are long-lived relative to the study period). Alternatively, relative abundance may be constant despite high rates of turnover of relatively short-lived individuals, in which case the ecological mechanisms responsible for stability are of considerable interest. What appear, superficially, to be rather stable abundance distributions may prove to be highly variable after long-term study. A classic example is provided by the Breckland plots studied by Watt (1981), in which a series of three different dominant species succeeded one another over the course of 44 years. Short-term study (3 or 4 years) would no doubt have concluded that any one of these three communities exhibited a rather stable relative abundance of species. However, the quadrats in this study were very small (only $10\,\mathrm{cm}^2$) and the results might be criticized as showing little more than the waxing and waning of individual genets of different plant species. Relative abundance studies should ideally be carried out in quadrats large enough to contain hundreds of genetic individuals.

Most relative abundance distributions are drawn up with data on mature plants. However, the structure of the distribution reflects the outcome of a *series* of population dynamics processes: seed immigration, germination, establishment, growth and mortality. Differential seed production between species, coupled with differential recruitment, growth and mortality, determine the final distribution of mature plants that we observe. The balance of relative performance may therefore fluctuate between species during the course of development; for example one species may have relatively high rates of seed immigration, another might have high germination, a third could exhibit high seedling survival, another show low rates of herbivore-induced mortality, a fifth might suffer less from shading as a mature plant, and so on.

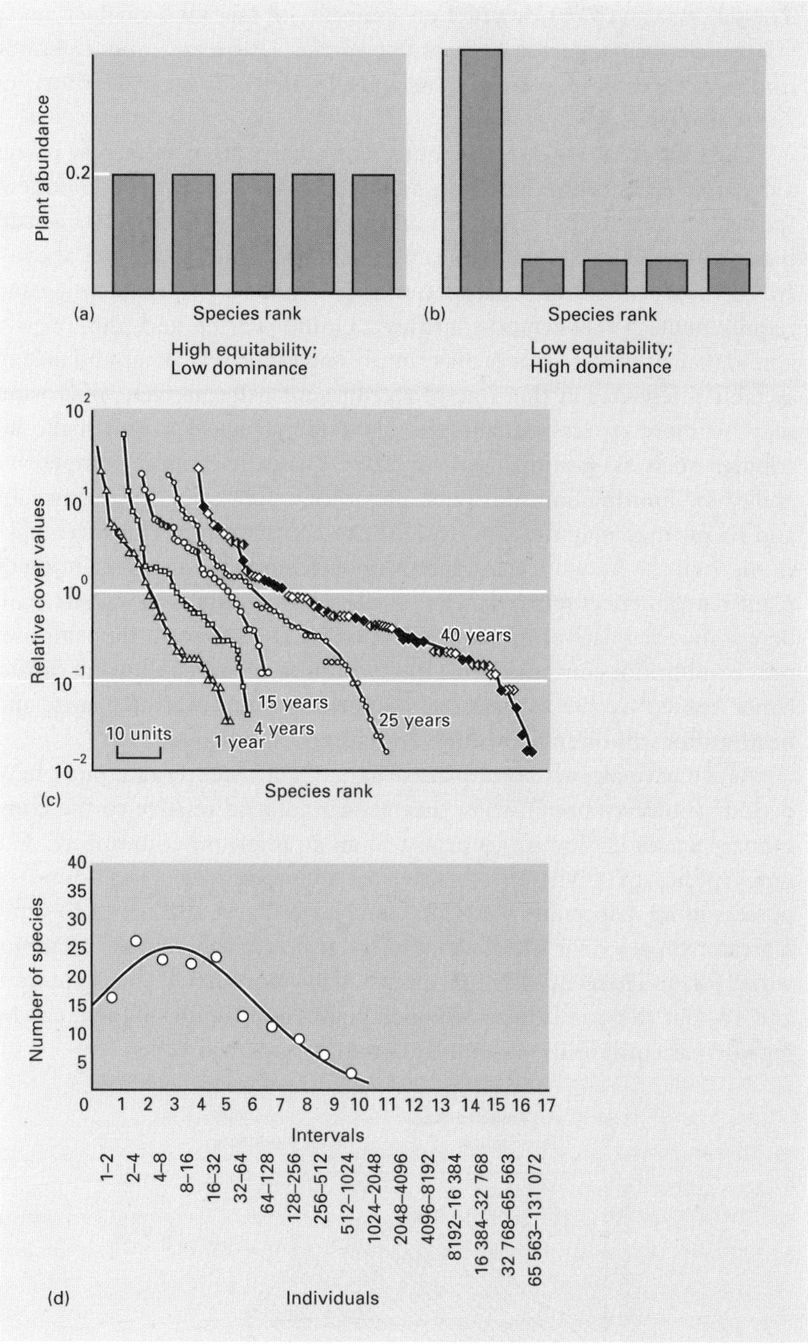

Fig. 14.8 (a,b) Equitability. Two hypothetical communities of five species differing in their degree of dominance. (c,d) Presentation of relative abundance data. (c) Rank abundance (or importance value) curves showing the relative abundance of plant species in old fields in five stages of abandonment in southern Illinois, USA, expressed as the percentage of total cover made up by each species, ranked from most abundant on the left, to least abundant on the right. Open symbols, herbs; half-open symbols, shrubs; closed symbols, trees. The slope of the rank abundance curve always decreases as species richness increases. (From Bazzaz 1975.) (d) Log normal distribution showing the frequency of species whose abundance lies in different 'octaves'. Relative abundance of diatom species in a sample taken from a stream in Pennsylvania, USA. (From Patrick 1973).

Torssell *et al.* (1975) describe an example of this kind of fluctuating relative performance for *Stylosanthes humilis* (Fabaceae) and *Digitaria ciliaris* (Poaceae) in annual grasslands in the Northern Territory of Australia (Fig. 14.9).

The long-term average distribution of relative abundance may reflect a dynamic equilibrium based on niche specialization of the component species, or it may be a non-equilibrium pattern reflecting the disturbance rate, species' longevities, site pre-emption, and so on (see above). In the first case, different species may have very specific microsite requirements ('regeneration niches'; Grubb 1977) and the relative abundance of mature plants may simply reflect the relative abundance of suitable microsites at the time of recruitment. Alternatively, microsites may be more or less equally abundant for all species, and niche attributes such as maximal growth rates, shade tolerance, competitive ability for limiting soil nutrients, phenology, longevity, pest resistance and so on may play the vital role. If the community is not in equilibrium, then the frequency and intensity of disturbance and the timing of plant reproduction relative to the availability of suitable microsites will determine the relative abundance of species. The species composition of non-equilibrium communities will tend not to recover following disturbance caused by the experimental addition or removal of plants, and nearby sites will often show different patterns of change.

As we have already seen, plants that are consistently rare must have periods (however brief) when they are advantaged relative to the commoner species if they are to persist in an equilibrium community. The rarer species of a Missouri prairie, for example, have been found to possess more dispersible seeds (Rabinowitz & Rapp 1981) and to show a greater capacity for interference in greenhouse experiments (Rabinowitz 1981). These matters are discussed in more detail in Chapters 8 and 15, but the question as to which plant communities, if any, can be regarded as equilibrium assemblages remains open to debate.

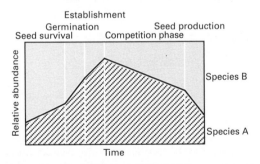

Fig. 14.9 Changes in relative abundance of two species from an annual pasture in a dry monsoonal climate in northern Australia. The legume *Stylosanthes humilis* (A) and the grass *Digitaria ciliaris* (B) show differing rates of loss in the seed, seedling, vegetative and reproductive stages of their life cycles. The legume shows higher rates of establishment, but the grass shows higher rates of seed production and seed survival. (After Torssell *et al.* 1975.)

14.6 Physical structure of plant communities

Plant communities exhibit a wide variety of three-dimensional structures and show spatial heterogeneity on many scales. The object of this section is to describe the kinds of vertical and horizontal structure that can be observed in different plant communities and to suggest how these patterns originate. The dynamics of the patterns are dealt with in Chapter 15.

14.6.1 Life-forms in plant communities

The great biomes of the world show a remarkable degree of *convergence* in their physiognomy, despite wide differences in the taxonomic affinities of their floras. Thus, experts aside, we would be hard-pressed to distinguish between the chaparral of California and Chile (Mooney 1977) or between the nutrient-poor heaths of South Africa and Australia (Milewski 1983). Arctic fell-fields look much like those in the Antarctic (Smith, R.I.L. 1984) and rain forest in Brazil resembles rain forest in South East Asia (Leigh 1975). This convergence of community physiognomy is vivid testimony of the importance of climatic factors as agents of natural selection. While there are numerous subtle differences in species richness and in the relative abundances of different forms, it is clear that there are broad trends in the dominant life-forms comprising plant communities under different climatic conditions.

Raunkiaer (1934) classified plants according to their adaptations for surviving the unfavourable season (winter cold or summer drought). These 'life-forms' (see Chapter 4) can be summarized into spectra describing the proportion of the total flora made up of different kinds of plants. Figure 14.10 shows the life-form spectra of five different climates, compared with the spectrum of the world's flora as a whole. Several points emerge from these comparisons.

1 Under ideal growth conditions (constant warmth and moisture) trees are dominant, simply because the competitive spoils go to the tallest individuals.

2 Where there is no unfavourable season, or the less favourable season is not too severe, then tree-like plants (phanerophytes) predominate in the flora as a whole.

3 In less equable climates, trees may still be the dominant plants in most communities, but the flora as a whole is made up predominantly of other life-forms (e.g. hemicryptophytes in northern temperate latitudes).

4 Where the summer is arid (as in deserts and Mediterranean climates), there is a preponderance of annual plants (which *avoid* drought by passing the dry period as dormant seeds) and geophytes (which avoid drought by die-back of their above-ground parts and survive by means of underground storage organs). Other desert plants that *tolerate* drought

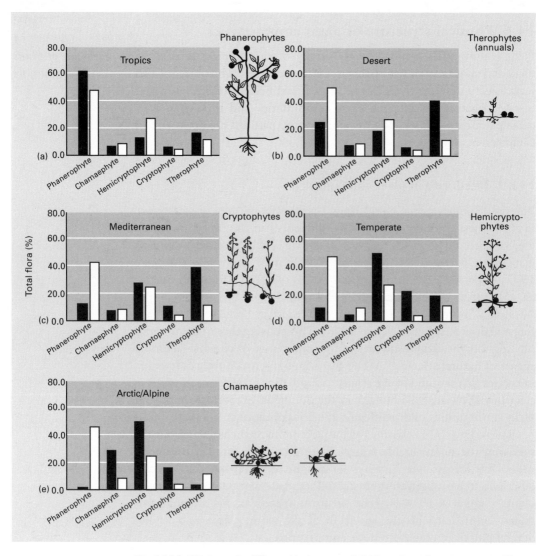

Fig. 14.10 Life-forms in different biome types. Solid bars show the percentages of five life-forms in a range of biomes. For comparison, the open bars show the composition of the world's flora as a whole (note that these figures are biased somewhat by the fact that most of the world's vascular plant species are tropical in origin). Each life-form reaches its peak representation in a different biome: (a) phanerophytes in the tropics; (b) annuals in deserts and Mediterranean ecosystems; (c) cryptophytes in Mediterranean and temperate systems; (d) hemicryptophytes in temperate ecosystems; and (e) chamaephytes in arctic and alpine habitats.

may appear to be dominant if the vegetation is surveyed during the dry season (e.g. xerophytic shrubs and stem succulents).

5 Where extreme cold and exposure characterize the unfavourable season, there is a shift towards cushion-forming plants and other species whose buds are held close to the soil surface (chamaephytes). Many arctic plants have their perennating buds protected by the dead leaf-bases of last year's shoots, because both the exposed aerial environment and the frozen soil are extremely inhospitable conditions for bud survival.

We should be careful not to read too much into the percentage make-up of entire floras. In terms of community dynamics it is much more important to know that trees dominate the vegetation than that hemicryptophytes are predominant in the flora. Again, the length of time a community had remained undisturbed and its history of human influence can have profound effects on the relative abundance of different life-forms (compare certain northern hemisphere Mediterranean communities, e.g. Californian chaparral, with their southern counterparts, e.g. South African fynbos; see Shimda & Whittaker 1984). Nevertheless, Raunkiaer's system does serve a valuable purpose in focusing our attention on the important (and as yet unanswered) question of why certain life-forms coexist in fairly predictable proportions in communities from quite different floristic regions.

14.6.2 Vertical structure of plant communities

The maximum height to which vegetation can grow is determined largely by climate. Wherever water is continuously available, and there is shelter from strong, drying winds, then tall trees are found. In the continuously warm, wet conditions of tropical rain forest, plant communities reach their pinnacle of vertical structuring, with a closed canopy at 25–45 m, a mosaic of light-gaps at differing heights, understorey species, regenerating canopy trees, vast buttressed trees emerging 20 m above the canopy top, festoons of lianas and epiphytes, and so on. Richards (1983) cautions against too broad an extrapolation of a multistrata, 'textbook' model of rain-forest structure. He points out that while single-dominant tropical rain forests do show clearly defined strata mixed forests usually do not, and that more important than the stratification of the trees is the boundary between the euphotic zone, in which the crowns are more or less fully exposed to sunlight, and the shaded zone (the undergrowth) beneath. Similarly, much of the vertical structuring is a result of gap formation and the dynamics of gap recolonization, rather than a consequence of the coexistence of specialized subcanopy or 'shrub layer' species beneath the crowns of the dominant trees (Clark & Clark 1992).

The world's tallest plants are found in the temperate rain forests of the west coast of North America, from northern California to the Olympic peninsula of Washington. Species like *Sequoia sempervirens* (Taxodiaceae) and *Pseudotsuga menziesii* (Pinaceae) reach extraordinary heights (over 80 m) and achieve massive girths (up to 20 m). Despite their great height, the vertical stratification of these communities does not approach the complexity of tropical rain forest. In areas of sporadic rainfall, the maximum height of the trees is reduced and the spacing between one tree and the next is increased. Vertical development of the woodland becomes progressively improverished as aridity increases, until the community passes into savannah with sparsely scattered, short trees surrounded by open grassland. Eventually, if rainfall is too low, fires too frequent or exposure too great, tree growth is precluded altogether. In such cases, vertical community development is restricted to dwarf woody shrubs (as in northern ericaceous heathlands), herbaceous or woody cushion plants (alpine and certain Arctic environments with prolonged snow cover), sclerophyllous scrub (in desert regions) or lichen heath (high latitude coastal communities). Another major factor limiting the vertical development of communities is feeding by vertebrate herbivores (see Chapter 13).

One of the most profound effects of vertical community structure is on total plant species richness. In tropical forests, for example, increased vertical complexity allows substantial increases in species richness by creating new habitats for plants like lianas (Putz 1984) and vascular epiphytes (Benzig 1983), growth forms that are rare or absent in other kinds of communities (see Chapter 4). The influence of vertical structure is also felt indirectly through its effects on microclimate (temperature and gas profiles, air-flow and turbulence, wind pollination; Wellington & Trimble 1984) and through effects on the structure of the animal community. There are close correlations, for example, between foliage height diversity and the number of bird species found in woodlands (MacArthur & MacArthur 1961) and between insect species richness and the vertical complexity of grasslands (Morris 1981).

14.6.3 Spatial structure of plant communities

Wherever we travel, we find that the environment alters from place to place, indeed often from step to step; correspondingly, the vegetation alters in both species composition and in the frequency of the component species. Figure 14.11 shows plant communities with two extreme spatial structures. In the first case there are distinct zones of different plant species, as we might see in the emergent vegetation surrounding shallow lakes. In the second, there are indistinct patterns, one grading imperceptibly into the other, as we might find in a transect up a forested mountainside. We need to determine the extent to which these spatial patterns in plant community structure reflect changes in abiotic conditions (soil nutrients, water depth, etc.) and to what extent the patterns are the result of biotic interactions (interspecific plant competition, seed dispersal, herbivorous animals, etc.). Also, to what extent do changing abiotic conditions alter the outcome of biotic interactions? For example, changing ratios of limiting soil nutrients may alter the relative competitive abilities of different plant species (see Chapter 8).

One of the most fundamental questions about the nature of plant communities concerns the way in which species respond to gradients in environmental conditions. Do whole recognizable *sets* of species appear

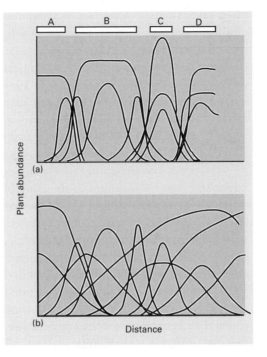

Fig. 14.11 Spatial structure of plant communities. Transects through two hypothetical plant communities: (a) where there are clear, discrete communities (A, B, C, D) separated by narrow intermediate zones known as ecotones (shown by the gaps between the open bars); (b) where there are no clear boundaries between communities, and plant species come and go along the transect more or less independently of one another. (After Whittaker 1975.)

and disappear at points along a gradient (as in Fig. 14.11a) or do species come and go more or less independently of one another along the gradient (Fig. 14.11b)? In the first case we would be in no doubt that there were distinct, recognizable communities. In the second case, one mix of species grades imperceptibly into another as we pass along the gradient; what we define as a community is therefore simply a matter of convenience (an identical community would not occur anywhere else). This distinction forms the basis for the long-running (and still unresolved) dispute between those who wish to classify and name discrete plant communities (the phytosociologists, e.g. Braun-Blanquet 1932; Becking 1957; Soó 1964–73) and those who see plant communities as continuously variable (the continuum school, e.g. Curtis & McIntosh 1951; Whittaker 1967). At this stage it is sufficient to note that, as in so many ecological disputes, it is the kind of system one studies that colours one's view of the ecological world as a whole.

In Europe, where humans have influenced all the plant communities for so long, and the scale of semi-natural vegetation is so limited, the attention of the ecologist has been focused on very abrupt changes in environmental conditions in very small patches of vegetation. For example, the sharp transition from a bracken-clad slope to a *Juncus*-dominated valley bottom, or the immediate change from heather moor to base-rich fen where a calcareous stream cuts through an acid peat-bog, define quite unequivocal plant communities where there is very little overlap in species composition between one and the other (Tansley 1939). In short, abrupt environmental transitions create distinctive plant communities. Even so, making these communities sufficiently uniform to permit the construction of a taxonomy requires the subjective elimination of a great many vegetation samples, including: (i) communities that are 'obviously unstable'; (ii) small fragments; (iii) ecotones; and (iv) mosaics (Poore 1955).

American ecologists, working on vast expanses of almost virgin forest, took a different view. They saw a world structured by rather smooth, gradual changes in environmental conditions. For example, the alluvial floodplain forest of the Mississippi stretches all the way from Minnesota to Louisiana; no ecologist would refer the forests of the upper and lower Mississippi to the same plant community, yet there is no place along the entire length where one can logically mark a boundary between them (Gleason 1926). In such a world it was impossible to erect a taxonomy of plant communities because no two sample areas contained the same species in the same relative abundances. These areas had to be described in terms of a continuum (Curtis 1959), and effort spent in trying to define fixed communities was effort essentially wasted (Davis 1981). Of course, the fact that vegetation changes as a continuum does not preclude our using sensible names to describe different communities. After all, electromagnetic radiation changes continuously in wavelength through the visible spectrum, but this does not stop us describing colours as red or blue. In the same way, the names we give to plant communities are labels of convenience rather than descriptions of rigidly structured, highly integrated, deterministic associations of plants.

Even where recognizable boundaries between communities do occur, their positions are not necessarily fixed. For example, Gleason (1927) describes the mosaic of prairie and alluvial floodplain forest found in Illinois. Prairie cannot invade forest because the majority of prairie species are intolerant of shade. On the other hand, the density of the prairie turf, the action of vertebrate herbivores and the high competitive ability of grasses for water preclude the invasion of prairie by tree seedlings. The comings and goings of forest and prairie therefore reflect the slow advance of the forest due to the shade cast by its leading, overhanging edge, abetted by the slow advance caused by erosion that gradually widens the river banks and makes more habitat available to forest. However, forest plants suffer much more severely from fire than do the hemicryptophytes of the prairie (see Chapter 4), so forest tends to be replaced by prairie in areas where fires are frequent.

Modern methods of multivariate analysis, ordination and classification are described by Gauch (1982). This proliferation of multivariate techniques for the analysis of spatial variation in plant community structure has not, however, led to great advances in our understanding of the processes underlying these patterns. As May (1985) observes, 'the wilderness of meticulous classification and ordination of plant communities, in which plant ecology has wandered for so long, began in the pursuit of answers to questions but then became an activity simply for

its own sake'. Generally, the analysis of multidimensional data sets will
be much more productive if it is carried out using *phylogenetically
independent contrasts* rather than multiple regression or standard multivariate techniques (Harvey & Pagel 1991; Harvey 1996).

[511]
*Chapter 14
Structure of Plant
Communities*

14.6.4 Allelopathy and spatial patterns

In addition to competing with one another through the exploitation of
shared resources or the physical pre-emption of space, plants may also
compete by chemical interference (Box 14.4). This is most likely to take
the form of suppression of the germination or growth of neighbouring
plants by secondary plant compounds (see Chapter 5), which are
leached, exuded or volatilized from the plant (so called 'allelochemicals'; see Rice 1984).

Some of the most pronounced patterns of zonation in California
annual grasslands are thought to be due to allelopathic interactions
between long-lived shrubs like *Salvia leucophylla* (Lamiaceae) and the
annual herbs and grasses (Muller & del Moral 1966). There is a bare
zone of up to 2 m around each *Salvia* plant, and a zone from 3 to 8 m
containing stunted plants of *Bromus mollis* (Poaceae), *Erodium cicutarium* (Geraniaceae) and *Festuca megalura* (Poaceae). Muller and del
Moral suggest that the zonation is initiated and perpetuated by volatile
terpenes, evolved into the air from the leaves of the *Salvia* during
periods of high temperature, adsorbed by the dry soil, and held there
until the early part of the following growing season. Seeds and seedlings
in contact with terpene-containing soils extract some of the terpenes by
solution in cutin, and the terpenes are transported into the cells via
phospholipids in the plasmodesmata.

As with all hypotheses of this kind, it is exceptionally difficult to
devise experiments which show unequivocally that the observed pattern
is due to allelopathic interference between plants, rather than due to
competition for water, light or nutrients, subtle soil interactions, or the
influence of herbivorous animals (Harper 1975). For example, Bartholomew (1970) suggested, on the basis of fencing experiments, that the
bare zones were caused by grazing animals rather than by allelochemicals. In one of the most thorough attempts to eliminate other possible
explanations, McPherson and Muller (1969) examined the causes of the
bare zones around the Californian shrub *Adenostoma fasciculatum* (Rosaceae). Fertilizer and shading experiments ruled out the likelihood of
competition for nutrients or light. Competition for water was dimissed
because *Adenostoma* has a very deep root system, and the phenology of
the excluded annual plants ensured that they were not subject to
drought (their germination was triggered by rainfall). *Adenostoma* produced no volatile terpenes, but toxins did leach from its leaves after each
rain. These accumulated in the soil and inhibited the germination of
annual plants. On fenced plots where the shrub was cut back to ground
level,more than 1000 seedlings m^{-2} appeared, compared with only

Box 14.4 Competition.

Competition is an interaction between two individual plants that reduces the fitness of one or both of them. Some ecologists insist that the word competition be reserved for mutually deleterious interactions ($-/-$) and that asymmetric interactions ($0/-$), where one party suffers little or no reduction in fitness, be called 'amensalisms'. In this book we use the one word, competition, to cover both kinds of interactions because: (i) a great many plant–plant interactions are asymmetric; and (ii) we often do not know the fitness implications of particular interactions in advance, so that calling them amensal rather begs the question.

Inter or intra?

Interactions between individuals of the *same* species are called intraspecific competition, while interactions between individuals of *different* species are called interspecific competition.

Interference or exploitation?

Exploitation occurs when plants compete by reducing the availability of shared, limiting resources such as light, water, soil nutrients or germination microsites. Exploitation may lead to more or less asymmetric competition (see below).

Interference occurs when fitness is reduced by essentially 'behavioural' mechanisms, which do not directly involve limiting, shared resources. For example, plants interfere when they compete via the production of allelopathic chemicals. Interference is usually strongly asymmetric.

Contest or scramble?

There is a continuum of exploitation competition from highly asymmetric (contest) to rather symmetric (scramble). In contest competition there are winners and losers; one individual gets all the resources it requires while another gets only the 'leftovers'. For example, plants contest for light, since the taller individual can pre-empt the incoming radiation. In scramble, the limited resources are divided equally between the competitors. Plants with similar root morphologies might scramble for soil nutrients. These two kinds of competition have different effects on population dynamics. If we plot the death rate (or some other measure of fitness loss) against the density of competing plants (on log axes), we obtain graphs of the form shown opposite.

continued on page 513

Box 14.4 *Continued.*

At low densities there is no competition at all, but above a certain threshold the rate of resource supply is insufficient for every plant to meet its full growth potential and competition begins. In scramble, all plants suffer equally, and in the limit no plants obtain sufficient resources to survive, so the death rate (k) is infinite (curve a). In pure contest, a *fixed number* of plants can be supported at a given rate of resource supply, so the death rate rises linearly with log density with a slope of 1.0 (curve b). Most real competitive interactions will fall between these extremes. We call the interactions scramble when their curves are close to (a) in shape, and contest when they are more like (b). In terms of dynamics, scramble competition is destabilizing compared with contest (see Begon *et al.* 1996, for details).

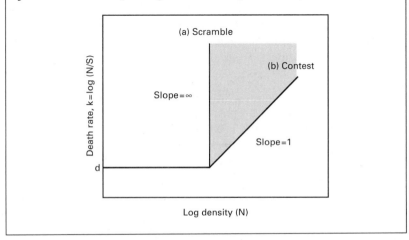

$40\,\text{m}^{-2}$ in the uncut controls. Small mammals did reduce recruitment beneath the shrubs (fencing increased seedling density from 40 to $70\,\text{m}^{-2}$), but the effect was small in comparison with the removal of the *Adenostoma* foliage and the consequent reduction in the input of toxic leachates.

The production of allelopathic compounds in living plants can be influenced by light quality (e.g. many plants produce less under glass), nutrient deficiency, drought, extremes of temperature, or by pathogen and herbivore attack (see Chapter 5). Allelochemicals are typically taken up by roots and translocated through xylem in the transpiration stream. In addition to volatilization, root exudation (Foy *et al.* 1971) and leaching from leaves (del Moral & Muller 1969), allelochemicals may move from plant to plant via natural root grafts (Graham & Bormann 1966), fungal bridges (Harley & Smith 1983) or via the haustoria of parasitic vascular plants (Rice 1984).

The modes of action of allelochemicals include:

1   inhibition of cell division in root meristems;

2   inhibition of oxidation of plant hormones like indoleacetic acid (IAA);

3   altered membrane permeability;

4   interference with mineral uptake (ash roots, for example, inhibit the uptake of labelled ^{32}P by oak roots);

5   reduced stomatal aperture leading to impaired photosynthetic rate;

6   interference with mitochondrial function;

7   reduced protein synthesis;

8   inhibition or stimulation of specific enzymes;

9   clogging of the xylem leading to impaired water balance;

10   various indirect effects, including inhibition of soil microorganisms, leading to reduced rates of decomposition, or altered chemical composition of the breakdown products (Rice 1984).

Subtle allelopathic effects are likely to occur in many (if not most) plant communities (e.g. Newman & Rovira 1975), and the experimental challenge lies in determining their importance in community dynamics in comparison with other ecological processes like exploitation competition, natural enemy attack, mutualistic relationships, and so on.

14.6.5 Quantitative methods for describing spatial patterns

Perhaps more energy has been devoted to this aspect of plant ecology than to any other (Goodall 1952, Pielou 1977; Grieg-Smith 1983). The spatial patterns described in these studies vary in scale from a few millimetres to many kilometres and are the result of very different ecological processes. On the smallest scales, the patterns reflect the interactions between individual plants (see Chapter 11) and are therefore the outcome (at least in part) of processes involved in plant population dynamics. Beyond the zone of immediate, first-order neighbours, direct plant–plant interactions are relatively unimportant (see Chapter 12), and pattern analyses on larger scales deal with either patterns of underlying soil conditions (water table, nutrient availability, heavy metal contamination) or the growth patterns of spreading, clonal plants (e.g. Kershaw 1963). Spatially variable biotic processes do occur on larger scales (e.g. the foraging of large vertebrate herbivores, see Chapter 13, or the activities of pollinating and seed-dispersing animals, see Chapter 6), but these are rarely addressed in traditional pattern analyses.

The desirable properties of a measure of plant spatial pattern include:

1   that the *degree* of non-randomness can be estimated (it is not enough simply to say that the pattern is not random);

2   that the measure is insensitive to the size of sampling unit chosen;

3   that the measure is easy to estimate from data and readily applied in theoretical models;

4   that the measure is not a function of plant density (so that density-independent mortality does not alter the estimate of pattern).

In fact, it is impossible to combine all these features in one measure. For instance, the parameter k of the negative binomial is a useful descriptor of the degree of aggregation, but the value of k is affected by density (see Elliot 1977 for a fully worked example). Similarly, quadrat methods are easy to apply in the field, but *all* fixed-area quadrats influence the kind of pattern they detect (see section 14.6.6). Lloyd's (1967) index of 'mean crowding' is useful because it is not sensitive to density-independent mortality, but it is cumbersome to use in theoretical models (Pielou 1977). Plotless (nearest-neighbour) methods do not suffer the disadvantages of fixed-area quadrats, but they only work well for plants with discrete individuals (like annuals and trees) and are difficult to incorporate in simple analytical models (Cottam & Curtis 1949; Goodall 1965; Byth & Ripley 1980). For a full discussion of the pros and cons of the different methods, see Grieg-Smith (1983) and Bartlett (1975). A full explanation of more advanced models of spatial patterning is given by Diggle (1983), and techniques for the statistical analysis of spatial data are clearly described by Upton and Fingleton (1985, 1989) and Cressie (1991).

Many of the early pattern analyses were attempts to correlate obvious surface features of the vegetation with underlying soil conditions (Knapp 1974). These studies suffered from the age-old problem that correlation does not imply causation, and it was always difficult to know whether or not the environmental factor they measured was the direct cause of the pattern. Many of the studies also failed to define a sensible null hypothesis; they did not ask what pattern would be expected if the hypothesized environmental factor were *not* operating. In other words, they had no 'null pattern' with which to compare their field measurements (Harvey *et al.*, 1983). All too often a random (Poisson) distribution was taken as the null hypothesis without any firm grounds for believing it to be appropriate. For example, soil conditions are unlikely to be uniform, with the consequence that plants are likely to be clumped in the 'better' patches. Similarly, plants might be aggregated due to limited seed dispersal around parents, clonal spread, and so on. Thus, finding that field data are not described by the Poisson distribution tells us nothing that we did not already know. On the other hand, tests of complete spatial randomness (CSR) can form a useful starting point, if only for the reason that when a pattern does not depart significantly from CSR, it scarcely merits any further formal statistical analysis (Diggle 1983). Modern methods of quantifying the degree of clumping in spatially explicit models are discussed below.

In discussing the use and usefulness of pattern analysis, it is well worth tracing the development of the controversy between those who believed that regular spacing in desert shrubs was indicative of competition for water (Woodell *et al.* 1969; Yeaton & Cody 1976) and those who challenged this view (Anderson 1971; Barbour 1973; Ebert & McMaster 1981). The acid test of this kind of hypothesis is a manipulative field experiment. For example, the water potentials of plants could be measured before and after their neighbours were removed. Such experiments demonstrate that some species (e.g. *Larrea tridentata* (Zygophyllaceae)) do compete for water and therefore benefit when their neighbours are cut down. Other species (e.g. *Ambrosia dumosa* (Asteraceae)) have only a brief growing season following rain, and water does not appear to be a limiting resource, despite the general aridity (Fonteyn & Mahall 1978, 1981). The important point is that it was the experiments and not the pattern analyses which provided the vital evidence for testing the hypothesis of competition between the individual shrubs.

14.6.6 Spatial patterns and quadrat size

Since we recognize spatial patterns in plant distribution on several natural scales (shoots within individuals, separate individuals, groups of individuals, etc.), it is not surprising that the size of quadrat we choose influences the kind of pattern we detect (Aberdeen 1958). For instance, if the quadrats are big enough for many individuals to fit within one quadrat, we should be able to detect clumping of individuals (caused, for example, by restricted seed dispersal about parent plants, or by underlying spatial heterogeneity in soils). If, on the other hand, the individual plants are about the same size as the quadrats (e.g. clonal plants of grasslands) then pattern analysis measures nothing more than the proportion of the habitat occupied (the analysis is essentially a description of frequency; see Box 14.3). If the quadrats are smaller than the individual plants (e.g. they may contain several to many shoots of the same clonal individual), then pattern analysis only tells us that shoots of the same individual tend to be clumped. This dependence of the pattern on quadrat size is illustrated in Fig. 14.12. The limitations of quadrat-based pattern analysis are so severe, and the alternative methods of nearest-neighbour analysis are so time-consuming and technically so difficult, that we should only undertake pattern analyses when they are absolutely necessary in order to answer very clearly formulated questions.

Quadrat size can also affect both the detection and the interpretation of interspecific association (Grieg-Smith 1983). If the quadrats are too small, then species will appear to be negatively associated simply because two individuals cannot occupy the same space. In medium-sized quadrats, large individuals can cause spurious positive associations between smaller plant species, since all smaller plants are constrained to grow in the gaps between the larger ones. Large quadrats may fail to detect genuine negative associations between species if the scale of spatial heterogeneity is small (i.e. one quadrat may contain several different patches, some with species A and some with B, even though A and B cannot coexist in any one kind of patch).

Quadrat size also influences the measurement of plant species rich-

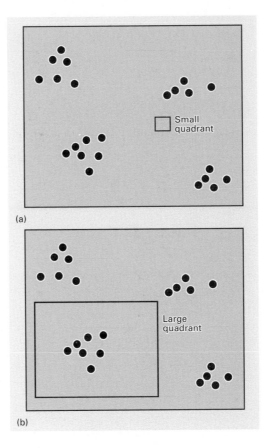

Fig. 14.12 Effect of quadrat size on pattern determination Varying the size of the sampling quadrat leads to interpretation of the same plant distribution in quite different ways: (a) small quadrats detect the pattern of plants as being aggregated; (b) large quadrats detect the same pattern as being regular.

ness (the standard quadrat area for work on vascular plant species richness is $1000 \, m^2$, but vegetation quadrats are typically much smaller than this, e.g. $16 \, m^2$; see above). There may be *apparent* changes in species richness when an experimental treatment leads to dramatic changes in the size of individual plants. For example, cessation of mowing leads to greatly increased size of a few individuals and thus to a reduction in the number of plants, and hence of species, found in small quadrats (as on Darwin's celebrated lawn (1859)). It is less widely appreciated that quadrat *shape* can also have a major impact on the assessment of plant species richness. Circular quadrats, with their minimal ratio of circumference to area, detect many fewer species than do rectangular quadrats and these, in turn, detect fewer than line transects (Fig. 14.13). The standard shape for plant species richness work is a rectangular quadrat with the long : short side ratio of $2.5 : 1$.

14.6.7 Spatial patterns reflecting temporal changes

A great many of the spatial patterns that we observe in plant communities today reflect processes of recovery from disturbances which occurred at different times in the past. Sometimes, these patterns are

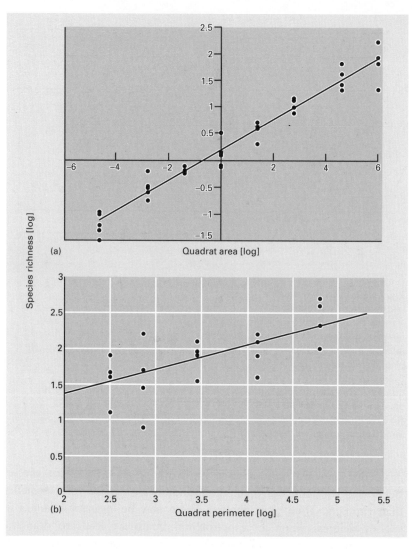

Fig. 14.13 Effects of quadrat size and shape on the measurement of species richness; data for tree species in Silwood Park. (a) A log–log plot (natural logs) of tree species richness versus quadrat area for quadrats of side 10 cm, 25 cm, 50 cm, 1 m, 2 m, 4 m, 10 m and 20 m. The slope of the regression line is 0.28 and the line accounts for 96% of the variance in ln(species richness). (b) For quadrats of a fixed area, increasing the length of the perimeter leads to the detection of greater species numbers. The data are for 9 m² quadrats measuring 3.75 × 2.4 m, 7.5 × 1.2 m, 15 × 0.6 m, 30 × 0.3 m and 60 × 0.15 m. The slope of the regression line is 0.34 and the line accounts for 54% of the variance in ln(species richness).

caused by localized, external forces like violent storms, landslides or lava-flows. When the disturbance does not kill the plants (as in many communities regularly affected by fire) the patterns may represent nothing more than stages in the *regrowth* of the dominant species. In other cases, the changes are driven by the life histories of the individual plants themselves, as they pass through their pioneer, building, mature and degenerate phases of growth (e.g. the 'stand cycles' described by Watt (1947, 1955) for *Calluna vulgaris* and *Pteridium aquilinum*). In these cases, there was a mosaic of patches, each of which cycled continually through a series of states, with adjacent patches cycling out of synchrony. Over a large area, a statistically stationary average is obtained, but local cycles persist indefinitely. Recently, the role of directional forces (e.g. the fact that the south-west-facing sector of a patch is sunnier than the north-east-facing sector) and memory (the history of long-lived individual plants) have been incorporated into mosaic models of forest dynamics (Remmert 1991; Wissel 1992; Hendry & McGlade 1995).

Aubreville (1971) suggested that spatial variation in species richness in tropical rain-forest trees is maintained by temporal variation in species composition at a given location. He saw the forest as a mosaic of patches at different stages in a cyclic succession. Hubbell and Foster (1986a,b) tested Aubreville's mosaic hypothesis of tropical regeneration for their Barro Colorado Island plot, and conclude that there is little evidence to support it. In very few cases was the probability of self-replacement significantly lower than of replacement by any other species. The *apparent* mosaic is simply a result of high tree species richness and the consequently low probability of replacement of a tree by an individual of its own kind.

The classic studies of succession (Cowles 1899; Clements 1928; Crocker & Major 1955) used observations on the structure of plant communities growing at different points in space (e.g. at different distances from the snout of a glacier, or from the front of the youngest sand dune) in order to *infer* the changes that occurred in plant communities through time. These, and other direct, long-term studies of successional changes, form the subject matter of the next section.

14.7 Succession

Succession is the process whereby one plant community changes into another. It involves the immigration and extinction of species, coupled with changes in the relative abundance of different plants. Succession represents community dynamics occurring on a time scale of the order of the lifespans of the dominant plants (in contrast to much slower, evolutionary changes, occurring over hundreds or thousands of generations, or the much more rapid seasonal or annual fluctuations in species' abundances). Succession occurs because, for each species, the probability of establishment changes through time, as both the abiotic environment (e.g. soil conditions, light intensity) and the biotic environment (e.g. the abundance of natural enemies, the nature and competitive ability of neighbouring plants) are altered. The degree to which these changes follow a predictable sequence determines the extent to which succession can be viewed as an orderly process. Some successions converge to rather uniform, predictable end-points, independent of the initial conditions. Others are non-converging or cyclic, or have alternative stable end-points, with their dynamics completely dominated by the history of disturbance and immigration. Thorough reviews of succession are to be found in Drury and Nisbet (1973), Connell and Slatyer (1977), Noble and Slatyer (1980) and Gray *et al.* (1986).

The great debate in the 1920s centred on whether the changes during succession were orderly and 'goal-directed' (headed towards an invariable climax community) or whether they were essentially random, with many possible end-points determined by initial conditions, accidents of immigration and by periodic, patchy episodes of disturbance (see above). The modern debate is no less intense, but concentrates on rather different processes. Nowadays, the role of chance in succession is universally admitted, and the central question concerns whether or not it is sensible to regard succession as a single process at all.

14.7.1 Interglacial cycles

As the polar ice-caps advanced and retreated during the Pleistocene, so individual plant species migrated southwards and northwards across the continents. As they migrated, so the composition of different plant communities altered (e.g. different North American tree species have advanced at average rates of between 100 and 400 m per year since the end of the last ice age; Davis 1981). During a 'typical' interglacial period we can recognize four broad phases of climate, soil and community development (Godwin 1975). Immediately following the retreat of the ice, there is an open community characterized by low temperatures and high exposure, supporting scattered arctic–alpine plants on soils of high pH. Next, as temperatures rise and leaching begins to reduce pH, a neutral grassland or scrub develops. By the height of the interglacial, weathering and leaching have further reduced the pH of the soil, and shade-casting, broad-leaved woodlands have developed on slightly acid, brown forest soils. As the ice readvances, declining temperatures coupled with continued leaching of already acid soils lead to the replacement of broad-leaved woodland by coniferous woodland and heathland on very acid, podsolized soils. Similar cycles can be reconstructed from fossil plant remains preserved in lake sediments throughout the north-temperate regions (Pennington 1969; Godwin 1975).

In regions that have not been glaciated at all during the Pleistocene (e.g. Australia), the process of leaching continues for millennia. Without the periodic input of nutrients that comes from glacial deposition, soils tend to be exceptionally low in nutrients such as phosphorus. These areas support unusual floras, and tend to undergo successions of a rather different kind (Noble & Slatyer 1980; Rice & Westoby 1983).

The impact of glacial cycles on the floristic composition of tropical plant communities is the subject of considerable debate (Sutton *et al.*

1983; Tallis 1991). It may be that the advancing ice sheets were a prime factor promoting the high species diversity of tropical rain-forest trees, because they forced large numbers of trees to coexist in small tropical refuges during the dry periods associated with the maximum advances of the glaciers. The trees may have evolved in isolation, only to be forced into cohabitation during later glacial episodes. In addition, it is likely that during longer periods of genetic isolation in fragmented forest refuges, speciation occurred that further increased species diversity. Far from representing long-term stable communities, therefore, it is possible that current tropical rain forests owe their high species richness to the fact that they have not remained stable long enough to cause the competitive exclusion of many of the tree species that were thrown together during the ice ages.

14.7.2 Primary succession

On 27 August 1883, telegrams reached Singapore from Jakarta, capital of the Dutch East Indies: 'During the night terrific detonations from Krakatoa (volcanic island, Straits of Sunda) audible as far as Soerakarta – ashes falling as far as Cheribon – flashes visible from here'. By the end of the eruption, only one-third of the entire island of Krakatau remained, and that was covered several metres deep in ash and pumice. New islands appeared where the sea had previously been 36 m deep. The explosion was heard 4652 km away and ash fell on ships 6076 km away. Giant waves, 40 m high, devastated everything in their path and caused the deaths of 36 417 people, washing away 165 coastal villages (Simkin & Fiske 1983).

No plant life at all survived the eruption, but the island was invaded by a series of plant species over the next 45 years (Fig 14.14; Docters van Leeuwen 1929). This kind of curve of species accumulation is

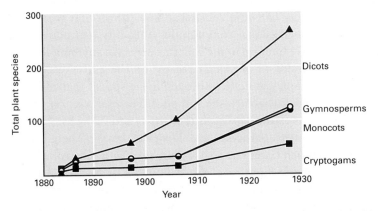

Fig. 14.14 Primary succession following the volcanic eruption that destroyed the island of Krakatau. Plant species richness increased rapidly over the first 40 years, associated with changes in the relative abundance of different life-forms (see text for details).

probably typical of the early stages of primary successions, and reflects a change in both species richness and life-form through time. The first colonists were cryptogams and the last were shade-loving forest species. No doubt the curve would reach an asymptote if followed for a longer period (Whittaker 1995).

The life history of the initial colonists of primary successions is determined largely by the nature of the substrate. On smooth bare rock, they will be lichens; on ribbed rock, bryophytes will appear first; on block scree the first plants might be trees and ferns; on gravel we find perennial herbs; on sandy substrates the colonists will be grasses, and so on (Grubb 1986). The dispersal capability and generation time of the plants alters the arrival sequence of species, and introduces a substantial random component into the succession (Fastie 1995).

Most primary successions (following glacial retreat, volcanic eruption, earthquake, sediment deposition or reduction of sea level) are associated with: (i) increasing soil nitrogen; and (ii) increasing height of the mature plants (leading to shading of low-growing species, Del Moral & Bliss 1993). The interaction between increasing shade and increasing soil nutrients often determines the order of species replacement. For example, the disappearance of heather *Calluna vulgaris* following heathland invasion by birch *Betula pubescens* can be almost wholly attributed to the increase in shade as the birch canopy closes. The grass *Deschampsia flexuosa* begins to increase as soon as the heather starts to decline, and eventually attains dominance of the ground layer under 35-year-old birch. As soil nutrient levels increase, so the competitive ability of *Deschampsia flexuosa* declines relative to *Agrostis capillaris* (Poaceae), with the result that under 45-year-old birch *Agrostis* begins to replace *Deschampsia* as the dominant (Hester 1988).

Perhaps the most important process is the accumulation of nitrogen in the soil. Mature ecosystems support soil nitrogen pools of between 5000 and 10 000 kg ha^{-1} in the topsoil (Table 14.1). During primary succession, this pool must be built up virtually from scratch. Experiments on a variety of virgin substrates have shown that woody plants cannot invade successional communities until nitrogen pools of 400–1200 kg ha^{-1} have built up in the soil; this may take anything from 20 to 100 years or more (Table 14.2). Many of the early vascular plant colonists are legumes and other nitrogen-fixing species that add substantial quantities of nitrogen to the soil pool. For example, on china clay waste, tree lupins *Lupinus arboreus* (Fabaceae) add 72 kg ha^{-1} year^{-1} and gorse *Ulex europaeus* (Fabaceae) add 27 kg ha^{-1} year^{-1}. These plants add nitrogen at a much faster rate than occurs via weathering (often negligible) or atmospheric input from rain (about 15 kg ha^{-1} year^{-1} (Marrs *et al.* 1983) but much higher than this in areas of intensive modern agriculture (e.g. The Netherlands) where atmospheric nitrogen input can exceed 75 kg ha^{-1} year^{-1}).

In a mature temperate ecosystem with a productivity of 10 t ha^{-1}

Table 14.1 Organic carbon and nitrogen storage in ecosystems classified according to Holdridge's system of life zones. (From Post *et al.* 1985.)

Life zone	Carbon (kg m^{-3})	Nitrogen (g m^{-3})	C/N ratio
Moist tundra	10.9	638.5	18.3
Wet tundra	20.7	1251.3	18.4
Rain tundra	36.6	2226.0	15.6
Boreal dry bush	10.2	631.0	16.0
Boreal moist forest	15.5	1034.1	16.0
Boreal wet forest	15.0	980.1	16.9
Boreal rain forest	32.2	1512.2	25.8
Cool temperate desert bush	9.9	779.8	12.5
Cool temperate steppe	13.3	1032.2	15.1
Cool temperate moist forest	12.0	626.1	22.5
Cool temperate wet forest	17.5	930.6	20.7
Cool temperate rain forest	20.3	1210.3	15.7
Warm temperate desert	1.4	106.5	15.3
Warm temperate thorn steppe	7.6	538.0	15.9
Warm temperate dry forest	8.3	644.8	14.0
Warm temperate moist forest	9.3	648.3	25.1
Warm temperate wet forest	26.8	2806.6	15.6
Subtropical desert bush	29.1	2282.6	12.7
Subtropical thorn woodland	5.4	379.3	14.3
Subtropical dry forest	11.5	1070.4	10.2
Subtropical moist forest	9.2	987.9	10.3
Subtropical wet forest	9.4	2853.4	3.3
Tropical thorn woodland	2.6	264.6	9.2
Tropical very dry forest	6.9	597.2	13.7
Tropical dry forest	10.2	885.9	13.3
Tropical moist forest	11.4	802.9	14.9
Tropical wet forest	15.0	655.2	30.2

year^{-1} (of which 1.5% is nitrogen), there must be an annual nitrogen uptake of about 150 kg ha^{-1}. This nitrogen supply is achieved by mineralization of the capital accumulated in the soil organic matter, and the rate of uptake is limited both by the size of the pool and by the rate of mineralization. It appears that young ecosystems function on a greatly reduced soil capital by virtue of high rates of mineralization coupled with a preponderance of nitrogen-fixing species (Marrs *et al.* 1983; Vitousek & Matson 1984).

14.7.3 Secondary succession

Secondary successions begin with a more or less mature soil containing a sizeable bank of seeds and vegetative propagules. One school of thought holds that secondary succession is nothing more than the expression of the life histories of the species already present at the outset. Thus short-lived plants with high relative growth rates flourish and achieve dominance first, followed by herbaceous perennials, then by rapidly growing, short-lived trees, followed eventually by slow-growing,

Table 14.2 Comparison of nitrogen content and compartmentation between major ecosystem pools in naturally colonized and reclaimed china clay wastes. n.d., not detectable. (From Marrs *et al.* 1983.)

Ecosystem	Age (years)	Total N (kg ha^{-1})	% Shoots	% Roots	% Litter	% Soil
Naturally colonized china clay wastes						
Pioneer						
Lupinus arboreus	16–18	291	37	1	6	56
Calluna vulgaris/Ulex europaeus	30–55	823	13	5	3	79
Intermediate						
Salix atrocinerea	40–76	981	8	18	6	68
Woodland						
Betula pendula/ Rhododendron ponticum/ Quercus robur	40–116	1770	30	3	0	67
Reclaimed china clay wastes						
Sand tips	3–84	211	11	59	n.d.	30
Mica dam walls	3–84	441	8	61	n.d.	31

long-lived trees. The observed 'succession' of species is nothing more than the replacement of small, short-lived plants by large, long-lived ones. This is known as the *initial floristic composition* model. At the other extreme is the model of *relay floristics*, which emphasizes a more or less strict sequence of plant species and stresses the importance of 'facilitation' (Egler 1954).

Facilitation is the process whereby species A paves the way for species B; the community cannot be invaded by species B until environmental conditions have been altered by the activities of species A. The most obvious examples of facilitation come from primary successions; for example, on freshly exposed glacial debris, mosses facilitate the invasion of grasses by providing them with roothold, shelter and moisture (Smith, R.I.L. 1984). On a rather different note, ericaceous plants like *Vaccinium* (Ericaceae) cannot enter a postglacial succession until substrate pH has been reduced from 8 to about 4.5 (Crocker & Major 1955). Rather more subtle examples occur when immigration of seed depends on the services of dispersing animals; for instance, primary successions on bare chalk are not invaded by bird-dispersed shrubs like *Crataegus* (Rosaceae) until woody plants like *Betula* have grown to sufficient height to provide perches for the birds (Finegan 1984). Whether these last two processes represent facilitation or simply 'enablement' is a moot point. In any case, it seems clear that genuine *interspecific* facilitation is the exception rather than the rule. Most examples of facilitation are a consequence of community-level processes affecting soils (increased nitrogen and organic matter, reduced pH) or light availability (increased shade), and are not the consequence of specific interactions between one plant species and another.

Between the extremes of initial floristic composition and relay floristics are the *tolerance* and *inhibition* models (Connell & Slatyer 1977). The tolerance model assumes that succession progresses by the replacement of early, fast-growing species by plants capable of regenerating in the conditions of depleted light and nutrient resources that these early species create (see Chapters 8 & 15). In contrast, the inhibition model assumes that early species simply inhibit the establishment of later species by site pre-emption. The longer-lived the initial colonists, the more slowly the succession proceeds. For instance, in South East Asia certain tropical forest successions following storm damage get 'stuck' after the development of a dense cover of ferns, and there is little change in the vegetation for a protracted period until the ferns are eventually shaded out (Whitmore 1985).

Other models of succession stress the importance of the proximity of seed parents, and the fact that immigrant species differ greatly from place to place, depending on what particular mature plants happen to grow nearby (see Chapter 15). Noble and Slatyer (1980) consider that certain 'vital attributes' of species determine their place in succession. These include their mode of arrival at a site, their persistence and their ability to establish at different stages of succession. These, in turn, can be related to the duration of their dormancy in the seed bank, their age at first fruiting and their longevity as adults (see also van der Valk 1981; Hils & Vankat 1982; Pickett 1982). In all secondary successions, the precise timing of the disturbance that initiates the succession is vital in determining the sequence of events that follows. For example, Keever's (1950) classic study showed the importance of the phenology of germination and early growth in relation to the time of year at which old fields are abandoned, in determining which plant species would dominate first, and how long that dominance would last.

Several quantitative descriptions of secondary succession have been attempted. The earliest models used multispecies versions of the Lotka–Volterra equations to address the question of what makes communities invasible (see Crawley 1986). Other models involved drawing up descriptive matrices of 'replacement probability' for individuals of each species in the community, based on the distribution of saplings of different species beneath the canopies of mature individuals of each tree species in the forest (Horn 1976; Usher 1986). More recent work uses spatially explicit simulation models to predict the fate of small patches of ground ('cells' the size of individual trees) as a function of the details of their neighbourhoods. These computationally intensive models involve explicit treatment of the birth, growth, reproductive performance and death of individual plants and take account of the limited dispersal of seeds around parent plants (see Chapter 15).

14.8 Models of spatial dynamics

The traditional approach in plant ecology was to view spatial patterns as a reflection of underlying heterogeneity in soil conditions and microclimate. The more exciting developments in recent years have concerned the study of pattern and non-randomness as emergent phenomena created by the interaction of ordering mechanisms and randomizing processes. The challenge is to work out the mechanism from an observation of the patterns, and to predict the patterns from the operation of known processes. What is becoming increasingly clear is that pattern and function are so tightly coupled that they need to be considered simultaneously.

The discovery of chaos has been one of the greatest intellectual revolutions of recent years (Lorenz 1963; May 1974; May & Oster 1976). The fact that chaotic systems show extreme sensitivity to initial conditions means that dynamic trajectories which start close together will diverge at an increasing rate as time goes by. Although chaos is deterministic, the long-term future of a chaotic system cannot be predicted because initial conditions can never be determined with sufficient precision, no matter how much is spent on data collection. While there is nothing to suggest that single-species plant populations show chaotic dynamics (they exhibit the wrong sort of density dependence, i.e. contest rather than overcompensating scramble competition; Watkinson 1980; Rees & Crawley 1991), there is ample scope for chaotic dynamics to occur in high-dimensional systems involving feedback between many interacting species with size, age and spatial structuring (Ruxton 1995). In the field, however, local extinction of low-density local populations and dispersal between asymmetrically fluctuating local populations will tend to reduce the incidence of chaotic dynamics (Hastings 1993), and unequivocal demonstrations of chaotic dynamics from field data would require such long time series for the calculation of their Lyapunov exponents that they are unlikely in the near future (Hastings *et al.* 1993, but see Hanski *et al.* 1993; Ellner & Turchin 1995).

In the past, it was believed that the complex behaviour of ecosystems required the invocation of complex external causes like environmental stochasticity, time-lags and age structure. Now we have the opposite problem; very simple chaotic systems produce intrinsically complex and intricate dynamics and spatial patterns (May & Oster 1976; Green 1991, May & Watts 1992). The notion that all ecological processes could be predicted, if only we had sufficient data, has been destroyed for ever. The behaviour of ecosystems is not predictable, even in principle. So why do we go on modelling?

Hand in hand with the discovery of chaos (the emergence of complexity from simple processes) has been the realization that complex mechanisms can produce simple ordered structures at a larger scale. The study of these emergent patterns has been greatly facilitated by the development of computationally intensive but conceptually very simple computer models for spatial dynamics. Broadly, the inclusion of space in models of population dynamics increases stability and fosters coexistence

[527]
Chapter 14
Structure of Plant
Communities

by allowing the possibility for the creation of refugia in which plants avoid consumption by their herbivores or suppression by superior competitors (see above). The different kinds of spatial models are introduced in the following sections and developed in Chapter 15.

14.8.1 Metapopulation models

The original metapopulation model was analysed by Levins & Culver (1971). Their great contribution was to realize that, instead of modelling the dynamics of the number of individuals in a large population, it was possible to gain important insights by considering the proportion of patches (or islands) that were occupied by the population and ignoring the details of precisely how many individuals occupied each patch. The important variables are the fraction of patches occupied, p, and the fraction of patches available for occupation, $(1-p)$. The occupied patches are assumed to produce migrants at a constant rate m, the migrants suffer mortality (the cost of migration) and the survivors settle at random over occupied and unoccupied patches alike. Occupied patches go extinct at a constant rate e. The rate of appearance of new occupied patches is proportional to the product of the number of migrants (mp) and the density of empty patches $(1-p)$. Thus:

$$\frac{dp}{dt} = -ep + mp(1-p) \qquad (14.3)$$

At equilibrium, the proportion of occupied patches is constant and the positive and negative terms on the right-hand side of Equation 14.3 are equal. Dividing both sides by p and rearranging gives the equilibrium proportion of occupied patches as

$$p^* = 1 - \frac{e}{m} \qquad (14.4)$$

The important point about this result is the existence of a threshold: unless $e/m < 1$ there can be no stable equilibrium (you cannot have negative patch occupancy). The migration rate must be large enough and the extinction rate low enough, otherwise the system of patches will decline to global extinction. Subsequent versions of the model included more realistic assumptions (e.g. smaller patches have higher extinction rates and produce fewer migrants) but the central insight is unchanged.

14.8.2 Patch models

One problem with the original Levins model of metapopulation dynamics was that by ignoring population dynamics within each patch it precluded the possibility of modelling local population extinction explicitly (e.g. when within-patch density fell below a threshold density); in Levins' model a patch with a large population was just as likely to go extinct as a patch with a small one.

Patch models did not contain any explicit spatial dimension, but consisted of identical patches containing subpopulations for most of the time, coupled together by limited migration from a common pool of dispersers (i.e. there was assumed to be total mixing and equal accessibility of all patches; Gadgil 1971; Holt 1984; Hanski 1989). Dispersal acts to prevent extinction, allows recolonization of empty patches and damps fluctuations in global population size (Fig. 14.15a).

The most important thing is that the dynamics in the various patches are out of phase; if all patches have low population density at the same time then the prospect of one patch rescuing another is greatly reduced (Crowley 1977; Ludwig & Levin 1991). This has clear and important implications for conservation biology, since catastrophes (e.g. harsh winters) often affect all the patches with a metapopulation simultaneously. The other finding of linked patch models was that the rate of dispersal between patches should be neither too great nor too small. If the migration was too great, then the system became just like one big

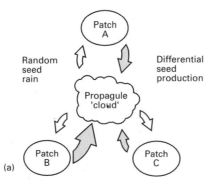

(a)

Fig. 14.15 Spatial models.
(a) Patch models are not spatially explicit. Patches A–C exchange seeds from a 'cloud' of propagules produced by each patch.
(b) Spatially explicit cellular automaton models predict the population in a cell i, j as a function of the previous occupants of i, j and of the neighbouring cells. When the neighbourhood is assumed to be four cells, the neighbours are $(i, j - 1)$, $(i, j + 1)$, $(i - 1, j)$ and $(i + 1, j)$; other models use neighbourhoods of six or eight cells (see Crawley & May 1987).

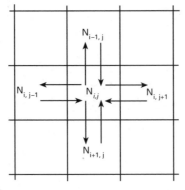

(b)

patch, and the stabilizing effects of subdivision were lost. If the migration was too little, then patches were not rescued at a sufficiently high rate, and the whole system drifted downwards towards global extinction.

Neither metapopulation nor patch models are spatially explicit; the immigration rate is not affected by the size or proximity of the source patch. Because the models are not spatially explicit, they are not capable of generating emergent spatial patterns. The definition of patches *imposes* a scale on the system. Ideally, we should like the heterogeneities to arise naturally and the scale of spatial patchiness to be determined by the dynamics.

14.8.3 Reaction diffusion models

In these models, space is treated as a one-dimensional (line) or two-dimensional (map) continuum, and changes in population size through space are modelled by partial differential equations. The *reaction* is the local population dynamic and the *diffusion* is the regional dispersal of the population (Levin & Segel 1976; Holmes *et al.* 1994).

The most interesting spatial aspects of reaction diffusion models concern the existence of waves. Travelling waves occur as a result of the transition between a low and a high equilibrium of a non-spatial model which arises as diffusion is added. The wave travels through space without changing shape, and the technique has been used to simulate the invasion of alien animals (e.g. grey squirrel; Okobu *et al.* 1989) and to model transient and persistent waves of herbivore populations (Lewis 1994). A second class, the kinematic waves, arise from the coupling of oscillators and become deformed through space and time (Murray 1989). The major problem with reaction diffusion models is that they become analytically intractable as soon as any realistic level of detail is added.

14.8.4 Cellular automata

In a cellular automaton, space is represented by a two-dimensional grid of cells, each big enough to hold just one plant (or one module of a plant, like the tiller of a grass). The state of the cell is discrete: it can be empty, it can be occupied by species 1, occupied by species 2, and so on. The model works on a series of simple rules which determine the various transitions that might occur, e.g. from empty to species 1, from species 2 to empty, and so on (Fig. 14.15b). These transition rules embody all the assumptions about the ecology of the system (relative competitive abilities, probabilities of establishment from seed, etc.). What makes cellular automata so useful is that it is very simple to make these rules spatially explicit (Wolfram 1986). What happens to a cell depends on what is happening in its neighbouring cells, and this is what makes cellular automata models so attractive for studying the dynamics of rooted vascular plants where neighbour effects are so influential (see Chapter 11).

The spatial behaviour of cellular automata models has been classified by Wolfram (1986) as falling into four categories: (i) spatial homogeneity, independent of initial conditions; (ii) stable or periodic spatial patterns that are strongly dependent on initial conditions; (iii) chaotic patterns and (iv) complicated localized patterns. The most curious spatial patterns are those arising from sophisticated self-organization, where regular structures emerge from initially random distributions and where complex patterns are created by extremely simple rules (e.g. the 'spatial swirlies' of Hassell *et al.* 1991 and the 'Persian carpets and lace doilies' of Nowak & May 1992).

An example of the use of a cellular automaton was by Crawley and May (1987) to demonstrate that plant species could coexist, even in the face of highly asymmetric competition. In this case, a perennial plant would always win in competition for a cell with an annual. However, perennial occupants of cells died with a probability d, so there was always some bare space for the annual to colonize, so long as seeds were available. Seeds from all of the annual plants in the previous generation were distributed at random over all cells of the automaton, occupied and unoccupied. The inferior competitor is able to persist so long as its potential rate of increase λ is greater than the reciprocal of the equilibrium proportion of empty space, E^*, available as a result of the relatively slow colonization ability of the superior competitor (persistence requires $\lambda > 1/E^*$). The analytical approximation to this result had been known for a long time (Skellam 1951). It would be straightforward to include more realistic assumptions like local seed dispersal and to incorporate more interesting assumptions about the biology of the seeds (e.g. smart seeds that only attempt to germinate in gaps; see Chapter 7). Models of this kind are increasingly used to study succession (Colasanti & Grime 1993; Hendry & McGlade 1995), fire (Hochberg *et al.* 1994) and pest outbreaks (Rand *et al.* 1995).

14.8.5 Coupled map lattice

The difference between a coupled map lattice (CML) and a cellular automaton is that instead of having a discrete state (e.g. occupied or not), the CML simulates the number of plants in each cell (Caswell & Etter 1993). For example, in a plant–herbivore CML there might be one set of equations to describe the number of plant species A in a cell, another to describe the number of plant species B and a third to describe the number of herbivores. The model works by allowing that the change in the number of occupants of the cell is affected by the numbers of competitors and herbivores in the cell in question and in a neighbourhood of (typically four, six or eight) adjacent cells. The *correlation length* is the separation distance in the lattice at which any two sites are uncorrelated in their dynamics (Moloney *et al.* 1992; Pacala & Tilman 1994). The *mean field approximation* is the dynamic behaviour predicted by the model once its spatially explicit elements have been taken away; the extent to which the predictions of the CML and the mean field approximation differ demonstrates the importance of the spatially explicit components of the model.

Since almost all spatially explicit models produce spatially aggregated output, there is little to be gained by demonstrating that the pattern is non-random. It is better to take a dynamic approach to variation in the level of clumping and plot a clumping index as a function of time, especially during transient dynamics (Hendry *et al.* 1996). Transience may persist for sufficiently long that it is the only relevant behaviour of the system (see Chapter 8). Two-species CML models have produced a new class of spatial dynamics known as chaotic Turing structures; these are patches that have stable boundaries but chaotic population dynamics within them (Solé & Bascompte 1993). The development of CML models in the context of plant coexistence is taken up by Pacala in Chapter 15.

14.9 Conclusions

We can recognize the following properties in communities.

1 The community is a contingent whole in reciprocal interaction with lower and higher level wholes, and not completely determined by them.

2 There are properties at the community level (like diversity, equitability, biomass, primary production, invasibility and the patterning of food webs) exhibiting patterns that are both striking and which show some kind of geographical regularity.

3 There are many configurations of populations that preserve the same qualitative properties of communities (e.g. species can be substituted, and communities persist despite their component populations constantly fluctuating).

4 The species of a community interact (plants with other plants, with their herbivores and pathogens, with their pollinators, dispersal agents and symbionts, with the enemies of their herbivores, and so on), so that the environment is structured from *interacting* biotic and abiotic factors.

5 The essential asymmetry of many interactions means that it is impossible for all species to be most abundant where their environment is most 'favourable' (Levins & Lewontin 1980).

6 The way a change in some physical character or genetic characteristic of a population affects the other populations in a community depends on the way the community is structured; thus if species *do* interact, then community structure determines the consequences of their interaction.

7 Spatially explicit models hold the key for a much more thorough understanding of plant community structure.

15: Dynamics of Plant Communities

Stephen W. Pacala

15.1 Introduction

Community ecologists rely on mathematical models primarily because humans have remarkably poor quantitative intuition, and so require models to determine the quantitative consequences of even simple assumptions. What will happen to coexisting plant species if we assume that they compete for a single limiting resource and that both the plants and resources are homogeneously distributed (e.g. there are no spatial effects)? The answer, though hardly surprising, is simply not obvious. The species capable of creating the lowest concentration of the limiting resource *in monoculture* will drive all the other species to extinction (the $R*$ result explained fully in Chapter 8).

Plant ecologists have now created a voluminous literature cataloguing and explaining 'if–then' statements derived from models. These models are usually designed to be as simple as possible rather than to facilitate direct empirical measurement of parameter values and functional forms. This is because the primary purpose of the models is to expose ideas, rather than to describe any one plant community.

The simplicity of the models has permitted a rich and rapid dialogue between empiricists and theoreticians. Field observations and experiments test the simple and qualitative predictions of models (e.g. Wedin & Tilman's (1993) test of the $R*$ result), and provide the empirical foundation for new models (e.g. the correlation between seed dormancy and dispersal that provides the rationale for some of the models in Chapter 7).

In this chapter, I first introduce the most widely studied class of simple mathematical models of plant communities, and then review the four primary hypotheses about the maintenance of plant biodiversity that have been derived from these. I also discuss a few empirical studies to show that each hypothesis has empirical support and to demonstrate some of the difficulties that are involved in testing hypotheses derived from simple models. Paradoxically, the very simplicity that facilitates interaction between model and experiment also leads to confusion over the interpretation of empirical findings and the applicability of theoretical results. Models of ideas are often based (by design) on such simple caricatures of nature that it is sometimes difficult to determine if a model should apply to any given field system.

For this reason, plant ecologists have also developed models of plant communities which it is both possible and practical to calibrate with empirical data (i.e. by estimating their parameters and functional forms). These models of natural systems place greater demands on the interaction of empiricists and modellers than do simple models of ideas. Typically, an intensive round of data-gathering and statistical estimation first defines the model. The model is then analysed to predict and understand the consequences of the parameter estimates. Finally, predictions and hypotheses emerging from the analysis are tested against independent data (see for example Pacala & Silander 1990; Pacala *et al.* 1996).

The primary disadvantage of this kind of work is that data-defined models tend to be more difficult to analyse than simple models of ideas. The problem is not so much the inevitable tendency to clutter a model with the particular attributes of one's favourite field system. Rather, it traces to two attributes of plants that are reflected in empirically calibrated models. Both dispersal distances and the spatial scale of resource uptake are surprisingly short for most plant species in nature. As a result, plants respond dramatically to stochastic variation in local spatial structure, often caused by the deaths of single individuals (e.g. canopy gaps in a forest). Local dispersal, in turn, ensures that a change in local fecundity and/or survivorship causes a further change in local spatial structure. In this way, the inevitable fine-scale randomness associated with birth, death and dispersal (demographic stochasticity; May 1973) can manifest itself in large-scale pattern. A random event, such as the death of a single individual, is first amplified by the responses of neighbours and then propagated to the local spatial pattern by short-distance dispersal. The change in local spatial pattern is, in turn, amplified by the responses of neighbours, the amplified response is again transmitted to the local spatial pattern, and so on. Population growth and spatial structure affect one another, and this interaction is stochastic at its heart.

Models that include these attributes of plant communities reside squarely within a corner of mathematics, known as non-linear spatial stochastic processes, which has been intractable until very recently. As a result, studies of calibrated models have been forced to rely primarily on computer simulations, and the literature on simple models of ideas is dominated by assumptions that effectively eliminate the interdependence of spatial pattern and dynamics.

In the second half of this chapter, I introduce models of plant communities that are designed to be calibrated. I then present a new theoretical result on the interdependence of spatial pattern and dynamics. This result relies on new mathematical methods (not presented here) and I include it in this chapter primarily to demonstrate the incompleteness of the current theory. Briefly, the result is that the interdependence of spatial pattern and dynamics leads to dramatically increased resilience of plant biodiversity to perturbation. I also show that published competition experiments support this 'spatial segregation hypothesis' and that the effect overturns what is perhaps the most central assumption of classical competition models.

15.2 Simple models of ideas

The majority of simple models of plant communities share a single conceptualization that traces to Skellam (1951). Imagine that the habitat is subdivided into an infinite number of discrete spatial cells and that each can contain at most a single adult plant. Also, let $N_i(t)$ be the fraction of cells occupied by species i at time t and $P_{ij}(t)$ be the rate, at time t, at which species i cells are captured by species j. Then, the change in the abundance of species i ($dN_i(t)/dt$) is:

$$\frac{dN_i(t)}{dt} \quad = \quad -\sum_{j=0}^{Q} P_{ji}(t) \quad + \quad \sum_{j=0}^{Q} P_{ij}(t) \tag{15.1}$$

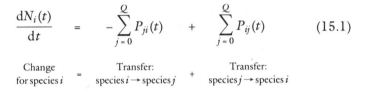

$$\underset{\substack{\text{Change}\\\text{for species } i}}{} = \underset{\substack{\text{Transfer:}\\\text{species } i \to \text{species } j}}{} + \underset{\substack{\text{Transfer:}\\\text{species } j \to \text{species } i}}{}$$

$i = 1,2,3,\ldots,Q$, where Q is the number of species and the subscript zero signifies the fraction of space unoccupied by any species: $N_0(t) = 1 - [N_1(t) + N_2(t) + \ldots + N_Q(t)]$.

As it stands, Equation 15.1 leaves the critical part of the ecological interactions among plant species unspecified. What determines the transfer of space from one plant species to another? The simplest assumption about the $P_{ij}(t)$ is made in the Markovian succession models of Horn (1975, 1976). Horn assumed that, during each small time period Δt, a constant fraction ($p_{ij}\Delta t$) of cells occupied by species j is transferred to species i. Thus, in Horn's models: $P_{ij}(t) = N_j(t)p_{ij}$.

Now, what does this mathematical assumption imply about the underlying biology? The most important biological implication is that a plant species can never be *recruitment-limited*. Recruitment limitation occurs if a species fails to colonize a site with propagules. Recruitment limitation is of particular interest to community ecologists because coexistence is usually possible only if each species is able to increase when rare. As a species' population size approaches zero, the number of offspring it produces also tends to zero, making recruitment limitation unavoidable. Horn's models do not include recruitment limitation because they assume that the rate at which a species captures new cells is independent of its population size. Thus, Horn's models are typically used to address problems other than questions of coexistence (e.g. the nature of a successional end-point).

To include recruitment limitation, we must construct models in which the $P_{ij}(t)$ depend on the population sizes of both species i and j.

Models of this sort have codified the four primary hypotheses explaining the coexistence of plant species.

15.2.1 Competition/colonization trade-off

Suppose that the plant species in a location are organized in a perfect linear competitive hierarchy, such that a species with a higher subscript will displace a species with a lower subscript from a cell as soon as the higher-subscript species colonizes the cell. Colonizing ability is assumed to trade off against competitive ability, with lower subscripts signifying greater capacity to colonize. Finally, disturbance is assumed to convert occupied cells into open space at random.

These assumptions may be formalized mathematically in terms of the $P_{ij}(t)$ in Equation 15.1:

$$P_{0j}(t) = DN_i(t)$$

$$\text{if } i > j: \quad P_{ij}(t) = C_i N_i(t) N_j(t) \tag{15.2}$$

$$\text{if } j \geq i: \quad P_{ij}(t) = 0$$

The first equation above says that disturbance converts occupied space to unoccupied space at rate D. The last equation states that a species never loses space to a species with lower competitive ability (signified by a lower subscript). The middle equation states that the rate of transfer of space to species i from empty space ($j = 0$), or from space occupied by a less competitive species ($j > 0$, $j < i$), is simply equal to the rate of colonization of the space by species i. The constant C_i in this expression gives the colonizing ability of the species. Because colonizing ability decreases as competitive ability increases: $C_1 > C_2 > \ldots > C_Q$. Finally, the right-hand side of the middle equation is correct only if dispersal is random and the mean dispersal distance is infinitely large. Like the rate of collision between two different types of molecules in an ideal gas, the rate of colonization is proportional to the product of the abundances of colonists and residents, but only if the spatial distribution is well mixed by large random movements.

The model given by Equations 15.1 and 15.2 has a long and illustrious history. Levins and Culver (1971) and Horn and MacArthur (1972) first showed that a competitively inferior 'fugitive species' could coexist with a superior competitor provided that the superior competitor is a sufficiently poor colonizer. Mathematically, the condition for coexistence of two species is: $C_1 > C_2 > D$. When this condition is met, the colonizing ability of species 1 allows it to persist on the fraction of habitat that the superior competitor is unable to exploit because of its severe recruitment limitation (low C_2) and the persistent disturbance. The result was generalized by Armstrong (1976), Hastings (1980), Shmida and Ellner (1984), Crawley and May (1987), Nee and May (1992), Tilman (1994) and Tilman *et al.* (1994). Most notably, Hast-

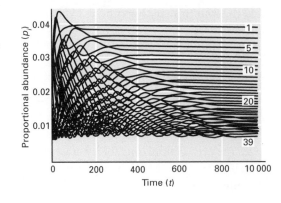

Fig. 15.1 Numerical solution of equations 1 and 2 from Tilman (1994) showing the coexistence of many species by the competition/colonization trade-off.

ings (1980) provided a recursive method showing that any number of species can coexist because of the competition/colonization trade-off and Tilman (1994) and Tilman *et al.* (1994) focused explicitly on the high diversity that can be maintained by the mechanism (see Fig. 15.1).

15.2.2 Resource partitioning

We now move from a purely hierarchical to a purely egalitarian view of community organization. Suppose that the environment is spatially variable and that each species is competitively superior under a different set of environmental conditions. The environmental variability could be either abiotic or biotic. Most abiotic heterogeneity is microclimatic or edaphic in origin: variation in local climate caused by topography and variation in water and nutrient availability caused both by climate and by changes in geological parent material of soils. Obviously, the relative competitive abilities of plant species change along gradients in the physical enviornment as is evidenced by the close correspondence among climate, soil type and vegetation (see Chapter 14).

Biotic heterogeneity, in contrast, is caused by organisms. The most common example of biotic heterogeneity is the gap created by the death of a single adult forest tree, which dramatically increases local light levels in the understorey and enhances the growth and survivorship of nearby saplings (Canham 1988). Because plants play such a prominent role in nutrient cycling, variation in the local spatial distribution and community composition of plants also causes biotic heterogeneity in below-ground resources. For example, the rate of decomposition of leaf litter is strongly affected by litter chemistry (i.e. the litter's lignin to nitrogen ratio; Mellilo *et al.* 1982; Pastor *et al.* 1984) and litter chemistry varies among plant species. Thus, changes in local community composition cause spatial variation in nitrogen mineralization rates. For this reason, the nitrogen mineralization rate is approximately twice as large underneath conifers as under adjacent hardwoods in Wisconsin forests (Pastor & Broschart 1990), and the presence of an individual maple (*Acer rubra* or *Acer sacharrum*) or ash (*Fraxinum americana,* Oleaceae) doubles the nitrogen mineralization rate directly beneath it in New England forests (Canham & Pacala 1995). Also, by gathering nutrients over larger distances than they drop leaf litter, arid-land shrubs create a grid-work of nutrient-rich pockets embedded in a nutrient-deprived matrix (Schlessinger *et al.* 1990). Finally, biotic heterogeneity sometimes is caused by organisms other than plants. For example, spatial patterns of herbivores (Crawley 1983; Pacala & Crawley 1992), pathogens (Dobson & Crawley 1994) and mutualists (e.g. mycorrhizal fungi or pollinators; Bertin 1989; Allen & Allen 1990) are notoriously variable, and may alter the relative competitive abilities of affected plants (Crawley 1983, 1989).

Now suppose that each plant species is competitively dominant under a fraction E_i (for species i) of environmental conditions and that competitive interactions occur only among the juveniles competing to fill a cell after the death of its former occupant (as before, deaths occur at rate D). Thus, if juveniles of all species are present in all cells, then each species will capture only those cells in which it is competitively dominant. For simplicity, assume first that adult plants have effectively infinite dispersal and fecundity so that juveniles of every species do reach every cell. Then, the terms on the right-hand side of Equation 15.1 become:

$$P_{0j}(t) = 0$$

$$\sum_{j=0}^{Q} P_{ji}(t) = DN_i \qquad (15.3)$$

$$\sum_{j=0}^{Q} P_{ij}(t) = DE_i$$

for $j = 0,1,2,\ldots,Q$. The first equation states that space never becomes empty (because the infinite fecundity fills it up as fast as it is created), the second states that a species loses space only when its adults die (unlike the previous model of the competition/colonization trade-off where competitive dominants could take space from competitive subordinants), and the third states that each species captures precisely the fraction of vacant cells in which it is competitively dominant (E_i).

The model consisting of Equations 15.1 and 15.3 is easy to solve. At equilibrium, the abundance of the ith species is simply E_i the fraction of space in which the species is competitively dominant. An unlimited diversity of species can coexist in the model, providing each has some set of environmental conditions in which it is the dominant species. This is the mechanism of coexistence in published models, ranging from models of plant species coexisting on heterogeneous soils (e.g. Levin 1974; Tilman 1982) to models in which conditions within each cell shift continuously over time (e.g. Commins & Nobel 1985; Pacala & Tilman

1994). In the former case, the habitat would be composed of a patch-work of monospecific stands, while in the latter, species would continuously track the spatial shifts in environmental conditions. I call these models of 'resource partitioning' because the species in them coexist by subdividing space on the basis of spatially variable environmental conditions, in a way analogous to the subdivision of prey on the basis of characteristics of the prey in the classical models of resource partitioning (the niche theory of MacArthur 1972 and May 1973).

With a bit more work, one may relax the assumption of infinite fecundity, while keeping the assumption of infinite dispersal. Suppose now that an individual produces juveniles at finite rate F. To persist, a species must have a colonizing ability (F) that is large enough to find, on average, those cells in which the species is a good competitor. Although the general necessary and sufficient condition for the persistence of a species at equilibrium is too complex to be presented here, a sufficient condition is simply: $E_i > 1/F$ (Pacala & Tilman 1994). Even with the restriction of finite fecundity, there is still no cap on the number of species that can coexist at equilibrium (Fig. 15.2; Hurtt & Pacala 1995).

15.2.3 Temporal partitioning: the storage effect

In the two models described above, conditions are assumed to be stochastic at small scale (at the level of a cell) but deterministic at large scale (over the whole ensemble of cells). For example, in the resource partitioning models, the *distribution* of environmental conditions is assumed to be temporally constant while conditions within an individual cell may change from one time to the next.

We now consider the alternative in which the environment varies temporally over the whole ensemble of cells, but is spatially uniform. Like spatial heterogeneity, temporal heterogeneity could be biotic (e.g. insect outbreaks) or abiotic (e.g. variation in weather).

For simplicity, assume that there are two species, each with death rate D, fecundity F, and infinite dispersal. Let Z_i be a random variable that designates which species is currently the competitive dominant. Let $Z_1 = 1$ and $Z_2 = 0$ for a fraction E_1 of the time and let $Z_1 = 0$ and $Z_2 = 1$ for the remaining time ($E_2 = 1 - E_1$). With these new definitions, the transition rates in Equation 15.1 are:

$$\sum_{j=0}^{Q} P_{ji}(t) = DN_i \tag{15.4}$$

$$\sum_{j=0}^{Q} P_{ij}(t) = [D(N_1 + N_2) + N_0]FZ_iN_i$$

Equations 15.1 and 15.4 describe a community in which the identity of

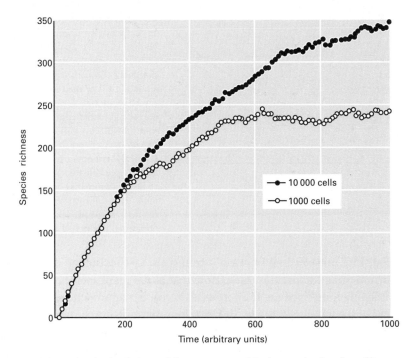

Fig. 15.2 Stochastic simulations of the resource partitioning mechanism from Hurtt and Pacala (1995), showing that many species can coexist by this mechanism (open symbols, 1000 cells; closed symbols, 10 000 cells; $F = 25$; $D = 0.01$; $\Delta t = 1.0$; dispersal is infinite). A new species was introduced during each time step for 1000 time steps. The environment varied randomly and independently from cell to cell and from time to time within each cell. Each species was the dominant competitor under a different set of environmental conditions (competitive rank was determined by the distance between a uniform random variable for each cell and time and a uniform random variable assigned permanently to each species). When an individual died, the successor was chosen as the highest ranked juvenile in the cell. All juveniles were replaced in each time step.

the competitive dominant changes randomly through time (see Section 14.4.2).

To simplify the mathematical treatment, suppose that F is large enough to make the fraction of empty space negligible and consider the discrete-time analogue of Equation 15.1 (replace $dN_i(t)/dt$ with $\Delta N_i(t)/\Delta t$). Then, the condition for the persistence of species i is given by Chesson's 'stochastic boundedness' criterion (Chesson 1983, 1984, 1985, 1986):

$$E_i > \frac{-\ln(1 - D\Delta t)}{\ln(1 - D\Delta t + DF\Delta t) - \ln(1 - D\Delta t)} \tag{15.5}$$

Equation 15.5 states that favourable environmental conditions for species i must occur more frequently than the threshold value given by the right-hand side (hereafter labelled E_i^*), if species i is to persist indefinitely. If $E_i < E_i^*$, then the species' gains during good years are not sufficient to balance its losses during bad years.

A plot of the right-hand side of Equation 15.5 versus the death rate

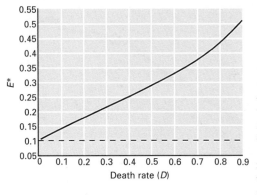

Fig. 15.3 Coexistence of species by temporal partitioning (Equation 15.5). To coexist, each of two species must be the dominant competitor more than a fraction E_i^* of the time. $\Delta t = 1$; $F = 10$. Note that E_i^* exceeds 0.5 on the right-hand edge of the plot shown. This indicates that species that are too short-lived (large D) cannot coexist, because it is not possible for each species to be the dominant competitor more than half the time as required by the model. The horizontal line is $1/F$.

D (Fig. 15.3) reveals two important things about coexistence caused by environmental fluctuations. First, note that as the plant species becomes increasingly long-lived (as D approaches zero), E_i^* approaches $1/F$ (the lower horizontal line in the figure). For long-lived plants, Equation 15.5 becomes approximately $E_i > 1/F$ which is the same condition required for persistence in a spatially heterogeneous environment (above). This shows the similarity between the mechanisms of spatial and temporal partitioning. To persist in a spatially heterogeneous environment, a species must disperse propagules from cells with favourable environmental conditions to other such cells. To persist in a temporally heterogeneous environment, a species must disperse *through time* from one period of favourable environmental conditions to another. If Equation 15.5 is true for both species in the model, then the species coexist because they partition the temporal variation in the environment.

Second, Fig. 15.3 illustrates the general result that only those species with overlapping generations can coexist by temporal partitioning (Chesson 1985). Species become annuals lacking dormant seed as $D\Delta t$ approaches 1 if Δt is taken to be 1 year. Because E_i^* approaches 1 as $D\Delta t$ approaches 1, short-lived species cannot coexist in the model. In general, coexistence of two annuals is impossible by the mechanism of temporal partitioning unless a fraction of seeds survives more than 1 year (see Chapter 7). Chesson (1985) labelled this mechanism the 'storage effect' because effects of favourable years are stored within a population (by increased population size) and then released over several years. If favourable years come frequently enough, then a population's stores allow it to weather the periods of decline.

15.2.4 Janzen–Connell hypothesis

In two classic papers, Janzen (1970) and Connell (1971) hypothesized that specialist herbivores, seed predators and pathogens could maintain unlimited species diversity of their host plants. Although the mechanism they proposed was later shown to be incorrect (Hubbell 1980), the hypothesis itself follows from another mechanism.

Suppose that adult trees harbour populations of specialist natural enemies and that these herbivores reduce the survivorship of conspecific juveniles growing beneath them. Let S be the survivorship beneath a conspecific and vS be the survivorship beneath a heterospecific, where $v \geqslant 1$. Assume that a fraction (m) of seeds have infinite dispersal, while the remaining fraction $(1 - m)$ fall within their mother's cell (this is as close as we can get to limited dispersal without creating the difficult interplay between spatial distributions and dynamics mentioned earlier). When an adult dies, a successor is chosen *at random* from among the juveniles present within the cell. As in the previous model, we assume that fecundity (F) is large enough and adult death rate (D) is small enough to make empty space negligible. Finally, to avoid the complication of age structure in Equation 15.1, we suppose the lifespan of adults for exceeds the life span of juveniles growing in the understorey.

If J_{ij} is the number of species i juveniles growing underneath an adult of species j, then $J_{ij} = mFN_i vS$ and $J_{ii} = mFN_i S + (1 - m)FS$. In words, the number of juveniles growing beneath a heterospecific adult (J_{ij} for $i \neq j$) is simply the number of dispersing species i seeds per cell (mFN_i) times the survivorship beneath a heterospecific (vS). The number of juveniles growing beneath a conspecific adult (J_{ii}) is the number of species i seeds that disperse to the cell from outside and survive ($MFN_i S$, note the survival rate is S instead of vS) plus the number of seeds produced within the cell which fail to disperse outside it ($(1 - m)FS$).

With these specifications, the $P_{ij}(t)$ in Equation 15.1 becomes:

$$P_{ij}(t) = \frac{J_{ij}}{\sum_{k=1}^{Q} J_{kj}} \qquad (15.6)$$

$$P_{ii}(t) = \frac{J_{ii}}{\sum_{k=1}^{Q} J_{ki}}$$

where $i = 1,2 \ldots ,Q$ and $j = 1,2, \ldots ,Q$. Here, the probability that a cell is taken over by species i on the death of its occupant is given by the probability that a randomly chosen juvenile in the cell is a species i juvenile. This probability is simply the number of species i juveniles in the cell divided by the total number of juveniles of all species in the cell.

If $v > 1$ and $m > 0$, then a rare species in Equations 15.1 and 15.6 has an advantage because juveniles of the rare species that disperse from their natal cell (the fraction m) have negligible probability of landing in a cell occupied by an adult of the same species. Thus, dispersing rare juveniles have higher average survivorship than dispersing common juveniles. For this reason, any number of species can coexist in Equations 15.1 and 15.6 so long as $v > 1$ and $m > 0$. Specialist herbivores, pathogens and seed predators might create a rare-species advantage that could maintain unlimited plant diversity.

15.3 Empirical tests

Each of the above four hypotheses about coexistence is supported by empirical studies. The mechanism of the Janzen–Connell hypothesis has been detected in empirical studies of tropical forests ranging from Queensland (Connell *et al.* 1984) to Panama (Condit *et al.* 1992). However, the effect is typically weak and detected in only a subset of the species. It is currently not known if other more powerful mechanisms obviate the Janzen–Connell effect. Similarly, a substantial number of studies show that the mechanism of the temporal partitioning hypothesis operates in natural communities. For example, germination rates of many plant species show temporal partitioning of annual variation in weather (see Chapter 2 of Harper 1977), desert ephemerals apparently specialize on years with different rainfall regimes (Chesson & Huntley 1989, 1993), and a subset of species in Minnesota grasslands increase dramatically in abundance during drought years (Tilman *et al.* 1994a). To my knowledge, however, no empirical study has shown definitively that fluctuations in recruitment are sufficient to maintain diversity in a natural system. The temporal partitioning hypothesis is particularly difficult to study because one must develop an understanding of each species' response to the full range of environmental conditions and also characterize the distribution of environmental fluctuations; this could take hundreds of years.

A comparison of evidence for the remaining two hypotheses is interesting, because both the resource partitioning hypothesis and the competition/colonization hypothesis purport to explain secondary successional diversity. The competition/colonization hypothesis explains early successional species as competitively inferior species that persist by repeatedly colonizing disturbed habitat before the late-successional good competitors arrive. For example, in a Minnesota grassland, species that allocate a large fraction of photosynthate to reproduction (colonizing ability) invade abandoned farmland before the arrival of species with low $R*$ values (the good competitors that allocate primarily to roots) (see Chapter 8). Tilman (1994) and Tilman *et al.* (1994b) argue that such trade-offs are unavoidable and may maintain the diversity of tropical forests.

In contrast, the resource partitioning hypothesis explains successional diversity in the same way as it explains changes in species composition across gradients in the physical environment. Each species is assumed to be a dominant competitor under a subset of conditions present in the habitat. In the case of zonation along a physical gradient, the dominant at each place along the gradient is explained as the best competitor under the environmental conditions at that place. An enormous literature supports this explanation (reviews in Whittaker 1975; Tilman 1982). In the case of secondary successional diversity, the heterogeneity in environmental conditions is biotic in origin rather than abiotic, but the explanation of coexistence is otherwise the same. For example in forests, early-successional species are assumed to be the dominant competitors under resource levels typical of gaps, while late-successional species are assumed to be the dominant competitors under closed canopy.

How are we to decide between these two explanations? The answer to this question is of more than academic interest. Tilman *et al.* (1994a,b) have shown that a low level of anthropogenic disturbance could destabilize a high-diversity community structured by the competition/ colonization mechanism, and cause the extinction of all late-successional dominants after a time delay. This prediction has alarming implications for the management of rain forests. The reason for the prediction is the extreme recruitment limitation necessary in competition/colonization models to prevent the good competitors from excluding the good colonists from all cells. In existing simple models, the recruitment limitation is so severe that the dominant competitors would occupy only a small fraction of the habitat even if all other species were removed. Because of their inability to colonize space, the late-successional species are perched at the brink of extinction, and the removal of habitat from these models (by increased anthropogenic disturbance) is enough to drive them extinct. In contrast, communities structured by resource partitioning could tolerate low levels of harvesting.

Again, how are we to decide between these explanations of diversity? The usual prescription is to perform an experiment, but there are two problems in this case. First, experimental tests are impractical in most natural systems because the vegetation is too long-lived. Obviously, one could focus only on model systems of short-lived species, but these are unlikely to be representative because they occur primarily in locations suffering unusually high levels of disturbance.

Second, models of ideas are typically false in many respects because they omit so much of nature's complexity (on purpose). Falsifying a simple model is thus both easy and unproductive; one wants a test that isolates the critical part of the explanation, but this is easier said than done. For example, Pacala *et al.* (1996) measured a suite of interspecific trade-offs among coexisting forest tree species in north-eastern North America. Early-successional species easily outgrow and overtop late-successional species under the high light levels in disturbed sites, but only late-successional species can survive the low light levels present in the understorey of undisturbed sites. This sounds very much like the resource partitioning hypothesis but does it falsify competition/

colonization models in which good colonizers are poorer competitors under all environmental conditions? Early-successional forest species also have longer dispersal than late-successional species and so might reach severely disturbed areas before late-successional species. Does this resurrect the competition/colonization hypothesis?

After analysing a model tailored specifically for this forest community, we eventually concluded that both mechanisms are present (and some others). Under natural disturbance regimes, successional diversity is maintained primarily by the resource partitioning mechanism. Natural disturbances in these forests typically do not kill all established saplings. As a result, the short dispersal of late-successional species does not significantly restrict their ability to colonize disturbed areas, and so the competition/colonization mechanism does not contribute significantly to successional diversity. Instead, early- and late-successional species partition the substantial variation in understorey light. Late-successional species are favoured in locations that had been previously dark enough to prevent the survivorship of early-successional saplings but light enough to permit the survivorship and growth of late-successional saplings, thus allowing a late-successional sapling to gain the large head start necessary to capture a gap once it forms. Early-successional species are favoured in areas that had been previously too dark to permit significant growth of late-successional saplings, because early-successional saplings can quickly overtop small late-successional saplings in gaps.

Although the competition/colonization mechanism does not control successional diversity, it does promote the diversity of old growth in the model. Old-growth species are differentiated by a growth/survivorship trade-off. Some late-successional species have high understorey survivorship but only moderate low-light growth rate, while others have rapid low-light growth but only moderate low-light survival. Species in the former category effectively colonize more gaps than species in the latter category, because their saplings are more likely to survive in the understorey until a gap forms. However, saplings in the latter category occasionally survive for a substantial period of time (by chance) and become large because they have relatively rapid low-light growth. These large saplings are competitively superior to the numerous, but comparatively small, saplings of slower growing species. Paradoxically, the good colonizers with high understorey survivorship are the late-successional dominants in the system (eastern hemlock, *Tsuga canadensis* (Pinaeceae) and American beech, *Fagus grandifolia*), while the good competitors with rapid low-light growth are the late-successional subdominants (yellow birch, *Betula alleghanensis* and sugar maple, *Acer saccharum*).

This directly contradicts the traditional view that late-successional dominants are superior competitors but poor colonizers, and illustrates the difficulty of mapping simple models of ideas onto the spatial and age-structured processes that govern most natural systems. Our analysis also required many more experiments with the tailored model than it would ever have been possible to complete in any natural system. This is partly because of the difficulty in establishing how the processes in the simple models relate to attributes of the system.

15.4 Models of natural systems

Perhaps the simplest way to formulate a model of a plant community is to specify the rules governing the performance (e.g. survivorship, fecundity and dispersal) of an individual plant throughout its life cycle. The performance rules could either be mechanistic (e.g. describe the uptake and renewal of resources) or phenomenological (e.g. growth and survivorship decreases with the local density of neighbours). One could then construct a computer model that includes each and every plant in the community, and use the performance rules to predict the fate of every individual through time (see Pacala 1986).

This approach is analogous to modelling a gas by simulating the motion of every atom in the gas. It has three advantages. The first is biological realism; models may be tailored to include the particulars important in a field system under study. Published examples include resource dynamics and age structure (Pacala *et al.* 1996), pathogens (Thrall & Antonovics 1995) and seed dormancy and spatially local interactions (Pacala and Silander 1990). The second advantage is ease of empirical calibration; it is comparatively easy to measure individual plants and so models may be designed so that it is both possible and practical to estimate functional forms and parameter values with empirical data (see Pacala & Silander 1990; Cain *et al.* 1995; Pacala *et al.* 1996). It is important to understand that not all individual-based models of plant communities are data-defined models. In particular, a large number of forest simulators derived from the JABOWA (Botkin *et al.* 1972) and FORET (Shugart & West 1977) models are individual based, but constructed from performance functions that are assumed rather than measured.

The third advantage is the capacity to extrapolate to spatio-temporal scales beyond the reach of any experimental programme. Although one cannot perform large-scale and long-term perturbation experiments in natural communities, one can do these experiments in modelled communities. Janis Antonovics calls the development of a data-defined model the 'taming of the system'. The idea is that once an empirical system is tame, then one can do with it what one likes. Obviously, experiments with a model are only as believable as the model. To increase confidence in a data-defined model, one should propagate the statistical uncertainty in parameter values and functional forms, to determine which large-scale and long-term predictions of the model are supported by the data that define it. For example, Pacala *et al.* (1996)

showed that parameter values in a forest model were defined precisely enough to determine the abundances of guilds of forest tree species at each point in succession, but not precisely enough to determine the relative abundances of species within each guild at each stage. One should also test predictions of the model against independent data. For example, the model in Thrall and Antonovics (1995) is capable of predicting the large-scale spatial association of pathogen resistance in a herbaceous weed with the abundance of a sexually transmitted fungus.

Balancing these advantages is a significant and important disadvantage. Data-defined models of plant communities are typically considerably more difficult to analyse *and understand* than are simple models of ideas. As indicated in section 15.1, the most important cause of this intractability is the finite spatial extent of interindividual interactions and dispersal in real plant communities. Note that each of the four simple models of ideas described in sections 15.2.1–15.2.4 assumes that dispersal is infinite. Without this assumption, the combination of finite dispersal and local competition (only within cells) would have rendered the models intractable.

How important is the interplay between spatial structure and dynamics caused by local interactions and finite dispersal? In the remainder of this chapter, I draw on new methods to answer this question for what is perhaps the most classical of competition models.

15.5 Spatial segregation hypothesis

Consider a community composed of two vegetatively reproducing plant species (say two grasses growing in a lawn). As in many such species, survivorship of new recruits is density independent (because of provisioning of a daughter ramet by its mother), but fecundity depends on local density. Thus, we assume that the mortality rate of an individual of either species is the constant μ, while its fecundity rate decreases with local density:

$$\text{fecundity rate} = f - \beta \, (\text{local density conspecifics})$$
$$- \gamma \, (\text{local density heterospecifics}) \quad (15.7)$$

Here, the constant f is the fecundity of a plant when growing alone, the constant β determines the strength of within-species effects of neighbours and the constant γ determines the strength of between-species effects of neighbours.

Two spatial processes must be specified to complete the model. First, suppose that the dispersal function for each species is a simple exponential decay. That is, the probability that a daughter ramet disperses a distance x from its mother is equal to $e^{-x/M}/M$, where the constant M is the mean dispersal distance. Dispersal is assumed to be equally likely in any direction. Second, suppose that effects of neighbours decrease with distance from the focal plant as in Weiner (1982). The local density of species i is assumed to be the sum, over all species i plants, of $e^{-x_j/\lambda}/Q$, where x_j is the distance between the focal plant and the jth individual, λ is the spatial scale of the competitive effect (the mean of the exponential function) and Q is a normalizing constant (equal to 2λ in a one-dimensional habitat and $2\pi\lambda$ in a two-dimensional habitat). The normalizing constant permits one to change the scale of density dependence (λ) without changing its strength (β or γ).

It is a simple matter to write a computer program that will simulate a community of plants governed by these rules. The program would begin with an initial number of individuals, each with a specified species identity and location (spatial coordinates). It would then compute the state of the community after the small time interval Δt in four steps.

1 Kill each individual with probability $\Delta t\mu$ (using a pseudo-random number generator).

2 Compute the local density of conspecifics about each individual by summing the distance-weights over all conspecific plants (for each plant sum the $e^{-x_j/\lambda}/Q$ where x_j is again the distance between the focal plant and the jth conspecific). Compute the local heterospecific densities by summing the distance-weights for the heterospecific neighbours.

3 Use Equation 15.7 to compute each individual's probability of giving birth (Δt times the fecundity rate for the individual) and have it give birth to a daughter with this probability.

4 Disperse each new daughter using the dispersal function (by drawing a pseudo-random distance and direction of dispersal for each daughter). By iterating this algorithm, the program will predict the state of the plant community at any time in the future.

Despite its simplicity, this model is very difficult to analyse and understand unless either dispersal or the spatial scale of competition (either M or λ) is infinitely large. Non-random spatial structure is produced in the model by two processes: finite dispersal generates within-species clumping, while spatially local competition works to separate individuals in space (because plants in clusters produce fewer progeny per capita). Infinite dispersal mixes up the community and destroys any non-random spatial structure, while infinitely large λ ensures that plant performance depends only on the global average population densities, no matter what spatial structure is present. With infinite M or λ, it is possible to show that the model converges to the familiar Lotka–Volterra competition equations (see Pacala 1986):

$$\frac{dN_1}{dt} = rN_1\left[1 - \frac{N_1}{K} - \alpha\frac{N_2}{K}\right]$$

$$\frac{dN_2}{dt} = rN_2\left[1 - \frac{N_2}{K} - \alpha\frac{N_1}{K}\right]$$

(15.8)

where the intrinsic rate of increase $r = f - \mu$ (birth rate minus death rate), the carrying capacity $K = (f - \mu)/\beta$ and the competition coefficient

$\alpha = \gamma/\beta$. Note that the two species would have species-specific r, K and α values if we had assumed species-specific f, μ, β and γ values. One can learn from any introductory ecology text that the two species in Equation 15.8 will coexist only if interspecific competition is less strong than intraspecific competition (if $\alpha < 1$). Moreover, the resilience of the community decreases as interspecific competition strengthens. In the limit as α approaches 1, the rate at which population sizes return to equilibrium after a perturbation approaches zero.

During the 1960s, the classical theory of competition was developed from the Lotka–Volterra competition equations. First, the additional assumption was added that (all else being equal) the strength of interspecific competition increases as competitors become more similar (the fundamental assumption of competition theory; Levins 1968; MacArthur 1972; May 1973). This, in turn, led to a formalization of Gause's competitive exclusion principle: species will not coexist if they are too similar. Similarity causes large competition coefficients and communities containing species with large competition coefficients cannot recover from perturbations (and may be destabilized altogether if α values get sufficiently large).

Classical competition theory was criticized heavily during the early 1980s (Connell 1980; Simberloff & Boecklin 1981; Strong & Simberloff 1981). This criticism centred on the lack of experimental field tests of the theory. Since that time, ecologists have responded with an enormous number of experimental measurements of the strength of interspecific competition in the field (see the review of experiments with plants in Law and Watkinson 1989 and Gurevitch 1992).

How does one measure the relative strength of competition in nature? The most common method is a removal experiment. One compares the response of species i after removing a portion of species j's density with its response after removing of a portion of its own density. If one performs this experiment on an equilibrium community governed by Equation 15.8, then the change in population size (dN_i/dt) following removal of a small number of heterospecifics divided by the change in population size following the removal of the same number of conspecifics is indeed the competition coefficient α.

Now, how much are these classical models and field methods affected by the interplay between dynamics and spatial distributions that occurs when the scales of dispersal and competition are both finite? Recently, mathematical methods have been developed that allow one to analyse problems of this sort (Durrett & Levin 1994a,b; Bolker & Pacala (in preparation); Pacala & Deutschman 1996). Unfortunately, the mathematics involved is too advanced to be presented here. The results that follow are derived using the methods in Bolker and Pacala manuscript.

If one performs a removal experiment on an equilibrium community govened by the rules in our computer model, then what is the relationship between the estimate of the competition coefficient obtained (α^{est})

and the competition coefficient that would be obtained were dispersal and/or the scale of competition infinite (α)? It is important to understand that α is still (with finite M and λ) the relevant measure of ecological similarity. As α approaches 1, the two species become equivalent because $\alpha = \gamma/\beta$. Neighbours of each species have equivalent effects on plant performance if $\alpha = 1$. In contrast, α^{est} is now the relevant measure of the strength of interspecific competition because α^{est} is a direct empirical summary of a species' population dynamic response to changes in conspecific and heterospecific density.

For simplicity, assume that the scale of competition is equal to the scale of dispersal ($M = \lambda$), that M is relatively large (but not infinite) and that the species are similar (so that α is near 1). I can generalize the result below to cases of unequal and/or small scales of competition and dispersal, species of any degree of similarity, species-specific values of all parameters and explicit resource competition. These generalizations increase the algebraic complexity of the result, but leave it qualitatively unchanged. The result is:

$$\alpha^{est} = \alpha \left[\frac{M\sqrt{1-\alpha} - w}{M\sqrt{1-\alpha} + w} \right] \qquad w = \frac{\beta}{f-\mu} \sqrt{\frac{2\mu}{f-\mu}} \qquad (15.9)$$

Equation 15.9 has three major implications. First, the measured strength of competition decreases with decreasing size of the ecological scales (mean dispersal distance M and spatial scale λ of competition) (Fig. 15.4). This effect is large. The strength of competition is reduced nearly tenfold by the reduction of M shown in Fig. 15.4. Note that the fact that α^{est} in Equation 15.9 may be negative for very small M is an artefact of the approximation (relatively large M) used to derive the simple formula. The strength of competition given by the general formula is always non-negative.

Second, α^{est} is a humped function of α (Fig. 15.5). If species are sufficiently similar, then the *measured strength of competition decreases as ecological similarity increases*. This is the precise opposite of the fundamental assumption of classical competition theory. It is easy to prove

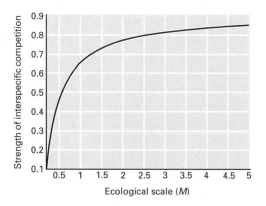

Fig. 15.4 Strength of competition, α^{est}, from Equation 15.9 as a function of the size of the ecological scales, M. Note that α^{est} approaches zero as the scales of dispersal and competition become short. $\alpha = 0.9$; $w = 0.05$.

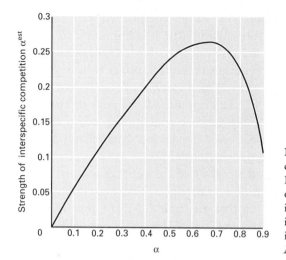

Fig. 15.5 Strength of competition, α^{est}, from Equation 15.9 as a function of the relative strength of interspecific and intraspecific interference, α. Note that α^{est} is a humped function of α. $M = 0.2; w = 0.05$.

(by differentiating α^{est} with respect to α) that the relationship between the strength of competition and ecological similarity is humped for all finite values of M and all positive values of w, and that the position of the hump increases as M increases. These analytical results are confirmed by simulated removal experiments within runs of the computer model. Figure 15.6a verifies that the measured strength of competition does indeed decrease as M decreases. Figure 15.6b confirms that α^{est} is a humped function of α and also shows the results of parallel experiments in a community with spatially global competition (effectively infinite M).

Third, α^{est} is an inverse measure of resilience simply because low α^{est} means that a decrease in the density of one species causes increased population growth primarily of that species rather than increased population growth of its competitor. This returns the relative abundance toward the pre-perturbation level. Thus, the final implications of Equation 15.9 are that resilience increases as the ecological scale (M) decreases and that resilience is *lowest* at intermediate levels of ecological similarity. Resilience actually increases as species become more similar to the right of the hump in Fig. 15.5. These implications are confirmed by runs of the computer model (Figs 15.7 and 15.8). Figure 15.7 shows that resilience increases as the scales of dispersal and competition decrease and Fig. 15.8 shows that resilience first decreases and then increases as ecological similarity increases.

Why are these counter-intuitive results obtained? The answer is that spatial distributions become increasingly interspecifically segregated as M decreases and as α increases. Recall that two 'forces' shape the spatial distributions. Local dispersal causes clustering, but only within-species clustering. Local competition pushes individuals apart regardless of what species they belong to. Collectively, these forces lead to intraspecific spatial aggregation and interspecific spatial segregation. This pattern is strongest if α is large and M is small because the interspecific

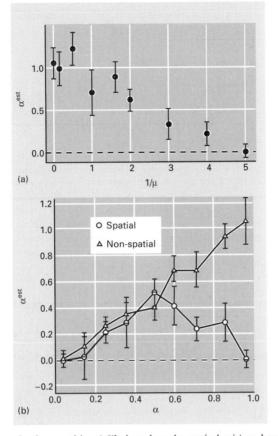

Fig. 15.6 Strength of competition (α^{cst} plotted on the vertical axis) as determined from simulated removal experiments in runs of the computer model. In every experiment, each of the two species was initiated at a population density of 5 and the model was iterated for 100 time steps (to approximate steady state). Then, 20% of individuals of either species was removed at random and the model was iterated for a further 0.2 time steps and recensused. The strength of competition was calculated as the average of the two species' change in population size following removal of heterospecific individuals divided by the change following removal of conspecifics. There were nine replicates for each set of parameter values. The mean of these nine is shown by a point and ± one standard error by the attached bars. (a) $\Delta t = 0.1; f = 3.2; \beta = 0.28; \gamma = 0.27; \mu = 0.4$; the one-dimensional habitat was 1000 units long and wrapped onto a torus. Note that the horizontal axis gives the *reciprocal* of the ecological scales M and that competition increases as M increases. (b) Circles depict runs with finite ecological scales ($M = 0.2$) while triangles depict runs with spatially global competition and infinite dispersal. Note that α^{cst} is a humped function of α as predicted by Equation 15.9. To vary α, we varied only γ and kept β constant at 0.28. All other parameter values were as in (a).

interference that leads to between-species segregation is then large, the dispersal that leads to intraspecific aggregation is then small and long dispersal events that break up spatial pattern are then rare. One measure of spatial segregation and aggregation is the spatial covariance. Suppose that we establish many pairs of small quadrats. The members of each pair are a distance x apart and the quadrats are small enough that they contain either one or no individuals. At equilibrium, the covariance

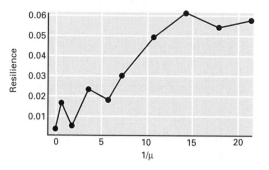

Fig. 15.7 Resilience as a function of spatial scale $(1/M)$ for the runs of the computer model in Fig. 15.6a. Let Υ_c be the change in population size of a species over a period of $t = 0.2$ after the removal of 20% of conspecifics, and let Υ_h be the corresponding change in abundance of the other species (following removal of the first species). The quantity $\Upsilon = \Upsilon_c - \Upsilon_h$ measures resilience. Large Υ_c and small Υ_h indicate that each species recovers quickly from the perturbation. Each point shows the mean of nine replicate runs in Fig. 15.6a. Note that resilience increases as the scales of competition and dispersal decrease.

between the population densities of the two species in these quadrats is approximately:

$$\text{interspecific covariance} = -\frac{N^*}{2M\sqrt{(1-\alpha)N^*/\mu}}\exp^{-\frac{x}{M}\sqrt{(1-\alpha)N^*/\mu}}$$

(15.10)

where N^* is the average density of individuals of either species in a quadrat. Equation 15.10 is obtained under the same conditions as Equation 15.9 (relatively large M and α near 1), but like Equation 15.9 is qualitatively correct in the general case. Note that species become more segregated as α increases and that the spatial extent of the segregation increases with M (Fig. 15.9).

The strength of competition (α^{est}) decreases as the scales of dispersal and interindividual interactions (M) decrease, simply because spatial segregation increases as M decreases. Together, interspecific spatial segregation and local competition ensure that the removal of an individual will increase primarily the performance of conspecific plants because most of the neighbours of the plants that are removed will be conspeci-

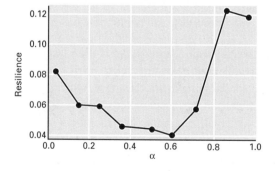

Fig. 15.8 Resilience as a function of degree of similarity (α) for the runs of the computer model in Fig. 15.6b. Resilience is measured as described in the caption of Fig. 15.7. Each point shows the mean of nine replicate runs in Fig. 15.6b. Note that resilience reaches a minimum for intermediate α.

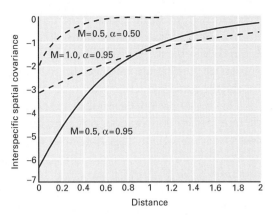

Fig. 15.9 The interspecific spatial covariance (Equation 15.10) as a function of spatial lag ($N^* = 0.5$; $\mu = 0.4$).

fics. This reduces interspecific competition and enhances resilience. The relationship between the strength of competition and the degree of ecological similarity is humped because an increase in α both increases the intensity of interference by heterospecific neighbours and decreases the likelihood that a plant has heterospecific neighbours. The hump occurs where the increasing spatial segregation overwhelms the increasing intensity of the interspecific interference.

The *spatial segregation hypothesis* states that finite dispersal and spatially local interactions lead to spatial structure that enhances ecological stability (resilience) and biodiversity. The magnitude of this effect is largest precisely where it is most critical, among those similar competitors which lack other mechanisms that reduce interspecific competition. Classical competition theory states that communities are stable because they are composed of dissimilar species. The spatial segregation hypothesis states that communities of similar plant species are stable (e.g. recover rapidly from perturbation) because the similarity of the species leads to stabilizing spatial structure.

15.6 Empirical evidence for the spatial segregation hypothesis

In an unpublished manuscript written in 1991, John Kelly and Christopher Tripler reviewed 302 field removal experiments published in *Ecology* over the previous decade. Their study divides the experiments into two classes: (i) plant-centred experiments in which neighbours were removed from the vicinity of a focal plant or a small cluster of plants, and (ii) plot-centred experiments in which quadrats were placed without reference to the locations of plants (usually at random). Although the two types of experiments involved similar plot sizes and fractions of biomass removed, statistically significant interspecific competition was detected in 72% of the plant-centred experiments (48 of 66 experiments) but in only 26% of the plot-centred experiments (60 of 234 experiments). The principal difference between the two types of experiments is that fine-scale spatial structure (within-quadrat spatial segregation) could separate plants from potential competitors in the plot-centred experiments but not in the plant-centred experiments. Thus, it appears that fine-scale interspecific spatial segregation was responsible for the threefold reduction in competition detected.

This result provides empirical support for both the prevalence and the potency of the competition-reducing spatial structure predicted by our simple model. It does not, however, demonstrate that the mechanisms in our simple model (local competition and dispersal) are responsible for the spatial structure in nature. An alternative to the mechanisms in the simple model is the possibility that habitats are spatially heterogeneous at fine scale and that plant species segregate on different habitat types (say because each species is the dominant competitor on a different habitat type). Obviously, it would be possible to test between these alternatives with field experiments.

During the last 10 years, I have been involved in three studies that estimated and tested models of plant communities in the field. Most recently, Pacala *et al.* (1996) calibrated a model of the nine dominant and subdominant tree species in transition oak–northern hardwood forests in north-eastern North America. The model is conceptually similar to the simple model described in this chapter, but is considerably more complex. It includes the mechanistic basis of the competition (which in this system is primarily competition for light), three spatial dimensions, and both age and size structure. Parameters and functional forms were estimated from field data and the model is capable of predicting a wide variety of natural phenomena including the rate and nature of succession, community composition and spatial distributions. Like the simple model, the forest model predicts spatial segregation among strategically similar species. In particular, eastern hemlock (*Tsuga canadensis*) and American beech (*Fagus grandifolia*) are predicted to form largely monospecific and spatially segregated patches of up to 1 ha in size. Such patches are easily observed in the field. With infinite spatial scales (analogous to infinite M in the simple model), the patches disappear, biodiversity collapses, succession progresses twice as fast and the forest supports only one-half the living biomass (Pacala & Deutschmann 1996). Thus, the interplay between spatial structure and dynamics in the forest model has a large effect on community and ecosystem dynamics and structure.

Of the remaining two studies involving calibrated models, one also showed a large effect of spatial processes (in a white clover–perennial grass assemblage; Cain *et al.* 1995), but the other showed no effect. The latter study was of an artificially depauperate community composed of two species of annual weeds on tilled land (Pacala & Silander 1990). Because of the large spatial scales of dispersal and competition in this community, dynamics were non-spatial like the dynamics of the simple model in the limit of large M.

To gain personal experience with the spatial segregation hypothesis, go outside to the nearest place where you can look at multispecies vegetation from above (any species-rich lawn will do). Estimate the abundance of a species in two ways: first by counting individuals close to randomly chosen plants of another species (say within a circle with radius equal to half average canopy diameter), and second by counting individuals within the same distance of randomly chosen points in space. If the first estimate is less than the second, then the species are spatially segregated. A good rule of thumb is that a bias of $n\%$ (where the first estimate is $n\%$ of the second) is equivalent to a $(100 - n)\%$ reduction in the strength of interspecific competition. I think you will find that reductions exceeding 50% are quite common.

15.7 Conclusions

Between 10 and 20 years ago, the widespread perception that ecologists lacked a mathematical theory of plant populations and communities led to repeated calls for the development of new models (Werner 1976; Antonovics & Levin 1980; Weiner & Conte 1981; Crawley 1986b). During the past decade, mathematical and empirical ecologists have collaborated on the development of a useful theory. Results from simple models of ideas are now routinely subjected to experimental tests that are the envy of animal ecologists (whose experimental subjects typically run away). Moreover, we have a growing body of empirically calibrated and tested models that allow one to extrapolate beyond the spatial and temporal scope of practical experiments. My own view is that we need to expend more effort in the area of data-defined models, both to sharpen the connections between our models and nature, and because the majority of phenomena of interest occur at spatial and temporal scales beyond the reach of field experiments.

The theory is probably most incomplete in its treatment of the interdependence of spatial structure and dynamics. The absence of models (other than computer simulators) that include local competition and finite dispersal is understandable because we have lacked the mathematical methods necessary to analyse these. The results developed here about the spatial segregation hypothesis show that even the oldest of chestnuts from competition models may be overturned by the effects of finite dispersal and local competition. The phenomenon is as large as it is unreported. The spatial structure that develops because of finite dispersal and local interactions drastically reduces the strength of competition (Fig. 15.4), enhances resilience and apparently decreases by a factor of two to three the detection of competition in field studies (see also Law & Watkinson 1989). Because new mathematical methods now permit one to analyse individual-based models with finite dispersal and local interactions, I suspect that this gap in the current theory will soon be closed.

16: Plants in Trophic Webs

James P. Grover and Robert D. Holt

It is obvious to all students of nature that a paramount question is: who eats whom? Answers are presented as food webs: networks of arrows and boxes. About 20 years ago, the initial, descriptive approach to food webs was replaced by a more abstract and analytical one (e.g. Pimm 1982); the current state of such studies is reviewed by Pimm *et al.* (1991) and Polis & Winemiller (1996) and a critique is provided by Paine (1988). Fascinating though it is, much of this literature is not directly useful to the plant ecologist. Very few food-web studies attempt to represent all species present in a habitat, and plants seem to suffer more from empirical myopia than any other organisms with the possible exception of decomposers. Often, plants are simply lumped into coarse categories – canopy trees versus grasses, phytoplankton versus benthic algae. Processes within these compartments that might influence the rest of the food web, such as the strong intraspecific and interspecific competition suffered by plants (Crawley 1990; see Chapter 15), may be obscured by this kind of aggregation. This competition usually takes place over abiotic resources: light, water and nutrients (Chapter 8). Nutrients are often supplied to plants mainly by decomposition and nutrient recycling, processes that are ignored in many food-web studies (but see DeAngelis 1992). Herbivores fare rather better than plants and decomposers in published food-web studies: many are identified to species, although coarse categories are often used, especially for invertebrates.

These and other deficiencies in the empirical basis of food-web research have stimulated new studies that attempt to catalogue organisms and processes more completely (e.g. Polis 1991). These efforts are an important way forward, but we take another tack in this chapter by adopting a theoretical approach that examines the properties of a small number of trophically connected populations. Rather than studying whole food webs, we focus on food-web 'modules' (Fig. 16.1), which allow us to represent important processes realistically. We emphasize the three key processes crucial for understanding plants: (i) herbivory; (ii) the dependence of plant growth on nutrient supply and recycling, and (iii) competition among plants for nutrients. This approach synthesizes the fundamental principles of population biology, first sketched by Hairston *et al.* (1960), with principles of ecosystem ecology concerning nutrient dynamics. Our presentation is graphical and intuitive; more

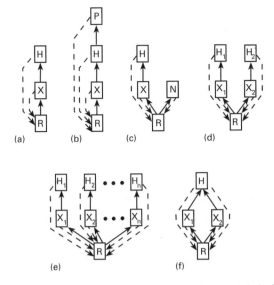

Fig. 16.1 Food-web modules: solid arrows show trophic flows and dashed arrows flows of recycled nutrient. (a) Simple food chain with an abiotic nutrient resource (R), a plant (X) and a herbivore (H). (b) Simple food chain with a nutrient, a plant, a herbivore and a predator (P). (c) Inedible plant module: of two plants, one is edible by the herbivore, and the other (N) is inedible; the two plants compete for the nutrient. (d) Module of two plants (X_1, X_2) competing for the same nutrient, in which each plant is eaten by its own specialist herbivore (H_1, H_2). (e) The specialist-herbivore module can be extended to an indefinite number of coexisting plants and herbivores, if certain conditions are met. (f) Module with a generalist herbivore feeding on two competing plants.

mathematically rigorous treatments are presented elsewhere (Grover 1994; Holt *et al.* 1994).

We begin with a simple food chain (Fig. 16.1a), based on an abiotic nutrient of concentration R, consumed by a plant of population density X, which is in turn consumed by a herbivore of population density H. Nutrient recycling occurs in two ways: (i) death of plants and herbivores, and (ii) incomplete assimilation of nutrient ingested during herbivory. Throughout, we assume that the microbial processes of decomposition responsible for nutrient recycling are rapid enough so that nutrient lost from populations appears effectively instantaneously in an available form (see Chapter 3). If the habitat is closed, or if it is an open system with inputs that balance outputs, then the total concentration of nutrient (S), both dissolved and bound in organisms, is constant in time. We also assume that the nutrient content (q) of organisms is constant. Therefore, whatever equations we write for the dynamics of the system components, a mass-balance constraint will be obeyed:

$$S = R + q_X X + q_H H \qquad (16.1)$$

To construct an equilibrium analysis of this simple food chain, we assume that herbivore growth rate increases with food supply and that herbivores experience a constant, density-independent loss rate, δ. This immediately fixes a plant density representing the food required by the herbivores to balance their losses, producing an equilibrium (Fig. 16.2a). We call this density $X^*_{(XH)}$, where the superscript * denotes dynamic equilibrium and the subscript (XH) denotes that the equilibrium pertains to the food-web module with plant X and herbivore H. Several factors influence the magnitude of $X^*_{(XH)}$. Among these is the herbivore's functional response, the rate at which an individual consumes plant material, which depends on several plant characters affecting attractiveness and ingestibility and on the herbivore's foraging behaviour. The magnitude of $X^*_{(XH)}$ also depends on the plant's nutritional value to herbivores. A 'good' food, one that is highly ingestible and nutritious, has a low value of $X^*_{(XH)}$, because the herbivore can balance a given loss rate with a low density of that food (Fig. 16.2a). With poor foods, a higher plant density is required for the herbivore to balance the same loss rate (Fig. 16.2a).

With the equilibrium plant density thus fixed, we then find the equilibrium nutrient concentration and herbivore density graphically, in the plane of nutrient concentration (R) versus herbivore density (H). The first tool we need is a zero net growth isocline, or simply isocline, describing the set of points (R,H) for which the plant population is at equilibrium. Suppose that plant growth rate increases with nutrient concentration and that plants experience a fixed, density-independent loss rate, ε. Then without herbivores, plants require a nutrient concentration $R^*_{(X)}$ (where the subscript (X) means a food-web module with only the plant present) to balance their losses and achieve an equilibrium (Fig. 16.2b; see Chapter 8). With herbivores present, plants suffer a higher loss rate and thus require a nutrient concentration higher than $R^*_{(X)}$. Therefore the plant isocline in the RH plane is a forward-sloping curve (Fig. 16.2c): above and to the left of the curve plant density decreases (too little nutrient and too many herbivores); below and to the right, plant density increases (few herbivores, plenty of nutrient). The more preferred a plant is by the herbivore, the greater its losses will be at a given herbivore density, and the more steeply the nutrient concentration required for equilibrium will increase with herbivore density. Therefore, a preferred plant has a lower isocline than a less preferred one (Fig. 16.2c), other things being equal.

The second tool we need comes from considering the mass balance of the nutrient. Setting X to its equilibrium value, and rearranging

Fig. 16.2 (*opposite*) Graphical analysis of a simple food chain with a nutrient, a plant and a herbivore. (a) Per capita growth rate of the herbivore population is an increasing function of plant density. At equilibrium with a given loss rate (δ, horizontal line), the herbivore requires a lower density of a good (i.e. nutritious) plant ($X^*_{(XH)}$, determined by the intersection of the growth curve with line δ). (b) Per capita growth rate of the plant population is an increasing function of nutrient concentration. In the absence of herbivores, plants experience a loss rate ε, which determines an equilibrium nutrient requirement, $R^*_{(X)}$. (c) Isocline of zero net growth for the plant population in the plane of nutrient concentration and herbivore density, given that the herbivore's net growth is

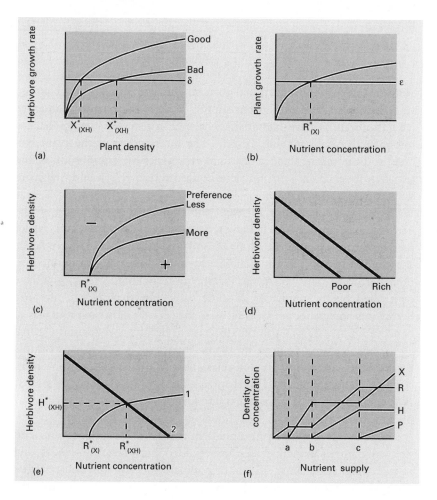

also zero. The plant isocline intersects the axis of nutrient concentration at $R^*_{(X)}$, the plant population's nutrient requirement in the absence of herbivory. The isocline of a plant more preferred by the herbivore lies below that of a less preferred plant. (d) Mass-balance constraints in the plane of nutrient concentration and herbivore density, given that the herbivore's net growth is zero. Nutrient-rich habitats have higher constraint lines than nutrient-poor habitats. (e) Equilibrium of the simple food chain is determined graphically by the intersection of the plant isocline with the mass-balance constraint. (f) Equilibrium densities of plants, herbivores and predators in simple food chains vary in a predictable way with the habitat's nutrient supply. In the poorest habitats, up to point a, there is insufficient nutrient to support plants or other populations, and nutrient concentration (R) simply increases with supply. From point a to point b, nutrient supply is high enough to support plants but not herbivores, and increasing nutrient leads to higher plant density (X), although nutrient concentration is controlled by plants. From point b to point c, nutrient supply is high enough to support herbivores, whose population density (H) increases with nutrient supply, as does nutrient concentration, while plant density is controlled by herbivores. In the richest habitats, above point c, nutrient supply is high enough to support predators, whose population density (P) increases with nutrient supply and who control herbivore density, while plant density also increases with nutrient supply; nutrient concentration is controlled by plants (Oksanen *et al.* 1981).

Equation 16.1, we get the equation of a straight line in the *XH* plane:

$$H = \frac{1}{q_H}(S - q_X X^*_{(XH)}) - \frac{1}{q_H}R \qquad (16.2)$$

whose elevation is proportional to total nutrient supply *S*, but whose slope is independent of nutrient supply. Therefore, rich and productive habitats (high *S*) have higher mass-balance constraints than poor and unproductive habitats (Fig. 16.2d). The intersection of the plant isocline and the mass-balance constraint represents the equilibrium of the two-step food chain, with nutrient concentration $R^*_{(XH)}$ and herbivore density $H^*_{(XH)}$ (Fig. 16.2e).

These two quantities increase with nutrient supply, and if the herbivore density in a rich habitat is sufficiently high we can expect predators to invade, giving a longer food chain (Fig. 16.1b). If predators are food-limited in the way that we previously imagined for herbivores, then herbivore density at equilibrium, now denoted $H^*_{(XHP)}$, will be controlled by the predators at some value less than $H^*_{(XH)}$. Because herbivores now suffer predation losses, their food requirement increases, hence $X^*_{(XHP)} > X^*_{(XH)}$, and the decrease in herbivory reduces the nutrient concentration required by plants, hence $R^*_{(XHP)} < R^*_{(XH)}$. Simple food chains, including longer ones, follow a pattern of 'top-down' control, as suggested by Hairston *et al.* (1960) and since elaborated (Oksanen *et al.* 1981). Each trophic level is alternately controlled by predation or competition for limiting resources. According to this theory, habitats arrayed along a gradient from nutrient-poor to nutrient-rich will contain an increasing number of trophic levels (Fig. 16.2f), and in every habitat dynamic control of population density propagates from the highest trophic level to the basal, a top-down process dubbed the 'trophic cascade' by Paine (1980). The theory of the trophic cascade has been enormously influential, especially in limnology. However, there are examples from aquatic systems (DeMelo *et al.* 1992) and terrestrial systems (Sih *et al.* 1985; Schoener 1989) where food-chain theory appears to be inadequate. One possible reason for this inadequacy is heterogeneity within the trophic level labelled 'plants' – up to this point, we have assumed that all plants are equal.

We explore the consequences of differences between plants later; first, we consider another factor that might obscure the patterns predicted by food-chain theory, namely its reliance on equilibrium analyses. In productive habitats, the plant–herbivore equilibrium can be unstable if the herbivore's growth rate saturates with plant density, as in Fig. 16.2a (Rosenzweig 1971). This is likely, since the ability of a herbivore to consume vegetation cannot increase indefinitely as plant density increases. In most models of two-step food chains, a stable limit cycle develops when the plant–herbivore equilibrium is unstable (DeAngelis 1992). With more trophic levels, there are more possible sources of instability (predator–prey interactions) and, moreover, a wide range of non-equilibrium dynamics can result from the response of food chains to environmental variability. The extent to which such dynamic instability contributes to observed variability in natural populations is unclear. Moreover, spatial heterogeneity, much of it created by plants themselves, can counter tendencies to instability in trophic interactions (Murdoch & Oaten 1975; Taylor 1990), allowing application of an equilibrium approach, though perhaps only at certain spatial scales.

Setting aside the question of non-equilibrium dynamics (see Chapter 14), not all kinds of plants are equally edible to herbivores. As a plant becomes less edible, its isocline in the RH plane rotates counter-clockwise (Fig. 16.2c). For the limiting case of a completely inedible plant (denoted N, Fig. 16.1c), the plant isocline becomes a vertical line (Fig. 16.3a), regardless of herbivore density; such a plant requires a nutrient concentration of $R^*_{(N)}$ to balance its losses. Under certain technical assumptions, the inedible plant (N), edible plant (X) and

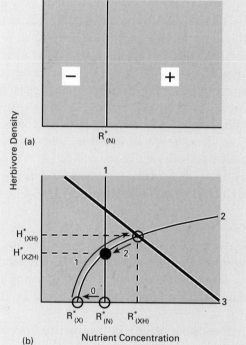

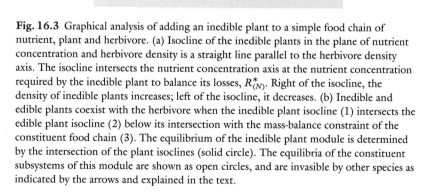

Fig. 16.3 Graphical analysis of adding an inedible plant to a simple food chain of nutrient, plant and herbivore. (a) Isocline of the inedible plants in the plane of nutrient concentration and herbivore density is a straight line parallel to the herbivore density axis. The isocline intersects the nutrient concentration axis at the nutrient concentration required by the inedible plant to balance its losses, $R^*_{(N)}$. Right of the isocline, the density of inedible plants increases; left of the isocline, it decreases. (b) Inedible and edible plants coexist with the herbivore when the inedible plant isocline (1) intersects the edible plant isocline (2) below its intersection with the mass-balance constraint of the constituent food chain (3). The equilibrium of the inedible plant module is determined by the intersection of the plant isoclines (solid circle). The equilibria of the constituent subsystems of this module are shown as open circles, and are invasible by other species as indicated by the arrows and explained in the text.

herbivore (H) will theoretically coexist if it is possible to construct a sequence of invasions leading to the full community (Hutson & Law 1985). At each invasion, we imagine that an infinitesimally small population of the invader is trying to increase, while residents are at equilibrium. Suppose that total nutrient supply exceeds the nutrient requirement of either plant growing alone (i.e. $S > R^*_{(X)}$ and $S > R^*_{(N)}$), so that either plant can invade an empty habitat. If the edible plant invades first, and then equilibrates at a density exceeding the herbivore's equilibrium food requirement, then the herbivore can also invade (pathway 1, Fig. 16.3b), eventually reaching the equilibrium density $H^*_{(XH)}$ (Figs 16.2e and 16.3b). This happens if nutrient supply is high enough. The inedible plant can now invade if the associated nutrient concentration, $R^*_{(XH)}$, exceeds its equilibrium nutrient requirement, $R^*_{(N)}$ (pathway 2, Fig. 16.3b). Verbally, this last condition says that the inedible plant is a better resource competitor than the edible plant, when the latter is burdened by herbivory. That is, a good resource competitor is one that reduces nutrient concentration to a low level, through its own consumption, thus denying the resource to other plants (this is the well-known 'R^* rule'; see Chapter 8).

However, the inedible plant cannot be too good a resource competitor. Suppose that $R^*_{(N)} < R^*_{(X)}$. Then, even when herbivores are absent, the inedible plant competitively excludes the edible plant, according to the R^* rule. Therefore, another condition for three-species coexistence is that the edible plant must be a better resource competitor than the inedible plant, in the absence of herbivory (this is the palatability/competitive ability trade-off; see Chapters 13 & 15). Then, if the habitat is dominated by the inedible plant, the edible plant can invade (pathway 0, Fig. 16.3b). The formal ordering relation

$$R^*_{(X)} < R^*_{(N)} < R^*_{(XH)} \qquad (16.3)$$

is an 'assembly rule', expressing conditions required for assembly of a food-web module by a sequence of invasions; Equation 16.3 can be interpreted as a generalization of the R^* rule applying to competitive dominance when plants compete for a nutrient in the absence of herbivores.

Of course, an inedible plant with no natural enemies is an unexploited resource. Suppose that a new herbivore arrives on the scene (immigrates or evolves), specializing on the inedible plant (Fig. 16.1d). This has the effect of increasing the equilibrium nutrient requirement of the previously inedible plant above its previous requirement, $R^*_{(N)}$, to compensate for losses to herbivory. So long as this increased requirement is not too large, all four species can coexist at equilibrium. The process of adding another inedible plant, followed by a specialist herbivore that eats it, can be continued indefinitely (Fig. 16.1e), so long as certain ordering relations are satisfied generalizing Equation 16.3 (Grover 1994). Verbally, each new species can invade if (i) plants invade in order of decreasing ability to compete for nutrient in the absence of herbivory, and (ii) each plant's herbivore invades before another plant is added. If a species violating these rules tries to invade, one of two things happens: either it is excluded from invasion by the residents, or it successfully invades but disrupts the relations allowing residents species to coexist, so that a cascade of extinctions take place. In the specialist-herbivore module, herbivores of the most competitive plants are 'keystone' species, whose role is crucial to community organization and which can be identified from the assembly rules. Releasing highly competitive plants from the burden of herbivory allows them to exclude other plants, and their herbivores.

Some insects may play the role we have sketched, of specialist herbivores that hobble a plant's competitiveness (Crawley 1989), but the world also has generalist herbivores and plenty of predators to eat them. Food-web theory is not yet able to specify assembly rules for communities with arbitrarily complex trophic connections. Nevertheless, we expect that more complex communities also have assembly rules that predict the outcome of species addition and deletion, and allow identification of keystone species. The assembly rules we have encountered so far, such as Equation 16.3, are based on quantities, such as $R^*_{(X)}$ and $R^*_{(XH)}$, that summarize a great deal of biology: the physiological and morphological characters determining plants' uptake of nutrients, growth and reproduction, their edibility and nutritional value to herbivores, herbivore foraging behaviour, life history and demography (see Chapter 13). We expect that more complex trophic webs will be characterized by similar critical quantities but that their assembly rules will become more complicated and contingent, as the foraging behaviours and ecological quirks of more and more herbivores and predators must be accounted for.

Taking a few steps in this direction, we now sketch some of the rules that apply to a food-web module with two plants and a generalist herbivore (Fig. 16.1f). In Fig. 16.2c, we drew isoclines for two plants differing in their preference to a herbivore. Likely causes of reduced herbivore preference are low nutritional value and morphological or chemical defences against herbivores (see Chapters 10 & 13). Such defences usually come at a cost of energy and nutrients (Chew & Rodman 1979, Bergelson & Purrington 1996), making the less preferred plant less competitive when herbivores are absent. Therefore, Fig. 16.2c would be more accurate if the preferred plant's isocline had a lower R intercept than the less preferred plant's, i.e. $R^*_{(1)} < R^*_{(2)}$, where the subscripts 1 and 2 represent preferred and less preferred plants, respectively. Then, since the isocline of the preferred plant has a lower slope than that of the less preferred plant, there will be an intersection of the two plants' isoclines, potentially representing an equilibrium in which they coexist (Fig. 16.4a–c). Either because of plant defences that reduce nutritional value, or because it is adaptive for herbivores to prefer

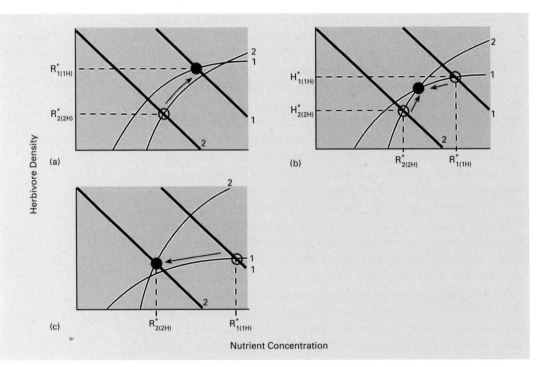

Fig. 16.4 Graphical analysis of the generalist herbivore module. In each graph, the isoclines and mass-balance constraints of simple food chains based on plants 1 and 2 are superimposed. (a) Nutrient-poor habitats. The intersection of the plant isoclines is above the highest of the two plants' mass-balance constraints. The equilibrium of the food chain containing the less preferred plant 2 (open circle) is invasible by the preferred plant (arrow), but the equilibrium of the food chain with the preferred plant 1 (solid circle) is not invasible by the less preferred plant 2. Plant 1 owes its victory to apparent competition. (b) Moderate habitats. The intersection of the plant isoclines falls between the two plants' mass-balance constraints (solid circle). The equilibria of the simple food chains (open circles) are each invasible by the missing plant (arrows) and coexistence occurs, due to a trade-off between resource and apparent competition. (c) Rich habitats. The intersection of the plant isoclines is below the lowest of the two plants' mass-balance constraints. The equilibrium of the constituent food chain containing the preferred plant 1 (open circle) is invasible by the less preferred plant (arrow), but the equilibrium of the food chain with the less preferred plant 2 (solid circle) is not invasible by the preferred plant 1. Plant 2 owes its victory to resource competition.

nutritious plants, plant 1 is likely to be a better food for the herbivore, so that less of it is required to support the herbivore than is required of plant 2 (Fig. 16.2a). That is $X^*_{1(1H)} < X^*_{2(2H)}$, where the simple food chains based on the two plants are distinguished by the subscripts $(1H)$ and $(2H)$. The nutrient cost of defences against herbivores could also result in a higher nutrient content for the less preferred plant '$(q_1 < q_2)$'. These two hypotheses imply that plant 1 has a higher mass-balance constraint than plant 2 (Equation 16.2).

If the habitat is relatively nutrient-poor, then plant 1's mass-balance constraint lies below the intersection of the plants' isoclines (Fig. 16.4a). Because the band of the *RH* plane lying between the two plants' mass-balance constraints represents the feasible space for a three-species equilibrium (Holt *et al.* 1994), both plants cannot coexist with the herbivore in a nutrient-poor habitat. Instead, when plant 1's food chain is at equilibrium, the corresponding point in the RH plane falls in a region where plant 2 has negative net growth; plant 1 excludes plant 2. However, when plant 2's food chain is at equilibrium, the corresponding point in the RH plane falls in a region where plant 1 has positive net growth; therefore, plant 1 invades. The competitive victory of plant 1 is signalled by the fact that it supports more herbivores at equilibrium than does plant 2 ($H^*_{1(1H)} > H^*_{2(2H)}$). The outcome of competition follows an 'H^* rule' because the generalist herbivore module involves apparent competition, the indirect interaction between the two plants mediated by the herbivore that eats them both (Holt 1977). In this type of competition, it is ability to support a generalist natural enemy that counts, rather than ability to depress resource availability.

For somewhat richer habitats, the band of feasible space lying between the plants' mass-balance constraints rises to encompass the intersection of the plant isoclines (Fig. 16.4b). Invasion analysis shows that when plant 1's food chain is at equilibrium, the corresponding point in the RH plane falls in a region where plant 2 has positive net growth; therefore, it invades. When plant 2's food chain is at equilibrium, the corresponding point in the RH plane falls in a region where plant 1 also has positive net growth; so plant 1 also invades. This mutual invasibility implies coexistence of the two plants that is mediated by the herbivore. Coexistence rests on the trade-offs and constraints dictating that plant defences are costly in terms of the resource over which competition occurs. The less defended, preferred plant supports more herbivores in its own food chain than does the less preferred plant ($H^*_{1(1H)} > H^*_{2(2H)}$). Balancing plant 1's advantage in apparent competition is plant 2's advantage in resource competition ($R^*_{2(2H)} < R^*_{1(1H)}$), which results from a reduced nutrient demand, achieved by reducing losses to herbivory through defence.

For relatively rich habitats, the band of feasible space lying between the plants' mass-balance constraints rises above the intersection of the plant isoclines (Fig. 16.4c). Coexistence of the two plants is thus no longer feasible. Invasion analysis shows that when plant 1's food chain is at equilibrium, the corresponding point in the RH plane falls in a region where plant 2 has positive net growth; therefore, plant 2 invades. However, when plant 2's food chain is at equilibrium, the corresponding point in the RH plant falls in a region where plant 1 has negative net growth; plane 2 excludes plant 1. The competitive victory of plant 2 is signalled by an R^* rule: at equilibrium, plant 2 depresses nutrient concentration to a lower level than plant 1 ($R^*_{2(2H)} < R^*_{1(1H)}$). Therefore, in rich habitats, with high herbivore densities, investment in defence against herbivory pays off, and it does so by reducing the population-level requirement for nutrient.

For this case (Fig. 16.4a–c), which we consider likely, herbivore-mediated coexistence of competing plants occurs in moderately rich habitats, where apparent competition trades off against resource competition. If we reverse the positions of either the isoclines or the mass balance-constraints in Fig. 16.4b, we get cases in which each of the constituent food chains in the generalist herbivore module can exclude the missing plant. The three-species equilibrium at the intersection of the plant isoclines is then unstable. This requires a constellation of plant and herbivore characters that seem unlikely to us, but we caution that so far we have implicitly assumed that the herbivore functional response is linear. With non-linear functional responses, graphical analysis is more complex, since the plant isoclines cannot be constrained to single curves in the *RH* plane. Moreover, we have not done justice to the full range of possibilities suggested in the rich literature on trade-offs and constraints in plants' abilities to compete for resources and resist herbivory (reviewed by Louda *et al.* 1990; Bryant *et al.* 1991; Pacala & Crawley 1992). With a broader view, some of the cases we dismiss here as unlikely may look more plausible.

It is premature to sketch any formal theory for more complex trophic architectures involving predators as well as plants and their herbivores. However, predation is always in some sense a dependent variable, erected on the foundation of plants which extract abiotic resources from the habitat. Therefore, plants will exert a number of effects that propagate from the bottom up (Price 1992). Once there are consumers present in a food web, however, the possibility always exists for strong top-down effects reminiscent of those suggested by food-chain theory (above). Top-down effects need not necessarily arise only from the top trophic level, but could now arise from lower within food webs (Menge & Sutherland 1976). Other effects reverberating up *and* down the trophic web arise from the interactions of population and nutrient dynamics (Bryant *et al.* 1991; DeAnglis 1992).

Up to this point we have treated plants and herbivores as if they were plankton, and have taken no account of the spatial heterogeneity that is universal in terrestrial systems (Oksanen *et al.* 1996). Much of this spatial heterogeneity is created by the plants themselves (Price 1992). Plants exist in patches composed of different species, and individual plants consist of various different tissues, each of which is potentially a patch of resource or a habitat for some animal. The spectrum of spatial variability presented by plants makes possible a wide range of indirect interactions, since plants provide both cover from, and cues to, predators that attack the animals feeding on or living on a plant (Price *et al.* 1980; Sih *et al.* 1985). Such spatial heterogeneity allows a diversity of consumers to rely on a single plant species (Hutchinson 1959; Lawton 1983). Because plants are fixed in single positions for long periods of time, relative to the generation times of many animals, animals can evolve spatial behaviours such as territoriality, habitat selection and migration (Holt 1987; Ostfeld 1992) in response to a mosaic of plants. The physical structure created by plants thus amplifies the possibility for idiosyncrasies of particular species to become important, and it may be a major cause of the diversity of terrestrial life (Hutchinson 1959; Price 1992; see Chapter 15).

Notwithstanding the complexities and contingencies that such considerations entail, ecologists need to develop a robust and predictive understanding of nature. If at all possible, our theory must retain both simplicity and generality. Simple advice on management of our threatened world is the most likely to be heeded. Much of what we present here is inspired by findings from the field of limnology. So is our conclusion. From studying lakes, we know that trophic dynamics are important, and that human intervention, whether at the top or the bottom of the trophic web, has strong, potentially disastrous effects. Despite the complexity of life within lakes, simple and general principles for their management are emerging, based largely on an understanding of trophic webs (but not without controversy; Carpenter & Kitchell 1992; DeMelo *et al.* 1992). To the extent that a lake is a microcosm (Forbes 1887), we may urgently hope for global progress.

17: Plants and Pollution

Mike Ashmore

17.1 Introduction

The impact of pollutants derived from human activity on vegetation has been recognized for many centuries. Some of the earliest descriptions of damage to vegetation from air pollution are found in the English diarist John Evelyn's book, published in 1661, in which he describes the difficulties of growing garden flowers and edible fruits in the coal-smoke polluted atmosphere of London (Evelyn 1661). Similarly, the idea that pollutants can be transported in the air over long distances is not a new one; thus at the end of the last century, Ibsen was writing in Norway that 'a sickening fog of smoke from British coal drops in a grimy pool upon the land'.

Over the present century the range of chemicals discharged into the environment from industry, transport, agriculture, waste disposal and many other human activities has increased dramatically. Once discharged, these compounds may be physically dispersed, through air movement, river currents or groundwater flow; furthermore, they may undergo chemical transformation, which may alter their potential environmental impact. The environmental impact of these chemicals will depend on their concentration in the environment, or on the dose received or accumulated by the target organism. At low doses, the environment may be contaminated but no adverse effects ensue; normally the chemical is only referred to as a pollutant if it has some environmental impact.

Pollutant problems may vary greatly in their spatial scales. Some are very local in character, with the environmental impact of the pollutant restricted to the immediate vicinity of, for example, a road, a factory or an old mine. Other problems are regional in character, as a result of the long-range transport of pollutants such as acid rain and tropospheric ozone (O_3). Finally, some are global in character, such as the accumulation of carbon dioxide (CO_2) and other greenhouse gases in the atmosphere of our planet, or the impact of changes in the O_3 layer on UV-B radiation levels.

Similarly, pollutant impacts may vary on different temporal scales. Some impacts, for example, are the result of an accidental release of large pollutant concentrations, which may cause an immediate impact, and from which there may be a slow and gradual recovery, while others are the result of an accumulation of pollutant deposition over years or even decades.

This chapter aims to provide an overview of the ways in which pollutants can impact the ecological processes described earlier in this book. Inevitably, it is not possible to provide a comprehensive account of the effects of the vast range of chemicals emitted into the environment by human activity. Instead, the examples used will mainly focus on the impacts of air pollutants and of heavy metals. Table 17.1 summarizes the major sources of these pollutants, their major impacts of relevance to plant ecology and the spatial scale of their impact, together with key references that give a more detailed account of these pollutants.

17.2 Effects on individual plants

Before considering the impacts of pollution on plant communities, it is essential to assess the ways in which individual plants respond to pollutants. The impact of a pollutant on any individual plant is complex and involves a large number of factors, which are summarized in Fig. 17.1 These may be divided as follows.

1 *The dose of the pollutant received*. This will partly depend on the concentration of the pollutant in the relevant medium, and the duration of exposure to it. However, pollutants may be present in the environment without actually being taken up by the plant. For example, the availability of heavy metals depends strongly on soil chemistry, while for many air pollutants plant uptake is limited by the stomatal conductance.

2 *Intrinsic factors*. There is large degree of variation between plant species in their sensitivity to particular pollutants; in addition there is often substantial genetic variation in response within species. This variation broadly relates to the ability of the plant to restrict pollutant uptake or, once it has been taken up, to detoxify, metabolize or sequester the pollutant. Within the same genotype, other factors such as plant age and growth stage may also influence response to pollutants. Finally, it is important to realize the dynamics of this response, with pollutant exposure frequently inducing biochemical, physiological or morphological responses that subsequently lead to a reduction in its adverse effects.

3 *Environmental factors*. Environmental conditions may modify plant responses to pollutants in a number of ways. First, they may modify pollutant dose; for example, soil water stress may lead to stomatal closure and thus to a reduced uptake of air pollutants, and reduced plant damage (e.g. Tingey & Hogsett 1985). Second, environmental factors may reduce the capacity of plants to detoxify and assimilate pollutants; for example, it is well established that sulphur dioxide (SO_2) is more phytotoxic when plants are grown at low temperatures or under low light conditions. This has been shown both in controlled fumigations (e.g. Jones & Mansfield 1982) and by field observations of direct

Table 17.1 Summary of pollutants considered in this chapter.

Pollutant	Major sources	Major impacts	Scale of effects	References
Sulphur dioxide (SO_2)	Power generation; industry; commercial and domestic heating	Visible foliar injury; reduced plant growth; elimination of lichens and bryophytes; forest decline	Local	UK TERG (1988), Ashmore *et al.* (1988), Wellburn (1988), Last & Watling (1991)
Nitrogen oxides (NO_x)	Power generation; transport	Altered plant growth; enhanced sensitivity to secondary stresses; eutrophication	Local	UK TERG (1988), Wellburn (1988), Last & Watling (1991)
Acid rain	Secondary pollutant formed from SO_2 and NO_x	Soil and freshwater acidification; forest decline	Regional	Ashmore *et al.* (1988), Adriano & Johnson (1989), Last & Watling (1991)
Ammonia (NH_3)	Intensive agriculture	Altered plant growth; enhanced sensitivity to secondary stress; soil acidification; eutrophication	Local	Pearson & Stewart (1993)
Heavy metals (lead, copper, cadmium, etc.)	Mining; smelting	Decreased root and mycorrhizal activity; reduced soil microbial activity	Local	Lepp (1981), Freedman (1989)
Ozone (O_3)	Secondary pollutant formed from NO_x and hydrocarbons	Visible foliar injury; reduced growth; forest decline	Regional	UK TERG (1988), Wellburn (1988), Ashmore & Bell (1991)
Carbon dioxide (CO_2)	Fossil fuel combustion; deforestation	Increased plant growth; altered water balance	Global	Rozema *et al.* (1993)

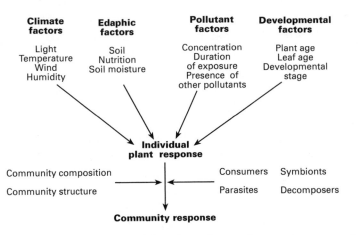

Fig. 17.1 Summary of factors influencing pollutant effects on plant communities.

forest damage (Makela *et al.* 1987). Third, exposure to pollutants may lead to changes in the morphology and/or physiology of the plant that make it more sensitive to environmental stresses; for example, increased deposition of sulphate or ammonium ions to conifer seedlings has been shown to increase their sensitivity to cold stress (Cape *et al.* 1991). Finally, pollutants seldom occur on their own, and the responses to pollutant combinations may be less than, greater than, or equal to the sum of their separate impacts; this has been shown for combinations of air pollutants (e.g. Ernst *et al.* 1985), for combinations of air pollutants and heavy metals and for elevated CO_2 levels in combination with air pollutants.

In general terms, four possibilities exist in terms of the impacts of pollutants on individual plants.

1   When the pollutant is present in sufficiently high concentrations, the plant may be completely destroyed.

2   At lower concentrations, the plant is able to survive, but its performance is adversely affected. For example, photosynthetic rates may be reduced, control of transpiration may be lost or root function impaired. These physiological effects will often lead to reduced biomass production or altered partitioning of resources between plant organs.

3   At yet lower concentrations, the physiology and growth of the plant may appear to be unaffected under optimal environmental conditions. However, the pollutant may cause subtle morphological or chemical changes that lead to altered tolerance of other environmental stresses, such as drought or frost, or altered susceptibility to pests and diseases.

4   Finally, in the case of certain pollutants, plant performance may be enhanced at appropriate concentrations. For example, certain toxic metals such as copper and iron are essential micronutrients for plant growth, while SO_2 can also stimulate plant growth at low concentrations, especially in soils that are sulphur deficient.

Pollutant exposure may affect not only vegetative growth but also reproductive performance. This may occur by impacts on carbon alloca-tion to reproductive organs but pollutants may also directly affect reproductive processes. Pollen germination has been shown to be sensi-tive to acid rain, SO_2 and O_3, often at relatively low concentrations (e.g. Mumford *et al.* 1972; Keller & Bada 1984; Percy 1986). Exposure of receptive stigmata to acidic rain has also been shown to reduce germina-tion following pollination, while there is also evidence of increased fruit abortion when the acid treatment followed pollination (Cox 1988). These effects on pollen germination may have implications for the vigour of progeny because of the decreased level of competition be-tween pollen (Cox 1992). Finally, pollution may have indirect effects on some species by reducing the numbers of insect pollinators; for exam-ple, there is evidence of reduced numbers of bees in conifer forests sprayed with insecticides, with an associated reduced fecundity in insect-pollinated plant species (Thaler & Plowright 1980).

17.3 Effects on species interactions

When a pollutant is present at a concentration that affects the physio-logy, growth or reproduction of individual plants, it is clear that the potential exists to influence the outcome of competition between species.

Several simple experiments with air pollutants such as SO_2 and O_3 have shown, as expected, that when a pollution-sensitive and pollution-resistant species (often a clover and grass species, respectively) are grown together, the presence of the pollutant shifts the competitive balance towards the pollution-resistant species. The precise mechanism of this response is not always clear, but one possibility is that allelopathic effects are involved. For example, Kochhar *et al.* (1980) found that leachate from fescue plants fumigated with O_3 inhibited nodulation of clover plants, but that from non-fumigated fescue plants did not.

While these simple experiments clearly demonstrate that pollutants can modify the outcome of plant competition, it is doubtful whether they provide much guidance as to how pollution modifies interactions between plant species in real communities. For example, Evans and Ashmore (1992) found that ambient air pollution at a rural site in southern Britain, where O_3 is the main pollutant, did not have the expected effects on the composition of an acid grassland community. At higher O_3 levels, the species composition shifted towards the O_3-sensitive forbs and away from the more resistant grass species. This result was explained by the observation that grass species were dominant in the community; when pollution levels were sufficient to reduce grass growth, the nominally more sensitive forbs could benefit from the increased light penetration through the canopy. Vertical gradients in O_3 through the canopy may also have been important, with lower concen-trations being found close to the ground. The importance of the vertical stratification of the plant community is also illustrated in other studies of responses to pollution stress. For example, Mclenahen (1978) examined deciduous forest stands in the Ohio Valley, USA, affected by different levels of industrial pollution and found a decline in density of the dominant tree species. This was accompanied by an increase in density in the lower canopy and in the shrub layer, presumably because of increased light penetration. Conversely, however, this increased density in the shrub layer was accompanied by a decreased density of the herb layer. Stratification may also be important below ground, since deep-rooting species may be able to survive surface soil contamination where shallow-rooted species cannot (e.g. Dawson & Nash 1980).

Increases in atmospheric CO_2 concentrations also have the potential to alter species composition, partly as a result of species differences in growth responses to changes in CO_2 concentration. For example, Arp *et al.* (1993) reported field studies which showed that the C_3 saltmarsh species *Scirpus olneyi* (Cyperaceae) showed a significant positive growth response to increasing CO_2 concentrations but that the C_4 species *Spartina anglica* (Poaceae) at the same location did not. In a mixed community of *Scirpus*, *Spartina* and *Distichlis spicata* (another C_4 grass), there was a larger increase in biomass of *Scirpus* in response to elevated CO_2, while that of *Spartina* declined, reflecting the influence of interspecific competition. The authors suggest that the increase in growth of *Scirpus* in pure stands is limited by intraspecific competition and self-shading, whereas in the mixed stand, where it is a relatively small component of species mix, it can expand further.

Another situation in which community composition may be influenced indirectly by changes in associated species, is in the case of epiphytic lichens. Many of these species are highly sensitive to direct effects of SO_2; they are often also highly sensitive to changes in bark chemistry. Where total acidic deposition is high, the buffering capacity of the bark may be a critical factor for lichen survival. For example, at UK sites with a high acidic deposition *Lobaria* spp. persist longer on *Fraxinus excelsior* (Oleaceae) than on *Quercus* spp. (Fagaceae) (Farmer *et al.* 1992), probably because of the higher cation status of its bark. Once bark has acidified, there may be a delayed response in terms of recovery of the lichen flora, especially on tree species, such as oaks, in which bark is not shed quickly (e.g. Bates *et al.* 1990). In The Netherlands, de Bakker (1989) reported that increased ammonia (NH_3) emissions have caused increased bark pH on oak species, but have also encouraged a more nitrophilic lichen flora.

There has been increasing evidence in recent years that exposure of vegetation to air pollution results in an altered performance of insect herbivores. At high concentrations, there may be direct effects on the insects themselves but, at the ambient pollution levels now found in most areas, the effect is probably mediated through chemical changes in the host plant. For SO_2 and nitrogen dioxide (NO_2), effects on amino acid composition may be particularly important, while for O_3 the response factors involved may be more complex. High NO_2 concentrations are probably a major factor in the high numbers of aphids, and other phytophagous insects, found alongside major roads (e.g. Fluckiger *et al.* 1978; Port & Thompson 1980). Increased CO_2 concentrations may reduce insect herbivore preformance (e.g. Fajer *et al.* 1991), possibly through increased C:N ratios or increased levels of phenolics and other defence compounds, although this may also be associated with increased levels of plant consumption to compensate for reduced nitrogen concentration (see Chapters 1 & 10).

Figure 17.2 summarizes the potential mechanisms through which air pollution can influence insect herbivore levels and hence plant performance (Bell *et al.* 1993), and serves to indicate some of the factors involved. It is interesting to note that increased insect damage can not only reduce control of transpiration but also increase pollutant penetration through the damaged cuticle (e.g. Warrington *et al.* 1989). Bell *et al.* (1993) provide a similar summary of pollutant interacting with plant pathogens; these organisms themselves very sensitive to pollutants such as SO_2, while chemical and biological changes on the leaf surface induced by pollutants may critically affect their performance.

Pollutant exposure may have an adverse effect on mycorrhizal associations, which are crucial to the performance, stress tolerance (including heavy metal tolerance) and competitive ability of many species. This

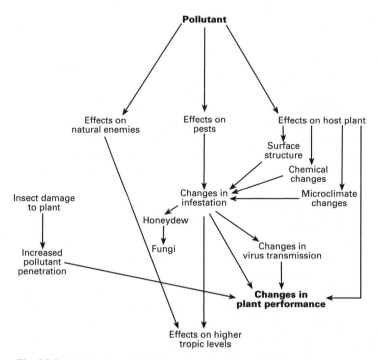

Fig. 17.2 Model of potential interactions involving air pollution, pests and host plants (from Bell *et al.* 1993).

may arise because of direct effects on some mycorrhizal types, by changes in soil chemistry induced by acid rain (e.g. Dighton & Skeffington 1987), possibly due to mobilization of aluminium, and by heavy metals. Alternatively, exposure to air pollutants above ground may influence mycorrhizal activity indirectly, by changes in carbon partitioning to the roots. For example, Norby *et al.* (1987) showed that exposure of *Pinus echinata* seedlings to elevated CO_2 levels resulted in increased fine root biomass and mycorrhizal density, as well as increased root carbon exudation, while Reich *et al.* (1985) reported a negative effect of O_3 exposure on mycorrhizal infection of oak seedlings.

17.4 Evolutionary responses

In competitive situations, it is likely that selection pressures will act in favour of individuals that are more resistant to a particular stress factor, and there is no reason that pollution should be an exception to this. Bradshaw and McNeilly (1991) identify three stages of population change in relation to a pollution stress. In the first, only the most sensitive genotypes are eliminated from the population. Next, all but the most resistant genotypes in the population are eliminated. Finally, the survivors interbreed, and among their progeny are even more resistant genotypes, which are then further selected. Which of these stages are actually reached will depend on a number of factors, including the level of selective pressure (pollution stress), the range of genetic variation in the population and the generation time.

The clearest evidence of these changes comes from studies of resistance to heavy metals. In many populations there are individuals that have the ability not only to exclude the toxic metal, but to bind or sequester it within the plant in sites where it is relatively innocuous. Many studies have demonstrated that individuals growing on sites that have been historically contaminated by heavy metals have this metal tolerance, but it is important to note that studies have shown that metal-tolerant populations can evolve within short periods of time. For example, Wu *et al.* (1975) studied copper tolerance of *Agrostis stolonifera* (Poaceae) populations of different ages around a copper refinery near Liverpool (Fig. 17.3). There was a marked shift towards copper tolerance within 4 years, and by 8 years all the most sensitive individuals had been eliminated from the population.

While many other examples could be cited of rapid evolution of tolerance to heavy metals and other pollutants, it is important to recognize that this phenomenon does not extend to all species. The rate of evolution will depend on a number of factors, of which the most important will be the degree of selective pressure, which will be related to the severity of the pollution stress, and the frequency of occurrence of tolerant individuals in the population. The latter appears to vary between species. For example, the frequency of occurrence of copper

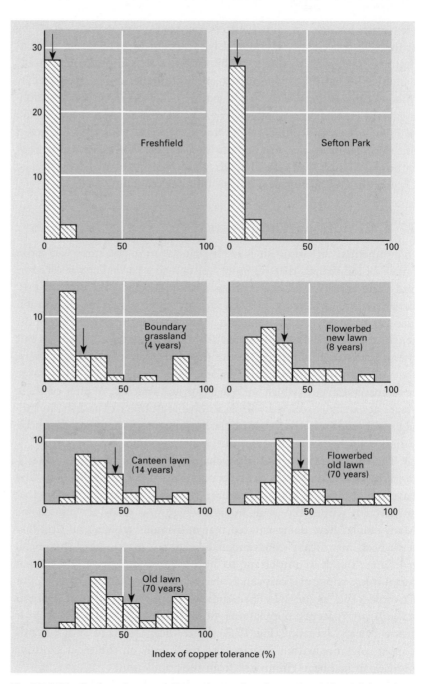

Fig. 17.3 Distribution of copper tolerance in samples of seven populations of *Agrostis stolonifera* (means denoted by arrows). The y axis indicates the number of individuals, from a sample of 30 tillers, with a given copper tolerance (0 being least tolerant and 100 being most tolerant). The Freshfield and Sefton Park sites are uncontaminated by copper or other metals, and sensitive individuals predominate. The other five sites are lawns of different ages at contaminated sites in the vicinity of the refinery (from Wu *et al.* (1975)).

tolerance among species found on old copper mines has been found to range from 0.01 to 0.16%; however, a number of species, which are not found on old copper mines, have a frequency of occurrence of zero (Bradshaw & McNeilly 1991). There is no obvious phylogenetic pattern in the occurrence of metal tolerance (Antonovics *et al.* 1971), although it does seem more common in some genera than others.

Old mine workings are often small in area and surrounded by non-contaminated areas. Thus, after a tolerant population has become established, it may be diluted by gene flow from pollen originating from non-tolerant plants in the surrounding area. One mechanism that appears in these situations is the development of breeding barriers between the two populations. There are several examples in the literature of the metal-tolerant plants flowering 1–4 weeks earlier than the surrounding vegetation, thus reducing substantially the rates of cross-pollination (e.g. McNeilly & Antonovics 1968).

Although there has been a long awareness among entomologists of the evolutionary effects of air pollution, as exemplified by the evolution of industrial melanism in moths and ladybirds (Kettlewell 1955), it was only in the 1970s that evolution of tolerance to air pollutants in plants was first demonstrated. In the USA, Taylor and Murdy (1975) showed that plants grown from seed of *Geranium carolinianum* collected close to a coal-fired power station were less sensitive to an acute dose of SO_2 than seed collected from unpolluted areas in the vicinity. In the UK, interest was first initiated by observations that commercial grass cultivars introduced into the Rossendale Valley of Lancashire, where SO_2 and particulate levels are high, performed poorly compared with local populations of the same species. Experimental fumigation of clones selected from the field showed that local ones were much more resistant to SO_2 than the introduced genotypes (Bell & Mudd 1976). Further work confirmed this observation, and extended it to a number of sites around Merseyside, where the degree of SO_2 tolerance in the population was shown to be related to the ambient SO_2 concentration (Horsman *et al.* 1979a).

Most studies of evolution of tolerance have involved pollutants that show strong gradients in concentration over short distances, and have involved herbaceous species. Where the pollutant is more widely dispersed, or generation times are longer, it is difficult to demonstrate the phenomenon, although there is no reason to suppose that it should not occur.

The consequences of evolution of pollution tolerance for overall genetic fitness are hard to evaluate, although there is some evidence that individuals from populations with a high resistance to SO_2 grow less well than control populations when grown in filtered air (Roose *et al.* 1982). There may be a physiological basis to such responses; for example, genotypes that are resistant to O_3 often have lower stomatal conductances, and hence may have reduced potential rates of photosynthesis. There may also be energy costs associated with the development of detoxification mechanisms. Such reduced capacity for photosynthesis and growth may have contributed to the rapid loss of tolerance to SO_2 in grass populations with falling ambient SO_2 levels reported by Wilson and Bell (1985).

In terms of the overall genetic variation within a population, there is little evidence on the impact of evolution of pollution tolerance. Key factors will be the strength of selective pressure and the frequency of resistant genotypes (Barrett & Bush 1991). Where, as with the mine populations, the frequency of resistant individuals is low and a large proportion of the population is eliminated, the effect on overall genetic variation in the population may be substantial. However, with less severe pollution levels, the effect may be much smaller. While some loss of genetic variation will result from pollution-induced stress, it is not clear how significant this is for the evolutionary potential of the affected populations.

17.5 Community-level effects

It is clear that when pollutants are present in sufficiently high concentrations, they can cause the complete elimination of all plant species. Around the Sudbury smelter in Ontario, for example, in the early 1970s, areas devoid of vegetation occurred up to 8 km distance. Up to 20–30 km from the smelter, species numbers and productivity in the overstorey were reduced; this was probably the result of the combined effects of SO_2 and heavy metal emissions from the smelter (Freedman & Hutchinson 1980). The effects on the understorey vegetation were less marked than those on the forest canopy. There was also evidence of acidification and heavy metal accumulation in lakes around the smelter, with adverse effects on the number of species of planktonic algae and macrophytes (Gorham & Gordon 1963). Over the last 20 years there has been a reduction in emissions of both SO_2 and heavy metals from the smelter, with some reinvasion evident, particularly of metal-tolerant ecotypes of certain grasses. Similar patterns of decreases in species diversity and productivity have been reported around other large point sources of pollutants (e.g. Gordon & Gorham 1963).

There is also evidence from historical records of changes in community composition as a result of pollution. For example, up to the early 1800s the dominant vegetation on the blanket bogs of the Peak District, in the southern Pennines of England, was *Sphagnum* mosses. However, with the onset of the Industrial Revolution in the surrounding valleys, pollutant levels in the region increased rapidly. The *Sphagnum* spp. began to disappear, and today the area is dominated by cotton-grass (*Eriophorum vaginatum* (Cyperaceae)). Overall, there has been a loss of species diversity, as species associated with the ombrotrophic *Sphagnum* spp., such as sundews (*Drosera* spp.) and marsh andromeda (*Andromeda polifolia* (Ericaceae)), have also disappeared (Lee *et al.* 1988). There is little doubt in this case that the change in community composition was a result of the direct elimination of the sensitive *Sphagnum* spp. by increased deposition of SO_2.

A more recent example of community change that may be linked to atmospheric pollution, but which is not due to direct adverse effects of these pollutants, comes from The Netherlands. Here, over recent decades, heathlands dominated by ericaceous shrubs, such as *Calluna vulgaris*, have been replaced in many areas by acid grassland communities, dominated by grasses like *Molinia caerulea* and *Deschampsia flexuosa* (Heil & Diemont 1983). This has been attributed to an increased nitrogen deposition, due primarily to increased emissions of NH_3 from intensive agriculture. Here, the primary mechanism postulated is a shift in the competitive balance between *Calluna* and the grass species; the latter are better able to respond to increased levels of nitrogen and thus can outcompete the *Calluna* (e.g. Heil & Brugginck 1987; see also Chapter 4). However, there is little doubt that the field situation is more complex than this, since once a mature *Calluna* canopy is established, invasion by grasses is unlikely, even when there are high levels of nitrogen input (Aerts *et al.* 1990). Thus it has been suggested that the increased nitrogen deposition also acts by increasing the sensitivity of *Calluna* to factors such as drought, cold stress and to attack by the heather beetle (Brunsting & Heil 1985), which will lead to canopy breakdown. Finally, it is likely that management practices are also significant; there have been reduced rates of grazing and sod-cutting in recent years, and these may have been important in maintaining the low nutrient status of the heathland communities (Marrs 1993). Thus in this case, community change may have resulted from a complex of factors including pollutant deposition, management practices, competitive shifts and the effects of climatic stress and insect herbivores.

A third area where community change as a result of pollutant stress has been intensively studied is in the San Bernardino mountains, which surround the city of Los Angeles. Here, an impact of O_3 pollution, arising primarily from the city below, began to be observed on the native forest community in the 1960s. The most dominant species of these mixed-conifer forests was *Pinus ponderosa*, but this also proved to be one of the most sensitive species to O_3. In many of these areas, *Pinus ponderosa* has suffered severe mortality. The loss of the dominant species has had substantial effects on the forest community (Miller 1989). In many areas, white fir or cedar species, which are more resistant to O_3, are becoming dominant, but at some higher elevation sites these and other conifer species do not regenerate naturally and the area may become dominated by shrubs. The complex pattern of changes in community composition following the loss of *Pinus ponderosa* is further complicated by the role of fire, which may favour other pine species rather than cedars or firs.

At a community level, the impact of pollutants on chemical, physical and biological processes in the soil may be of great significance. Thus, in soils in which buffering is dominated by cation exchange, increased deposition of acidic ions, such as sulphate and nitrate, may lead to exchange and leaching of base cations, such as calcium and magnesium, down the soil profile and beyond the rooting zone (Ulrich 1989). In more acidic soils, buffering is dominated by aluminium exchange, and acid deposition can increase levels of available aluminium. This aluminium can be directly toxic to fine roots (e.g. Neitzke & Runge 1985) or restrict uptake of other mineral nutrients (e.g. Bengtsson *et al.* 1988).

Soil acidification may also increase the availability of other metals. Such metals can have direct impacts on rates of decomposition because of their toxic effects on soil microorganisms, and there is some evidence that SO_2 may have similar effects (Ineson & Wookey 1988). There is clear evidence of these effects in the vicinity of metal smelters (e.g. Tyler 1975), but away from these large sources the evidence of real effects on decomposition and nutrient cycling is less convincing.

There is strong evidence of acidification and base-cation depletion of forest soils in western Europe over the past three decades (e.g. Hallbäcken & Tamm 1986). Over a similar period, there has been a decline in health of major forest species such as beech and Norway spruce over many areas of western Europe. The causes of these declines are complex, but they are often associated with mineral nutrient deficiencies, as shown by experiments in which symptoms have been temporarily reversed by soil fertilization. It is probable that these result from a combination of soil acidification, with effects in the canopy, such as direct leaching of base cations and direct effects in the canopy especially a growth stimulation by nitrate and ammonium deposition that increases the demand for mineral nutrients (Oren *et al.* 1988). Thus any analysis of the causes of forest decline needs to consider atmospheric, edaphic and biological factors influencing nutrient cycling, as well as their interaction with other stress factors and with changes in forest management.

Soil acidification and eutrophication associated with increased acid and nitrogen deposition may also have consequences for woodland ground flora composition. For example, both Falkengren-Grerup (1986) in southern Sweden and Thimonier *et al.* (1994) in north-east France have reported an increase in species numbers in deciduous forests over the last two to three decades. As might be expected, some basophilic or neutrophilic species had decreased in frequency or been lost, and there had been an increase in acidophilic and nitrophilic species.

17.6 Concluding remarks

This chapter summarizes some of key aspects of pollutant impacts on vegetation at an ecological level. Although the discussion focuses, for the sake of clarity, on a limited range of pollutants, many of the same principles and ideas will apply in the case of other pollutants and other ecological situations. While in the past many situations arose in which high concentrations of pollutants were the dominant factor causing changes in the structure and function of local plant community, most of the situations of concern today tend to involve more widely dispersed pollutants at lower concentrations, the effects of which may only become apparent over many years or even decades. In such cases, the pollutant is no longer the single dominant factor, but one of a range of interacting biological, climatic and edaphic factors that may influence plant communities. If we are fully to understand the role of pollution in such situations, it is essential that we gain further understanding of the ways in which pollution can interact with these other factors and place our analysis of pollutant impacts more clearly within an ecological framework.

18: Climate Change and Vegetation

J. Philip Grime

18.1 Introduction

The present century is distinguished from its predecessors in being the first in which the activities of humans have begun to seriously affect the functioning and indeed the viability of our planet as a stable support system for human populations and for the present complement of ecosystems. Evidence of a rapid rise in global concentrations of atmospheric CO_2 is incontrovertible (Fig. 1.27, Keeling *et al.* 1982) and there is strong evidence (Wigley & Jones 1981) that this is causing an enhanced 'greenhouse effect' in which the extra CO_2, together with other pollutant gases (methane, chlorofluorocarbons, nitrous oxide and other trace gases), is acting as an increasingly effective radiation trap and bringing about a progressive rise in global mean temperature.

The vegetation covering the land surface of the earth is not immune to these changes and a current task for plant ecologists is to determine to what extent plant populations are already responding to change and to predict the future course of vegetation development. This research is essential if we are to estimate the repercussions on agriculture and to plan an effective response that safeguards food supplies. It is equally important, however, to examine the feedback mechanisms whereby vegetation changes wrought by climate may themselves influence, and perhaps even accelerate, future changes in climate. Here, it is salutary to remember that the evaporative surfaces provided by vegetation are a potent controller of atmospheric humidity and rainfall patterns. It is also important to recognize that the carbon currently stored in the living and dead components of terrestrial ecosystems is a significant contributor to the global carbon budget; it is therefore a matter of some concern to know whether this carbon pool will expand or contract in future. A 'worst case' scenario might be envisaged in which changes in temperature and moisture supplies resulted in a net increase in the rate of oxidation of the terrestrial carbon store and an acceleration of greenhouse warming.

Faced by such an uncertain future, land managers, conservationists policy-makers and concerned individuals are already pressing plant ecologists for assessments of how the vegetation in various parts of the world will respond to climate change. Given the current discrepancies between different meteorological models and the necessity to take into account other factors likely to affect vegetation (forest clearance, agriculture, urbanization), these demands for diagnosis and prediction are suddenly placing plant ecology in a position that is simultaneously both challenging (can we really work at this scale?) and invigorating (Gosh! what we are doing *is* important!). In these unprecedented circumstances, two responses are emerging from the plant ecology research community. The first is immediate and addressed to policy-makers and consists of 'best guesses' of the future, based on our present state of knowledge. In this chapter, only an outline description of this activity will be presented. The second response, which is the main focus of this chapter, is to embark upon new long-term projects designed to resolve present uncertainties, refine current predictions and test them rigorously.

18.2 Importance of land use

At the present time, the most potent forces for change acting on vegetation are the effects of land use. These arise from the direct effects of an expanding human population (habitat destruction by agriculture, forestry, industry, human settlements, overgrazing) and indirect effects (eutrophication through ground water and atmospheric pollutants, and phytotoxicity resulting from aerial and soil contamination). Reviewed on a world scale, the most consistent effect of these phenomena is the inexorable replacement of mature, often species-rich ecosystems by early-successional states in which the vegetation is composed of recently established and potentially short-lived species. This process has two important implications for studies that seek to predict the impacts of climate change. The first is the notion that vegetation is already experiencing such radical processes of change that impacts of future climates are perhaps most appropriately analysed as a fine tuning of the rates and trajectories of changes which are already well advanced. The second implication arises from the reduction in average lifespan in vegetation subjected to modern forms of intensive or disruptive land use. The higher rates of population turnover characteristic of the vegetation of disturbed and intensively exploited landscapes create conditions in which the plant cover is likely to respond more quickly to climate change either by permitting invasions and extinctions or by allowing rapid genetic changes within existing populations. Hence, land use is an essential factor in any calculations of the direction and rate of vegetation responses to climate change.

18.3 Current predictions

18.3.1 World vegetation patterns

Many ecologists (e.g. Raunkiaer 1934; Holdridge 1947) have recognized the correlations between world climate zones and the structural characteristics of the dominant vegetation present within them (see

Chapter 14). More recently, efforts have been made to replace mere correlations with a functional understanding in which both plant morphology (Box 1981) and physiology (Woodward 1987) are used to predict what kind of vegetation will occur under specific climatic conditions. This approach brings the obvious advantage that such predictions can be tested by comparison with the patterns of vegetation observed in the real world. Such a test is an essential preliminary to any attempt to predict patterns of vegetation in the climates of the future. Another method of validating current models of the climatic control of vegetation zones is to refer to the palaeoecological record. In recent years, geographically extensive databases of pollen deposits have been constructed for both Europe and North America (e.g. Huntley & Birks 1983) allowing a quite detailed mapping of the dynamic changes in forest cover that were associated with past climatic changes. These analyses reveal major shifts in the distribution of tree species and indicate remarkable constancy in the physiological tolerance of individual species. In some cases (e.g. Davis *et al.* 1986) it has even proved possible to estimate the rates at which particular tree species responded to climate change by seedling encroachment across continents.

Unfortunately our ability to explain the present array of world vegetation types and to predict some past distributions does not guarantee a sure basis for prediction of future patterns. Here, it is important to recognize that all existing models refer to circumstances where an equilibrium is allowed to establish between vegetation and climate. As already explained, such a scenario is of declining relevance to a rapidly expanding proportion of the land surface. Many areas of agricultural and industrial landscape now exist in a state of continuous disturbance and truncated succession. Habitat fragmentation in heavily populated or exploited landscapes has reduced dramatically the capacity of late-successional species to disperse to new habitats. At the same time, however, many other species, with human assistance, are likely to achieve unprecedented rates of invasion as a consequence of rapid colonization of disturbed habitats, deliberate introductions and passive transport of seeds in earth-moving operations. In these circumstances, past assemblages and events may be of little value in predicting plant communities of the future.

18.3.2 Regional vegetation patterns

Attempts to predict the impact of climate change at a local scale are difficult at present due to the lack of reliable, regional definition in models of future climates. In the British Isles, for example, the task is particularly hazardous because the region lies at a meeting point of climatic zones and in consequence has notoriously inconsistent weather. These uncertainties are compensated to some extent by the fact that in Britain there has been a strong tradition of detailed recording of both climate and plant species distributions. The flora contains contrasted geographical elements (Lousley 1953; Perring & Walters 1962), most species have been mapped and many have sharply defined boundaries that coincide with climatic patterns. Prediction is also assisted by the existence of several classical studies in which plant ecologists have investigated the mechanisms by which the present climate limits the distribution of particular species (Pigott 1968; Davison 1977; Carter & Prince 1981; Pigott & Huntley 1981). Using these quite rich sources of information and making the broad assumption that climate change in Britain will be characterized by a modest elevation in mean temperature and an increased frequency of extreme climatic events (Houghton *et al.* 1990), it has been found possible to devise some tentative predictions of effects on British vegetation (Grime & Callaghan 1988). A summary of these predictions is given in Box 18.1.

Box 18.1 Some predictions of the response of British vegetation to global warming.

1   In southern England, particularly on dry soils, the growing season will shift significantly towards the autumn, winter and spring, with a quiescent phase often occurring in summer. In northern Britain, lengthening of the summer growing season will occur.

2   Conspicuous changes will occur in the relative abundance of various constituents in familiar vegetation types. These will be most obvious in southern Britain. They will vary locally according to habitat factors and vegetation management, and will be predictable from phenological data and nuclear DNA amounts (for an explanation of the predictive value of DNA amount see Box 18.3). In general, we expect grasses and geophytes with high nuclear DNA amounts to remain important in drier habitats but where moisture supply is sufficient to sustain summer growth progressive dominance by species with low DNA amounts seems likely. In southern Britain bryophytes may be expected to decline in abundance at well-drained sites in grassland, scrub and woodland. The current bimodal seasonality in bryophyte growth (autumn and spring) is predicted to be replaced at many southern sites by unimodal (winter) patterns, and lichens may become more abundant locally.

3   Warmer conditions and lengthening of the summer will promote a generally higher level of flowering, seed and spore production and a greater incidence of insect cross-pollination in British vegetation. This may have profound long-term effects, especially in northern Britain and at higher altitudes where effective sexual reproduction is currently a rare event in many populations.

continued on p. 586

Box 18.1 *Continued.*

4 Many of the trees, shrubs and herbaceous plants (e.g. *Tilia cordata*, *Euonymus europaeus* (Celastraceae), *Cirsium acaulon* (Asteraceae)) that currently reach their northern limit in England or southern Scotland will begin to expand northwards. This will arise mainly through increased seed viability, particularly in exceptionally warm years but other factors will be important in particular species, e.g. climate warming will lead to an expansion in the area of desiccated open vegetation available for colonization by ephemerals such as *Hordeum murinum* and *Anisantha sterilis* (both Poaceae).

5 Species currently restricted to south-facing slopes in northern Britain will begin to colonize other aspects.

6 Expansion to higher altitudes will begin to occur in many lowland species.

7 Plants of northern distribution in Europe may be expected to retreat from the southern extremities of their ranges in Britain. This is likely to be most evident in shallow-rooted, calcifuge species (e.g. *Deschampsia flexuosa* (Poaceae), *Galium saxatile* (Rubiaceae), *Viola lutea*) and will be a discontinuous process involving vegetation and soil changes, accelerated by drought mortalities, fire and soil erosion in exceptionally warm years.

8 For the reasons described under 7, northern species may retreat from lower altitudes, at localities in southern Britain and at more northerly sites there will be a tendency for populations to show reduced vigour on south-facing slopes.

9 The advent of a warmer climate will create the obvious possibility of the incursion into Britain of plants that flourish under Mediterranean conditions. Many of these plants have already established 'toeholds' in the south or survive locally in coastal and urban areas (Perring & Walters 1962). The scale of invasion is likely to depend crucially on future policies of land use in Britain.

10 Climate change, operating in isolation from other modifying factors, is unlikely to lead to widespread extinctions of common species from the British flora. Nevertheless, many rare plants may become even more precarious through their limited abilities for genetic change or effective migration. Particularly threatened are species such as *Polemonium caeruleum* in which small populations now occupy cool, continuously damp refugia on isolated steep north-facing slopes.

18.4 Current research

As illustrated in Box 18.1, by using existing sources of information plant ecologists can sometimes offer suggestions about the potential of climate change to influence vegetation. Often, however, our scope for prediction is limited by an uneven distribution of past research effort, such that we know a great deal about a small number of plants of economic importance (crop plants, forage grasses and trees of commercial forests) whilst the dominant plants of some of the world's most extensive ecosystems have received scant attention. Although some of these gaps in our knowledge will be filled by new research, it is inevitable that only a small proportion of terrestrial species can be the subject of intensive study. This situation necessitates careful planning if we are to develop a useful comparative database of plant responses to climatic factors. Here a crucial problem is to determine whether, for the purposes of analysing and predicting community and ecosystem responses, plants (and animals) can be classified into a relatively small number of universal functional types. This adds new urgency to an old debate (MacLeod 1894; Ramenskii 1938; Hutchinson 1959; MacArthur & Wilson 1967; Grime 1974; Southwood 1977; Noble & Slatyer 1979; Whittaker & Goodman 1979), resolution of which would mark a significant step in the development of ecology as a predictive science, quite apart from its usefulness to climate change studies. In this chapter only passing reference will be made to specific theories relating to plant functional types. The main objective will be to describe a protocol in which the search for functional types is part of the process whereby predictions of vegetation responses to climate change can be developed and tested.

18.4.1 A research protocol

Figure 18.1 summarizes a set of procedures designed to formulate, test and refine predictions of vegetation response to climate change. The scheme is generic in the sense that it could be applied both at the coarse scale required for global vegetation models or in a more detailed investigation of a regional flora.

The starting point in the protocol of Fig. 18.1 consists of a screening process in which standardized data are collected on each of the species under investigation. As we see later, the range of measurements with the

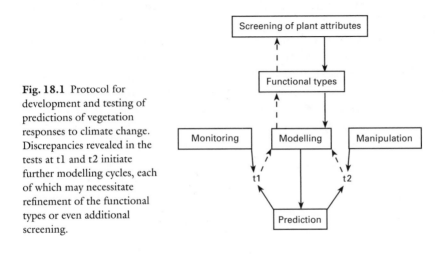

Fig. 18.1 Protocol for development and testing of predictions of vegetation responses to climate change. Discrepancies revealed in the tests at t1 and t2 initiate further modelling cycles, each of which may necessitate refinement of the functional types or even additional screening.

theoretical potential to inform prediction is extremely wide and requires careful adjustment to specific objectives. Review of the screening outputs then permits recognition of plant functional types that, in return, are incorporated into a predictive model of the vegetation under investigation. The lower part of Fig. 18.1 summarizes two complementary sets of procedures that can be used to test model predictions. The first compares predictions of the responses of populations or communities of plants against data collected by direct monitoring of events in natural conditions. The second matches predictions against the results of manipulative experiments in which controlled impacts of ecological factors simulating climate change are applied to natural vegetation or to plant communities synthesized in growth rooms or experimental plots. As indicated in Fig. 18.1, the logical pathway associated with the use of monitored data is the same as that involving manipulation experiments. In both cases discrepancies between prediction and reality drive further cycles, each consisting of further screening, refinement of prediction and tests of prediction.

Each step in the protocol will now be considered in greater detail and two examples (Boxes 18.3 and 18.4) are given to illustrate the two pathways of Fig. 18.1.

18.4.2 Screening of plant attributes

Very different measurements have been made, some in the field others in the laboratory, in previous attempts to define plant functional types. By describing demographic patterns, population biologists have often sought to define functional types (e.g. Deevey 1947; Whittaker & Goodman 1979) and there is no doubt that knowledge relating to the spatial and temporal distribution of reproduction, mortality, dispersal and dormancy can contribute vital information to a functional typology. Some population biologists (e.g. Harper 1982) have argued until recently that demography, particularly when allied to studies of plant morphology, provides an adequate basis for interpretation and prediction of vegetation processes. However, this school of thought has been notably silent on the topic of climate change where the need for physiological information is indisputable. There is now widespread agreement that for this purpose plant functional types must be defined by reference to both demographic criteria *and* those features of life history, physiology and biochemistry that determine responsiveness to climatic factors. In Box 18.2 an attempt has been made to identify some of the plant attributes most likely to condition responses to climate change.

18.4.3 Formal searches for plant functional types

In order to use screening outputs as a reliable basis for recognizing functional types it is essential that large numbers of species are involved

---

Box 18.2 Some plant attributes likely to influence plant response to climate change.

Affecting the direction of response

1 Photosynthetic pathway (C_3, C_4, CAM; see Chapter 1)
2 Temperature optimum for growth
3 Nuclear DNA amount (see Box 18.3)
4 Water-use efficiency (see Chapter 2)
5 Desiccation tolerance
6 Rooting depth
7 Freezing tolerance

Affecting the rate of response (see Chapter 4)

8 Life history and lifespan
9 Reproductive rate
10 Seed dispersal
11 Seed persistence
12 Nature and extent of mycorrhizal infection (see Chapter 3)

(Grime 1965; Clutton-Brock & Harvey 1979; Keddy 1992) and that steps are taken to distinguish between adaptive specializations and phylogenetic constraints (Stebbins 1971; Hodgson & Mackey 1986; Givnish 1987; Harvey & Pagel 1992). Greatest rigour is ensured in circumstances where many plant attributes have been screened and statistical tests can be performed (Grime *et al.* 1987) to confirm (or refute) the existence of recurring sets of traits, associated with distinctive ecologies. Research on this large scale is in its infancy but significant progress has been achieved at research centres in Australia, Britain, Canada and The Netherlands and a laboratory manual of methods used in the formal search for functional types has been published (Hendry & Grime 1993). Some procedures examine basic features of anatomy, morphology, physiology and biochemistry at different stages of development (seed, seedling, established plant); others measure survivorship and growth responses under particular environmental stresses. Further sets of procedures are designed with specific relevance to impacts of land use and climate change (temperature, moisture supply, CO_2 concentration) and efforts have been made to measure attributes relevant to competitive interactions and the potential to dominate plant communities. The majority of these procedures are conducted over a relatively short period of time, under standard conditions on material of known and consistent genetic origin. However, some important attributes are not amenable to this approach and for the present we must rely on data collected from the field or reported in the literature. Examples in this category are seed persistence in the soil and the nature and extent of mycorrhizal infections. However, extreme caution must be applied in using these sources of data, particularly where they allow contamination of the database by information of weaker genetic definition.

Box 18.3 Genome size as a predictor of plant community responses to climate at Bibury.

From large-scale screening operations (e.g. Stebbins 1956; Bennett & Smith 1976; Levin & Funderburg 1979) and various other investigations (Hartsema 1961; Bennett 1971, 1976; Grime & Mowforth 1982; Grime 1983; Grime *et al.* 1985), it has been established that there is more than a thousand-fold variation in nuclear DNA amount in vascular plants and that differences in DNA amount in cool temperate regions such as the British Isles coincide with differences in the timing of shoot growth. An illustration of such differences is provided in Fig. 18.2 which, in particular, shows the inability of plants with small genomes to grow rapidly in the early spring. It has been suggested (Grime & Mowforth 1982) that this limitation is imposed by the inhibitory effect of low temperatures on cell division. It is also hypothesized that plants of higher DNA amount, cell size and length of the cell cycle (the three attributes are inextricably linked) are less inhibited because much of their spring growth is a process of expansion of large cells divided and stored in an unexpanded state during preceding warmer seasons or weather. One of several implications of this theory for predictions of the effects of climate change is that the greater dependence of plants with small genomes on warm temperatures will make them more responsive to year-to-year variation in climate. In Fig. 18.3 this hypothesis is tested by measuring the variance in total shoot biomass over the duration of the Bibury experiment in three groups of species classified by genome size. The results show a progressive increase of variance in shoot yield with decreasing DNA amount. This supports the prediction in Box 18.1 that in temperate floras global warming will stimulate an expansion in plants with small genomes.

18.4.4 Monitoring of vegetation responses to climate

As shown in the left-hand side of Fig. 18.1 one of the ways in which some predictions of vegetation responses to climate change can be tested is by reference to long-term studies in which fluctuations in the abundance of species can be compared with year-to-year variation in climate. Reliable monitoring studies are a scarce resource and there are only a small number of cases (e.g. Leps *et al.* 1982; Willis *et al.* 1993) where they have been used to test predictions of plant community response to climatic factors. A remarkable example is provided by the Bibury plots, which are located on a road verge in southern England where they have been monitored continuously since 1958 by Professor A.J. Willis. Each year in the second week of July the shoot biomass of each species has been estimated non-destructively in replicated plots and the data have been calibrated at intervals by comparison with harvested samples. By this method it has been possible to construct a record of the species composition of the community over a 39-year period and to begin to use the results to test predictions of the response of constituent species and functional types with year-to-year variation in climate. An example of the use of Bibury data in hypotheses testing is provided in Box 18.3, which explores the predictive value of nuclear DNA amounts.

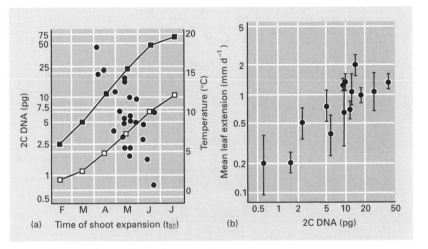

Fig. 18.2 (a) Relationship between quantity of nuclear DNA (●) and the time of shoot expansion in 24 plant species commonly found in the Sheffield region. Temperature is average for each month of daily minima (□) and maxima (■) in air temperature 1.5 m above ground. (b) Relationship between quantity of nuclear DNA and leaf extension from 25 March to 5 April in 14 grassland species coexisting in the same turf; vertical lines represent 95% confidence limits.

18.4.5 Manipulative experiments

Of all the potential effects of climate change on vegetation, those arising from the direct effects of elevated CO_2 concentrations on plant growth have attracted most attention. This is not only because a continuing rise in level of atmospheric CO_2 well into the next century is common to all current meteorological predictions but also because, as explained in Chapter 1, it is well known that differences between plants in photosynthetic pathway (C_3, C_4 and CAM) may cause them to respond to elevated CO_2 to different extents and with different water-use efficiency.

However, following the arguments of Grime (1989) and Korner (1993) it may be important not to overemphasize the importance of direct effects of CO_2 on photosynthesis. Ecological responses to elevating CO_2 will also depend on all those processes that lie 'downstream' from photosynthesis and translate photosynthate into plant tissue. This means that for a sustained stimulation of carbon assimilation to occur, other potentially limiting resources (water, mineral nutrients) must remain available to the plant.

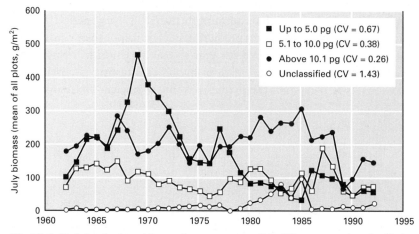

Fig. 18.3 Variation in shoot biomass in components of the Bibury vegetation classified with respect to nuclear DNA amount.

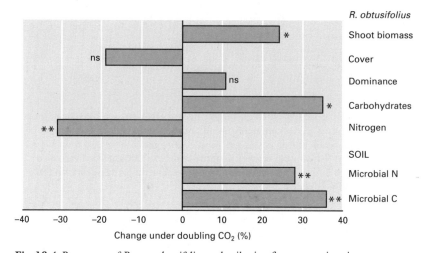

Fig. 18.4 Responses of *Rumex obtusifolius* and soil microflora grown in microcosms to a doubling of atmospheric CO_2 (700 vpm) as compared with controls at 350 vpm. Vegetation was allowed to develop for 84 days by natural recruitment from seed banks in soils removed from a tall herb community in Derbyshire and placed in microcosms (six replicates per treatment) in cabinets without nutrient addition. Shoot biomass was measured as mg dry weight; cover as number of touches in a point–quadrat analysis; dominance as biomass of *R. obtusifolius*: total community biomass; carbohydrates (starch + glucose + sucrose) as mg g^{-1} fresh weight; nitrogen as mg g^{-1} dry weight of fully expanded young leaves; microbial C and N as mg g^{-1} dry soil. (ns = non significant, * = $P < 0.05$, ** = $P < 0.01$, *** = $P < 0.001$.)

As in the case of other facets of climate change, it is essential that any predictions of responses to elevated CO_2 based on the laboratory responses of individual plants are eventually tested at community or ecosystem level following a protocol such as that outlined in Fig. 18.1. Experiments that subject ecosystems to elevated levels of CO_2 for substantial periods of time are technically difficult and expensive. However, as illustrated in the example in Box 18.4, they are an essential component of the research if we are not to run the risk of failing to recognize important factors and interactions that are not included in laboratory investigations with single species.

Box 18.4 Response of native British herbs to elevated CO_2.

Screening of growth responses has provided an extensive database on the responses of native British plants to elevated CO_2 (Hunt *et al.* 1991, 1993). From these data it is evident that there are large species-specific differences in response to elevated CO_2 varying from zero in some ephemerals and slow-growing perennials to very large increases in dry-matter production in some species. Among the latter, potentially large, fast-growing, clonal perennials of productive habitats (e.g. *Chamaenerion angustifolium* (Onagraceae) and *Urtica dioica*) are conspicuous, prompting the hypothesis that responsiveness to elevated CO_2 may be dependent on the existence of strong carbon sinks. Here it is interesting to note that fast-growing clonal herbs are prominent among the life-forms identified as currently expanding in abundance in the British flora (Grime *et al.* 1988). In previous attempts to explain this phenomenon (e.g. Hodgson 1989), emphasis has been placed on changes in land use and eutrophication by mineral fertilizers and atmospheric deposition of nitrogen. The results of the screening of responses to CO_2 suggest that a stimulatory effect of elevating CO_2 could be contributing to the observed shift in functional types.

Following the protocol of the right-hand side of Fig. 18.1, experiments have been conducted to determine whether responses to CO_2 detected by screening individual plants can be used to predict functional responses at the community level. Figure 18.4 presents results from one such experiment in which the effect of elevated CO_2 (twice current ambient level) has been examined on the structure of an early-successional herbaceous community that had been allowed to regenerate from natural seed banks on disturbed fertile soil transferred to the laboratory. The results of these manipulative experiments with natural communities (Diaz *et al.* 1993) are strongly at variance with the screening data in that the promotory effects of CO_2 were not confirmed in potentially responsive species. It is particularly interesting to note

continued on p. 594

Box 18.4 *Continued.*

that under a doubling of CO_2 concentration the early-successional, fast-growing and potentially large *Rumex obtusifolius* (Polygonaceae), although attaining a higher shoot biomass, did not increase its relative abundance in the community and showed marked symptoms of leaf stunting, reduced levels of foliar nitrogen and carbohydrate accumulation. These results suggest that elevated CO_2 causes an increase in substrate release into the soil leading to mineral nutrient sequestration by the expanded microflora and a consequent nutritional limitation on plant growth.

Whilst these results do not invalidate the patterns of response evident in the screening data, they are a reminder of the complexities that distinguish the screening laboratory from a multispecies ecosystem. They suggest that sink strength is not the only trait accounting for CO_2 responsiveness. Other characteristics, such as species' nutritional requirements and the nature of their interaction with the microbial flora may be important in predicting responses to high CO_2 at the community level.

18.5 Conclusions

If the current efforts to predict the future impacts of climate change on vegetation are to mature from speculation to science they will need to include both modelling and experimental tests. In scale and technical content these activities are extending ecologists far beyond the 'normal' ambit of their activities and bringing them into new alliances with a wide range of other specialists. Quite apart from the need to utilize complex datasets such as that at Bibury or to grapple with the complexities of CO_2 effects on ecosystems, the sudden promotion of climate change to the forefront of the ecologist's research agenda is also acting as a rude reminder of a large number of unsolved ecological problems and gaps in our knowledge that, for too long, have escaped our attention. A satisfying botanical approach to climate change will take the enquiring mind back to almost all of the preceding chapters of this book.

19: Biodiversity

Michael J. Crawley

Say you want to convince your father-in-law to get involved in conservation – in rescuing biodiversity. How would you start? Would you tell him about the genes for disease resistance in wild relatives of crop plants? Would you mention the probable existence of undiscovered, valuable pharmaceuticals, talk of tropical rain forests and their rates of conversion, or describe a personal experience of nature that still brings tears to your eyes or goose bumps to your skin? That is, would you appeal to his intelligence or his emotions? (Michael E. Soulé 1988)

19.1 Introduction

Biodiversity is the totality of hereditary variation in life-forms, across all levels of biological organization, from genes and chromosomes within individual species to the array of species themselves and finally, at the highest level, the living communities of ecosystems such as forests and lakes (Wilson 1994). Out of the myriad possibilities, one slice of biodiversity might be the chromosomes and genes within an offshore island population of the orchid species *Dactylorhiza majalis*. Another would be the diversity of orchid species within the British Isles, and still another would be the richness of the entire vascular plant flora of Britain. However it is measured, biodiversity is strongly scale dependent (see Chapter 14).

In 1979, E.O. Wilson invented the term *biophilia* to describe the inborn affinity human beings have for other forms of life, an affiliation evoked, according to circumstance, by pleasure, or a sense of security, or awe, or even fascination blended with revulsion (Wilson 1994). Gordon Orians (1980) even goes so far as to suggest that humans share a similar conception of the ideal landscape: a viewpoint atop a prominence, close to a lake or seaside, surrounded by park-like terrain with scattered, flat-topped trees. That is to say, a landscape just like the savannah of tropical East Africa where humanity evolved for several million years. Open, so that the approach of enemies can be detected. Trees to climb if pursued. Water in which to catch fish. And so on. Whether or not you buy this argument, it is clear that people do have deep-rooted feelings about landscape and biological diversity, and that this diversity is seriously threatened by the present-day activities of humans through habitat destruction, pollution and climate change. Unfortunately, statis-

[595]

Table 19.1 Species richness of various plant groups. Compiled from various sources.

Plant group		No. of species
Bryophyta	Mosses, liverworts, hornworts	16 600
Psilophyta	Psilopsids	9
Lycopodiophyta	Lycophytes	1 275
Equisetophyta	Horsetails	15
Filicophyta	Ferns	10 000
Gymnospermae	Gymnosperms	529
Dicotyledonae	Dicots	170 000
Monocotyledonae	Monocots	50 000
Total		248 428

tics about extinction don't change peoples' behaviour, and we have neither the time nor the resources to convert sufficient people to biophilia to make a difference (Wilson 1984). Sustained political pressure is the only course of action likely to make a significant long-term difference. Politicians need to know that paying a price for biodiversity is a vote-winning issue. As the old saying goes: 'think global, act local'.

19.2 The number of plant species

Compared to the insect species, which are counted in their millions, and the nematodes in their tens of millions, the number of plant species on earth seems positively modest (Tables 19.1 & 19.2). Few vascular plant species are described as new to science each year (hundreds rather than thousands) and the total inventory of the world's living plant species is probably not much more than 250 000. The rate of appearance of new species by speciation is trivially low, and these days new species tend to come about as a result of hybridization between a native plant and an alien species or between two alien species (see below). In contrast, the rate of extinction, caused mainly by direct destruction of habitats by human beings, is accelerating at an alarming rate. Raven (1988) esti-

Table 19.2 Number of seed plant species in various regions of the world. Compiled from various sources.

Biogeographic unit	No. of species	Area (1000 km^2)	Species (per 1000 km^2)
Tropical Africa	30 000	20 000	1.5
West Tropical Africa	7 300	4 500	1.6
Sudan	3 200	2 505	1.28
Southern Africa	22 977	2 573	8.93
Cape Floral Kingdom	8 504	90	94.49
Brazil	40 000	8 456	4.73
Peninsular India	20 000	4 885	4.09
Australia	25 000	7 716	3.24
Eastern N. America	4 425	3 238	1.37

mates that vascular plant species are being lost at an average rate of about five species *per day* and the argues that this will rise to about 10 species lost per day during the early decades of the twenty-first century.

There are several clear, broad patterns in plant species richness (Fig. 19.1). Plant species richness declines with increasing latitude, and

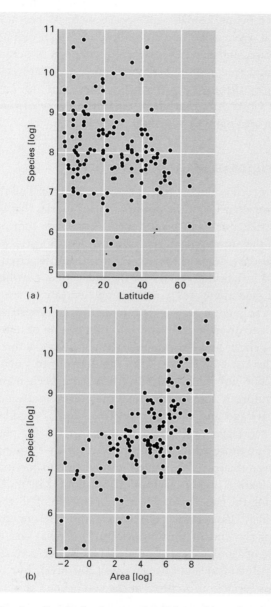

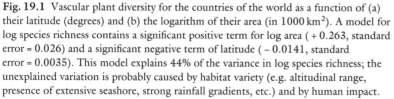

Fig. 19.1 Vascular plant diversity for the countries of the world as a function of (a) their latitude (degrees) and (b) the logarithm of their area (in $1000\,km^2$). A model for log species richness contains a significant positive term for log area (+ 0.263, standard error = 0.026) and a significant negative term of latitude (− 0.0141, standard error = 0.0035). This model explains 44% of the variance in log species richness; the unexplained variation is probably caused by habitat variety (e.g. altitudinal range, presence of extensive seashore, strong rainfall gradients, etc.) and by human impact.

at a given latitude it declines with increasing altitude. At a given latitude, larger and more populous countries support more plant species, chiefly because of their greater variety of habitats (see Chapter 14). In regions that have been glaciated in the recent past, plant richness is higher on continents where the mountain chains run north to south (e.g. North America) than on those where they run east to west (e.g. Europe; see below). Small-scale species richness per unit area (alpha diversity) at a given latitude and altitude is much more variable from place to place, and depends on local conditions of geology, drainage, microclimate, seed source, herbivory, etc. (see Chapters 8, 14 & 15). For example, there are more species in $14\,km^2$ of La Selva Forest Reserve in Costa Rica than there are in the whole of the British Isles, which is an area more than 17 000 times as great.

19.3 Origins of plant biodiversity

The fossil record is patchy, partial and biased, but it's all that we've got. Probably less than 1% of the terrestrial plant species that ever lived have left any fossilized remains behind them. The single continent of Pangea permitted the deveopment of relatively few floristic provinces, but continental drift, leading to the production of dispersed land masses, allowed the isolation necessary for the independent evolution of terrestrial floras, and there was a massive net increase in species numbers. However, the biggest factors in the increase in terrestrial plants came through niche diversification and the colonization of new habitats (e.g. mountains, coasts and deserts, outside the primordial forest).

Over the last 400 million years the diversity of vascular plants has increased, but not monotonically (there have been mass extinctions) and not uniformly across taxa (the ferns peaked in diversity about 250 million years ago and the gymnosperms about 70 million years ago; Fig. 19.2; see Signor 1990). Extinction patterns for plants might be different from those exhibited by the better preserved animal groups like marine invertebrates. The biggest of the five major mass extinction events of marine organisms took place about 245 million years ago (mya) during the late Permian, when about 80% of all genera and 54% of all families disappeared. The second most severe mass extinction occurred at the end of the Ordovician (440 mya) and accounted for about 22% of families of marine taxa. Others occurred during the late Devonian (350 mya) and late Triassic (200 mya), but the most celebrated mass extinction occurred at the end of the Cretaceous (66 mya); although it accounted for a mere 15% of marine families it spelled the demise of such charismatic animals as the dinosaurs, plesiosaurus and pterosaurs (Erwin *et al.* 1987).

The Permian extinction appears to have been a long drawn out affair, lasting about 7 million years during the formation of the supercontinent of Pangea, at which time there was massive global change in

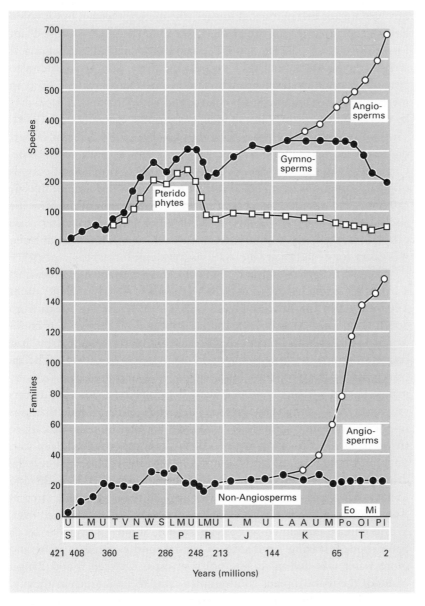

Fig. 19.2 Evolutionary history of vascular plant measured along the *x* axis in millions of years ago (mya). S, Silurian; D, Devonian; E, Eocene; P, Permian; J, Jurassic; K, Cretaceous; T, Tertiary. (a) Number of plant species, showing the exponential rise of the angiosperms after 100 mya. (b) Number of families showing the explosive increase of the angiosperms. (After Signor 1990.)

climate and volcanism. In contrast, the Ordovician and Cretaceous extinctions probably occurred over much shorter periods of time. The Ordovician was associated with three major episodes of glaciation during a period of only 500 000 years (the Hirnantian glaciation). Controversy still surrounds the cause of the Cretaceous mass extinction. The most popular (and certainly the most spectacular) explanation involves a giant extraterrestrial impact (the meteor theory), but it could have been caused by climate change over a much longer period. Plant fossils do not, in general, show the same sudden mass extinctions as seen in the earlier animal fossil record; the extinction periods tend to last longer and the peak extinction rates do not coincide. The twin mechanisms of gradual climate change and competitive exclusion may be more influential than any sudden catastrophic events (Knoll 1984). The end of Cretaceous catastrophe did, however, have a major impact on terrestrial vegetation with perhaps as many as 75% of the late-Cretaceous plant species going extinct. Two other, more recent periods of plant extinctions have been documented: one during the late Eocene (50 mya), and a second, more local event at the junction of the Miocene and Quaternary periods (Niklas *et al.* 1983; Knoll 1986). After mass extinctions, there is presumably tremendous competitor release, making it is easier for evolutionary innovations (new rare forms) to prosper.

The gymnosperm- and fern-dominated forests roamed by dinosaurs were rather similar at the time of the break-up of Gondwanaland about 140 mya, when the Atlantic Ocean opened up, separating Africa from South America. The first angiosperms appeared 113 mya when continental break-up was already well advanced and the angiosperms had become dominant in many terrestrial ecosystems by 80 mya. The replacement of gymnosperms by flowering plants and of dinosaurs by mammals was virtually complete by 65 mya. The next 60 million years saw spectacular and relatively rapid evolution in a climate that was uniformly warm and sea levels were high (there were no polar ice-caps to tie up fresh water during this period). Weathering produced deep soils on a uniformly forested landscape. By 35 mya the climate had become drier and cooler, and forest break-up had begun. Between 25 and 15 mya Antarctica separated from South America, the cold circum-Antarctic current was set up and the continent began to freeze over. World climate continued to deteriorate and, about 12 mya, sea levels had dropped, leaving large areas of coastal sand exposed as more and more water was tied up in the polar ice-caps. Between 5 and 2 mya, many of the world's plant communities came to look much as they do today. Many of the present angiosperm species proliferated during the last 1.5 million years, during which time glacial periods of about 100 000 years of cold alternated with brief 10 000-year interglacial (greenhouse) periods when temperatures were about 5 °C higher. Glacial climates arrived over periods of just a few hundred years as temperatures dropped and sea level fell. There have been about 10 of these cycles, each of which allowed an interaction between fragmentation and rapid climate to create the geographic isolation that is a prerequisite for speciation. The last glacial period lasted from roughly 75 000 to 12 000 years ago, during which period sea levels were low and sea shores were 50–200 km further out to sea than at present in various parts of the world.

19.4 Postglacial changes in plant biodiversity

Much has been written about the precise locations of the tropical refugia in which plant diversity was preserved (some would say enhanced) during the most recent ice ages (Tallis 1991). Those high-latitude countries like Britain and Canada that were almost completely stripped bare of plants by the glaciers have quite different histories of recolonization. In North America, the mountain ranges run north–south, so the migration of plant species towards their warm, southern refugia as the ice advanced was unimpeded; likewise their return, after the ice had melted, was unimpaired (Fig. 19.3). In contrast, great mountain ranges cut off the lines of retreat of the north-west European flora in the face of the advancing ice. High mountains stretched east–west from the Pyrenees, through the Alps to the Carpathians. This is correlated with present-day differences in forest tree species diversity. In Great Britain, the climax forest might contain five species of tree, whereas in New England it would contain at least 10 tree species. In unglaciated islands at similar latitudes in Japan, the tree flora is even richer, with as many as 20 species. This implies that the tree flora of north-east North America is unsaturated in species and that the north-west European tree flora is highly unsaturated. Recent theoretical modelling for north-east American forests corroborate this prediction (see Chapter 15). The picture of postglacial dispersal dynamics, reconstructed by Davis (1994) and others, had had a major impact on the way we think about the structure of plant communities. Far from being organic 'entities', which moved to and fro as identifiable units with the coming and going of the ice-sheets, we have come to see communities as serendipitous reassortments of species, thrown together by chance and by differences in their dispersal rates (i.e. to Gleason's rather than Clements' view of community; see Chapter 14).

19.5 Current geographical distribution of biodiversity

About 65% of the world's 250 000 flowering plant species are found in the tropics (Raven 1976). The world's flora is divided into six plant kingdoms (Boreal, Palaeotropic, Neotropic, Australasian, Cape and Antarctic; Fig. 19.4). The smallest of the these, the Cape floral kingdom, is about the size of Portugal (90 000 km^2) and yet has 8600 species of which 5800 are endemic and more than 1400 are Red Data Book species (critically rare, endangered or vulnerable) and at least 29 are known already to be extinct. The whole of tropical Africa, which is 235 times the size has only 3.5 times this number of species. The Cape Peninsula, an area roughly the size of London, supports more plant species (2285) than the entire British Isles (1515).

The flora is distributed over the world's 12 major biomes, showing substantial convergence of physiognomy but radical differences in

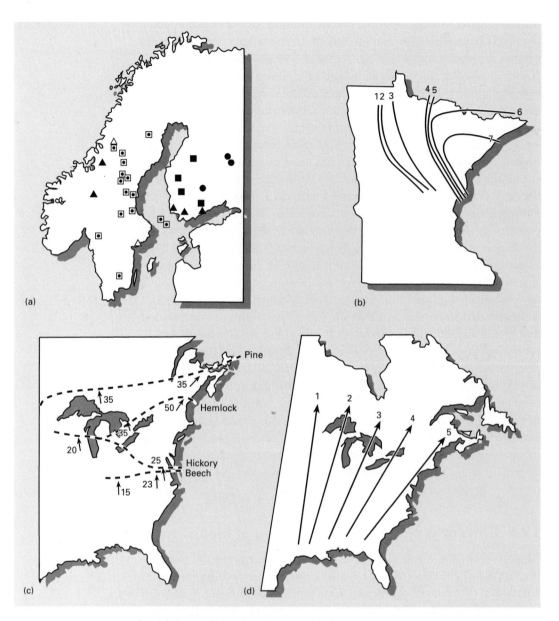

Fig. 19.3 Postglacial migration of trees in Europe and North America. (a) *Picea abies* in Scandinavia: large circles, 4000–5000 years ago; solid squares, 3000–4000 years ago; dotted squares, 1700–3000 years ago; triangles, 1000–1700 years ago; small circles, more recent than AD 1000. (b) Migration of *Pinus strobus* westwards across northern Minnesota: numbers indicate the position of the front in thousands of years ago. (c) Northern limits of three trees 8000 years ago and their rates of northward migration in km per 100 years over the previous 2000–4000 years. (d) Migration in km per century for five different tracks in eastern North America: across tracks 1–5, *Carya* migrated at 19.3, 9.6, 15.4, 7.6 and 7.4 km per century whereas *Populus* showed the opposite pattern, migrating at 16.7, 22.3, 27.6, 29.7 and 35.0 km per century (references in Tallis 1991).

species composition from one continent to another. Travelling north or south from the equator we might pass (with judicious minor detours) through the following biomes: (i) tropical rain forest; (ii) tropical deciduous forest; (iii) tropical savannah; (iv) tropical thorn scrub; (v) desert; (vi) sclerophyllous Mediterranean scrub (maquis and chaparral); (vii) steppe; (viii) mountain tops; (ix) temperate deciduous forest; (x) temperate grassland; (xi) boreal conifer forest (taiga); and (xii) tundra. At each step we are likely to observe reductions in both alpha and beta diversities (see Chapter 14).

19.5.1 Biodiversity hot-spots

The species-rich areas are far from being randomly distributed over the world's land surface. They are aggregated into what have come to be called hot-spots; some of these reflect ancient refugia, others are centres of speciation, others reflect unusual geology or extreme isolation

The International Union for the Conservation of Nature (IUCN) recognizes three kinds of high-diversity sites:

1 botanically rich sites that can be recognized geographically, like Mt Kinabalu in Borneo;

2 geographical regions with high species richness and/or endemism such as the Atlas Mountains in Morocco; and

3 vegetation types and floristic provinces that are exceptionally rich in plant species like the Amazonian rain forests or the mallee of south-western Australia.

Myers (1990) reckons that about 20% of the world's flora (about 50 000 endemic species) are contained in just 18 hot-spots in lowland tropical forest, montane tropical forest and Mediterranean regions (see Table 19.3).

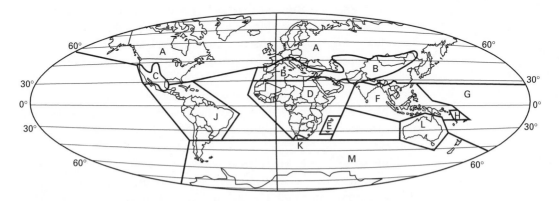

Fig. 19.4 The floral realms. Holarctic kingdom: A, Boreal subkingdom; B, Tethyan subkingdom; C, Madrean subkingdom. Palaeotropical kingdom: D, African; E, Madagascan; F, Indo-Malesian; G, Polynesian; H, Neocaledonian subkingdoms. J, Neotropical kingdom. K, Cape kingdom. L, Australian kingdom. M, Antarctic kingdom. (After Takhatjan 1969.)

Table 19.3 Hot-spots of regional plant biodiversity. Compiled from data in Davis *et al.* 1994.

Region	Endemic species
Cape, South Africa	6 000
Upland western Amazonia	5 000
Atlantic coastal Brazil	5 000
Madagascar	4 900
Philippines	3 700
Borneo, north	3 500
Himalaya, east	3 500
Australia, south-west	2 830
Ecuador, west	2 500
Columbia	2 500
Malaysia	2 400
California	2 140
India, western Ghats	1 600
Chile, central	1 450
New Caledonia	1 400
Tanzania, E. mountains	535
Sri Lanka, south-west	500
Ivory Coast, south-west	200
Total	49 655

The world record for alpha diversity appears to have been held by a 50×20 m plot in heavily grazed Mediterranean scrub dominated by annual grasses and legumes, lying on the road from Jerusalem to the Dead Sea in Israel (more than 250 species; R.H. Whittaker, personal communication). One of the main reasons that this vegetation sample was so rich in species is that the individual plants were so tiny. Many of the annual plant species in these arid, goat-grazed communities are less than 5 cm tall, so total numbers of individual plants per 0.1-ha sample area are staggering. Suppose that each tiny Mediterranean annual were to grow in a square of side 1 cm, then there would be 10 million individuals in the sample area. In contrast, suppose that the Costa Rican rain-forest species occupied 1 m squares, then we would find only 1000 individuals per sample area (in fact, tropical trees would tend to be packed much less tightly than this, and so total population densities would be even lower). A change of four orders of magnitude in total plant density precludes any meaningful comparison of total species richness.

Thus while it is essential to make comparisons of species richness from sample areas of similar sizes, it is also important to compare samples containing roughly equal numbers of individuals. In practice, therefore, we should restrict our comparisons to plants of similar stature and life history. It makes much more sense to compare the species richness of trees in 0.1-ha plots in Japan and England than it does to compare trees in Costa Rica with annual herbs in Israel (Table 19.4).

Table 19.4 Mean numbers of species per 0.1-ha sample area (non-Mediterranean sites include only data for woody plants over 2.5 cm in diameter at breast height). Compiled from various sources by Mooney (1988).

Sample area	Mean no. of species
Dry tropical forest	
Costa Rica upland, Guanacaste	41
Costa Rica riparian, Guanacaste	64
Venezuelan Llanos, Calabozo	41
Venezuelan coastal, Boca de Uchire	67
Moist tropical forest	
Panama Canal Zone, Curundu	88
Brazil, Manaus	91
Panama Canal Zone, Madden Forest	125
Wet tropical forest	
Panama Canal Zone, pipeline road	151
Ecuador, Rio Palenque	118
Costa Rica, near La Selva	236
Temperate zone	
Missouri, Babler State Park	21
Temperate zone	
Australia, forests and woodlands	48
Tennessee, Great Smoky Mountains	25
Oregon, Siskiyou Mountains	26
Arizona, Santa Catalina Mountains	21
Colorado, Rocky Mountain National Park	32
Mediterranean zone	
Israel, grazed woodlands	136
Israel, open shrubland	139
Israel, closed shrubland	35
California, grazed woodlands	64
California, closed shrubland	24
Chile, open shrubland	108
Australia, heath	65
South Africa, fynbos	75

19.5.2 Cape floral kingdom of South Africa

No one really understands why the Cape flora is quite as rich as it is, and there are certainly no watertight explanations for the stunning diversity exhibited by some of the genera (e.g. the Cape supports 526 of the world's 740 *Erica* species, 96 of 160 *Gladiolus* (Iridaceae) and 69 of the 112 species of *Protea*. 'Taxonomic uncertainty is the hallmark of experienced fynbos botanists.' In terms of species richness per genus there is nowhere on earth to match it; for example, an area of just 240 km^2 in the Kogelberg supports 98 species of *Erica*. Fynbos alpha diversity is about 65 species per 1000 m^2 and changes rather little from place to place. This is not outstandingly high by world standards (see Table 19.3; compare with the Park Grass plots, Rothamsted, which support about 50 species in an equivalent area, or a British chalk grassland, which might support 80 species). The beta diversity of fynbos is extraordinarily high. As one goes along an environmental gradient (e.g. up a mountainside) there may be complete species turnover (i.e. beta diversity of 1.0) within just a few hundred metres. It is the gamma diversity of fynbos vegetation, however, that is truly remarkable and unparalleled anywhere else in the world. For example, Kogelberg and Cape Peninsula have similar soil and climate and are separated only by the sandy Cape Flats, but they have just 50% of their plant species in common. Fynbos is a shrubland characterized by tall protea shrubs with large leaves (proteoids), heath-like shrubs (ericoids), wiry reed-like plants (restioids) and bulbs (geophytes), growing in areas of low soil nutrient, high summer drought, recurrent bush fire and exposure to high winds. Many fynbos species are rare, with populations numbering less than 50 individuals in total (Cowling 1992). Another biome peculiar to South Africa is succulent karoo, a sparsely vegetated plant community with scattered dwarf shrubs and a great diversity of succulent plants ('mesembs') growing in dry areas (rain < 250 mm per year) on relatively rich soils and not prone to fires. It supports the richest succulent flora in the world with more than 2000 species of Mesembryanthemaceae, crassulas, euphorbias, stapeliads, haworthias and aloes, in addition to winter-flowering geophytes and a rich flora of spring annuals.

19.5.3 Island floras

There are two sorts of island endemism. In the first case, islands may act as refugia for formerly widespread species that are now extinct on the continent. In the second, isolated islands may be rich in endemic species as a result of *in situ* speciation and adaptive radiation from a small number of initial colonist species (the initial colonists were presumably widespread and abundant on the continent, and also good at dispersing). Remote islands tend to have a high percentage of locally evolved endemics, i.e. species derived, sometimes by spectacular adaptive radiation, from single original colonists (e.g. *Cyanea* (Campanulaceae) and *Cyrtandra* (Gesneriaceae) on Hawaii have more than 50 species each; *Echium* (Boraginaceae), *Picris* (Asteraceae) and *Aeonium* (Crassulaceae) on the Canary Islands), see Plate 13, facing p. 366. Less isolated islands have a smaller proportion of endemics, and many of these are relics, i.e. plants that formerly had much wider distributions (e.g. *Lactoris fernandeziana* on Juan Fernandez, *Dirachma socotrana* on Socotra and *Degeneria vitiensis* on Fiji, all are representatives of monotypic plant families). On the island of Crete, on the tall inaccessible cliffs, safe from the ubiquitous and voracious goats, plant species (chasmophytes) survive that were once widespread on the mainland in the Balkans. In the Lesser Antilles (the arc of islands form Anguilla to Grenada) there are 327 endemics, but only 107 of these are restricted to a single group of islands; 60 of the endemics are found on five or more islands (Synge 1992).

Gigantism is a characteristic feature of plants on islands. Representatives of genera that are lowly herbs on the mainland have evolved to become trees on islands; a well-known example is the daisy tree *Senecio redivivus* (Asteraceae) on the island of St Helena.

Because remote islands lack specialist pollinators like long-tongued bees (see Chapter 6) their floras tend to be dominated by plants showing adaptations to generalized pollination (small, greenish or white, unspectacular flowers). One of the most curious features of island plant biology is the almost complete absence of incompatibility systems and the unusually high representation of dioecious (separate-sexed) plant species. This suggests that there are significant costs to inbreeding for plants on islands, and implies that it may be easier to evolve dioecism rather than incompatibility systems on islands (see Barrett 1996).

The principal ecological processes associated with plants on islands are summarized by Vitousek (1988):

1   the plants tend to exhibit reduced competitive ability as a result of founder effects and genetic bottlenecks (see Chapter 6);

2   small populations on islands are capable of maintaining little genetic diversity;

3   this low genetic diversity implies a relative lack of adaptability to change;

4   there has been a loss of resistance to predators, pathogens and herbivores (especially mammalian herbivores to which many remote island populations were never exposed);

5   loss of coevolved organisms, or failure of potential mutualists ever to arrive;

6   lack of disturbances like fire in recent evolutionary history;

7   intensive exploitation by humans; and

8   most importantly, introduction of ungulates by ancient mariners as an insurance against shipwreck (many oceanic islands of the tropics and subtropics support feral populations of goat, pig, sheep, cattle and horses, with devastating impact on local endemic plant communities).

Current extinction rates on islands are poorly documented in most places, but depressingly well documented in others. For example, the following plant extinctions have been recorded from the islands of the Hawaiian archipelago: Lycopodiaceae (1 species); Aspleniaceae (3 species); Blechnaceae (1 species); Ophioglossaceae (1 species); Amaranthaceae (2 species); Campanulaceae (31 species, mainly *Cyanea* and *Delissea*); Caryophyllaceae (7 species); Asteraceae (13 species); Euphorbiaceae (1 species); Gesneriaceae (8 species); Lamiaceae (15 species); Malvaceae (4 species); Myrsinaceae (1 species); Piperaceae (1 species);

Primulaceae (1 species); Rubiaceae (1 species); Rutaceae (9 species); Violaceae (1 species); Poaceae (7 species). These were all caused directly or indirectly by Europeans, either by direct habitat destruction for agriculture or urbanization, or by the ravages of introduced ungulates (mainly feral pigs and goats; Vitousek *et al.* 1987).

19.6 Variation in plant biodiversity within the British Isles

The flora of the British Isles is better known than any other in the world. There are more active botanists in England than there are plant species for them to study. As a result of this extraordinary asymmetry, the flora has been mapped and then mapped again, with each mapping scheme at a finer spatial resolution than the one before it. The first scheme (Perring & Walters 1962) mapped the entire vascular plant flora of 1515 native species at a scale of 10×10 km squares (a total of about 3600 squares for the British Isles as a whole); amateur botanists spent the next 20 years 'square-bashing', filling in species that were overlooked during the original survey. This means that nowadays any white space on these distribution maps can be reliably regarded as indicating that a species is missing, rather than just unrecorded. Subsequent schemes (like county floras compiled during the 1970s and 1980s) used tetrads (2×2 km squares), which improved the resolution 25-fold, while the current generation of floras use 1×1 km squares (an amazing 100-fold increase in resolution compared with the original scheme). Remote sensing methods are capable of mapping variation in plant communities at a patch size of about 20×20 m, so the process has clearly still got a long way to go.

In Britain, *scarce* species are defined as those that occur in less than 100 of the 10-km squares and *rare* species as those found in 15 or fewer squares. The first map in Fig. 19.5 shows the number of scarce species found per 10-km square since systematic recording began in the early 1800s while the second shows the number of scarce species found in recent years (since 1970). The maps show four important patterns. First, and most obviously, there are several extremely well-defined botanical hot-spots, in which large numbers of scarce species are aggregated. These include the New Forest, Hampshire (ancient woodland), the Lizard, Cornwall (coastal serpentine grassland), Dungeness, Kent (coastal shingle), Craven, Yorkshire (upland limestone grassland) and Ben Lawers and surrounding mountains in the Central Highlands of Scotland (alpine and subalpine calcareous mica-schists). Each of these areas boasts more than 20 scarce plant species in several adjacent 10-km squares. Second, there are not many places where there are no scarce species at all to be found; the few such areas are concentrated in regions with intensive arable agriculture (e.g. the East Midlands) or heavy industrial development (e.g. the central lowlands of Scotland or south

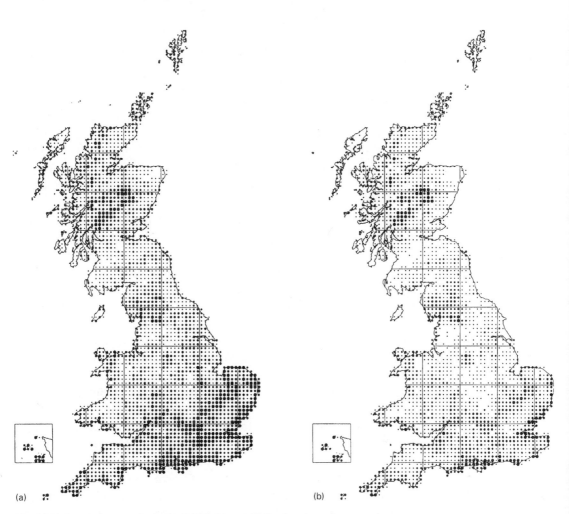

Fig. 19.5 Scarce plant species in the British Isles are defined as those that occur in more than 15 but less than 101 10-km squares. (After Stewart *et al*. 1994.) (a) Number of scarce species per 10-km square since recording began, with dots of increasing size representing 1–4, 5–9, 10–19 and 20+ scarce species per square. (b) Scarce species per square recorded since 1970.

Lancashire). Third, there has been a marked decline in the distribution of scarce species; the majority of the 10-km squares in the post-1970 map have substantially fewer scarce species than they had in earlier years. Centres of pronounced loss of scarce species include the coast of north Norfolk, the Norfolk Broads, the Thames Valley, metropolitan London, arable Lincolnshire and the chalk of the central Chilterns. In the few cases where there are more scarce species on the recent map, this reflects nothing more than increased recording effort. Fourth, the scarce species are geographically distinct, so that the gamma diversity of scarce species is high across different hot-spots (for example, the scarce species on Ben Lawers are totally different from those on the Lizard; the scarce plants on the limestone of Westmoreland are quite distinct from those on the chalk of Kent). This means that you cannot conserve more than a small fraction of scarce species by concentrating resources on a few 'flagship' hot-spots.

Recent changes in the distribution of plant species within the British Isles have been, and continue to be, profound. Perring (1994) described the loss of native species as 'extensive and alarming'. All 113 of the British counties have experienced a decline in the richness of scarce species; the average number of scarce species per county was 55 in the earlier years of this century but this has declined to an average of 37 in the post-1970 period. The average loss of scarce species of 35% masks wide regional variations; for example, the minimum loss was 6% in Shetland and the maximum loss was 73% in Cheshire. It is noteworthy that the proportional loss of scarce species is significantly lower in counties where the original scarce species richness was higher, indicating that scarce species are being lost most rapidly from areas which were already low in species diversity (Stewart *et al.* 1994). Analysis of traits of increasing and decreasing plant species suggests that most of the changes result from direct human intervention in habitat destruction or creation and that the typical declining species is a stress-tolerator while the typical increaser is a fast-growing weed of eutrophic soils (Hodgson 1986).

The reasons for increases in plant distribution include: range expansion by recently arrived aliens (*Epilobium ciliatum* (Onagraceae)); habitat creation (motorways and the spread of halophytes like *Cochlearia danica* (Brassicaceae)); changing fashions in agriculture (*Brassica napus* subsp. *oleifera*); deliberate sowing (*Onobrychis viciifolia* (Fabaceae) on motorway verges); climate change (escape of *Conyza sumatrensis* (Asteraceae) from London's 'heat island'); changed grazing practice (*Pteridium aquilinum* (Dennstaedtiaceae)). Declines are caused by direct habitat destruction (building sites, roads, drainage of wetlands, dredging of canals), agricultural practices (fertilizer pollution, herbicides, hedgerow removal, ploughing ancient grasslands), clear-felling woodlands, atmospheric pollution (nitrates and acidifying ions), eutrophication of water courses, reductions in alien seed introduction, municipal tidying-up (frequent or ill-timed mowing of rough grassy areas), amenity management, landfill, repointing of stonework on ruins and walls, and so on. Many of these changes have important practical and policy implications, and knowledge of the relative importance of different processes in affecting the distribution and abundance of native and alien species will be of considerable value in strategic planning, and will broaden a debate that is currently focused on climate-change scenarios (i.e. models based on temperature and CO_2 concentration).

In their study of British biodiversity, Prendergast *et al.* (1993) defined hot-spots for various taxa (butterflies, birds, dragonflies, liverworts and aquatic vascular plants) as being the top 5% of record-containing 10-km squares. The number of species ranged from 137 in the case of liverworts to 37 for dragonflies, and the number of hot-spots from 107 in the case of liverworts to 135 for aquatic plants. Their study showed that neither of two key assumptions about conservation planning was true for these taxa: (i) habitats that were rich for one taxon were not necessarily rich for others; and (ii) many rare species did not occur in, and hence could not benefit from the conservation of, hotspots. Thus, using one flagship group of organisms like birds or orchids to define areas of conservation priority is not an effective means of preserving biodiversity in other taxa (see Prendergast & Eversham 1995).

19.7 Threats to biodiversity

19.7.1 Species loss in Britain

The principal causes of species loss are listed in Table 19.5 in the context of their relative importance to loss of scarce species in the British Isles. Different relative importance would certainly be given to each cause in different parts of the world, but the increasing role of direct human habitat destruction means that there is likely to be a considerable convergence to British patterns over the years to come. Processes vastly more important in other parts of the world include clear-felling or primary forest (see below; there is virtually no primary forest left in Britain) and grubbing and grazing by alien vertebrate herbivores (feral populations of pigs, goats, sheep, cattle and horses pose a major threat to plant diversity, especially on oceanic islands).

19.7.2 Species loss in tropical environments

The main causes of tropical forest destruction are catalogued by Spears (1988):

1 increasing pressure of human population and the need for additional land for cultivation;
2 patterns of land ownership that force peasant families and landless people into forests and marginal land;
3 commercial agriculture like plantation crops such as palm oil, rubber, coconut, etc.;
4 commercial logging and forest road building, opening up previously inaccessible forest to cultivation and fuelwood harvesting.
The scale of the problem of forest loss is shown in Fig. 19.6.

Some solutions could include: (i) provide the people living in and next to the forest edge with attractive alternatives to further forest encroachment; (ii) intensification of forest management and improved policing of illegal timber extraction and wildlife poaching; (iii) planting new fuelwood and timber plantations; and (iv) creation of extensive forest reserves. It is possible that 2–4 years cultivation on a plot

Table 19.5 Causes of decline in scarce native species in the British Isles, ranked by the percentage of all cases of decline attributed to a given cause. The decline of a species is often attributed to several causes simultaneously (e.g. fertilizer, herbicide and reseeding of grassland). Note that too little grazing is responsible for more species decline than too much grazing. Note also that fertilizer application is apparently a greater menace than herbicide spraying. Action (like drainage or ploughing) and inaction (like scrub encroachment or neglect of coppicing) can be equally damaging to species richness. Other causes of decline, listed in fewer than 1% of cases, include cessation of reed cutting, saltmarsh succession, acidification, trawling and cockle fishing, invasion by *Spartina anglica*, flailing of verge vegetation, rotary mowing of churchyards, decline of village greens, flooding for reservoirs, land fill, intentional eradication of toxic plants, overgrowth by alien plants, fish farming, the switch from spring to autumn cereal cultivation, liming, reseeding old pasture, stopping sand movement in dunes, hedgerow removal, hybridization with other species, decline in seabird roosts, tarmac surfacing of roads, peat extraction, collection for horticulture, recreational boat traffic, shingle removal, *Urtica* invasion following Dutch elm disease, water extraction, skiing and abseiling on cliffs. (From Crawley 1996 after Stewart *et al.* 1994.)

Cause of species decline	Percentage of citations
Reduced grazing leading to overgrowth by rank vegetation	10
Fertilizers	9
Drainage	7.9
Scrub encroachment	7.4
Herbicides	6.4
Ploughing ancient grasslands	5.9
Overgrazing	4.9
Afforestation with conifers	4.1
Poor forest management (e.g. failure to cut coppice)	3.3
Seaside development (bungalows, car parks, etc.)	3.1
Coastal defence construction	2.8
Botanical collecting	2.6
River works (e.g. straightening, bank grading)	2.6
Eutrophication of freshwater	2.3
Pond filling	2.3
Quarrying and gravel extraction	2.3
Trampling by visitors and off-road vehicle use	2.1
Heather burning	1.8
Felling ancient woodland	1.5
Lowland heath destruction	1.3
Tidying-up waste places	1.3
Small population size	1.3
Destruction of limestone pavement	1.3
Other causes of decline (see legend)	12.5

followed by 10–20 years of fallow can be sustained without fertility loss and soil erosion, and may even promote biodiversity by providing a mosaic of habitat patches. However, more extensive clearance and shorter fallowing periods leads to irreversible decline. In the surviving forests, commercial logging and illegal tree extraction, opening up new areas to slash-and-burn human populations by the creation of logging roads, pose an enormous threat to plant diversity. The conversion of natural dry areas to irrigated agriculture poses direct problems of plant loss and indirect problems of increased salination and lowered water

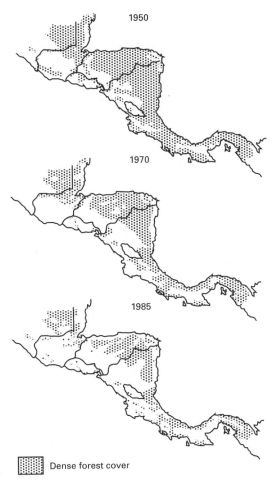

1950

1970

1985

Fig. 19.6 Forest destruction in Central America over the period 1950–1985 (Leonard 1987). The situation is substantially worse in 1995 although detailed maps are not yet published. The forest cover indicated does not include coastal mangrove forests and open pine savanna.

Dense forest cover

tables. Fire has been used by humans in Africa for at least 150 000 years, but misuse of fire can favour spread of flammable invasive shrubs like *Chromolaena odorata* (Asteraceae). Fuelwood harvesting, and bark stripping from medicinal plants like *Warburgia salutaris* (Canellaceae), has almost eliminated the scrub woodland from much of semi-arid Africa. Toxic run-off from mineral extraction has laid waste to huge areas of valley land. Well-intentioned measures like the drilling of deep wells or the control of insect pests like tsetse open up areas to human settlement with a cascade of subsequent impacts, all harmful to plant biodiversity: encroachment of invasive alien plants like *Maesopsis eminii* (Rhamnaceae) in mountains of Tanzania; destruction by cattle of bamboo communities; logging for charcoal production; overgrazing by goats; fragmentation of habitats; fuel for local brick-making; badly managed burning for grazing management (e.g. dry-season fires); building poles; establishment of villages inside forest reserves, especially after the construction of forest roads; dams (e.g. in Semlike and Rutshuru, Zaire); little if any forest management; logging often followed by human settlement; refugees; siltation of rivers and destruction of riverine vegetation following mining; timber poaching by illegal pit-sawyers is highly selective of trees like *Aningueria robusta* (Sapotaceae) and *Milicia excelsa* (Moraceae) and leads to severe depletion of these tree populations; high elephant densities in some places leads to increase in grass (and hence more fires, more animal ingress and hence even more grass); well-sinking, predator removal and improved veterinary care all lead to increased livestock numbers; stabilized political boundaries leading to reduced nomadic movement; plant-collecting for bulbs and succulents in Karoo. Of the 232 Red Data Book plants in the Cape Floristic Region of South Africa, 84 are threatened by invasive plant species, agriculture threatens a further 42, urbanization 21, fire 21, grazing 10, collecting 10 and mining 10 more species.

19.7.3 Urbanization

Huge losses of natural and semi-natural habitats are associated with urban pressures from human population growth (e.g. in tropical countries) or resulting from changes in family structure (e.g. in Britain where there is a huge demand for greenfield housing land to accommodate elderly people who traditionally have lived in nuclear families). Disposal of solid wastes takes up large areas, many of which (e.g. abandoned quarries and ponds) are local diversity hot-spots. Road-building still goes on apace, despite the universal appreciation amongst traffic experts that current rates of energy expenditure and pollution cannot be sustained. Telling people not to use their cars is seen by politicians as a guaranteed vote-loser, as is increasing taxes on energy profligacy (did you realise that aviation fuel was tax free worldwide?). Investment in fast, efficient, convenient public transport has yet to be accepted as a political imperative.

Other human activities have big impacts on plant diversity. These are most starkly clear at the seaside. As sea level rises with global warming, there will be more and more pressure to erect coastal defences. Coastal development for tourism has ruined large areas that were formerly rich in plants (e.g. the southern and eastern Mediterranean coasts of Spain). Agriculture and forestry destroy pristine habitats, and agrochemicals have effects on plants a long way away from the point of application. Most people are aware of the problems that can arise from the indiscriminate use of herbicides, but fewer people realize the enormous loss in plant diversity that is associated with fertilizer pollution and terrestrial eutrophication (see Fig. 14.5). The general loss of species from British hedgerows is due in large part of fertilizer pollution, but an important component comes from atmospheric inputs of nitrogen (including atmospheric ammonia), which can be more than $60 \, \text{kg ha}^{-1} \, \text{year}^{-1}$ in intensive agricultural areas. Acid rain and the deposition of particulate sulphates cause substantial loss of plant diversity in areas that lie downwind of industrial centres; the higher the chimneys are built, the further away from the source the effects are felt, and the more difficult it becomes to compensate for any problems (e.g. coal-burning power stations in England cause pollution problems in Norway on the other side of the North Sea).

In other parts of the world, recreational interest in hunting means that ungulate populations are often allowed to reach such high densities that they threaten plant biodiversity (e.g. deer in the forests of the USA, pigs in the forests of Hawaii). Changes in fire regimes can have massive impacts on plant diversity (see Chapter 4) and overgrazing by domestic livestock is a massive problem in arid parts of the developing world (see Chapter 13).

Others problems representing varying degrees of threat to plant species richness in different parts of the world include water extraction; hydroelectric and water-storage schemes; industrialization and mineral extraction (e.g. limestone extraction for cement-making in treeless Crete); plant-collecting for festivals and traditional (e.g. medicinal) uses or for horticulture (cycads, cacti, orchids, etc.); disappearance of traditional land management with depopulation and loss of rural skills; drainage (e.g. loss of 400 km^2 of wetlands in the Danube delta, Romania); alien plant pathogens; peat extraction for horticulture; and, tragically, the vast local tides of human migration as refugees move out of war zones.

19.7.4 Enforcement of conservation legislation

As this chapter is being written, a magnificent avenue of oak trees *Quercus robur* planted in Windsor Great Park in 1720 during the reign of Queen Anne, and incorporating one 600-year-old tree and several individuals over 300 years from the former parkland, are threatened with felling on the instructions of the Duke of Edinburgh. More than 60 trees were felled on 4 September 1995 before the protesters got themselves organized and built tree houses in several of the remaining trees. The Windor oaks are symbolic of a classic conflict of interest between the large landowner, who sees himself as an enlightened, long-term custodian of the landscape, and the protesters, who see the landowner as an insensitive, uncaring vandal. The conflict epitomizes the 'environmental neatness' versus 'let nature get on with it' viewpoints. The Crown Estate wanted to create a brand new avenue of young oaks, exhibiting military precision in their straight lines and even spacing, which they argued would be an imposing landscape feature for hundreds of years to come. The protesters' response was that there is no way of knowing that oak will prosper in the altered climate 200 years from now, so why destroy a resource that is presently at its peak of biological interest (e.g. one of the trees to be felled was host to a Red Data Book fungus *Phellinus robustus* and to several internationally rare species of beetles). The argument goes that if we cannot protect our plant biodiversity in an area like wealthy, well-informed south-eastern England, then what hope is there elsewhere? Public pressure following extensive media coverage was such that the plan to fell the trees was abandoned. It is clear, however, that without direct action, the plan would have gone ahead.

19.8 Alien plants

The definitions of native and alien species and the varying degrees of naturalization of alien plants are presented in Box 19.1. The key point is that alien plants are species moved directly or indirectly by people well beyond the original geographic range in which they evolved. In the process of translocation, alien plants are often freed from their pathogens and herbivores, which are left behind in the country of origin. Two classic examples of herbivore release are Monterey pine *Pinus radiata* (Pinaceae) and rubber *Hevea brasiliensis* (Euphorbiaceae). In its native home on the Monterey peninsula in California, *Pinus radiata* is a scrubby often sickly little tree, but growing in plantations in New Zealand it is a straight-stemmed giant. Rubber simply cannot be grown as a commercial plantation crop in its native Brazil because of pest problems, but it flourishes under plantation conditions in South East Asia.

Alien plants need to be considered in two groups: (i) those that have naturalized following *intentional* introduction (crop plants, timber trees, garden ornamentals); and (ii) accidental and unintentional introductions. In the case of intentional introductions, there is no reason to suppose that there would be any obvious relationship between attributes of the plant's native geography and its success as an alien. For unintentional introductions, however, it is clear that abundant and widespread species are more likely to find their way on board ship in a foreign port than rare or local species, and that plants growing wild on the dockside are more likely to be picked up than plants growing in remote, mountain-top habitats. Thus, the sample of species picked up is highly biased and much of it is preadapted to live in human-disturbed habitats. Thus, because the probability of establishment depends on the rate of introduction of propagules, it is not surprising that alien plant species richness is positively correlated with human transport centres (docks, cities, railways, arterial roads) and negatively correlated with the degree of isolation of a habitat within the new, introduced range (e.g. alien plant species richness is low in mountain areas and on uninhabited islands; see Fig. 19.7 and Plate 14, facing p. 366). In Britain, we are fortunate to suffer very little from our alien plants, but in many parts of the tropics and subtropics alien plants are the biggest single threat to plant conservation in nature reserves. It is noteworthy that most of the serious problems are caused by intentionally introduced plants rather than by accidentally introduced species (see below).

Box 19.1 Alien plants: a glossary of terms.

Aliens non-indigenous species moved intentionally or unintentionally
by human beings from one country and released into another;
much of the movement of plant species occurred during the
opening up of worldwide maritime trade routes during the period
1500–1750

Naturalized alien plant species that have formed self-replacing
populations based on recruitment from seed or spread of
vegetative fragments

Natives the indigenous flora, which reached the country unaided by
humans, plus anthropogenic species (archaeophytes) that may have
been introduced unintentionally prior to the Roman invasion of
Britain

Casuals alien plant species that do not form self-replacing populations
and rely on repeated reintroduction for persistence (includes many
frost-susceptible tropical and subtropical plant species that can
flower but generally do not set seed in temperate environments)

Archaeophytes the house-sparrows of the botanical world; species
living in artificial habitats, and found wherever human habitation
and cultivation extends; cosmopolitan temperate weeds like *Poa
annua, Capsella bursa-pastoris* and *Plantago major*. Considered as
native species but possibly introduced before 1500

Climatic matching the ability of a plant to increase when rare
requires that the abiotic environment is conducive (the nature of
the unfavourable season is particularly important, including
extremes of high and low temperature and extremes of drought)

Hemerophytes species introduced intentionally by people (contrast
xenophytes)

Herbivore release the extraordinary vigour of some alien plant
species is attributable, at least in part, to the plants having left their
specialist invertebrate herbivores and fungal pathogens behind in
their native environments. For example, as an alien in South
Africa, *Banksia ericifolia* (Proteaceae) produces 16 500 seeds at 8
years of age, whereas in its native habitat near Sydney in Australia
it takes 25 years to produce a seed crop less than 10% this size.
Many familiar Eurasian weeds grow much taller as aliens in the
New World

Invasible communities all plant communities are invasible, but some
are much more invasible than others (Fig. 19.7). Analysis of data
on the number of alien plant species in different habitats needs to
control for differences in the rate of introduction of propagules,
i.e. probability of seed from a given native habitat being
transported to the new country (a function of geographic

continued on p. 618

> **Box 19.1** *Continued.*
>
> isolation, distribution, abundance, viability of transport stages, etc.) and the probability of the plant becoming established in its new home (disturbance regime, transport activity, isolation and extent of target habitat, percentage ground cover, etc.)
>
> **Invasive species** the ability to 'increase when rare' (the invasion criterion) is a property exhibited by *all species* in their native habitats; i.e. all plant species are invasive under certain environmental conditions. This being the case, it makes no sense to search for traits of 'invasive' and 'non-invasive' species. The study of invasions requires an approach involving the plant species, the abiotic and biotic environments of the target habitat, the disturbance regime, and the interactions between these and the pattern of temporal variability (e.g. year-to-year changes in weather, and in the densities of competitors and natural enemies)
>
> **Neophytes** naturalized aliens introduced since 1500 (contrast with archaeophytes)
>
> **Xenophytes** unintentionally introduced aliens

19.8.1 Notions of invasive and non-invasive species

Perhaps the most fundamental concept in population dynamics is the invasion criterion, which states that every persistent species must exhibit the ability to increase when rare. Technically, this requires that

$$\frac{dN}{dt} > 0 \tag{19.1}$$

when population density, N, is small. All theoretical models of species coexistence have the invasion criterion at their heart. Without it, species would simply drift inexorably to extinction, as one catastrophe after another (drought, flood, frost or hurricane) drove their population density ever downward. Thus, *all* persistent species (that is to say, all species recognized as being an integral part of a given plant community) exhibit the ability to increase when rare. All species pass the invasion criterion. This being so, it makes no sense to talk about invasive species and non-invasive species. All species are invasive, given the right combination of environmental conditions. If all species are invasive then it must follow that it is pointless to search for traits that will separate invasive from non-invasive species. The *competitive ability* of a plant is not a genotype-specific trait; it depends on the environmental setting and the specific identity of the other competitors. In order to understand the population behaviour of an invading plant genotype we need a knowledge of the following: (i) the abiotic environment; (ii) the biotic environment (resources, competitors, natural enemies and mutualists);

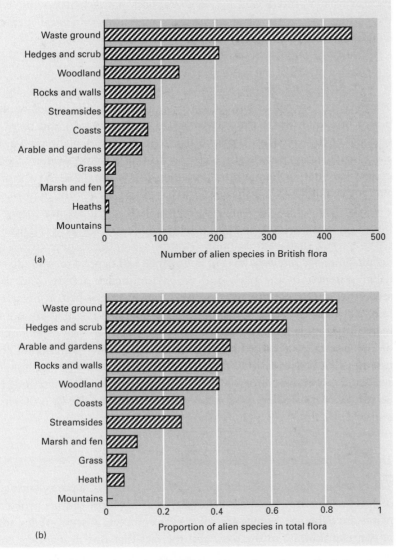

Fig. 19.7 Alien plant species in the British Isles: (a) numbers of naturalized alien plant species in 10 principal habitat groups; (b) proportion of the total flora made up by alien species in 10 habitats. (From Crawley 1997.)

(iii) the interaction between biotic and abiotic environments; and most importantly (vi) the year. One year differs from another to such an extent that a genotype may fail miserably for several years in a row before prospering mightily in a year when conditions are just right. It is salutary to remember Sod's Law of Experiments: in a 3-year study, the important things are bound to happen in the fourth year. Clearly, then, there can be no single-factor explanation of invasiveness; the minimal description of an invasion should be four-dimensional (genotype by abiotic environment by biotic environment by year). Indeed, the model should really be five-dimensional because it is almost inevitable that there will be an interaction between year and biotic environment (e.g. the performance of the plant's principal competitors or natural enemies will differ from year to year).

19.8.2 Problem plants

The problem of defining weeds is legendary ('a plant in the wrong place', 'a plant whose virtues have yet to be discovered') and there is seldom much agreement between different groups of plant biologists about whether any particular species is a weed or not (one person's weed is another person's herbarium specimen; Willamson 1993). To short-circuit these difficulties, I shall define what I mean by a problem plant. It will have the following two characteristics: (i) the plant has passed a threshold level of abundance; (ii) someone is concerned. You will notice straight away that both defining features are entirely subjective. The definition of the *threshold* level of abundance will depend on an individual's awareness of the plant and its environmental setting, and will reflect the degree of tolerance (or intolerance) of the person involved (e.g. the mere fact that a plant was alien might mean that for some nature-reserve managers the threshold of tolerance is zero). The nature of the *concern* that caused the person to define the existence of the threshold, and to set its level where they did, relies on a set of economic, aesthetic, moral and cultural judgements. None of these has anything to do with ecological science. Problem plants are not defined by ecological traits.

19.8.3 What are the problem plants?

The problem plants of Britain are, by world standards, a pretty tame set of species. We are extremely fortunate that our most serious problem plants are as benign as they are. Given the complete absence of any sort of plant quarantine in the past, and the fact that British mariners and botanical explorers brought back to Britain virtually every plant species that they could lay their hands on (plus countless other accidental introductions that they didn't lay their hands on), we have indeed escaped lightly. We have a lot to thank the frosty British winter for.

As to numbers, it appears that about 5% of the plants which can grow in Britain have become naturalized, of the naturalized plant species about 6% have become so widely naturalized that they behave for all intents and purposes as if they were native plant species, and of these about 22% have become problem plants in one habitat or another (see Tables 19.6 & 19.7). That is to say, of the 23 550 plant species that we know can grow in the wild in Britain, only 0.1% has become a problem. It is difficult to do this sort of calculation for other countries, because there is so little information on what plant species were imported but did not become established. However, two things are quite

Table 19.6 Quantitative make-up of the alien flora of the British Isles. The challenge is to understand how the filters operate at each level in the hierarchy. For example, of all those species that produce seedlings in British gardens (the self-sowers) what is it that distinguishes those that go on to become thoroughly naturalized from those that remain as mere casuals?

Hierarchy	No. of species
World	250 000
Brought to Britain intentionally	200 000
(mainly as herbarium specimens to kew, etc.)	
Brought to Britain unintentionally	20 000
Grown intentionally in Britain	
(outdoors and under glass)	
Botanic gardens	25 000
Commercially available garden plants	14 000
Grown outdoors in Britain	
Botanic gardens	16 000
Private gardens	10 000
Native flora	1 515
Non-cultivated aliens	6 000
British flora	
Native	1 515
Naturalized aliens	1 169
Widely naturalized aliens	68
Problem alien plants	15

clear: (i) more alien plant species are regarded as being problem plants in other countries; and (ii) the problems caused by alien plants are much more severe than they are in the British Isles (see Crawley 1997 for details).

British problem plants come in a range of life histories and infest a variety of different plant communities (Fig. 19.7). It is noteworthy that the only annual plants on the list are problems are arable agriculture. There are no alien annual plants that are problems in British natural habitats (again, we are fortunate here, since there are some pernicious annual problem plants in other countries, like the highly invasive annual grass *Bromus tectorum* in the north-western USA that in Britain, is a pathetic little thing, restricted to a tiny area in East Anglia). The problems increase with some of the stout herbaceous perennials, which can dominate river banks, railway embankments and derelict urban land, to the complete local exclusion of native vegetation. Giant hogweed has the added problem of inflicting on small children a vicious photosensitized rash. The real problem plants, however, are the trees and shrubs. Most British readers will be familiar with the problem of *Rhododendron ponticum* (Ericaceae), especially in wet upland areas like Snowdonia National Park where it runs rampant through native woodland and out onto the open heath. More insidious, and much less well publicized, is the invasion of oak woodlands of the highest conservation value (e.g. Windsor Great Park, a World Heritage site) by Turkey oak

Quercus cerris. This south-east European invader regenerates freely on light, acid soils and can outcompete the native dominant *Quercus robur* by the neat evolutionary trick of playing host to the sexual generation of an insect which, in its agamic generation, seeks out and destroys the acorns of native *Quercus robur* (the animal in question is a host-alternating, cynipid gall wasp from Balkan Europe called the knopper gall insect *Andricus quercuscalicis*; Hails & Crawley 1989). Trees are much more problematic than other plants by dint of their sheer size and longevity, and because of their ability to dominate ecosystem processes

Table 19.7 Problem plants in the British Isles. These alien plant species have become so thoroughly naturalized that in their ecological behaviour they are virtually indistinguishable from native plant species.

Habitat	Problem plants	Threat to native biota	Nature of the problem
Woodland	*Acer pseudoplatanus*	?	Ubiquitous, but not obviously problematical to native tree diversity
	Quercus cerris	Y	Replaces *Quercus robur* regeneration in parts of south-east England
	Rododendron ponticum	?	Can be a problem in Ireland, in wet, western oak woodlands
	Allium triquetrum	N	Forms monocultures on the floor of deciduous plantation woodland in south-west Britain; also on walls and banks
Open water	*Elodea canadensis*	N	Less common than formerly, when it choked several canals, impeding navigation
	E. nuttallii	N	More recently arrived, and reaching lower peak abundance
	Crassula helmsii	Y	Capable of forming solid mats on bare mud or when fully submerged
Banks	*Impatiens glandulifera*	?	An annual plant forming thickets of soft stems on muddy stream banks
	Heracleum mantegazzianum	N	Reaches peak abundance on gravel spits and banks in fast-flowing rivers
	Fallopia japonica	Y	Impenetrable thickets on river banks, railways and industrial waste ground. Spreads by root and stem fragments, not by seed
Arable	*Avena fatua*	N	Not a problem outside arable fields
	Veronica persica	N	Uncommon outside arable fields
Gardens	*Aegopodium podagraria*	N	Pestilential garden weed, and a minor component of lowland plantation woodland
	Epilobium ciliatum	N	Minor garden weed, commonest on urban waste ground
Heath	*Rhododendron ponticum*	Y	Forms monocultures on wet upland heath to the exclusion of all other plants
	Campylopus introflexus	?	Forms an unbroken carpet over the surface of the peat. Possible impaired regeneration of heathland plants from seed

Y, yes (native plant diversity reduced, at least locally); N, no; ?, case yet to be demonstrated (the precautionary principle would council caution).

like primary production, litter quality, nutrient cycling, microclimate and associated biodivesity.

19.8.4 Problem plants in other countries

Many of the oak woodland nature reserves in central European countries have been completely robbed of their conservation value after having been overrun by the false acacia *Robinia pseudoacacia* (Fabaceae) introduced originally from North America for use as a windbreak and shelterbelt.

The island of Hawaii boasts more than its share of problem aliens (Vitousek *et al.* 1987; Stone *et al.* 1992).

1 Species that short-circuit entire primary successions by recruiting directly onto bare lava (e.g. the nitrogen-fixing tree *Myrica faya*).

2 Vines that smother native trees (e.g. *Passiflora mollissima* and *Rubus argutus* (Rosaceae), which can overrun forests of *Acacia koa* (Fabaceae) or *Metrosideros polymorpha* (Myrtaceae)).

3 Thicket-forming, evergreen woody species that exclude all native plants (e.g. *Psidium cattleianum* (Myrtaceae), *Leucaena polycephala* (Fabaceae), *Ulex europaeus* (Fabaceae) and many others).

4 Trees that replace native canopy dominants in forest reserves (e.g. *Spathodea campanulata* (Bignonaceae) and various melastomes including *Miconia calvescens* (Melastomaceae)).

5 Entire alien plant communities, reassembled in matched climatic zones (e.g. Scottish grasslands on the cool, moist, upper slopes of Mona Loa with *Anthoxanthum, Holcus, Dactylis* (all Poaceae) *Hypochoeris* (Asteraceae), *Ulex* (Fabaceae), etc. all in their characteristic 'Scottish' relative abundances).

6 Infiltration of flammable species along paths, roads and powerline rights-of-way that bring fires into the heart of previously unburned natural habitats (e.g. fuel-rich grasses like *Pennisetum setaceum* and *Andropogon virginicus*.

7 Replacement of the native forest understorey (e.g. *Hedychium gardnerianum* (Zingiberaceae), *Lantana camera* (Verbenaceae) or the Australian tree fern *Cyathea cooper*.

The seeds of many Hawaiian alien plants are dispersed deep into the heart of natural habitats by alien, fruit-feeding birds. Recruitment conditions favouring regeneration of the aliens are provided by feral populations of alien ungulates (e.g. principally pigs and goats but there are feral wild populations of cattle, sheep and horses as well).

In South Africa, about 750 fynbos plant species are threatened by the spread of alien plants. In coastal lowlands, invasive Australian wattles like rooikrans *Acacia cyclops* (Fabaceae) and Port Jackson willow *Acacia saligna* (7600 km^2 of sandy coastal lowland, planted intentionally to stabilize dunes in 1850) possess a seed bank and fire-stimulated germination and thrive best in disturbed areas. Streams and rivers are invaded by black wattle *Acacia mearnsii*, blackwood *Acacia melanoxylon*, stinkbean *Paraserianthes lophantha* (all from Australia) and *Sesbania punicea* (Fabaceae, from Argentina and Uruguay); mountain areas by three pines, three hakeas and wattles, cluster, alleppo and Monterey pine (silky hakea *Hakea sericea* (hedge and fuel), cluster pine *Pinus pinaster* and long-leaf wattle *Acacia longifolia* are the worst problems).

What's to be done? Of the approximately 4600 species of alien plants on Hawaii, more than 700 reproduce in the wild and 86 are considered to be serious threats to native ecosystems (Smith 1985; Stone *et al.* 1992; Crawley *et al.* 1996). Serious conficts of interest have arisen between local people (e.g. pig hunters on Hawaii) and conservationists (often Caucasians who don't live full-time on the islands) who want to get rid of the alien plants and animals. Given the difficulty of the steep, rugged volcanic terrain, and the high cost of labour, the only realistic prospect for the control of the 100 or so problem alien plant species on Hawaii is biological control. There have been some notable successes already (e.g. *Opuntia megacantha* (Cactaceae) by *Cactoblastis cactorum* and *Dactylopius opuntiae*, and *Ageratina riparia* (Asteraceae) controlled by the fungus *Entyloma ageratinae*) but much needs to be done. For example, the entire family Melastomaceae, including such pernicious weeds as *Clidemia hirta*, *Tibouchina herbaceae* and *Melastoma candidum*, has been targeted for biocontrol. Biological weed control is the only practical, affordable, long-term (and potentially permanent) solution to alien weed invasions of nature reserves, particularly in tropical environments. Despite its excellent safety record, biocontrol of weeds in no panacea; the process of exploration and careful screening is relatively slow, and the eventual success rate is substantially less than 50% (Crawley 1989c). Nevertheless, once established, biocontrol is the ideal pest management for nature reserves, because it is cheap, permanent and highly selective.

19.8.5 Overview of problem plants

Problem plants from various parts of the world are shown in Table 19.8.

19.8.5.1 Perennial herbs

Many examples exist in wetlands and arid rangelands; several apparently tame British plants make aggressive invaders in other countries (e.g. purple loosestrife *Lythrum salicaria* in North American wetlands and the stoloniferous *Pilosella officinarum* (Asteraceae) in pastures in New Zealand overgrazed by rabbits and sheep).

19.8.5.2 Perennial grasses

These replace native grasses in grazed rangeland and pose a serious fire

Table 19.8 Problem plants in various parts of the world.

Country	Growth form	Problem plant
Hawaii	Tree	*Spathodea*
	Cactus	*Opuntia*
	Shrub	*Melastoma*
	Crop	*Pisidium*
	Vine	*Passiflora*
	Grass	*Andropogon*
South Africa	Pine	*Pinus radiata, Pinus halepensis*
	Acacia	*Acacia longifolia*
	Protea	*Hakea sericea*
New Zealand	Pine	*Pinus radiata*
	Shrub	*Ulex europaeus, Cytisus scoparius*
	Herb	*Pilosella officinarum*

risk to native trees in forest ecosystems. Perennial grasses often pose a double fire hazard: (i) they produce a much greater bulk of flammable material than was produced by the native plants and therefore cause more intense fires; and (ii) the timing of their flammability is different (e.g. they stay green further into the dry season) so that fires occur at different times of year, with further damaging consequences to native vegetation. We are extremely fortunate in Britain that we have so few problem perennial grasses (people whose flower beds are infested by gardeners' *Phalaris arundinaria* garters Picta may disagree).

19.8.5.3 Vines

These are very serious problems in subtropical and torpical woodland and wayside habitats (e.g. kudzu vine *Pueraria thunbergiana* (Fabaceae) a Chinese alien widely naturalized in subtropical south-eastern USA). We are lucky in Britain that the mile-a-minute vine (Russian vine) *Fallopia baldschuanica* (Polygonaceae) does not spread by seed, otherwise many of our natural scrublands would be quickly overrun.

19.8.5.4 Shrubs

These are very serious pests in most climates; in places like New Zealand, sheep pastures are overrun by one or more of gorse *Ulex europaeus*, broom *Cytisus scoparius* or bramble *Rubus fruticosus*. Several tropical crop shrubs and small trees are devastating invaders (e.g. the guavas *Psidium* spp. on Pacific islands exclude native vegetation in extensive, impenetrable monocultures).

19.8.5.5 Trees

These are potentially disastrous for nature conservation and watershed management. Alien invasive trees are the canopy dominants in many parts of the tropics, especially on islands. Some of the most highly valued conservation areas in the world are seriously jeopardized by invasive trees. For example, the South African fynbos has one of the richest floras on earth and is threatened by a suite of trees including pines from America (*Pinus radiata*) and the Mediterranean (*Pinus halepensis*), as well as a host of woody plants from Australia including proteas (*Hakea, Grevillea*) and legumes (several species of *Acacia*). British woodland is presently threatened by sycamore *Acer pseudoplatanus* and Turkey oak *Quercus cerris*, and is likely to be threatened in the near future (especially with climate warming) by tree of heaven *Ailanthus altissima* (Simaronbaceae), false acacia *Robinia pseudoacacia* and box elder *Acer negundo*.

19.9 Plant conservation

The traditionally accepted reasons why we should conserve biodiversity are as follows: (i) ethical and moral obligations; (ii) aesthetic and recreational potential; (iii) ecological impact; and (iv) economic importance. The difficulty is that none of these is a big vote-winner, and all of the arguments supporting them are currently somewhat fragile. Issues of ethics and aesthetics cut little ice with a subsistence farmer in the Third World. The ecological impact of biodiversity loss is almost wholly speculative (see Chapters 15 & 16). The first large-scale experiments on the ecosystem-level impact of variations in plant species richness are currently under way in Europe and the USA. The economic argument is hard to sustain in the face of the fact that the loss of 90 or so species from the US flora has had no discernible economic consequence (Norton 1988).

The accidental discovery of leaf-cutter ants *Atta* that reject the leaves of a particular leguminous tree *Hymenaea coubaril* on Barro Colorado island in the Panama Canal led to the discovery of a valuable pesticide. The reason the ants do this is because the leaves are fungicidal and kill the species of fungus employed by the ants in their fungus garden to break down leaves and provide the insects with food. Subsequent laboratory work discovered a potent new fungicide in the *Hymenaea* leaves, a terpene-caryophyllene epoxide, which turned out to be toxic to a wide range of pathogenic fungi (Hubbell *et al.* 1983). The argument that the unexplored pharmaceutical riches of the tropical forests represent a capital that it would be economic madness to squander is often repeated. However, those mostly likely to profit from such discoveries are multinational chemical companies rather than the people who live in the forest, and it is on the behaviour of these local people that the future of the forest ultimately depends.

On a more positive note, African ecosystems are not as fragile or as vulnerable as popularly believed. Throughout their evolution they have been subjected to enormous environmental pressures (including hunter-gatherers using fire for at least the last 5000 years). Biotic diversity is not linked to the distribution of elephant, rhino, hippo, giraffe and other charismatic magaherbivores. 'The massive investment in conservation campaigns directed at these species does more for the souls of the donors and the egos of the elephant experts than it does for biotic diversity, which is centred on less exciting communities of montane forests, Mediterranean heathlands, wetlands, lakes and rivers' (Huntley 1988). Real biodiversity hot-spots are seldom evident from vegetation maps (the standard starting point for biotic inventory). Sadly, however knowledge of biotic conservation needs and priorities far exceeds the ability of African governments to implement conservation action plans.

19.9.1 Parks and nature reserves

Most people would agree that the best way to preserve biodiversity is to buy up large areas of pristine habitat and to manage them as nature reserves whose prime objective is conservation. But what proportion of threatened plant species occur within the boundaries of protected areas? It varies. In New Zealand, about 70% are in permanent protected areas managed to benefit the biota, while 11% of threatened species have no effective protection in the wild. In Australia, the figure is about 50%. In Spain, however, only 5% of nationally threatened plants occur in national parks. In many developing countries the picture is much worse; often none of the threatened plants has effective *in situ* protection. The money to buy the necessary reserves will have to come from developed countries, and political action by voters in developed countries is the only answer.

19.9.2 Habitat restoration

In addition to the preservation of pristine habitats, we need to understand how to restore the vast numbers of habitats that have been degraded to various degrees by human activity. Some excellent case histories involve the recreation of species-rich prairie in Wisconsin (Jordan 1988), coastal wetlands in Caifornia (Zedler 1988), restoration of degraded land in the Amazon basin (Uhl 1988) and of logged coniferous forest in Michigan (Cairns 1988). It is becoming clear that damaged ecosystems can recover rather rapidly under sensitive management, in such a way that biological diversity is vastly improved, but it does appear that restoration is often more difficult in tropical compared with temperate habitats (Cairns 1988). There is much to be said for benign neglect as a management policy, and many an urban nature reserve in Britain has simply been derelict urban wasteland, reclassified after 30 years of natural succession. Practical things that can be done include short- or long-term propagation and reintroduction programmes involving relocation and transplantation, fostering and cross-fostering, artificial cross-pollination, germination and seedling establishment. However, reintroduction programmes are not favoured by all botanists; there are those who will argue vehemently against the loss of original native genotypes and who will point out the possible problems of using introduced genotypes ('genetic pollution'; see Chapter 6).

19.9.3 Botanic gardens

Kew Gardens in London, along with its outstation at Wakehurst Place in Sussex, is one of the world's greatest living plant collections. They have 85 000 living accessions between them, representing 33 400 taxa (mainly species, but some subspecies and horticultural varieties; certainly more than 25 000 species). This represents about one-tenth of the world's flora maintained in a single institution. Of these living accessions, about half are grown under glass for all or part of the year, but the remainder are grown in the open air, exposed to the rigours of the English winter (P. Cribb, personal communication).

Some people may try to convince you that it is impossible to conserve the world's flora by keeping plants in cultivation and that habitat protection is the only sensible course of action. But think about it. Suppose that there was the political will to provide the funding to set up institutions like Kew Gardens in 10 new locations throughout the tropics, and to train the people to staff them. That would be enough, in principle, to rescue at least some genotypes of the entire world flora. The task is clearly not impossible. Propagation of endangered wild plants is much less difficult than is commonly believed. What is lacking are people willing and able to do the work. Again, it requires political action by people like you who are both knowledgeable and interested.

19.9.4 Gene banks

Seeds stored under carefully controlled conditions may be valuable as an adjunct to other conservation measures, but they are certainly not a realistic substitute for protected areas and botanic gardens. Their maintenance costs are high, and many of the high-tech methods of gene preservation like tissue cultures and meristem storage are simply not appropriate for use in underdeveloped countries.

19.10 Food plant conservation

We all rely on plants for food and for a wide range of other important direct and indirect functions (direct uses like fuel, timber, paper, ornamental plants, medicines, perfumes, pesticides, dyes, cosmetics, plastics, oils, polishes, etc.; indirect uses like shelter, watershed and coastal protection, erosion control, sewage treatment and water purification). Of the worlds 250 000 plants species only 3000 have been used directly for human food and just 200 species have been domesticated (Zohary &

Hopf 1988). Of these, just 15 species (rice, maize, wheat, sorghum, rye, yam, cassava, millet, barley, sugarcane, sugarbeet, groundnut, soybean, broad bean, banana) are crops of major economic importance, providing 95% of the plant calories consumed directly by humans (which for most of the world's people, of course, is almost all of their calories consumed). The extremes are great. In one part of rural Peru, people regularly eat 193 species of plants, of which 120 are collected exclusively from the wild. In Britain, we eat meat and two veg (about 20 species in total, based on wheat, potato and rice as staples). In developed countries, commerical plant breeders make sure that they conserve genetic diversity for their future use. In the developing world, however, loss of landraces and loss of semi-natural habitats where the wild relatives of crop plants are to be found is occurring at an alarming rate (Williams 1988). Recent losses of biodiversity in crop genotypes has been brought about by increased mechanization, fertilizer input and pesticide use. In India, for example, where rice was traditionally grown from at least 3000 varieties, more than 75% of rice is now cultivated from fewer than 10 varieties (Rhoades 1991), and in Indonesia 74% of current rice varieties are descended from a single maternal parent (Hargrove *et al.* 1988). In the USA 50% of the wheat crop comes from nine varieties, 75% of the potato crop from four varieties and 50% of the cotton crop from just three varieties. Catastrophic crop failures attributable to extreme genetic uniformity include wheat in the former USSR, maize in the USA, rice in Indonesia and citrus in Florida.

The choice timber species are inevitably the first species to be selectively logged from rich tropical forests, often illegally by 'tree poachers' and by unlicensed local people who see pit-sawing as their only potential source of external income. Given that population densities of many of the tree species are often very low to begin with, selective logging by people can easily cause local extinction over large areas of forest; only trees in the most remote and inaccessible parts are likely to survive. This process will have disproportionately great effects on the loss of within-species genetic diversity in rare, long-lived species like timber trees (see Chapter 6).

19.11 Economics of plant conservation

Most attempts to put a value on the conservation of plant biodiversity use cost–benefit analysis and trade-offs. The mainstream economic approach is doggedly non-judgemental about people's preferences: what the individual *wants* is presumed to be good for that individual. Biodiversity is treated in a way that is utilitarian (i.e. things count to the extent that people want them), anthropocentric (humans are assigning the values) and instrumentalist (i.e. biota is regarded as an instrument for human satisfaction; see Randall (1988) for a full discussion of these terms). The major functions of plants in the landscape (food production, watershed protection, erosion control, shelter, timber, fuel) can be costed with relative ease. Other aspects (like the future value of undiscovered pharmaceuticals or DNA sequences for use in genetic engineering) are much more speculative. Some important components are so subjective that there will always be debate about their relevance (e.g. the value of environmental peace and beauty, the scientific value of relatively undisturbed ecosystems). Environmental economists use a number of indirect methods for costing these things. The commonest is the *contingent valuation method* (CVM), which aims to detect peoples' preferences for environmental quality by discovering what people are willing to pay (WTP) for increments in environmental quality or what they are willing to accept (WTA) in compensation for forgoing these benefits (interestingly, WTP is often greater than WTA). The method is called 'contingent' because it depends upon individuals' perceptions of a mass of uncontrolled background information. To many environmentalists, however, the subjectivity of these measures undermines their credibility completely. For example, the notion of what a person with no income and no material possessions is 'willing to pay' for conservation verges on the odious. Another method (especially useful for assessing the value of sites for ecotourism) is the *travel cost method*. This simply quantifies how much rich people are willing to spend in order to visit a particular destination. It, too, is contingent, since an avid birder is going to spend vastly more money to visit a remote island to see the last bird on their 'life-list' compared with someone who likes to spend their vacation visiting art galleries in capital cities. The technique of *hedonic prices* arrives at the value of non-marketed environmental services by subtraction; the value of soil fertility, scenic beauty and clear air are obtained by subtracting known economic costs (like land, property, wages) from the actual total costs in areas which differ in environmental quality (Groombridge 1992; Swanson 1995).

The depths of the folly of using traditional economics to underpin biological conservation is shown in two classic studies. The first is the universality of human self-interest and the dire consequences that this will inevitably have for shared natural resources (the 'Tragedy of the Commons'; see Box 19.2). The second is the way that compound interest rates and opportunity costs conspire to undermine the value of biological resources. In a masterful analysis, Lande *et al.* (1994) demonstrated that when the economic discount rate of future harvests exceeds a critical value related to the population growth rate, the strategy that maximizes the present value of cumulative harvest is immediate extinction (liquidation) of the population. Given an unstable equilibrium at small population size (Allee effects, see Chapter 14), the critical discount rate is even lower in more realistic, stochastic models than in their equivalent deterministic forms. These two fundamental insights argue against the use of economic discounting in the development of optimal strategies for sustainable use of biological resources.

Box 19.2 The Tragedy of the Commons.

The following extract is taken from one of the most influential essays ever written in environmental science. Garrett Hardin was a leading figure in the environmental movement of the 1960s; a rare combination of first-rate ecologist and brilliant communicator.

The tragedy of the commons develops in this way. Picture a pasture open to all. It is to be expected that each herdsman will try to keep as many cattle as possible on the commons. Such an arrangement may work reasonably satisfactorily for centuries because of tribal wars, poaching and disease which keep the numbers of both men and beast well below the carrying capacity of the land. Finally, however, comes the day of reckoning, that is the day when the long-desired goal of social stability becomes a reality. At this point, the inherent logic of the commons remorselessly generate tragedy.

As a rational being, each herdsman tries to maximise his gain. Explicitly or implicitly, more or less consciously he asks 'What is the utility *to me* of adding one more animal to the herd?' This utility has one negative and one positive component.

(1) The positive component is a function of the increment of one animal. Since the herdsman receives all the proceeds from the sale of the additional animal the positive utility is nearly $+1$.
(2) The negative component is a function of the additional overgrazing created by one more animal. Since, however, the effects of overgrazing are shared by all the herdsmen, the negative utility for any decision-making herdsman is only a fraction of -1.

Adding together the component partial utilities, the rational herdsman concludes that the only sensible course for him to pursue is to add another animal to his herd. And another . . . But this is the conclusion reached by each and every rational herdsman sharing the commons. Therein is the tragedy. Each man is locked into a system that compels him to increase his herd without limit–in a world that is limited. Ruin is the destination toward which all men rush, each pursuing his own best interest in a society that believes in the freedom of the commons. Freedom in a commons brings ruin to all. Garret Hardin (1968)

19.12 Conclusions

Biodiversity is being lost at a staggering rate through habitat destruction and degradation of surviving habitat fragments. Although much of the attention tends to focus on loss of tropical biodiversity, there is no cause for complacency in the developed countries. In the contiguous USA, for example, an estimated 90 species have been driven to extinction in the last 200 years, and hundreds more are in imminent danger of extinction (Ayensu & DeFilipps 1978). The resources to do much about the loss of diversity are almost all in the wrong place. For example, of the world's 1500 botanic gardens, 532 are in Europe; Africa only has 82 and South America just 66. The Biodiversity Convention at Rio made a significant step towards improving the prospect for world conservation with its five key points:

1   ensuring that use of biological resources is sustainable;

2   funding for conservation and rational use in developing countries;

3   sharing the costs and benefits between developed and developing countries;

4   providing economic and legal conditions for technology transfer; and

5   ensuring a fair share of the benefits of biotechnology research that arise from the conservation of biological diversity and exploitation of developing world genetics resources.

But in the end, it all comes down to you. Get active. Do things. Believe me, it makes a difference.

References

Abbott, H.G. & Quink, T.F. (1970) Ecology of eastern white pine seed caches made by small forest animals. *Ecology* **51**, 271–278.

Aberdeen, J.E.C. (1958) The effect of quadrat size, plant size and plant distribution on frequency estimates in plant ecology. *Australian Journal of Botany* **6**, 47–59.

Abrahamson, W.G. & Weis, A.G. (1986) The nutritional ecology of arthropod gall makers. In *The Nutritional Ecology of Insects, Mites and Spiders* (Eds F. Slansky & J.G. Rodriguez), pp. 235–258. John Wiley & Sons, New York.

Abrahamson, W.G., McCrea, K.D., Whitwell, A.J. & Vernieri, L.A. (1991) The role of phenolic compounds in goldenrod ball gall resistance and formation. *Biochemical Systematics and Ecology* **19**, 615–622.

Abrams, P. (1976) Environmental variability and niche overlap. *Mathematical Bioscience* **28**, 357–372.

Abrams, P.A. (1994) The fallacies of ratio-dependent predation. *Ecology* **75**, 1842–1850.

Adler, F.R. & Karban, R. (1994) Defended fortresses or moving targets? Another model of inducible defences inspired by military metaphors. *American Naturalist* **144**, 813–832.

Adriano, D.C. & Johnson, A.H. (Eds) (1989) *Acidic Precipitation, Vol. 2. Biological and Ecological Effects*. Springer-Verlag, New York.

Aerts, R. (1995) The advantages of being evergreen. *Trends in Ecology and Evolution* **10**, 402–407.

Aerts, R., Berendse, F., De Caluwe, H. & Schmitz, M. (1990) Competition in heathland along an experimental gradient of nutrient availability. *Oikos* **57**, 310–318.

Agrios, G.N. (1988) *Plant Pathology*. Academic Press, Orlando.

Aguilera, M.O. & Lauenroth, W.K. (1995) Influence of gap disturbances and types of microsites on seedling establishment in *Bouteloua gracilis*. *Journal of Ecology* **83**, 87–97.

Ahrens, U., Lorenz, K.H. & Seemuller, E. (1993) Genetic diversity among mycoplasma like organisms associated with stone fruit diseases. *Molecular Plant Microbe Interactions* **6**, 686–691.

Akhteruzzaman, M. (1991) *Insect herbivory and the growth of* Quercus robur. Unpublished PhD thesis, University of London.

Alexander, H.M. & Antonovics, J. (1988) Disease spread and population dynamics of anther smut infection of Silene alba caused by the fungus *Ustilago violaceae*. *Journal of Ecology* **76**, 91–104.

Alexander, H.M. & Burdon, J.J. (1984) The effect of disease induced by *Albugo candida* (white rust) and *Peronospera parasitica* (downy mildew) on the survival and reproduction of *Capsella bursa-pastoris* (Shepherd's purse). *Oecologia* **64**, 314–318.

Alexander, I.J. (1983) The significance of ectomycorrhizas in the nitrogen cycle. In *Nitrogen as an Ecological Factor* (Eds J.A. Lee, S. McNeill & I.H. Rorison), pp. 69–93. Blackwell Scientific Publications, Oxford.

Allen, E.B. & Allen, M.F. (1990) The mediation of competition by mycorrhizae in successional and patchy environments. In *Perspectives on Plant Competition* (Eds J. Grace & D. Tilman). Academic Press, New York.

Allen, M.F. (1991) *The Ecology of Mycorrhizae*. Cambridge University Press, Cambridge.

Allred, W.S., Gaud, W.S. & States, J.S. (1994) Effects of herbivory by Abert squirrels (*Sciurus aberti*) on cone crops of Ponderosa pine. *Journal of Mammalogy* **75**, 700–703.

Al-Mufti, M.M., Sydes, C.L., Furness, S.B., Grime, J.P. & Band, S.R. (1977) A quantitative analysis of shoot phenology and dominance in herbaceous vegetation. *Journal of Ecology* **65**, 759–791.

Althawadi, A.B. & Grace, J. (1986) Water use by the desert cucurbit *Citrullus colocynthis* (L.) Schrad. *Oecologia* **70**, 475–480.

Alvarez-Buylla, E.R. (1994) Density dependence and patch dynamics in tropical rain forests: matrix models and applications to a tree species. *American Naturalist* **143**, 155–191.

Alward, R.D. & Joern, A. (1993) Plasticity and overcompensation in grass responses to herbivory. *Oecologia* **95**, 358–364.

Amor-Prats, D. & Harborne, J.B. (1993) Allelochemical effects of ergoline alkaloids from *Ipomoea parasitica* on *Heliothis virescens*. *Chemoecology* **4**, 55–61.

Anderson, D.J. (1971) Pattern in desert perennials. *Journal of Ecology* **59**, 555–560.

Anderson, R.M. & May, R.M. (1979) Population biology of infectious diseases: Parts 1 & 2. *Nature* **280**, 361–367; 455–461.

Anderson, R.M. & May, R.M. (1991) *Infectious Diseases of Humans: Dynamics and Control*. Oxford University Press, Oxford.

Andrew, C.S. (1978) Legumes and acid soil. In *Limitations and Potentials for Biological Nitrogen Fixation in the Tropics* (Eds J. Dubereiner, R. Burris & A. Hollaender), pp. 135–160. Plenum Press, New York.

Antonovics, J. (1978) The population genetics of mixtures. In *Plant Relations in Pastures* (Ed. J.R. Wilson), pp. 233–253. CSIRO, Melbourne.

Antonovics, J. (1992) Towards community genetics. In *Plant Resistance to Herbivores and Pathogens: Ecology, Evolution and Genetics* (Eds R.S. Fritz & E.L. Simms), pp. 426–449. University of Chicago Press, Chicago.

Antonovics, J. & Levin, D.A. (1980) The ecological and genetical consequences of density dependent regulation in plants. *Annual Review of Ecology and Systematics* 11, 411–452.

Antonovics, J. & Thrall, P.H. (1994) Cost of resistance and the maintenance of genetic polymorphism in host–pathogen systems. *Proceedings of the Royal Society of London Series B* 257, 105–110.

Antonovics, J., Bradshaw, A.D. & Turner, R.G. (1971) Heavy metal tolerance in plants. *Advances in Ecological Research* 7, 1–85.

Antonovics, J., Iwasa, Y. & Hassell, M.P. (1995) A generalized model of parasitoid, venereal, and vector-based transmission processes. *American Naturalist* 145, 661–675.

Appanah, S. (1993) Mass flowering of dipterocarp forests in the aseasonal tropics. *Journal of Biosciences* 18, 457–474.

Appel, H.M. (1993) Phenolics in ecological interactions: the importance of oxidation. *Journal of Chemical Ecology* 19, 1521–1552.

Arenovski, A.L. & Howes, B.L. (1992) Lacunal allocation and gas transport capacity in the salt marsh grass *Spartina alterniflora*. *Oecologia* 90, 316–322.

Armbruster, W.S. (1993) Evolution of plant pollination systems: hypotheses and tests with the neotropical vine *Dalechampia*. *Evolution* 47, 1480–1505.

Armesto, J.J. & Pickett, S.T.A. (1985) Experiments on disturbance in old-field plant communities: impact on species richness and abundance. *Ecology* 66, 230–240.

Armstrong, D.P. & Westoby, M. (1993) Seedlings from large seeds tolerate defoliation better: a test using phylogenetically independent contrasts. *Ecology* 74, 1092–1110.

Armstrong, R.A. (1976) Fugitive species: experiments with fungi and some theoretical considerations. *Ecology* 57, 953–963.

Armstrong, R.A. & McGehee, R. (1976a) Coexistence of two competitors on one resource. *Journal of Theoretical Biology* 56, 499–502.

Armstrong, R.A. & McGehee, R. (1976b) Coexistence of species competing for shared resources. *Theoretical Population Biology* 9, 317–328.

Armstrong, R.A. & McGehee, R. (1980) Competitive exclusion. *American Naturalist* 115, 151–170.

Arnitzen, F.K., Visser, J.H.M. & Hoogendoorn, J. (1994) The effect of the potato cyst nematode *Globodera pallida* on *in vitro* root growth of potato genotypes, differing in tolerance. *Annals of Applied Biology* 124, 59–64.

Arp, W.J., Drake, B.G., Pockman, W.T., Curtis, P.S. & Whigham, D.F. (1993) Interaction between C3 and C4 salt marsh plant species during four years of exposure to elevated atmospheric CO_2. *Vegetatio* 104/105, 133–144.

Arrhenius, O. (1921) Species and area. *Journal of Ecology* 9, 95–99.

Ashenden, T.W. & Mansfield, T.A. (1978) Extreme pollution sensitivity of grasses when SO_2 and NO_2 are present in the atmosphere together. *Nature* 273, 142–143.

Ashman, T. (1992) The relative importance of inbreeding and maternal sex in determining progeny fitness in *Sidalcea oregana* ssp. *spicata*, a gynodioecious plant. *Evolution* 46, 1862–1874.

Ashman, T.L. (1994) A dynamic perspective on the physiological cost of reproduction in plants. *American Naturalist* 144, 300–316.

Ashman, T. & Stanton, M. (1991) Seasonal variation in pollination dynamics of sexual dimorphic *Sidalcea oregana* ssp. *spicata* (Malvaceae). *Ecology* 72, 993–1003.

Ashmore, M.R. & Bell, J.N.B. (1991) The role of ozone in global change. *Annals of Botany* 67, 39–48.

Ashmore, M.R., Bell, J.N.B. & Garretty, C. (1988) *Acid Rain and Britain's Natural Ecosystems*. Imperial College Centre for Environmental Technology, London.

Asquith, T.N., Uhlig, J., Mehensha, H., Putman, L., Carlson, D.M. & Butler, L. (1987) Binding of condensed tannins to salivary proline-rich glycoproteins: the role of carbohydrates. *Journal of Agricultural and Food Chemistry* 35, 331–334.

Atkinson, W.D. & Shorrocks, B. (1981) Competition on a divided and ephemeral resource: a simulation model. *Journal of Animal Ecology* 50, 461–471.

Atlan, A., Gouyon, P.H., Fournial, T., Pomente, D. & Couvet, D. (1992) Sex allocation in an hermaphroditic plant – the case of gynodioecy in *Thymus vulgaris* L. *Journal of Evolutionary Biology* 5, 189–203.

Aubreville, A. (1971) Regenerative patterns in the closed forest of the Ivory Coast. In *World Vegetation Types* (Ed. S.R. Eyre), pp. 41–55. Columbia University Press, New York.

Auerbach, M.J. & Strong, D.R. (1981) Nutritional ecology of *Heliconia* herbivores: experiments with plant fertilisation and alternative hosts. *Ecological Monographs* 51, 63–83.

Augspurger, C.K. (1980) Mass-flowering of a tropical shrub (*Hybanthus prunifolius*): influence on pollinator attraction and movement. *Evolution* 34, 475–488.

Augspurger, C.K. (1981) Reproductive synchrony of a tropical shrub: experimental studies on effects of pollinators and seed predators on *Hybanthus prunifolius* (Violaceae). *Ecology* 62, 775–788.

Augspurger, C.K. (1983a) Offspring recruitment around tropical trees: changes in cohort distance with time. *Oikos* 40, 189–196.

Augspurger, C.K. (1983b) Seed dispersal of the tropical trees, *Platypodium elegans*, and the escape of its seedlings from fungal pathogens. *Journal of Ecology* 71, 759–772.

Augspurger, C.K. (1986) Morphology and dispersal potential of wind-dispersed diaspores of neotropical trees. *American Journal of Botany* 73, 353–363.

Augspurger, C.K. (1988) Impact of pathogens on natural plant populations. In *Plant Population Ecology* (Eds A.J. Davy, M.J. Hutchings & A.R. Watkinson), pp. 413–433. Blackwell Scientific Publications, Oxford.

Augspurger, C.K. (1989) Morphology and aerodynamics of wind-dispersed legumes. In *Advances in Legume Biology* (Eds C.H. Stirton & J.L. Zarucchi), pp. 451–466. Missouri Botanical Garden, St Louis.

Augspurger, C.K. & Franson, S.E. (1986) Wind dispersal of artificial fruits varying in mass, area, and morphology. *Ecology* 68, 27–42.

Augspurger, C.K. & Kelly, C.K. (1984) Pathogen mortality of tropical tree seedlings: experimental studies of the effects of dispersal distance, seedling density, and light conditions. *Oecologia* 61, 211–217.

Ayal, Y. (1994) Time lags in insect response to plant productivity – significance for plant insect interactions in deserts. *Ecological Entomology* 19, 207–214.

Ayres, M.P. (1993) Plant defense, herbivory, and climate change. In *Biotic Interactions and Global Change* (Eds P.M. Kareiva, J.G. Kingsolves & R.B. Huey), pp. 75–94. Sinauer, Sunderland, Massachusetts.

Ayres, M.P. & MacLean, S.F. (1987) Development of birch leaves and the growth-energetics of *Epirrata autumnata* (Geometridae). *Ecology* 68, 558–569.

Baas, W.J. (1989) Secondary plant compounds, their ecological significance and consequences for the carbon budget: introduction of the carbon/nutrient cycle theory. In *Causes and Consequences of Variation in Growth Rate and Production of Higher Plants* (Eds H. Lambers, M.L. Cambridge, H. Konings & T.L. Pons), pp. 313–340. SBP Academic Publishing, The Hague.

Bach, C.E. (1980) Effects of plant density and diversity on the population dynamics of a specialist herbivore, the striped cucumber beetle, *Acalymma vittata* (Fab.). *Ecology* 61, 1515–1530.

Bach, C.E. (1994) Effects of a specialist herbivore (*Altica subplicata*) on *Salix cordata* and sand dune succession. *Ecological Monographs* 64, 423–445.

Baker, G.A. & O'Dowd, D.J. (1982) Effects of parent plant density on the production of achene type in the annual *Hypochoeris glabra*. *Journal of Ecology* 70, 201–215.

Baker, H.G. (1972) Seed weight in relation to environmental conditions in California. *Ecology* 53, 997–1010.

Baldwin, I.T. (1988) The alkaloidal responses of wild tobacco to real and simulated herbivory. *Oecologia* 77, 378–381.

Baldwin, I.T. (1989) The mechanism of damaged-induced alkaloids in wild tobacco. *Journal of Chemical Ecology* 151, 1661–1680.

Baldwin, I.T. (1991) Damage induced alkaloids in wild tobacco. In *Phytochemical Induction by Herbivores* (Eds D.W. Tallamy & M.J. Raupp), pp. 46–47. Wiley-Interscience, New York.

Baldwin, I.T. & Ohnmeiss, T.E. (1994) Swords into plowshares? *Nicotiana sylvestris* does not use nicotine as a nitrogen source under nitrogen-limited growth. *Oecologia* 98, 385–392.

Baldwin, I.T., Schultz, J.C. & Ward, D. (1987) Patterns and sources of leaf tannin variation in yellow birch (*Betula alleghemiensis*) and sugar maple (*Acer saccharum*). *Journal of Chemical Ecology* 13, 1069–1078.

Baldwin, I.T., Sims, C.L. & Kean, S.E. (1990) The reproductive consequences associated with inducible alkaloidai responses in wild tobacco. *Ecology* 71, 252–262.

Baldwin, I.T., Oesch, R.C., Merhige, P.M. & Hayes, K. (1993) Damage-induced root nitrogen metabolism in *Nicotiana sylvestris*: testing C/N predictions for alkaloid production. *Journal of Chemical Ecology* 19, 3029–3043.

Baldwin, I.T., Karb, M.J. & Ohnmeiss, T.E. (1994) Allocation of ^{15}N from nitrate to nicotine: production and turnover of an induced defence. *Ecology* 75, 1705–1713.

Ballardie, R.T. & Whelan, R.J. (1986) Masting, seed dispersal and seed predation in the cycad *Macrozamia communis*. *Oecologia* 70, 100–105.

Ballaré, C.L., Scopel, A.L., Ghersa, C.M. & Sánchez, R.A. (1987) The population ecology of *Datura ferox* in soybean crops. A simulation approach incorporating seed dispersal. *Agriculture, Ecosystems and Environment* 19, 177–188.

Baltensweiler, W., Benz, G., Bovey, P. & Delucchi, V. (1977) Dynamics of larch budmoth populations. *Annual Review of Ecology and Systematics* 22, 79–100.

Barber, J. & Baker, N.J. (Eds) (1985) Photosynthesis and the Environmental. Volume 6. Elsevier, Amsterdam.

Barbosa, P. & Krischik, V.A. (1987) Influence of alkaloids on feeding preference of eastern deciduous forest trees by the gypsy moth *Lymantria dispar*. *American Naturalist* 130, 53–69.

Barbosa, P. & Schultz, J.C. (1987) *Insect Outbreaks*. Academic Press, New York.

Barbour, M.G. (1973) Desert dogma re-examined: root/shoot production and plant spacing. *American Midland Naturalist* 89, 41–57.

Barkham, J.P. & Hance, C.E. (1982) Population dynamics of the wild daffodil (*Narcissus pseudonarcissus*). III. Implications of a computer model of 1000 years of population change. *Journal of Ecology* 70, 323–344.

Barnea, A., Yom-Tov, Y. & Friedman, J. (1991) Does ingestion by birds affect seed germination? *Functional Ecology* 5, 394–402.

Barrett, J.A. (1988) Frequency dependent selection in plant fungal interactions. *Philosophical Transactions of the Royal Society of London Series B* 319, 473–483.

Barrett, S.C.H. (1996) The reproductive biology and genetics of island plants. *Philosophical Transactions of the Royal Society of London Series B* 351, 725–733.

Barrett, S.C.H. & Bush, E.J. (1991) Population processes in plants and the evolution of resistance to gaseous air pollutants. In *Ecological Genetics and Air Pollution* (Eds G.E. Taylor Jr, L.F. Pitelka & M.T. Clegg), pp. 137–165. Springer-Verlag, New York.

Barrett, S.C.H. & Kohn, J.R. (1991) Genetic and evolutionary consequences of small population size in plants: implications for conservation. In *Genetics and Conservation of Rare Plants* (Eds D.A. Falk & K.E. Holsinger), pp. 3–30. Oxford University Press, New York.

Barrett, S.C.H., Harder, L.D. & Cole, W.W. (1994) Effects of flower number and position on self fertilization in experimental populations of *Eichhornia paniculata* (Pontederiaceae). *Functional Ecology* 8, 526–535.

Barth, F. (1985) *Insects and Flowers*. Princeton University Press, Princeton.

Bartholomew, B. (1970) Bare zones between California shrub and grassland communities: the role of animals. *Science* 170, 1210–1212.

Bartlett, M.S. (1975) *The Statistical Analysis of Spatial Pattern*. Chapman & Hall, London.

Barton, A.M. (1993) Factors controlling plant distributions – drought, competition, and fire in montane pines in Arizona. *Ecological Monographs* 63, 367–397.

Basey, J.M. & Jenkins, S.H. (1993) Production of chemical defences in relation to plant growth rate. *Oikos* 68, 323–328.

Baskin, J.M. & Baskin, C.C. (1989) Physiology of dormancy and germination in relation to seed bank ecology. In *Ecology of Soil Seed Banks* (Eds M.A. Leck, V.T. Parker & R.L. Simpson), pp. 53–66. Academic Press, London.

Bateman, A.J. (1948) Intra-sexual selection in *Drosophila*. *Heredity* 2, 349–368.

Bates, G.H. (1935) The vegetation of footpaths, sidewalks, cart tracks and gateways. *Journal of Ecology* 23, 470–487.

Bates, J.W., Bell, J.N.B. & Farmer, A.M. (1990) Epiphyte recolonization of oaks along a gradient of air pollution in south-east England, 1979–1990. *Environmental Pollution* 68, 81–99.

Bauer, H., Nagele, M., Comploj, M., Galler, V., Mair, M. & Unterpertinger, E. (1994) Photosynthesis in cold acclimated leaves of plants with various degrees of freezing tolerance. *Physiologia Plantarum* 91, 403–412.

Bawa, K.S. (1980) Evolution of dioecy in flowering plants. *Annual Review of Ecology and Systematics* 11, 15–39.

Bawa, K.S. (1983) Patterns of flowering in tropical plants. In *Handbook of Experimental Pollution Biology* (Eds C.E. Jones & R.J. Little), pp. 394–410. Van Nostrand Reinhold, New York.

Baylis, G.T.S. (1975) The magnolioid mycorrhiza and mycotrophy in root systems derived from it. In *Endomycorrhizas* (Eds F.E. Sanders, B. Mosse & P.B. Tinker), pp. 373–389. Academic Press, London.

Bazzalo, H.E.M., Del Pero Martinez, M.A. & Caso, O.H. (1985) Phenolic compounds in stems of sunflower plants inoculated with *Sclerotinia sclerotiorum* and their inhibitory effects on the fungus. *Phytopathology Zeitschrift* 112, 322–332.

Bazzaz, F.A. (1975) Plant species diversity in old field successional ecosystems in southern Illinois. *Ecology* 56, 485–488.

Bazzaz, F.A. & Harper, J.L. (1976) Relationship between plant weight and numbers in mixed population of *Sinapis alba* (L.) Rabenh. and *Lepidium sativum* L. *Journal of Applied Ecology* 13, 211–216.

Bazzaz, F.A., Levin, D.A. & Schmierbach, M.R. (1982) Differential survival of genetic variants in crowded populations of *Phlox*. *Journal of Applied Ecology* 19, 891–900.

Bazzaz, F.A., Chiariello, N.R., Coley, P.D. & Pitelka, L.F. (1987) Allocating resources to reproduction and defense. *Bioscience* 37, 58–67.

Beart, J.E., Lilley, T.H. & Haslam, E. (1985) Plant polyphenols – secondary metabolism and chemical defence: some observations. *Phytochemistry* 24, 33–38.

Beattie, A.J. & Culver, D.C. (1979) Neighborhood size in *Viola*. *Evolution* 33, 1226–1229.

Becking, R.W. (1957) The Zürich–Montpellier School of Phytosociology. *Botanical Review* 23, 411–488.

Beehler, B. (1983) Frugivory and polygamy in birds of paradise. *Auk* 100, 1–12.

Begon, M. (1984) Density and individual fitness: asymmetric competition. In *Evolutionary Ecology* (Ed. B. Shorrocks), pp. 179–194. Blackwell Scientific Publications, Oxford.

Begon, M. & Mortimer, M. & Thompson, D. (1995b) *Population Ecology: a Unified Study of Animals and Plants*, 3rd edn. Blackwell Science, Oxford.

Begon, M.E., Harper, J.L. & Townsend, C.R. (1995) *Ecology: Individuals, Populations and Communities*, 3rd edn. Blackwell Science, Oxford.

Bell, A.D., Roberts, D. & Smith, A. (1979) Branching patterns: the simulation of plant architecture. *Journal of Theoretical Biology* 81, 351–375.

Bell, C.M. & Harestad, A.S. (1987). Efficacy of pine oil as repellent to wildlife. *Journal of Chemical Ecology* 13, 1409–1417.

Bell, G. (1985) On the function of flowers. *Proceedings of the Royal Society of London Series B* 224, 223–265.

Bell, J.N.B. & Mudd, C.H. (1976) Sulphur dioxide in plants – a case study of *Lolium perenne*. In *Effects of Air Pollution on Plants* (Ed. T.A. Mansfield), pp. 88–103. Cambridge University Press, Cambridge.

Bell, J.N.B., McNeill, S., Houlden, G., Brown, V.C. & Mansfield, P.J. (1993) Atmospheric change: effect on plant pests and diseases. *Parasitology* 106, 811–824.

Belovsky, G.E. & Joern, A. (1995) Regulation of rangeland grasshoppers: differing dominant mechanisms in space and time. In *Population Dynamics: New Approaches and Synthesis* (Eds N. Cappuccino & P.W. Price), pp. 359–386. Academic Press, San Diego.

Belsky, A.J. (1992) Effects of grazing, competition, dis-

[636]

turbance and fire on species composition and diversity in grassland communities. *Journal of Vegetation Science* **3**, 187–200.

Belsky, A.J., Carson, W.P., Jensen, C.L. & Fox, G.A. (1993) Over-compensation by plants – herbivore optimization or red herring. *Evolutionary Ecology* **7**, 109–121.

Bendel, R.B., Higgins, S.S., Teberg, J.E. & Pyke, D.A. (1989) Comparison of skewness coefficients, coefficient of variation, and Gini coefficient as inequality measures within populations. *Oecologia* **78**, 394–400.

Benedict, J.B. (1984) Rates of tree-island migration, Colorado Rocky Mountains, USA. *Ecology* **65**, 820–823.

Benedict, J.H. & Hatfield, J.L. (1988) Influence of temperature-induced stress on host plant suitability to insects. In *Plant Stress–Insect Interactions* (Ed. E.A. Heinrichs), pp. 139–166. John Wiley & Sons, New York.

Bengtsson, B., Asp, H., Jensen, P. & Berggren, D. (1988) Influence of aluminium on phosphate and calcium uptake in beech (*Fagus sylvatica*) grown in nutrient solution and soil solution. *Physiologia Plantarum* **74**, 299–305.

Bennet, R.N. & Wallsgrove, R.M. (1994) Secondary metabolites in plant defence mechanisms. *New Phytologist* **127**, 617–633.

Bennett, K.D. (1983) Postglacial population expansion of forest trees in Norfolk, UK. *Nature* **303**, 164–167.

Bennett, M.D. (1971) The duration of meiosis. *Proceedings of the Royal Society of London Series B* **178**, 277–299.

Bennett, M.D. (1976) DNA amount, latitude and crop plant distribution. *Environmental and Experimental Botany* **16**, 93–108.

Bennett, M.D. & Smith, J.P. (1976) Nuclear DNA amounts in angiosperms. *Philosophical Transactions of the Royal Society of London Series B* **274**, 227–274.

Benzing, D.H. (1983) Vascular epiphytes: a survey with special reference to their interactions with other organisms. In *Tropical Rain Forest: Ecology and Management* (Eds S.L. Sutton, T.C. Whitmore & A.C. Chadwick), pp. 11–24. Blackwell Scientific Publications, Oxford.

Berenbaum, M. (1978) Toxicity of a furanocoumarin to armyworms: a case of biosynthetic escape from insect herbivores. *Science* **201**, 532–534.

Berenbaum, M.R. (1992) Coumarins. In *Herbivores: their Interaction with Secondary Plant Metabolites*, 2nd edn (Eds G.A. Rosenthal & M.R. Berenbaum), Vol. 1, pp. 221–250. Academic Press, San Diego.

Berenbaum, M.R. (1995) The chemistry of defense: theory and practice. *Proceedings of the National Academy of Sciences USA* **92**, 2–8.

Berenbaum, M.R. & Feeny, P. (1981) Toxicity of angular furanocoumarins to swallowtail butterflies: escalation in a coevolutionary arms race? *Science* **212**, 927–929.

Berenbaum, M.R. & Neal, J. (1985) Synergism between myristicin and xanthotoxin, a naturally co-occurring plant toxicant. *Journal of Chemical Ecology* **11**, 1349–1358.

Berenbaum, M.R. & Zangerl, A.R. (1988) Stalemates in the coevolutionary arms race: syntheses, synergisms and sundry other sins. In *Chemical Mediation of Coevolution* (Ed. K.C. Spencer), pp. 113–132. American Institute of Biological Sciences, Washington.

Berenbaum, M.R. & Zangerl, A.R. (1992) Qualification of chemical coevolution. *In Plant Resistance to Herbivores and Pathogens: Ecology, Evolution and Genetics* (Eds R.S. Fritz & E.L. Simms), pp. 69–90. University of Chicago Press, Chicago.

Berendse, F. (1985) The effect of grazing on the outcome of competition between plant species with different nutrient requirements. *Oikos* **44**, 35–39.

Berendse, F. (1994) Competition between plant populations at low and high nutrient. *Oikos* **71**, 253–260.

Berendse, F. & Elberse, W.T. (1990) Competition and nutrient availability in heathland and grassland ecosystems. In *Perspectives on Plant Competition* (Eds J.B. Grace & D. Tilman), pp. 93–116. Academic Press, San Diego.

Bergelson, J. (1990) Life after death – site pre-emption by the remains of *Poa annua*. *Ecology* **71**, 2157–2165.

Bergelson, J.M. & Crawley, M.J. (1992a) Overcompensation in response to mammalian herbivory: the disadvantage of being eaten. *American Naturalist* **139**, 870–882.

Bergelson, J. & Crawley, M.J. (1992b) The effects of grazers on the performance of individuals and populations of scarlet gilia, *Ipomopsis aggregata*. *Oecologia* **90**, 435–444.

Bergelson, J.M. & Lawton, J.H. (1988) Does foliage damage influence predation on the insect herbivores of birch? *Ecology* **69**, 434–445.

Bergelson, J. & Purrington, C.B. (1996) Surveying the costs of resistance in plants. *American Naturalist* (in press).

Bergelson, J.M., Fowler, S. & Hartley, S. (1986) The effects of foliage damage on case bearing moth larvae, *Coleophora serratella*, feeding on birch. *Ecological Entomology* **11**, 241–250.

Bergelson, J., Juenger, T. & Crawley, M.J. (1996) Regrowth following herbivory in *Ipomopsis aggregata*: compensation but not overcompensation. *American Naturalist* (in press).

Bergström, G. (1991) Chemical ecology of terpenoid and other fragrances of angiosperm flowers. In *Ecological Chemistry and Biochemistry of Plant Terpenoids* (Eds J.B. Harborne & F.A. Tomas-Barberan), pp. 287–296. Clarendon Press, Oxford.

Bergstrom, R. & Danell, K. (1995) Effects of simulated summer browsing by moose on leaf and shoot biomass of birch, *Betula pendula*. *Oikos* **72**, 132–138.

Berkowitz, A.R., Canham, C.D. & Kelly, V.R. (1995) Competition vs facilitation of tree seedling growth and survival in early successional communities. *Ecology* **76**, 1156–1168.

Bernays, E.A. (1981) Plant tannins and insect

herbivores: an appraisal. *Ecological Entomology* **6**, 353–360.

Bernays, E.A. (1990) Plant secondary compounds deterrent but not toxic to the grass specialist acridid *Locusta migratoria*: implications for the evolution of graminivory. *Entomologia Experimentatia Applicata* **54**, 53–56.

Bernays, E.A. (1991) Evolution of insect morphology in relation to plants. *Philosophical Transactions of the Royal Society of London Series B* **333**, 257–264.

Bernays, E.A. & Chapman, R.F. (1977) Deterrent chemicals as a basis of oligophagy in *Locusta migratoria* (L). *Ecological Entomology* **2**, 1–18.

Bernays, E.A. & Chapman, R.F. (1978) Plant chemistry and acridoid feeding behaviour. In *Biochemical Aspects of Plant and Animal Coevolution* (Ed. J.B. Harborne), pp. 99–142. Academic Press, London.

Bernays, E.A. & Chapman, R.F. (1995) *Host-plant Selection by Phytophagous Insects*. Chapman & Hall, New York.

Bernays, E.A. & Graham, M. (1988) On the evolution of host specificity in phytophagous arthropods. *Ecology* **69**, 886–892.

Bernays, E.A. & Woodhead, S. (1982) Plant phenols utilised as nutrients by a phytophagous insect. *Science* **216**, 201–203.

Bernays, E.A., Cooper-Driver, G. & Bilgener, M. (1989) Herbivores and tannins. *Advances in Ecological Research* **19**, 263–302.

Berner, R.A. (1991). A model for atmospheric CO_2 over Phanerozoic time. *American Journal of Science* **291**, 339–376.

Berninger, F., Mencuccini, M., Nikinmaa, E., Grace, J. & Hari, P. (1995) Evaporative demand determines branchiness of Scots pine. *Oecologia* **102**, 164–168.

Berry, J. & Björkman, O. (1980). Photosynthetic response and adaption to temperature in higher plants. *Annual Review of Plant Physiology* **31**, 491–543.

Berryman, A.A. (1991) Population theory: an essential ingredient in population prediction, management, and policy making. *American Entomologist* **37**, 138–142.

Berryman, A.A., Dennis, B., Raffa, K.F. & Stenseth, N.C. (1985) Evolution of optimal group attack, with particular reference to bark beetles (Coleoptera, Scolytidae). *Ecology* **66**, 898–903.

Berryman, A.A., Gutierrez, A.P. & Arditi, R. (1995) Credible, parsimonious and useful predator–prey models. *Ecology* **76**, 1980–1985.

Bertin, R.I. (1988) Paternity in plants. In *Plant Reproductive Ecology* (Eds J.L. Doust & L.L. Doust), pp. 30–59. Oxford University Press, New York.

Bertin, R.I. (1989) Pollination biology. In *Plant–Animal Interactions* (Ed. W.G. Abrahamson), pp. 23–83. McGraw-Hill, New York.

Bertin, R.I. & Newman, C.M. (1993) Dichogamy in angiosperms. *Botanical Review* **59**, 112–152.

Bewley, J.D. (1979) Physiological aspects of desiccation tolerance. *Annual Review of Plant Physiology* **30**, 195–238.

Bewley, J.D. & Black, M. (1982) *Physiology and Biochemistry of Seeds*, Vol. 2. Springer-Verlag, Berlin.

Biere, A. (1995) Genotypic and plastic variation in plant size – effects on fecundity and allocation patterns in *Lychnis flos cuculi* along a gradient of natural soil fertility. *Journal of Ecology* **83**, 629–642.

Bierzychudek, P. (1981) Pollinator limitation of plant reproductive effort. *American Naturalist* **117**, 838–840.

Bierzychudek, P. (1982) The demography of Jack-in-the-pulpit, a forest perennial that changes sex. *Ecological Monographs* **52**, 335–351.

Bierzychudek, P. (1984) Assessing optimal life histories in a fluctuating environment. The evolution of sex-changing by Jack-in-the-pulpit. *American Naturalist* **123**, 829–840.

Bilbrough, C.J. & Richards, J.H. (1993) Growth of sagebrush and bitterbrush following simulated winter browsing – mechanism of tolerance. *Ecology* **74**, 481–492.

Bink, F.A. (1986) Acid stress in *Rumex hydrolapathum* (Polygonaceae) and its influence on the phytophage *Lycaena dispar* (Lepidoptera; Lycaenidae). *Oecologia* **70**, 447–451.

Bishnoi, N.R., Dua, A., Gupta, V.K. & Sawhney, S.K. (1993) Effect of chromium on seed germination, seedling growth and yield of peas. *Agriculture Ecosystems and Environment* **47**, 47–57.

Bishop, G.F. & Davy, A.J. (1984) Significance of rabbits for the population regulation of *Hieracium pilosella* in Breckland. *Journal of Ecology* **72**, 273–284.

Bishop, G.F., Davy, A.J. & Jefferies, R.L. (1978) Demography of *Hieracium pilosella* in a Breck grassland. *Journal of Ecology* **66**, 615–629.

Björkman, C.S., Larsson, S. & Gref, R. (1991) Effects of nitrogen fertilisation on pine needle chemistry and sawfly performance. *Oecologia* **86**, 202–209.

Björkman, O. (1968) Further studies on differentiation of photosynthetic properties in sun and shade ecotypes of *Solidago virgaurea*. *Physiologia Plantarum* **21**, 84–99.

Björkman, O., Boardman, N.K., Anderson, J.M., Thorne, S.W., Goodchild, D.J. & Pyliotis, N.A. (1972) Effect of light intensity during growth of *Atriplex patula* on the capacity of photosynthetic reactions, chloroplast components and structure. *Carnegie Institution Year Book* **71**, 115–135.

Black, J.N. (1957) The early vegetative growth of three strains of subterranean clover (*Trifolium subterraneum* L). in relation to size of seed. *Australian Journal of Agricultural Research* **8**, 1–14.

Black, J.N. (1959) Seed size in herbage legumes. *Herbage Abstracts* **29**, 235–241.

Black, J.N. (1964) An analysis of the potential production of swards of subterranean clover (*Trifolium subterraneum* L.) at Adelaide, South Australia. *Journal of Applied Ecology* **1**, 3–18.

Bliss, L.C. (1956) A comparison of plant development in micro-environments of arctic and alpine tundras. *Ecological Monographs* **26**, 303–307.

Bliss, L.C. (1985) Alpine. In *Physiological Ecology of North American Plant Communities* (Eds B.F. Chabot & H.A. Mooney), pp. 41–65. Chapman & Hall, New York.

Blom-Zandstra, M. & Jupijn, G.L. (1987) A computer-controlled multi-titration system to study transpiration, OH⁻ efflux and nitrate uptake by intact lettuce plants (*Lactuca sativa* L.) under different environmental conditions. *Plant Cell and Environment* 10, 545–550.

Blossey, B. (1995) Impact of *Galerucella pusilla* Duft. and *G. calmariensis* L. (Coleoptera: Chrysomelidae) on field populations of purple loosestrife (*Lythrum salicaria* L.). In *Proceedings of the VIII International Symposium on the Biological Control of Weeds* (Eds E.S. Delfosse & R.R. Scott). CSIRO, Melbourne.

Boardman, N.K. (1977) Comparative photosynthesis of sun and shade plants. *Annual Review of Plant Physiology* 28, 355–377.

Bodnaryk, R.P. (1991) Development profile of sinalbin in mustard seedlings and its relationship to insect resistance. *Journal of Chemical Ecology* 17, 1543–1556.

Boecklen, W.J., Mopper, S. & Price, P.W. (1994) Sex biased herbivory in arroyo willow – are there general patterns among herbivores? *Oikos* 71, 267–272.

Boerner, R.E.J. (1984) Foliar nutrient dynamics and nutrient use efficiency of four deciduous tree species in relation to site fertility. *Journal of Applied Ecology* 21, 1029–1040.

Bogler, D.J., Neff, J.L. & Simpson, B.B. (1995) Multiple origins of the yucca–yucca moth association. *Proceedings of the National Academy of Sciences USA* 92, 6864–6867.

Bolker, B.M. & Pacala, S.W. Understanding the ecological implications of spatial pattern formation using ensemble models (submitted).

Bonan, G.B. (1991) Density effects on the size structure of annual plant populations: an indication of neighbourhood competition. *Annals of Botany* 68, 341–347.

Bond, G. (1976) The results of the IBP survey of root nodule formation in non-leguminous angiosperms. In *Symbiotic Nitrogen Fixation in Plants* (Ed. P.S. Nutman), pp. 443–474. Cambridge University Press, Cambridge.

Bond, W.J. (1985) Canopy-stored seed reserves (serotiny) in Cape Proteaceae. *South African Journal of Botany* 51, 181–186.

Bond, W.J. (1995) Assessing the risk of plants extinction due to pollinator and dispersal failure. In *Extinction Rates* (Eds J.H. Lawton & R.M. May), pp. 131–146. Oxford University Press, Oxford.

Bond, W.J. & Midgley, J.J. (1995) Kill thy neighbour – an individualistic argument for the evolution of flammability. *Oikos* 73, 79–85.

Borghetti, M., Edwards, W.R.N., Grace, J., Jarvis, P.G. & Raschi, A. (1991) The refilling of embolized xylem in *Pinus sylvestris* L. *Plant, Cell and Environment* 14, 357–369.

Borghetti, M., Grace, J. & Raschi, A. (Eds) (1993) *Water Transport in Plants under Climatic Stress*. Cambridge University Press, Cambridge.

Bossema, I. (1968) Recovery of acorns in the European jay (*Garrulus g. glandarius* L.). *Verhandelingen van der Koninklijke Vlaamse Academie voor Wetenschappen* 71, 1–5.

Bossema, I. (1979) Jays and oaks: an eco-ethological study of a symbiosis. *Behaviour* 70, 1–118.

Bosy, J.L. & Reader, R.J. (1995) Mechanisms underlying the suppression of forb seedling emergence by grass (*Poa pratensis*) litter. *Functional Ecology* 9, 635–639.

Botkin, D.R., Janak, J.F. & Willis, J.R. (1972) Some ecological consequences of a computer model of forest growth. *Journal of Ecology* 60, 849–873.

Boucher, D.H. (Ed.) (1985) *The Biology of Mutualism: Ecology and Evolution*. Croom Helm, London.

Boutin, J.P., Provot, M. & Roux, L. (1981) Effect of cycloheximide and renewal of phosphorus supply on surface acid phosphatase activity of phosphorus deficient tomato roots. *Physiologia Plantarum* 51, 353–360.

Bouwmeester, H.J. (1990) *The effect of environmental conditions on the seasonal dormancy and germination of weed seeds*. Unpublished PhD thesis, University of Wageningen.

Bowen, G.D. (1980) Mycorrhizal roles in tropical plants and ecosystems. In *Tropical Mycorrhiza Research* (Ed. P. Mikola), pp. 165–190. Clarendon Press, Oxford.

Bowers, M.A. (1993) Influence of herbivorous mammals on an old field plant community years 1–4 after disturbance. *Oikos* 67, 129–141.

Bowers, M.D. & Stamp, N.E. (1993) Effects of plant age, genotype, and herbivory on *Plantago* performance and chemistry. *Ecology* 74, 1778–1791.

Box, E.O. (1981) *Macroclimate and Plant Forms: An Introduction to Predictive Modelling in Phytogeography*. Dr. W. Junk, The Hague.

Boyce, M.S. (1992) Population viability analysis. *Annual Review of Ecology and Systematics* 23, 481–506.

Boyer, J.S. (1985) Water transport. *Annual Review of Plant Physiology* 36, 473–516.

Boyer, J.S. (1995) Biochemical and biophysical aspects of water deficits and the predisposition to disease. *Annual Review of Phytopathology* 33, 251–274.

Braakhekke, W.G. (1980) On coexistence: a causal approach to diversity and stability of grassland vegetation. *Verslagen van Lanabouwkundige Onderzoekingen* 902, 1–164.

Bradford, D.F. & Smith, C.C. (1977) Seed predation and seed number in *Schellea* palm fruits. *Ecology* 58, 667–673.

Bradshaw, A.D. (1973) Environment and phenotypic plasticity. *Brookhaven Symposium in Biology* 25, 75–94.

Bradshaw, A.D. (1976) Pollution and evolution. In *Effects of Air Pollutants on Plants* (Ed. T.A. Mansfield), pp. 135–159. Cambridge University Press, Cambridge.

Bradshaw, A.D. & Chadwick, M.J. (1980) *The Restoration of Land*. Blackwell Scientific Publications, Oxford.

Bradshaw, A.D. & McNeilly, T. (1981) *Evolution and Pollution*. Arnold, London.

Bradshaw, A.D. & McNeilly, T. (1991) Evolution in relation to environmental stress. In *Ecological Genetics and Air Pollution* (Eds G.E. Taylor Jr, L.F. Pitelka & M.T. Clegg), pp. 11–31. Springer-Verlag, New York.

Bradstock, R.A. & Myerscough, P.J. (1981) Fire effects on seed release and the emergence and establishment in *Banksia ericifolia* L. *Australian Journal of Botany* **29**, 521–531.

Brandt, R.N. & Lamb, R.J. (1994) Importance of tolerance and growth rate in the resistance of oilseed rapes and mustards to flea beetles, *Phyllotreta cruciferae* (Goeze) (Coleoptera, Chrysomelidae). *Canadian Journal of Plant Science* **74**, 169–176.

Brattsten, L.B. (1992) Metabolic defenses against plant allelochemicals. In *Herbivores: their Interaction with Secondary Plant Metabolites*, 2nd edn (Eds G.A Rosenthal & M.R. Berenbaum), pp. 176–242. Academic Press, San Diego.

Braun-Blanquet, J. (1932) *Plant Sociology: the Study of Plant Communities*. McGraw-Hill, New York.

Brewer, J.S. & Platt, W.J. (1994) Effects of fire season and herbivory on reproductive success in a clonal forb, *Pityopsis graminifolia*. *Journal of Ecology* **82**, 665–675.

Briggs, M.A. & Schultz, J.C. (1990) Chemical defense production in *Lotus corniculatus* L. II. Trade-offs among growth, reproduction and defense. *Oecologia* **83**, 32–37.

Brodbeck, B. & Strong, D. (1987) Amino acid nutrition of herbivorous insects and stress to host plants. In *Insect Outbreaks* (Eds P. Barbosa & J.C. Schultz), pp. 347–364. Academic Press, New York.

Brokaw, N.V.L. (1982) The definition of treefall gap and its effect on measures of forest dynamics. *Biotropica* **14**, 158–160.

Brokaw, N.V.L. (1985) Gap-phase regeneration in a tropical forest. *Ecology* **66**, 682–687.

Bronstein, J.L. & McKey, D. (1989) The comparative biology of figs. *Experientia* **45**, 601–604.

Brooks, R.R., Baker, A.J.M. & Malaisse, F. (1992) Copper flowers. *Research and Exploration* **8**, 338–351.

Brown, D.G. & Weis, A.E. (1995) Direct and indirect effects of prior grazing of goldenrod upon the performance of a leaf beetle. *Ecology* **76**, 426–436.

Brown, G. (1995) The effects of lead and zinc on the distribution of plant species at former mining areas of Western Europe. *Flora* **190**, 243–249.

Brown, J.H. & Heske, E.J. (1990) Control of a desert-grassland transition by a keystone rodent guild. *Science* **250**, 1705–1707.

Brown, J.H. & Munger, J.C. (1985) Experimental manipulation of a desert rodent community – food addition and species removal. *Ecology* **66**, 1545–1563.

Brown, J.H., Davidson, D.W., Munger, J.C. & Inouye, R.S. (1986) Experimental community ecology: the desert granivore system. *Community Ecology* (Eds J. Diamond & T. Case), pp. 41–61. Harper & Row, New York.

Brown, J.R. & Stuth, J.W. (1993) How herbivory affects grazing tolerant and sensitive grasses in a central Texas grassland – integrating plant response across hierarchical levels. *Oikos* **67**, 291–298.

Brown, N.A.C., Kotze, G. & Botha, P.A. (1993) The promotion of seed germination of Cape *Erica* species by plant-derived smoke. *Seed Science and Technology* **21**, 573–580.

Brown, V.K. & Gange, A.C. (1992) Secondary plant succession – how is it modified by insect herbivory? *Vegetatio* **101**, 3–13.

Brunet, J. & Charlesworth, D. (1995) Floral sex allocation in sequentially blooming plants. *Evolution* **49**, 70–79.

Brunsting, A.M.H. & Heil, G.W. (1985) The role of nutrients in the interactions between a herbivorous beetle and some competing plant species in heathlands. *Oikos* **44**, 23–26.

Bryant, J.P. & Kuropat, P.J. (1980) Selection of winter forage by subarctic browsing vertebrates: the role of plant chemistry. *Annual Review of Ecology and Systematics* **11**, 261–285.

Bryant, J.P., Chapin, F.S. & Klein, D.R. (1983) Carbon/nutrient balance of boreal plants in relation to vertebrate herbivory. *Oikos* **40**, 357–368.

Bryant, J.P., Clausen, T.P., Reichardt, P.B., McCarthy, M.C. & Werner, R.A. (1987) Effects of nitrogen fertilisation upon the secondary chemistry and nutritional value of quacking aspen (*Populus tremuloides* Michx.) leaves for the large aspen tortrix (*Choristoneura conflicta* (Walker)). *Oecologia* **73**, 513–517.

Bryant, J.P., Kuropat, P.J., Cooper, S.M. & Owen-Smith, N. (1989) Resource availability hypothesis of plant antiherbivore defence tested in a South African savanna ecosystem. *Nature* **340**, 227–229.

Bryant, J.P., Provenza, F.D., Pastor, J., Reichardt, P.B., Clausen, T.P. & du Toit, J.T. (1991) Interactions between woody plants and browsing mammals mediated by secondary metabolites. *Annual Review of Ecology and Systematics* **22**, 431–446.

Bryant, J.P., Clausen, T.P. & Werner, R.A. (1993) Effects of mineral nutrition on delayed inducible resistance in Alaskan paper birch. *Ecology* **74**, 2072–2084.

Buchmann, S.L. (1983) Buzz pollination in angiosperms. In *Handbook of Experimental Pollination Biology* (Eds C.E. Jones & R.J. Little), pp. 73–113. Van Nostrand Reinhold, New York.

Bucking, H. & Heyser, W. (1994) The effect of ectomycorrhizal fungi on Zn uptake and distribution in seedlings of *Pinus sylvestris* L. *Plant and Soil* **167**, 203–212.

Bucyanayandi, J.D., Bergeron, J.M., Soucie, J., Thomas, D.W. & Jean, Y. (1992) Differences in nutritional quality between herbaceous plants and bark of conifers as winter food for the vole *Microtus pennsylvanicus*. *Journal of Applied Ecology* **29**, 371–377.

Bullock, J.M., Clear Hill, B. & Silvertown, J. (1994a) Demography of *Cirsium vulgare* in a grazing experiment. *Journal of Ecology* **82**, 101–111.

Bullock, J.M., Hill, B.C., Dale, M.P. & Silvertown, J. (1994b) An experimental study of the effects of sheep grazing on vegetation change in a species poor grassland and the role of seedling recruitment into gaps. *Journal of Applied Ecology* **31**, 493–507.

Bullock, S.H. (1994) Wind pollination of neotropical dioecious trees. *Biotropica* **26**, 172–179.

Bulmer, M.G. (1984) Delayed germination of seeds: Cohen's model revisited. *Theoretical Population Biology* **26**, 376–377.

Bulmer, M.G. (1985) Selection for iteroparity in a variable environment. *American Naturalist* **126**, 63–71.

Bu'Lock, J.D. (1980) Mycotoxins as secondary metabolites. In *The Biosynthesis of Mycotoxins* (Ed. P.S. Steyn), pp. 1–16. Academic Press, New York.

Bultman, T.L. & Faeth, S.H. (1986) Experimental evidence for intraspecific competition in a lepidopteran leaf miner. *Ecology* **67**, 442–448.

Burd, M. (1994a) A probabilistic analysis of pollinator behavior and seed production in *Lobelia deckenii*. *Ecology* **75**, 1635–1646.

Burd, M. (1994b) Bateman principle and plant reproduction – the role of pollen limitation in fruit and seed set. *Botanical Review* **60**, 83–139.

Burd, M. & Head, G. (1992) Phenological aspects of male and female function in hermaphroditic plants. *American Naturalist* **140**, 305–324.

Burdon, J.J. (1987) *Diseases and Plant Population Biology*. Cambridge University Press, Cambridge.

Burdon, J.J. (1993) The structure of pathogen populations in natural plant communities. *Annual Review of Phytopathology* **31**, 305–323.

Burdon, J.J. (1994) The distribution and origin of genes for race specific resistance to *Melampsora lini* in *Linum marginale*. *Evolution* **48**, 1564–1575.

Burdon, J.J. & Chilvers, G.A. (1994) Demographic changes and the development of competition in a native Australian eucalypt forest invaded by exotic pines. *Oecologia* **97**, 419–423.

Burdon, J.J. & Thompson, J.N. (1995) Changed patterns of resistance in a population of *Linum marginale* attacked by the rust pathogen *Melampsora lini*. *Journal of Ecology* **83**, 199–206.

Burdon, J.J., Marshall, D.R. & Brown, A.H.D. (1983) Demographic and genetic changes in populations of *Echium plantagineum*. *Journal of Ecology* **71**, 667–679.

Burdon, J.J., Groves, R.H., Kaye, P.E. & Spear, S.S. (1984) Competition in mixtures of susceptible and resistant genotypes of *Chondrilla juncea* differentially infested with rust. *Oecologia* **64**, 199–203.

Burdon, J.J., Wenström, A., Ericson, L., Müller, W.J. & Morton, R. (1992) Density-dependent mortality in *Pinus sylvestris* caused by the snow blight pathogen *Phacidium infestans*. *Oecologia* **90**, 74–79.

Burdon, J.J. Wennstrom, A., Muller W.J. & Ericson, L. (1994) Spatial patterning in young stands of *Pinus sylvestris* in relation to mortality caused by the snow blight pathogen *Phacidium infestans*. *Oikos* **71**, 130–136.

Burger, J.C. & Louda, S.M. (1994) Indirect versus direct effects of grasses on growth of a cactus (*Opuntia fragilis*) – insect herbivory versus competition. *Oecologia* **99**, 79–87.

Burgess, T.M. & Webster, R. (1980) Optimal interpolation and isarithmic mapping of soil properties. II. Block Kriging. *Journal of Soil Science* **31**, 333–341.

Burnett, W.C., Jones, S.B. & Mabry, T.J. (1978) The role of sesquiterpene lactones in plant–animal coevolution. In *Biochemical Aspects of Plant and Animal Coevolution* (Ed. J.B. Harborne), pp. 233–257. Academic Press, London.

Burrows, F.M. (1986) The aerial motion of seeds, fruits, spores and pollen. In *Seed Dispersal* (Ed. D.R. Murray), pp. 1–48. Academic Press, Sydney.

Bustamante, R.H., Branch, G.M. & Eekhout, S. (1996) Maintenance of an exceptional intertidal grazer biomass in South Africa—subsidy by subtidal kelps. *Ecology* **76**, 2314–2329.

Butcher, R.E. (1983) *Studies on interference between weeds and peas*. PhD thesis, University of East Anglia.

Butler, L.G., Rogler, J.C., Mehansha, H. & Carlson, D.M. (1986) Dietary effects of tannins. In *Plant Flavonoids in Biology and Medicine* (Eds M. Elliott, V. Cody & J.B. Harborne), pp. 141–157. Alan R. Liss, New York.

Byers, D.L. & Meagher, T.R. (1992) Mate availability in small populations of plant species with homomorphic sporophytic self-incompatibility. *Heredity* **68**, 353–359.

Byers, J.A., Lanne, B.S. & Lofquist, J. (1989) Host tree unsuitability recognised by pine shoot beetles in flight. *Experientia* **45**, 489–492.

Byrne, M.M. & Levey, D.J. (1993) Removal of seeds from frugivore defecations by ants in a Costa Rican rain forest. *Vegetatio* **107/108**, 363–374.

Byth, K. & Ripley, B.D. (1980) On sampling spatial patterns by distance methods. *Biometrics* **36**, 279–284.

Cachorro, P., Olmos, E., Ortiz, A. & Cerda, A. (1995) Salinity induced changes in the structure and ultrastructure of bean root cells. *Biologia Plantarum* **37**, 273–283.

Cain, M.L., Pacala, S.W., Silander, J.A. Jr & Fortin, M.J. (1995) Neighborhood models of clonal growth in the white clover, *Trifolium repens*. *American Naturalist* **145**, 888–917.

Cain, S.A. (1947) Characteristics of natural areas and factors in their development. *Ecological Monographs* **17**, 185–200.

Cairns, J. (1988) Increasing diversity by restoring damaged ecosystems. In *Biodiversity* (Eds E.O. Wilson & F.M. Peter), pp. 333–343. National Academy of Sciences, Washington, DC.

Caldwell, M.M., Eissenstat, D.M., Richards, J.H. & Allen, M.F. (1985) Competition for phosphorus: differential uptake from dual isotope-labelled soil interspaces between shrub and grass. *Science* **229**, 384–386.

Callaghan, T.V. & Collins, N.J. (1981) Life cycles, population dynamics and the growth of tundra plants. In *Tundra Ecosystems: a Comparative Analysis* (Eds L.C. Bliss, O.W. Heal & J.J. Moore), pp. 257–284. Cambridge University Press, Cambridge.

Cammell, M.E. & Way, M.J. (1983) Aphid pests. In *The Faba Bean* (Vicia faba *L.*). *A Basis for Improvement* (Ed. P.D. Hebblethwaite), pp. 315–346. Butterworth, London.

Campbell, B.D., Grime, J.P. & Mackey, J.M. (1991) A trade-off between scale and precision in resource foraging. *Oecologia* **87**, 532–538.

Campbell, D. (1985) Pollen and gene dispersal: the influence of competition for pollination. *Evolution* **39**, 418–431.

Campbell, M.M. & Ellis, B.E. (1992) Fungal elicitor-mediated responses in pine cell cultures: purification and characterisation of phenylalanine ammonia-lyase. *Plant Physiology* **98**, 62–70.

Canham, C.D. (1988) Growth and architecture of shade-tolerant trees: response to canopy gaps. *Ecology* **69**, 786–795.

Canham, C.D. & Pacala, S.W. (1995) Linking tree population dynamics and forest ecosystem processes. In *Species Effects on Ecosystems* (Eds D. Jones & J. Lawton), pp. 84–94. Chapman & Hall, London.

Cannell, M.G.R., Rothery, P. & Ford, E.D. (1984) Competition within stands of *Picea sitchensis* and *Pinus contorta*. *Annals of Botany* **53**, 349–362.

Canny, M.J. (1990) What becomes of the transpiration stream? *New Phytologist* **114**, 341–368.

Canny, M.J. (1995) Apoplastic water and solute movement: new rules for an old space. *Annual Review of Plant Physiology and Molecular Biology* **46**, 215–236.

Cantlon, J.E. (1969) The stability of natural populations and their sensitivity to technology. *Brookhaven Symposium in Biology* **22**, 197–205.

Cape, J.N., Leith, I.D., Fowler, D., *et al.* (1991) Sulphate and ammonium in mist impair the frost hardening of red spruce seedlings. *New Phytologist* **118**, 119–126.

Caporn, S.J.M., Risager, M. & Lee, J.A. (1994) Effect of nitrogen supply on frost hardiness in *Calluna vulgaris* (L.) Hull. *New Phytologist* **128**, 461–468.

Caraco, T. & Kelly, C.K. (1991) On the adaptive value of physiological integration in plants. *Ecology* **72**, 81–93.

Carey, P.D., Fitter, A.H. & Watkinson, A.R. (1992) A field study using the fungicide benomyl to investigate the effect of mycorrhizal fungi on plant fitness. *Oecologia* **90**, 350–355.

Carey, P.D., Watkinson, A.R. & Gerard, F.F.O. (1995) The determinants of the distribution and abundance of the winter annual grass *Vulpia ciliata* subsp. *ambigua*. *Journal of Ecology* **83**, 177–187.

Carne, P.B. (1969) On the population dynamics of the eucalypt-defoliating sawfly *Perga affinis affinis* Kirby (Hymenoptera). *Australian Journal of Zoology* **17**, 113–141.

Carney, H.J., Richerson, P.J., Goldman, C.R. and Richards, R.C. (1988) Seasonal phytoplankton demographic processes and experiments on interspecific

competition. *Ecology* **69**, 664–678.

Carpenter, F.L. (1983) Pollination energetics in avian communities: simple concepts and complex realities. In *Handbook of Experimental Pollination Biology* (Eds C.E. Jones & R.J. Little), pp. 215–234. Van Nostrand Reinhold, New York.

Carpenter, S.R. & Kitchell, J.F. (1992) Trophic cascade and biomanipulation: interface of research and management – a reply to the comment by DeMelo *et al. Limnology and Oceanography* **37**, 208–213.

Carrasco, A., Boudet, A.M. & Maringo, S. (1978) Enhanced resistance of tomato plants to *Fusarium* by controlled stimulation of their natural phenolic production. *Physiological Plant Pathology* **12**, 225–232.

Carroll, C.R. & Hoffman, C.A. (1980) Chemical feeding deterrent mobilised in response to insect herbivory and counteradaptation by *Epilachna tredecimnotata*. *Science* **209**, 414–416.

Carter, A.J. & Robinson, E.R. (1993) Genetic structure of a population of the clonal grass *Setaria incrassata*. *Biological Journal of the Linnean Society* **48**, 55–62.

Carter, R.N. & Prince, S.D. (1981) Epidemic models used to explain biogeographical distribution limits. *Nature* **293**, 644–645.

Carver, T.L.W., Robbins, M.P. & Zeyen, R.J. (1991) Effects of two PAL inhibitors on the susceptibility and localised autofluorescent host cell responses of oat leaves attacked by *Erysiphe graminis* DC 39. *Physiological and Molecular Pathology* **39**, 269–287.

Caswell, H. (1989) *Matrix Population Models*. Sinauer, Sunderland, Massachusetts.

Caswell, H. & Etter, R.J. (1993) Ecological interactions in patchy environments: from patch-occupancy models to cellular automata. In *Patch Dynamics* (Eds S.A. Levin, T.M. Powell & J.H. Steel), pp. 93–108. Springer-Verlag, Berlin.

Cates, R.G. (1975) The interface between slugs and wild ginger: some evolutionary aspects *Ecology* **56**, 391–400.

Cates, R.G. & Zou, J. (1990) Douglas fir (*Pseudotsuga menziesii*) population variation in terpene chemistry and its role in budworm (*Choristoneura occidentalis* Freeman) dynamics. In *Population Dynamics of Forest Insects* (Eds A.D. Watt, S.R. Leather, M.D. Hunter & N.A.C. Kidd), pp. 169–182. Intercept, Andover.

Cebrian, J. & Duarte, C.M. (1994) The dependence of herbivory on growth rate in natural plant communities. *Functional Ecology* **8**, 518–525.

Cerling, T.E., Quade, J., Wang, Y. & Bowman, J.R. (1989) Carbon isotopes in soils and palaeosols as ecologic and palaeoecologic indicators. *Nature* **341**, 138–139.

Chabot, B.F. & Hicks, D.J. (1982) The ecology of leaf life spans. *Annual Review of Ecology and Systematics* **13**, 229–259.

Chaboudez, P. & Burdon, J.J. (1995) Frequency dependent selection in a wild plant pathogen system. *Oecologia* **102**, 490–493.

Candler, G.E. & Anderson, J.W. (1976) Studies of the

nutrition and growth of *Drosera* spp. with reference to the carnivorous habit. *New Phytologist* **76**, 129–141.

Chapin, D.M. (1995) Physiological and morphological attributes of 2 colonizing plant species on Mount St. Helens. *American Midland Naturalist* **133**, 76–87.

Chapin, F.S. (1980) The mineral nutrition of wild plants. *Annual Review of Ecology and Systematics* **11**, 233–260.

Chapin, F.S. & Shaver, G.R. (1985) Individualistic growth response of tundra plant species to environmental manipulations in the field. *Ecology* **66**, 564–576.

Chapin, F.S., McGraw, J.B. & Shaver, G.R. (1989) Competition causes regular spacing of alder in Alaskan shrub tundra. *Oecologia* **79**, 412–416.

Chapman, V.J. (1977) *Wet Coastal Ecosystems. Ecosystems of the World, Vol. I.* Elsevier, Amsterdam.

Charles-Dominique, P., Atramentowicz, M., Charles-Dominique, M. *et al.* (1981) Les mammiferes frugivores abroricoles nocturnes d'une foret Guyanaise: inter-relations plantes–animaux. *Terre et Vie* **35**, 341–435.

Charlesworth, B. (1980) *Evolution in Age-structured Populations.* Cambridge University Press, Cambridge.

Charlesworth, D. & Charlesworth, B. (1978) A model for the evolution of dioecy and gynodioecy. *American Naturalist* **112**, 975–997.

Charlesworth, D. & Charlesworth, B. (1979) The evolutionary genetics of sexual systems in flowering plants. *Proceedings of the Royal Society of London Series B* **205**, 513–530.

Charlesworth, D. & Charlesworth, B. (1981) Allocations of resources to male and female functions in hermaphrodites. *Biological Journal of the Linnean Society* **15**, 57–74.

Charlesworth, D. & Charlesworth, B. (1987) Inbreeding depression and its evolutionary consequences. *Annual Review of Ecology and Systematics* **18**, 237–268.

Charlesworth, D., Morgan, M.T. & Charlesworth, B. (1993) Mutation accumulation in finite outbreeding and inbreeding populations. *Genetical Research* **61**, 39–56.

Charnov, E.L. (1979) Simultaneous hermaphroditism and sexual selection. *Proceedings of the National Academy of Sciences USA* **76**, 2480–2484.

Charnov, E.L. (1980) Sex allocation and local mate competition in barnacles. *Marine Biology Letters* **1**, 269–272.

Charnov, E.L. (1982a) Parent–offspring conflict over reproductive effort. *American Naturalist* **119**, 736–737.

Charnov, E.L. (1982b) *The Theory of Sex Allocation.* Princeton University Press, Princeton.

Charnov, E.L. (1984) The behavioural ecology of plants. In *Behavioural Ecology: an Evolutionary Approach* (Eds J.R. Krebs & N.B. Davies), pp. 362–379. Blackwell Scientific Publications, Oxford.

Charnov, E.L. (1986) An optimization principle for sex allocation in a temporally varying environment. *Heredity* **56**, 119–121.

Charnov, E.L. & Bull, J.J. (1986) Sex allocation, pollinator attraction and fruit dispersal in cosexual plants. *Journal of Theoretical Biology* **118**, 321–325.

Charnov, E.L., Maynard Smith, J. & Bull, J.J. (1977) Why be an hermaphrodite? *Nature* **263**, 125–126.

Chen, J., Chang, C.J. & Jarret, R.L. (1992) DNA probes as molecular markers to monitor the seasonal occurrence of walnut witches broom mycoplasma like organism. *Plant Disease* **76**, 1116–1119.

Cheplick, G.P. (1992) Sibling competition in plants. *Journal of Ecology* **80**, 567–575.

Chesson, P.L. (1983) Coexistence of competitors in a stochastic environment: the storage effect. In *Population Biology* (Eds H.I. Freedman & C. Strobeck), pp. 188–198. Springer-Verlag, New York.

Chesson, P.L. (1984) The storage effect in stochastic population models. In *Mathematical Ecology: Miramare-Trieste Proceedings* (Eds S.A. Levin & T.G. Hallam), pp. 76–89. Springer-Verlag, New York.

Chesson, P.L. (1985) Coexistence of competitors in spatially and temporally varying environments: a look at the combined effects of different sorts of variability. *Theoretical Population Biology* **28**, 263–287.

Chesson, P.L. (1986) Environmental variation and the coexistence of species. In *Community Ecology* (Eds J. Diamond & T.J. Case), pp. 240–256. Harper & Row, New York.

Chesson, P.L. (1988) Interactions between environment and competition: how fluctuations mediate coexistence and competitive exclusion. In *Community Ecology* (Ed. A. Hastings), pp. 51–71. Springer-Verlag, London.

Chesson, P. & Huntley, N. (1989) Short-term instabilities and long-term community dynamics. *Trends in Ecology and Evolution* **4**, 292–298.

Chesson, P. & Huntley, N. (1993) Hierarchies of temporal variation and species diversity. *Plant Species Biology* **8**, 195–298.

Chew, F.S. & Rodman, J.E. (1979) Plant resources for chemical defense. In *Herbivores: Their Interaction with Secondary Plant Metabolites* (Eds G.A. Rosenthal & D.H. Janzen), pp. 271–307. Academic Press, New York.

Chiariello, N., Hickman, J.C. & Mooney, H.A. (1982) Endomycorrhizal role for interspecific transfer of phosphorus in a community of annual plants. *Science* **217**, 941–943.

Chiariello, N.R., Field, C.B. & Mooney, H.A. (1987) Midday wilting in a tropical pioneer tree. *Functional Ecology* **1**, 3–11.

Chin, H.F. & Roberts, E.H. (1980) *Recalcitrant Crop Seeds.* Tropical Press, Kuala Lumpur, Malaysia.

Chippendale, H.G. & Milton, W.E.J. (1934) On the viable seeds present in the soil beneath pastures. *Journal of Ecology* **22**, 508–531.

Christensen, K.M. & Whitham, T.G. (1991) Indirect herbivore mediation of avian seed dispersal in Pinyon pine. *Ecology* **72**, 534–542.

Christie, E.K. & Moorby, J. (1975) Physiological responses of arid grasses. I. The influence of phosphorus supply on growth and phosphorus absorption. *Australian Journal of Agricultural Research* **26**, 423–436.

Clark, D.A. & Clark, D.B. (1984) Spacing dynamics of a tropical rain forest tree: evaluation of the Janzen–Connell model. *American Naturalist* **124**, 769–788.

Clark, D.A. & Clark, D.B. (1992) Life-history diversity of canopy and emergent trees in a neotropical rainforest. *Ecological Monographs* **62**, 315–344.

Clark, J.R. & Benforado, J. (Eds) (1981) *Wetlands of Bottomland Hardwood Forests*. Elsevier, New York.

Clark, J.S. (1989) Ecological disturbance as a renewal process: theory and application to fire history. *Oikos* **56**, 17–30.

Clark, J.S. (1990) Fire and climate change during the last 750 yr in northwestern Minnesota. *Ecological Monographs* **60**, 135–159.

Clarke, A.E. & Newbigin, E. (1993) Molecular aspects of self-incompatibility in flowering plants. *Annual Review of Genetics* **27**, 257–279.

Clarkson, D.T. & Hanson, J.B. (1980) The mineral nutrition of higher plants. *Annual Review of Plant Physiology* **31**, 239–298.

Clarkson, D.T. & Lüttge, U. (1991) Mineral nutrition: inducible and repressible nutrient transport systems. *Progress in Botany* **52**, 61–83.

Clausen, J., Keck, D.D. & Hiesey, W.M. (1948) *Experimental Studies on the Nature of Species. III. Environmental Responses of Climatic Races of* Achillea. Carnegie Institute Washington, Publication 581.

Clay, K. (1990) The impact of parasitic and mutualistic fungi on competitive interactions among plants. In *Perspectives on Plant Competition* (Eds J. Grace & D. Tilman), pp. 391–412. Academic Press, New York.

Clay, K. (1993) The ecology and evolution of endophytes. *Agriculture Ecosystems and Environment* **44**, 39–64.

Clay, K., Marks, S. & Cheplick, G.P. (1993) Effects of insect herbivory and fungal endophyte infection on competitive interactions among grasses. *Ecology* **74**, 1767–1777.

Clements, F.E. (1916) *Plant Succession. An Analysis of the Development of Vegetation*. Carnegie Institute Washington, Publication 242.

Clements, F.E. (1928) *Plant Succession and Indicators*. H.W. Wilson, New York.

Clutton-Brock, T.H. & Harvey, P.H. (1979) Comparison and adaptation. *Proceedings of the Royal Society of London Series B* **205**, 547–565.

Clymo, R.S. (1962) An experimental approach to part of the calcicole problem. *Journal of Ecology* **50**, 701–731.

Clymo, R.S. (1973) The growth of *Sphagnum*: some effects of environment. *Journal of Ecology* **61**, 849–869.

Cody, M.L. (1966) A general theory of clutch size. *Evolution* **20**, 174–184.

Cody, M.L. (1984) Branching patterns in columnar cacti. In *Being Alive on Land* (Eds N.S. Margaris, M. Arianoustou-Faraggitaki & W.C. Oechel), pp. 201–233. Junk, The Hague.

Cohen, D. (1966) Optimizing reproduction in a randomly varying environment. *Journal of Theoretical Biology* **12**, 119–129.

Cohen, D. (1968) A general model of optimal reproduction in a randomly varying environment. *Journal of Ecology* **56**, 219–228.

Cohen, D. (1971) Maximizing final yield when growth is limited by time or by limiting resources. *Journal of Theoretical Biology* **33**, 299–307.

Colasanti, R.L. & Grime, J.P. (1993) Resource dynamics and vegetation processes – a deterministic model using 2-dimensional cellular automata. *Functional Ecology* **7**, 169–176.

Cole, D.N. (1995) Experimental trampling of vegetation. 2. Predictors of resistance and resilience. *Journal of Applied Ecology* **32**, 215–224.

Coleman, D.C., Coe, C.V. & Elliot, E.T. (1984) Decomposition, organic matter turnover and nutrient dynamics in agro-ecosystems. In *Agricultural Ecosystems* (Eds R. Lowrance, B.R. Stinner & G.T. House), pp. 83–104. John Wiley & Sons, New York.

Coleman, J.S. & Jones, C.G. (1988) Plant stress and insect performance: eastern cottonwood, ozone and a leaf beetle. *Oecologia* **76**, 57–61.

Coleman, J.S. & Jones, C.G. (1991) A phytocentric perspective of phytochemical induction by herbivores. In *Phytochemical Induction by Herbivores* (Eds D.W. Tallamy & M.J. Raupp), pp. 3–46. Wiley-Interscience, New York.

Coleman, J.S. & Leonard, A.S. (1995) Why it matters where on a leaf a folivore feeds. *Oecologia* **101**, 324–328.

Coleman, J.S., McConnaughay, K.D.M. & Bazzaz, F.A. (1993) Elevated CO_2 and plant nitrogen use – is reduced tissue nitrogen concentration size dependent? *Oecologia* **93**, 195–200.

Coley, P.D. (1983) Herbivory and defensive characteristics of tree species in a lowland tropical forest. *Ecological Monographs* **53**, 209–233.

Coley, P.D. (1987) Interspecific variation in plant antiherbivore properties: the role of habitat quality and rate of disturbance. *New Phytologist* **106**, 251–263.

Coley, P.D. (1988) Effects of plant growth rate and leaf lifetime on the amount and type of anti-herbivore defense. *Oecologia* **74**, 531–536.

Coley, P.D. (1993) Gap size and plant defences. *Trends in Ecology and Evolution* **8**, 1–2.

Coley, P.D., Bryant, J.P. & Chapin, F.S. (1985) Resource availability and plant antiherbivore defense. *Science* **230**, 895–899.

Collins, S.L., Glenn, S.M. & Gibson, D.J. (1995) Experimental analysis of intermediate disturbance and initial floristic composition – decoupling cause and effect. *Ecology* **76**, 486–492.

Comins, H.N. & Noble, I.R. (1985) Dispersal, variability and transient, niches: species coexistence in a uniformly variable environment. *American Naturalist* **126**, 706–723.

Condit, R., Hubbell, S.P. & Foster, R.B. (1992) Recruitment near conspecific adults and the maintenance of tree and shrub diversity in a neotropical forest. *American Naturalist* **140**, 261–286.

Condit, R., Hubbell, S.P. & Foster, R.B. (1994) Density dependence in two understorey tree species in a neotropical forest. *Ecology* **75**, 671–680.

Condit, R., Hubbell, S.P. & Foster, R.B. (1995) Mortality rates of 205 neotropical tree and shrub species and the impact of a severe drought. *Ecological Monographs* **65**, 419–439.

Conn, E.E. (1981) *The Biochemistry of Plants. Vol. 7. Secondary Plant Products* (Eds P.K. Stumpf & E.E. Conn) Academic Press, New York.

Connell, J.H. (1971) On the role of natural enemies in preventing competitive exclusion in some marine animals and in rain forest trees. In *Dynamics of Populations* (Eds B.J. den Boer & G.R. Gradwell), pp. 298–310. Centre for Agricultural Publishing and Documentation, Wageningen.

Connell, J.H. (1978) Diversity in tropical rainforests and coral reefs. *Science* **199**, 1302–1310.

Connell, J.H. (1979) Tropical rain forests and coral reefs as open non-equilibrium systems. In *Population Dynamics: 20th Symposium of the British Ecological Society* (Eds R.M. Anderson, B.D. Turner & L.R. Taylor), pp. 141–163. Blackwell Scientific Publications, Oxford.

Connell, J.H. (1980) Diversity and the coevolution of competitors, or the ghost of competition past. *Oikos* **35**, 131–138.

Connell, J.H. (1983) On the prevalence and relative importance of interspecific competition: evidence from field experiments. *American Naturalist* **122**, 661–696.

Connell, J.H. & Slatyer, R.O. (1977) Mechanisms of succession in natural communities and their role in community stability and organization. *American Naturalist* **111**, 1119–1144.

Connell, J.H., Tracey, J.G. & Webb, L.J. (1984) Compensatory recruitment, growth, and mortality as factors maintaining rain forest tree diversity. *Ecological Monographs* **54**, 141–164.

Connor, E.F. & McCoy, E.D. (1979) The statistics and biology of the species–area relationship. *American Naturalist* **113**, 791–833.

Constable, G.A. & Rawson, H.M. (1980) Carbon production and utilization in cotton: inferences from a carbon budget. *Australian Journal of Plant Physiology* **7**, 539–553.

Cook, C.E., Whichard, L.P., Wall, M.E., *et al.* (1972) Germination stimulants. II. The structure of strigol – a potent seed germination stimulant for witchweed (*Striga lutea* Lour.). *Journal of the American Chemical Society* **94**, 6198–6199.

Cook, R. (1980) The biology of seeds in the soil. In *Demography and Evolution in Plant Populations* (Ed. O.T. Solbrig), pp. 107–129. Blackwell Scientific Publications, Oxford.

Cook, R.E. (1988) Growth in *Medeola virginiana* clones. I. Field observations. *American Journal of Botany* **75**, 725–731.

Cook, R., Evans, D.R., Williams, T.A. & Mizen, K.A. (1992) The effect of stem nematode on establishment and early yields of white clover. *Annals of Applied Biology* **120**, 83–94.

Cook, S.P. & Hain, F.P. (1988) Toxicity of host monoterpenes to *Dendroctonus frontalis* and *Ips calligraphus*. *Journal of Entomological Science* **23**, 287–292.

Cooper, W.S. (1926) The fundamentals of vegetation change. *Ecology* **7**, 391–413.

Cornell, H.V. (1983) Why and how gall wasps form galls: Cynipids as genetic engineers? *Antenna* **7**, 53–58.

Costich, D.E. (1995) Gender specialization across a climatic gradient: experimental comparison of monoecious and dioecious *Ecballium*. *Ecology* **76**, 1036–1050.

Costich, D.E. & Galan, F. (1988) The ecology of the monoecious and dioecious subspecies of *Ecballium elaterium* (L.) A. Rich. (Cucurbitaceae). I. Geographic distribution and its relationship to climatic conditions in Spain. *Lagascalia* **15** (Extra), 697–710.

Costich, D.E. & Meagher, T.R. (1992) Genetic variation in *Ecballium elaterium* (Cucurbitaceae): breeding system and geographic distribution. *Journal of Evolutionary Biology* **5**, 589–601.

Cottam, D.A. (1986) The effects of slug-grazing on *Trifolium repens* and *Dactylis glomerata* in monoculture and mixed swards. *Oikos* **47**, 275–279.

Cottam, G. & Curtis, J.T. (1949) A method for making rapid surveys of woodlands, by means of pairs of randomly selected trees. *Ecology* **30**, 101–104.

Courtney, S. (1988) If it's not coevolution, it must be predation? *Ecology* **69**, 910–911.

Courtney, S.P. & Manzur, M.I. (1985) Fruiting and fitness in *Crataegus monogyna* – the effects of frugivores and seed predators. *Oikos* **44**, 398–406.

Cousens, R. (1995) Can we determine the intrinsic dynamics of real plant populations? *Functional Ecology* **9**, 15–20.

Cowan, I.R. (1977) Stomatal behaviour and environment. *Advancements in Botanical Research* **4**, 117–223.

Cowles, H.C. (1899) The ecological relations of the vegetation on the sand dunes of Lake Michigan. *Botanical Gazette* **27**, 95–117; 167–202; 281–308; 361–391.

Cowling, R.M. (1992) *The Ecology of Fynbos: Nutrients, Fire and Diversity*. Oxford University Press, Oxford.

Cox, P.A. (1985) Islands and dioecism: insights from the reproductive ecology of *Pandanus tectorius* in Polynesia. In *Studies in Plant Demography* (Ed. J. White), pp. 359–372. Academic Press, London.

Cox, R.M. (1988) Sensitivity of forest plant reproduction to long-range transported air pollutants: the effects of wet-deposited acidity and copper on the reproduction of *Populus tremuloides*. *New Phytologist* **110**, 33–38.

Cox, R.M. (1992) Air pollution effects on plant reproductive processes and possible consequences to their population biology. In *Air Pollution Effects on Biodiver-*

sity (Eds J.R. Barker & D.T. Tingey), pp. 131–158. Van Nostrand Reinhold, New York.

Craig, T.P., Price, P. & Itami, J.K. (1986) Resource regulation by a stem-galling sawfly on the arroyo willow. *Ecology* **67**, 419–425.

Craig, T.P., Itami, J.K. & Price, P. (1990) Intraspecific competition and facilitation by a shoot-galling sawfly. *Journal of Animal Ecology* **59**, 147–159.

Cramer, M.D., Schierholt, A., Wang, Y.Z. & Lips, S.H. (1995) The influence of salinity on the utilization of root anaplerotic carbon and nitrogen metabolism in tomato seedlings. *Journal of Experimental Botany* **46**, 1569–1577.

Crawley, M.J. (1983) *Herbivory: the Dynamics of Animal–Plant Interactions*. Blackwell Scientific Publications, Oxford.

Crawley, M.J. (1985) Reduction of oak fecundity by low density herbivore populations. *Nature* **314**, 163–164.

Crawley, M.J. (1986a) The population biology of invaders. *Philosophical Transactions of the Royal Society of London Series B* **314**, 711–731.

Crawley, M.J. (1986b) The structure of plant communities. In *Plant Ecology* (Ed. M.J. Crawley), pp. 1–50. Blackwell Scientific Publications, Oxford.

Crawley, M.J. (1987a) Benevolent herbivores? *Trends in Ecology and Evolution* **2**, 167–168.

Crawley, M.J. (1987b) What makes a community invasible? In *Colonization, Succession and Stability* (Eds A.J. Gray, M.J. Crawley & P.J. Edwards), pp. 429–453. Blackwell Scientific Publications, Oxford.

Crawley, M.J. (1988) Herbivores and plant population dynamics. In *Plant Population Ecology* (Eds A.J. Davy, M.J. Hutchings & A.R. Watkinson), pp. 367–392. Blackwell Scientific Publications, Oxford.

Crawley, M.J. (1989a) Insect herbivores and plant population dynamics. *Annual Review of Entomology* **34**, 531–564.

Crawley, M.J. (1989b) The relative importance of vertebrate and invertebrate herbivores in plant population dynamics. In *Insect–Plant Interactions* (Ed. E.A. Bernays), Vol. 1, pp. 45–71. CRC Press, Boca Raton.

Crawley, M.J. (1989c) The successes and failures of weed biocontrol using insects. *Biocontrol News and Information* **10**, 213–222.

Crawley, M.J. (1990a) The population dynamics of plants. *Philosophical Transactions of the Royal Society of London Series B* **330**, 125–140.

Crawley, M.J. (1990b) Rabbit grazing, plant competition and seedling recruitment in acid grassland. *Journal of Applied Ecology* **27**, 803–820.

Crawley, M.J. (1992a) Population dynamics. In *Natural Enemies: the Population Biology of Predators, Parasites and Diseases* (Ed. M.J. Crawley), pp. 40–89. Blackwell Scientific Publications, Oxford.

Crawley, M.J. (1992b) Seed predators and population dynamics. In *Seeds: the Ecology of Regeneration in Plant Communities* (Ed. M. Fenner), pp. 157–191. CAB International, Wallingford.

Crawley, M.J. (1993a) *GLIM for Ecologists*. Blackwell Scientific Publications, Oxford.

Crawley, M.J. (1993b) On the consequences of herbivory. *Evolutionary Ecology* **7**, 124–125.

Crawley, M.J. (1997) *Aliens: the Ecology of Non-indigenous Plants*. Oxford University Press, Oxford.

Crawley, M.J. & Akhteruzzaman, M. (1988) Individual variation in the phenology of oak trees and its consequences for herbivorous insects. *Functional Ecology* **2**, 409–415.

Crawley, M.J. & Brown, S.L. (1995) Seed limitation and the dynamics of feral oilseed rape on the M25 motorway. *Proceedings of the Royal Society of London Series B* **259**, 49–54.

Crawley, M.J. & Gillman, M.P. (1989) Population dynamics of cinnabar moth and ragwort in grassland. *Journal of Animal Ecology* **58**, 1035–1050.

Crawley, M.J. & Krebs, J.R. (1992) Foraging theory. In *Natural Enemies: the Population Biology of Predators, Parasites and Diseases* (Ed. M.J. Crawley), pp. 90–114. Blackwell Scientific Publications, Oxford.

Crawley, M.J. & Long, C.R. (1995) Alternate bearing, predator satiation and seedling recruitment in *Quercus robur* L. *Journal of Ecology* **83**, 683–696.

Crawley, M.J. & May, R.M. (1987) Population dynamics and plant community structure: competition between annuals and perennials. *Journal of Theoretical Biology* **125**, 475–489.

Crawley, M.J. & Nachapong, M. (1985) The establishment of seedlings from primary and regrowth seeds of ragwort (*Senecio jacobaea*). *Journal of Ecology* **73**, 255–261.

Crawley, M.J. & Rees, M. (1996) The Balance of Plant Populations. In *Frontiers of Population Ecology* (Eds R.B. Floyd, A.W. Sheppard & P.J. De Bono), pp. 165–196. CSIRO, Melbourne.

Crawley, M.J. & Weiner, J. (1991) Plant size variation and vertebrate herbivory: winter wheat grazed by rabbits. *Journal of Applied Ecology* **28**, 154–172.

Crawley, M.J., Hails, R.S., Rees, M., Kohn, D. & Buxton, J. (1993) Ecology of transgenic oilseed rape in natural habitats. *Nature* **363**, 620–623.

Crawley, M.J., Harvey, P.H. & Purvis, A. (1996) Comparative ecology of the native and alien floras of the British Isles. *Philosophical Transactions of the Royal Society of London Series B* (in press).

Crespi, M., Vereecke, D., Temmerman, W., Vanmontagu, M. & Desomer, J. (1994) The FAS operon of *Rhodococcus fascians* encodes new genes required for efficient fasciation of host plants. *Journal of Bacteriology* **176**, 2492–2501.

Cressie, N.A.C. (1991) *Statistics for Spatial Data*. Wiley-Interscience, New York.

Cresswell, E.G. & Grime, J.P. (1981) Induction of a light requirement during seed development and its ecological consequences. *Nature* **291**, 583–585.

Cresswell, J.E. & Robertson, A.W. (1994) Discrimination by pollen collecting bumblebees among differentially rewarding flowers of an alpine wild flower,

Campanula rotundifolia (Campanulaceae). *Oikos* **59**, 304–308.

Crisp, M.D. & Lange, R.T. (1976) Age structure, distribution and survival under grazing of the arid-zone shrub *Acacia burkittii*. *Oikos* **27**, 86–92.

Crocker, R.L. & Major, J. (1955) Soil development in relation to vegetation and surface age at Glacier Bay, Alaska. *Journal of Ecology* **43**, 427–448.

Crome, F.H.J. (1975) The ecology of fruit pigeons in tropical northern Queensland. *Australian Wildlife Research* **2**, 155–185.

Crow, J.F. & Kimura, M. (1970) *An Introduction to Population Genetics Theory*. W.H. Freeman, New York.

Crowley, D.E., Wang, Y.C., Reid, C.P.P. & Szaniszlo, P.J. (1991) Mechanisms of iron acquisition from siderophores by micro-organisms and plants. *Plant and Soil* **130**, 179–198.

Crowley, P.H. (1977) Spatially distributed stochasticity and the constancy of ecosystems. *Bulletin of Mathematical Biology* **39**, 157–166.

Cruden, R.W. (1976) Intraspecific variation in pollen–ovule ratios and nectar secretion – preliminary evidence of ecotypic adaptation. *Annals of the Missouri Botanical Garden* **63**, 277–289.

Cruden, R.W. (1977) Pollen–ovule ratios: a conservative indicator of the breeding systems in the flowering plants. *Evolution* **31**, 32–46.

Cruden, R.W. & Jensen, K.G. (1979) Viscin threads, pollination efficiency and low pollen–ovule ratios. *American Journal of Botany* **66**, 875–879.

Cruden, R.W. & Lyon, D.L. (1985) Patterns of biomass allocation to male and female functions in plants with different mating systems. *Oecologia* **66**, 299–306.

Cullen, J.M. & Groves, R.H. (1977) The population biology of *Chondrilla juncea* L. in Australia. *Proceedings of the Ecological Society of Australia* **10**, 121–134.

Curtis, J.T. (1959) *The Vegetation of Wisconsin. An Ordination of Plant Communities*. University of Wisconsin Press, Madison.

Curtis, J.T. & McIntosh, R.P. (1951) An upland forest continuum in the prairie–forest border region of Wisconsin. *Ecology* **32**, 476–496.

Cyr, H. & Pace, M.L. (1993) Magnitude and patterns of herbivory in aquatic and terrestrial ecosystems. *Nature* **361**, 148–150.

Dale, M.P. & Causton, D.R. (1992) The ecophysiology of *Veronica chamaedrys, V. montana* and *V. officinalis*. 2. The interaction of irradiance and water regime. *Journal of Ecology* **80**, 493–504.

Damgaard, C. & Abbott, R.J. (1995) Positive correlations between selfing rate and pollen ovule ratio within plant populations. *Evolution* **49**, 214–217.

Damman, H. (1987) Leaf quality and enemy avoidance by the larvae of a moth. *Ecology* **68**, 88–97.

Danell, K., Hjalten, J., Ericson, L. & Elmqvist, T. (1991) Vole feeding on male and female willow shoots along a gradient of plant productivity. *Oikos* **62**, 145–152.

Danell, K., Bergstrom, R. & Iedenius, L. (1994) Effects of large mammalian browsers on architecture, biomass, and nutrients of woody plants. *Journal of Mammalogy* **75**, 833–844.

Darwin, C. (1859) *On the Origin of Species by Means of Natural Selection*. John Murray, London.

Darwin, C. (1876) *The Effects of Cross- and Self-fertilization in the Vegetable Kingdom*. John Murray, London.

Da Silva, H.R., de Britto-Pereira, M.C. & Caramaschi, U. (1989) Frugivory and seed dispersal by *Hyla truncata*, a neotropical treefrog. *Copeia* 1989, 781–783.

Davidar, P. (1983) Birds and neotropical mistletoes: effects on seedling recruitment. *Oecologia* **60**, 271–273.

Davidar, P. & Morton, E.S. (1986) The relationship between fruit crop sizes and fruit removal rates by birds. *Ecology* **67**, 262–265.

Davidson, D.W. (1993) The effects of herbivory and granivory on terrestrial plant succession. *Oikos* **68**, 23–35.

Davidson, D.W. & Morton, S.R. (1981) Competition for dispersal in ant-dispersed plants. *Science* **213**, 1259–1261.

Davidson, D.W., Inouye, R.S., Brown, J.H. (1984) Granivory in a desert ecosystem: Experimental evidence for indirect facilitation of ants by rodents. *Ecology* **65**, 1780–1786.

Davies, W.J. & Zhang, J.H. (1991) Root signals and the regulation of growth and development of plants in drying soil. *Annual Review of Plant Physiology and Plant Molecular Biology* **42**, 55–76.

Davis, M.B. (1981) Quaternary history and the stability of plant communities. In *Forest Succession: Concepts and Application* (Eds D.C. West, H.H. Shugart & D.B. Botkin), pp. 132–153. Springer-Verlag, Berlin.

Davis, M.B. (1988) Invasions of forest communities during the holocene: beech and hemlock in the Great Lakes region. In *Colonization, Succession and Stability* (Eds A.J. Gray, M.J. Crawley & P.J. Edwards), pp. 373–393. Blackwell Scientific Publications, Oxford.

Davis, M.B. (1994) Ecology and paleoecology begin to merge. *Trends in Ecology and Evolution* **9**, 357–358.

Davis, M.B., Woods, K.D., Webb, S.L. & Futyma, R.B. (1986) Dispersal versus climate: expansion of *Fagus* and *Tsuga* into the upper Great Lakes region. *Vegetatio* **67**, 93–103.

Davis, M.B., Sugita, S., Calcot, R.R., Ferrari, J.B. & Frelich, L.E. (1994) Historical development of alternate communities in a hemlock-hardwood forest in northern Michigan, USA. In *Large-scale Ecology and Conservation Biology* (Eds P.J. Edwards, R.M. May & N.R. Webb), pp. 19–39. Blackwell Scientific Publications, Oxford.

Davis, S.D., Heywood, V.H. & Hamilton, A.C. (1994) *Centres of Plant Diversity* (3 volumes). IUCN Publications, Cambridge.

Davis-Carter, J.G. & Shuman, L.M. (1993) Influence of texture and pH of kaolinitic soils on zinc fractions and zinc uptake by peanuts. *Soil Science* **155**, 376–384.

Davison, A.D. (1977) The ecology of *Hordeum murinum* L. 3: some effects of adverse climate. *Journal of Ecology* **65**, 523–530.

Davy, A.J. & Taylor, K. (1974) Seasonal patterns of nitrogen availability in contrasting soils in the Chiltern Hills. *Journal of Ecology* **62**, 793–807.

Dawson, J.L. & Nash, T.H. III (1980) Effects of air pollution from copper smelters on a desert grassland community. *Environmental and Experimental Botany* **20**, 61–62.

Dawson, T.E. (1990) Spatial and physiological overlap of three co-occurring alpine willows. *Functional Ecology* **4**, 13–25.

Day, F.P. & Monk, C.D. (1974) Vegetation patterns on a southern Appalachian watershed. *Ecology* **55**, 1064–1074.

De Angelis, D.L. (1992) *Dynamics of Nutrient Cycling and Food Webs*. Chapman & Hall, London.

De Angelis, D.L. & Huston, M.A. (1993) Further considerations on the debate over herbivore optimization theory. *Ecological Applications* **3**, 30–31.

De Angelis, D.L., Travis, C.C. & Post, W.M. (1979) Persistence and stability of seed-dispersed species in a patchy environment. *Theoretical Population Biology* **16**, 107–125.

Debacke, P. (1988) Population dynamics of some broad-leaved weeds in cereal. I. Relation between standing vegetation and soil seed bank. *Weed Research* **28**, 251–263.

De Bakker, A.J. (1989) Effects of ammonia emission on epiphytic lichen vegetation. *Acta Botanica Neerlandica* **38**, 337–342.

Declerk-Floate, R. & Price, P.W. (1994) Impact of a bud galling midge on bud populations of *Salix exigua*. *Oikos* **70**, 253–260.

Deevey, E.S. (1947) Life tables for natural populations of animals. *Quarterly Review of Biology* **22**, 283–314.

de Kroon, H., Plaisir, A., van Groenedaal, J. & Caswell, H. (1986) Elasticity: the relative contribution of demographic parameters to population growth rate. *Ecology* **67**, 1427–1431.

Delesalle, V.A. & Mooreside, P.D. (1995) Estimating the costs of allocation to male and female functions in a monoecious cucurbit, *Lagenaria siceraria*. *Oecologia* **102**, 9–16.

Del Moral, R. & Bliss, L.C. (1993) Mechanisms of primary succession – insights resulting from the eruption of Mount St. Helens. *Advances in Ecological Research* **24**, 1–66.

Del Moral, R. & Muller, C.H. (1969) Fog drip: a mechanism of toxin transport from *Eucalyptus globosus*. *Bulletin of the Torrey Botanical Club* **96**, 467–475.

Delph, L.F. & Meagher, T.R. (1995) Sexual dimorphism masks life history trade-offs in the dioecious plant *Silene latifolia*. *Ecology* **76**, 775–785.

DeMelo, R., France, R. & McQueen, D.J. (1992) Biomanipulation: hit or myth? *Limnology and Oceanography* **37**, 192–207.

Dement, W.A. & Mooney, H.A. (1974) Seasonal variation in the production of tannins and cyanogenic glycosides in the chaparral shrub *Heteroneless arbutifolia*. *Oecologia* **15**, 65–70.

Demmig-Adams, B. & Adams III, W.V. (1992) Photoprotection and other responses of plants to high light stress. *Annual Review of Plant Physiology and Molecular Biology* **43**, 599–626.

Dennill, G.B., Donnelly, D. & Chown, S.L. (1993) Expansion of host plant range of a biocontrol agent *Trichilogaster acaciaelongifoliae* (Pteromalidae) released against the weed *Acacia longifolia* in South Africa. *Agriculture Ecosystems and Environment* **43**, 1–10.

Denno, R.F. & McClure, M.S. (1983) *Variable Plants and Herbivores in Natural and Managed Systems*. Academic Press, New York.

Denno, R.F., McClure, M.S. & Ott, J.R. (1995) Interspecific interactions in phytophagous insects: competition reexamined and resurrected. *Annual Review of Entomology* **40**, 297–331.

De Rooi-Van der Goes, P.C.E.M., Vanderputten, W.H. & Peters, B.A.M. (1995) Effects of sand deposition on the interaction between *Ammophila arenaria*, plant parasitic nematodes, and pathogenic fungi. *Canadian Journal of Botany* **73**, 1141–1150.

De Steven, D. (1982) Seed production and seed mortality in a temperate witch-hazel forest shrub (*Hamamelis virginiana*). *Journal of Ecology* **70**, 437–443.

Dethier, V.G. (1941) Chemical factors determining the choice of food plants by *Papilo* larvae. *American Naturalist* **75**, 61–73.

Devlin, B. & Ellstrand, N.C. (1990a) Male and female fertility variation in wild radish, a hermaphrodite. *American Naturalist* **136**, 87–107.

Devlin, B. & Ellstrand, N.C. (1990b) The development and application of a refined method for estimating gene flow from angiosperm paternity analysis. *Evolution* **44**, 248–259.

Devlin, B., Roeder, K. & Ellstrand, N.C. (1988) Fractional paternity assignment – theoretical development and comparison to other methods. *Theoretical and Applied Genetics* **76**, 369–380.

Devlin, B., Clegg, J. & Ellstrand, N.C. (1992) The effect of flower production on male reproductive success in wild radish populations. *Evolution* **46**, 1030–1042.

Dewar, R.C. (1995) An empirical model for stomatal conductance in terms of guard cell function: theoretical paper. *Plant, Cell and Environment* **18**, 365–372.

Díaz, S., Grime, J.P., Harris, J. & McPherson, E. (1993) Rising atmospheric carbon dioxide: evidence of a feedback mechanism limiting plant response. *Nature* (in review).

Dicke, M., Baarlen, P. van, Wessels, R. & Dijkman, H. (1993) Herbivory induces systemic production of plant volatiles that attract predators of the herbivore: extraction of endogenous elicitor. *Journal of Chemical Ecology* **19**, 581–599.

Diggle, P.J. (1983) *Statistical Analysis of Spatial Point Patterns*. Academic Press, London.

Dighton, J. & Skeffington, R.A. (1987) Effects of artificial acid precipitation on the mycorrhizas of Scots pine seedlings. *New Phytologist* **107**, 191–202.

Dinerstein, E. & Wemmer, C.M. (1988) Fruits *Rhinoceros* eat: megafaunal seed dispersal on a South Asian flood plain. *Ecology* **69**, 1768–1775.

Dirr, M.A., Barker, A.V. & Maynard, D.M. (1973) Extraction of nitrate reductase from leaves of Ericaceae. *Phytochemistry* **12**, 1261–1264.

Dixon, A.F.G. (1971a) The role of aphids in wood formation. I. The effect of the sycamore aphid, *Drepanosiphum platanoidis* (Schr.) (Aphididae), on the growth of sycamore, *Acer pseudoplatanus* (L.). *Journal of Applied Ecology* **8**, 165–179.

Dixon, A.F.G. (1971b) The role of aphids in wood formation. II. The effect of the lime aphid, *Eucallipterus tiliae* L. (Aphididae), on the growth of lime *Tilia × vulgaris* Hayne. *Journal of Ecology* **8**, 393–399.

Dixon, R.A. (1986) The phytoalexin response: elicitation, signalling and control of host gene expression. *Biological Reviews* **61**, 239–291.

Doak, D.F. (1992) Life time impacts of herbivory for a perennial plant. *Ecology* **73**, 2086–2099.

Dobkin, D.S. (1984) Flowering patterns of long-lived *Heliconia* inflorescences: implications for visiting and resident nectarivores. *Oecologia* **64**, 245–254.

Dobson, A. & Crawley, W. (1994) Pathogens and the structure of plant communities. *Trends in Ecology and Evolution* **9**, 393–398.

Docters van Leeuwen, W.M. (1929) Krakatau's new flora. *Proceedings of the Fourth Pacific Science Congress (Batavia) Part 2*, pp. 56–71g.

Dodd, M., Silvertown, J., McConway, K., Potts, J. & Crawley, M. (1995) Community stability – a 60-year record of trends and outbreaks of species in the Park Grass Experiment. *Journal of Ecology* **83**, 277–285.

Dohmen, G.P., McNeill, S. & Bell, J.N.B. (1984) Air pollution increases *Aphis fabae* pest potential. *Nature* **307**, 52–53.

Dominguez, C.A. (1995) Genetic conflicts-of-interest in plants. *Trends in Ecology and Evolution* **10**, 412–416.

Dong, M. (1995) Morphological responses to local light conditions in clonal herbs from contrasting habitats, and their modification due to physiological integration. *Oecologia* **101**, 282–288.

Doussard, D.E. & Denno, R.F. (1994) Host range of generalist caterpillars: trenching permits feeding on plants with secretory canals. *Ecology* **75**, 69–78.

Doyle, C.J., Cousens, R. & Moss, C.R. (1986) A model of the economics of controlling *Alopecurus myosuroides* Huds. *Crop Protection* **5**, 143–150.

Dressler, R.L. (1981) *The Orchids: Natural History and Classification*. Harvard University Press, Cambridge, Massachusetts.

Drew, M.C. (1975) Comparison of the effects of a localized supply of phosphate, nitrate, ammonium and potassium on the growth of the seminal root system, and the shoot, in barley. *New Phytologist* **75**, 479–490.

Drury, W.H. & Nisbet, I.C. (1973) Succession. *Journal of the Arnold Aboretum* **54**, 331–368.

Dudley, S.A. & Schmitt, J. (1995) Genetic differentiation in morphological responses to simulated foliage shade between populations of *Impatiens capensis* from open and woodland sites. *Functional Ecology* **9**, 655–666.

Dudt, J.F. & Shure, D.J. (1994) The influence of light and nutrients on foliar phenolics and insect herbivory. *Ecology* **75**, 86–98.

Duffey, S.S. & Felton, G.W. (1989) Plant enzymes in resistance to insects. In *Biocatalysis in Agricultural Biotechnology* (Eds J.R. Whitaker & P.E. Sonnet), pp. 289–313. American Chemical Society, Washington DC.

Duggan, A.E. (1985) Pre-dispersal seed predation by *Anthocharis cardamines* (Pieridae) in the population dynamics of the perennial *Cardamine pratensis* (Brassicaceae). *Oikos* **44**, 99–106.

Dunham, R.J. & Nye, P.H. (1974) The influence of soil water content on the uptake of ions by roots. II. Chloride uptake and concentration gradients in soil. *Journal of Applied Ecology* **11**, 581–595.

Durrett, R. & Levin, S.A. (1994a) The importance of being discrete (and spatial). *Theoretical Population Biology* **46**, 363–394.

Durrett, R. & Levin, S.A. (1994b) Stochastic spatial models: a user's guide to ecological applications. *Philosophical Transactions of the Royal Society of London Series B* **343**, 329–350.

Dussourd, D.E. & Eisner, T. (1987) Vein-cutting behaviour: insect counterplay to the later defence of plants. *Science* **237**, 898–901.

Dyer, M.I. (1975) The effects of redwing blackbirds (*Agelaius phoeniceus* L.) on biomass production of corn grains (*Zea mays* L.). *Journal of Applied Ecology* **12**, 719–726.

Eastop, V.F. (1973) Deductions from the present day host plants of aphids and related insects. *Symposium Royal Entomological Society of London* **6**, 157–178.

Ebert, T.A. & McMaster, G.S. (1981) Regular patterns of desert shrubs – a sampling artifact. *Journal of Ecology* **69**, 559–564.

Eckert, C.G. & Barrett, S.C.H. (1994) Tristyly, self-compatibility and floral variation in *Decodon verticillatus* (Lythraceae). *Biological Journal of the Linnean Society* **53**, 1–30.

Eckert, C.G. & Barrett, S.C.H. (1995) Style morph ratios in *Tristylous decodon verticillatus* (Lythraceae) – selection vs. historical contingency. *Ecology* **76**, 1051–1066.

Edees, E.S. & Newton, A. (1988) *Brambles of the British Isles*. Ray Society, London.

Edenius, L. (1993) Browsing by moose on Scots pine in relation to plant resource availability. *Ecology* **74**, 2261–2269.

Edenius, L., Danell, K. & Nyquist, H. (1995) Effects of simulated moose browsing on growth, mortality, and fecundity in Scots pine – relations to plant productivity. *Canadian Journal of Forest Research* **25**, 529–535.

Edmunds, G.F. & Alstad, D.N. (1982) Effects of air pollutants on insect populations. *Annual Review of Entomology* **27**, 369–384.

Edwards, P.B. & Wanjura, W.J. (1989) Eucalypt-feeding insects bite off more than they can chew: sabotage of induced defenses? *Oikos* **54**, 246–248.

Edwards, P.J. & Wratten, S.D. (1983) Wound-induced defenses in plants and their consequences for patterns in insect grazing. *Oecologia* **59**, 88–93.

Edwards, P.J., Wratten, S.D. & Parker, E.A. (1992) The ecological significance of rapid wound-induced changes in plants – insect grazing and plant competition. *Oecologia* **91**, 266–272.

Egler, F.E. (1947) Arid southeast Oahu vegetation, Hawaii. *Ecological Monographs* **17**, 383–435.

Egler, F.E. (1954) Vegetation science concepts. I. Initial floristic composition – a factor in old-field vegetation development. *Vegetatio* **4**, 412–417.

Egley, G.H. & Duke, S.O. (1985) Physiology of weed seed dormancy and germination. In *Weed Physiology, Vol. 1, Reproduction and Ecophysiology* (Ed. S.O. Duke), pp. 27–64. CRC Press, Boca Raton.

Ehleringer, J.R. (1984) Intraspecific competitive effects on water relations, growth and reproduction in *Encelia farinosa*. *Oecologia* **63**, 153–158.

Ehleringer, J.R. & Björkman, O. (1978) Pubescence and leaf spectral characteristics in a desert shrub, *Encelia farinosa*. *Oecologia* **36**, 151–162.

Ehleringer, J.R. & Cooper, T.A. (1988) Correlations between carbon isotope ratio and microhabitat in desert plants. *Oecologia* **76**, 562–566.

Ehleringer, J.R. & Monson, R.K. (1993) Evolutionary and ecological aspects of photosynthetic pathway variation. *Annual Review of Ecology and Systematics* **24**, 411–439.

Ehleringer, J.R. & Mooney, H.A. (1978) Leaf hairs: effects on physiological activity and adaptive value to a desert shrub. *Oecologia* **37**, 183–200.

Ehleringer, J.R., Sage, R.F., Flanagan, L.B. & Pearcy, R.W. (1991) Climate change and the evolution of C_4 photosynthesis. *Trends in Ecology and Evolution* **6**, 95–99.

Ehleringer, J.R., Phillips, S.L. & Comstock, J.P. (1992) Seasonal variation in the carbon isotopic composition of desert plants. *Functional Ecology* **6**, 396–404.

Ehrlén, J. (1995) Demography of the perennial herb *Lathyrus vernus*. 1. Herbivory and individual performance. *Journal of Ecology* **83**, 287–295.

Ehrlén, J. & Eriksson, O. (1995) Pollen limitation and population growth in a herbaceous perennial legume. *Ecology* **76**, 652–656.

Ehrlich, P.R. & Raven, P.H. (1964) Butterflies and plants: a study in coevolution. *Evolution* **18**, 586–608.

Eldredge, N. (1995) *Reinventing Darwin: the Great Evolutionary Debate*. Weidenfeld & Nicolson, London.

Ellenberg, H. (1953) Physiologisches und ökologisches Verhalten derselben Pflanzenarten. *Bericht der Deutschene Botanischen Gesellschaft* **65**, 351–361.

Elliot, J.M. (1977) *Some Methods for the Statistical Analysis of Samples of Benthic Invertebrates*, 2nd edn. Freshwater Biological Association Scientific Publication No. 25.

Elliott, S. & Loudon, A. (1987) Effects of monoterpene odours on food selection by red deer calves. *Journal of Chemical Ecology* **13**, 1343–1350.

Ellis, W.M., Keymer, R.J. & Jones, D.A. (1977) On the polymorphism of cyanogenesis in *Lotus corniculatus* L. VIII. Ecological studies in Anglesey. *Heredity* **39**, 45–65.

Ellison, A.M. & Rainowitz, D. (1989) Effects of plant morphology and emergence time on size hierarchy formation in experimental populations of two varieties of cultivated peas (*Pisum sativum*). *American Journal of Botany* **76**, 427–436.

Ellner, S. (1986) Germination dimorphism and parent offspring conflict in seed germination. *Journal of Theoretical Biology* **123**, 173–185.

Ellner, S. & Shmida, A. (1981) Why are adaptations for long-range seed dispersal rare in desert plants? *Oecologia* **51**, 133–144.

Ellner, S. & Turchin, P. (1995) Chaos in a noisy world: new methods and evidence from time series analysis. *American Naturalist* **145**, 343–375.

Ellstrand, N.C. (1992) Gene flow by pollen – implications for plant conservation genetics. *Oikos* **63**, 77–86.

Ellstrand, N.C. & Ellam, D.R. (1993) Population genetic consequences of small population size – implications for plant conservation. *Annual Review of Ecology and Systematics* **24**, 217–242.

Ellstrand, N.C. & Roose, M.L. (1987) Patterns of genotypic diversity in clonal plant species. *American Journal of Botany* **74**, 123–131.

van Emden, H.F. (1972) Aphids as phytochemists. In *Phytochemical Ecology* (Ed. J.B. Harborne), pp. 25–44. Academic Press, London.

Emmingham, W.H. (1982) Ecological indexes as a means of evaluating climate, species distribution and primary production. In *Analysis of Coniferous Forest Ecosystems in the Western United States* (Ed. R.L. Edmonds). Hutchinson, Ross Publishing Company, Stroudsburg, Pennsylvania.

English-Loeb, G.M. (1990) Plant drought stress and outbreaks of spider mites: a field test. *Ecology* **71**, 1401–1411.

English-Loeb, G.M. & Karban, R. (1992) Consequences of variation in flowering phenology for seed head herbivory and reproductive success in *Erigeron glaucus*. *Oecologia* **89**, 588–595.

Enright, N.J. & Watson, A.D. (1992) Population dynamics of the nikau palm, *Rhopalostylis sapida* (Wendl. et Drude), in a temperate forest remnant near Auckland, New Zealand. *New Zealand Journal of Botany* **30**, 29–43.

Enright, N.J., Franco, M. & Silvertown, J. (1995) Comparing plant life histories using elasticity analysis – the importance of life span and the number of life cycle stages. *Oecologia* **104**, 79–84.

Epling, C., Lewis, H. & Bell, F.M. (1960) The breeding

group and seed storage: a study in population dynamics. *Evolution* **14**, 238–255.

Ericson, L., Elmqvist, T., Jakobsson, K., Danell, K. & Salomonson, A. (1992) Age structure of boreal willows and fluctuations in herbivore populations. *Proceedings of the Royal Society of Edinburgh Series B* **98**, 75–89.

Ericsson, T. (1995) Growth and shoot–root ratio of seedlings in relation to nutrient availability. *Plant and Soil* **169**, 205–214.

Eriksson, O. (1989) Seedling dynamics and life histories in clonal plants. *Oikos* **55**, 231–238.

Eriksson, O. (1993) Dynamics of genets in clonal plants. *Trends in Ecology and Evolution* **8**, 313–316.

Eriksson, O. & Jerling, L. (1990) Hierarchical selection and risk spreading in clonal plants. In *Clonal Growth in Plants: Regulation and Function* (Eds J. van Groenendael & H. de Kroon), pp. 79–94. SPB Academic Publishing, The Hague.

Ernst, W.H.O., Tonneijck, N.E.C. & Pasman, R.J.M. (1985) Ecotypic response of *Silene cucubalus* to air pollutants (SO₂, O₃) *Journal of Plant Physiology* **118**, 439–450.

Erwin, D.H., Valentine, J.W. & Sepkoski, J.J. (1987) A comparative study of diversification rates: the early Paleozoic versus the Mesozoic. *Evolution* **41**, 1177–1186.

Esau, K. (1965) *Plant Anatomy*, 2nd ed. John Wiley & Sons, New York.

Escarre, J. & Thompson, J.D. (1991) The effects of successional habitat variation and time of flowering on seed production in *Rumex acetosella*. *Journal of Ecology* **79**, 1099–1112.

Estiarte, M., Filella, I., Serra, J. & Penuelas, J. (1994) Effects of nutrient and water stress on leaf phenolic content of peppers and susceptibility to generalist herbivore *Helicoverpa armigera* (Hubner). *Oecologia* **99**, 387–391.

Etherington, J.R. (1982) *Environment and Plant Ecology*, 2nd edn. John Wiley & Sons, Chichester.

Evans, A.S. & Cabin, R.J. (1995) Can dormancy affect the evolution of post germination traits – the case of *Lesquerella fendleri*. *Ecology* **76**, 344–356.

Evans G.C. (1972) *The Quantitative Analysis of Plant Growth*. Blackwell Scientific Publications, Oxford.

Evans, J.R. (1983) *Photosynthesis and nitrogen partitioning in leaves of* T. aestivum *and related species*. PhD thesis, Australian National University, Canberra.

Evans, P.A. & Ashmore, M.R. (1992) The effects of ambient air on a semi-natural grassland community. *Agriculture, Ecosystems and Environment* **38**, 91–97.

Evelyn, J. (1661) *Fumifugium – or the Inconvenience of the Aere and Smoake of London Dissipated*. Second Reprint (1972). National Society for Clean Air, Brighton.

Ewens, W.J., Brockwell, P.J., Gani, J.M. & Resnick, S.I. (1987) Minimum viable population size in the presence of catastrophes. In *Viable Populations for Conservation* (Ed. M.E. Soulé), pp. 59–68. Cambridge University Press, Cambridge.

Facelli, J.M. (1994) Multiple indirect effects of plant litter affect the establishment of woody seedlings in old fields. *Ecology* **75**, 1727–1735.

Faegri, K. & van der Pijl, L. (1979) *The Principles of Pollination Ecology*, 3rd edn. Pergamon Press, New York.

Faeth, S.H. (1985) Host leaf selection by leaf miners: interactions among three trophic levels. *Ecology* **66**, 870–875.

Faeth, S.H. (1986) Indirect interactions between temporarily separated herbivores mediated by the host plant. *Ecology* **67**, 474–494.

Faeth, S.H. (1987) Community structure and folivorous insect outbreaks: the role of vertical and horizontal interactions. In *Insect Outbreaks: Ecological and Evolutionary Perspectives* (Eds P. Barbosa & J.C. Schultz), pp. 135–172. Academic Press, New York.

Faeth, S.H. (1988) Plant-mediated interactions between seasonal herbivores: enough for evolution or coevolution? In *Chemical Mediation of Coevolution* (Ed. K.C. Spencer), pp. 391–414. American Institute of Biological Sciences, Washington, D.C.

Faeth, S.H. (1992) Interspecific and intraspecific interactions via plant responses to folivory – an experimental field test. *Ecology* **73**, 1802–1813.

Fagouri, M., Gay, C.W. & Banner, R.L. (1995) Factors affecting survival of perennial wheatgrasses seeded during drought in Morocco. *Arid Soil Research and Rehabilitation* **9**, 51–62.

Fajer, E.D., Bowers, M.D. & Bazzaz, F.A. (1991) The effect of enriched CO₂ atmospheres on the buckeye butterfly, *Junonia coenia*. *Ecology* **72**, 751–754.

Falconer, D.S. (1989) *Introduction to Quantitative Genetics*, 3rd edn. Longman, London.

Falkengren-Grerup, U. (1989) Soil acidification and its impact on ground vegetation. *Ambio* **18**, 179–183.

Falkowski, P.G. (1995) Ironing out what controls primary production in the nutrient-rich waters of the open ocean. *Global Change Biology* **1**, 161–164.

Farmer, A.M., Bates, J.W. & Bell, J.N.B. (1992) Ecophysiological effects of acid rain on bryophytes and lichens. In *Bryophytes and Lichens in a Changing Environment* (Eds J.W. Bates & A.M. Farmer), pp. 284–313. Clarendon Press, Oxford.

Farnsworth, K.D. & Niklas, K.J. (1995) Theories of optimization, form and function in branching architecture in plants. *Functional Ecology* **9**, 355–363.

Farquhar, G.D. & Sharkey, T.D. (1982) Stomatal conductance and photosynthesis. *Annual Review of Plant Physiology* **33**, 317–345.

Farquhar, G.D., Ehleringer, J.R. & Hubick, K.T. (1989) Carbon isotope discrimination and photosynthesis. *Annual Review of Plant Physiology and Molecular Biology* **40**, 503–537.

Farris, M.A. & Lechowicz, M.J. (1990) Functional interactions among traits that determine reproductive success in a native annual plant. *Ecology* **71**, 548–557.

Fastie, C.L. (1995) Causes and ecosystem consequences of multiple pathways of primary succession at Glacier Bay, Alaska. *Ecology* **76**, 1899–1916.

Feeny, P. (1969) Inhibitory effects of oak leaf tannins on the hydrolysis of proteins by trypsin. *Phytochemistry* 8, 2116–2126.

Feeny, P. (1970) Seasonal changes in oak leaf tannins and nutrients as a cause of spring feeding by winter moth caterpillars. *Ecology* 51, 565–581.

Feeny, P. (1976) Plant apparency and chemical defense. In *Biochemical Interactions Between Plants and Insects* (Eds J.W. Wallace & R.L. Mansell), pp. 1–40. Plenum Press, New York.

Feeny, P. (1990) Theories of Plant-chemical defence: a brief historical summary. *Symposia Biologica Hungarica* 89, 163–175.

Feeny, P. (1992) The evolution of chemical ecology: contributions from the study of herbivorous insects. In *Herbivores: their Interaction with Secondary Plant Metabolites*, 2nd edn (Eds G.A. Rosenthal & M.R. Berenbaum), pp. 1–44. Academic Press, San Diego.

Feinsinger, P. (1983) Coevolution and pollination. In *Coevolution* (Eds D.J. Futuyma & M. Slatkin), pp. 282–310. Sinauer, Sunderland, Massachusetts.

Felsenstein, J. (1974) The evolutionary advantage of recombination. *Genetics* 78, 737–756.

Felton, G.W. & Duffey S.S. (1991a) Protective action of midgut catalase in lepidopteran larvae against oxidative plant defenses. *Journal of Chemical Ecology* 17, 1715–1732.

Felton, G.W. & Duffey, S.S. (1991b) Reassessment of the role of gut alkalinity and detergency in insect herbivory. *Journal of Chemical Ecology* 17, 1821–1836.

Fenner, M. (1978) Susceptibility to shade in seedlings of colonising and closed turf species. *New Phytologist* 81, 739–744.

Fenner, M. (1983) Relationships between seed weight, ash content and seedling growth in twenty-four species of Compositae. *New Phytologist* 95, 697–706.

Fenner, M. (1985) *Seed Ecology*. Chapman & Hall, London.

Fenner, M. (1991a) The effects of the parental environment on seed germinability. *Seed Science Research* 1, 75–84.

Fenner, M. (1991b) Irregular seed crops in forest trees. *Quarterly Journal of Forestry* 85, 166–172.

Fenner, M. (Ed.) (1992) *Seeds: the Ecology of Regeneration in Plant Communities*. CAB International, Wallingford.

Fernandes, G.W. & Price, P.W. (1992) The adaptive significance of insect gall distribution – survivorship of species in xeric and mesic habitats. *Oecologia* 90, 14–20.

Fialho, R.F. (1990) Seed dispersal by a lizard and a treefrog – effect of dispersal site on seed survivorship. *Biotropica* 22, 423–424.

Field, C. & Mooney, H.A. (1986) The photosynthesis-nitrogen relationship in wild plants. In *On the Economy of Plant Form and Function* (Ed. T.J. Givnish), pp. 25–55. Cambridge University Press, Cambridge.

Finch, C.E. & Rose, M.R. (1995) Hormones and the physiological architecture of life history evolution. *Quarterly Review of Biology* 70, 1–52.

Fineblum, W.L. & Rausher, M.D. (1995) Tradeoff between resistance and tolerance to herbivore damage in a morning glory. *Nature* 377, 517–519.

Finegan, B. (1984) Forest succession. *Nature* 312, 109–114.

Firbank, L.G. & Mortimer, A.M. (1985) Weed–crop interaction and the optimal timing of weed control. Proceedings of the British Crop Protection Conference – Weeds, pp. 879–887. BCPC, Farnham.

Firbank, L.G. & Watkinson, A.R. (1985) On the analysis of competition within two-species mixtures of plants. *Journal of Applied Ecology* 22, 503–517.

Firbank, L.G. & Watkinson, A.R. (1986) Modelling the population dynamics of an arable weed and its effects upon crop yield. *Journal of Applied Ecology* 23, 147–159.

Firbank, L.G. & Watkinson, A.R. (1990) On the effects of competition: from monocultures to mixtures. In *Perspectives on Plant Competition* (Eds J.B. Grace & D. Tilman), pp. 165–192. Academic Press, San Diego.

Firbank, L.G., Mortimer, A.M. & Putwain, P.D. (1985) *Bromus sterilis* in winter wheat: a test of a predictive population model. *Aspects of Applied Biology* 9, 59–66.

Firn, R.D. & Jones, C.G. (1995) Plants may talk but can they hear? *Trends in Ecology and Evolution* 10, 9.

Fischer, K.E. & Chapman, C.A. (1993) Frugivores and fruit syndromes – differences in patterns at the genus and species level. *Oikos* 66, 472–482.

Fisher, B.L., Howe, H.F. & Wright, S.J. (1991) Survival and growth of *Virola surinamensis* yearlings: water augmentation in gap and understory. *Oecologia* 86, 292–297.

Fisher, R.A. (1930) *The Genetical Theory of Natural Selection*. Clarendon Press, Oxford.

Fitter, A.H. (1977) Influence of mycorrhizal infection on competition for phosphorus and potassium by two grasses. *New Phytologist* 79, 119–125.

Fitter, A.H. (1985) Functional significance of root morphology and root system architecture. In *Ecological Interactions in Soil* (Eds A.H. Fitter, D. Atkinson D.J. Read & M.B. Usher), pp. 87–106. Blackwell Scientific Publication, Oxford.

Fitter, A.H. (1990) The role and ecological significance of vesicular-arbuscular mycorrhizas in temperate ecosystems. *Agriculture, Ecosystems and Environment* 29, 137–151.

Fitter, A.H. (1991a) Costs and benefits of mycorrhizal infection. *Experientia* 47, 350–355.

Fitter, A.H. (1994) Architecture and biomass allocation as components of the plastic response of root systems to soil heterogeneity. In *Exploitation of Environmental Heterogeneity by Plants* (Eds M.M. Caldwell & R.W. Pearcy), pp. 305–323. Academic Press, New York.

Fitter, A.H. (1995) The characteristics of root systems. In *Plant Roots: the Hidden Half* (Eds Y. Waisel, A. Eshel & U. Kafkafi), pp. 3–24. Marcel Dekker, New York.

Fitter, A.H. & Garbaye, J. (1994) Interactions between

mycorrhizal fungi and other soil organisms. *Plant and Soil* **159**, 123–132.

Fitter, A.H. & Hay, R.K.M. (1981) *Environmental Physiology of Plants*. Academic Press, London.

Fitter, A.H. & Hay, R.K.M. (1987) *Environmental Physiology of Plants*, 2nd edn. Academic Press, London.

Fitter, A.H. & Peat, H.J. (1994) The Ecological Flora Database. *Journal of Ecology* **82**, 415–425.

Fitter, A.H., Stickland, T.R., Harvey, M.L. & Wilson, G.W. (1991) Architectural analysis of plant root systems. I. Architectural correlates of exploitation efficiency. *New Phytologist* **118**, 375–382.

Flach, B.M.T., Eller, B.M. & Egli, A. (1995) Transpiration and water uptake of *Senecio medley woodii* and *Aloe jucunda* under changing environmental conditions – measurements with a potometric water budget meter. *Journal of Experimental Botany* **46**, 1615–1624.

Fleming, T.H. (1988) *The Short-Tailed Fruit Bat: a Study in Plant–Animal Interactions*. University of Chicago Press, Chicago.

Fleming, T.H. & Estrada, A. (Eds) (1993) *Frugivory and seed dispersal: Ecological and Evolutionary Aspects*. Kluwer Academic, Dordrecht.

Flint, L.H. & McAllister, E.D. (1937) Wavelengths of radiation in the visible spectrum promoting the germination of light sensitive lettuce seeds. *Smithsonian Miscellaneous Collections* **96**, 1–8.

Flor, H.H. (1956) The complementary genic systems in flax and flax rust. *Advances in Genetics* **8**, 29–54.

Flower-Ellis, J.G.K. & Persson, H. (1980) Investigation of structural properties and dynamics of Scots pine stands. *Ecological Bulletin (Stockholm)* **32**, 125–138.

Fluckiger, W., Oertli, J.J. & Baltensweiler, W. (1978) Observations on aphid infestation on hawthorn in the vicinity of a motorway. *Naturwissenschaften* **65**, 654–655.

Fogden, M.P.L. (1972) The seasonality and population dynamics of equatorial forest birds in Sarawak. *Ibis* **114**, 307–343.

Fogel, R. (1980) Mycorrhizae and nutrient cycling in natural forest ecosystems. *New Phytologist* **86**, 199–212.

Foggo, A., Speight, M.R. & Gregoire, J.C. (1994) Root disturbance of common ash, *Fraxinus excelsior* (Oleaceae), leads to reduced foliar toughness and increased feeding by a folivorous weevil, *Stereonychus fraxini* (Coleoptera, Curculionidae). *Ecological Entomology* **19**, 344–348.

Foley, W.J., Lassak, E.V. & Brophy, J. (1987) Digestion and absorption of *Eucalyptus* essential oils in greater glider and brushtail possum. *Journal of Chemical Ecology* **13**, 2115–2130.

Fonteyn, P.J. & Mahall, B.E. (1978) Competition among desert perennials. *Nature* **275**, 544–545.

Fonteyn, P.J. & Mahall, B.E. (1981) An experimental analysis of structure in a desert plant community. *Journal of Ecology* **69**, 883–896.

Forbes, S.A. (1887) The lake as a microcosm. *Bulletin of the Peoria Scientific Association* 77–87. Reprinted in *Foundations of Ecology* (Eds L.R. Real & J.H. Brown), 1991, pp. 14–27. University of Chicago Press, Chicago.

Ford, E.D. (1975) Competition and stand structure in some even-aged plant monocultures. *Journal of Ecology* **63**, 311–333.

Forget, P.M. (1992) Regeneration ecology of *Eperua grandiflora* (Caesalpiniaceae), a large-seeded tree in French Guiana. *Biotropica* **24**, 146–156.

Foster, R.B. (1982) The seasonal rhythm of fruit fall on Barro Colorado Island. In *The Ecology of a Tropical Forest: Seasonal Rhythms and Long-term Changes* (Eds E.G. Leigh, A.S. Rand & D.S. Windsor), pp. 151–172. Smithsonian Press, Washington.

Foster, S.A. (1986) On the adaptive value of large seeds for tropical moist forest trees: a review and synthesis. *Botanical Review* **52**, 260–299.

Fowler, N. (1981) Competition and coexistence in a North Carolina grassland. II. Effects of the experimental removal of species. *Journal of Ecology* **69**, 843–854.

Fowler, N.L. (1984) The role of germination date, spatial arrangement and neighbourhood effects in competitive interactions in *Linum*. *Journal of Ecology* **72**, 307–318.

Fowler, N.L. (1986) Density-dependent population regulation in a Texas grassland. *Ecology* **67**, 545–554.

Fowler, S.V. & Lawton, J.H. (1985) Rapidly induced defenses and talking trees: the Devil's advocate position. *American Naturalist* **126**, 181–195.

Fowler, S.V. & MacGarvin, M. (1986) The effects of leaf damage on the performance of insect herbivores on birch, *Betula pubescens*. *Journal of Animal Ecology* **55**, 565–574.

Fox, L.R. (1981) Defense and dynamics in plant–herbivore systems. *American Zoologist* **21**, 853–864.

Fox, L.R. (1988) Diffuse coevolution within complex communities. *Ecology* **69**, 906–907.

Fox, L.R. & Macaulay, B.J. (1977) Insect grazing on *Eucalyptus* in response to variation in leaf tannin and nitrogen. *Oecologia* **29**, 145–162.

Foy, C.L., Hurtt, W. & Hale, M.G. (1971) Root exudation and plant growth regulators. In *Biochemical Interactions Among Plants* (Eds US National Committee for IBP), pp. 75–85. National Academy of Sciences, Washington, DC.

Fraenkel, G.S. (1959) The raison d'être of secondary plant substances. *Science* **129**, 1466–1470.

Frank, S.A. (1990) Sex allocation theory of birds and mammals. *Annual Review of Ecology and Systematics* **21**, 13–55.

Frank, S.A. (1991) Ecological and genetic models of host pathogen coevolution. *Heredity* **67**, 73–83.

Frankel, O.H. (1974) Genetic conservation: our evolutionary responsibility. *Genetics* **78**, 53–65.

Frankel, O.H. & Soule, M.A. (1981) *Conservation and Evolution*. Cambridge University Press, Cambridge.

Frankel, O.H., Brown, A.H.D & Burdon, J.J. (1995) *The Conservation of Plant Biodiversity*. Cambridge University Press, Cambridge.

Frankel, R. & Galun, E. (1977) *Pollination Mechanisms, Reproduction and Plant Breeding.* Springer-Verlag, Berlin.

Frankie, G.W. & Haber, W.A. (1983) Why bees move among mass-flowering neotropical trees. In *Handbook of Experimental Pollination Biology* (Eds C.E. Jones & R.J. Little), pp. 360–372. Van Nostrand Reinhold New York.

Franklin, I.A. (1980) Evolutionary change in small populations. In *Conservation Biology: an Evolutionary-Ecological Perspective* (Eds M.E. Soulé & B.A. Wilcox), pp. 135–150. Sinauer, Sunderland, Massachusetts.

Freedman, B. (1989) *Environmental Ecology.* Academic Press, London.

Freedman, B. & Hutchinson, T.C. (1980) Long-term effects of smelter pollution at Sudbury Ontario on forest community composition *Canadian Journal of Botany* **58**, 2123–2140.

Freeman, D.C., Klikoff, L.G. & Harper, K.T. (1976) Differential resource utilisation by the sexes of dioecious plants. *Science* **193**, 597–599.

Freeman, D.C., Harper, K.T. & Charnov, E.L. (1980) Sex change in plants: old and new observations and new hypotheses. *Oecologia* **47**, 222–232.

French, K. & Westoby, M. (1992) Removal of vertebrate-dispersed fruits in vegetation on fertile and infertile soils. *Oecologia* **91**, 447–454.

Friedli, H., Lötscher, H., Oeschger, H., Siegenthaler, U. & Stauffer, B. (1986) Ice core record of the $^{13}C/^{12}C$ ratio of atmospheric CO_2 in the past two centuries. *Nature* **324**, 237–238.

Frischknecht, P.M., Dufek, J.V. & Baumann, T.W. (1986) Purine alkaloid formation in buds and developing leaflets of *Coffea arabica*: expression of an optimal defence strategy. *Phytochemistry* **25**, 613–616.

Fritz, R.S. & Simms, E.L. (Eds) (1992) *Plant Resistance to Herbivores and Pathogens: Ecology, Evolution and Genetics.* University of Chicago Press, Chicago.

Fritz, R.S., Gaud, W.S., Sacchi, C.F. & Price, P.W. (1987) Patterns of intra- and interspecific association of gall-forming sawflies in relation to shot size on their willow host plant. *Oecologia* **73**, 159–169.

Fuller, W., Hance, C.E. & Hutchings, M.J. (1983) Within-season fluctuations in mean fruit weight in *Leontodon hispidus* L. *Annals of Botany* **51**, 545–549.

Futuyma, D.J. (1983a) Evolutionary interactions among herbivorous insects and plants. In *Coevolution* (Eds D.J. Futuyma & M. Slatkin), pp. 207–231. Sinauer, Sunderland, Massachusetts.

Futuyma, D.J. (1983b) Selective factors in the evolution of host choice by phytophagous insects. In *Herbivorous Insects: Host Seeking Behaviour and Mechanisms* (Ed. S. Ahmad), pp. 227–244. Academic Press, New York.

Futuyma, D.J. & Mayer, G.C. (1980) Non-allopatric speciation in animals. *Systematic Zoology* **29**, 254–271.

Futuyma, D.J. & Wasserman, S.S. (1980) Resource concentration and herbivory in oak forests. *Science* **210**, 920–922.

Gadd, G.M. (1993) Interactions of fungi with toxic metals. *New Phytologist* **124**, 25–60.

Gadgil, M. (1971) Dispersal: population consequences and evolution. *Ecology* **52**, 253–261.

Gadgil, M.D. & Solbrig, O.T. (1972) The concept of *r*- and *K*-selection: evidence from wild flowers and some theoretical considerations. *American Naturalist* **106**, 14–31.

Galen, C., Zimmer, K.A. & Newport, M.E. (1987) Pollination in floral scent morphs of *Polemonium viscosum* – a mechanism for disruptive selection on flower size. *Evolution* **41**, 599–606.

Ganders, F.R. (1979) The biology of heterostyly. *New Zealand Journal of Botany* **17**, 607–636.

Gange, A.C. & Brown, V.K. (1989) Insect herbivory affects size variation in plant populations. *Oikos* **56**, 351–356.

Gange, A.C., Brown, V.K. & Farmer, L.M. (1990) A test of mycorrhizal benefit in an early successional plant community. *New Phytologist* **115**, 85–91.

Gange, A.C., Brown, V.K. & Sinclair, G.S. (1993) Vesicular-arbuscular mycorrhizal fungi: a determinant of plant community structure in early succession. *Functional Ecology* **7**, 616–622.

Garwood, N.C. (1983) Seed germination in a seasonal tropical forest in Panama: a community study. *Ecological Monographs* **53**, 159–181.

Garwood, N.C. (1989) Tropical soil seed banks: a review. In *Ecology of Soil Seed Banks* (Eds M.A. Leck, V.T. Parker & R.L. Simpson), pp. 149–209. Academic Press, London.

Garwood, N.C., Janos, D.P. & Brokaw, N. (1979) Earthquake-caused landslides: a major disturbance to tropical forests. *Science* **205**, 997–999.

Gasser, U.G., Dahlgren, R.A., Ludwig, C. & Lauchli, A.E. (1995) Release kinetics of surface associated Mn and Ni in serpentinitic soils – pH effects. *Soil Science* **160**, 273–280.

Gatsuk, E., Smirnova, O.V., Vorontzova, L.I., Zaugolnova, L.B. & Zhukova, L.A. (1980) Age-states of plants of various growth forms: a review. *Journal of Ecology* **68**, 675–696.

Gauch, H.G. (1982) *Multivariate Analysis in Community Ecology.* Cambridge University Press, Cambridge.

Gauthier, G., Hughes, R.J., Reed, A., Beaulieu, J. & Rochefort, L. (1995) Effect of grazing by greater snow geese on the production of graminoids at an Arctic site (Bylot Island, NWT, Canada). *Journal of Ecology* **83**, 653–664.

Gautier-Hion, A. & Maisels, F. (1994) Mutualism between a leguminous tree and large African monkeys as pollinators. *Behavioral Ecology and Sociobiology* **34**, 203–210.

Gautier-Hion, A., Duplantier, J.M., Quris, R. *et al.* (1985) Fruit characters as a basis of fruit choice and seed dispersal in a tropical forest vertebrate community. *Oecologia* **65**, 324–337.

Gebauer, R.L.E., Reynolds, J.F. & Tenhunen, J.D. (1995) Growth and allocation of the arctic sedges *Eriophorum angustifolium* and *Eriophorum vaginatum* –

effects of variable soil oxygen and nutrient availability. *Oecologia* **104**, 330–339.

Geber, M.A. (1989) Interplay of morphology and development on size inequality: a *Polygonum* greenhouse study. *Ecological Monographs* **59**, 267–288.

Gedge, K.E. & Maun, M.A. (1994) Compensatory response of 2 dune annuals to simulated browsing and fruit predation. *Journal of Vegetation Science* **5**, 99–108.

Geritz, S.A.H. (1995) Evolutionarily stable seed polymorphism and small-scale spatial variation in seedling density. *American Naturalist* **146**, 685–707.

Geritz, S.A.H., de Jong, T.J. & Klinkhamer, P.G.L. (1984) The efficacy of dispersal in relation to safe-site area and seed production. *Oecologia* **62**, 219–221.

Gershenzon, J. (1984) Changes in the level of plant secondary metabolites under water and nutrient stress. In *Phytochemical Adaptations to Stress* (Eds B.N. Timmerman, C. Steerlink & E.A. Loewus), pp. 273–321. Plenum Press, New York.

Gershenzon, J. (1994) The cost of plant chemical defence against herbivory: a biochemical perspective. In *Insect–Plant Interactions* (Ed. E.A. Bernays), Vol. 5, pp. 105–173. CRC Press, Boca Raton.

Gershenzon, J. & Croteau, R. (1992) Terpenoids. In *Herbivores: their Interaction with Secondary Plant Metabolites*, 2nd edn (Eds G.A. Rosenthal & M.R. Berenbaum), pp. 165–209. Academic Press, San Diego.

Gholz, H.L. (1982) Environmental limits on aboveground net primary productivity, leaf area, and biomass in vegetation zones of the Pacific Northwest. *Ecology* **63**, 469–481.

Gibbens, R.P., Havstad, K.M., Billheimer, D.D. & Herbel, C.H. (1993) Creosote bush vegetation after 50 years of lagomorph exclusion. *Oecologia* **94**, 210–217.

Giblin-Davis, R.M., Busey, P. & Center, B.J. (1992) Dynamics of *Belonolaimus longicaudatus* parasitism on a susceptible St. Augustine grass host. *Journal of Nematology* **24**, 432–437.

Gill, D.E. (1986) Individual plants as genetic mosaics: ecological organisms versus evolutionary individuals. In *Plant Ecology* (Ed. M.J. Crawley), pp. 321–343. Blackwell Scientific Publications, Oxford.

Gill, D.E. (1989) Fruiting failure, pollinator inefficiency, and speciation in orchids. In *Speciation and its Consequences* (Eds D. Otte & J.A. Endler), pp. 458–481. Sinauer, Sunderland, Massachusetts.

Gill, R.M.A. (1992) A review of damage by mammals in north temperate forests. 2. Small mammals. *Forestry* **65**, 281–308.

Gillman, M.P. & Crawley, M.J. (1990) The cost of sexual reproduction in ragwort (*Senecio jacobaea* L.). *Functional Ecology* **4**, 585–589.

Gillman, M., Bullock, J.M., Silvertown, J. & Clear Hill, B. (1993) A density-dependent model of *Cirsium vulgare* population dynamics using field-estimated parameter values. *Oecologia* **96**, 282–289.

Gilpin, M.E. (1991) The genetic effective size of a metapopulation. *Biological Journal of the Linnean Society* **42**, 165–175.

Gilpin, M. & Hanski, I. (Eds) (1991) *Metapopulation Dynamics: Empirical and Theoretical Investigations*. Academic Press, San Diego.

Givnish, T.J. (1978) On the adaptive significance of compound leaves, with particular reference to tropical trees. In *Tropical Trees as Living Systems* (Eds P.B. Tomlinson & H. Zimmermann), pp. 351–380. Cambridge University Press, Cambridge.

Givnish, T.J. (1979) On the adaptive significance of leaf form. In *Topics in Plant Population Biology* (Eds O.T. Solbrig, S. Jain, G.B. Johnson & P.H. Raven), pp. 375–407. Macmillan, London.

Givnish, T.J. (1980) Ecological constraints on the evolution of breeding systems in seed plants: dioecy and dispersal in gymnosperms. *Evolution* **34**, 959–972.

Givnish, T.J. (1982) On the adaptive significance of leaf height in forest herbs. *American Naturalist* **120**, 353–381.

Givnish, T.J. (1987) Comparative studies of leaf form: assessing the relative roles of selective pressures and phylogenetic constraints. *New Phytologist* **106** (suppl.), 131–160.

Gleason, H.A. (1917) The structure and development of the plant association. *Bulletin of the Torrey Botanical Club* **44**, 463–481.

Gleason, H.A. (1926) The individualistic concept of plant association. *Bulletin of the Torrey Botanical Club* **53**, 7–26.

Gleason, H.A. (1927) Further views on the succession concept. *Ecology* **8**, 299–326.

Gleaves, J.T. (1973) Gene flow mediated by wind borne pollen. *Heredity* **31**, 355–366.

Gleeson, S. & Tilman, D. (1990) Allocation and the transient dynamics of succession on poor soils. *Ecology* **71**, 1144–1155.

Glitzenstein, J.S., Platt, W.J. & Streng, D.R. (1995) Effects of fire regime and habitat on tree dynamics in North Florida longleaf pine savannas. *Ecological Monographs* **65**, 441–476.

Glyphis, J.P. & Puttick, G.M. (1989) Phenolics, nutrition and insect herbivory in some garrigue and maquis plant species. *Oecologia* **78**, 259–263.

Godfray, H.C.J. (1995) Evolutionary theory of parent–offspring conflict. *Nature* **376**, 133–138.

Godwin, H. (1975) *The History of the British Flora. A Factural for Phytogeography*, 2nd edn. Cambridge University Press, Cambridge.

Goldberg, D.E. (1985) Effects of soil pH, competition, and seed predation on the distributions of two tree species. *Ecology* **66**, 503–511.

Goldberg, D.E. (1987) Neighbourhood competition in an old-field plant community. *Ecology* **68**, 1211–1223.

Goldberg, D.E. & Barton, A.M. (1992) Patterns and consequences of interspecific competition in natural communities – a review of field experiments with plants. *American Naturalist* **139**, 771–801.

Goldberg, D.E. & Werner, P.A. (1983) Equivalence of competitors in plant communities: a null hypothesis and a field experimental approach. *American Journal of Botany* **70**, 1098–1104.

Gollan, T., Turner, N.C. & Schulze, E.D. (1985) The responses of stomata and leaf gas exchange to vapour pressure deficits and soil water contents in the sclerophyllous woody species *Nerium oleander*. *Oecologia* **65**, 356–362.

Gomez, J.M. & Zamoram, R. (1994) Top-down effects in a tritrophic system: parasitoids enhance plant fitness. *Ecology* **75**, 1023–1030.

Gonzalez-Andujar, J.L. & Ferandez-Quintanilla, C. (1991) Modelling the population dynamics of *Avena sterilis* under dry-land cereal cropping systems. *Journal of Applied Ecology* **28**, 16–27.

Goodall, D.W. (1952) Quantitative aspects of plant distribution. *Biological Reviews* **27**, 194–245.

Goodall, D.W. (1965) Plotess tests of interspecific association. *Journal of Ecology* **53**, 197–210.

Gordon, A.G. & Gorham, E. (1963) Ecological aspects of air pollution form an iron sintering plant at Wawa, Ontario. *Canadian Journal of Botany* **41**, 1063–1078.

Gorham, E. & Gordon, R.J. (1963) Some effects of smelter pollution on aquatic vegetation near Sudbury, Ontario. *Canadian Journal of Botany* **41**, 371–378.

Gorski, T. (1975) Germination of seeds in the shadow of plants. *Physiologia Plantarum* **34**, 342–346.

Gorski, T., Gorska, K. & Nowicki, J. (1977) Germination of seeds of various herbaceous species under leaf canopy. *Flora Batava* **166**, 249–259.

Gottlieb, L.D. (1977) Genotypic similarity of large and small individuals in a natural population of the annual plant *Stephanomeria exigua* ssp. *coronaria* (Compositae). *Journal of Ecology* **65**, 127–134.

Goudey, J.S., Saini, H.S. & Spencer, M.S. (1988) Role of nitrogen in regulating germination of *Sinapis arvensis* L. (wild mustard). *Plant, Cell and Environment* **11**, 9–12.

Gould, F. (1983) Genetics of plant–herbivore systems: interactions between applied and basic study. In *Variable Plants and Herbivores in Natural and Managed Systems* (Eds R. Denno & B. McClure), pp. 599–653. Academic Press, New York.

Gould, F. (1988a) Evolutionary biology and genetically engineered crops. *BioScience* **38**, 26–33.

Gould, F. (1988b) Genetics of pairwise and multispecies plant–herbivore coevolution. In *Chemical Mediation of Coevolution* (Ed. K.C. Spencer), pp. 13–55. Academic Press, New York.

Goulding, M. (1980) *The Fishes and the Forest: Explorations in Amazonian Natural History*. University of California Press, Berkeley.

Gowing, D.J.G., Jones, H.G. & Davies, W.J. (1993) Xylem transported abscisic acid: the relative importance of its mass and its concentration in the control of stomatal aperture. *Plant, Cell and Environment* **16**, 453–459.

Grace, J. (1977) *Plant Response to Wind*. Academic Press, London.

Grace, J. (1993a) Refilling of embolised xylem. In *Water*

Transport in Plants under Climatic Stress* (Eds M. Borghetti, J. Grace & A. Raschi), pp. 52–62. Cambridge University Press, Cambridge.

Grace, J. (1993b) Consequences of xylem cavitation for plant water deficits. In *Water Deficits, Plant Responses from Cell to Community* (Eds J.A. Smith & H. Griffiths), pp. 109–128. Bios Scientific Publishers, Oxford.

Grace, J.B. & Tilman, D. (Eds) (1990) *Perspectives on Plant Competition*. Academic Press, San Diego.

Grace, J., Ford, E.D. & Jarvis, P.G (Eds) (1981) *Plants and their Atmospheric Environment*. Blackwell Scientific Publications, Oxford.

Gracia, J.A., Reeleder, R.D. & Belair, G. (1991) Interactions between *Phythium tracheiphilum*, *Meloidogyne hapla* and *Pratylenchus penetrans* on lettuce. *Phytoprotection* **72**, 105–114.

Graham, B.F. & Bormann, F.H. (1966) Natural root grafts. *Botanical Review* **32**, 255–292.

Grant, M.C. & Mitton, J.B. (1979) Elevational gradients in adult sex ratios and sexual differentiation in vegetative growth rates of *Populus tremuloides* Michx. *Evolution* **33**, 914–918.

Grant, V. (1958) The regulation of recombination in plants. *Cold Spring Harbor Symposium on Quantitative Biology* **23**, 337–363.

Grant, V. (1981) *Plant Speciation*, 2nd edn. Columbia University Press, New York.

Grant, V. & Grant, K. (1965) *Flower Pollination in the Phlox Family*. Columbia University Press, New York.

Gray, A.J., Crawley, M.J. & Edwards, P.J. (Eds) (1986) *Colonization, Succession and Stability*. Blackwell Scientific Publications, Oxford.

Grayer, R. & Harborne, J.B. (1994) A survey of antifungal compounds from higher plants 1982–1993. *Phytochemistry* **37**, 19–42.

Greathead, D.J. & Greathead, A.H. (1992) Biological control of insect pests by insect parasitoids and predators: the BIOCAT database. *Biocontrol News and Information* **13**, 61–80.

Green, D.M. (1991) Chaos, fractals and non-linear dynamics in evolution and phylogeny. *Trends in Ecology and Evolution* **6**, 333–337.

Greenway, H. & Munns, R. (1980) Mechanisms of salt tolerance in non-halophytes. *Annual Review of Plant Physiology* **31**, 149–190.

Greig-Smith, P. (1983) *Quantitative Plant Ecology*, 3rd edn. Blackwell Scientific Publications, Oxford.

Griffith, G.W. & Hedger, J.N. (1994) The breeding biology of biotypes of the witches broom pathogen of cocoa, *Crinipellis perniciosa*. *Heredity* **72**, 278–289.

Griffiths, B. & Robinson, D. (1992) Root induced nitrogen mineralization – a nitrogen-balance model. *Plant and Soil* **139**, 253–263.

Griffiths, H.M., Sinclair, W.A., Davis, R.E. *et al.* (1994) Characterisation of mycoplasma like organisms from *Fraxinus*, *Syringa*, and associated plants from geographically diverse sites. *Phytopathology* **84**, 119–126.

Grime, J.P. (1965) Comparative experiments as a key to the ecology of flowering plants. *Ecology* **45**, 513–515.

Grime, J.P. (1973) Competitive exclusion in herbaceous vegetation. *Nature* **242**, 344–347.

Grime, J.P. (1974) Vegetation classification by reference to strategies. *Nature* **250**, 26–31.

Grime, J.P. (1979) *Plant Strategies and Vegetation Processes.* John Wiley & Sons, Chichester.

Grime, J.P. (1983) Prediction of weed and crop response to climate based upon measurements of nuclear DNA content. In *Aspects of Applied Biology. 4: Influence of Environmental Factors on Herbicide Performance and Crop and Weed Biology.* National Vegetable Research Station, Wellesbourne.

Grime, J.P. (1988) The C-S-R model of primary plant strategies – origins, implications and tests. In *Plant Evolutionary Biology* (Eds L.D. Gottlieb & S.K. Jain), pp. 371–393. Chapman & Hall, London.

Grime, J.P. (1989) Ecological effects of climate change on plant populations and vegetation composition with particular reference to the British flora. In *Climatic change and plant genetic resources* (Eds Jackson M., Ford-Lloyd, B.V., Parry, M.L.), pp. 40–60. Belhaven Press, London.

Grime, J.P. (1993) A comment on Silvertown, Franco and McConway (1992) *Functional Ecology* **7**, 380.

Grime, J.P. & Callaghan, T.V. (1988) Direct and indirect effects of climatic change on species, ecosystems and processes of conversation and amenity interest. Contract Report to the Department of the Environment, London.

Grime, J.P. & Curtis, A.V. (1976) The interaction of drought and mineral nutrient stress in calcareous grassland. *Journal of Ecology* **64**, 975–988.

Grime, J.P. & Hodgson, J.G. (1969) An investigation of the ecological significance of lime-chlorosis by means of large-scale comparative experiments. In *Ecological Aspects of the Mineral Nutrition of Plants* (Ed. I.H. Rorison), pp. 67–100. Blackwell Scientific Publications, Oxford.

Grime, J.P. & Mowforth, M.A. (1982) Variation in genome size – an ecological interpretation. *Nature* **299**, 151–153.

Grime, J.P., Mason, G., Curtis, A.V. *et al.* (1981) A comparative study of germination characteristics in a local flora. *Journal of Ecology* **69**, 1017–1059.

Grime, J.P., Shacklock, J.M.L. & Band, S.R. (1985) Nuclear DNA contents, shoot phenology and species coexistence in a limestone grassland community. *New Phytologist* **100**, 435–444.

Grime, J.P., Hunt, R. & Kryzanowski, W.J. (1987) Evolutionary physiological ecology of plants. In *Evolutionary Physiological Ecology* (Ed. P. Calow), pp. 105–126. Cambridge University Press, Cambridge.

Grime, J.P., Hodgson, J.G. & Hunt, R. (1988) *Comparative Plant Ecology: a Functional Approach to Common British Species.* Unwin Hyman, London.

Groombridge, B. (1992) *Global Biodiversity: Status of the Earth's Living Resources.* Chapman & Hall, London.

Gross, K.L. (1980) Colonization of *Verbascum thapsus* (Mullein) in an old field in Michigan: the effects of vegetation. *Journal of Ecology* **68**, 919–928.

Gross, K.L. (1981) Predictions of fate from rosette size in four 'biennial' plant species: *Verbascum thapsus, Oenothera biennis, Daucus carota* and *Tragopogon dubius. Oecologia* **48**, 209–213.

Gross, R.S. & Werner, P.A. (1983) Relationship among flowering phenology, insect visitors, and seed-set of individuals: experimental studies on four co-occurring species of goldenrod (*Solidago*: Compositae). *Ecological Monographs* **53**, 95–117.

Grover, J.P. (1988) Dynamics of competition in a variable environment: experiments with two diatom species. *Ecology* **69**, 408–417.

Grover, J.P. (1989) Effects of Si : P supply ratio, supply variability, and selective grazing in the plankton: an experiment with a natural algal and protistan assemblage. *Limnology and Oceanography* **34**, 349–367.

Grover, J.P. (1990) Resource competition in a variable environment: phytoplankon growing according to Monod's model. *American Naturalist* **136**, 771–789.

Grover, J.P. (1991) Non-steady state dynamics of algal population growth: experiments with two chlorophytes. *Journal of Phycology* **27**, 70–79.

Grover, J.P. (1994) Assembly rules for communities of nutrient-limited plants and specialist herbivores. *American Naturalist* **143**, 258–282.

Grubb, P.J. (1977) The maintenance of species-richness in plant communities: the importance of the regeneration niche. *Biological Reviews* **52**, 107–145.

Grubb, P.J. (1986) The ecology of establishment. In *Ecology and Landscape Design* (Eds A.D. Bradshaw, D.A. Goode & E. Thorpe). Blackwell Scientific Publications, Oxford.

Grubb, P.J. (1992) A positive distrust in simplicity – lessons from plant defenses and from competition among plants and among animals. *Journal of Ecology* **80**, 585–610.

Grubb, P.J., Green, H.E. & Merrifield, R.C.J. (1969) The ecology of chalk heath: its relevance to the clacicole-calcifuge and soil acidification problem. *Journal of Ecology* **57**, 175–211.

Grubb, P.J., Kelly, D. & Mitchley, J. (1982) The control of relative abundance in communities of herbaceous plants. In *The Plant Community as a Working Mechanism* (Ed. E.I. Newman), pp. 79–97. Blackwell Scientific Publications, Oxford.

Gulmon, S.L. & Mooney, H.A. (1986) Costs of defense on plant productivity. In *On the Economy of Plant Form and Function* (Ed. T.J. Givnish), pp. 681–698. Cambridge University Press, Cambridge.

Gulmon, S.L., Rundel, P.W., Ehleringer, J.R. & Mooney, H.A. (1979) Spatial relationships and competition in a Chilean desert cactus. *Oecologia* **44**, 40–43.

Gundersen, D.E., Lee, I.M., Rehner, S.A., Davis, R.E. & Kingsbury, D.T. (1994) Phylogeny of mycoplasma like organisms (Phytoplasmas) – a basis for their classification. *Journal of Bacteriology* **176**, 5244–5254.

Gurevitch, J. (1992) A meta-analysis of competition in

field experiments. *American Naturalists* **140**, 539–572.

Gurevitch, J., Wilson, P., Stone, J.L., Teese, P. & Stoutenburgh, R.J. (1990) Competition among old-field perennials at different levels of soil fertility and available space. *Journal of Ecology* **78**, 727–744.

Gutterman, Y. (1973) Differences in progeny due to day length and hormone treatments of the mother plant. In *Seed Ecology* (Ed. W. Heydecker), pp. 59–80. Butterworth, London.

Gutterman, Y. (1974) The influence of the photoperiodic regime and red–far red light treatments of *Portulaca oleracea* L. plants on the germinability of their seeds. *Oecologia* **17**, 27–38.

Gutterman, Y. (1977) Influence of environmental conditions and hormonal treatments of the mother plants during seed maturation, on the germination of their seed. In *Advances in Plant Reproductive Physiology* (Ed. C.P. Malik), pp. 288–294. Kalyani Publishers, New Dehli.

Gutterman, Y. (1992) Maternal effects on seeds during development. In *Seeds: the Ecology of Regeneration in Plant Communities* (Ed. M. Fenner), pp. 27–59. CAB International, Wallingford.

Guy, R.D., Reid, D.M. & Krouse, H.R. (1980) Shifts in carbon isotope ratios of two C_3 halophytes under natural and artificial conditions. *Oecologia* **44**, 241–247.

Habte, M. & Soedarjo, M. (1995) Limitation of vesicular arbuscular mycorrhizal activity in *Leucaena leucocephala* by Ca insufficiency in an acid Mn rich oxisol. *Mycorrhiza* **5**, 387–394.

Hacker, S.D. & Bertness, M.D. (1995) Morphological and physiological consequences of a positive plant interaction. *Ecology* **76**, 2165–2175.

Hackett, C. (1965) Ecological aspects of the nutrition of *Deschampsia flexuosa* (L.) Trin. II. The effects of Al, Ca, Fe, K, Mn, P and pH on the growth of seedlings and established plants. *Journal of Ecology* **53**, 315–333.

Hadas, A. (1982) Seed–soil contact and germination. In *The Physiology and Biochemistry of Seed Development, Dormancy and Germination* (Ed. A.A. Khan), pp. 507–527. Elsevier Biomedical Press, Amsterdam.

Häggstrom, H. & Larsson, S. (1995) Slow larval growth on a suboptimal willow results in high predation mortality in the leaf beetle *Galerucella lineala*. *Oecologia* **104**, 308–315.

Hahlbrock, K., Lamb, C.J., Purwin, C., Ebel, J., Fautz, E. & Schäfer, E. (1981) Rapid response of suspension-cultured parsley cells to the elicitor from *Phytophthora megasperma* var. *soje*. *Plant Physiology* **67**, 768–773.

Haig, D. & Westoby, M. (1988) Inclusive fitness, seed resources, and maternal care. In *Plant Reproductive Ecology: Patterns and Strategies* (Eds J. Lovett Doust & L. Lovett Doust), pp. 60–79. Oxford University Press, Oxford.

Hails, R.S. & Crawley, M.J. (1991) The population dynamics of an alien insect *Andricus quercuscalicis* (Hymenoptera: Cynipidae). *Journal of Animal Ecology* **60**, 545–562.

Hain, R., Reif, H., Krause, E. *et al.* (1993) Disease resistance results from foreign phytoalexin expression in a novel plant. *Nature* **361**, 153–156.

Hainsworth, J.M. & Aylmore, L.A.G. (1986) Water extraction by single plant roots. *Soil Science Society of America Journal* **50**, 841–848.

Hairston, N.G. (1989) *Ecological Experiments: Purpose, Design and Execution*. Cambridge University Press, Cambridge.

Hairston, N.G., Smith, F.E. & Slobodkin, L.B. (1960) Community structure, population control, and competition. *American Naturalist* **94**, 421–425.

Haldane, J.B.S. (1949) Disease and evolution. *La Ricerca Scientifica* **19**, 1–11.

Hallbäcken, L. & Tamm, C.O. (1986) Changes in soil acidity from 1927 to 1982–84 in a forested area of south-west Sweden. *Scandinavian Journal of Forest Research* **1**, 219–232.

Hallé, F., Oldemann, R.A.A. & Tomlinson, P.B. (1978) *Tropical Trees and Forests: an Architectural Analysis*. Springer-Verlag, Berlin.

Hamilton, W.D. (1964) The genetical evolution of social behaviour. I & II. *Journal of Theoretical Biology* **7**, 1–52.

Hamilton, W.D. (1966) The moulding of senescence. *Journal of Theoretical Biology* **12**, 12–45.

Hamilton, W.D. (1967) Extraordinary sex ratios. *Science* **156**, 477–488.

Hamilton, W.D. (1972) Altruism and related phenomena, mainly in social insects. *Annual Review of Ecology and Systematics* **3**, 193–232.

Hamilton, W.D. (1979) Wingless and fighting males in fig wasps and other insects. In *Reproductive Competition and Sexual Selection in Insects* (Eds M.S. Blum & N.A. Blum), pp. 167–220. Academic Press, London.

Hamilton, W.D. (1980) Sex versus non-sex parasite. *Oikos* **35**, 282–290.

Hamilton, W.D. & May, R.M. (1977) Dispersal in stable habitats. *Nature* **269**, 578–581.

Han, K. & Lincoln, D.E. (1994) The evolution of carbon allocation to plant secondary metabolites: a genetic analysis of cost in *Diplacus aurantiacus*. *Evolution* **48**, 1550–1563.

Handel, S.N., Fisch, S.B. & Schatz, G.E. (1981) Ants disperse a majority of herbs in a mesic forest community in New York State. *Bulletin of the Torrey Botanical Club* **108**, 430–437.

Hanhimäki, S., Senn, J. & Haukioja, E. (1995) The convergence in growth of foliage-chewing insect species on individual mountain trees. *Journal of Animal Ecology* **64**, 543–552.

Hanley, M.E., Fenner, M. & Edwards, P.J. (1995) An experimental field study of the effects of mollusk grazing on seedling recruitment and survival in grassland. *Journal of Ecology* **83**, 621–627.

Hansen, S.R. & Hubbell, S.P. (1980) Single-nutrient microbial competition: qualitative agreement between experimental and theoretically forecast outcomes. *Science* **207**, 1491–1493.

Hanski, I. (1983) Coexistence of competitors in patchy environments. *Ecology* **64**, 493–500.

Hanski, I. (1989) Metapopulation dynamics – does it help to have more of the same? *Trends in Ecology and Evolution* **4**, 113–114.

Hanski, I. (1991) Single-species metapopulation dynamics: concepts, models and observations. *Biological Journal of the Linnean Society* **42**, 17–38.

Hanski, I., Hansson, L. & Hettonen, H. (1991) Specialist predators, generalist predators, and the microtine rodent cycle. *Journal of Animal Ecology* **60**; 353–367.

Hanski, I., Turchin, P., Korpimaki, E. & Henttoren, H. (1993) Population oscillations of borel rodents: regulation by mustelid predators leads to chaos. *Nature* **363**, 232–235.

Hansson, L. (1994) Bark consumption by voles in relation to geographical origin of tree species. *Scandinavian Journal of Forest Research* **9**, 288–296.

Hara, T. (1984) A stochastic model and the moment dynamics of the growth and size distribution in plant populations. *Journal of Theoretical Biology* **109**, 173–190.

Hara, T., Kimura, M. & Kikuzawa, K. (1991) Growth patterns of tree height and stem diameter in populations of *Abies veitchii*, *A. mariesii* and *Betula ermanii*. *Journal of Ecology* **79**, 1085–1098.

Harborne, J.B. (1982) *Introduction to Ecological Biochemistry*. Academic Press, London.

Harborne, J.B. (1993) *Introduction to Ecological Biochemistry*, 4th edn. Academic Press, London.

Harborne, J.B. (1994) *The Flavonoids: Advances in Research since 1986*. Chapman & Hall, London.

Harborne, J.B. & Tomas-Barberan, F.A. (1991) *Ecological Chemistry and Biochemistry of Plant Terpenoids*. Clarendon Press, Oxford.

Harborne, J.B., Ingham, J.L., King, L. & Payne, M. (1976) The isopentenylisoflavone luteone as a pre-infectional antifungal agent in the genus *Lupinus*. *Phytochemistry* **15**, 1485–1488.

Hardin, G. (1968) *Exploring New Ethics for Survival. The Voyage of the Spaceship Beagle*. Viking, London.

Hardner, C.M. & Potts, B.M. (1995) Inbreeding depression and changes in variation after selfing in *Eucalyptus globulus* ssp. *globulus*. *Silvae Genetica* **44**, 46–54.

Hargrove, T.R., Cabanilla, V.L. & Coffman, W.R. (1988) Twenty years of rice breeding. *BioScience* **38**, 675–681.

Harlan, H.V. & Martini, M.L. (1938) The effect of natural selection on a mixture of barley varieties. *Journal of Agricultural Research* **57**, 189–199.

Harley, J.L. & Smith, S.E. (1983) *Mycorrhizal Symbioses*. Academic Press, London.

Harper, J.L. (1957) The ecological significance of dormancy and its importance in weed control. In *Proceedings of the IV International Congress on Crop Protection, Hamburg*, pp. 415–420.

Harper, J.L. (1961) Approaches to the study of plant competition. In *Mechanisms in Biological Competition* (Ed. F.L. Milthorpe), pp. 1–39. Symposium of the Society for Experimental Biology, **15**. Cambridge University Press, Cambridge.

Harper, J.L. (1967) A Darwinian approach to plant ecology. *Journal of Ecology* **55**, 247–270.

Harper, J.L. (1969) *The role of predation in vegetational diversity*. Brookhaven Symposia in Biology **22**, 48–62.

Harper, J.L. (1975) Review of 'Allelopathy' by E.L. Rice. *Quarterly Review of Biology* **50**, 493–495.

Harper, J.L. (1977) *Population Biology of Plants*. Academic Press, London.

Harper, J.L. (1981) The concept of population in modular organisms. In *Theoretical Ecology: Principles and Applications*, 2nd edn (Ed. R.M. May), pp. 53–77. Blackwell Scientific Publications, Oxford.

Harper, J.L. (1982) After description. In *The Plant Community as a Working Mechanism* (Ed. E.I. Newman), pp. 11–25. Blackwell Scientific Publications, Oxford.

Harper, J.L., Lovell, P.H. & Moore, K.G. (1970) The shapes and sizes of seeds. *Annual Review of Ecology and Systematics* **1**, 327–356.

Harris, P., Wilkinson, A.T.S., Thompson, L.S. & Neary, M. (1978) Interaction between the cinnabar moth, *Tyria jacobaeae* L. (Lep.: Arctiidae) and ragwort, *Senecio jacobaea* L. (Compositae) in Canada. In *Proceedings of the 4th International Symposium on the Biological Control of Weeds*, pp. 174–180.

Harrison, N.A., Richardson, P.A., Kramer, J.B. & Tsai, J.H. (1994) Detection of the mycoplasma like organism associated with lethal yellowing disease of palms in Florida by polymerase chain reaction. *Plant Pathology* **43**, 998–1008.

Harrison, S. (1994) Metapopulations and conservation. In *Large-scale Ecology and Conservation Biology* (Eds P.J. Edwards, R.M. May & N.R. Webb), pp. 111–128. Blackwell Scientific Publications, Oxford.

Harrison, S. (1995) Lack of strong induced or maternal effects in tussock moths (*Orgyia vestusta*) on bush lupine (*Lupinus arboreus*). *Oecologia* **103**, 343–348.

Harrison, S. & Karban, R. (1986) Effects of an early-season folivorous moth on the success of a later-season species, mediated by a change in the quality of a shared host, *Lupinus arboreus* Sims. *Oecologia* **69**, 354–359.

Harrison, S. & Maron, J.L. (1995) Impacts of defoliation by tussock moths (*Orgyia vestusta*) on the growth and reproduction of bush lupine (*Lupinus arboreus*). *Ecological Entomology* **20**, 223–229.

Hart, R. (1977) Why are biennials so few? *American Naturalist* **111**, 792–799.

Hartgerink, A.P. & Bazzaz, F.A. (1984) Seedling-scale environmental heterogeneity influences individual fitness and population structure. *Ecology* **65**, 198–206.

Hartley, S.E. (1988) The inhibition of phenolic biosynthesis in damaged and undamaged birch foliage and its effect on insect herbivores. *Oecologia* **76**, 95–70.

Hartley, S.E. & Firn, R.D. (1989) Phenolic biosynthesis, leaf damage, and insect herbivory in birch (*Betula pendula*). *Journal of Chemical Ecology* **15**, 275–283.

Hartley, S.E. & Lawton, J.H. (1987) The effects of different types of damage on the chemistry of birch foliage and the responses of birch-feeding insects. *Oecologia* **74**, 432–437.

Hartley, S.E. & Lawton, J.H. (1991) Biochemical aspects and significance of the rapidly induced accumulation of phenolics in birch foliage. In *Phytochemical Induction by Herbivores* (Eds D.W. Tallamy & M.J. Raupp), pp. 105–132. John Wiley & Sons, New York.

Hartley, S.E. & Lawton, J.H. (1992) Host-plant manipulation by gall insects: a test of the nutrition hypothesis. *Journal of Animal Ecology* **61**, 113–119.

Hartley, S.E., Nelson, K. & Gorman, M. (1995) The effect of fertiliser and shading on plant chemical composition and palatability to Orkney voles, *Microtis arvalis orcadensis. Oikos* **72**, 79–87.

Hartmann, T., Ehmke, A., Sander, H., van Borsted, K., Adolph, R. & Toppel, G. (1989) Metabolic integration of the pyrrolizidine alkaloids in respect to their function in chemical protection of *Senecio vulgaris. Planta Medica* **55**, 218–219.

Hartnett, D.C., Hetrick, B.A.D., Wilson, G.T.W. & Gibson, D.J. (1993) Mycorrhizal influence on intra- and interspecific neighbourhood interactions among co-occurring prairie grasses. *Journal of Ecology* **81**, 787–795.

Hartsema, A.M. (1961) Influence of temperature on flower formation and flowering of bulbous and tuberous plants. In *Handbuch der Pflanzenphysiologie. 16. Ansenfaktoren in Wachstum und Entwicklung* (Ed. W. Ruhland), pp. 123–167. Springer-Verlag, Berlin.

Hartvigsen, G. & McNaughton, S.J. (1995) Trade-off between height and relative growth-rate in a dominant grass from the Serengeti ecosystem. *Oecologia* **102**, 273–276.

Hartvigsen, G., Wait, D.A. & Coleman, J.S. (1995) Tri-trophic interactions influenced by resource availability: predator effects on plant performance depend on plant resources. *Oikos* **74**, 463–468.

Harvey, P.H. (1996) Phylogenies for ecologists. *Journal of Animal Ecology* **65**, 255–263.

Harvey, P.H. & Pagel, M.D. (1991) *The Comparative Method in Evolutionary Biology.* Oxford University Press, Oxford.

Harvey, P.H., Colwell, R.K., Silvertown, J.W. & May, R.M. (1983) Null models in ecology. *Annual Review of Ecology and Systematics* **14**, 189–211.

Harvey, P.H., Read, A.F. & Nee, S. (1995) Why ecologists need to be phylogenetically challenged. *Journal of Ecology* **83**, 535–536.

Haslam, E. (1986) Secondary metabolism – fact and fiction. *Natural Products Reports* **3**, 217–249.

Haslam, E. (1988) Plant polyphenols (syn. vegetable tannins) and chemical defense – a reappraisal. *Journal of Chemical Ecology* **14**, 1780–1805.

Haslam, E. (1989) *Plant Polyphenols: Vegetable Tannins Revisited.* Cambridge University Press, Cambridge.

Hassell, M.P. & May, R.M. (1985) From individual behaviour to population dynamics. In *Behavioural Ecology* (Eds R. Sibly & R. Smith), pp. 3–32. Blackwell Scientific Publications, Oxford.

Hassell, M.P., Lawton, J.H. & May, R.M. (1976) Patterns of dynamical behaviour in single-species populations. *Journal of Animal Ecology* **45**, 471–486.

Hassell, M.P., Comins, H.N. & May, R.M. (1991) Spatial structure and chaos in insect population dynamics. *Nature* **353**, 255–258.

Hastings, A. (1980) Disturbance, coexistence, history, and competition for space. *Theoretical Population Biology* **18**, 363–373.

Hastings, A. (1993) Complex interactions between dispersal and dynamics – lessons from coupled logistic equations. *Ecology* **74**, 1362–1372.

Hastings, A., Hom, C.L., Ellner, S., Turchin, P. & Godfray, H.C.J. (1993) Chaos in ecology – is mother nature a strange attractor? *Annual Review of Ecology and Systematics* **24**, 1–33.

Hatcher, P.E., Paul, N.D., Ayres, P.G. & Whittaker, J.B. (1994) The effect of an insect herbivore and a rust fungus individually, and combined in sequence, on the growth of 2 *Rumex* species. *New Phytologist* **128**, 71–78.

Haukioja, E., Niemela, L., Iso-Iivari, L., Ojala, H. & Aro, E. (1978) Birch leaves as a resource for herbivores. Variation in the suitability of leaves. *Report of Kevo Subarctic Research Station* **14**, 5–12.

Hawksworth, D.L. (1973) Mapping studies. In *Air Pollution and Lichens* (Eds B.W. Ferry, M.S. Baddeley & D.L. Hawksworth), pp. 38–76. Athlone Press, London.

Hay, M.E. (1981) Herbivory, algal distribution, and the maintenance of between-habitat diversity on a tropical fringing reef. *American Naturalist* **118**, 520–540.

Hay, M.E. & Fenical, W. (1988) Marine plant–herbivore interactions: the ecology of chemical defense. *Annual Review of Ecology and Systematics* **19**, 111–145.

Hay, M.E., Kappel, Q.E. & Fenical, W. (1994) Synergisms in plant defenses against herbivores: interactions of chemistry, calcification, and plant quality. *Ecology* **75**, 1714–1726.

Hedrick, P.W. (1983) Recombination and directional selection. *Nature* **302**, 727.

Heide, O.M., Junttila, O. & Samuelsen, R.T. (1976) Seed germination and bolting in red beet as affected by parent plant environment. *Physiologia Plantarum* **36**, 343–349.

Heil, B.G.W. & Diemont, W.M. (1988) Raised nutrient levels change heathland into grassland. *Vegetatio* **53**, 113–120.

Heil, G.W. & Brugginck, M. (1987) Competition for nutrients between *Calluna vulgaris* and *Molinia caerulea* (L.) Moench. *Oecologia* **73**, 105–107.

Heinrich, B. & Collins, S.L. (1983) Caterpillar leaf damage, and the game of hide and seek with birds. *Ecology* **64**, 592–602.

Hemming, J.D.C. & Lindroth, R.L. (1995) Intraspecific variation in aspen phytochemistry: effects on performance of gypsy moths and forest tent caterpillars. *Oecologia* **103**, 79–88.

Hendrix, S.D. (1979) Compensatory reproduction in a biennial herb following insect defloration. *Oecologia* **42**, 107–118.

Hendrix, S.D. & Sun, I-F. (1989) Inter- and intraspecific variation in seed mass in seven species of umbellifer. *New Phytologist* **112**, 445–451.

Hendry, G.A.F. & Grime, J.P. (1993) *Methods in Comparative Plant Ecology*. Chapman & Hall, London.

Hendry, R.J. & McGlade, J.M. (1995) The role of memory in ecological systems. *Proceedings of the Royal Society of London Series B* **259**, 153–159.

Hendry, R.J., McGlade, J.M. & Weiner, J. (1996) A coupled map lattice model of the growth of plant monocultures. *Ecological Modelling* **84**, 81–90.

Herms, D.A. & Mattson, W.J. (1992) The dilemma of plants – to grow or defend? *Quarterly Review of Biology* **67**, 283–335.

Herms, D.A. & Mattson, W.J. (1994) Plant growth and defense. *Trends in Ecology and Evolution* **9**, 488.

Herre, E.A. (1987) Optimality, plasticity and selective regime in fig-wasp sex ratios. *Nature* **329**, 627–629.

Herrera, C.M. (1985) Determinants of plant–animal coevolution: the case of mutualistic vertebrate seed dispersal systems. *Oikos* **44**, 132–141.

Herrera, C.M. (1993) Selection on floral morphology and environmental determinants of fecundity in a hawk moth pollinated violet. *Ecological Monographs* **63**, 251–275.

Herrera, C.M. (1995) Plant–vertebrate seed dispersal systems in the Mediterranean: ecological, evolutionary, and historical determinants. *Annual Review of Ecology and Systematics* **26**, 705–727.

Heske, E.J., Brown, J.H. & Mistry, S. (1994) Long term experimental study of a Chihuahuan Desert rodent community – 13 years of competition. *Ecology* **75**, 438–445.

Hester, A.J. (1988) Vegetation succession under birch: the effect of shading and fertilizer treatments on the growth and competitive ability of *Deschampsia flexuosa*. In *Ecological Change in the Uplands* (Eds M.B. Usher & D.B.A. Thompson), pp. 71–74. Blackwell Scientific Publications, Oxford.

Hetrick, B.A.D., Wilson, G.W.T. & Figge, D.A.H. (1994) The influence of mycorrhizal symbiosis and fertilizer amendments on establishment of vegetation in heavy metal mine spoil. *Environmental Pollution* **86**, 171–179.

Hett, J.M. (1971) A dynamic analysis of age in sugar maple seedlings. *Ecology* **52**, 1071–1074.

Heywood, J.S. (1986) The effect of plant size variation on genetic drift in populations of annuals. *American Naturalist* **127**, 851–861.

Hik, D.S. & Jefferies, R.J. (1990) Increases in the net above ground productivity of a salt-marsh forage grass – a test of the predictions of the herbivore optimization model. *Journal of Ecology* **78**, 180–195.

Hill, M.O., Evans, D.F. & Bell, S.A. (1992) Long-term effects of excluding sheep from hill pastures in North Wales. *Journal of Ecology* **80**, 1–13.

Hils, M.H. & Vankat, J.L. (1982) Species removals from a first-year old field plant community. *Ecology* **63**, 705–711.

Hjalten, J., Danell, K. & Ericson, L. (1993) Effects of simulated herbivory and intraspecific competition on the compensatory ability of birches. *Ecology* **74**, 1136–1142.

Hladik, C.M. (1967) Surface relative du tractus digestif de quelques primates. Morphologie des villosites intestinales et correlations avec la regime alimentaire. *Mammalia* **31**, 120–147.

Hobbs, R.J. & Mooney, H.A. (1995) Spatial and temporal variability in California annual grassland – results from a long term study. *Journal of Vegetation Science* **6**, 43–56.

Hochberg, M.E., Menaut, J.C. & Gignoux, J. (1994) Influences of tree biology and fire in the spatial structure of the West African savanna. *Journal of Ecology* **82**, 217–226.

Hodgson, J.G. (1986) Commonness and rarity in plants with special reference to the Sheffield flora. *Biological Conservation* **36**, 199–314.

Hodgson, J.G. (1989) What is happening to the British flora? An investigation of commonness and rarity. *Plants Today* **2**, 26–32.

Hodgson, J.G. & Mackey, J.M.L. (1986) The ecological specialisation of dicotyledonous families within a local flora: some factors constraining optimisation of seed size and their possible evolutionary significance. *New Phytologist* **104**, 479–515.

Hoffmann, J.H. (1990) Interactions between three weevil species in the biological control of *Sesbania punicea* (Fabaceae): the role of simulation models in evaluation. *Agriculture, Ecosystems and Environment* **32**, 77–87.

Hoffmann, J.H. & Moran, V.C. (1992) Oviposition patterns and the supplementary role of a seed feeding weevil, *Rhyssomatus marginatus* (Coleoptera, Curculionidae), and the biological control of a perennial leguminous weed, *Sesbania punicea*. *Bulletin of Entomological Research* **82**, 343–347.

Holdridge, L.R. (1947) Determination of world plant formations from simple climatic data. *Science* **105**, 367–368.

Holliday, N.J. (1977) Population ecology of the winter moth (*Operophtera brumata*) on apple in relation to larval dispersal and time of budburst. *Journal of Applied Ecology* **14**, 803–814.

Hollinger, D.Y. (1983) *Photosynthesis, water and herbivory in co-occurring deciduous and evergreen oaks*. PhD thesis, Stanford University.

Hollinger, D.Y. (1989) Canopy organization and foliage photosynthetic capacity in a broad-leaved evergreen montane forest. *Functional Ecology* **3**, 53–62.

Holm, L.G., Plucknett, D.L., Pancho, J.V. & Herberger, J.P. (1977) *The World's Worst Weeds: Distribution and Biology*. University of Hawaii Press, Honolulu.

Holmes, E.E., Lewis, M.A., Banks, J.E. & Veit, R.R. (1994) Partial differential equations in ecology: spatial interactions and population dynamics. *Ecology* **75**, 17–29.

Holmes, W. (Ed.) (1980) *Grass: its Production and Utilization*. Blackwell Scientific Publications, Oxford.

Holt, R.D. (1977) Predation, apparent competition, and the structure of prey communities. *Theoretical Population Biology* **12**, 197–229.

Holt, R.D. (1984) Spatial heterogeneity, indirect interactions, and the coexistence of prey species. *American Naturalist* **124**, 377–406.

Holt, R.D. (1987) Prey communities in patchy environments. *Oikos* **50**, 276–290.

Holt, R.D., Grover, J.P. & Tilman, G.D. (1994) Simple rules for interspecific dominance in systems with exploitative and apparent competition. *American Naturalist* **144**, 741–771.

Honig, M.A., Linder, H.P. & Hond, W.J. (1992) Efficacy of wind pollination – pollen load size and natural microgametophyte populations in wind-pollinated *Staberoha banksii* (Restionaceae). *American Journal of Botany* **79**, 443–448.

Honkanen, T., Haukioja, E. & Suomela, J. (1994) Effects of simulated defoliation and debudding on needle and shoot growth in Scots pine (*Pinus sylvestris*) – implications of plant source sink relationships for plant herbivore studies. *Functional Ecology* **8**, 631–639.

Hook, P.B., Lauenroth, W.K. & Burke, I.C. (1994) Spatial patterns of roots in a semi-arid grassland – abundance of canopy opening and regeneration gaps. *Journal of Ecology* **82**, 485–494.

Horn, H.S. (1971) *The Adaptive Geometry of Trees*. Princeton University Press, Princeton.

Horn, H.S. (1975) Markovian properties of forest succession. In *Ecology and Evolution of Communities* (Eds M. Cody & J. Diamond), pp. 196–211. Belknap Press, Cambridge, Massachusetts.

Horn, H.S. (1976) Successions. In *Theoretical Ecology* (Ed. R.M. May), pp. 187–204. Blackwell Scientific Publications, Oxford.

Horn, H.S. & MacArthur, R.H. (1972) Competition among fugitive species in a harlequin environment. *Ecology* **53**, 749–752.

Horsman, D.C., Roberts, T.M. & Bradshaw, A.D. (1978) Evolution of sulphur dioxide tolerance in perennial ryegrass. *Nature* **276**, 493–494.

Horvitz, C.C. & Schemske, D.W. (1988) A test of the pollinator limitation hypothesis for a neotropical herb. *Ecology* **69**, 200–206.

Horvitz, C.C. & Schemske, D.W. (1994) Effects of dispersers, gaps, and predators on dormancy and seedling emergence in a tropical herb. *Ecology* **75**, 1949–1958.

Houghton, J.T., Jenkins, G.J. & Ephraums, J.J. (Eds) (1990) *Climate Change: the IPCC Scientific Assessment*. Cambridge University Press, Cambridge.

Howard-Williams, C. (1970) The ecology of *Becium homblei* in central Africa with special reference to metalliferous soils. *Journal of Ecology* **58**, 745–763.

Howarth, C.J. & Ougham, J.H. (1993) Tansley review. 51. Gene expression under temperature stress. *New Phytologist* **125**, 1–26.

Howe, H.F. (1983) Annual variation in a neotropical seed-dispersal system. In *Tropical Rain Forest: Ecology and Management* (Eds S.L. Sutton, T.C. Whitmore & A.C. Chadwick), pp. 211–227. Blackwell Scientific Publications, Oxford.

Howe, H.F. (1984) Constraints on the evolution of mutualisms. *American Naturalist* **123**, 764–777.

Howe, H.F. (1985) Gomphothere fruits: a critique. *American Naturalist* **125**, 853–865.

Howe, H.F. (1986) Seed dispersal by fruit-eating birds and mammals. In *Seed Dispersal* (Ed. D.R. Murray), pp. 123–189. Academic Press, Sydney.

Howe, H.F. (1989) Scatter- and clump-dispersal and seedling demography: hypothesis and implications. *Oecologia* **79**, 417–426.

Howe, H.F. (1990) Survival and growth of juvenile *Virola surinamensis* in Panama: effects of herbivory and canopy closure. *Journal of Tropical Ecology* **6**, 259–280.

Howe, H.F. (1993a) Specialized and generalized dispersal systems: where does 'The Paradigm' stand? *Vegetatio* **107/108**, 3–14.

Howe, H.F. (1993b) Aspects of variation in a neotropical seed dispersal system. *Vegetatio* **107/108**, 149–162.

Howe, H.F. & Smallwood, J. (1982) Ecology of seed dispersal. *Annual Review of Ecology and Systematics* **13**, 201–228.

Howe, H.F. & Vande Kerckhove, G.A. (1980) Nutmeg dispersal by tropical birds. *Science* **210**, 925–927.

Howe, H.F. & Vande Kerckhove, G.A. (1981) Removal of wild nutmeg (*Virola surinamensis*) crops by birds. *Ecology* **62**, 1093–1106.

Howe, H.F. & Westley, L.C. (1988) *Ecological Relationships of Plants and Animals*. Oxford University Press, New York.

Hsu, S.B., Hubbell, S.P. & Waltman, P. (1977) A mathematical theory for single-nutrient competition in continuous cultures of microorganisms. *SIAM Journal of Applied Mathematics* **32**, 366–384.

Huang, B.R., Johnson, J.W., Nesmith, D.S. & Bridges, D.G. (1995) Nutrient accumulation and distribution of wheat genotypes in response to waterlogging and nutrient supply. *Plant and Soil* **173**, 47–54.

Hubbell, S.P. (1979) Tree dispersion, abundance and diversity in a tropical dry forest. *Science* **203**, 1299–1309.

Hubbell, S.P. (1980) Seed predation and the coexistence of tree species in tropical forests. *Oikos* **35**, 214–229.

Hubbell, S.P. & Foster, R.B. (1986a) Biology, chance and history, and the structure of tropical tree communities. In *Community Ecology* (Eds J.M. Diamond & T.J. Case), pp. 314–324. Harper & Row, New York.

Hubbell, S.P. & Foster, R.B. (1986b) The spatial context of regeneration in a neotropical forest. In *Colonization,*

Succession and Stability (Eds A.J. Gray, M.J. Crawley & P.J. Edwards). Blackwell Scientific Publications, Oxford.

Hubbell, S.P. & Foster, R.B. (1986c) Canopy gaps and the dynamics of a neotropical forest. In *Plant Ecology* (Ed. M.J. Crawley), pp. 77–96. Blackwell Scientific Publications, Oxford.

Hubbell, S.P. & Werner, P.A. (1979) On measuring the intrinsic rate of increase of populations with heterogeneous life histories. *American Naturalist* 113, 277–293.

Hubbell, S.P., Wiemer, D.F. & Adejare, A. (1983) An anti-fungal terpenoid defends a neotropical tree (*Hymenaea*) against attack by fungus-growing ants (*Atta*). *Oecologia* 60, 321–327.

Hubbell, S.P., Condit, R. & Foster, R.B. (1990) Presence and absence of density dependence in a neotropical tree community. *Philosophical Transactions of the Royal Society of London Series B* 330, 269–281.

Hughes, M.K., Lepp, N.W. & Phipps, D.A. (1980) Aerial heavy metal pollution and terrestrial ecosystems. *Advances in Ecological Research* 11, 217–327.

Hughes, T.P. (1984) Population dynamics based on individual size rather than age: a general model with a reef coral example. *American Naturalist* 123, 778–795.

Hulme, P.E. (1994) Seedling herbivory in grasslands – relative impact of vertebrate and invertebrate herbivores. *Journal of Ecology* 82, 873–880.

Hulme, P.E. (1996) Herbivores and the performance of grassland plants – a comparison of arthropod, mollusk and rodent herbivory. *Journal of Ecology* 84, 43–51.

Hunt, R., Hand, D.W., Hannah, M.A. & Neal, A.M. (1991) Response to CO_2 enrichment in 27 herbaceous species. *Functional Ecology* 5, 410–421.

Hunt, R. *et al.* (1993) Further responses to CO_2 enrichment in British herbaceous species. *Functional Ecology* (submitted).

Hunter, A.F. (1993) Gypsy moth population sizes and the window of opportunity in Spring. *Oikos* 68, 531–538.

Hunter, M.D. (1992) A variable insect–plant interaction: the relationship between tree budburst phenology and population levels of insect herbivores among trees. *Ecological Entomology* 17, 91–95.

Hunter, M.D. (1996) Incorporating variation in plant chemistry into a spatially-explicit ecology of phytophagous insects. In *Forests and Insects*. 18th Symposium of the Royal Entomological Society, Chapman & Hall, London.

Hunter, M.D. & Price, P.W. (1992) Playing chutes and ladders: heterogeneity and the relative roles of bottom-up and top-down forces in natural communities. *Ecology* 73, 724–732.

Hunter, M.D. & Schultz, J.C. (1995) Fertilisation mitigates chemical induction and herbivore responses within damaged oak trees. *Ecology* 76, 1226–1232.

Hunter, M.D. & West, C. (1990) Variation in the effects of spring defoliation on the late season phytophagous

insects of *Quercus robur*. In *Population Dynamics of Forest Insects* (Eds A.D. Watt, S.R. Leather, M.D. Hunter & N.A.C. Kidd), pp. 123–136. Intercept, Andover.

Hunter, M.D. & Willmer, P.G. (1989) The potential for interspecific competition between two abundant defoliators on oak: leaf damage and habitat quality. *Ecological Entomology* 14, 267–277.

Hunter, M.D., Ohgushi, T. & Price, P.W. (Eds) (1992) *Effects of Resource Distribution on Animal–Plant Interactions*. Academic Press, San Diego.

Hunter, M.D., Hall, L.A. & Schultz, J.C. (1994) Evaluation of resistance to tufted apple bud moth (Lepidoptera: Tortricidae) within and among apple cultivars. *Environmental Entomology* 23, 282–291.

Huntley, B.J. (1988) Conserving and monitoring biotic diversity: some African examples. In *Biodiversity* (Eds E.O. Wilson & F.M. Peter), pp. 248–260. National Academy of Sciences, Washington, DC.

Huntley, B. & Birks, H.J.B. (1983) *An Atlas of Past and Present Pollen Maps for Europe: 0–13 000 years ago*. Cambridge University Press, Cambridge.

Huntly, N.J. (1991) Herbivores and the dynamics of communities and ecosystems. *Annual Review of Ecology and Systematics* 22, 477–503.

Huntly, N. & Reichman, O.J. (1994) Effects of subterranean mammalian herbivores on vegetation. *Journal of Mammalogy* 75, 852–859.

Hurlbert, S.H. (1984) Pseudo-replication and the design of ecological field experiments. *Ecological Monographs* 54, 187–211.

Hurtt, G.C. & Pacala, S.W. (1995) The consequences of recruitment limitation: reconciling chance, history, and competitive differences between plants. *Journal of Theoretical Biology* 176, 1–12.

Huston, M. (1979) A general hypothesis of species diversity. *American Naturalist* 113, 81–101.

Huston, M. (1986) Size bimodality in plant populations: an alternative hypothesis. *Ecology* 67, 265–269.

Hutchings, M.J. (1985) Plant population biology. In *Methods in Plant Ecology*, 2nd edn (Eds P.D. Moore & S.B. Chapman), pp. 377–435. Blackwell Scientific Publications, Oxford.

Hutchings, M.J. (1987) The population biology of the early spider orchid, *Ophrys sphegodes* Mill. I. A demographic study from 1975 to 1984. *Journal of Ecology* 75, 711–727.

Hutchings, M.J. & Bradbury, I.K. (1986) Ecological perspectives on clonal perennial plants, *BioScience* 36, 178–182.

Hutchings, M.J. & de Kroon, H. (1994) Foraging in plants: the role of morphological plasticity in resource acquisition. *Advances in Ecological Research* 25, 159–238.

Hutchinson, G.E. (1951) Copepodology for the ornithologist. *Ecology* 32, 571–577.

Hutchinson, G.E. (1957) Concluding remarks. *Cold Spring Harbor Symposium on Quantitative Biology* 22, 415–457.

Hutchinson, G.E. (1959) Homage to Santa Rosalia or why are there so many kinds of animals? *American Naturalist* **93**, 145–159.

Hutson, V. & Law, R. (1985) Permanent coexistence in general models of three interacting species. *Journal of Mathematical Biology* **21**, 285–298.

Iason, G.R. & Hester, A.J. (1993) The response of heather to shade and nutrients: predictions of the carbon/nutrient balance hypothesis. *Journal of Ecology* **81**, 75–80.

Ineson, P. & Wookey, P.A. (1988) Effects of sulphur dioxide on forest litter decomposition and nutrient release. In *Air Pollution and Ecosystems* (Ed. P. Mathy), pp. 254–260. D. Reidel, Dordrecht.

Inoue, K., Maki, M. & Masuda, M. (1996) Evolution of *Campanula* flowers in relation to insect pollinators on islands. In *Floral Biology: Studies of Floral Evolution in Animal-pollinated Plants* (Eds D.G. Lloyd & S.C.H. Barrett), pp. 377–400. Chapman & Hall, New York.

Inouye, R.S. (1980) Density-dependent germination response by seeds of desert annuals. *Oecologia* **46**, 235–238.

Irvine, A.K. & Armstrong, J.E. (1990) Beetle pollination in tropical forests of Australia. In *Reproduction Ecology of Tropical Forest Plants* (Eds K.S. Bawa & M. Hadley), pp. 135–150. Parthenon, Carnforth.

Islam, Z. & Crawley, M.J. (1983) Compensation and regrowth in ragwort (*Senecio jacobaea*) attacked by cinnabar moths (*Tyria jacobaeae*). *Journal of Ecology* **71**, 829–843.

Isman, M.B. & Duffey, S.S. (1982) Toxicity of tomato phenolic-compounds to the tomato fruitworm, *Heliothis zea*. *Entomologia Experimentatia Applicata* **31**, 370–376.

Iwasa, Y. & Roughgarden, J. (1984) Shoot/root balance of plants: optimal growth of a system with many vegetative organs. *Theoretical Population Biology* **25**, 78–105.

Jackson, G. (1991) *The effects of ozone and nitrogen dioxide on cereal/aphid interactions.* PhD thesis, University of London.

Jackson, G.E., Irvine, J., Grace, J. & Khalil, A.A.M. (1995a) Abscisic acid concentration and fluxes in droughted conifer saplings. *Plant, Cell and Environment* **18**, 13–22.

Jackson, G.E., Irvine, J. & Grace, J. (1995b) The vulnerability to xylem cavitation of Scots pine and Sitka spruce saplings during water stress. *Tree Physiology*.

Jackson, G.E., Irvine, J. & Grace, J. (1995c) Xylem cavitation in two mature Scots pine forests growing in a wet and dry area of Britain. *Plant, Cell and Environment*.

Jackson, J.B.C., Buss, L.W. & Cook, R.E. (Eds) (1985) *Population Biology and Evolution of Clonal Organisms.* Yale University Press, New Haven.

Jackson, M.B. & Drew, M.C. (1984) Effects of flooding on growth and metabolism of herbaceous plants. In *Flooding and Plant Growth* (Ed. T.T. Kozlowski), pp. 47–128. Academic Press, London.

Jackson, M.B. & Caldwell, M.M. (1993) Geostatistical patterns of soil heterogeneity around individual perennial plants. *Journal of Ecology* **81**, 682–692.

Jager, H.J. & Grill, D. (1975) Einfluss von SO_2 und HF auf freie Aminosauren der Fichte. (*Picea abies* (L.) Karsten). *European Journal of Forest Pathology* **5**, 279–286.

Jaindl, R.G., Doescher, P., Miller, R.F. & Eddleman, L.E. (1994) Persistence of Idaho fescue on degraded rangelands – adaptation to defoliation or tolerance. *Journal of Range Management* **47**, 54–59.

James, C.D., Hoffman, M.T., Lightfoot, D.C., Forbes, G.S. & Whitford, W.G. (1993) Pollination ecology of yucca elata – an experimental study of a mutualistic association. *Oecologia* **93**, 512–517.

James, L.F., Keeler, R.F., Bailey, E.M., Cheek, P.R. & Hegarty, M.P. (eds) (1992) Poisonous plants. Proceedings of the Third International Symposium, Iowa State University Press, Ames.

James, R.R., McEvoy, P.B. & Cox, C.S. (1992) Combining the cinnabar moth (*Tyria jacobaeae*) and the ragwort flea beetle (*Longitarsus jacobaeae*) for control of ragwort (*Senecio jacobaeae*) – an experimental analysis. *Journal of Applied Ecology* **29**, 589–596.

Janson, C.H. (1983) Adaptation of fruit morphology to dispersal agents in a neotropical forest. *Science* **219**, 187–189.

Janzen, D.H. (1966) Coevolution of mutualism between ants and acacias in Central America. *Evolution* **20**, 249–275.

Janzen, D.H. (1970) Herbivores and the number of tree species in tropical forests. *American Naturalist* **104**, 501–508.

Janzen, D.H. (1971) Escape of *Cassia grandis* L. beans from predators in time and space. *Ecology* **52**, 964–979.

Janzen, D.H. (1976) Why bamboo wait so long to flower. *Annual Review of Ecology and Systematics* **7**, 347–391.

Janzen, D.H. (1977) What are dandelions and aphids? *American Naturalist* **111**, 586–589.

Janzen, D.H. (1978) Complications in interpreting the chemical defences of trees against tropical arboreal plant-eating vertebrates. In *The Ecology of Arboreal Folivores* (Ed. C.F. Montgomery), pp. 73–84. Smithsonian Institution, Washington DC.

Janzen, D.H. (1978) Seeding patterns of tropical trees. In *Tropical Trees as Living Systems* (Eds P.B. Tomlinson & M.H. Zimmermann), pp. 83–128. Cambridge University Press, Cambridge.

Janzen, D.H. (1979) New horizons in the biology of plant defenses. In *Herbivores: their Interaction with Secondary Plant Metabolites* (Eds. G.A. Rosenthal & D.H. Janzen), pp. 331–350. Academic Press, New York.

Janzen, D.H. (1980) When is it coevolution? *Evolution* **34**, 611–612.

Janzen, D.H. (1982) Removal of seeds from horse dung by tropical rodents: influence of habitat and amount of dung. *Ecology* **63**, 1887–1900.

Janzen, D.H. (1983) Seed and pollen dispersal by animals: convergence in the ecology of contamination and sloppy harvest. *Biological Journal of the Linnean Society* **20**, 103–113.

Janzen, D.H. (1984) Dispersal of small seeds by big herbivores – foliage is the fruit. *American Naturalist* **123**, 338–353.

Janzen, D.H. (1985) The natural history of mutualisms. In *The Biology of Mutualism* (Ed. D.H. Boucher), pp. 40–99. Croom Helm, London.

Janzen, D.H. (1988) On the broadening of insect–plant research. *Ecology* **69**, 905.

Janzen, D.H. & Martin, P. (1982) Neotropical anachronisms: what the gomphotheres ate. *Science* **215**, 19–27.

Jarosz, A.M. & Burdon, J.J. (1992) Host pathogen interactions in natural populations of *Linum marginale* and *Melampsora lini*. 3. Influence of pathogen epidemics on host survivorship and flower production. *Oecologia* **89**, 53–61.

Jarosz, A.M. & Davelos, A.L. (1995) Effects of disease in wild plant populations and the evolution of pathogen aggressiveness. *New Phytologist* **129**, 371–387.

Jarvis, P.G. & Jarvis, M.S. (1964) Growth rates of woody plants. *Physiologia Plantarum* **17**, 654–666.

Jefferies, R.L. & Gottlieb, L.D. (1983) Genetic variation within and between populations of the asexual plant *Puccinellia* (hybrid) *phryganodes* (Trin.) Scribner and Merr. *Canadian Journal of Botany* **61**, 774–779.

Jefferies, R.L., Davy, A.J. & Rudmik, T. (1981) Population biology of the salt marsh annual *Salicornia europaea* agg. *Journal of Ecology* **69**, 17–32.

Jefferies, R.L., Klein, D.R. & Shaver, G.R. (1994) Vertebrate herbivores and northern plant-communities – reciprocal influences and responses. *Oikos* **71**, 193–206.

Jelinski, D.E. & Cheliak, W.M. (1992) Genetic diversity and spatial subdivision of *Populus tremuloides* (Salicaceae) in a heterogeneous landscape. *American Journal of Botany* **79**, 728–736.

Jensen, T.S. (1985) Seed–predator interactions of European beech, *Fagus sylvatica*, and forest rodents, *Clethrionomys glareolus* and *Apodemus flavicollis*. *Oikos* **44**, 149–156.

Jermy, T. (1976) Insect host plant relationships – coevolution or sequential evolution? *Symposia Biologica Hungarica* **16**, 109–113.

Jermy, T. (1984) Evolution of insect/host plant interactions. *American Naturalist* **124**, 609–630.

Jermy, T. (1993) Evolution of insect–plant relationships – a devil's advocate approach. *Entomologia Experimentatia Applicata* **66**, 3–12.

Jing, S.W. & Coley, P.D. (1990) Dioecy and herbivory: the effect of growth rate on plant defense in *Acer negundo*. *Oikos* **58**, 369–377.

Joenje, W. (1978) *Plant colonization and succession on embanked sandflats: a case study in the Lauwerszeepolder*. PhD thesis, University of Groningen.

Johnson, E.A., Miyanishi, K. & Kleb, H. (1994) The hazards of interpretation of static age structures as shown by stand reconstructions in *Pinus contorta–Picea engelmanii* forest. *Journal of Ecology* **82**, 923–931.

Johnson, I.R. & Parsons, A.J. (1985) A theoretical analysis of grass growth under grazing. *Journal of Theoretical Biology* **112**, 345–367.

Johnson, S.G., Delph, L.F. & Elderkin, C.L. (1995) The effect of petal size manipulation on pollen removal, seed set, and insect visitor behavior in *Campanula americana*. *Oecologia* **102**, 174–179.

Johnston, M.A. (1992) *Rabbit grazing and the dynamics of plant communities*. Unpublished PhD thesis, University of London.

Joly, C.A. & Brandle, R. (1995) Fermentation and adenylate metabolism of *Hedychium coronarium* (Zingiberaceae) and *Acorus calamus* L. (Araceae) under hypoxia and anoxia. *Functional Ecology* **9**, 505–510.

Jones, C.G. (1983) Phytochemical variation, colonisation, and insect communities: the case of bracken fern. In *Variable Plants and Herbivores in Natural and Managed Systems* (Eds R.F. Denno & M.S. McClure), pp. 513–549. Academic Press, New York.

Jones, C.G. (1984) Microorganisms as mediators of plant resource exploitation by insect herbivores. In *A New Ecology: Novel Approaches to Interactive Systems* (Eds P.W. Price, C.N. Slobodchikoff & W.S. Gaud), pp. 51–84. John Wiley & Sons, New York.

Jones, C.G. & Coleman, J.S. (1991) Plant stress and insect herbivory: toward an integrated perspective. In *Responses of Plants to Multiple Stresses* (Eds H.A. Mooney, W.E. Winner & E.J. Pell), pp. 249–280, Academic Press, New York.

Jones, C.G. & Firn, R.D. (1979) Resistance of *Pteridium aquilinium* to attack by non-adapted phytophagous insects. *Biochemical Systematics and Ecology* **7**, 95–101.

Jones, C.G. & Firn, R.D. (1991) On the evolution of plant secondary chemical diversity. *Philosophical Transactions of the Royal Society of London Series B* **333**, 273–280.

Jones, C.G. & Lawton, J.H. (1991) Plant chemistry and insect species richness of British umbellifers. *Journal of Animal Ecology* **60**, 767–778.

Jones, C.G., Hopper, R.F., Coleman, J.S. & Krischik, V.A. (1993) Control of systemically induced herbivore resistance by plant vascular architecture. *Oecologia* **93**, 452–456.

Jones, C.G., Coleman, J.S. & Findlay, S. (1994) Effects of ozone on interactions between plants, consumers and decomposers. In *Plant Responses to the Gaseous Environment* (Eds R.G. Alscher & A. Wellburn), pp. 339–363. Chapman & Hall, London.

Jones, D.A. (1988) Cyanogenesis in animal–plant interactions. In *Cyanide Compounds in Biology* (Eds D. Evered & S. Harnett), pp. 151–157. John Wiley & Sons, Chichester.

Jones, H.G. (1992) *Plants and Microclimate*. Cambridge University Press, Cambridge.

[665]

Jones, K.C. & Klocke, J.A. (1987) Aphid feeding deterrency of ellagitannins, their phenolic derivatives. *Entomologia Experimentatia Applicata* **44**, 229–232.

Jones, L.H. (1961) Aluminium uptake and toxicity in plants. *Plant and Soil* **13**, 297–310.

Jones, M.G. (1933) Grassland management and its influence on the sward. *Empire Journal of Experimental Agriculture* **1**, 43–57; 122–128; 223–234; 361–367.

Jones, T. & Mansfield, T.A. (1982) The effect of SO₂ on growth and development of seedlings of *Phleum pratense* under different light and temperature environments. *Environmental Pollution, Series A* **27**, 57–71.

Jones, T.H., Hassell, M.P. & Godfray, H.C.J. (1997) Host–multiparasitoid interactions. In *BES Symposium on Multitrophic Interactions in Terrestrial Systems* (Eds A. Gange & V.K. Brown), Blackwell Science, Oxford.

de Jong, T.J., Klinkhamer, P.G.L. & Metz, J.A.J. (1987) Selection for biennial life histories. *Vegetatio* **70**, 149–156.

Joost, R.A. (1995) *Acremonium* in fescue and rye grass – boon or bane – a review. *Journal of Animal Science* **73**, 881–888.

Jordan, D.N. & Smith, W.K. (1995) Radiation frost susceptibility and the association between sky exposure and leaf size. *Oecologia* **103**, 43–48.

Jordan, W.R. (1988) Ecological restoration: reflections on a half century of experience at the University of Wisconsin-Madison arboretum. In *Biodiversity* (Ed. E.O. Wilson), pp. 311–316. National Academy of Sciences, Washington, DC.

Jordano, P. & Herrera, C.M. (1981) The frugivorous diet of blackcap populations *Sylvia atricapilla* wintering in southern Spain. *Ibis* **123**, 502–507.

Joseph, G., Kelsey, R.G., Moldenke, A.F., Miller, J.C., Berry, R.E. & Wernz, J.G. (1993) Effects of nitrogen and Douglas-fir allelochemicals on development of the gypsy moth, *Lymantria dispar. Journal of Chemical Ecology* **19**, 1245–1263.

Jump, B.A. & Woodward, S. (1994) Histology of witches brooms on *Betula pubescens. European Journal of Forest Pathology* **24**, 229–237.

Kachi, N. & Hirose, T. (1983) Bolting induction in *Oenothera erythrosepala* Borbas in relation to rosette size, vernalization and photoperiod. *Oecologia* **60**, 6–9.

Kachi, N. & Hirose, T. (1985) Population dynamics of *Oenothera glazioviana* in a sand-dune system with special reference to the adaptive significance of size-dependent reproduction. *Journal of Ecology* **73**, 887–901.

Kadmon, R. (1995) Plant competition along soil moisture gradients: a field experiment with the desert annual *Stipa capensis. Journal of Ecology* **83**, 253–262.

Kadmon, R. & Shmida, A. (1992) Departure rules used by bees foraging for nectar – a field-test. *Evolutionary Ecology* **6**, 142–151.

Kadmon, R., Shmida, A. & Selten, R. (1991) Within-plant foraging behavior of bees and its relationship to nectar distribution in *Anchusa strigosa. Israel Journal of Botany* **40**, 283–294.

Kalisz, S. (1991) Experimental determination of seed bank age structure in the winter annual *Collinsia verna. Ecology* **72**, 575–585.

Kamenetsky, R. & Gutterman, Y. (1994) Life cycles and delay of seed dispersal in some geophytes inhibiting the Negev Desert Highlands of Israel. *Journal of Arid Environments* **27**, 337–345.

Karasov, W.H. & Levey, D.J. (1990) Digestive system trade-offs and adaptations of frugivorous passerine birds. *Physiological Zoology* **63**, 1248–1270.

Karban, R. (1980) Periodical cicada nymphs impose periodical oak tree wood accumulation. *Nature* **287**, 326–327.

Karban, R. (1986) Interspecific competition between folivorous insects on *Erigeron glaucus. Ecology* **67**, 1063–1072.

Karban, R. (1989) Community organisation of *Erigeron glaucus* folivores: effects of competition, predation and host plant. *Ecology* **70**, 1028–1039.

Karban, R. (1993a) Induced resistance and plant density of a native shrub, *Gossypium thurberi*, affects its herbivores. *Ecology* **74**, 1–8.

Karban, R. (1993b) Costs and benefits of induced resistance and plant density for a native shrub, *Gossypium thurberi. Ecology* **74**, 9–19.

Karban, R. (1997) BES Symposium on Multitrophic Interactions in Terrestrial Systems (Eds A. Gange & V.K. Brown), pp. 199 and 299. Blackwell Science, Oxford.

Karban, R. & Baldwin, I.T. (1997) *Induced Responses to Herbivory.* Chapman & Hall, London (in press).

Karban, R. & Myers, J.H. (1989) Induced plant responses to herbivory. *Annual Review of Ecology and Systematics* **20**, 331–348.

Karban, R. & Niiho, C. (1995) Induced resistance and susceptibility to herbivory: plant memory and altered plant development. *Ecology* **76**, 1220–1225.

Karban, R. & Strauss, S.Y. (1993) Effects of herbivores on growth and reproduction of their perennial host, *Erigeron glaucus. Ecology* **74**, 39–46.

Karban, R., Adamchek, R. & Schnathorst, W.C. (1987) Induced resistance and interspecific competition between spider mites and a vascular wilt fungus. *Science* **235**, 678–680.

Karlin, S. & Lessard, S. (1986) *Theoretical Studies on Sex Ratio Evolution.* Princeton University Press, Princeton.

Karron, J.D. (1987) A comparison of levels of genetic polymorphism and self-compatibility in geographically restricted and widespread plant congeners. *Evolutionary Ecology* **1**, 47–58.

Karron, J.D., Linhart, Y.B., Chaulk, C.A. & Robertson, C.A. (1988) Genetic structure of populations of geographically restricted and widespread species of *Astragalus* (Fabaceae). *American Journal of Botany* **75**, 1114–1119.

Kato, M. (1994) Alteration of bottom-up and top-down regulation in a natural population of an agromyzid leafminer, *Chromatomyia suikazurae. Oecologia* 97, 9–16.

Kay, Q.O.N. (1982) Intraspecific discrimination by pollinators and its role in evolution. In *Pollination and Evolution* (Eds J.A. Armstrong, J.M. Powell & A.J. Richards), pp. 9–28. Royal Botanic Gardens, Sydney.

Kearsley, M.J. & Whitham, T.G. (1989) Developmental changes in resistance to herbivory: implications for individuals and populations. *Ecology* 70, 422–434.

Keating, S.T., Yendol, W.G. & Schultz, J.C. (1988) Relationship between susceptibility of gypsy moth larvae (Lepidoptera: Lymantriidae) to a baculovirus and host plant foliage constituents. *Environmental Entomology* 17, 952–958.

Keating, S.T., Hunter, M.D. & Schultz, J.C. (1990) Leaf phenolic inhibition of the gypsy moth nuclear polyhedrosis virus: the role of polyhedral inclusion body aggregation. *Journal of Chemical Ecology* 16, 1445–1457.

Keddy, P.A. (1981) Experimental demography of the sand-dune annual, *Cakile edentula*, growing along an environmental gradient in Nova Scotia. *Journal of Ecology* 69, 615–630.

Keddy, P.A. (1982) Population ecology on an environmental gradient: *Cakile edentula* on a sand dune. *Oecologia* 52, 348–355.

Keddy, P.A. (1989) *Competition*. Chapman & Hall, London.

Keddy, P.A. (1990) Competitive hierarchies and centrifugal organization in plant communities. In *Perspectives on Plant Competition* (Eds J.B. Grace & D. Tilman), pp. 265–290. Academic Press, San Diego.

Keddy, P. (1992) A pragmatic approach to functional ecology. *Functional Ecology* 6, 621–626.

Keeley, J.E. (1979) Population differentiation along a flood frequency gradient: physiological adaptation to flooding in *Nyssa sylvatica. Ecological Monographs* 49, 98–108.

Keeling, C.D., Bacastow, R.B. & Whorf, T.P. (1982) Measurements of the concentration of carbon dioxide at Mauna Loa Observatory, Hawaii. In *Carbon Dioxide Review: 1982* (Ed. W.C. Clark), pp. 377–385. Oxford University Press, New York.

Keeling, C.D., Bacastow, R.B., Carter, A.F. *et al.* (1989) A three-dimensional model of atmospheric CO_2 transport based on observed winds: 1. Analysis of observational data. In *Aspects of Climate Variability in the Pacific and the Western Americas* (Ed. D.H. Peterson), pp. 165–235. Geophysical Monograph vol. 55. American Geophysical Union, Washington DC.

Keever, C. (1950) Causes of succession on old fields of the Piedmont, North Carolina. *Ecological Monographs* 20, 229–250.

Kefeli, V.I. & Dashek, W.V. (1984) Non-hormonal stimulators and inhibitors of plant growth and development. *Biological Reviews* 59, 273–288.

Keller, T. & Beda, H. (1984) Effects of SO_2 on the germination of conifer pollen. *Environmental Pollution Series A* 33, 237–243.

Kelly, D. (1985) Why are biennials so maligned? *American Naturalist* 125, 473–479.

Kemp, P.R. & Williams III, G.J. (1980) A physiological basis of niche separation between *Agropyron smithii* (C_3) and *Bouteloua gracilis* (C_4). *Ecology* 61, 846–858.

Kenkel, N.C. (1988) Pattern of self-thinning in jack pine: testing the random mortality hypothesis. *Ecology* 69, 1017–1024.

Kenkel, N.C., Hoskins, J.A. & Hoskins, W.D. (1989) Local competition in a naturally established jack pine stand. *Canadian Journal of Botany* 67, 2630–2635.

Kerley, G.I.H., Tiver, F. & Whitford, W.G. (1993) Herbivory of clonal population – cattle browsing affects reproduction and population structure of *Yucca elata. Oecologia* 93, 12–17.

Kerner, A. (1894) *The Natural History of Plants. Their Forms, Growth, Reproduction and Distribution*. Blackie & Son, London.

Kerr, A. (1987) The impact of molecular genetics on plant pathology. *Annual Review of Phytopathology* 25, 87–110.

Kershaw, K.A. (1963) Pattern in vegetation and its causality. *Ecology* 44, 377–388.

Kerslake, J. & Hartley, S.E. (1997) Phenology of winter moth feeding on common heather: effects of source population and experimental manipulation of hatch dates. *Journal of Animal Ecology* (in press).

Kettlewell, H.B.D. (1955) Selection experiments on industrial melanism in the Lepidoptera. *Heredity* 9, 323–342.

Kevan, P.G. (1983) Floral colors through the insect eye: what they are and what they mean. In *Handbook of Experimental Pollination Biology* (Eds C.E. Jones & R.J. Little), pp. 3–30. Van Nostrand Reinhold, New York.

Kevan, P.G. & Baker, H.G. (1983) Insects as flower visitors and pollinators. *Annual Review of Entomology* 28, 407–453.

Khalil, A.A.M. & Grace, J. (1992) Acclimation to drought in *Acer pseudoplatanus* seedlings. *Journal of Experimental Botany* 257, 1591–1602.

Khalil, A.A.M. & Grace, J. (1993) Does xylem ABA control the stomatal behaviour of water-stressed sycamore (*Acer pseudoplatanus* L.) seedlings? *Journal of Experimental Botany* 44, 1127–1134.

Khan, A.H., Ashraf, M.Y., Naqvi, S.S.M., Khanzada, B. & Ali, M. (1995) Growth, ion and solute contents of sorghum grown under NaCl and Na_2SO_4 salinity stress. *Acta Physiologiae Plantarum* 17, 261–268.

Khan, M.B. & Harborne, J.B. (1991) Induced alkaloid defence in *Atropa acuminata* in response to mechanical and herbivore leaf damage. *Chemoecology* 1, 77–81.

Khan, M.L. & Tripathi, R.S. (1991) Seedling survival and growth of early and late successional tree species as affected by insect herbivory and pathogen attack in subtropical humid forest stands of north-east India. *International Journal of Ecology* 12, 569–579.

Kiang, Y.T. (1972) Pollination study in a natural population of *Mimulus guttatus. Evolution* 26, 308–310.

Kidd, N.A.C., Lewis, G.B. & Howell, C.A. (1985) An

association between two species of pine aphid, *Schizolachnus pineti* and *Eulachnus agilis*. *Ecological Entomology* **10**, 427–432.

Kimmerer, T.W. & Potter, D.A. (1987) Nutritional quality of specific tissues and selective feeding by a specialist leafminer. *Oecologia* **71**, 548–551.

Kimura, M. & Crow, J.F. (1964) The number of alleles that can be maintained in a finite population. *Genetics* **49**, 725–738.

King, R.W., Wardlaw, I.F. & Evans, L.T. (1967) Effects of assimilate utilization and photosynthetic rate in wheat. *Planta* **77**, 261–276.

King, T.J. (1975) Inhibition of seed germination under leaf canopies in *Arenaria serpyllifolia*, *Veronica arvensis* and *Cerastium holosteoides*. *New Phytologist* **75**, 87–90.

Klinkhamer, P.G.L., Jong, T.D. de & Meelis, E. (1987) Delay of flowering in the 'biennial' *Cirsium vulgare*: size effects and devernalization. *Oikos* **49**, 303–308.

Klocke, J.A. & Chan, B.G. (1982) Effects of cotton condensed tannin on feeding and digestion in *Heliothus zea*. *Journal of Insect Physiology* **28**, 85–91.

Klocke, J.A., Van Wagenen, B. & Balandrin, M.F. (1986) The ellagitannin geraniin and its hydrolysis products isolated as insect growth inhibitors from semi-arid land plants. *Phytochemistry* **25**, 85–91.

Knapp, R. (Ed.) (1974) *Vegetation Dynamics*. Junk, The Hague.

Knogge, W., Kombrink, E., Schmelzer, E. & Halbrock, K. (1987) Occurrence of phytoalexins and other putative defense-related substances in uninfected plants. *Planta* **171**, 279–287.

Knoll, A.H. (1984) Patterns of extinction in the fossil record of vascular plants. In *Extinctions* (Ed. M. Nitecki), pp. 21–67. University of Chicago Press, Chicago.

Knoll, A.H. (1986) Patterns of change in plant communities through geologic times. In *Community Ecology* (Eds J. Diamond & T.J. Case), pp. 125–141. Harper & Row, New York.

Knowles, P. & Grant, M.C. (1983) Age and size structure analyses of Engelmann spruce, Ponderosa pine, Lodgepole pine and Limber pine in Colorado. *Ecology* **64**, 1–9.

Knox, P.B. (1967) Apomixis: seasonal and population differences in a grass. *Science* **157**, 325–326.

Knox, R.G., Peet, R.K. & Christensen, N.L. (1989) Population changes in loblolly pine stands: changes in skewness and size inequality. *Ecology* **70**, 1153–1166.

Kochhar, M., Blum, U. & Reinert, R.A. (1980) Effects of O_3 and(or) fescue on ladino clover: interactions. *Canadian Journal of Botany* **58**, 241–249.

Koide, R.T., Huenneke, L.F., Hamburg, S.P. & Mooney, H.A. (1988) Effects of applications of fungicide, phosphorus and nitrogen on the structure and productivity of an annual serpentine plant community. *Functional Ecology* **2**, 335–344.

Körner, C.H. (1993) CO_2 fertilization: The great uncertainty in future vegetation development. In *Vegetation Dynamics and Global Change*, (Eds A.M. Solomon,

H.H. Shugart), pp. 53–70. Chapman & Hall, London.

Körner, C.H. (1994) Leaf diffusive conductance in the major vegetation types of the globe. In *Ecophysiology of Photosynthesis* (Eds E.-D. Schulze & M.M. Caldwell), pp. 463–490. Springer-Verlag, Berlin.

Körner, C.H., Scheel, J.A. & Bauer, H. (1979) Maximum leaf diffusive conductance in vascular plants. *Photosynthetica* **13**, 45–83.

Kosuge, T. (1969) The role of phenolics in host response to infection. *Annual Review of Phytopathology* **7**, 195–222.

Koyama, H. & Kira, T. (1956) Intraspecific competition among higher plants. VIII. Frequency distribution of individual plant weight as affected by the interaction between plants. *Journal of the Institute of Polytechnics, Osaka City University* **7**, 73–94.

Koziol, M.J. & Whatley, F.R. (Eds) (1984) *Gaseous Air Pollutants and Plant Metabolism*. Butterworth, London.

Kozlowski, T.T. (1972) *Water Deficits and Plant Growth*. Academic Press, London.

Kozlowski, T.T. (1981) *Water Deficits and Plant Growth VI. Woody Plant Communities*. Academic Press, New York.

Kozlowski, T.T. (Ed.) (1984) *Flooding and Plant Growth*, Academic Press, London.

Kozlowski, T.T. (Ed.) (1992) Carbohydrate sources and sinks in woody plants. *Botanical Review* **58**, 107–222.

Kozlowski, T.T. & Ahlgren, C.E. (Eds) (1974) *Fire and Ecosystems*. Academic Press, New York.

Kramer, P.J. (1980) Drought, stress, and the origin of adaptations. In *Adaptation of Plants to Water and High Temperature Stress* (Eds N.C. Turner & P.J. Kramer), pp. 7–20. John Wiley & Sons, New York.

Kramer, P.J. (1983) *Water Relations of Plants*. Academic Press, New York.

Kramer, P.J. & Boyer, J. (1995) *Water Relations of Plants and Soils*. Academic Press, San Diego.

Krannitz, P.G., Aarssen, L.A. & Dow, J.M. (1991) The effect of genetically based differences in seed size on seedling survival in *Arabidopsis thaliana* (Brassicaceae). *American Journal of Botany* **78**, 446–450.

Krefting, L.W. & Roe, E.I. (1949) The role of some birds and mammals in seed germination. *Ecological Monographs* **19**, 269–286.

Krieger, R.I., Feeny, P.P. & Wilkinson, C.F. (1971) Detoxification enzymes in the guts of caterpillars; an evolutionary answer to plant defense? *Science* **172**, 1578–1581.

Kruger, F.J. (1982) Prescribing fire frequencies in Cape fynbos in relation to plant demography. In *Dynamics and Management of Mediterranean-type Ecosystems* (Eds C.E. Conrad & W.C. Oechel), pp. 483–489. *USDA Forest Service General Technical Report PSW* 58.

Kuc, J. (1972) Phytoalexins. *Annual Review of Phytopathology* **10**, 207–232.

Kuc, J. (1982) Plant immunisation – mechanisms and practical implications. In *Active Defense Mechanisms in*

Plants (Eds R.K.S. Wood & E. Tjamos), pp. 157–178. Plenum Press, New York.

Kunin, W.E. (1993) Sex and the single mustard – population-density and pollinator behaviour effects on seed-set. *Ecology* **74**, 2145–2160.

Lambers, H. (1993) Rising CO₂, secondary plant metabolism, plant–herbivore interactions and litter decomposition. *Vegetatio* **104/105**, 263–271.

Lande, R. (1993) Risks of population extinction from demographic and environmental stochasticity and random catastrophes. *American Naturalist* **142**, 911–927.

Lande, R. & Barrowclough, G.F. (1987) Effective population size, genetic variation and their use in population management. In *Viable Populations of Conservation* (Ed. M.E. Soulé), pp. 87–123. Cambridge University Press, Cambridge.

Lande, R., Engen, S. & Saether, B.E. (1994) Optimal harvesting, economic discounting and extinction risk in fluctuating populations. *Nature* **372**, 88–90.

Landsburg, J. & Ohmart, C. (1989) Levels of insect defoliation in forests – patterns and concepts. *Trends in Ecology and Evolution* **4**, 96–100.

Lange, O.L. (1959) Untersuchungen über Wärmehaushalt und Hitzeresistenz mauretanischer Wüsten- und Savannenpflanzen. *Flora (Jena)* **147**, 595–651.

Lange, O.L. & Kappen, L. (1972) Photosynthesis of lichens from Antartica. *Antarctic Research Series* **20**, 80–95. American Geophysical Union.

Lange, O., Schulze, E.D., Evanari, M., Kappen, L. & Buschbom, U. (1974) The temperature-related photosynthetic capacity of plants under desert conditions. I. Seasonal changes of the photosynthetic response to temperature. *Oecologia* **17**, 91–110.

Lange, O.L., Meyer, A., Zellinger, H. & Heber, U. (1994) Photosynthesis and water relations of lichen soil crusts: field measurements in the coastal fog zone of the Namib Desert. *Functional Ecology* **8**, 253–264.

Larcher, W. (1995) *Physiological Plant Ecology*. Springer-Verlag, Berlin.

Larcher, W. & Bauer, H. (1981) Ecological significance of resistance to low temperature. In *Encyclopedia of Plant Physiology. New Series, Vol. 12A Physiological Plant Ecology 1* (Eds O.L. Lange, P.S. Nobel, C.B. Osmond & H. Ziegler), pp. 403–437. Springer-Verlag, Berlin.

Larson, D. (1991) *Dispersal ecology of phainopeplas and mistletoes*. PhD dissertation, University of Illinois at Chicago.

Larsson, S. (1989) Stressful times for the plant stress–insect performance hypothesis. *Oikos* **56**, 277–283.

Larsson, S., Wiren, A., Lundgren, L. & Ericsson, T. (1985) Effects of light and nutrient stress on leaf phenolic chemistry in *Salix dasyclados* and susceptibility to *Galerucella lineola* (Coleoptera). *Oikos* **45**, 205–210.

Larsson, S., Wiren, A., Lundgren, L. & Ericsson, T. (1986) Effects of light and nutrient stress on leaf phenolic chemistry in *Salix dasyclados* and susceptibility to *Galerucella lineola* (Coleoptera). *Oikos* **47**, 205–210.

Last, F.T. & Watling, R. (1991) Acidic deposition: its nature and impacts. *Proceedings of the Royal Society of Edinburgh Section* B, **97**, 343.

Latch, G.C.M. (1994) Influence of agremonium endophytes on perennial grass improvement. *New Zealand Journal of Agricultural Research* **37**, 311–318.

Laurie, S., Bradbury, M. & Stewart, G.R. (1994) Relationships between leaf temperature, compatible solutes and anti-transpirant treatment in some desert plants. *Plant Science* **100**, 147–156.

Law, R. (1975) *Colonization and the evolution of life histories in* Poa annua. PhD thesis, University of Liverpool.

Law, R. (1981) The dynamics of a colonizing population of *Poa annua*. *Ecology* **62**, 1267–1277.

Law, R. (1983) A model for the dynamics of a plant population containing individuals classified by age and size. *Ecology* **64**, 224–230.

Law, R. (1988) Some ecological properties of intimate mutualisms involving plants. In *Plant Population Ecology* (Eds A.J. Davy, M.J. Hutchings & A.R. Watkinson), pp. 315–342. Blackwell Scientific Publications, Oxford.

Law, R. & Watkinson, A.R. (1989) Competition. In *Ecological Concepts* (Ed. J.M. Cherrett), pp. 243–284. Blackwell Scientific Publications, Oxford.

Law, R., McLellan, A. & Mahdi, A.-K.S. (1993) Spatio-temporal processes in a calcareous grassland. *Plant Species Biology* **8**, 175–193.

Laws, R.M., Parker, I.S.C. & Johnstone, R.C.B. (1975) *Elephants and their Habitat*. Clarendon Press, Oxford.

Lawton, J.H. (1976) The structure of the arthropod community on bracken. *Botanical Journal of the Linnean Society* **73**, 187–216.

Lawton, J.H. (1982) Vacant niches and unsaturated communities: a comparison of bracken herbivores at sites on two continents. *Journal of Animal Ecology* **51**, 573–595.

Lawton, J.H. (1983) Plant architecture and the diversity of phytophagous insects. *Annual Review of Entomology* **28**, 23–39.

Lawton, J.H. (1987) Food-shortage in the midst of apparent plenty: the case for birch feeding insects. In *Proceedings of the Third World European Congress of Entomology* (Ed. H.W. Velthuis), pp. 219–228. Nederlandse Entomolgische Verening, Amsterdam.

Lawton, J.H. & McNeill, S. (1979) Between the devil and the deep blue sea: on the problem of being a herbivore. In *Population Dynamics* (Eds K. Anderson, B. Turner & L.R. Taylor), pp. 223–245. Blackwell Scientific Publications, Oxford.

Leather, S.R. & Barbour, D.A. (1987) Associations between soil type, lodgepole pine (*Pinus contorta*) provenance, and the abundance of the pine beauty moth, *Panolis flammea*. *Journal of Applied Ecology* **24**, 945–951.

Leck, M.A., Parker, V.T. & Simpson, R.L. (Eds) (1989) *Ecology of Soil Seed Banks*. Academic Press, London.

Lederau, M. (1995) Plant growth and defense: reply to Herms and Mattson. *Trends in Ecology and Evolution* **10**, 39.

Lederau, M., Litvak, M. & Monson, R. (1994) Plant chemical defence: monoterpenes and the growth–differentiation balance hypothesis. *Trends in Ecology and Evolution* **9**, 58–61.

Lee, J.A., Press, M.C., Studholme, C. & Woodin, S.J. (1988) Effects of acidic deposition on wetlands. In *Acid Rain and Britain's Natural Ecosystems* (Eds M.R. Ashmore, J.N.B. Bell & C. Garretty), pp. 27–37. Imperial College Centre for Environmental Technology, London.

Lee, J.H., Hubel, A. & Schoffl, F. (1995) Derepression of the activity of genetically engineered heat shock factor causes constitutive synthesis of heat shock proteins and increased thermo-tolerance in transgenic *Arabidopsis*. *Plant Journal* **8**, 603–612.

Lefkovitch, L.P. (1965) The study of population growth in organisms grouped by stages. *Biometrics* **21**, 1–18.

Legg, C.J., Maltby, E. & Proctor, M.C.F. (1992) The ecology of severe moorland fire on the North York moors – seed distribution and seedling establishment of *Calluna vulgaris*. *Journal of Ecology* **80**, 737–752.

Lehman, J.T. (1982) Microscale patchiness of nutrients in plankton communities. *Science* **216**, 729–730.

Lehmann, E. (1909) Zur Keimungsphysiologie und biologie von *Ranunculus sceleratus* L. und einingen anderan Samen. *Berichte der Deutsche Botanische Gesellschaft* **27**, 476–494.

Lehtila, K. & Syrjanen, K. (1995a) Positive effects of pollination on subsequent size, reproduction, and survival of *Primula veris*. *Ecology* **76**, 1084–1098.

Lehtila, K. & Syrjanen, K. (1995b) Compensatory responses of 2 *Melampyrum* species after damage. *Functional Ecology* **9**, 511–517.

Leigh, E.G. (1975) Structure and climate in tropical rain forest. *Annual Review of Ecology and Systematics* **6**, 67–86.

Leigh, E.G. (Ed.) (1982) *The Ecology of a Tropical Forest: Seasonal Rhythms and Long-term Changes*. Smithsonian Institution Press, Washington, DC.

Leighton, M. (1993) Modeling dietary selectivity by Bornean orangutans – evidence for integration of multiple criteria in fruit selection. *International Journal of Primatology* **14**, 257–313.

Leishman, M.R. & Westoby, M. (1994) Hypotheses on seed size: tests using the semiarid flora of western New South Wales, Australia. *American Naturalist* **143**, 890–906.

Leishman, M.R., Westoby, M. & Jurado, E. (1995) Correlates of seed size variation – a comparison among 5 temperate floras. *Journal of Ecology* **83**, 517–529.

Leon, J.A. (1985) Germination strategies. In *Evolution: Essays in Honour of John Maynard Smith* (Eds P.J. Greenwood, P.H. Harvey & M. Slatkin), pp. 129–142. Cambridge University Press, Cambridge.

Leonard, H.J. (1987) *Natural Resources and Economic Development in Central America: a Regional Environmental Profile*. Transaction Books, Oxford.

Lepp, N.W. (1981) *Metals in the Environment, Vol. 2. Effect of Heavy Metal Pollution on Plants*. Applied Science Publishers, London.

Leps, J., Osbornova-Kosinova, J. & Rejmanek, M. (1982) Community stability, complexity and species life-history strategies. *Vegetatio* **50**, 53–63.

Lertzman, K.P. & Gass, C.L. (1983) Alternative methods of pollen transfer. In *Handbook of Experimental Pollination Biology* (Eds C.E. Jones & R.J. Little), pp. 474–489. Van Nostrand Reinhold, New York.

Lesica, P. & Shelly, J.S. (1995) Effects of reproductive mode on demography and life history in *Arabis fecunda* (Brassicaceae). *American Journal of Botany* **82**, 752–762.

Leslie, P.H. (1945) On the use of matrices in certain population mathematics. *Biometrika* **33**, 183–212.

Leslie, P.H. (1948) Some further notes on the use of matrices in population mathematics. *Biometrika* **35**, 213–245.

Leuning, R. (1995) A critical appraisal of a combined stomatal photosynthesis model for C3 plants: a theoretical paper. *Plant, Cell and Environment* **18**, 339–356.

Leverich, W.J. & Levin, D.A. (1979) Age-specific survivorship and reproduction in *Phlox drummondii*. *American Naturalists* **113**, 881–903.

Levey, D.J. (1986) Methods of seed processing by birds and seed deposition patterns. In *Frugivores and Seed Dispersal* (Eds A. Estrada & T.H. Fleming), pp. 147–158. Dr W. Junk, Dordrecht.

Levey, D.J. & Byrne, M.M. (1993) Complex ant plant interactions – rain forest ants as secondary dispersers and post dispersal seed predators. *Ecology* **74**, 1802–1812.

Levey, D.J. & Grajal, A. (1991) Evolutionary implications of fruit-processing limitations in cedar waxwings. *American Naturalist* **138**, 171–189.

Levin, B.R., Stewart, F.M. & Chao, L. (1977) Resource-limited growth, competition, and predation: a model and experimental studies with bacteria and bacteriophage. *American Naturalist* **111**, 3–24.

Levin, D.A. (1975) Gametophytic selection in *Phlox*, In: *Gamete Competition in Plants and Animals* (Ed. D.L. Mulcahy), pp. 207–217. North-Holland Publishing Company, Oxford.

Levin, D.A. (1983) An immigration–hybridization episode in *Phlox*. *Evolution* **37**, 575–582.

Levin, D.A. (1988) The paternity pools of plants. *American Naturalist* **132**, 309–317.

Levin, D.A. & Funderburg, S.W. (1979) Genome size in angiosperms: temperate versus tropical species. *American Naturalist* **114**, 784–795.

Levin, D.A. & Kerster, H.W. (1974) Gene flow in seed plants. *Evolutionary Biology* **7**, 139–220.

Levin, S.A. (1974) Dispersion and population interactions. *American Naturalist* **108**, 207–228.

Levin, S.A. & Segel, L.A. (1976) Hypothesis for the origin of plankton patchiness. *Nature* **259**, 659.

Levins, R. (1968) *Evolution in Changing Environments.* Princeton University Press, Princeton.

Levins, R. (1969) Dormancy as an adaptive strategy. In: *Dormancy and Survival* (Ed. H.W. Woolhouse) **23**, 1–10. Cambridge University Press, Cambridge.

Levins, R. (1979) Coexistence in a variable environment. *American Naturalist* **114**, 765–783.

Levins, R. & Culver, D. (1971) Regional coexistence of species and competition between rare species. *Proceedings of the National Academy of Sciences USA* **68**, 1246–1248.

Levins, R. & Lewontin, R. (1980) Dialectics and reductionism in ecology. In *Conceptual Issues in Ecology* (Ed. E. Saarinen), pp. 107–138. Reidel, London.

Levitt, J. (1972) *Responses of Plants to Environmental Stresses.* Academic Press, London.

Levitt, J. (1978) An overview of freezing injury and survival, and its inter-relationships with other stresses. In *Plant Cold Hardiness and Freezing Stress* (Eds P.H. Li & A. Sakai). Academic Press, London.

Lewis, A.C. (1979) Feeding preference for diseased and wilted sunflower in the grasshopper *Malanopus differentialis*. *Entomologia Experimentalis et Applicata* **26**, 202–207.

Lewis, M.A. (1994) Spatial coupling of plant and herbivore dynamics – the contribution of herbivore dispersal to transient and persistent waves of damage. *Theoretical Population Biology* **45**, 277–312.

Leyval, C., Singh, B.R. & Joner, E.J. (1995) Occurrence and infectivity of arbuscular mycorrhizal fungi in some Norwegian soils influenced by heavy metals and soil properties. *Water Air and Soil Pollution* **84**, 201–216.

Li, B. (1995) *Studies of weed–crop competition.* PhD thesis, University of East Anglia.

Li, M., Lieberman, M. & Lieberman, D. (1996) Seedling demography in undisturbed tropical wet forest in Costa Rica. In *Ecology of Tropical Forest Tree Seedlings* (Ed. M.D. Swaine). UNESCO/Parthenon, Paris/Carnforth.

Liddle, M.J., Budd, C.S.J. & Hutchings, M.J. (1982) Population dynamics and neighbourhood effects in establishing swards of *Festuca rubra* L. *Oikos* **38**, 52–59.

Lidon, F.C. & Henriques, F.S. (1992) Copper toxicity in rice – diagnostic criteria and effect on tissue Mn and Fe. *Soil Science* **154**, 130–135.

Lieberman, D., Hall, J.B., Swaine, M.D. & Lieberman, M. (1979) Seed dispersal by baboons in the Shai Hills, Ghana. *Ecology* **60**, 65–75.

Lincoln, D.E. (1993) The influence of plant carbon dioxide and nutrient supply on susceptibility to insect herbivores. *Vegetatio* **104/105**, 273–280.

Lincoln, D.E. & Mooney, H.A. (1984) Herbivory on *Diplacus aurantiacus* shrubs in sun and shade. *Oecologia* **64**, 173–176.

Lincoln, D.E., Newton, T.S., Ehrlich, P.R. & Williams, K.S. (1982) Coevolution of the checkerspot butterfly *Euphydras chalcedona* and its larval food plant *Diplacus aurantiacus*: larvae response to protein and leaf resin. *Oecologia* **52**, 216–223.

Lincoln, D.E., Fajer, E.D. & Johnson, R.H. (1993) Plant insect herbivore interactions in elevated CO_2 environments. *Trends in Ecology and Evolution* **8**, 64–68.

Linder, S. & Axelsson, B. (1982) Changes in carbon uptake and allocation patterns as a result of irrigation and fertilisation in a young *Pinus sylvestris* stand. In *Carbon Uptake and Allocation in Subalpine Ecosystems as a Key to Management* (Ed. R.H. Waring), pp. 38–44. Oregon State University, Corvallis.

Linhart, Y.B. (1976) Density-dependent seed germination strategies in colonising versus non-colonising species. *Journal of Ecology* **64**, 375–380.

Linhart, Y.B. & Thompson, J.D. (1995) Terpene-based selective herbivory by *Helix aspersa* (Mollusca) on *Thymus vulgaris* (Labiatae). *Oecologia* **102**, 126–132.

Liu, S., Norris, D.M., Hartwig, E.E. & Xu, M. (1992) Inducible phytoalexins in juvenile soybean genotypes predict soybean looper resistance in the fully developed plants. *Plant Physiology* **100**, 1479–1485.

Lloyd, D.G. (1974) Theoretical sex ratios in dioecious and gynodioecious angiosperms. *Heredity* **32**, 11–34.

Lloyd, D.G. (1982) Selection of combined versus separate sexes in seed plants. *American Naturalist* **120**, 571–585.

Lloyd, D.G. (1984) Gender allocations in outcrossing cosexual plants. In *Perspectives on Plant Population Ecology* (Eds R. Dirzo & J. Sarukhán), pp. 277–300. Sinauer, Sunderland, Massachusetts.

Lloyd, D.G. & Barrett, S. (1996) *Floral Biology.* Chapman & Hall, New York.

Lloyd, D.G. & Bawa, K.S. (1984) Modification of the gender of seed plants in varying conditions. *Evolutionary Biology* **17**, 255–338.

Lloyd, D.G. & Webb, C.J. (1977) Secondary sex characters in plants. *Botanical Review* **43**, 177–216.

Lloyd, M. (1967) Mean crowding. *Journal of Animal Ecology* **36**, 1–30.

Logan, D.C. & Stewart, G.R. (1992) Germination of the seeds of parasitic angiosperms. *Seed Science Review* **2**, 179–190.

Lokesha, R., Hedge, S.G., Uma Shaanker, R. & Ganeshaiah, J.N. (1992) Dispersal mode as a selective force in shaping the chemical composition of seeds. *American Naturalist* **140**, 520–525.

Lonsdale, W.M. (1989) Interpreting seed survivorship curves. *Oikos* **52**, 361–364.

Lonsdale, W.M. (1990) The self-thinning rule: dead or alive? *Ecology* **71**, 1373–1388.

Lonsdale, W.M. (1993) Rates of spread of an invading species – *Mimosa pigra* in northern Australia. *Journal of Ecology* **81**, 513–521.

Lorenz, E.M. (1963) Deterministic non-periodic flow. *Journal of Atmospheric Sciences* **20**, 130–141.

Lotka, A.J. (1925) *Elements of Physical Biology.* Williams & Wilkins, Baltimore.

Louda, S.M. (1982) Limitation of the recruitment of the shrub *Haplopappus squarrosus* (Asteraceae) by flower- and seed-feeding insects. *Journal of Ecology* **70**, 43–53.

[671]

Louda, S.M. (1983) Seed predation and seedling mortality in the recruitment of a shrub *Haplopappus venetus* (Asteraceae), along a climatic gradient. *Ecology* **64**, 511–521.

Louda, S.M. & Collinge, S.K. (1992) Plant resistance to insect herbivores – field test of the environmental stress hypothesis. *Ecology* **73**, 153–169.

Louda, S.M. & Mole, S. (1991) Glucosinolates: chemistry and ecology. In *Herbivores: their Interaction with Secondary Plant Metabolites*, 2nd edn (Eds G.A. Rosenthal & M.R. Berenbaum), Vol. 1, pp. 124–164. Academic Press, San Diego.

Louda, S.M. & Potvin, M.A. (1995) Effect of inflorescence feeding insects on the demography and lifetime fitness of a native plant. *Ecology* **76**, 229–245.

Louda, S.M., Ferris, M.A. & Blaa, M.J. (1987) Variation in methylglucosinolate and insect damage to *Cleome serrulata* along a natural soil moisture gradient. *Journal of Chemical Ecology* **13**, 569–582.

Louda, S.M., Keeler, K.H. & Holt, R.D. (1990) Herbivore influences on plant performance and competitive interactions. In *Perspectives on Plant Competition* (Eds J.B. Grace & D. Tilman), pp. 413–444. Academic Press, New York.

Lousley, J.E. (Ed.) (1953) *The Changing Flora of Britain*. Buncle, Arbroath.

Loveless, M.D. & Hamrick, J.L. (1984) Ecological determinants of genetic structure in plant populations. *Annual Review of Ecology and Systematics* **15**, 65–95.

Lovett Doust, L. (1981) Population dynamics and local specialization in a clonal perennial (*Ranunculus repens*). I. The dynamics of ramets in contrasting habitats. *Journal of Ecology* **69**, 743–755.

Lovric, L.M. & Lovric, A.Z. (1984) Morpho-anatomical syndromes in phyto-indicators of extreme stormy habitats in the north-eastern Adriatic. In *Being Alive on Land* (Eds N.S. Margaris, M. Arianoustou-Faraggitaki & W.C. Oechel), pp. 41–49. Junk, The Hague.

Lowenberg, G.J. (1994) Effects of floral herbivory on maternal reproduction in *Sanicula arctopoides* (Apiaceae). *Ecology* **75**, 359–369.

Lowman, M.D. & Box, J.R. (1983) Variation in leaf toughness and phenolic content among five species of Australian rain forest trees. *Australian Journal of Ecology* **5**, 31–35.

Lubchenco, J. (1978) Plant species diversity in marine intertidal community: importance of herbivore food preference and algal competitive abilities. *American Naturalist* **112**, 23–39.

Ludwig, D. & Levin, S.A. (1991) Evolutionary stability of plant communities and the maintenance of multiple dispersal types. *Theoretical Population Biology* **40**, 285–307.

Lynch, S.P. & Martin, R.A. (1987) Cardenolide content and TLC profiles on Monarch butterflies and their larval host-plant milkweed *Asclepias viridis* in Northwest Louisiana. *Journal of Chemical Ecology* **13**, 47–70.

Lyons, E.E., Miller, D. & Meagher, T.R. (1994) Evolutionary dynamics of sex ratio and gender dimorphism in *Silene latifolia*. 1. Environmental effects. *Journal of Heredity* **85**, 196–203.

van der Maarel, E. & Leertouwer, J. (1967) Variation in vegetation and species diversity along a local environmental gradient. *Acta Botanica Neerlandica* **16**, 211–221.

MacArthur, R.H. (1972) *Geographical Ecology. Patterns in the Distribution of Species*. Harper & Row, New York.

MacArthur, R.H. & MacArthur, J. (1961) On bird species diversity. *Ecology* **42**, 594–598.

MacArthur, R.H. & Wilson, E.O. (1967) *The Theory of Island Biogeography*. Princeton University Press, Princeton.

McCanny, S.J., Keddy, P.A., Arnason, T.J., Gaudet, C.L., Moore, D.R.J. & Shipley, B. (1990) Fertility and the food quality of wetland plants: a test of the resource availability hypothesis. *Oikos* **59**, 373–381.

McClure, M.S. (1983) Competition between herbivores and increased resource heterogeneity. In *Variable Plants and Herbivores in Natural and Managed Systems*. (Eds R.F. Denno & M.S. McClure), pp. 125–153. Academic Press, New York.

McConnaughay, K.D.M. & Bazzaz, F.A. (1987) The relationship between gap size and performance of several colonizing annuals. *Ecology* **68**, 411–416.

McCullough, D.G. & Kulman, H.M. (1991) Effects of nitrogen fertilisation on young jack pine (*Pinus banksiana*) and on its suitability as a host for jack budworm (*Choristoneura pinus pinus*) (Lepidoptera: Tortricidae). *Canadian Journal of Forest Research* **21**, 1447–1458.

MacDonald, N. & Watkinson, A.R. (1981) Models of an annual plant population with a seedbank. *Journal of Theoretical Biology* **93**, 643–653.

Macevicz, S. & Oster, G. (1976) Modelling social insect populations. II. Optimal reproductive strategies in annual eusocial insect colonies. *Behavioral Ecology and Sociobiology* **1**, 265–282.

McEvoy, P.B. & Rudd, N.T. (1993) Effects of vegetation disturbances on biological control of tansy ragwort, *Senecio jacabaea*. *Ecological Applications* **3**, 682–698.

McGonigle, T.P. (1988) A numerical analysis of published field trials with vesicular arbuscular mycorrhizal fungi. *Functional Ecology* **2**, 473–478.

McGuiness, K.A. (1984) Equations and explanations in the study of species area curves. *Biological Reviews* **59**, 423–440.

McInnes, P.F., Naiman, R.J., Pastor, J. & Cohen, Y. (1992) Effects of moose browsing on vegetation and litter of the boreal forest, Isle Royale, Michigan, USA. *Ecology* **73**, 2059–2075.

McIntyre, S., Lavorel, S. & Tremont, R.M. (1995) Plant life history attributes – their relationship to disturbance responses in herbaceous vegetation. *Journal of Ecology* **83**, 31–44.

Mack, R.N. (1981) Invasion of *Bromus tectorum* L. into western North America: an ecological chronicle. *Agro-Ecosystems* **7**, 145–165.

Mack, R.N. & Harper, J.L. (1977) Interference in dune

annuals: spatial pattern and neighbourhood effects. *Journal of Ecology* **65**, 345–363.

Mack, R.N. & Pyke, D.A. (1983) The demography of *Bromus tectorum*: variation in time and space. *Journal of Ecology* **71**, 69–93.

Mack, R.N. & Pyke, D.A. (1984) The demography of *Bromus tectorum*: the role of micro-climate, grazing and disease. *Journal of Ecology* **72**, 731–748.

McKane, R.B., Grigal, D.F. & Russelle, M. (1990) Spatio-temporal differences in ^{15}N uptake and the organization of an old-field plant community. *Ecology* **71**, 1126–1132.

McKay, A.C., Ophel, K.M., Reardon, T.B. & Gooden, J.M. (1993) Livestock deaths associated with *Clavibacter toxicus anguina* sp. infection in seed heads of *Agrostis avenacea* and *Polypogon monspeliensis*. *Plant Disease* **77**, 635–641.

McKee, K.L. (1995) Mangrove species distribution and propagule predation in Belize – an exception to the dominance predation hypothesis. *Biotropica* **27**, 334–345.

McKey, D. (1979) The distribution of secondary compounds within plants. In *Herbivores: their Interaction with Secondary Plant Metabolites* (Eds G.A. Rosenthal & D.H. Janzen), pp. 55–133. Academic Press, London.

Mclenahen, J.R. (1978) Community changes in a deciduous forest exposed to air pollution. *Canadian Journal of Forest Research* **8**, 432–438.

MacLeod, J. (1894) Over de bevruchting der bloemen in het Kempisch gedeelte van Vlaanderen. *Deel II. Bot. Jaarboek Dodonaea* **6**, 119–511.

McMahon, T. (1975) The mechanical design of trees. *Scientific American* **233**, 92–102.

McNaughton, S.J. (1968) Autotoxic feedback in relation to germination and seedling growth in *Typha latifolia*. *Ecology* **49**, 367–369.

McNaughton, S.J. (1979) Grazing as an optimization process: grass–ungulate relationships in the Serengeti. *American Naturalist* **113**, 691–703.

McNaughton, S.J. (1983) Compensatory growth as a response to herbivory. *Oikos* **40**, 329–336.

McNaughton, S.J. (1986) On plants and herbivores. *American Naturalist* **128**, 765–770.

McNaughton, S.J. (1992) Laboratory simulated grazing – interactive effects of defoliation and canopy closure on Serengeti grasses. *Ecology* **73**, 170–182.

McNaughton, S.J. (1993) Grasses and grazers, science and management. *Ecological Applications* **3**, 17–20.

McNeill, S. & Southwood, T.R.E. (1978) The role of nitrogen in the development of insect/plant relationships. In *Biochemical Aspects of Plant and Animal Co-evolution* (Ed. J.B. Harbone), pp. 77–98. Academic Press, London.

McNeilly, T. & Antonovics, J. (1968) Evolution in closely adjacent plant populations. IV. Barriers to gene flow. *Heredity* **23**, 205–218.

McPartlan, H.C. & Dale, P.J. (1994) An assessment of gene-transfer by pollen from field-grown transgenic potatoes to non-transgenic potatoes and related species. *Transgenic Research* **3**, 216–225.

McPherson, G.R. (1993) Effects of herbivory and herb interference on oak establishment in a semiarid temperate savanna. *Journal of Vegetation Science* **4**, 687–692.

McPherson, J.K. & Muller, C.H. (1969) Allelopathic effects of *Adenostoma fasciculatum* 'chamise', in the Californian chaparral. *Ecological Monographs* **39**, 177–198.

Maekawa, T. (1924) On the phenomenon of sex transition in *Arisaema japonica*. *Journal of the College of Agriculture, Hokkaido Imperial University* **13**, 217–305.

Maher, T.F. & Shepherd, R.F. (1992) Mortality and height growth losses of coniferous seedlings damaged by the black army cutworm. *Canadian Journal of Forest Research* **22**, 1364–1370.

Maiorana, V.C. (1981) Herbivory in sun and shade. *Biological Journal of the Linnean Society* **15**, 151–156.

Major, E.J. (1990) Water stress in Sitka spruce and its effect on the green spruce aphid *Elatobium abietinum*. In *Population Dynamics of Forest Insects* (Eds A.D. Watt, S.R. Leather, M.D. Hunter & N.A.C. Kidd), pp. 85–94. Intercept, Andover.

Makela, A., Materna, J. & Schopp, W. (1987) Direct effects of sulfur on forests in Europe – a regional model of risk. Working Paper 87–577, International Institute for Applied Systems Analysis, Laxenburg, Austria.

Mallik, A.U., Hobbs, R.J. & Legg, C.J. (1984) Seed dynamics in *Calluna–Arctostaphylos* heath in northeastern Scotland. *Journal of Ecology* **72**, 855–871.

Malloch, D.W., Pirozynski, K.A. & Raven, P.H. (1980) Ecological and evolutionary significance of mycorrhizal symbioses in vascular plants (a review). *Proceedings of the National Academy of Sciences USA* **77**, 2113–2118.

Malo, J.E. & Suarez, F. (1995) Cattle dung and the fate of *Biserrula pelecinus* (Leguminosae) in a Mediterranean pasture – seed dispersal, germination and recruitment. *Botanical Journal of the Linnean Society* **118**, 139–148.

Mamolos, A.P., Veresoglou, D.S. & Barbayiannis, N. (1995) Plant species abundance and tissue concentrations of limiting nutrients in low nutrient grasslands – a test of competition theory. *Journal of Ecology* **83**, 485–495.

Manasse, R.S. & Howe, H.F. (1983) Competition for dispersal agents among tropical trees: influences of neighbors. *Oecologia* **59**, 185–190.

Manicacci, D. & Barrett, S.C.H. (1995) Stamen elongation, pollen size, and siring ability in tristylous *Eichhornia paniculata* (Pontederiaceae). *American Journal of Botany* **82**, 1381–1389.

Mann, J. (1978) *Secondary Metabolism*. Clarendon Press, Oxford.

Marcone, C., Firrao, G., Ragozzino, A. & Locci, R. (1994) Detection of MLOs in declining alder trees in southern Italy and their characterization by RFLP analysis. *European Journal of Forest Pathology* **24**, 217–228.

Marks, P.L. (1974) The role of pin cherry (*Prunus pennsylvanica* L.) in the maintenance of stability in northern hardwood ecosystems. *Ecological Monographs* **44**, 73–88.

Marquis, R.J. (1984) Leaf herbivores decreased fitness of a tropical plant. *Science* **226**, 537–539.

Marquis, R.J. (1992) A bite is a bite is a bite? Constraints on response to folivory in *Piper arieianum* (Piperaceae). *Ecology* **73**, 143–152.

Marquis, R.J. & Whelan, C.J. (1994) Insectivorous birds increase growth of white oak through consumption of leaf-chewing insects. *Ecology* **75**, 2007–2014.

Marrs, R.H. (1993) An assessment of change in *Calluna* heathlands in Breckland, eastern England, between 1983 and 1991. *Biological Conservation* **65**, 133–139.

Marrs, R.H., Roberts, R.D., Skeffington, R.A. & Bradshaw, A.D. (1983) Nitrogen and the development of ecosystems. In *Nitrogen as an Ecological Factor* (Eds J.A. Lee, S. McNeill & I.H. Rorison), pp. 113–136. Blackwell Scientific Publications, Oxford.

Marschner, H., Römheld, V., Horst, W.J. & Martin, P. (1986) Root-induced changes in the rhizosphere: importance for the mineral nutrition of plants. *Zeitschrift für Pflanzenernährung und Bodenkunde* **149**, 441–456.

Marschner, H., Treeby, M. & Romheld, V. (1989) Role of root-induced changes in the rhizosphere for iron acquisition in higher plants. *Zeitschrift für Pflanzenernährung und Bodenkunde* **152**, 197–204.

Marshall, D.R. & Jain, S.K. (1969) Interference in pure and mixed populations of *Avena fatua* and *A. barbata*. *Journal of Ecology* **57**, 251–270.

Martens, S.N. & Boyd, R.S. (1994) The ecological significance of nickel hyperaccumulation: a plant chemical defence. *Oecologia* **98**, 379–384.

Martin, J.S., Martin, M.M. & Bernays, E.A. (1987) Failure of tannic acid to inhibit digestion or reduce digestibility of plant protein in gut fluids of insect herbivores: implication for theories of plant defense. *Journal of Chemical Ecology* **13**, 605–621.

Martin, M.M. & Martin, J.S. (1984) Surfactants: their role in preventing the precipitation of proteins in insect guts. *Oecologia* **61**, 342–345.

Martinez del Rio, C. & Restrepo, C. (1993) Ecological and behavioral consequences of digestion in frugivorous animals. *Vegetatio* **107/108**, 205–216.

Martinez del Rio, C., Baker, H.G. & Baker, I. (1992) Ecological and evolutionary implications of digestive processes: bird preferences and the sugar constituents of floral nectar and fruit pulp. *Experientia* **48**, 544–551.

Martínez-Ramos, M. & Soto-Castro, A. (1993) Seed rain and advanced regeneration in a tropical rain forest. *Vegetatio* **107/108**, 299–318.

Martínez-Ramos, M., Sarukhán, J. & Piñero, D. (1988) The demography of tropical trees in the context of forest gap dynamics. In *Plant Population Ecology* (Eds A.J. Davy, M.J. Hutchings & A.R. Watkinson), pp. 293–313. Blackwell Scientific Publications, Oxford.

Masaki, T., Kominami, Y. & Nakashizuka, T. (1994) Spatial and seasonal patterns of seed dissemination of *Cornus controversa* in a temperate forest. *Ecology* **75**, 1903–1910.

Mason, H.L. (1947) Evolution in certain floristic associations in western North America. *Ecological Monographs* **17**, 201–210.

Masters, G.J. & Brown, V.K. (1992) Plant mediated interactions between two spatially separated insects. *Functional Ecology* **6**, 175–179.

Masters, G.J. & Brown, V.K. (1997) Host–plant mediated interactions between spatially separated herbivores: effects on community structure. In *BES Symposium on Multitrophic Interactions in Terrestrial Systems*. Blackwell Science, Oxford.

Matches, A.G. (1992) Plant response to grazing – a review. *Journal of Production Agriculture* **5**, 1–7.

Mateille, T. (1994) Biology of the plant nematode relationship – physiological changes and the defense mechanism of plants. *Nematologica* **40**, 276–311.

Matson, P.A. & Hunter, M.D. (1992) The relative contributions of top-down and bottom-up forces in population and community ecology. *Ecology* **73**, 723.

Matthews, N.J. & Flegg, J.J.M. (1981) Seeds, buds and bullfinches. In *Pests, Pathogens and Vegetation* (Ed. J.M. Thresh), pp. 375–383. Pitman, London.

Mattson, W.J. Jr (1980) Herbivory in relation to plant nitrogen content. *Annual Review of Ecology and Systematics* **11**, 119–161.

Mattson, W.J. & Haack, R.A. (1987) The role of drought stress in provoking outbreaks of phytophagous insects. In *Insect Outbreaks* (Eds P. Barbosa & J.C. Schultz), pp. 365–407. Academic Press, New York.

Mattson, W.J. & Scriber, J.M. (1987) Nutritional ecology of insect folivores of woody plants: nitrogen, water, fiber, and mineral considerations. In *Nutritional Ecology of Insects, Mites, Spiders and Related Invertebrates* (Eds F. Slansky & J.G. Rodriguez), pp. 105–146. John Wiley & Sons, New York.

Maurer, R., Seemuller, E. & Sinclair, W.A. (1993) Genetic relatedness of mycoplasma like organisms affecting elm, alder, and ash in Europe and North America. *Phytopathology* **83**, 971–976.

Maurice, S., Charlesworth, D., Desfeux, C., Couvet, D. & Gouyon, P.H. (1993) The evolution of gender in hermaphrodites of gynodioecious populations with nucleocytoplasmic male sterility. *Proceedings of the Royal Society of London Series B* **251**, 253–261.

Maurice, S., Belhassen, E., Couvet, D. & Gouyon, P.H. (1994) Evolution of dioecy – can nuclear cytoplasmic interactions select for maleness? *Heredity* **73**, 346–354.

Mauricio, R., Bowers, M.D. & Bazzaz, F.A. (1993) Pattern of leaf damage affects fitness of the annual plant *Raphanus sativus* (Brassicaceae). *Ecology* **74**, 2066–2071.

Maxwell, F.G. & Jennings, P.R. (Eds) (1980) *Breeding Plants Resistant to Insects*. John Wiley & Sons, New York.

May, R.M. (1973) *Stability and Complexity in Model Ecosystems*. Princeton University Press, Princeton.

May, R.M. (1974) Biological populations with non-overlapping generations: stable points, stable cycles and chaos. *Science* **186**, 645–647.

May, R.M. (1975) Patterns of species abundance and diversity. In *Ecology and Evolution of Communities* (Eds M.L. Cody & J.M. Diamond), pp. 81–120. Harvard University Press, Cambridge, Massachusetts.

May, R.M. (1981b) Models for two interacting populations. In *Theoretical Ecology* (Ed. R.M. May), pp. 78–104. Blackwell Scientific Publications, Oxford.

May, R.M. (Ed.) (1981a) *Theoretical Ecology. Principles and Applications*, 2nd edn. Blackwell Scientific Publications, Oxford.

May, R.M. (1985) Evolutionary ecology and John Maynard Smith. In *Evolution: Essays in Honour of John Maynard Smith* (Eds P.J. Greenwood, P.H. Harvey & M. Slatkin), pp. 107–116. Cambridge University Press, Cambridge.

May, R.M. & MacArthur, R.H. (1972) Niche overlap as a function of environmental variability. *Proceedings of the National Academy of Sciences USA* **69**, 1109–1113.

May, R.M. & Oster, G.F. (1976) Bifurcations and dynamic complexity in simple ecological models. *American Naturalist* **110**, 573–599.

May, R.M. & Watts, C.H. (1992) The dynamics of predator–prey and resource–harvester systems. In *Natural Enemies: the Population Biology of Predators, Parasites and Diseases* (Ed. M.J. Crawley), pp. 431–457. Blackwell Scientific Publications, Oxford.

Maynard Smith, J. (1966) Sympatric speciation. *American Naturalist* **100**, 637–650.

Maynard Smith, J. (1972) *On Evolution*. Edinburgh University Press, Edinburgh.

Maynard Smith, J. (1978) *The Evolution of Sex*. Cambridge University Press, Cambridge.

Maynard Smith, J. (1982) *Evolution and the Theory of Games*. Cambridge University Press, Cambridge.

Maynard Smith, J. (1989) *Evolutionary Genetics*. Oxford University Press, Oxford.

Mazer, S.J. (1989) Ecological, taxonomic and life-history correlates of seed mass among Indiana dune angiosperms. *Ecological Monographs* **59**, 153–175.

Meagher, T.R. (1980) Population biology of *Chamaelirium luteum*, a dioecious lily. I. Spatial distribution of males and females. *Evolution* **34**, 1127–1137.

Meagher, T.R. (1986) Analysis of paternity within a natural population of *Chamaelirium luteum*. I. Identification of most likely male parents. *American Naturalist* **128**, 199–215.

Meagher, T.R. (1992) The quantitative genetics of sexual dimorphism in *Silene latifolia* (Caryophyllaceae). I. Genetic variation. *Evolution* **46**, 445–457.

Meagher, T.R. & Thompson, E. (1987) Analysis of parentage for naturally established seedlings of *Cha-maelirium luteum* (Liliaceae). *Ecology* **68**, 803–812.

Meharg, A.A. (1994) Integrated tolerance mechanisms – constitutive and adaptive plant responses to elevated metal concentrations in the environment. *Plant Cell and Environment* **17**, 989–993.

Mehrhoff, L.A. (1989) The dynamics of declining populations of an endangered orchid, *Isotria medeoloides*. *Ecology* **70**, 783–786.

Meidner, H. & Mansfield, T.A. (1968) *Physiology of Stomata*. McGraw-Hill, London.

Melillo, J.M., Aber, J.D. & Muratove, J.F. (1982) Nitrogen and lignin control of hardwood leaf litter decomposition dynamics. *Ecology* **63**, 621–626.

Mencuccini, M. & Grace, J. (1995) Climate influences the leaf area–sapwood relationship in Scots pine (*Pinus sylvestris* L.). *Tree Physiology* **15**, 1–10.

Mencuccini, M. & Grace, J. (1996) Hydraulic architecture parameters in a Scots pine chronosequence (Thetford, UK). *Plant, Cell and Environment*.

Menge, B.A. (1976) Organization of the New England rocky intertidal community: role of predation, competition, and environmental heterogeneity. *Ecological Monographs* **46**, 355–393.

Menge, B.A. & Sutherland, J.P. (1976) Species diversity gradients: synthesis of the roles of predation, competition, and temporal heterogeneity. *American Naturalist* **110**, 351–369.

Menges, E.S. (1990) Population viability analysis for an endangered plant. *Conservation Biology* **4**, 52–62.

Menges, E.S. (1991) Seed germination percentage increases with population size in a fragmented prairie species. *Conservation Biology* **5**, 158–164.

Mercer, C.F. (1994) Plant parasitic nematodes in New Zealand. *New Zealand Journal of Zoology* **21**, 57–65.

Mesmar, M.N. & Jaber, K. (1991) The toxic effect of lead on seed germination, growth, chlorophyll and protein contents of wheat and lens. *Acta Biologica Hungarica* **42**, 331–344.

Meyer, G.A. (1993) A comparison of the impacts of leaf feeding and sap feeding insects on growth and allocation of goldenrod. *Ecology* **74**, 1101–1116.

Meyer, G.A. & Montgomery, M.E. (1987) Relationships between leaf age and the food quality of cottonwood foliage for the gypsy moth, *Lymantria dispar*. *Oecologia* **72**, 527–532.

Meyer, G.A. & Root, R.B. (1993) Effects of herbivorous insects and soil fertility on reproduction of goldenrod. *Ecology* **74**, 1117–1128.

Mihaliak, C.A., Couvet, D. & Lincoln, D.E. (1989) Genetic and environmental contributions to variation in leaf mono- and sesquiterpenes of *Heterotheca subaxillaris*. *Biochemical Systematics and Ecology* **17**, 529–533.

Milburn, J.A. & Johnson, R.P.C. (1966) The conduction of sap. II. Detection of vibrations produced by sap cavitation in *Ricinus* xylem. *Planta* **69**, 43–52.

Milchunas, D.G., Lauenroth, W.K. & Chapman, P.L. (1992) Plant competition, abiotic, and long term and short term effects of large herbivores on demography of opportunistic species in a semi-arid grassland. *Oecologia* 92, 520–531.

Milchunas, D.G., Varnamkhasti, A.S., Lauenroth, W.K. & Goetz, H. (1995) Forage quality in relation to long term grazing history, current year defoliation, and water resource. *Oecologia* 101, 366–374.

Milewski, A.V. (1983) A comparison of ecosystems in Mediterranean Australia and Southern Africa: nutrient-poor sites at the Barrens and the Caledon Coast. *Annual Review of Ecology and Systematics* 14, 57–76.

Miller, K.R. (1979) The photosynthetic membrane. *Scientific American* 21, 102–113.

Miller, P.R. (1989) Concept of forest decline in relation to western US forests. In *Air Pollution's Toll on Forests and Crops* (Eds J.J. Mackenzie & M.T. El-Ashry), pp. 75–112. Yale University Press, New Haven.

Miller, T.E. & Weiner, J. (1989) Local density variation may mimic effects of asymmetric competition on plant size variability. *Ecology* 70, 1188–1191.

Miller, T.E. & Werner, P.A. (1987) Competitive effects and responses between plant species in a first-year old-field community. *Ecology* 68, 1201–1210.

Milton, S.J. (1995) Effects of rain, sheep and tephritid flies on seed production of 2 arid Karoo shrubs in South Africa. *Journal of Applied Ecology* 32, 137–144.

Milton, W. (1940) The effect of manuring, grazing and cutting on the yield, botanical and chemical composition of natural hill pastures. *Journal of Ecology* 28, 326–356.

Moermond, T.C. & Denslow, J.S. (1985) Neotropical avian frugivores: patterns of behavior, morphology and nutrition with consequences for fruit selection. In *Neotropical Ornithology* (Eds P.A. Buckley, M.S. Foster, E.S. Morton, R.S. Ridgely & N.G. Smith), pp. 865–897. *Ornithological Monographs* Vol. 36, American Ornithologists Union, Allen Press, Lawrence, Kansas.

Moesta, P. & Griesbach, H. (1982) L-2-Aminooxy-3-phenylpropionic acid inhibits phytoalexin accumulation in soybean with concomitant loss of resistance against *Phytophthora megasperma* f.sp. *glycinae*. *Physiological Plant Pathology* 21, 65–70.

Mogie, M. (1992) *The Evolution of Asexual Reproduction in Plants*. Chapman & Hall, London.

Mohan, K.S. & Pillai, G.B. (1993) Biological control of *Oryctes rhinoceros* (L.) using an Indian isolate of *Oryctes baculovirus*. *Insect Sciences and its Application* 14, 551–558.

Mohler, C.L., Marks, P.L. & Sprugel, D.G. (1978) Stand structure and allometry of trees during self-thinning of pure stands. *Journal of Ecology* 66, 599–614.

Moles, S. (1994) Trade-offs and constraints in plant herbivore defense theory – a life history perspective. *Oikos* 71, 3–12.

Mole, S. & Waterman, P.G. (1985) Stimulatory effects of tannins and cholic acid on tryptic hydrolysis of proteins: ecological implications. *Journal of Chemical Ecology* 9, 1323–1331.

Mole, S., Ross, J.A.M. & Waterman, P.G. (1988) Light-induced variation in phenolic levels in foliage of rain-forest plants. *Journal of Chemical Ecology* 14, 1–21.

Mole, S., Butler, L.G. & Iason, G. (1990) Defence against dietary tannin in herbivores: a survey for proline rich salivary proteins in mammals. *Biochemical Systematics and Ecology* 18, 287–293.

Mølgaard, P. (1986) Food plant preferences by slugs and snails: a simple method to evaluate the relative palatability of the food plants. *Biochemical Systematics and Ecology* 14, 113–121.

Moloney, K.A., Levin, S.A., Chiarello, N.R. & Buttel, L. (1992) Pattern and scale in a serpentine grassland. *Theoretical Population Biology* 41, 257–276.

Monson, R.K. (1989) On the evolutionary pathways resulting in C_4 photosynthesis and Crassulacean acid metabolism (CAM). *Advances in Ecological Research* 19, 57–110.

Monson, R.K., Harley, P.C., Litvak, M.E. *et al.* (1994) Environmental and developmental controls over the seasonal pattern of isoprene emission from aspen leaves. *Oecologia* 99, 260–270.

Montalvo, J., Casado, M.A., Levassor, C. & Pineda, F.D. (1993) Species diversity patterns in Mediterranean grasslands. *Journal of Vegetation Science* 4, 213–222.

Monteith, J.L. (1995) A reinterpretation of stomatal responses to humidity. *Plant, Cell and Environment* 18, 357–364.

Monteith, J.L. & Unsworth, M.H. (1990) *Principles of Environmental Physics*. Arnold, London.

Montgomery, M.E. (1982) Life-cycle nitrogen budget for the gypsy moth, *Lymantria dispar*, reared on an artificial diet. *Journal of Insect Physiology* 28, 437–442.

Mooney, H.A. (Ed.) (1977) *Convergent Evolution in Chile and California: Mediterranean Climate Ecosystems*. Dowden, Hutchinson & Ross, Stroudsburg, Pennsylvania.

Mooney, H.A. (1988) Lessons from Mediterranean–Chilean regions. In *Biodiversity* (Ed. Wilson, F.O.), pp. 157–165. National Academy Press, Washington DC.

Mooney, H.A. & Conrad, C.E. (Eds) (1977) *Environmental Consequences of Fire and Fuel Management in Mediterranean Ecosystems*. General Technical Reports of the USDA Forest Service (Washington) GTR-WO-3.

Mooney, H.A. & Godron, M. (Eds) (1983) *Disturbance and Ecosystems*. Springer-Verlag, Berlin.

Mooney, H.A. & Gulmon, S.L. (1982) Constraints on leaf structure and function in reference to herbivory. *BioScience* 32, 198–206.

Mooney, H.A. & Koch, G.W. (1994) The impact of rising CO_2 concentrations on the terrestrial biosphere. *Ambio* 23, 74–76.

Mooney, H.A., Björkman, O., Ehleringer, J. & Berry, J. (1976) Photosynthetic capacity of *in situ* Death Valley plants. *Carnegie Institution Year Book* 75, 410–413.

Mooney, H.A., Bonnicksen, T.M., Christensen, N.L., Lothan, J.E. & Reiners, W.A. (Eds) (1981) *Fire Regimes and Ecotype Properties*. USDA Forest Service General Technical Reports (Washington) GTR-WO-26.

Mooney, H.A., Field, C., Williams, W.E., Berry, J.A. & Björkman, O. (1983). Photosynthetic characteristics of plants of a Californian cool coastal environmental. *Oecologia* **57**, 38–42.

Moore, R.P. (Ed.) (1985) *Handbook on Tetrazolium Testing*. International Seed Testing Association, Zurich.

Mopper, S. & Simberloff, D. (1995) Differential herbivory in an oak population: the role of plant phenology and insect performance. *Evolution* **76**, 1233–1241.

Mor, M. & Spiegel, Y. (1993) Infection of *Narcissus* roots by *Aphelenchoides subtenuis*. *Journal of Nematology* **25**, 476–479.

Moravec, J. (1973) The determination of the minimal area of phytocenoses. *Folia Geobotanica et Phytotaxonomica* **8**, 23–47.

Morgan, D.C. & Smith, H. (1979) A systematic relationship between phytochrome-induced development and species habitat, for plants grown in simulated natural radiation. *Planta* **145**, 253–258.

Morinaga, T. (1926) The effect of alternating temperatures upon the germination of seeds. *American Journal of Botany* **13**, 141–158.

Morris, M.G. (1981) Responses of grassland invertebrates to management by cutting. III. Adverse effects on Auchenorhyncha. *Journal of Applied Ecology* **18**, 107–123.

Morrison, D.A., Cary, G.J., Pengelly, S.M. *et al.* (1995) Effects of fire frequency on plant species composition of sandstone communities in the Sydney region – inter fire interval and time since fire. *Australian Journal of Ecology* **20**, 239–247.

Morrow, P.A. & Fox, L.R. (1980) Effects of variation in *Eucalyptus* essential oil yield on insect growth and grazing damage. *Oecologia* **45**, 209–219.

Mortimer, A.M. (1987) Contributions of plant population dynamics to understanding early succession. In *Colonization, Succession and Stability* (Eds A.J. Gray, M.J. Crawley & P.J. Edwards), pp. 57–80. Blackwell Scientific Publications, Oxford.

Motten, A.F. & Antonovics, J. (1992) Determinants of outcrossing rate in a predominantly self-fertilizing weed, *Datura stramonium* (Solanaceae). *American Journal of Botany* **79**, 419–427.

Mulcahy, D.L., Curtis, P.S. & Snow, A.A. (1983) Pollen competition in a natural population. In *Handbook of Experimental Pollination Biology* (Eds C.E. Jones & R.J. Little), pp. 330–337. Van Nostrand Reinhold, New York.

Mulder, C.P.H. & Harmsen, R. (1995) The effect of muskox herbivory on growth and reproduction in an arctic legume. *Arctic and Alpine Research* **27**, 44–53.

Muller, C.H. & del Moral, R. (1966) Soil toxicity induced by terpenes from *Salvia leucophylla*. *Bulletin of the Torrey Botanical Club* **93**, 130–137.

Muller-Scharer, H. (1991) The impact of root herbivory as a function of plant density and competition survival, growth and fecundity of *Centaurea maculosa* in field plots. *Journal of Applied Ecology* **28**, 759–776.

Mullick, D. (1977) The non-specific nature of defence in bark and wood during wounding, insect and pathogen attack. *Recent Advances in Phytochemistry* **11**, 395–441.

Mumford, R.A., Lipke, H., Loufer, D.A. & Feder, W.A. (1972) Ozone-induced changes in corn pollen. *Environmental Science and Technology* **6**, 427–430.

Murdoch, A.J. & Ellis, R.H. (1992) Longevity, viability and dormancy. In *Seeds: the Ecology of Regeneration in Plant Communities* (Ed. M. Fenner), pp. 193–229. CAB International, Wallingford.

Murdoch, W.W. & Oaten, A. (1975) Predation and population stability. *Advances in Ecological Research* **9**, 2–131.

Murray, D.R. (1986) Seed dispersal by water. In *Seed Dispersal* (Ed. D.R. Murray), pp. 49–86. Academic Press, Sydney.

Murray, K.G., Russell, S., Picone, C.M., Winnett-Murray, K., Sherwood, W. & Kuhlmann, M.L.. (1994) Fruit laxatives and seed passage rates in frugivores: consequences for plant reproductive success. *Ecology* **75**, 989–994.

Murray, M.G. & Brown, D. (1993) Niche separation of grazing ungulates in the Serengeti – an experimental test. *Journal of Animal Ecology* **62**, 380–389.

Murton, R.K. (1971) The significance of a specific search image in the feeding behaviour of the wood pigeon. *Behaviour* **40**, 10–42.

Muzika, R.M. & Pregitzer, K.S. (1992) Effect of nitrogen fertilisation on leaf phenolic production of grand fir seedlings. *Trees* **6**, 241–244.

Myers, J.H. (1987) Nutrient availability and the deployment of mechanical defences in grazed plants: a new experimental approach to the optimal defense theory. *Oikos* **49**, 350–351.

Naidoo, G. & Mundree, S.G. (1993) Relationship between morphological and physiological responses to waterlogging and salinity in *Sporobolus virginicus* (L.) Kunth. *Oecologia* **93**, 360–366.

Nakashima, K. & Murata, N. (1993) Destructive plant diseases caused by mycoplasma like organisms in Asia. *Outlook on Agriculture* **22**, 53–58.

Naveh, Z. (1975) The evolutionary significance of fire in the Mediterranean region. *Vegetatio* **29**, 199–208.

Nee, S. & May, R.M. (1992) Dynamics of metapopulations: habitat destruction and competitive coexistence. *Journal of Animal Ecology* **61**, 37–40.

Neitzke, M. & Runge, M. (1985). Keimlings- und Jungpflanzenentwicklung der Buche (*Fagus sylvatica* L.) Abhängigkeit vom Al/Ca Verhältnis des Bodenextraktes. *Flora* **177**, 239–249.

Nettancourt, D. de (1977) *Incompatibility in Angiosperms*. Springer-Verlag, New York.

Neuvonen, S. & Haukioja, E. (1991) The effects of inducible resistance in host foliage on birch-feeding herbivores. In *Phytochemical Induction by Herbivores* (Eds D.W. Tallamy & M.J. Raupp), pp. 277–292. John Wiley & Sons, New York.

Newman E.I. (1974) Root and soil water relations. In

The Plant Root and its Environment (Ed. E.W. Carson), pp. 363–440. University Press of Virginia, Charlottesville.

Newman, E.I. (1985) The rhizosphere: carbon sources and microbial populations. In *Ecological Interactions in Soil, Plants, Microbes and Animals* (Eds A.H. Fitter, D. Atkinson, D.J. Read & M.B. Usher), pp. 107–121. Blackwell Scientific Publications, Oxford.

Newman, E.I. & Rovira, A.D. (1975) Allelopathy among some British grassland species. *Journal of Ecology* 63, 727–737.

Newsham, K.K., Fitter, A.H. & Watkinson, A.R. (1994) Root pathogenic and arbuscular mycorrhizal fungi determine fecundity of asymptomatic plants in the field. *Journal of Ecology* 82, 805–814.

Newsham, K.K., Fitter, A.H. & Watkinson, A.R. (1995a) Multi-functionality and biodiversity in arbuscular mycorrhizas. *Trends in Ecology and Evolution* 10, 407–411.

Newsham, K.K., Fitter, A.H. & Watkinson, A.R. (1995b) Arbuscular mycorrhiza protect an annual grass from root pathogenic fungi in the field. *Journal of Ecology* 83, 991–1000.

Newsham, K.K., Watkinson, A.R., West, H.M., & Fitter, A.H. (1995c) Symbiotic fungi determine plant community structure: changes in a lichen-rich plant community induced by fungicide application. *Functional Ecology* 9, 442–447.

Nichols-Orians, C.M. (1991) The effects of light on foliar chemistry, growth and susceptibility of seedlings of a canopy tree to an attine ant. *Oecologia* 87, 552–560.

Nichols-Orians, C.M. & Schultz, J.C. (1989) Leaf toughness affects leaf harvesting by the leafcutter ant, *Atta cephalotes* (L.) (Hymenoptera: Formicidae). *Biotropica* 21, 80–83.

Niinemets, U. & Kull, K. (1994) Leaf weight per area and leaf size of 85 Estonian woody species in relation to shade tolerance and light availability. *Forest Ecology and Mangement* 70, 1–10.

Niklas, K.J. (1988) Biophysical limitations on plant form and evolution. In *Plant Evolutionary Biology* (Eds L.D. Gottlieb & S.K. Jain), pp. 185–220. Chapman & Hall, London.

Niklas, K.J. (1994) Morphological evolution through complex domains of fitness. *Proceedings of the National Academy of Sciences USA* 91, 6772–6779.

Niklas, K.J., Tiffney, B.H. & Knoll, A.H. (1983) Patterns in vascular plant diversification. *Nature* 303, 614–616.

Nilsson, S.G. & Wastljung, U. (1987) Seed predation and cross pollination in mast seeding beech (*Fagus sylvatica*) patches. *Ecology* 68, 260–265.

Noble, P. (1991) *Physicochemical and Environmental Plant Physiology*. Academic Press, New York.

Noble, I.R. & Slatyer, R.O. (1980) The use of vital attributes to predict successional changes in plant communities subject to recurrent disturbances. *Vegetatio* 43, 5–21.

Noble, J.C., Bell, A.D. & Harper, J.L. (1979) The population biology of plants with clonal growth. I. The morphology and structural demography of *Carex arenaria*. *Journal of Ecology* 67, 983–1008.

Norby, R.J., O'Neill, E.G., Hood, W.G. & Luxmoore, R.J. (1987) Carbon allocation, root exudation and mycorrhizal colonisation. *Tree Physiology* 3, 203–210.

Norby, R.J., Gunderson, C.A., Wullschleger, S.D., O'Neill, E.G. & McCracken, M.K. (1992) Productivity and compensatory responses of yellow-poplar trees in elevated CO_2. *Nature* 357, 322–324.

Northup, R.R., Yu, Z., Dahlgren, R.A. & Vogt, K.A. (1995) Polyphenol control of nitrogen release from pine litter. *Nature* 377, 227–229.

Norton, B. (1988) Commodity, amenity, and morality: the limits of quantification in valuing biodiversity. In *Biodiversity* (Eds E.O. Wilson & F.M. Peter), pp. 200–205. National Academy of Sciences, Washington, DC.

Norton, D.A. & Kelly, D. (1988) Mast seeding over 33 years by *Dacrydium cupressinum* Lam. (rimu) (Podocarpaceae) in New Zealand: the importance of economies of scale. *Functional Ecology* 2, 399–408.

Novak, S.J., Mack, R.N. & Soltis, P.S. (1993) Genetic variation in *Bromus tectorum* (Poaceae) – introduction dynamics in North America. *Canadian Journal of Botany* 71, 1441–1448.

Nowak, M.A. & May, R.M. (1992) Evolutionary games and spatial chaos. *Nature* 359, 826–829.

Nowak, M.A. & May, R.M. (1994) Superinfection and the evolution of parasite virulence. *Proceedings of the Royal Society of London Series B* 255, 81–89.

Noy Meir, I. (1993) Compensating growth of grazed plants and its relevance to the use of rangelands. *Ecological Applications* 3, 32–34.

Numata, M. (1989) The ecological characteristics of aliens in natural and semi-natural stands. *Memoirs of Shukutoku University* 23, 23–35.

Nunney, L. & Campbell, K.A. (1993) Assessing minimum viable population size: demography meets population genetics. *Trends in Ecology and Evolution* 8, 234–239.

Nybom, H. & Schaal, B.A. (1990) DNA 'fingerprints' reveal genotypic distributions in natural populations of blackberries and raspberries (*Rubus*, Rosaceae). *American Journal of Botany* 77, 883–888.

Nye P.H. & Tinker P.B. (1977) *Solute Movement in the Soil-Root System*. Blackwell Scientific Publications, Oxford.

Oba, G. (1994) Responses of *Indigofera spinosa* to simulated herbivory in a semi-desert of North West Kenya. *International Journal of Ecology* 15, 105–117.

Obeid, M., Machin, D. & Harper, J.L. (1967) Influence of density on plant to plant variations in fiber flax, *Linum usitatissimum*. *Crop Science* 7, 471–473.

O'Brien, W.J. (1974) The dynamics of nutrient limitation of phytoplankton algae: a model reconsidered. *Ecology* 55, 135–141.

O'Dowd, D.J. & Hay, M.E. (1980) Mutualism between harvester ants and a desert ephemeral: seed escape from rodents. *Ecology* 61, 531–540.

[678]

Oesterheld, M. (1992) Effect of defoliation intensity on above ground and below ground relative growth rates. *Oecologia* **92**, 313–316.

Ogden, J. (1970) Plant population structure and productivity. *Proceedings of the New Zealand Ecological Society* **17**, 1–9.

Ogden, J. (1985) Past, present and future: studies on the population dynamics of some long-lived trees. In *Studies on Plant Demography* (Ed. J. White), pp. 3–16. Academic Press, London.

Ohgushi, T. (1992) Resource limitation on insect herbivore populations. In *Effects of Resource Distribution on Animal–Plant Interactions* (Eds M.D. Hunter, T. Ohgushi & P.W. Price), pp. 199–241. Academic Press, San Diego.

Ohmart, C.P., Stewart, L.G. & Thomas, J.R. (1985) Effects of food quality, particularly concentrations of *Eucalyptus blakelyi* foliage on the growth of *Paropsis atomaria* larvae nitrogen (Coleoptera: Chrysomelidae). *Oecologia* **65**, 543–549.

Ohnmeiss, T.E. & Baldwin, I.T. (1994) The allometry of nitrogen allocation to growth and an inducible defense under nitrogen-limited growth. *Ecology* **75**, 995–1002.

Ohsaki, N. & Sato, Y. (1994) Food plant choice of *Pieris* butterflies as a trade-off between parasitoid avoidance and quality of plants. *Ecology* **75**, 59–68.

Oksanen, L. (1983) Trophic exploitation and arctic phytomass patterns. *American Naturalist* **122**, 45–52.

Oksanen, L. (1988) Ecosystem organization: mutualism and cybernetics or plain Darwinian struggle for existence? *American Naturalist* **131**, 424–444.

Oksanen, L. *et al.* (1997) Outlines of food webs in a low arctic tundra landscape in relation to three theories of trophic dynamics. In *BES Symposium on Multitrophic Interactions in Terrestrial Ecosystems* (Eds A. Gange & V.K. Brown). Blackwell Science, Oxford.

Oksanen, L., Fretwell, S.D., Arruda, J. & Niemela, P. (1981) Exploitation ecosystems in gradients of primary productivity. *American Naturalist* **118**, 240–261.

Okubo, A. & Levin, S. (1989) A theoretical framework for data analysis of wind dispersal of seeds and pollen. *Ecology* **70**, 329–338.

Okubo, A., Maini, P.K., Williamson, M.H. & Murray, J.D. (1989) On the spatial spread of the grey squirrel in Britain. *Proceedings of the Royal Society of London Series B* **238**, 113–125.

Olsen, C. (1923) Studies on the hydrogen ion concentration of the soil and its significance to the vegetation, especially to the natural distribution of plants. *Compte rendu des travaux du Laboratoire de Carlsberg* **15**, 1–166.

Olsen, C. (1953) The significance of concentration for the rate of ion absorption by higher plants in water culture. IV. The influence of hydrogen ion concentration. *Physiologia Plantarum* **6**, 848.

Onuf, C.P., Teal, J.M. & Valiela, I. (1977) Interactions of nutrients, plant growth and herbivory in a mangrove ecosystem. *Ecology* **58**, 514–526.

Oostermeijer, J.G.B., Altenburg, R.G.M. & Dennijs, H.C.M. (1995) Effects of outcrossing distance and selfing on fitness components in the rare *Gentiana pneumonanthe* (Gentianaceae). *Acta Botanica Neerlandica* **44**, 257–268.

Oostermeijer, J.G.B., van't Veer, R. & den Nijs, J.C.M. (1994) Population structure of the long-lived perennial *Gentiana pneumonanthe* in relation to vegetation and management in the Netherlands. *Journal of Applied Ecology* **31**, 428–438.

Oren, R., Schulze, E.D., Werk, K.S. & Meyer, J. (1988) Performance of two *Picea abies* (L.) Karst. stands at different stages of decline. VII. Nutrient relations and growth. *Oecologia* **77**, 163–173.

Orians, G.H. (1980) Habitat selection: general theory and applications to human behaviour. In *The Evolution of Human Social Behaviour* (Ed. J.S. Lockard), pp. 49–66. Elsevier North Holland, New York.

Orians, G.H. & Solbrig, O.T. (1977) A cost–income model of leaves and roots with special reference to arid and semi-arid areas. *American Naturalist* **111**, 677–690.

Orive, M.E. (1995) Senescence in organisms with clonal reproduction and complex life histories. *American Naturalist* **145**, 90–108.

Ornduff, R. (1969) Reproductive biology in relation to systematics. *Taxon* **18**, 121–133.

Osmond, C.B., Winter, K. & Ziegler, H. (1981) Functional significance of different pathways of CO_2 fixation in photosynthesis. In *Encyclopedia of Plant Physiology. New Series*, Vol. 12B (Eds A. Pirson & M.H. Zimmerman), pp. 480–547. Springer-Verlag, Berlin.

Ostfeld, R.S. (1992) Small-mammal herbivores in a patchy environment: individual strategies and population responses. In *Effects of Resource Distribution on Animal–Plant Interactions* (Eds M.D. Hunter, T. Ohgushi & P.W. Price), pp. 43–74. Academic Press, New York.

Pacala, S.W. (1986a) Neighborhood models of plant population dynamics: 2. Multispecies models of annuals. *Theoretical Population Biology* **29**, 262–292.

Pacala, S.W. (1986b) Neighborhood models of plant population dynamics. 4. Single-species and multispecies models of annuals with dormant seeds. *American Naturalist* **128**, 859–878.

Pacala, S.W. & Crawley, M.J. (1992) Herbivores and plant diversity. *American Naturalist* **140**, 243–260.

Pacala, S.W. & Deutschman, D.J. (1996) Details that matter: the spatial structure of individual trees maintains forest ecosystem function. *Oikos* **74**, 357–365.

Pacala, S.W. & Roughgarden, J. (1982) Spatial heterogeneity and interspecific competition. *Theoretical Population Biology* **21**, 92–113.

Pacala, S.W. & Silander, J.A. (1985) Neighborhood models of plant population dynamics. I. Single-species models of annuals. *American Naturalist* **125**, 385–411.

Pacala, S.W. & Silander, J.A. Jr (1990) Field tests of neighborhood population dynamic models of two annual weed species. *Ecological Monographs* **60**, 113–134.

Pacala, S.W. & Tilman, D. (1994) Limiting similarity in mechanistic and spatial models of plant competition in heterogeneous environments. *American Naturalist* **143**, 222–257.

Pacala, S.W., Canham, C.D., Saponara, J., Silander, J.A., Kobe, R.K. & Ribbens, E. (1996) Forest models defined by field measurements. II. Estimation, error analysis and dynamics. *Ecological Monographs* **66**, 1–44.

Paces, T. (1985) Sources of acidification in Central Europe estimated from elemental budgets in small basins. *Nature* **315**, 31–36.

Paige, K.N. & Whitham, T.G. (1985) Individual and population shifts in flower color by scarlet gilia: a mechanism for pollination tracking. *Science* **227**, 315–317.

Paige, K.N. & Whitham, T.G. (1987) Overcompensation in response to mammalian herbivory: the advantage of being eaten. *American Naturalist* **129**, 419–428.

Paine, R.T. (1966) Food web complexity and species diversity. *American Naturalist* **100**, 65–75.

Paine, R.T. (1969) A note on trophic complexity and community stability. *American Naturalist* **103**, 91–93.

Paine, R.T. (1980) Food webs: linkage, interaction strength and community infrastructure. *Journal of Animal Ecology* **49**, 667–685.

Paine, R.T. (1988) Food webs: road maps of interactions or grist for theoretical development? *Ecology* **69**, 1648–1654.

Pake, C.E. & Venable, D.L. (1995) Is coexistence of Sonoran desert annuals mediated by temporal variability in reproductive success? *Ecology* **76**, 246–261.

Palit, S., Sharma, A. & Talukder, G. (1994) Effects of cobalt on plants. *Botanical Review* **60**, 149–181.

Palmblad, I.G. (1968) Competition in experimental studies on populations of weeds with emphasis on the regulation of population size. *Ecology* **49**, 26–34.

Paltridge, F.W. & Denholm, J.V. (1974) Plant yield and the switch from vegetative to reproductive growth. *Journal of Theoretical Biology* **44**, 23–34.

Pandey, C.B. & Singh, J.S. (1992) Influence of rainfall and grazing on herbage dynamics in a seasonally dry tropical savanna. *Vegetatio* **102**, 107–124.

Pantone, D.J. (1995) Replacement series analysis of the competitive interaction between a weed and a crop as influenced by a plant parasitic nematode. *Fundamental and Applied Nematology* **18**, 93–97.

Parker, A.J. & Peet, R.K. (1984) Size and age structure of coniferous forests. *Ecology* **65**, 1685–1689.

Parker, I.M., Nakamura, R.R. & Schemske, D.W. (1995) Reproductive allocation and the fitness consequences of selfing in 2 sympatric species of *Epilobium* (Onagraceae) with contrasting mating systems. *American Journal of Botany* **82**, 1007–1016.

Parker, M.A. & Salzman, A.G. (1985) Herbivore exclosure and competitor removal: effects on juvenile survivorship and growth in the shrub *Gutierrezia microcephala*. *Journal of Ecology* **73**, 909–913.

Parkhurst, D.F. & Loucks, O.L. (1972) Optimal leaf size in relation to environment. *Journal of Ecology* **60**, 505–537.

Pastor, J. & Broschart, M. (1990) The spatial pattern of a northern conifer–hardwood landscape. *Landscape Ecology* **4**, 55–68.

Pastor, J., Aber, J.D., McClaugherty, C.A. & Melillo, J.M. (1984) Aboveground production and N and P cycling along a nitrogen mineralization gradient on Blackhawk Island, Wisconsin. *Ecology* **65**, 256–268.

Paterniani, E. & Short, A.C. (1974) Effective maize pollen dispersal in the field. *Euphytica* **23**, 129–134.

Patrick, R. (1973) Uses of algae, especially diatoms, in the assessment of water quality. *American Society for Testing and Materials, Special Technical Publication* **528**, 76–95.

Pearcy, R.W. & Ehleringer, J. (1984) Comparative ecophysiology of C_3 and C_4 plants. *Plant, Cell and Environment* **7**, 1–13.

Pearson, J. & Stewart, G.R. (1993) The deposition of ammonia and its effects on plants. *New Phytologist* **125**, 283–305.

Peart, D.R. (1982) *Experimental analysis of succession in a grassland at Sea Ranch, California*. PhD thesis, University of California, Davis.

Peat, H.J. & Fitter, A.H. (1993) The distribution of arbuscular mycorrhizas in the British flora. *New Phytologist* **125**, 845–854.

Pedersen, B. (1995) An evolutionary theory of clonal senescence. *Theoretical Population Biology* **47**, 292–320.

Pellmyr, O. & Huth, C.J. (1994) Evolutionary stability of mutualism between yuccas and yucca moths. *Nature* **372**, 257–260.

Pennings, S.C. & Callaway, R.M. (1992) Salt marsh plant zonation – the relative importance of competition and physical factors. *Ecology* **73**, 681–690.

Pennington, W. (1969) *The History of British Vegetation*. English Universities Press, London.

Percy, K. (1986) The effects of a simulated acid rain on germinative capacity, growth and morphology of forest tree seedlings. *New Phytologist* **104**, 473–484.

Perring, F.H. (1994) Foreword. In *Scarce Plants in Britain* (Eds A. Stewart, D.A. Pearman & C.D. Preston), p. 8. JNCC, Peterborough.

Perring, F.H. & Walters, S.M. (1962) *Atlas of the British Flora*. Thomas Nelson, London.

Perrins, J., Fitter, A. & Williamson, M. (1993) Population biology and rates of invasion of 3 introduced *Impatiens* species in the British Isles. *Journal of Biogeography* **20**, 33–44.

Petanidou, T., Dennijs, J.C.M. & Oostermeijer, J.G.B. (1995) Pollination ecology and constraints on seed set of the rare perennial *Gentiana cruciata* L. in the Netherlands. *Acta Botanica Neerlandica* **44**, 55–74.

Peterken, G.F. (1981) *Woodland Conservation and Management*. Chapman & Hall, London.

Peters, N.K. & Verma, D.P.S. (1990) Phenolic compounds as regulators of gene expression in plant–microbe interactions. *Molecular Plant–Microbe Interactions* **3**, 4–8.

Petraitis, P.S., Latham, R.E. & Niesenbaum, R.A. (1989) The maintenance of species diversity by disturbance. *Quarterly Review of Biology* **64**, 393–418.

Philips, D.L. & McMahon, J.A. (1981) Competition and spacing patterns in desert shrubs. *Journal of Ecology* **69**, 97–115.

Pianka, E.R. (1970) On r- and K-selection. *American Naturalist* **104**, 592–597.

Pichersky, E., Raguso, R.A., Lewinsohn, E. & Croteau, R. (1994) Floral scent production in *Clarkia* (Onagracea). *Plant Physiology* **106**, 1533–1540.

Pickett, S.T.A. (1982) Population patterns through twenty years of old field succession. *Vegetatio* **49**, 15–59.

Pickett, S.T.A. & White, P.S. (Eds) (1985) *The Ecology of Natural Disturbance and Patch Dynamics*. Academic Press, New York.

Pielou, E.C. (1961) Segregation and symmetry in two-species populations as studied by nearest-neighbour relationships. *Journal of Ecology* **49**, 255–269.

Pielou, E.C. (1977) *Mathematical Ecology*. John Wiley & Sons, New York.

Pierce, S.M., Esler, K. & Cowling, R.M. (1995) Smoke-induced germination of succulents (Mesembryanthemaceae) from fire-prone and fire-free habitats in South Africa. *Oecologia* **102**, 520–522.

Pigott, C.D. (1968) Biological flora of the British Isles: *Cirsium acaulon*. *Journal of Ecology* **56**, 597–612.

Pigott, C.D. (1983) Regeneration of oak–birch woodland following exclusion of sheep. *Journal of Ecology* **71**, 629–646.

Pigott, C.D. & Huntley, J.P. (1981) Factors controlling the distribution of *Tilia cordata* at the northern limit of its geographical range. 3. Nature and cause of seed sterility. *New Phytologist* **87**, 817–839.

van der Pijl, L. (1972) *Principles of Dispersal in Higher Plants*, 2nd edn. Springer-Verlag, Berlin.

Pilson, D. (1992) Aphid distribution and the evolution of goldenrod resistance. *Evolution* **46**, 1358–1372.

Pimm, S.L. (1982) *Food Webs*. Chapman & Hall, New York.

Pimm, S.L., Lawton, J.H. & Cohen, J.E. (1991) Food web patterns and their consequences. *Nature* **350**, 669–674.

Piñero, D., Martínez-Ramos, M. & Sarukhán, J. (1984) A population model of *Astrocaryum mexicanum* and a sensitivity analysis of its finite rate of increase. *Journal of Ecology* **72**, 977–991.

Pitelka, L.F. & Ashmun, J.W. (1985) Physiology and integration of ramets in clonal plants. In *Population Biology and Evolution of Clonal Organisms* (Eds J.B.C. Jackson, L.W. Buss & R.E. Cook), pp. 399–435. Yale University Press, New Haven.

Pitelka, L.F., Thayer, M.E. & Hansen, S.B. (1983) Variation in achene weight in *Aster acuminatus*. *Canadian Journal of Botany* **61**, 1415–1420.

Pitt, M.D. & Heady, H.F. (1978) Responses of annual vegetation to temperature and rainfall patterns in northern California. *Ecology* **59**, 336–350.

Platenkamp, G.A.J. & Shaw, G.S. (1993) Environmental and genetic maternal effects on seed characters in *Nemophila menziesii*. *Evolution* **47**, 540–555.

Platt, W. & Weis, I.M. (1977) Resource partitioning and competition within a guild of fugitive prairie plants. *American Naturalist* **111**, 479–513.

Plumptre, A.J. (1994) The effects of trampling damage by herbivores on the vegetation of the Parc National des Volcans, Rwanda. *African Journal of Ecology* **32**, 115–129.

Polis, G.A. (1991) Complex trophic interactions in deserts: an empirical critique of food web theory. *American Naturalist* **138**, 123–155.

Polis, G.A. and Winemiller, K.O. (1996) *Food webs: integration of Patterns & Dynamics*. Chapman & Hall, London.

Ponquett, R.T., Smith, M.T. & Ross, G. (1992) Lipid autoxidation and seed ageing: Putative relationships between seed longevity and lipid stability. *Seed Science Research* **2**, 51–54.

Pons, T.L. (1989) Breaking of seed dormancy as a gap detection mechanism. *Annals of Botany* **63**, 139–143.

Pons, T.L. (1991) Induction of dark-dormancy in seeds: its importance for the seed bank in the soil. *Functional Ecology* **5**, 669–675.

Pons, T.L., van der Werf, A. & Lambers, H. (1994) Photosynthetic nitrogen use efficiency of inherently slow- and fast-growing species: possible explanations for observed differences. In *A Whole Plant Perspective On Carbon–Nitrogen Interactions*, (Eds J. Roy & E. Garnier), pp. 61–77. SPB Academic Publishing, The Hague.

Poore, M.E.D. (1955) The use of phytosociological methods in ecological investigations. III. Practical application. *Journal of Ecology* **43**, 606–651.

Port, G.R. & Thompson, J.R. (1980) Outbreaks of insect herbivores on plants along motorways in the United Kingdom. *Journal of Applied Ecology* **17**, 649–656.

Portnoy, S. & Willson, M.F. (1993) Seed dispersal curves – behaviour of the tail of the distribution. *Evolutionary Ecology* **7**, 25–44.

Possingham, H.P., Comins, H.N. & Noble, I.R. (1995) The fire and flammability niches in plant communities. *Journal of Theoretical Biology* **174**, 97–108.

Post, W.M., Pastor, J., Zinke, P.J. & Stangenberger, A.G. (1985) Global patterns of soil nitrogen storage. *Nature* **317**, 613–616.

Potter, A. & Kimmerer, T.W. (1986) Seasonal allocation of defence investment in *Ilex opaca* Alton and constraints on a specialist leafminer. *Oecologia* **69**, 217–224.

Powell, J.A. (1992) Inter-relationships of yuccas and yucca moths. *Trends in Ecology and Evolution* **7**, 10–15.

Power, M.E. (1992) Top-down and bottom-up forces in

food webs: do plants have primacy? *Ecology* 73, 724–746.

Prendergast, J.R. & Eversham, B.C. (1995) Butterfly diversity in southern Britain – hotspot losses since 1930. *Biological Conservation* 72, 109–114.

Prendergast, J.R., Quinn, R.M., Lawton, J.H., Eversham, B.C. & Gibbons, D.W. (1993) Rare species, the coincidence of diversity hotspots and conservation strategies. *Nature* 365, 335–337.

Preston, F.W. (1962) The canonical distribution of commonness and rarity. Part I. *Ecology* 43, 185–215.

Prestridge, R.A. (1997) A catch 22: the utilization of endophytic fungi of pest management. In *BES Symposium on Multitrophic Interactions in Terrestrial Systems* (Eds A. Gange & V.K. Brown) Blackwell Science, Oxford.

Price, E.A.C. & Hutchings, M.J. (1992) The causes and developmental effects of integration and independence between different parts of *Glechoma hederacea* clones. *Oikos* 63, 376–386.

Price, M.V. & Wasser, N.M. (1979) Pollen dispersal and optimal outcrossing in *Delphinium nelsoni*. *Nature* 277, 294–297.

Price, P.W. (1987) The role of natural enemies in insect populations. In *Insect Outbreaks* (Eds P. Barbosa & J.C. Schultz), pp. 287–312. Academic Press, New York.

Price, P.W. (1988) Inversely density-dependent parasitism: the role of plant refuges for hosts. *Journal of Animal Ecology* 57, 89–96.

Price, P.W. (1989) Clonal development of coyote willow, *Salix exigua* (Salicaceae) and attack by the shoot-galling sawfly, *Euura exiguae* (Hymenoptera: Tenthredinidae). *Environmental Entomology* 18, 16–68.

Price, P.W. (1991) The plant vigour hypothesis and herbivore attack. *Oikos* 62, 244–251.

Price, P.W. (1992) Plant resources as the mechanistic basis for insect herbivore population dynamics. In *Effects of Resource Distribution on Animal–Plant Interactions* (Eds M.D. Hunter, T. Ohgushi & P.W. Price), pp. 139–173. Academic Press, San Diego.

Price, P.W., Bouton, C.E., Gross, P., McPheron, B.A., Thompson, J.N. & Weis, A.E. (1980) Interactions among three trophic levels: influence of plants on interactions between insect herbivores and natural enemies. *Annual Review of Ecology and Systematics* 11, 41–65.

Price, P.W., Waring, G.L., Julkunen-Tiitto, R., Tahvanainen, J., Mooney, H.A. & Craig, T.P. (1989) Carbon-nutrient balance hypothesis in within-species phytochemical variation of *Salix lasiolepis. Journal of Chemical Ecology* 15, 1117–1131.

Price, P.W., Cobb, N., Craig, T.P. *et al.* (1990) Insect herbivore population dynamics on trees and shrubs: new approaches relevant to latent and eruptive species and life table development. In *Insect–Plant Interactions* (Ed. E.A. Bernays), Vol. 2, pp. 1–38. CRC Press, Boca Raton.

Primack, R.B. (1985) Longevity of individual flowers. *Annual Review of Ecology and Systematics* 16, 15–37.

Primack, R.B. & Hall, P. (1990) Costs of reproduction in the pink lady's slipper orchid: a four-year experimental study. *American Naturalist* 136, 638–656.

Prins, A.H., Nell, H.W. & Klinkhamer, P.G.L. (1992) Size dependent root herbivory on *Cynoglossum officinale. Oikos* 65, 409–413.

Probert, R.J. & Longley, P.L. (1989) Recalcitrant seed storage physiology in three aquatic grasses (*Zizania palustris, Spartine anglica* and *Porteresia coarctata*). *Annals of Botany* 63, 53–63.

Proctor, J. & Woodell, S.R.J. (1975) The ecology of serpentine soils. *Advances in Ecological Research* 9, 256–366.

Proctor, M. & Yeo, P. (1972) *The Pollination of Flowers.* Taplinger, New York.

Proksch, P., Wray, V., Isman, M.B. & Rahans, I. (1990) Ontogenetic variation of biologically active natural products in *Ageratina adenophora. Phytochemistry* 29, 453–458.

Putz, F.E. (1984) The natural history of lianes on Barro Colorado Island, Panama. *Ecology* 65, 1713–1724.

Pyke, G.H. (1978) Optimal foraging in bumblebees and coevolution with their plants. *Oecologia* 36, 181–193.

Pysek, P. (1992) Seasonal changes in response of *Senecio ovatus* to grazing by the chrysomelid beetles *Chrysomela speciosissima. Oecologia* 91, 596–628.

Queller, D.C. (1989) The evolution of eusociality – reproductive head starts of workers. *Proceedings of the National Academy of Sciences USA* 86, 3224–3226.

Quinn, J.A., Mowrey, D.P., Emanuele, S.M. & Whalley, R.D.B. (1994) The foliage is the fruit hypothesis: *Buchloe dactyloides* (Poaceae) and the short grass prairie of North America. *American Journal of Botany* 81, 1545–1554.

Quiring, D.T. (1993) Influence of intra-tree variation in time of budburst of white spruce on herbivory and the behaviour and survivorship of *Zeiraphera canadensis. Ecological Engineering* 18, 353–364.

Quist, M.E. (1995) Reversibility of damages to forest floor plants by episodes of elevated hydrogen ion and aluminum ion concentrations in the soil solution. *Plant and Soil* 176, 297–305.

Rabinowitz, D. (1979) Bimodal distributions of seedling weight in relation to density of *Festuca paradoxa* Desv. *Nature* 277, 297–298.

Rabinowitz, D. (1981) Seven forms of rarity. In *The Biological Aspects of Rare Plant Conservation* (Ed. H. Synge), pp. 205–217. John Wiley & Sons, New York.

Rabinowitz, D. & Rapp, J.K. (1980) Seed rain in a North American tall grass prairie. *Journal of Applied Ecology* 17, 793–802.

Rabinowitz, D. & Rapp, J.K. (1981) Dispersal abilities of seven sparse and common grasses from a Missouri prairie. *American Journal of Botany* 68, 616–624.

Rabotnov, T.A. (1978) On coenopopulations of plants reproducing by seeds. In *Structure and Functioning of Plant Populations* (Eds A.H.J. Freysen & J.W. Woldendorp), pp. 1–26. North-Holland Publishing Company, Amsterdam.

Ramenskii, L.G. (1938) *Introduction to the Geobotanical Study of Complex Vegetations*. Selkzgiz, Moscow.

Rand, D.A., Keeling, M. & Wilson, H.B. (1995) Invasion, stability and evolution to criticality in spatially extended, artificial host–pathogen ecologies. *Proceedings of the Royal Society of London Series B* **259**, 55–63.

Randall, A. (1988) What mainstream economists have to say about the value of biodiversity. In *Biodiversity* (Eds E.O. Wilson & F.M. Peter), pp. 217–223. National Academy of Sciences, Washington, DC.

Ranwell, D.S. (1972) *Ecology of Salt Marshes and Sand Dunes*. Chapman & Hall, London.

Rao, M.V., Hale, B.A. & Ormond, D.P. (1995) Amelioration of ozone-induced oxidative damage in wheat plants grown under high carbon dioxide. *Plant Physiology* **109**, 421–432.

Rapport, D.J. (1971) An optimization model of food selection. *American Naturalist* **105**, 575–578.

Rathcke, B. (1983) Competition and facilitation among plants for pollination. In *Pollination Biology* (Ed. L. Real), pp. 305–329. Academic Press, New York.

Raunkiaer, C. (1934) *The Life Forms of Plants and Statistical Plant Geography*. Clarendon Press, Oxford.

Raupp, M.J. & Denno, R.F. (1983) Leaf age as a predictor of herbivore distribution and abundance. In *Variable Plants and Herbivores in Natural and Managed Systems* (Eds R.F. Denno & M.S. McClure), pp. 91–118. Academic Press, New York.

Rausher, M.D. (1988) Is coevolution dead? *Ecology* **69**, 898–901.

Rausher, M.D. & Feeny, P. (1980) Herbivory, plant density, and plant reproductive success: the effect of *Battus philenor* on *Aristolochia reticulata*. *Ecology* **61**, 905–917.

Rausher, M.D., Iwao, K., Simms, E.L., Ohsaki, N. & Hall, D. (1993) Induced resistance in *Ipomoea purpurea*. *Ecology* **74**, 20–29.

Raven, J. (1995) The early evolution of land plants: aquatic ancestors and atmospheric interactions. *Botanical Journal of Scotland* **47**, 151–175.

Raven, P.H. (1988) Our diminishing tropical forests. In *Biodiversity* (Eds E.O. Wilson & F.M. Peter), pp. 119–122. National Academy of Sciences, Washington, DC.

Raven, P.H., Evert, R.F. & Eichhorn, S.E. (1992) *Biology of Plants*, 5th edn. Worth Publishers, New York.

Read, D.J. (1991) Mycorrhizas in ecosystems. *Experientia* **47**, 376–391.

Read, D.J., Francis, R. & Finlay, R.D. (1985) Mycorrhizal mycelia and nutrient cycling in plant communities. In *Ecological Interactions in Soil, Plants, Microbes and Animals* (Eds A.H. Fitter, D. Atkinson, D.J. Read & M.B. Usher), pp. 193–217. Blackwell Scientific Publications, Oxford.

Real, E.A. & Ellner, S. (1992) Life-history evolution in stochastic environments – a graphical mean-variance approach. *Ecology* **73**, 1227–1236.

Rees, M. (1993a) Null models and dispersal distributions: comments on an article by Caley. *American Naturalist* **141**, 812–815.

Rees, M. (1993b) Trade-offs among dispersal strategies in British plants. *Nature* **336**, 150–152.

Rees, M. (1994) Delayed germination of seeds: a look at the effects of adult longevity, the timing of reproduction and population age/stage structure. *American Naturalist* **144**, 43–64.

Rees, M. (1995a) Community structure in sand dune annuals: is seed weight a key quantity? *Journal of Ecology* **83**, 857–864.

Rees, M. (1995b) EC-PC comparative analyses? *Journal of Ecology* **83**, 891–892.

Rees, M. & Brown, V.K. (1991) The effects of established plants on recruitment in the annual forb *Sinapis arvensis*. *Oecologia* **87**, 58–62.

Rees, M. & Brown, V.K. (1992) Interactions between invertebrate herbivores and plant competition. *Journal of Ecology* **80**, 353–360.

Rees, M. & Crawley, M.J. (1989) Growth, reproduction and population dynamics. *Functional Ecology* **3**, 645–653.

Rees, M. & Crawley, M.J. (1991) Do plant populations cycle? *Functional Ecology* **5**, 580–582.

Rees, M. & Long, M.J. (1992) Germination biology and the ecology of annual plants. *American Naturalist* **139**, 484–508.

Rees, M. & Long, M.J. (1993) The analysis and interpretation of seedling recruitment curves. *American Naturalist* **141**, 233–262.

Rees, M., Grubb, P.J. & Kelly, D. (1996) Quantifying the impact of competition and spatial heterogeneity on the structure and dynamics of a four-species guild of winter annuals. *American Naturalist* **147**, 1–32.

Rees, S.B. & Harborne, J.B. (1985) The role of sesquiterpene lactones and phenolics in the chemical defence of the chicory plant. *Phytochemistry* **24**, 2225–2231.

Reese, J.C. & Beck, S.D. (1978) Inter-relationships of nutritional indices and dietary moisture in the black cutworm, *Agrotis ipsilon*, digestive efficiency. *Journal of Insect Physiology* **24**, 473–479.

Regal, P.J. (1977) Ecology and evolution of flowering plant dominance. *Science* **196**, 622–629.

Regal, P.J. (1982) Pollination by wind and animals. *Annual Review of Ecology and Systematics* **13**, 497–524.

Reich, P.B., Schoettle, A.W., Stroo, H.F., Troiano, J. & Amundsen, R.G. (1985) Effects of O_3, SO_2, and simulated acid rain on mycorrhizal infection of northern red oak. *Canadian Journal of Botany* **63**, 2049–2055.

Reichardt, P.B., Bryant, J.P., Clausen, T.P. & Wieland, G.D. (1984) Defence of winter-dormant Alaska paper birch against snowshoe hare. *Oecologia* **65**, 58–69.

Reichardt, P.B., Chapin, F.S. III, Bryant, J.P., Mattes, B.R. & Clausen, T.P. (1991) Carbon/nutrient balance as a predictor of plant defense in Alaskan balsam poplar: potential importance of metabolite turnover. *Oecologia* **88**, 401–406.

Reid, N. (1986) Pollination and seed dispersal of mistletoes (Loranthaceae) by birds in southern Australia. In *The Dynamic Partnership: Birds and Plants in Southern Australia* (Eds H.A. Ford & D.C. Paton), pp. 64–173. Government Printer, South Australia.

Reid, N. (1989) Dispersal of a mistletoe by a honeyeater and a flowerpecker: components of seed dispersal quality. *Ecology* **70**, 137–145.

Remmert, H. (1991) The mosaic-cycle concept of ecosystems – an overview. In *The Mosaic-Cycle Concept of Ecosystems* (Ed. H. Remmert), pp. 1–21. Springer-Verlag, New York.

Reuther, R. (1992) Arsenic introduced into a littoral fresh water model ecosystem. *Science of the Total Environment* **115**, 219–237.

Rhoades, D.F. (1983) Herbivore population dynamics and plant chemistry. In *Variable Plants and Herbivores in Natural and Managed Systems* (Eds R.F. Denno & M.S. McClure), pp. 155–204. Academic Press, New York.

Rhoades, D.F. (1985) Offensive–defensive interactions between herbivores and plants: their relevance in herbivore population dynamics and ecological theory. *American Naturalist* **125**, 205–238.

Rhoades, D.F. & Cates, R.G. (1976) Toward a general theory of plant anti-herbivore chemistry. In *Biochemical Interaction Between Plants and Insects* (Eds J.W. Wallace & R.L. Mansell), pp. 168–213. Plenum Press, New York.

Rhoades, R.E. (1991) The world's food supply at risk. *National Geographic* **179**, 74–103.

Rice, B. & Westoby, M. (1983) Plant species richness at the 0.1 hectare scale in Australian vegetation compared to other continents. *Vegetatio* **52**, 129–140.

Rice, B. & Westoby, M. (1986) Evidence against the hypothesis that ant-dispersed seeds reach nutrient-enriched microsites. *Ecology* **67**, 1270–1274.

Rice, E.L. (1983) *Allelopathy*, 2nd edn. Academic Press, London.

Rice, K.J. (1985) Responses of *Erodium* to varying microsites: the role of germination cueing. *Ecology* **66**, 1651–1657.

Rice, R.L., Lincoln, D.E. & Langenheim, J. (1978) Palatability of monoterpenoid compositional types of *Satureja douglasii* to a generalist mollusc *Ariolimax dolichophallus*. *Biochemical Systematics and Ecology* **6**, 45–53.

Richards, A.J. (1972) The *Taraxacum* flora of the British Isles. *Watsonia* **9**, 1–141.

Richards, A.J. (1986) *Plant Breeding Systems*. George Allen & Unwin, London.

Richards, M.B., Stock, W.D. & Cowling, R.M. (1995) Water relations of seedlings and adults of two fynbos *Protea* species in relation to their distribution patterns. *Functional Ecology* **9**, 575–583.

Richards, P.W. (1983) The three dimensional structure of tropical rain forest. In *Tropical Rain Forest: Ecology and Management* (Eds S.L. Sutton, T.C. Whitmore & A.C. Chadwick), pp. 3–10. Blackwell Scientific Publications, Oxford.

Richter, J.P. (1970) *The Notebooks of Leonardo da Vinci (1452–1519), Compiled and Edited from the Original Manuscripts*. Dover, New York.

Ricklefs, R.E. (1977) Environmental heterogeneity and plant species diversity: a hypothesis. *American Naturalist* **111**, 376–381.

Riemer, J. & Whittaker, J.B. (1989) Air pollution and insect herbivores: observed interactions and possible mechanisms. In *Insect–Plant Interactions* (Ed. E.A. Bernays), Vol. 1, pp. 73–105. CRC Press, Boca Raton.

Risch, S. (1980) The population dynamics of several herbivorous beetles in tropical agro-ecosystem: the effect of intercropping corn, beans and squash in Costa Rica. *Journal of Applied Ecology* **17**, 593–612.

Ritchie, M.E. & Tilman, D. (1992) Interspecific competition among grasshoppers and their effect on plant abundance in experimental field environments. *Oecologia* **89**, 524–532.

Ritland, K. & Jain, S. (1984) The comparative life histories of two annual *Limnanthes* species in a temporally variable environment. *American Naturalist* **124**, 656–679.

Ritz, K. & Newman, E.I. (1984) Movement of ^{32}P between intact grassland plants of the same age. *Oikos* **43**, 138–142.

Roberts, E.H. (1988) Temperature and seed germination. In *Plants and Temperature* (Eds S.P. Long & F.I. Woodward), pp. 109–132. Company of Biologists, Cambridge.

Roberts, E.H. & Smith, R.D. (1977) Dormancy and the pentose phosphate pathway. In *The Physiology and Biochemistry of Seed Dormancy* (Ed. A.A. Khan), pp. 385–411. Elsevier Biomedical Press, Amsterdam.

Roberts, H.A. (1962) Studies on the weeds of vegetable crops. II. Effects of six years of cropping on the weed seeds in the soil. *Journal of Ecology* **50**, 803–813.

Roberts, H.A. (1964) Emergence and longevity in cultivated soil of seeds of some annual weeds. *Weed Research* **4**, 296–307.

Roberts, H.A. (1979) Periodicity of seedling emergence and seed survival in some Umbelliferae. *Journal of Applied Ecology* **16**, 195–201.

Roberts, H.A. (1981) Seed banks in soils. In *Advances in Applied Biology 6* (Ed. T.H. Coaker), pp. 1–55. Academic Press, London.

Roberts, H.A. (1986) Seed persistence in soil and seasonal emergence in plant species from different habitats. *Journal of Applied Ecology* **23**, 639–656.

Roberts, H.A. & Boddrell, J.E. (1983) Seed survival and periodicity of seedling emergence in eight species of Cruciferae. *Annals of Applied Biology* **103**, 301–309.

Robert, H.A. & Dawkins, P.A. (1967) Effect of cultivation on the numbers of viable weed seeds in soil. *Weed Research* **7**, 290–301.

Roberts, H.A. & Neilson, J.E. (1980) Seed survival and periodicity of seedling emergence in some species of *Atriplex, Chenopodium, Polygonum* and *Rumex*. *Annals of Applied Biology* **94**, 111–120.

Roberts, J., Osvaldo, M.R.C. & de Agular, L.F. (1990)

Stomatal and boundary layer conductances in an Amazonian *terra firme* rain forest. *Journal of Applied Ecology* 27, 336–353.

Roberts, R.J. & Morton, R. (1985) Biomass of larval Scarabaeidae (Coleoptera) in relation to grazing pressures in temperate, sown pastures. *Journal of Applied Ecology* 22, 863–874.

Robinson, D. (1994) The response of plants to non-uniform supplies of nutrients. *New Phytologist* 127, 635–674.

Robinson, D. & Rorison, I.H. (1983) A comparison between the responses of Lolium perenne L., Holcus lanatus L. and Deschampsia flexuosa L. Trin. to a localized supply of nitrogen. *New Phytologist* 94, 263–273.

Rodwell J.S. (1991–1997) British Plant Communities (5 volumes). Cambridge University Press, Cambridge.

Roeder, K., Devlin, B. & Lindsay, B.D. (1989) Application of maximum likelihood methods to population genetic data for the estimation of individual fertilities. *Biometrics* 45, 363–379.

Roehrig, N.E. & Capinera, J.L. (1983) Behavioural and developmental responses of range caterpillar larvae. *Hemileca oliviae*, to condensed tannin. *Journal of Insect Physiology* 29, 901–906.

Roland, J. & Myers, J.M. (1987) Improved insect performance from host-plant defoliation: winter moth on oak and apple. *Ecological Entomology* 12, 409–414.

Romme, W.H., Knight, D.H. & Yavitt, J.B. (1986) Mountain pine beetle outbreaks in the Rocky Mountains – regulators of primary productivity. *American Naturalist* 127, 484–494.

Room, P.M. & Thomas, P.A. (1983) Nitrogen and establishment of a beetle for biological control of the floating weed *Salvinia* in Papua New Guinea. *Journal of Applied Ecology* 22, 139–156.

Roose, M.L., Bradshaw, A.D. & Roberts, T.M. (1982) Evolution of resistance to gaseous air pollutants. In *Effects of Gaseous Air Pollution in Agriculture and Horticulture* (Eds M.H. Unsworth & D.P. Ormrod), pp. 379–409. Butterworth Scientific, London.

Root, R.B. (1973) Organization of a plant–arthropod association in simple and diverse habitats: the fauna of collards (*Brassica oleracea*). *Ecological Monographs* 43, 95–124.

Rosenthal, G.A. & Berenbaum, M.R. (Eds) (1992) *Herbivores: their Interaction with Secondary Plant Metabolites*, 2nd edn. Academic Press, San Diego.

Rosenthal, G.A. & Janzen, D.H. (Eds) (1979) *Herbivores: their Interaction with Secondary Plant Metabolites*. Academic Press, New York.

Rosenzweig, M.L. (1971) Paradox of enrichment: destabilization of exploitation ecosystems in ecological time. *Science* 171, 385–387.

Ross, M.A. & Harper, J.L. (1972) Occupation of biological space during seedling establishment. *Journal of Ecology* 60, 77–88.

Roth, I. (1984) *Stratification of Tropical Forests as seen in Leaf Structure*. Junk, The Hague.

Rothhaupt, K.O. (1988) Mechanistic resource competition theory applied to laboratory experiments with zooplankton. *Nature* 333, 660–662.

Roubik, D.W. (1993) Direct costs of forest reproduction, bee-cycling and the efficiency of pollination modes. *Journal of Biosciences* 18, 537–552.

Roughgarden, J. (1976) Resource partitioning among competing species: a coevolutionary approach. *Theoretical Population Biology* 9, 388–424.

Rowell-Rahier, M. & Pasteels, J.M. (1992) Third trophic level influences of plant allelochemicals. In *Herbivores: Their Interaction with Secondary Plant Metabolites*, 2nd edn. (Eds G.A. Rosenthal & M.R. Berenbaum), Vol. 2, pp. 243–277. Academic Press, San Diego.

Rowland, A.J., Borland, A.M. & Lea, P.J. (1988) Changes in amino acids, amides and proteins in response to air pollutants. In *Air Pollution and Plant Metabolism* (Eds S. Schulte-Hostende, N.M. Darrall, L.W. Blank & A.R. Wellburn), pp. 189–221. Elsevier Applied Science, London.

Roy, J. & Bergeron, J.M. (1990) Role of phenolics of coniferous trees as deterrents against debarking behaviour of meadow voles. *Journal of Chemical Ecology* 16, 801–808.

Rozema, J., Lambers, H., van der Gejn, S.C. & Cambridge, M.L. (Eds) (1993) *CO_2 and the Biosphere. Vegetatio* 104/105. Kluwer Academic Publishers, Dordrecht.

Runge, M. (1983) Physiology and ecology of nitrogen nutrition. In *Physiological Plant Ecology III. Encyclopedia of Plant Physiology* (Eds O.L. Lange, P.S. Nobel, C.B. Osmond & H. Ziegler), pp. 163–200. Springer-Verlag, Berlin.

Ruxton, G.D. (1995) Temporal scales and the occurrence of chaos in coupled populations. *Trends in Evolution and Ecology* 10, 141–143.

Ryan, C.A. (1974) Assay and biochemical properties of the proteinase inhibitor-inducing factor: a wound hormone. *Plant Physiology* 54, 328–332.

Ryan, C.A., Bishop, P.D., Graham, J.S., Broadway, R.M. & Duffey, S.S. (1986) Plant and fungal cell-wall fragments activate expression of proteinase-inhibitor genes for plant defense. *Journal of Chemical Ecology* 12, 1025–1036.

Ryser, P. & Lambers, H. (1995) Root and leaf attributes accounting for the performance of fast growing and slow growing grasses at different nutrient supply. *Plant and Soil* 170, 251–265.

Sagers, C.L. (1992) Manipulation of host plant quality: herbivores keep leaves in the dark. *Functional Ecology* 6, 741–743.

Sagers, C.L. & Coley, P.D. (1995) Benefits and costs of defense in a neotropical shrub. *Ecology* 76, 1835–1843.

Saini, H.S., Bassi, P.K. & Spencer, M.S. (1985) Seed germination of *Chenopodium album* L.: further evidence for the dependence of the effects of growth regulators on nitrate availability. *Plant, Cell and Environment* 8, 707–711.

Sakai, A.K. (1990) Sex ratios of red maple (*Acer rubrum*)

populations in northern lower Michigan. *Ecology* **71**, 571–580.

Sakai, A.K., Wagner, W.L., Ferguson, D.M. & Herbst, D.R. (1995) Origins of dioecy in the Hawaiian flora. *Ecology* **76**, 2517–2529.

Salisbury, E.J. (1921) The significance of the calcicolous habit. *Journal of Ecology* **8**, 202–215.

Salisbury, E.J. (1942) *The Reproductive Capacity of Plants: Studies in Quantitative Biology*. G. Bell & Sons, London.

Salisbury, E.J. (1974) Seed mass in relation to environment. *Proceedings of the Royal Society of London Series B* **186**, 83–88.

Salzman, A.G. (1985) Habitat selection in a clonal plant. *Science* **228**, 603–604.

Samson, D.A. & Werk, K.S. (1986) Size-dependent effects in the analysis of reproductive effort in plants. *American Naturalist* **127**, 667–680.

Samson, D.A., Philippi, T.E. & Davidson, D.W. (1992) Granivory and competition as determinants of annual plant diversity in the Chihuahuan Desert. *Oikos* **65**, 61–80.

Sarathchandra, S.U., Dimenna, M.E., Burch, G. *et al.* (1995) Effects of plant parasitic nematodes and rhizosphere micro-organisms on the growth of white clover (*Trifolium repens* L.) and perennial rye grass (*Lolium perenne* L.). *Soil Biology and Biochemistry* **27**, 9–16.

Sarukhán, J. (1974) Studies on plant demography: *Ranunculus repens* L., *R. bulbosus* L. and *R. acris* L. II. Reproductive strategies and seed population dynamics. *Journal of Ecology* **62**, 151–177.

Sarukhán, J.L. & Harper, J.L. (1973) Studies on plant demography: *Ranunculus repens* L., *R. bulbosus* L. and *R. acris* L. I. Population flux and survivorship. *Journal of Ecology* **61**, 675–716.

Sarukhán, J., Martínez-Ramos, M. & Piñero, D. (1984) The analysis of demographic variability at the individual level and its population consequences. In: *Perspectives on Plant Population Ecology* (Eds R. Dirzo & J. Sarukhán), pp. 83–106. Sinauer, Sunderland, Massachusetts.

Schaeffer, W.M. & Gadgil, M.D. (1975) Selection for optimal life histories in plants. In *The Ecology and Evolution of Communities* (Eds M. Cody & J. Diamond), pp. 142–157. Harvard University Press, Cambridge, Massachusetts.

Schaffner, J.H. (1922) Control of sexual state in *Arisaema triphyllum* and *A. draconitum*. *American Journal of Botany* **9**, 72–78.

Schaal, B.A. (1980) Reproductive capacity and seed size in *Lupinus texensis*. *American Journal of Botany* **67**, 703–709.

Scheffler, J.A., Parkinson, R. & Dale, P.J. (1993) Frequency and distance of pollen dispersal from transgenic oilseed rape (*Brassica napus*). *Transgenic Research* **2**, 356–364.

Scheiner, S.M. (1987) Size and fecundity hierarchies in an herbaceous perennial. *Oecologia* **74**, 128–132.

Schemske, D.W. & Horvitz, C. (1984) Variation among floral visitors in pollination ability: a precondition for mutualism specialization. *Science* **225**, 519–521.

Schemske, D.W. & Horvitz, C. (1988) Plant–animal interactions and fruit production in a neotropical herb: a path analysis. *Ecology* **69**, 1128–1137.

Schierenbeck, K.A., Mack, R.N. & Sharitz, R.R. (1994) Effects of herbivory on growth and biomass allocation in native and introduced species of *Lonicera*. *Ecology* **75**, 1661–1672.

Schlessinger, W.H., Reynolds, J.F., Cunningham, G.L., *et al.* (1990) Biological feedbacks in global desertification. *Science* **247**, 1043–1048.

Schlessman, M.A. (1988) Gender diphasy ('sex choice'). In *Plant Reproductive Ecology* (Eds J.L. Doust & L.L. Doust), pp. 139–156. Oxford University Press, New York.

Schlichting, C.D., Stephenson, A.G., Small, L.E. & Winsor, J.A. (1990) Pollen loads and progeny vigor in *Cucurbita pepo*: the next generation. *Evolution* **44**, 1358–1372.

Schmid, B. (1994) Effects of genetic diversity in experimental stands of *Solidago altissima*: evidence for the potential role of pathogens as selective agents in plant populations. *Journal of Ecology* **82**, 165–175.

Schmitt, J. (1993) Reaction norms of morphological and life history traits to light availability in *Impatiens capensis*. *Evolution* **47**, 1654–1668.

Schmitt, J., Eccleston, J. & Ehrhardt, D.W. (1987) Density-dependent flowering phenology, outcrossing and reproduction in *Impatiens capensis*. *Oecologia* **72**, 341–347.

Schoen, D.J. (1982) Genetic variation and the breeding system of *Gilia achilleifolia*. *Evolution* **36**, 361–370.

Schoen, D.J. & Stewart, S.C. (1987) Variation in male fertilities and pairwise mating probabilities in *Picea glauca*. *Genetics* **116**, 141–152.

Schoener, T.W. (1983) Field experiments on interspecific competition. *American Naturalist* **122**, 240–285.

Schoener, T.W. (1985) Some comments on Connell's and my reviews of field experiments on interspecific competition. *American Naturalist* **125**, 730–740.

Schoener, T.W. (1989) Food webs from the large to the small. *Ecology* **70**, 1559–1589.

Scholander, P.F., Hammel, M.T., Bradstreet, E.D. & Hemmingsen, E.A. (1965) Sap pressure in vascular plants. *Science* **148**, 339–346.

Schultz, J.C. (1983) Habitat selection and foraging tactics of caterpillars in heterogeneous trees. In *Variable Plants and Herbivores in Natural and Managed Systems* (Eds R.F. Denno & M.S. McClure), pp. 61–90. Academic Press, New York.

Schultz, J.C. (1988) Many factors influence the evolution of herbivore diets, but plant chemistry is central. *Ecology* **68**, 896–897.

Schultz, J.C. (1992) Factoring natural enemies into plant tissue availability to herbivores. In *Effects of Resource Distribution on Animal–Plant Interactions* (Eds M.D. Hunter, T. Ohgushi & P.W. Price), pp. 175–197. Academic Press, San Diego.

Schultz, J.C., Nothnagle, P.J. & Baldwin, I.T. (1982) Seasonal and individual variation in leaf quality of two northern hardwood tree species. *American Journal of Botany* **69**, 755–759.

Schultz, J.C., Hunter, M.D. & Appel, H.M. (1992) Antimicrobial activity of polyphenols mediates plant–herbivore interactions. In *Plant Polyphenols: Biogenesis, Chemical Properties and Significance* (Ed. R.W. Hemingway), pp. 621–637. Plenum Press, New York.

Schulze, E.D. (1986) Carbon dioxide and water vapor exchange in response to drought in the atmosphere and in the soil. *Annual Review of Plant Physiology* **37**, 247–274.

Schulze, E.D. & Hall, A.E. (1982) Stomatal control of water loss. In *Encyclopedia of Plant Physiology*. Physiological Plant Ecology (Eds O.L. Lange, P.S. Nobel, C.B. Osmond, H. Ziegler). Vol. 12B, pp. 181–230. Springer, Berlin.

Schulze, E.D., Hall, A.E., Lange, O.L. & Walz, H. (1982) A portable steady-state porometer for measuring the carbon dioxide and water vapour exchanges of leaves under natural conditions. *Oecologia* **53**, 141–145.

Schulze, E.D., Kelliher, F.M., Lloyd, D. & Leuning, R. (1994) Relationships among maximum stomatal conductance, ecosystem surface conductance, carbon assimilation rate, and plant nitrogen nutrition: a global ecology scaling exercise. *Annual Review of Ecology and Systematics* **25**, 629–660.

Schupp, E.W. (1992) The Janzen–Connell model for tropical tree diversity – population implications and the importance of spatial scale. *American Naturalist* **140**, 526–530.

Schupp, E.W. (1993) Quantity, quality, and the effectiveness of seed dispersal. *Vegetatio* **107/108**, 15–29.

Scriber, J.M. (1977) Limiting effects of low leaf-water content on the nitrogen utilisation, energy budget, and larval growth of *Hyalophora cecropia* (Lepidoptera: Saturniidae). *Oecologia* **28**, 269–287.

Scriber, J.M. (1979) Effects of leaf-water supplementation upon post-ingestive nutritional indices of forb-, shrub-, vine-, and tree-feeding Lepidoptera. *Entomologia Experimentata Applicata* **25**, 240–252.

Scriber, J.M. & Feeny, P. (1979) Growth of herbivorous caterpillars in relation to feeding specialisation and to the growth form of their food plants. *Ecology* **60**, 829–850.

Scriber, J.M. & Slansky, F. Jr (1981) The nutritional ecology of immature insects. *Annual Review of Entomology* **26**, 183–211.

Seastedt, T.R., Ramundo, R.A. & Hayes, D.C. (1988) Maximization of densities of soil animals by foliage herbivory – empirical evidence, graphical and conceptual models. *Oikos* **51**, 243–248.

Seaward, M.R.D. (Ed.) (1977) *Lichen Ecology*. Academic Press, London.

Seiber, J.N., Lee, S.M. & Benson, J.M. (1984) Chemical characteristics and ecological significance of cardenolides in *Asclepias* species. In *Isopentenoids in Plants: Biochemistry and Function* (Eds W.D. Nes, G. Fuller & L.S. Tsai), pp. 563–588. Marcel Dekker, New York.

Seigler, D.S. (1981) Secondary metabolites and plant systematics. In *The Biochemistry of Plants. Vol. 7. Secondary Plant Products* (Eds P.K. Stumpf & E.E. Conn), pp. 139–176. Academic Press, New York.

Seigler, D.S. (1991) Cyanide and cyanogenic glycosides. In *Herbivores: their Interaction with Secondary Metabolites*, 2nd edn (Eds G.A. Rosenthal & M.R. Berenbaum), Vol. 1, pp. 35–78. Academic Press, San Diego.

Seigler, D. & Price, P.W. (1976) Secondary compounds in plants: primary functions. *American Naturalist* **110**, 101–105.

Seldal, T., Andersen, K.J. & Hogstedt, G. (1994) Grazing-induced proteinase-inhibitors – a possible cause for lemming population-cycles. *Oikos* **70**, 3–11.

Seliskar, D.M. (1995) Coastal dune restoration – a strategy for alleviating dieout of *Ammophila breviligulata*. *Restoration Ecology* **3**, 54–60.

Senn, J., Hanhimaki, S. & Haukioja, E. (1992) Among-tree variation in leaf phenology and morphology and its correlation with insect performance in the mountain birch. *Oikos* **63**, 215–222.

Sestak, Z. (Ed.) (1985) *Photosynthesis during Leaf Development*. T:VS 11, Junk. The Hague.

Shabel, A.B. & Peart, D.R. (1994) Effects of competition, herbivory and substrate disturbance on growth and size structure in pin cherry (*Prunus pensylvanica* L.) seedlings. *Oecologia* **98**, 150–158.

Shaffer, M.L. (1981) Minimum population sizes for species conservation. *Bioscience* **31**, 131–134.

Sharitz, R.R. & McCormick, J.F. (1973) Population dynamics of two competing annual species. *Ecology* **54**, 723–740.

Sharkey, T.D. & Singass, E.L. (1995) Why plants emit isoprene. *Nature* **34**, 769.

Shaw, R.F. & Mohler, J.D. (1953) The selective significance of the sex ratio. *American Naturalist* **87**, 337–342.

Shea, M.M., Dixon, P.M. & Sharitz, R.R. (1993) Size differences, sex ratio, and spatial distribution of male and female water tupelo, *Nyssa aquatica* (Nyssaceae). *American Journal of Botany* **80**, 26–30.

Shibata, M. & Nakashizuka, T. (1995) Seed and seedling demography of 4 co-occurring *Carpinus* species in a temperate deciduous forest. *Ecology* **76**, 1099–1108.

Shimda, A. & Ellner, S. (1984) Coexistence of plant species with similar niches. *Vegetatio* **58**, 29–55.

Shmida, A. & Whittaker, R.H. (1984) Convergence and non-convergence of Mediterranean type communities in the Old and the New World. In *Being Alive on Land* (Eds N.S. Margaris, M. Arianoustou-Faraggitati & W.C. Oechel), pp. 5–11. Junk, The Hague.

Shugart, H.H. & West, D.C. (1977) Development of an Appalachian deciduous forest succession model and its application to the assessment of the impact of the chestnut blight. *Journal of Environmental Management* **5**, 161–179.

Shumway, S.W. (1995) Physiological integration among

clonal ramets during invasion of disturbance patches in a New England salt marsh. *Annals of Botany* **76**, 225–233.

Shure, D.J. & Wilson, L.A. (1993) Patch-size effects on plant phenolics in successional openings of the southern Appalachians. *Ecology* **74**, 55–67.

Shykoff, J.A. & Bucheli, E. (1995) Pollinator visitation patterns, floral rewards and the probability of transmission of *Microbotryum violaceum*, a venereal disease of plants. *Journal of Ecology* **83**, 189–198.

Sibly, R. & Calow, P. (1982) Asexual reproduction in protozoa and invertebrates. *Journal of Theoretical Biology* **96**, 401–424.

Sibly, R. & Calow, P. (1983) An integrated approach to life-cycle evolution using selective landscapes. *Journal of Theoretical Biology* **102**, 527–547.

Signor, P.W. (1990) The geologic history of diversity. *Annual Review of Ecology and Systematics* **21**, 509–539.

Sih, A., Crowley, P., McPeek, M., Petranka, J. & Strohmeier, K. (1985) Predation, competition, and prey communities: a review of field experiments. *Annual Review of Ecology and Systematics* **16**, 269–311.

Silander, J.A. (1985) Microevolution in clonal plants. In *Population Biology and Evolution of Clonal Organisms* (Eds J.B.C. Jackson, L.W. Buss & R.E. Cook), pp. 107–152. Yale University Press, New Haven.

Silander, J.A. & Antonovics, J. (1982) Analysis of interspecific interactions in a coastal plant community – a perturbation approach. *Nature* **298**, 557–560.

Silander, J.A. & Pacala, S.W. (1990) The application of plant population dynamic models to understanding plant competition. In *Perspectives on Plant Competition* (Eds J.B. Grace & D. Tilman), pp. 67–91. Academic Press, San Diego.

Silkstone, B.E. (1987) The consequences of leaf damage for subsequent insect grazing on birch (*Betula* spp.). *Oecologia* **74**, 149–152.

Silverman, P., Seskar, M., Kanter, D., Métraux, J. & Raskin, I. (1995) Salicylic acid in rice: biosynthesis, conjugation, and possible role. *Plant Physiology* **108**, 633–639.

Silvertown, J.W. (1980) The dynamics of a grassland ecosystem: botanical equilibrium in the Park Grass Experiment. *Journal of Applied Ecology* **17**, 491–504.

Silvertown, J.W. (1981) Seed size, life span, and germination date as coadapted features of plant life history. *American Naturalist* **118**, 860–864.

Silvertown, J.W. (1983) Why are biennials sometimes not so few? *American Naturalist* **121**, 448–453.

Silvertown, J.W. (1985) When plants play the field. In *Evolution: Essays in Honour of John Maynard Smith* (Eds P.J. Greenwood, P.H. Harvey & M. Slatkin), pp. 143–153. Cambridge University Press, Cambridge.

Silvertown, J. (1991) Modularity, reproductive thresholds and plant population dynamics. *Functional Ecology* **5**, 577–582.

Silvertown, J. & Franco, M. (1993) Plant demography and habitat: a comparative approach. *Plant Species Biology* **8**, 67–73.

Silvertown, J.W. & Lovett Doust, J. (1993) *Introduction to Plant Population Biology*. Blackwell Scientific Publications, Oxford.

Silvertown, J. & Smith, B. (1989) Germination and population structure of spear thistle *Cirsium vulgare* in relation to experimentally controlled sheep grazing. *Oecologia* **81**, 369–373.

Silvertown, J., Franco, M. & McConway, K. (1993a) The eternal triangle – an attempt at reconciliation. *Functional Ecology* **7**, 380–381.

Silvertown, J., Franco, M., Pisanty, I. & Mendoza, A. (1993b) Comparative plant demography – relative importance of life-cycle components to the finite rate of increase in woody and herbaceous species. *Journal of Ecology* **81**, 465–476.

Silvertown, J., Dodd, M.E., McConway, K., Potts, J. & Crawley, M.J. (1994) Rainfall, biomass variation and community composition in the Park Grass Experiment. *Ecology* **75**, 2430–2437.

Simberloff, D. & Boecklen, W. (1981) Santa Rosalia reconsidered: size ratios and competition. *Evolution* **35**, 1206–1228.

Simkin, T. & Fiske, R.S. (1983) *Krakatau 1883. The Volcanic Eruption and its Effects*. Smithsonian Institute Press, Washington, DC.

Simms, E.L. (1992) Costs of resistance to herbivory. In *Plant Resistance to Herbivores and Pathogens: Ecology, Evolution and Genetics* (Eds R.S. Fritz & E.L. Simms), pp. 392–425. University of Chicago Press, Chicago.

Simms, E.L. & Vision, T.J. (1995) Pathogenic-induced systemic resistance in *Ipomoea purpurea*. *Oecologia* **102**, 494–500.

Simpson, B.B. & Neff, J.L. (1983) Evolution and diversity of floral rewards. In *Handbook of Experimental Pollination Biology* (Eds C.E. Jones & R.J. Little), pp. 142–159. Van Nostrand Reinhold, New York.

Sims, T.L. (1993) Genetic regulation of self incompatibility. *Critical Reviews in Plant Sciences* **12**, 129–167.

Sinclair, A.R.E. (1975) The resource limitation of trophic levels in tropical grassland ecosystems. *Journal of Animal Ecology* **44**, 497–520.

Sinclair, A.R.E., Jagia, M.K. & Anderson, R.J. (1988) Camphor from juvenile white spruce as an antifeedant for snowshoe hares. *Journal of Chemical Ecology* **14**, 1505–1514.

Sinclair, W.A., Griffiths, H.M. & Treshow, M. (1994) Ash yellows in velvet ash in Zion National Park, Utah – high incidence but low impact. *Plant Disease* **78**, 486–490.

Skarpe, C. (1991) Impact of grazing in savanna ecosystems. *Ambio* **20**, 351–356.

Skellam, J.G. (1951) Random dispersal in theoretical populations. *Biometrika* **38**, 196–218.

Slansky, F. Jr & Feeny, P. (1977) Stabilization of the rate of nitrogen accumulation by larvae of the cabbage butterfly on wild and cultivated plants. *Ecological Monographs* **47**, 209–228.

Slansky, F. Jr & Rodriguez, J.G. (Eds) (1987) *Nutritional*

Ecology of Insects, Mites, Spiders and Related Invertebrates. John Wiley & Sons, New York.

Slatkin, M. (1974) Competition and regional coexistence. *Ecology* **55**, 128–134.

Slatyer, R.O. (1967) *Plant–Water Relationships.* Academic Press, London.

Smallwood, P.D. (1993) Web-site tenure in the long-jawed spider – is it risk sensitive foraging, or conspecific interactions? *Ecology* **74**, 1826–1835.

Smedley, M.P., Dawson, T.E., Comstock, J.P., Donovan, L.A., Sherrill, D.E., Cook, C.S. & Ehleringer, J.R. (1991) Seasonal carbon isotope discrimination in a grassland community. *Oecologia* **85**, 314–320.

Smith, A.M. (1994) Xylem transport and the negative pressure sustained by water. *Annals of Botany* **74**, 647–651.

Smith, B.H. (1983) Demography of *Floerkea proserpinacoides*, a forest-floor annual. III. Dynamics of seed and seedling populations. *Journal of Ecology* **71**, 413–425.

Smith, B.H. (1984) The optimal design of a herbaceous body. *American Naturalist* **123**, 197–211.

Smith, C.C. (1975) The coevolution of plants and seed predators. In *Coevolution of Animals and Plants* (Eds L.E. Gilbert & P.H. Raven), pp. 53–77. University of Texas Press, Texas.

Smith, C.C. & Follmer, D. (1972) Food preferences of squirrels. *Ecology* **53**, 82–91.

Smith, C.C. & Fretwell, S.D. (1974) The optimal balance between size and number of offspring. *American Naturalist* **108**, 499–506.

Smith, C.C., Hamrick, J.L. & Kramer, C.L. (1990) The advantage of mast years for wind pollination. *American Naturalist* **136**, 154–166.

Smith, C.W. (1985) Impact of alien plants on Hawaii's native biota. In *Hawaii's Terrestrial Ecosystems: Preservation and Management* (Eds C.P. Stone & J.M. Scott), pp. 180–250. University of Hawaii Press, Honolulu.

Smith, R.I.L. (1984) Terrestrial plant biology of the sub-Antarctic and Antarctic. In *Antarctic Biology* (Ed. R.M. Laws), Vol. 1, pp. 61–162. Academic Press, London.

Smouse, P.E. & Meagher, T.R. (1994) Genetic analysis of male reproductive contributions in *Chamaelirium luteum* (L.) Gray (Liliaceae). *Genetics* **136**, 313–322.

Snow, D.W. (1981) Tropical frugivorous birds and their food plants: a world survey. *Biotropica* **13**, 1–14.

Snyder, M.A. (1993) Interactions between Aberts squirrel and Ponderosa pine – the relationship between selective herbivory and host plant fitness. *American Naturalist* **141**, 866–879.

Soane, I.D. & Watkinson, A.R. (1979) Clonal variation in populations of *Ranunculus repens*. *New Phytologist* **82**, 557–573.

Sobrado, M.A., Grace, J. & Jarvis, P.G. (1992) The limits to xylem embolism recovery in *Pinus sylvestris* L. *Journal of Experimental Botany* **43**, 831–836.

Sokal, R.R. & Rohlf, J.E. (1981) *Biometry*, 2nd edn. Freeman, San Francisco.

Solbrig, O.T. (1981) Studies on the population biology of the genus *Viola*. II. The effect of plant size on fitness in *Viola sororia*. *Evolution* **35**, 1080–1093.

Solbrig, O.T., Sarandón, R. & Bossert, W. (1988) A density-dependent growth model of a perennial herb, *Viola fimbriatula*. *American Naturalist* **131**, 385–400.

Solé, R.V. & Bascompte, J. (1993) Chaotic Turing structures. *Physics Letters A* **179**, 325–331.

Sommer, U. (1984) The paradox of the plankton: fluctuations of phosphorus availability maintain diversity of phytoplankton in flow-through cultures. *Limnology and Oceanography* **29**, 633–636.

Sommer, U. (1985) Comparison between steady state and non-steady state competition: experiments with natural phytoplankton. *Limnology and Oceanography* **30**, 335–346.

Soó, R. (1964–1973) *A magyar flóra és vegetáció rendszertani-növényföldrajzi kézikönyue*, Vols 1–5. Akadémian Kiadó, Budapest.

Sorensen, A.E. (1986) Seed dispersal by adhesion. *Annual Review of Ecology and Systematics* **17**, 443–464.

Sorensen, T. (1958) Sexual chromosome aberrants in apomictic Taraxaca. *Botaniska Tidsskrift* **54**, 1–22.

Sork, V.L. (1985) Germination response in a large-seeded neotropical tree species, *Gustavia superba* (Lecythidaceae). *Biotropica* **17**, 130–136.

Sork, V.L. (1987) Effects of predation and light on seedling establishment in *Gustavia superba*. *Ecology* **68**, 1341–1350.

Sork, V.L., Bramble, J. & Sexton, O. (1993) Ecology of mast-fruiting in three species of North American deciduous oaks. *Ecology* **74**, 528–541.

Soulé, M.E. (1988) Mind in the biosphere; mind of the biosphere. In *Biodiversity* (Eds E.O. Wilson & F.M. Peter), pp. 465–469. National Academy of Sciences, Washington, DC.

Soumela, J. & Ayres, M.P. (1994) Within-tree and among-tree variation in leaf characteristics of mountain birch and its applications for herbivory. *Oikos* **70**, 212–222.

Southon, I.W. & Buckingham, J. (1989) *Dictionary of Alkaloids*. Chapman & Hall, London.

Southwood, T.R.E. (1973) The insect/plant relationship – an evolutionary perspective. In *Insect Plant Relationships* (Ed. H.F. van Emden), pp. 3–30. John Wiley & Sons, London.

Southwood, T.R.E. (1977) Habitat, the templet for ecological strategies? *Journal of Animal Ecology* **46**, 337–365.

Spalinger, D.E. & Hobbs, N.T. (1992) Mechanisms of foraging in mammalian herbivores: new models of functional response. *American Naturalist* **140**, 325–348.

Spears, J. (1988) Preserving biological diversity in the tropical forests of the Asian region. In *Biodiversity* (Eds E.O. Wilson & F.M. Peter), pp. 393–402. National Academy of Sciences, Washington, DC.

Spencer, K.C. (1988) Chemical mediation of coevolution in the *Passiflora-Heliconius*. In *Chemical Mediation of Coevolution* (Ed. K.C. Spencer). American Institute of Biological Sciences, Washington, DC.

Sperry, J.S., Holbrook, N.M., Zimmermann, M.H. & Tyree, M.T. (1987) Spring filling of xylem vessels in wild grapevine. *Plant Physiology* **83**, 414–417.

Sperry, J.S., Nichols, K.L., Sullivan, J.E.M. & Eastlack, S.E. (1994) Xylem embolism in ring porous, diffuse porous, and coniferous trees of Northern Utah and interior Alaska. *Ecology* **75**, 1736–1752.

Sporne, K.R. (1965) *The Morphology of Gymnosperms: the Structure and Evolution of Primitive Seed-plants.* Hutchinson University Library, London.

Sprent, J.I. (1983) Adaptive variation in legume nodule physiology resulting from host–rhizobial interactions. In *Nitrogen as an Ecological Factor* (Eds J.A. Lee, S. McNeill & I.H. Rorison), pp. 29–42. Blackwell Scientific Publications, Oxford.

Sprugel, D.G. (1976) Dynamic structure of wave-generated *Abies balsamea* forests in the north-eastern United States. *Journal of Ecology* **64**, 889–911.

Sprugel, D.G. & Bormann, F.H. (1981) Natural disturbance and the steady state in high-altitude balsam fir forests. *Science* **211**, 390–393.

Stafford, H.A. (1981) Compartmentation in natural product biosynthesis by multienzyme complexes. In *The Biochemistry of Plants Vol. 7. Secondary Plant Products* (Eds P.K. Stumpf & E.E. Conn), pp. 118–138. Academic Press, New York.

Stalfelt, M.G. (1972) *Plant Ecology: Plants, the Soil and Man.* Translated by M.S. Jarvis & P.G. Jarvis. Longman, London.

Stalter, R. & Serrao, J. (1983) The impact of defoliation by gypsy moths on the oak forest at Greenbrook Sanctuary, New Jersey. *Bulletin of the Torrey Botanical Club* **110**, 526–529.

Standell, C. & Clements, R.O. (1994) Influence of grass seed rate, herbicide (Bentazone), molluscicide (Methiocarb) and insecticide (Triazophos) on white clover (*Trifolium repens*) establishment, herbage yield and sward botanical composition. *Crop Protection* **13**, 429–432.

Staniforth, R.J. & Cavers, P.B. (1979) Field and laboratory germination responses of achenes of *Polygonum lapathifolium, P. pensylvanicum* and *P. persicaria. Canadian Journal of Botany* **57**, 877–885.

Stanton, M.L. (1984) Seed variation in wild radish: effect of seed size on components of seedling and adult fitness. *Ecology* **65**, 1105–1112.

Stanton, M.L., Snow, A.A. & Handel, S.N. (1986) Floral evolution: attractiveness to pollinators increases male fitness. *Science* **232**, 1625–1627.

Stanton, M.L., Young, H.J., Ellstrand, N.C. & Clegg, J.M. (1991) Consequences of floral variation for male and female reproduction in experimental populations of wild radish, *Raphanus sativus* L. *Evolution* **45**, 268–280.

Stapanian, M.A. & Smith, C.C. (1978) A model for scatter-hoarding: coevolution of fox squirrels and black walnuts. *Ecology* **59**, 884–896.

Stapanian, M.A. & Smith, C.C. (1984) Density dependent survival of scatter-hoarded nuts – an experimental approach. *Ecology* **65**, 1387–1396.

Stapleton, A.E. & Walbot, V. (1994) Flavonoids can protect maize DNA from the induction of ultraviolet radiation damage. *Plant Physiology* **105**, 881–889.

Stark, R.W. (1965) Recent trends in forest entomology. *Annual Review of Entomology* **10**, 303–325.

Stearns, S.C. (1976) Life-history tactics: a review of ideas. *Quarterly Review of Biology* **51**, 3–47.

Stearns, S.C. (1977) The evolution of life history traits: a critique of the theory and a review of the data. *Annual Review of Ecology and Systematics* **8**, 145–171.

Stearns, S.C. (1989) Trade-offs in life-history evolution. *Functional Ecology* **3**, 259–268.

Stearns, S.C. & Crandall, R.E. (1981) Bet-hedging and persistence as adaptation of colonizers. In *Evolution Today. Proceedings of 2nd International Congress on Systematics and Evolution 1980*, pp. 371–384. Vancouver, BC.

Stebbins, G.L. (1956) Cytogenetics and evolution of the grass family. *American Journal of Botany* **43**, 890–905.

Stebbins, G.L. (1971) *Chromosomal Evolution in Higher Plants.* Edward Arnold, London.

Steenbergh, W.F. & Lowe, C.H. (1969) Critical factors during the first years of life of the saguaro (*Cereus giganteus*) at Saguaro National Monument. *Ecology* **50**, 825–834.

Steenbergh, W.F. & Lowe, C.H. (1977) *Ecology of the Saguaro: II. Reproduction, Germination, Establishment, Growth, and Survival of the Young Plant.* National Park Service Scientific Monograph Series, No. 8, Washington, D.C.

Steinger, T. & Muller-Scharer, H. (1992) Physiological and growth responses of *Centaurea maculosa* (Asteraceae) to root herbivory under varying levels of interspecific plant competition and soil nitrogen availability. *Oecologia* **91**, 141–149.

Stephenson, A.G. (1981) Flower and fruit abortion: proximate causes and ultimate function. *Annual Review of Ecology and Systematics* **12**, 253–281.

Stephenson, A.G. & Winsor, J.A. (1986) *Lotus corniculatus* regulates offspring quality through selective fruit abortion. *Evolution* **40**, 453–458.

Stern, A.C., Boubel, R.W., Turner, D.B. & Fox, D.L. (1984) *Fundamentals of Air Pollution*, 2nd edn. Academic Press, New York.

Sterner R.W. (1995) Elemental stoichiometry of species in ecosystems. In *Linking Species and Ecosystems* (Eds C.G. Jones & J.H. Lawton), pp. 240–252. Chapman & Hall, New York.

Stewart, A., Pearman, D.A. & Preston, C.D. (1994) *Scarce Plants in Britain.* JNCC, Peterborough.

Stiles, E.W. (1980) Patterns of fruit presentation and seed dispersal in bird-disseminated woody plants in the eastern deciduous forest. *American Naturalist* **116**, 670–688.

Stitt, M. (1991) Rising CO_2 levels and their potential significance for carbon flow in photosynthetic cells. *Plant, Cell and Environment* **14**, 741–762.

Stocker, G.C. & Irvine, A.K. (1983) Seed dispersal by cassowaries (*Casuarius casuarius*) in northern Queensland's rainforests. *Biotropica* **15**, 170–176.

Stone C.P., Smith, C.W. & Tunison, J.T. (1992) *Alien Plant Invasions in Native Ecosystems of Hawaii: Management and Research*. University of Hawaii Press, Honolulu.

Strauss, S.Y. & Karban, R. (1994) The significance of outcrossing in an intimate plant herbivore relationship. 1. Does outcrossing provide an escape from herbivores adapted to the parent plant? *Evolution* **48**, 454–464.

Strong, D.R. & Simberloff, D. (1981) Straining of gnats and swallowing ratios: character displacement. *Evolution* **35**, 810–812.

Strong, D.R., Lawton, J.H. & Southwood, T.R.E. (1984) *Insects on Plants. Community Patterns and Mechanisms*. Blackwell Scientific Publications, Oxford.

Strong, D.R., Maron, J.L., Connors, P.G. Whipple, A., Harrison, S. & Jefferies, R.L. (1995) High mortality, fluctuation in numbers, and heavy subterranean insect herbivory in bush lupine, *Lupinus arboreus*. *Oecologia* **104**, 85–92.

Suighara, G. (1981) $S = CA^z$ $z = 1/4$: a reply to Connor and McCoy. *American Naturalist* **117**, 790–793.

Suighara, G. (1984) Graph theory, homology and food webs. *Proceedings of Symposia in Applied Mathematics* **30**, 83–101.

Summerhayes, V.S. (1951) *Wild Orchids of Britain*. Collins New Naturalist, London.

Sun, D. & Liddle, M.J. (1993) Trampling resistance, stem flexibility and leaf strength in 9 Australian grasses and herbs. *Biological Conservation* **65**, 35–41.

Sunderland, N. (1960) Germination of the seeds of angiospermous root parasites. In *The Biology of Weeds* (Ed. J.L. Harper), pp. 83–93. Symposium of the British Ecological Society. Blackwell Scientific Publications, Oxford.

Sutherland, W.J. & Stillman, R.A. (1988) The foraging tactics of plants. *Oikos* **52**, 239–244.

Sutton, S.L., Whitmore, T.C. & Chadwick, A.C. (Eds) (1983) *Tropical Rain Forest: Ecology and Management*. Blackwell Scientific Publications, Oxford.

Swain, T. (1977) Secondary compounds as protection agents. *Annual Review of Plant Physiology* **28**, 479–501.

Swain, T. (1979) Tannins and lignins. In *Herbivores: their Interaction with Secondary Plant Metabolites* (Eds G.A. Rosenthal & D.H. Janzen) pp. 657–718. Academic Press, San Deigo.

Swaine, M.D. & Whitmore, T.C. (1988) On the definition of ecological species groups in tropical rain forests. *Vegetatio* **75**, 81–86.

Swanson, T.M. (1995) The Economics and Ecology of Biodiversity. Cambridge University Press, Cambridge.

Sydes, C.L. & Grime, J.P. (1984) A comparative study of root development using a simulated rock crevice. *Journal of Ecology* **72**, 937–946.

Symonides, E. (1983) Population size regulation as a result of intra-population interactions. I. Effect of density on the survival and development of individuals of *Erophila verna* (L.). *Ekologia Polska* **31**, 839–881.

Symonides, E. (1988) Population dynamics of annual plants. In *Plant Population Ecology* (Eds A.J. Davy, M.J. Hutchings & A.R. Watkinson), pp. 221–248. Blackwell Scientific Publications, Oxford.

Symonides, E., Silvertown, J. & Andreasen, V. (1986) Population cycles caused by overcompensating density-dependence in an annual plant. *Oecologia* **71**, 156–158.

Synge, H. (1992) Plants on oceanic islands. In *Global Biodiversity: Status of the Earth's Living Resources* (Ed. B. Groombridge), pp. 147–150. Chapman & Hall, London.

Tahvanainen, J. (1983) The relationship between flea beetles and their cruciferous host plants: the role of plant and habitat characteristics. *Oikos* **40**, 433–437.

Takhtajan, A. (1969) *Flowering Plants – Origin and Dispersal*. Translated by C. Jeffrey. Oliver & Boyd, Edinburgh.

Tallamy, D.W. & Raupp, M.J. (1991) *Phytochemical Induction by Herbivores*. John Wiley & Sons, New York.

Tallis, J.H. (1991) *Plant Community History*. Chapman & Hall, London.

Tamm, C.O. (1972) Survival and flowering of perennial herbs. III. The behaviour of *Primula veris* on permanent plots. *Oikos*, **23**, 159–166.

Tamm, C.O. (1991) Behaviour of some orchid populations in a changing environment: observations on permanent plots, 1943–1990. In *Population Ecology of Terrestrial Orchids* (Eds T.C.E. Wells & J.H. Willems), pp. 1–13. SPB Academic Publishing, The Hague.

Tan, C.S. & Buttery, B.R. (1995) Determination of the water use of 2 pairs of soybean isolines differing in stomatal frequency using a heat balance stem flow gauge. *Canadian Journal of Plant Science* **75**, 99–103.

Tang, W. (1989) Seed dispersal in the cycad *Zamia pumila* in Florida. *Canadian Journal of Botany* **67**, 2066–2070.

Tansley, A.G. (1935) The use and abuse of vegetational concepts and terms. *Ecology* **16**, 284–307.

Tansley, A.G. (1939) *The British Islands and their Vegetation*. Cambridge University Press, Cambridge.

Tapper, P.-G. (1992) Irregular fruiting in *Fraxinus excelsior*. *Journal of Vegetation Science* **3**, 41–46.

Taylor, A.D. (1990) Metapopulations, dispersal, and predator–prey dynamics: an overview. *Ecology* **71**, 429–433.

Taylor, G.E. Jr & Murdy, W.H. (1975) Population differentiation of an annual plant species, *Geranium carolinianum*, in response to sulfur dioxide. *Botanical Gazette* **136**, 212–215.

Tenhunen, J.D., Lange, O.L., Pereira, J.S., Losch, R. & Catarino, F. (1981) Midday stomatal closure in *Arbutus unedo* leaves. In *Components of Productivity of Mediterranean-climate Regions – Basic and Applied Aspects* (Eds N.S. Margaris & H.A. Mooney), pp. 61–69. Junk, The Hague.

Terborgh, J. (1986) Community aspects of frugivory in tropical forests. In *Frugivory and Seed Dispersal* (Eds A. Estrada & T.H. Fleming), pp. 371–384. Dr W. Junk, Boston.

Terborgh, J., Losos, E., Riley, M.P. & Riley, B. (1993) Predation by vertebrates and invertebrates on the seeds of five canopy tree species of an Amazonian forest. *Vegetatio* 107/108, 375–386.

Thaler, C.E. & Plowright, R.C. (1980) The effect of aerial insecticide spraying for spruce budworm control on the fecundity of entomophilous plants in New Brunswick. *Canadian Journal of Botany* 58, 2022–2027.

Thimonier, A., Dupouey, J.L., Bost, F. & Becker, M. (1994) Simultaneous eutrophication and acidification of a forest ecosystem in North-East France. *New Phytologist* 126, 533–539.

Thom, E.R., Prestidge, R.A., Wildermoth, D.D., Taylor, M.J. & Marshall, S.L. (1992) Effect of hessian fly (*Mayetiola destructor*) on production and persistence of prairie grass (*Bromus willdenowii*) when rotationally grazed by dairy cows. *New Zealand Journal of Agricultural Research* 35, 75–82.

Thomas, S.C. & Weiner, J. (1989) Growth, death and size distribution change in an *Impatiens pallida* population. *Journal of Ecology* 77, 524–536.

Thompson, F. (1918) *The Mistress of Vision*. Douglas Pepler, Ditchling.

Thompson, J.N. (1984) Variation among individual seed masses in *Lomatium grayi* (Umbelliferae) under controlled conditions: magnitude and partitioning of the variance. *Ecology* 66, 1608–1616.

Thompson, J.N. (1988) Coevolution and alternative hypotheses on insect/plant interactions. *Ecology* 68, 893–895.

Thompson, J.N. (1989) Concepts of coevolution. *Trends in Ecology and Evolution* 4, 179–183.

Thompson, J.N. (1994) *The Coevolutionary Process*. University of Chicago Press, Chicago.

Thompson, J.N. & Burdon, J.J. (1992) Gene-for-gene coevolution between plants and parasites. *Nature* 360, 121–125.

Thompson, J.N. & Willson, M.F. (1979) Evolution of temperate fruit/bird interactions: phenological strategies. *Evolution* 33, 973–982.

Thompson, K. (1986) Small-scale heterogeneity in the seed bank of an acidic grassland. *Journal of Ecology* 74, 733–738.

Thompson, K. (1992) The functional ecology of seed banks. In *Seeds: the Ecology of Regeneration in Plant Communities* (Ed. M. Fenner), pp. 231–258. CAB International, Wallingford.

Thompson, K. & Grime, J.P. (1979) Seasonal variation in the seed banks of herbaceous species in ten contrasting habitats. *Journal of Ecology* 67, 893–921.

Thompson, K. & Grime, J.P. (1983) A comparative study of germination responses to diurnally-fluctuating temperatures. *Journal of Applied Ecology* 20, 141–156.

Thompson, K. & Rabinowitz, D. (1989) Do big plants have big seeds? *American Naturalist* 133, 722–728.

Thompson, K., Grime, J.P. & Mason, G. (1977) Seed germination in response to fluctuating temperatures. *Nature* 267, 147–149.

Thompson, L. & Harper, J.L. (1988) The effect of grasses on the quality of transmitted radiation and its influence on the growth of white clover *Trifolium repens*. *Oecologia* 75, 343–347.

Thomson, J.D. & Barrett, S.C.H. (1981) Temporal variation of gender in *Aralia hispida* Vent. (Araliaceae). *Evolution* 35, 1094–1107.

Thrall, P. & Antonovics, J. (1995) Theoretical and empirical studies of meta-populations: population and genetic dynamics of the *Silene–Ustilago* system. *Canadian Journal of Botany* 73 (Suppl. 1), S1249–S1258.

Tieszen, L.L., Senyimba, M.M., Imbamba, S.K., & Troughton, J.H. (1979) The distribution of C_3 and C_4 grasses and carbon isotope discrimination along an altitudinal and moisture gradient in Kenya. *Oecologia* 37, 337–350.

Tilman, D. (1976) Ecological competition between algae: experimental confirmation of resource-based competition theory. *Science* 192, 463–465.

Tilman, D. (1977) Resource competition between planktonic algae: an experimental and theoretical approach. *Ecology* 58, 338–348.

Tilman, D. (1980) Resources: a graphical–mechanistic approach to competition and predation. *American Naturalist* 116, 362–393.

Tilman, D. (1982) *Resource Competition and Community Structure*. Princeton University Press, Princeton.

Tilman, D. (1985) The resource ratio hypothesis of succession. *American Naturalist* 125, 827–852.

Tilman, D. (1987) Secondary succession and the pattern of plant dominance along experimental nitrogen gradients. *Ecological Monographs* 57, 189–214.

Tilman, D. (1988) *Plant Strategies and the Dynamics and Structure of Plant Communities*. Princeton University Press, Princeton.

Tilman, D. (1990a) Mechanisms of plant competition for nutrients: the elements of a predictive theory of competition. In *Perspectives on Plant Competition* (Eds J.B. Grace & D. Tilman), pp. 117–141. Academic Press, San Diego.

Tilman, D. (1990b) Constraints and tradeoffs: toward a predictive theory of competition and succession. *Oikos* 58, 3–15.

Tilman, D. (1994) Competition and biodiversity in spatially structured habitats. *Ecology* 75, 2–16.

Tilman, D. (1994) Competition and biodiversity in spatially structured habitats. *Ecology* 75, 2–16.

Tilman, D. & Downing, J.A. (1994) Biodiversity and stability in grasslands. *Nature* 367, 363–365.

Tilman, D. & Pacala, S. (1993) The maintenance of species richness in plant communities. In *Species Diversity in Ecological Communities*. University of Chicago Press, Chicago.

Tilman, D. & Wedin, D. (1991a) Plant traits and resource reduction for five grasses growing on a nitrogen gradient. *Ecology* **72**, 685–700.

Tilman, D. & Wedin, D. (1991b) Dynamics of nitrogen competition between successional grasses. *Ecology* **72**, 1038–1049.

Tilman, D., Mattson, M. & Langer, S. (1981) Competition and nutrient kinetics along a temperature gradient: an experimental test of a mechanistic approach to niche theory. *Limnology and Oceanography* **26**, 1020–1033.

Tilman, D., Kilham, S.S. & Kilham, P. (1982) Phytoplankton community ecology: the role of limiting nutrients. *Annual Review of Ecology and Systematics* **13**, 349–372.

Tilman, D., Downing, J.A. & Wedin, D.A. (1994a) Does diversity beget stability? *Nature* **371**, 113–114.

Tilman, D., May, R.M., Lehman, C.L. & Nowak, M.A. (1994b) Habitat destruction and the extinction debt. *Nature* **371**, 65–66.

Timmons, A.M. *et al.* (1996) Risks from transgenic crops. *Nature* **380**, 487.

Tingey, D.T. & Hogsett, W.E. (1985) Water stress reduces ozone injury via a stomatal mechanism. *Plant Physiology* **77**, 944–947.

Tolvanen, A., Laine, K., Pakonen, T., Saari, E. & Havas, P. (1994) Responses to harvesting intensity in a clonal dwarf shrub, the bilberry (*Vaccinium myrtillus* L.). *Vegetatio* **110**, 163–169.

Torssell, B.W.R., Rose, C.W. & Cunningham, B.R. (1975) Population dynamics of an annual pasture in a dry monsoonal climate. *Proceedings of the Ecological Society of Australia* **9**, 157–162.

Trame, A.M., Coddington, A.J. & Paige, K.N. (1995) Field and genetic studies testing optimal outcrossing in *Agave schottii*, a long-lived clonal plant. *Oecologia* **104**, 93–100.

Tran, V.N. & Cavanagh, A.K. (1984) Structural aspects of seed dormancy. In *Seed Physiology, Vol. 2, Germination and Reserve Mobilization* (Ed. D.R. Murray), pp. 1–44. Academic Press, London.

Traveset, A., Willson, M.F. & Gaither, J.C. (1995) Avoidance by birds of insect-infested fruits of *Vaccinium ovalifolium*. *Oikos* **73**, 381–386.

Trenbath, B.R. (1993) Intercropping for the management of pests and diseases. *Field Crops Research* **34**, 381–405.

Treshow, M. (Ed.) (1984) *Air Pollution and Plant Life.* John Wiley & Sons, Chichester.

Trivers, R.L. (1971) The evolution of reciprocal altruism. *Quarterly Review of Biology* **46**, 35–57.

Trivers, R.L. (1974) Parent–offspring conflict. *American Zoologist* **14**, 249–264.

Trivers, R.L. & Willard, D.E. (1973) Natural selection of parental ability to vary the sex ratio of offspring. *Science* **179**, 90–91.

Troelstra, S.R., Wagenaar, R. & Smant, W. (1995) Nitrogen utilization by plant species from acid heathland soils. 1. Comparison between nitrate and ammonium nutrition at constant low pH. 2. Growth and shoot/root partitioning of NO_3 assimilation at constant low pH and varying NO_3^-/NH_4^+ ratio. *Journal of Experimental Botany* **46**, 1103–1112; 1113–1121.

Trumble, J.T., Kolodnyhirsch, D.M. & Ting, I.P. (1993) Plant compensation for arthropod herbivory. *Annual Review of Entomology* **38**, 93–119.

Tuba, Z., Lichtenthaler, H.K., Csintalan, Z. & Pocs, T. (1993) Regreening of desiccated leaves of the poikilochlorophyllous *Xerophyta scabrida* upon rehydration. *Journal of Plant Physiology* **142**, 103–108.

Tuomi, J. (1992) Towards integration in plant defense theories. *Trends in Ecology and Evolution* **7**, 365–367.

Tuomi, J., Niemela, P., Hankioja, E., Siren, S. & Neuvonen, S. (1984) Nutrient stress: an explanation for plant antiherbivore responses to defoliation. *Oecologia* **61**, 208–210.

Tuomi, J., Fagerstrom, T. & Niemela, P. (1991) Carbon allocation, phenotypic plasticity, and induced defenses. In *Phytochemical Induction by Herbivores* (Eds D.W. Tallamy & M.J. Raupp), pp. 85–104. John Wiley & Sons, New York.

Tuomi, J., Nilsson, P. & Astrom, M. (1994) Plant compensatory responses – bud dormancy as an adaptation to herbivory. *Ecology* **75**, 1429–1436.

Turkington, R. (1983) Leaf and flower demography of *Trifolium repens*. I. Growth in mixture with grasses. *New Phytologist* **93**, 599–616.

Turlings, T.C.L., Tumlinson, J.H. & Lewis, W.J. (1990) Exploitation of herbivore-induced plant odors by host-seeking parasitic wasps. *Science* **250**, 1251–1253.

Turner, C.L., Seastedt, T.R. & Dyer, M.I. (1993) Maximization of above ground grassland production – the role of defoliation frequency, intensity, and history, *Ecological Applications* **3**, 175–186.

Turner, M.D. & Rabinowitz, D. (1983) Factors suppressing frequency distributions of plant mass: the absence of dominance and suppression in competing monocultures of *Festuca paradoxa*. *Ecology* **64**, 469–475.

Turner, M.G., Hargrove, W.W., Gardner, R.H. & Romme, W.H. (1994) Effects of fire on landscape heterogeneity in Yellowstone National Park, Wyoming. *Journal of Vegetation Science* **5**, 731–742.

Turner, N.C. (1988) Measurement of plant water status by the pressure chamber technique. *Irrigation Science* **9**, 289–308.

Turner, R.M., Alcorn, S.M., Olin, G. & Booth, J.A. (1966) The influence of shade, soil, and water on saguaro seedling establishment. *Botanical Gazette* **127**, 95–102.

Turrill, W.B. (1929) *The Plant Life of the Balkan Peninsula.* Clarendon Press, Oxford.

Tyler, G. (1974) Heavy metal pollution and soil enzymatic activity. *Plant and Soil* **41**, 303–311.

Tyler, G (1994) A new approach to understanding the calcifuge habit of plants. *Annals of Botany* **73**, 327–330.

Tyler, G. & Strom, L. (1995) Differing organic acid exudation pattern explains calcifuge and acidifuge behavior of plants. *Annals of Botany* **75**, 75–78.

Tyre, A.J. & Addicott, J.F. (1993) Facultative non-mutualistic behavior by an obligate mutualist – cheating by yucca moths. *Oecologia* **94**, 173–175.

Tyree, M.T. & Sperry, J.S. (1989) Vulnerability of xylem to cavitation and embolism. *Annual Review of Plant Physiology and Molecular Biology* **40**, 19–38.

Uhl, C. (1988) Restoration of degraded lands in the Amazon Basin. In *Biodiversity* (Eds E.O. Wilson & F.M. Peter), pp. 326–332. National Academy of Sciences, Washington, DC.

UK TERG (1988) *The Effects of Acid Deposition on the Terrestrial Environment in the United Kingdom*. First Report of the United Kingdom Terrestrial Effects Review Group. HMSO, London.

Ulrich, B. (1989) Effects of acidic precipitation on forest ecosystems in Europe. In *Acidic Precipitation, Vol. 2. Biological and Ecological Effects* (Eds D.C. Adriano & A.H. Johnson), pp. 189–272. Springer-Verlag, New York.

Uma Shaanker, R., Ganeshaiah, K.N. & Bawa, K.S. (1988) Parent–offspring conflict, sibling rivalry, and brood size patterns in plants. *Annual Review of Ecology and Systematics* **19**, 177–205.

Underwood, A.J. & Denley, E.J. (1984) Paradigms, explanations and generalizations in models for the structure of intertidal communities on rocky shores. In *Ecological Communities: Conceptual Issues and Evidence* (Eds D.R. Strong, D. Simberloff, L.G. Abele & A.B. Thistle), pp. 151–180. Princeton University Press, Princeton.

Ungar, I.A. (1987) Population biology of halophyte seeds. *Botanical Review* **53**, 301–334.

Ungar, I.A., Benner, D.K. & McGraw, D.C. (1979) The distribution and growth of *Salicornia europaea* on an inland salt pan. *Ecology* **60**, 329–336.

Upton, G.J.C. & Fingleton, B. (1985) *Spatial Data Analysis by Example. Vol. 1. Point Patterns, and Quantitative Data*. John Wiley & Sons, Chichester.

Upton, G.J.G. & Fingleton, B. (1989) *Spatial Data Analysis by Example. Vol. 2. Categorical and Directional Data*. John Wiley & Sons, Chichester.

Usher, M.B. (1986) Modelling successional processes in ecosystems. In *Colonization, Succession and Stability* (Eds A.J. Gray, M.J. Crawley & P.J. Edwards). Blackwell Scientific Publications, Oxford.

Uyenoyama, M.K. (1995) A generalized least squares estimate for the origin of sporophytic self-incompatibility. *Genetics* **139**, 975–992.

van der Valk, A.G. (1981) Succession in wetlands: a Gleasonian approach. *Ecology* **62**, 688–696.

Valledares, G.R. & Hartley, S.E. (1994) Effects of scale on detecting interactions between Coleophora and Eriocrania leaf-miners. *Ecological Entomology* **19**, 257–262.

Van Breemen, N. (1985) Acidification and decline of Central European forest. *Nature* **315**, 16.

Van den Honert, T.H. (1948) Water transport in plants as a catenary process. *Discussions of the Faraday Society* **3**, 146–153.

Van der Meijden, E. & van der Veen, C.A.M. (1996) Tritrophic metapopulation dynamics: a case study of ragwort, cinnabar moth and the parasitoid *Cotesia popularis*. In *Metapopulation Dynamics: Ecology, Genetics and Evolution* (Eds I.A. Hanski & M.E. Gilpin), Academic Press, San Diego.

Van der Putten, W.H., van Dijk, C. & Peters, B.A.M. (1993) Plant-specific soil-borne diseases contribute to succession in fore dune vegetation. *Nature* **362**: 53–56.

Vander Wall, S.B. (1990) *Food Hoarding in Animals*. University of Chicago Press, Chicago.

Vander Wall, S.B. (1994) Seed fate pathways of antelope bitterbrush: dispersal by seed-caching yellow pine chipmunks. *Ecology* **75**, 1911–1926.

Van Gardingen, P.R. & Grace, J. (1991) Plants and Wind. *Advances in Botanical Research* **18**, 192–254.

Van Gardingen, P.R., Jeffree, C.E. & Grace, J. (1989) Variation in stomatal aperture in leaves of *Avena fatua* L. observed by low-temperature scanning electron microscopy. *Plant, Cell & Environment* **12**, 887–897.

Van Groenendael, J.M. & Slim, P. (1988) The contrasting dynamics of two populations of *Plantago lanceolata* classified by age and size. *Journal of Ecology* **76**, 585–599.

Van Tooren, B.F. & Pons, T.L. (1988) Effects of temperature and light on the gemination in chalk grassland species. *Functional Ecology* **2**, 303–311.

Vázquez-Yanes, C. & Orozco-Segovia, A. (1984) Ecophysiology of seed germination in tropical humid forests of the world: a review. In *Physiological Ecology of Plants in the Wet Tropics* (Eds E. Medina, H.A. Mooney & C. Vázquez-Yanes), pp. 37–50. Junk, The Hague.

Venable, D.L. (1992) Size–number trade-offs and the variation of seed size with plant resource status. *American Naturalist* **140**, 287–304.

Venable, D.L. & Brown, J.S. (1988) The selective interactions of dispersal, dormancy, and seed size as adaptations for reducing risk in variable environments. *American Naturalist* **131**, 360–384.

Venable, D.L. & Lawlor, L. (1980) Delayed germination and dispersal in desert annuals: escape in space and time. *Oecologia* **46**, 272–282.

Venable, D.L., Pake, C.E. & Caprio, A.C. (1993) Diversity and coexistence of sonoran desert winter annuals. *Plant Species Biology* **8**, 207–216.

Vet, L.E.M. & Dicke, M. (1992) Ecology of infochemical use by natural enemies in a tritrophic context. *Annual Review of Entomology* **37**, 141–172.

Vicari, M. & Bazely, D.R. (1993) Do grasses fight back? The case for antiherbivore defences. *Trends in Ecology and Evolution* **8**, 137–141.

Vickery, J.A., Watkinson, A.R. & Sutherland, W.J. (1994) The solution to the Brent goose problem – an economic analysis. *Journal of Applied Ecology* **31**, 371–382.

Vickery, J.A., Sutherland, W.J., Watkinson, A.R., Lane, S.J. & Rowcliffe, J.M. (1995) Habitat switching by dark-bellied Brent geese *Branta bernicla* (L) in relation to food depletion. *Oecologia* **103**, 499–508.

Vinton, M.A. & Hartnett, D.C. (1992) Effects of bison grazing on *Andropogon gerardii* and *Panicum virgatum* in burned and unburned tall grass prairie. *Oecologia* **90**, 374–382.

Vitousek, P.M. (1982) Nutrient cycling and nutrient use efficiency. *American Naturalist* **119**, 553–572.

Vitousek, P. (1988) Diversity and biological invasions of oceanic islands. In *Biodiversity* (Eds E.O. Wilson & F.M. Peter), pp. 181–189. National Academy of Sciences, Washington, DC.

Vitousek, P.M. & Matson, P.A. (1984) Mechanisms of nitrogen retention in forest ecosystems: a field experiment. *Science* **255**, 51–52.

Vitousek, P.M. & Matson, P.A. (1985) Disturbance, nitrogen availability, and nitrogen losses in an intensively managed loblolly pine plantation. *Ecology* **66**, 1360–1376.

Vitousek, P.M., Walker, L.R., Whiteacre, L.D., Mueller-Dombois, D. & Matson, P.A. (1987) Biological invasion by *Myrica faya* alters ecosystem development in Hawaii. *Science* **238**, 802–804.

Volterra, V. (1926) Variation and fluctuations of the number of individuals in animal species living together. In *Animal Ecology* (Ed. R.N. Chapman), pp. 409–448. McGraw-Hill, London.

Von Wiren, N., Morel, J.L., Guckert, A., Romheld, V. & Marschner, H. (1993) Influence of microorganisms on iron acquisition in maize. *Soil Biology and Biochemistry* **25**, 371–376.

Vrain, T.C. (1993) Pathogenicity of *Pratylenchus penetrans* to American ginseng (*Panax quinquefolium*) and to kiwi (*Actinidia chinensis*). *Canadian Journal of Plant Science* **73**, 907–912.

Vrieling, K. & van Wijk, C.A.M. (1994) Estimating costs and benefits of the pyrrolizidine alkaloids of *Senecio jacobaea* under natural conditions. *Oikos* **70**, 449–454.

Wacquant, J.P. & Picard, J.B. (1992) Nutritional differentiation among populations of the Mediterranean shrub *Dittrichia viscosa* (Asteraceae) in siliceous and calcareous habitats. *Oecologia* **92**, 14–22.

Wainwright, S.J. (1984) Adaptations of plants to flooding with salt water. In *Flooding and Plant Growth* (Ed. T.T. Kozlowski), pp. 295–343. Academic Press, London.

Waite, S. & Hutchings, M.J. (1978) The effects of sowing density, salinity and substrate upon the germination of seeds of *Plantago coronopus* L. *New Phytologist* **81**, 341–348.

Waite, S. & Hutchings, M.J. (1979) A comparative study of establishment of *Plantago coronopus* L. from seeds sown randomly and in clumps. *New Phytologist* **82**, 575–583.

Walker, T.W. & Syers, J.K. (1976) The fate of phosphorous during pedogenesis. *Geoderma* **15**, 1–19.

Wallace, J.P. & O'Hop, J. (1985) Life on a fast pad: waterlily leaf beetle impact on water lilies. *Ecology* **66**, 1534–1544.

Wallace, L.L. & Macko, S.A. (1993) Nutrient acquisition by clipped plants as a measure of competitive success – the effects of compensation. *Functional Ecology* **7**, 326–331.

Waller, D.M. (1979) Models of mast fruiting in trees. *Journal of Theoretical Biology* **80**, 223–232.

Waller, D.M. (1986) The dynamics of growth and form. In *Plant Ecology* (Ed. M.J. Crawley), pp. 291–320. Blackwell Scientific Publications, Oxford.

Waller, D.M. (1993) How does mast-fruiting get started? *Trends in Ecology and Evolution* **8**, 122–123.

Waller, D.M. & Steingraeber, D. (1986) Branching and modular growth: theoretical models and empirical patterns. In *The Population Biology and Evolution of Clonal Organisms* (Eds J.B.C. Jackson, L.W. Buss & R.E. Cook). Yale University Press, New Haven.

Waloff, B. & Richards, O.W. (1977) The effect of insect fauna on growth, mortality and natality of broom, *Sarothamnus scoparius*. *Journal of Applied Ecology* **14**, 787–798.

Wandera, J.L., Richards, J.H. & Mueller, R.J. (1992) The relationships between relative growth rate, meristematic potential and compensatory growth of semi-arid land shrubs. *Oecologia* **90**, 391–398.

Ware, A.B. & Compton, S.G. (1994) Responses of fig wasps to host plant volatile cues. *Journal of Chemical Ecology* **20**, 785–802.

Waring, C.L. & Cobb, N.S. (1992) The impact of plant stress on herbivore population dynamics. In *Insect–Plant Interactions* (Ed. E.A. Bernays), Vol. 4, pp. 168–215. CRC Press, Boca Raton.

Waring, R.H. & Running, S.W. (1978) Sapwood water storage: its contribution to transpiration and effect upon water and conductance through the stems of old-growth Douglas fir. *Plant, Cell and Environment* **1**, 131–140.

Waring, R.H., Schroeder, P.E. & Oren, R. (1982) Application of the pipe model theory to predict canopy leaf area. *Canadian Journal of Forest Research* **12**, 556–560.

Waring, R.H., McDonald, A.J.S., Larsson, S. *et al.* (1985) Differences in chemical composition of plants grown at constant relative growth rates with stable mineral nutrition. *Oecologia* **66**, 157–160.

Warington, K. (1936) The effect of constant and fluctuating temperatures on the germination of weed seeds in arable soils. *Journal of Ecology* **24**, 185–204.

Warming, E. (1909) *Oecology of Plants. An Introduction to the Study of Plant Communities*. Clarendon Press, Oxford.

Warrington, S., Cottam, D.A. & Whittaker, J.B. (1989) Effects of insect damage on photosynthesis, transpiration and SO₂ uptake by sycamore. *Oecologia* **80**, 136–139.

Waser, N.M. (1978) Competition for hummingbird pollination and sequential flowering in two Colorado wildflowers. *Ecology* **59**, 934–944.

Waser, N.M. & Price, M.V. (1983) Optimal and actual outcrossing in plants, and the nature of plant-poll-

[695]

inator interaction. In *Handbook of Experimental Pollination Biology* (Eds C.E. Jones & R.J. Little), pp. 341–359. Van Nostrand Reinhold, New York.

Waterman, P.G. (1992) Roles for secondary metabolites in plants. In *Secondary Metabolites: their Functions and Evolution* (Eds D.J. Chadwick & J. Whekan), pp. 255–275. John Wiley & Sons, Chichester.

Waterman, P.G. & Mole, S. (1989) Extrinsic factors influencing production of secondary metabolites in plants. In *Insect–Plant Interactions* (Ed. E.A. Bernays), pp. 107–134. CRC Press, Boca Raton.

Waterman, P.G. & Mole, S. (1994) *Analysis of Phenolic Plant Metabolites*. Blackwell Science, Oxford.

Watkinson, A.R. (1978a) The demography of a sand dune annual: *Vulpia fasciculata*. II. The dynamics of seed populations. *Journal of Ecology* **66**, 35–44.

Watkinson, A.R. (1978b) The demography of a sand dune annual, *Vulpia fasciculata*. III. The dispersal of seeds. *Journal of Ecology* **66**, 483–498.

Watkinson, A.R. (1980) Density-dependence in single species populations of plants. *Journal of Theoretical Biology* **83**, 345–357.

Watkinson, A.R. (1981a) Interference in pure and mixed populations of *Agrostemma githago*. *Journal of Applied Ecology* **18**, 967–976.

Watkinson, A.R. (1981b) The population ecology of winter annuals. In *The Biological Aspects of Rare Plant Conservation* (Ed. H. Synge), pp. 253–264. John Wiley & Sons, Chichester.

Watkinson, A.R. (1985a) On the abundance of plants along an environmental gradient. *Journal of Ecology* **73**, 569–578.

Watkinson, A.R. (1985b) Plant responses to crowding. In *Studies on Plant Demography: a Festschrift for John L. Harper* (Ed. J. White), pp. 275–289. Academic Press, London.

Watkinson, A.R. (1986) Plant population dynamics. In *Plant Ecology* (Ed. M.J. Crawley), pp. 137–184. Blackwell Scientific Publications, Oxford.

Watkinson, A.R. (1990) The population dynamics of *Vulpia fasciculata*: a nine year study. *Journal of Ecology* **78**, 196–209.

Watkinson, A.R. & Davy, A.J. (1985) Population biology of salt marsh and sand dune annuals. *Vegetatio* **62**, 487–497.

Watkinson, A.R. & Harper, J.L. (1978) The demography of a sand dune annual: *Vulpia fasciculata*. 1. The natural regulation of populations. *Journal of Ecology* **66**, 15–33.

Watkinson, A.R. & Powell, J.C. (1993) Seedling recruitment and the maintenance of clonal diversity in plant populations: a computer simulation of *Ranunculus repens*. *Journal of Ecology* **81**, 707–717.

Watkinson, A.R. & White, J. (1986) Some life-history consequences of modular construction in plants. *Philosophical Transactions of the Royal Society of London Series B* **313**, 31–52.

Watkinson, A.R. Lonsdale, W.M. & Andrew, M.H. (1989) Modelling the population dynamics of an annual plant *Sorghum intrans* in the wet–dry tropics. *Journal of Ecology* **77**, 162–181.

Watkinson, A.R. Lintell-Smith, G., Newsham, K.K. & Rowcliffe, J.M. (1993) Population interactions and the determinants of population size. *Plant Species Biology* **8**, 149–158.

Watson, M.A. & Casper, B.B. (1984) Morphogenetic constraints on patterns of carbon distribution in plants. *Annual Review of Ecology and Systematics* **15**, 233–258.

Watt, A.D. (1990) The consequences of natural, stress-induced and damage-induced differences in tree foliage on the population dynamics of the pine beauty moth. In *Population Dynamics of Forest Insects* (Eds A.D. Watt, S.R. Leather, M.D. Hunter & N.A.C. Kidd), pp. 157–168. Intercept, Andover.

Watt, A.D. (1994) The relevance of the stress hypothesis to insect feeding on tree foliage. In *Individuals, Populations and Patterns in Ecology* (Eds S.R. Leatherman, A.D. Watt, N.J. Mills & K.F.A. Walters), pp. 73–85. Intercept, Andover.

Watt, A.D. & MacFarlane, A.M. (1991) Does damage-mediated intergeneration conflict occur in the beech leaf mining weevil? *Oikos* **63**, 171–174.

Watt, A.S. (1947) Pattern and process in the plant community. *Journal of Ecology* **35**, 1–22.

Watt, A.S. (1955) Bracken versus heather, a study in plant sociology. *Journal of Ecology* **43**, 490–506.

Watt, A.S. (1971) Factors controlling the floristic composition of some plant communities in Breckland. In *The Scientific Management of Animal and Plant Communities for Conservation* (Eds E. Duffey & A.S. Watt), pp. 137–152. Blackwell Scientific Publications, Oxford.

Watt, A.S. (1981) A comparison of grazed and ungrazed grassland in East Anglian Breckland. *Journal of Ecology* **69**, 499–508; 509–536.

Weaver, S.E. & Cavers, P.B. (1980) Reproductive effort in two perennial weed species in different habitats. *Journal of Applied Ecology* **17**, 505–513.

Wedin, D. & Tilman, D. (1993) Competition among grasses along a nitrogen gradient: initial conditions and mechanisms of competition. *Ecological Monographs* **63**, 199–229.

Weiblen, G.D. & Thompson, J.D. (1995) Seed dispersal in *Erythronium grandiflorum* (Liliaceae). *Oecologia* **182**, 211–219.

Weiner, J. (1982) A neighborhood model of annual plant interference. *Ecology* **63**, 1237–1241.

Weiner, J. (1986) How competition for light and nutrients affects size variability in *Ipomoea tricolor* populations. *Ecology* **67**, 1425–1427.

Weiner, J. (1988) Variation in the performance of individuals in plant populations. In *Plant Population Ecology* (Eds A.J. Davy, M.J. Hutchings & A.R. Watkinson), pp. 59–81. Blackwell Scientific Publications, Oxford.

Weiner, J. (1990) Asymmetric competition in plant populations. *Trends in Ecology and Evolution* **5**, 360–364.

Weiner, J. (1993) Competition, herbivory and plant size variability: *Hypochaeris radicata* grazed by snails (*Helix aspersa*). *Functional Ecology* 7, 47–53.

Weiner, J. & Conte, P.T. (1981) Dispersal and neighborhood effects in an annual plant competition model. *Ecological Modelling* 13, 131–147.

Weiner, J. & Solbrig, O.T. (1984) The meaning and measurement of size hierarchies in plant populations. *Oecologia* 61, 334–336.

Weiner, J. & Thomas, S.C. (1986) Size variability and competition in plant monocultures. *Oikos* 47, 211–222.

Weiner, J., Berntson, G.M. & Thomas, S.C. (1990a) Competition and growth form in a woodland annual. *Journal of Ecology* 78, 459–469.

Weiner, J., Mallory, E.B. & Kennedy, C. (1990b) Growth and viability in crowded and uncrowded populations of dwarf marigolds (*Tagetes patula*). *Annals of Botany* 65, 513–524.

Weis, A.E., Abrahamson, W.G. & Andersen, M.C. (1992) Variable selection on *Eurostas* gall size. 1. The extent and nature of variation in phenotypic selection. *Evolution* 46, 1674–1697.

Wellburn, A. (1988) *Air Pollution and Acid Rain: the Biological Impact*. Longman Scientific, New York.

Weller, D.E. (1987) A re-evaluation of the – 3/2 power rule of plant self-thinning. *Ecological Monographs* 57, 23–43.

Weller, S.J., Ohm, H.W., Patterson, F.L., Foster, J.E. & Taylor, P.L. (1991) Genetics of resistance of CI 15160 Durum wheat to biotype D of hessian fly. *Crop Science* 31, 1163–1168.

Wellings, P.W. & Dixon, A.F.G. (1987) The role of weather and natural enemies in determining aphid outbreaks. In *Insect Outbreaks* (Eds P. Barbosa & J.C. Schultz), pp. 314–346. Academic Press, New York.

Wellington, W.G. & Trimble, R.M. (1984) Weather. In *Ecological Entomology* (Eds C.B. Huffaker & R.L. Rabb), pp. 399–425. John Wiley & Sons, New York.

Wells, T.C.E. (1981) Population ecology of terrestrial orchids. In *The Biological Aspects of Rare Plant Conservation* (Ed. H. Synge), pp. 281–295. John Wiley & Sons, Chichester.

Went, F.W. (1953) The effect of temperature on plant growth. *Annual Review of Plant Physiology* 4, 347–362.

Went, F.W., Juhren, G. & Juhren, M.C. (1952) Fire and biotic factors affecting germination. *Ecology* 33, 351–364.

Werner, P.A. (1976) Ecology of plant populations in seasonal environments. *Systematic Botany* 1, 246–268.

Werner, P.A. (1977) Colonization success of a biennial plant species: experimental field studies of species cohabitation and replacement. *Ecology* 58, 840–849.

Werner, P.A. & Caswell, H. (1977) Population growth rates and age versus stage distribution models for teasel (*Dipsacus sylvestris* Huds.). *Ecology* 58, 1103–1111.

Wesson, G. & Wareing, P.F. (1969a) The role of light in the germination of naturally occurring populations of buried weed seeds. *Journal of Experimental Botany* 20, 402–413.

Wesson, G. & Wareing, P.F. (1969b) The induction of light sensitivity in weed seeds by burial. *Journal of Experimental Botany* 20, 414–425.

West, C. (1985) Factors underlying the late-seasonal appearance of the lepidopterous leaf mining guild on oak. *Ecological Entomology* 10, 111–120.

West, H.M., Fitter, A.H. & Watkinson, A.R. (1993) The influence of three biocides on the fungal associates of the roots of *Vulpia ciliata* ssp. *ambigua* under natural conditions. *Journal of Ecology* 81, 345–350.

Weste, G. & Ashton, D.H. (1994) Regeneration and survival of indigenous dry sclerophyll species in the Brisbane Ranges, Victoria, after *Phytophthora cinnamomi* epidemic. *Australian Journal of Botany* 42, 239–253.

Westley, L.C. (1993) The effect of inflorescence bud removal on tuber production in *Helianthus tuberosus* L. (Asteraceae). *Ecology* 74, 2136–2144.

Westoby, M. (1982) Frequency distributions of plant size during competitive growth of stands: the operation of distribution-modifying functions. *Annals of Botany* 50, 733–735.

Westoby, M., Jurado, E. & Leishman, M. (1992) Comparative evolutionary ecology of seed size. *Trends in Ecology and Evolution* 7, 368–372.

Wheelwright, N.T. (1985) Fruit size, gape width and the diets of fruit-eating birds. *Ecology* 66, 808–818.

Wheelwright, N.T. (1993) Fruit size in a tropical tree species – variation, preference by birds, and heritability. *Vegetatio* 108, 163–174.

Wheelwright, N.T. & Bruneau, A. (1992) Population sex-ratios and spatial-distribution of *Ocotea tenera* (Lauraceae) trees in a subtropical forest. *Journal of Ecology* 80, 425–432.

Wheelwright, N.T. & Janson, C.H. (1985) Colors of fruit displays of bird-dispersed plants in two tropical forests. *American Naturalist* 126, 777–799.

Whipps, J.M. (1984) Environmental factors affecting the loss of carbon from the roots of wheat and barley seedlings. *Journal of Experimental Botany* 35, 767–773.

White, J. (1985) The thinning rule and its application to mixtures of plant populations. In *Studies on Plant Demography* (Ed. J. White), pp. 291–309. Academic Press, London.

White, J. & Harper, J.L. (1970) Correlated changes in plant size and number in populations. *Journal of Ecology* 58, 467–485.

White, P.S. (1984) The architecture of Devil's Walking Stick, *Aralia spinosa* L. (Araliaceae). *Journal of the Arnold Arboretum* 65, 403–418.

White, T.C.R. (1970) Some aspects of life history, host selection, dispersal, and oviposition of adult *Cardiaspina densitexta* (Homoptera: Psyllidae). *Australian Journal of Zoology* 18, 105–117.

White, T.C.R. (1974) A hypothesis to explain outbreaks of looper caterpillars, with special reference to populations of *Selido suavis* in a plantation of *Pinus radiata* in New Zealand. *Oecologia* **16**, 279–301.

White, T.C.R. (1984) The abundance of invertebrate herbivores in relation to the availability of nitrogen in stressed food plants. *Oecologia* **63**, 90–105.

White, T.C.R. (1993) *The Inadequate Environment. Nitrogen and the Abundance of Animals*. Springer-Verlag, Berlin.

Whitehouse, H.L.K. (1950) Multiple-allelomorph incompatibility of pollen and style in the evolution of the angiosperms. *Annals of Botany* **14**, 198–216.

Whitham, T.C. (1980) The theory of habitat selection: examined and extended using *Pemphigus* aphids. *American Naturalist* **115**, 449–466.

Whitham, T.G. (1983) Host manipulation of parasites: within-plant variation as a defense against rapidly evolving pests. In *Variable Plants and Herbivores in Natural and Managed Systems* (Eds R.F. Denno & M.S. McClure), pp. 15–38. Academic Press, New York.

Whitham, T.G. & Mopper, S. (1985) Chronic herbivory: impacts on architecture and sex expression in pinyon pine. *Science* **228**, 1089–1091.

Whitmore, T.C. (1985) Forest succession. *Nature* **315**, 692.

Whitney, G. (1986) Relation of Michigan's presettlement pine forests to substrate and disturbance history. *Ecology* **67**, 1548–1559.

Whittaker, J.B. (1982) The effect of grazing by a chrysomelid beetle, *Gastrophysa viridula* on growth and survival of *Rumex crispus* on a shingle bank. *Journal of Ecology* **70**, 291–296.

Whittaker, R.H. (1956) Vegetation of the Great Smoky Mountains. *Ecological Monographs* **26**, 1–80.

Whittaker, R.H. (1967) Gradient analysis of vegetation. *Biological Reviews* **42**, 207–264.

Whittaker, R.H. (1975) *Communities and Ecosystems*, 2nd edn. Collier MacMillan, London.

Whittaker, R.H. & Goodman, D. (1979) Classifying species according to their demographic strategy. 1: Population fluctuations and environmental heterogeneity. *American Naturalist* **113**, 185–200.

Whittaker, R.J. (1995) Disturbed island ecology. *Trends in Ecology and Evolution* **10**, 421–425.

Wiebes, J.T. (1979) Figs and their insect pollinators. *Annual Review of Ecology and Systematics* **10**, 1–12.

Wiens, D. (1984) Ovule survivorship, brood size, life history, breeding systems, and reproductive success in plants. *Oecologia* **64**, 47–53.

Wiens, D., Calvin, C.L., Wilson, C.A., Davern, C.I., Frank, D. & Seavey, S.R. (1987) Reproductive success, spontaneous embryo abortion, and genetic load in flowering plants. *Oecologia* **71**, 501–509.

Wigley, T.M.L. & Jones, P.D. (1981) Detecting CO_2-induced climate change. *Nature* **292**, 205–208.

Williams, G.C. (1957) Pleiotropy, natural selection, and the evolution of senescence. *Evolution* **11**, 398–411.

Williams, G.C. (1966) *Adaptation and Natural Selection*. Princeton University Press, Princeton.

Williams, G.C. (1975) *Sex and Evolution*. Princeton University Press, Princeton.

Williams, J.T. (1988) Identifying and protecting the origins of our food plants. In *Biodiversity* (Eds E.O. Wilson & F.M. Peter), pp. 240–247. National Academy of Sciences, Washington, DC.

Williams, K.S. & Myers, J.H. (1984) Previous herbivore attack of red alder may improve food quality for fall web-worm larvae. *Oecologia* **63**, 166–170.

Williams, R.D. & Hoogland, R.E. (1982) The effects of naturally occurring phenolic compounds on seed germination. *Weed Science* **30**, 206–210.

Williams, S.L. (1995) Surf grass (*Phyllospadix torreyi*) reproduction – reproductive phenology, resource allocation, and male rarity. *Ecology* **76**, 1953–1970.

Williamson, M.H. (1972) *The Analysis of Biological Populations*. Arnold, London.

Willis, A.J., Dunnett, N.P., Hunt, R. & Grime, J.P. (1995) Does Gulf Stream position affect vegetation dynamics in Western Europe? *Oikos* **73**, 408–410.

Willson, M.F. (1993) Dispersal mode, seed shadows and colonization patterns. *Vegetatio* **108**, 261–280.

Willson, M.F. & Burley, N. (1983) *Mate Choice in Plants: Tactics, Mechanisms and Consequences*. Princeton University Press, Princeton.

Wilson, D.E. & Jansen, D.H. (1972) Predation on *Scheelea* palm seeds by bruchid beetles: seed density and distance from the parent palm. *Ecology* **53**, 954–959.

Wilson, E.O. (1984) *Biophilia*. Harvard University Press, Cambridge, Massachusetts.

Wilson, E.O. (1994) *Naturalist*. Island Press, Washington.

Wilson, G.R. & Bell, J.N.B. (1985) Studies on the tolerance to sulphur dioxide of grass populations in polluted areas. III. Investigations on the rate of development of tolerance. *New Phytologist* **100**, 63–77.

Wilson, S.D. & Keddy, P.A. (1986a) Species competitive ability and position along a natural stress/disturbance gradient. *Ecology* **67**, 1236–1242.

Wilson, S.D. & Keddy, P.A. (1986b) Measuring diffuse competition along an environmental gradient: results from a shoreline plant community. *American Naturalist* **127**, 862–869.

Wilson, S.D. & Tilman, D. (1991) Components of plant competition along an experimental gradient of nitrogen availability. *Ecology* **72**, 1050–1065.

Wilson, S.D. & Tilman, D. (1993) Plant competition and resource availability in response to disturbance and fertilization. *Ecology* **74**, 599–611.

Windle, P.N. & Franz, E.H. (1979) Plant population structure and aphid parasitism in barley monocultures and mixtures. *Journal of Applied Ecology* **16**, 259–268.

Wink, M., Hofer, A., Bilfinger, M., Englert, E., Martin, M. & Schneider, D. (1993) Geese and dietary allelochemicals – food palatability and geophagy. *Chemoecology* **4**, 93–107.

Winsor, J.A., Davis, L.E. & Stephenson, A.G. (1987) The relationship between pollen load and fruit maturation and the effect of pollen load on offspring vigor in

Cucurbita pepo. American Naturalist **129**, 643–656.

Wissel, C. (1992) Modelling the mosaic cycle of a middle European beech forest. *Ecological Modelling* **63**, 29–43.

Wolfram, S. (1986) *Theory and Applications of Cellular Automata*. World Scientific, Singapore.

Woodell, S.R.J., Mooney, H.A. & Hill, A.J. (1969) The behaviour of *Larrea divaricata* (creosote bush) in response to rainfall in California. *Journal of Ecology* **57**, 37–44.

Woodhead, S. (1981) Environmental and biotic factors affecting the phenolic content of different cultivars of *Sorghum bicolor. Journal of Chemical Ecology* **7**, 1035–1047.

Woodward, F.I. (1987) *Climate and Plant Distribution*. Cambridge University Press, Cambridge.

Woodward, F.I. (1992) Predicting plant responses to global environmental change. *New Phytologist* **122**, 239–251.

Wrangham, R.W., Chapman, C.A. & Chapman, L.J. (1994) Seed dispersal by forest chimpanzees in Uganda. *Journal of Tropical Ecology* **10**, 355–368.

Wright, S. (1952) The theoretical variance within and among subdivisions of a population that is in a steady state. *Genetics* **37**, 312–321.

Wu, L., Bradshaw, A.D. & Thurman, D.A. (1975) The potential for evolution of heavy metal tolerance in plants. 3. The rapid evolution of copper tolerance in *Agrostis stolonifera. Heredity* **34**, 165–187.

Yahara, T. & Oyama, K. (1993) Effects of virus infection on demographic traits of an agamospermous population of *Eupatorium chinense* (Asteraceae). *Oecologia* **96**, 310–315.

Yakir, D. & Yechieli, Y. (1995) Plant invasion of newly exposed hypersaline Dead sea shores. *Nature* **374**, 803–805.

Yeaton, R.I. & Cody, M.L. (1976) Competition and spacing in plant communities: the northern Mohave desert. *Journal of Ecology* **64**, 689–696.

Yeaton, R.I., Travis, J. & Gilinsky, E. (1977) Competition and spacing in plant communities: the Arizona upland association. *Journal of Ecology* **65**, 587–595.

Yoda, K., Kira, T., Ogawa, H. & Hozumi, K. (1963) Self-thinning in overcrowded pure stands under cultivated and natural conditions. *Journal of Biology, Osaka City University* **14**, 107–129.

Yost, R.S., Behara, G. & Fox, R.L. (1982) Geostatistical analysis of soil chemical properties of large land areas. I. Semi-variograms. *Soil Science Society of America Journal* **46**, 1028–1032.

Young, H.J. & Stanton, M.L. (1990) Influences of floral variation on pollen removal and seed production in wild radish. *Ecology* **71**, 536–547.

Young, H.J. & Young, T.P. (1992) Alternative outcomes of natural and experimental high pollen loads. *Ecology* **73**, 639–647.

Zakaria, S. (1989) *The influence of previous insect feeding on the rate of damage of birch tree leaves*. Unpublished PhD thesis, University of London.

Zedler, J.B. (1988) Restoring diversity in salt marshes: can we do it? In *Biodiversity* (Eds E.O. Wilson & F.M. Peter), pp. 317–325. National Academy of Sciences, Washington, DC.

Zedler, P.H. (1995) Are some plants born to burn? *Trends in Ecology and Evolution* **10**, 393–395.

Zeide, B. (1987) Analysis of the −3/2 power law of self-thinning. *Forest Science* **33**, 517–537.

Zhang, J. & Davies, W.J. (1989) Sequential response of whole plant water relations to prolonged soil drying and the involvement of xylem ABA in the regulation of stomatal behaviour in sunflower plants. *New Phytologist* **113**, 167–174.

Zimmerman, M. (1982) The effect of nectar production on neighborhood size. *Oecologia* **52**, 104–108.

Zimmermann, M.H. (1983) *Xylem Structure and the Ascent of Sap*. Springer-Verlag, Berlin.

Zohary, D. & Hopf, M. (1988) *Domestication of Plants in the Old World*. Clarendon Press, Oxford.

Zohary, M. (1950) Evolutionary trends in the fruiting head of Compositae. *Evolution* **4**, 103–109.

Zucker, W.V. (1982) How aphids choose leaves: the role of phenolics in host selection by a galling aphid. *Ecology* **63**, 927–981.

Index

Page references to figures are in italic, tables are in bold and boxes are followed by B, e.g. 89B

[705]